DIFFERENTIAL EQUATIONS

GRAPHICS • MODELS • DATA

DIFFERENTIAL EQUATIONS

GRAPHICS • MODELS • DATA

David Lomen
David Lovelock

University of Arizona

JOHN WILEY & SONS, INC.

New York • Chichester • Weinheim • Brisbane • Singapore • Toronto

MATHEMATICS EDITOR Barbara Holland
MARKETING MANAGER Katherine Hepburn
PRODUCTION EDITOR Ken Santor
COVER DESIGNER Karin Kincheloe
TEXT DESIGNER Nancy Field
COVER ILLUSTRATION Roy Wiemann
ILLUSTRATION EDITOR Sigmund Malinowski
ILLUSTRATION STUDIO Wellington Studios

Library of Congress Cataloging-in-Publication Data

Lomen, David O., (date)
Differential equations : graphics, models, data / by David Lomen, David Lovelock.
p. cm.
Includes index.
ISBN 0-471-07648-1 (alk. paper)
ISBN 13: 978-0-471-07648-3
1. Differential equations. I. Lovelock, David, (date).
II. Title.
QA371.L645 1999
515'.35–DC21 98-14494
CIP

10 9 8 7 6 5 4 3 2

PREFACE

We have written this book to enable everyone to be an active participant in their own learning. Our objective is to introduce Ordinary Differential Equations (ODEs) in a manner that invites students to think, experiment, and comprehend. We want students to see ODEs as a logical extension of calculus – not only to understand them, but also to view them as a natural tool for investigating many aspects of science and engineering. We encourage students to explore differential equations by employing technology as an instrument to check, discover, and interpret the behavior of solutions. Our intent is to provide lively reading with compelling applications, projects, and experiments that supply students with opportunities to explore the relationship between a differential equation and the process it models.

Differential Equations Are a Generalization of Indefinite Integrals

- Like indefinite integrals, if the occasional differential equation can be solved in terms of familiar functions, we should solve it. Thus, the book provides analytical solutions of differential equations.
- Like indefinite integrals, most differential equations cannot be solved in terms of familiar functions. When this occurs, we try to understand the solution by qualitative means (monotonicity, concavity, symmetry, singularities, existence and uniqueness of solutions) and by quantitative means (numerical solutions).

How do we know the information we obtain about a differential equation is correct? The simple answer is to use everything we have at our disposal – analytical solutions, quantitative calculations, and qualitative arguments. We treat differential equations from a global point of view, subject to many different mathematical procedures.

Visualization and the Rule of Four

We cover and use graphical and numerical techniques early in the book because they are essential whenever analytical solutions cannot be found in terms of familiar functions, or when their properties are obscure because of the form of the analytical solution. We emphasize visual exploration of solutions via slope fields, direction fields, phase plane solutions, and other graphical representations. Whenever appropriate we also implement a "rule of four," treating topics from numerical, graphical, analytical, and descriptive viewpoints. Each of these viewpoints brings its unique contribution and perspective to the solution of an ODE.

The Role of Real Data and Modeling

Most of the topics in this book are problem or data driven. We use data sets to develop differential equations, to obtain values of parameters in differential equations, and to check the accuracy of mathematical models. We give careful attention to mathematical modeling, solution techniques, and interpretation of results. Different disciplines view the data-modeling process in different ways. In some disciplines, the model is constructed based on established principles, and then the data set is used to test the model. In other disciplines, the data set is used to construct the model empirically, and then the model is used to predict subsequent behavior. This book deals with both approaches. Nevertheless, the emphasis of the book is on differential equations, not modeling.

Many of the data sets used in the book are available from the website **http: //www.math.arizona.edu/~dsl /data.htm**, which has links to other data sources.

Technology

The rapid development and increased use of technology has required a closer scrutiny of the content of the traditional differential equations course. Because modern technology can quickly draw slope fields, graph complicated functions, and display data with just a few keystrokes, we no longer need to avoid subject matter which requires such things. Yet this same technology is capable of giving incomplete, misleading, or incorrect solutions, so it is important to look at all solutions from more than one point of view. This book does just that, equipping students with confidence because they fully understand the solution by using graphical and numerical means in addition to the traditional analytical ones.

We feel that the appropriate use of technology is vital to using differential equations as a tool. Thus, we expect students to have access to mathematical software or to a calculator that draws slope fields or direction fields, generates numerical solutions, and simultaneously displays data and graphs of functions. With these tools students become active participants in the learning process and they are encouraged to think, experiment, and comprehend. By these means they understand the origins, qualitative behavior, and interpretations of solutions of ODEs.

Key Changes in This Edition

We streamlined examples and exposition, making the book easier to teach and learn from. The development of standard techniques has been condensed while comparisons between nonlinear and linear equations, autonomous and nonautonomous equations have been expanded. We put additional emphasis on the interplay between first order systems and second order equations and treated spring-mass systems, pendulums, and electric circuits in a parallel fashion.

We added new material on nonlinearity, chaos, and qualitative reasoning and included many more interesting and realistic examples and exercises. More emphasis was placed on having students understand the reason for each step in the development as well as their importance and usefulness. Some of this occurs under the new heading "Where We Are Going — And Why" at the start of each chapter, where we explain why the subject of the chapter is important.

We added pedagogical devices to promote clarity, understanding, ease of use, and importance of ideas. For example, we:

- included more mathematical models where obtaining an explicit solution is not possible (or, if it is possible, gives a solution which is impossible to graph directly) yet the solution behavior is easy to determine by graphical techniques.
- moved many tables and figures into the margin to make the book much more readable.
- rearranged exercises so similar ones are together.
- put "key words" in the margin to help students keep the strategies in mind and added "subheadings" for the same reason.
- included the answers to most of the odd-numbered problems.

Prerequisites

The prerequisite for this book is a course in single variable calculus, from either a traditional or a reform approach. Linear algebra and multivariable calculus are not required. Sections 6.2 and 9.5 use elementary properties of 2 by 2 matrices. This book is appropriate for any major, although many of the applications are from engineering and the sciences (physical, biological, and social).

Suggested Syllabus

This book may be covered completely in a two quarter course, but contains more material than may be covered in a one semester course. The following are suggestions for a one semester course:

- Chapters 1 through 7, 8.1 through 8.4, 9, and 10.1 through 10.4.
- Chapters 1 through 7, 8.1 through 8.4, 11, and 12.

As far as selecting material from Chapters 1 through 7 is concerned, you should note the following: Sections 1.4, 2.6, 3.4, 3.5, 4.5, 4.6, and 5.3 are not extensively used in later sections, and so may be selectively omitted.

Some comments on the dependence of Chapters 8 through 12 follow, assuming that you have covered Chapters 1 through 7, and Sections 8.1 through 8.4, in the manner previously described.

- Sections 8.5 through 8.8 are not critical for later sections.
- Chapter 9 follows from Chapter 7. It does not use any material from Chapter 8.
- Chapter 10 follows from Section 9.4. It is not used in later chapters.
- Chapter 11 follows from Section 8.4. It does not use any material from Chapters 9 and 10. It is not used in later chapters.
- Chapter 12 follows from Section 8.4.

Additional Materials

- A Solutions Manual for students in which all odd-numbered exercises are solved.
- An Instructor's Resource Manual containing detailed suggestions together with additional materials, exercises, experiments, data sets, and projects.
- Free DOS-based computer programs developed for this differential equations course which may be downloaded from the website **http: //www.math.arizona.edu/ software/uasoft.html** or by using ftp from **software@ math.arizona.edu**.

ACKNOWLEDGMENTS

We would like to thank the following for their constructive ideas and valuable assistance:

Students (past and present, local and global):

Alicia Acevedo, Gabriel Aldaz, Rupesh Amin, David Anthony, Kate Baird, John Beardslee, Mark Biedrzycki, David Brokl, Erik Brown, Jesse Cameron, Ken Cardell, Danny Carillo, Elizabeth Cheney, Haiqian Cheng, Steve Connor, Paul Deneke, Shelley DeVere, Scott Downs, Mason Dykstra, Melisa Enrico, Melissa Erickson, Jason Figueroa, Andy Folkening, Bruce Gungle, Lisa Hanley, David Harman, Scott Hayes, Steve Hiller, Shih-Chieh Huang, Ralph Huarto, Janelle Ivie, Susan Kern, Stephen Koester, Tapia Kolunsarka, Alaric Lebaron, Clay Lines, Jennifer Lowe, Rachael Ludwick, John Magras, Danielle Manuszak, Elizabeth McBride, Terri McSweeney, David Megaw, Thanh Nguyen, Becki Norris, Michael Oddy, Chad Parkhill, Lesley Perg, Thai Phan, Tim Prescott, Paul Richards, Doug Robinson, Arturo Ruiz, Joe Scionti, Ian Scott, Patrick Shipman, Shane Sickafoose, Chris Sinclair, Joel Statkevicus, Cathy Steffens, Michelle Switala, Anna Szidarovszky, Katherine Tydersley, Michael Van Zeeland, Dale Williams, Tracy Wood, Jue Wu, Herbert Yee, Peter Yip, Weijun Zhu, and Richard Ziehmer.

Colleagues and others:

Bruce Bayly, Melanie Bell, Jim Clark, Jim Cushing, Sam Evens, Hermann Flaschka, David Gay, Rick Greenberg, Larry Grove, Tom Hallam, Brink Harrison, Shandelle Henson, Dean Hickerson, Barbara Holland, Simone Jacobsen, John Kessler, Donna Krawczyk, Ira Lackow, Kathy Lackow, Connie Lomen, Cynthia Lovelock, Denise Lovelock, Fiona Lovelock, Jim Mark, Michael Mayersohn, Sean McHaney, Lang Moore, Bill Mueller, Charlie Nafzigger, Alan Newell, Javier Osante Vazquez, John Palmer, Terry Passott, Yuri Pinelis, John Robson, Ken Santor, Pam Sharratt, Steve Sheldon, Stephen Shipman, Fred Stevensen, David Sutherland, Hal Tharp, Mark Torgerson, Doug Ulmer, Maciej Wojtkowski, David Yabi, and Chris Yetman.

Reviewers:

Frank J. Avenoso, Nassau Community College; Edgar M. Chandler, Paradise Valley Community College; Robert Cole, The Evergreen State College; Michael R. Colvin, California Polytechnic State University; Wade Ellis, Jr., West Valley College;

William D. Emerson, Metropolitan State College of Denver; Jeffrey P. Igo, Henry Ford Community College; Kenneth R. Kellum, San Jose State University; Reza Malek- Madani, United States Naval Academy; Joe A. Marlin, North Carolina State University; Anthony V. Phillips, SUNY at Stony Brook; Roger Pinkham, Stevens Institute of Technology; Zwi Reznick, Fresno City College; Asok Sen, Purdue University; Paul Schembari, East Stroudsburg University; and Robert L. Wheeler, Virginia Technological Institute & State College.

We also acknowledge the National Science Foundation for their support. The Instrument and Laboratory Improvement grants, DUE94-50953 and DUE 94-55970, equipped two electronic classrooms so that each student had access to a computer at all times during class. NSF grants USE90-53431 and USE90-54181 allowed us to develop software to use in these classrooms. These grants provided a means for meaningfully incorporating graphical and numerical methods into classroom discussions and greatly aided the graphical and numerical presentations of this book. Finally, NSF grant DUE90-53799 supplied funds for mathematics education students and teachers to work through the book and suggest improvements.

TO THE STUDENT FROM OTHER STUDENTS

by David Brokl, Jennifer Lowe, Ian Scott, and Michael Van Zeeland, students at the University of Arizona

This book is different! Most math books and math classes follow the simple formula: (1) present an abstract definition or theorem; (2) attempt to clarify this definition or theorem with examples and dialogue; and then (3) assign exercises which can be done by repeating the examples in the book with changed numbers. This book often works contrary to that formula in that graphical and numerical examples are presented first, and from them some general conclusions are drawn. Sometimes sections start with a data set or real world example, often about situations we have encountered in our other science and engineering classes. This is great, as it makes the book much more interesting to read.

As you are working through these examples, keep your eyes open for connections to past results. These are there, and when you finish, there are some new conclusions which await you. Sometimes there are clues as to what to look for in the "Where We Are Going — and Why" at the start of each chapter.

To help understand these general conclusions, we would first read through a new section as if reading a story, looking for main ideas, but not dwelling on them. Then we would go back through the section carefully with pencil and paper (and often a computer or graphing calculator) at hand to make sure the details made sense. After each example, we stopped to examine the logic and details of the analysis. In many places of the text there were questions asking why something was true (these are contained in parentheses). We found it very worthwhile to figure out the answer before going on. The words "Comments about" appeared at various places in the book. These were not simply idle comments, but either emphasized an important point or made useful remarks about a prior result. Be sure to read and think about these "Comments"! As you work through each section, pay close attention to every single graph. In other math books, graphs may seem to be incidental, but in this book the graphs are central to understanding the ideas. A computer, or graphing calculator, was also crucial to understanding the material. We would question what we saw on the screen and guess what would happen if we changed one of the inputs. We would also make sure that what we saw agreed with our analytical work. Computers can be misleading, so we questioned things which did not make sense. We found that answering these questions led to a much better understanding of the material. It was a great feeling to discover that we could explain why the computer results were misleading.

Most of the exercises cannot be done by simply looking at the examples in the book without reading the section. However, by working through the material in the book you will be able to recognize similarities between the exercises and the examples and see the proper approach for the exercise. If part of a section is confusing, look ahead to a "How to" box — which contains concise explanations — or to related definitions, or maybe to the "What Have We Learned?" summary at the end of the chapter.

Finally, find a friend (or friends) to study with. These people should be able to trade ideas with you and discuss alternative approaches to solving problems. Having several points of view is often a big help in solving some of the more challenging problems. Someone else can often easily spot the place where your thinking or analysis is incorrect. Of course, explaining things to each other is an effective way to really understand the material. We really enjoyed using this book, and think you will also.

CONTENTS

CHAPTER 8 SECOND ORDER LINEAR DIFFERENTIAL EQUATIONS – QUALITATIVE AND QUANTITATIVE ASPECTS 315

CHAPTER 9 LINEAR AUTONOMOUS SYSTEMS 379

CHAPTER 10 NONLINEAR AUTONOMOUS SYSTEMS 315

CHAPTER 11 USING LAPLACE TRANSFORMS 505

CHAPTER 1

BASIC CONCEPTS

Where We Are Going – and Why

We will develop ways of analyzing ordinary differential equations that take full advantage of the power of calculus and technology. We do this by treating topics from graphical, numerical, and analytical points of view. Because each of these points of view can at times give incomplete information, we always need to compare our results for consistency. The strategies presented in this book will empower you to correctly analyze solutions of a differential equation, even when those solutions are not obtainable as analytical expressions.

In this chapter, we introduce these three points of view by considering differential equations of the form

$$\frac{dy}{dx} = g(x).$$

We start with a brief introduction, definitions, and examples, which make use of prior knowledge of antiderivatives. Then we spend the next two sections illustrating graphical techniques, including slope fields and isoclines. These techniques often let us discover many properties of the solution of a specific differential equation by simply analyzing the differential equation from a graphical point of view. We end this chapter with a discussion of Taylor series, which are especially useful when solutions are given by integrals that do not have simple antiderivatives.

The purpose of this chapter is to illustrate graphical, numerical, and analytical approaches in the familiar setting of antiderivatives. We want to make sure you have a firm foundation in these approaches so you can quickly grasp the new ideas in subsequent chapters.

1.1 SIMPLE DIFFERENTIAL EQUATIONS AND EXPLICIT SOLUTIONS

Ever since the ideas of calculus were developed by Newton, Leibniz, and others in the seventeenth century, people have been using differential equations to describe many phenomena that touch our lives. Differential equations are the most common mathematical tool used for the precise formulation of the laws of nature and other phenomena described by a relationship between a function and its derivatives. In this book you will see many examples of such relationships.

The simple definition of a differential equation is an equation that involves a derivative. Thus, many differential equations are solved in beginning calculus courses, perhaps without anyone stating it.

Two Basic Examples

We start with a familiar example.

EXAMPLE 1.1 *Parabolas*

Consider the problem of finding the most general antiderivative of the function x. If we call this antiderivative $y(x)$, then the derivative of y is x; that is, y satisfies the differential equation

$$\frac{dy}{dx} = x. \tag{1.1}$$

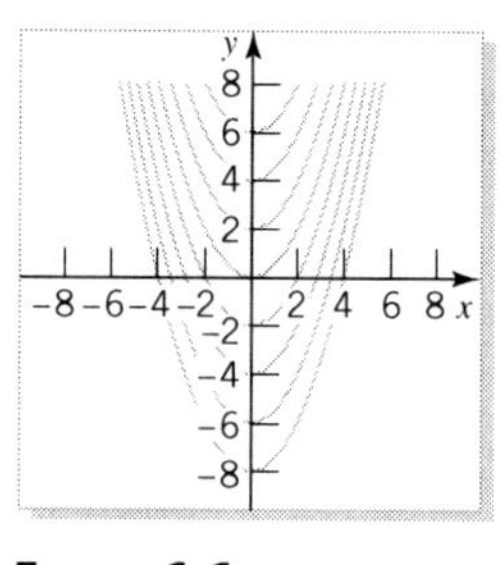

FIGURE 1.1
Some solutions of $dy/dx = x$

Because any antiderivative of x may be written as

$$y(x) = \frac{1}{2}x^2 + C, \tag{1.2}$$

where C is an arbitrary constant, it seems reasonable to call (1.2) solutions of our differential equation (1.1). Because of the arbitrary constant, we have an infinite number of solutions, a different one for each choice of C (collectively called a family of solutions). Some of these solutions are graphed in Figure 1.1 (upward opening parabolas) where we notice that the role of the arbitrary constant is to determine the vertical position. All solutions of this differential equation have the same general shape, and any two solutions will differ from each other by a vertical translation. Thus, two different solutions will not intersect.[1]

Vertical translation

We now look at another familiar example, one to which we will return frequently in this chapter.

EXAMPLE 1.2 *The Natural Logarithm Function*

We consider the problem of finding the most general antiderivative, $y(x)$, of the function $1/x$ — that is, finding solutions of the differential equation

$$\frac{dy}{dx} = \frac{1}{x}. \tag{1.3}$$

[1] If two functions $y_1(x)$ and $y_2(x)$ have a point x_0 in common, so that $y_1(x_0) = y_2(x_0)$, we say that $y_1(x)$ and $y_2(x)$ intersect, touch, or cross, at $x = x_0$.

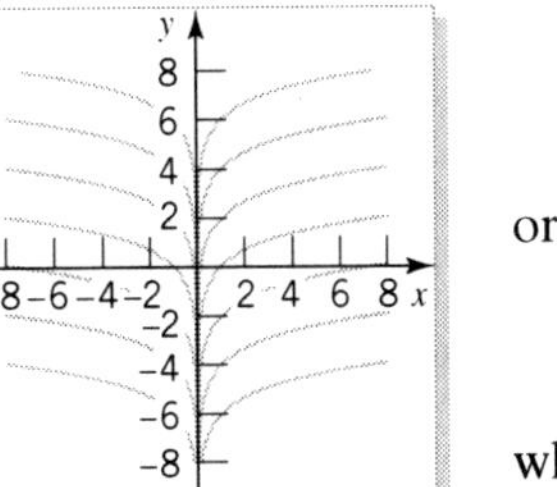

FIGURE 1.2
Some solutions of $dy/dx = 1/x$

Because all antiderivatives of $1/x$ may be written as

$$y(x) = \ln|x| + C, \tag{1.4}$$

or

$$y(x) = \begin{cases} \ln x + C & \text{if } x > 0, \\ \ln(-x) + C & \text{if } x < 0, \end{cases} \tag{1.5}$$

where C is an arbitrary constant, these are our solutions. Again, because of the arbitrary constant, we have a family of solutions. Some of these solutions are graphed in Figure 1.2, where we notice again that the role of the arbitrary constant is to determine the vertical position. If $x > 0$ all solutions of this differential equation have the same general shape, and any two solutions will differ from each other by a vertical translation. The same is true for $x < 0$.

Vertical translation

Definitions and Comments

These are two of many examples of differential equations covered in calculus. All problems where we found the indefinite integral (or antiderivative) of a function, $g(x)$, could have been formulated as finding $y(x)$ as a solution of

$$\frac{dy}{dx} = g(x). \tag{1.6}$$

The solutions of (1.6) all have the form

$$y(x) = \int g(x)\,dx + C, \tag{1.7}$$

where $\int g(x)\,dx$ is any specific antiderivative of $g(x)$.[2] The arbitrary constant C indicates that we have an infinite number of solutions, related to each other by a vertical translation.

Vertical translation

Before developing any methods for finding solutions of differential equations, we give some formal definitions which will be helpful later when we consider more complicated situations.

Differential equations such as (1.1), (1.3), and (1.6) are called first order differential equations, because the first derivative is the highest one that occurs in each equation. Thus we have

◆ *Definition 1.1:* **A FIRST ORDER ORDINARY DIFFERENTIAL EQUATION is an equation that involves at most the first derivative of an unknown function. If y, the unknown function, is a function of x, then we write the first order differential equation as**

$$\frac{dy}{dx} = g(x, y), \tag{1.8}$$

where $g(x, y)$ is a given function of the two variables x and y.[3] ◆

[2] In calculus it is customary to have the symbol $\int g(x)\,dx$ include the arbitrary constant C. Here we add the constant explicitly to emphasize geometrical ideas.

[3] Even though all our examples in this chapter are of the form $dy/dx = g(x)$, we use $dy/dx = g(x, y)$ in all our definitions so that they also apply to subsequent chapters.

Comments about First Order Differential Equations

- The right-hand side of (1.8) may contain x and y explicitly, for example, $x^2 + y^2$. However, in this chapter we will consider the case where $g(x, y)$ is a function of x alone.
- If y is a function of x, then x is called the independent variable and y is called the dependent variable.

We previously noted that (1.2), (1.4), and (1.7) are solutions of (1.1), (1.3), and (1.6), respectively. We know this because if we differentiate these functions and substitute the result into the proper differential equation, we obtain an identity. These solutions are called explicit because they have the dependent variable, y, given solely in terms of the independent variable, x. This prompts our second definition.

◆ *Definition 1.2:* **An EXPLICIT SOLUTION of the first order ordinary differential equation**

$$\frac{dy}{dx} = g(x, y) \tag{1.9}$$

is any function $y = y(x)$, with a derivative in some interval $a < x < b$, that identically satisfies the differential equation (1.9). ◆

Comments about Explicit Solutions

- Because an explicit solution has a derivative in the interval $a < x < b$, it must be continuous in that interval. (Why?) An explicit solution can never have a vertical tangent. (Why?)
- An explicit solution may contain an arbitrary constant. If it does, we have an infinite number of solutions, called a FAMILY OF EXPLICIT SOLUTIONS. If it does not contain an arbitrary constant, we have a PARTICULAR EXPLICIT SOLUTION. Often particular explicit solutions are just called particular solutions.
- The graph of a particular solution is called a SOLUTION CURVE of the differential equation.

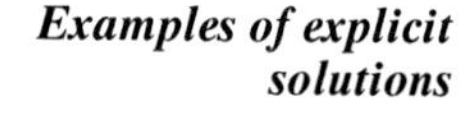
Examples of explicit solutions

The explicit solutions of the three differential equations mentioned so far all contain an arbitrary constant, so they are families of explicit solutions. This constant may be determined if we know the value of the solution at some specific value of x.

Thus, on the one hand, if we specify that the solution (1.4), $y(x) = \ln|x| + C$, of the differential equation (1.3) must pass through the point $(-1, 0)$, the value of the constant C must satisfy $0 = \ln|-1| + C$, so $C = 0$. This appears to give the particular solution $\ln|x|$. However, the graph of $\ln|x|$ consists of two disconnected branches (one with $x < 0$, and the other with $x > 0$) and so is not continuous, whereas any particular solution must be continuous. Because our initial point $(-1, 0)$ is given on the left branch of $\ln|x|$, the particular solution that passes through $(-1, 0)$ is $\ln|x|$ on the interval $-\infty < x < 0$; that is, $\ln(-x)$. Its graph is shown in Figure 1.3.

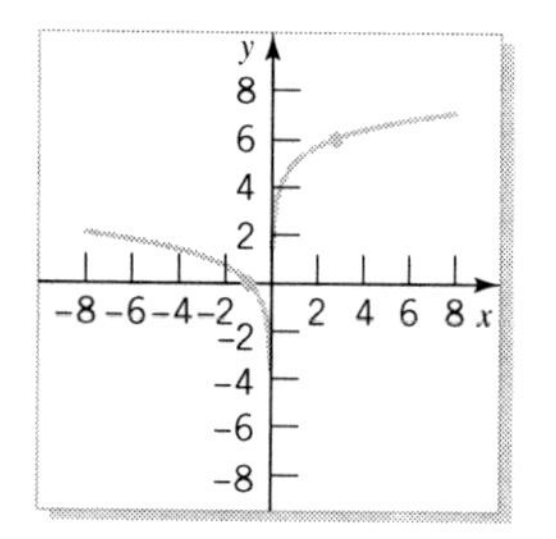

FIGURE 1.3
Particular solutions of $dy/dx = 1/x$ through $(-1, 0)$ and $(e, 6)$

On the other hand, if we specify that the solution of (1.3) is to pass through the point $(e, 6)$, the value of C must satisfy $6 = \ln|e| + C$, so $C = 5$. With similar reasoning to that just used, the particular solution passing through $(e, 6)$ is $\ln|x| + 5$

on the interval $0 < x < \infty$; that is, $\ln x + 5$. Its graph is also shown in Figure 1.3. If we look at Figure 1.2 through these eyes, we see that it represents 14 particular solutions, not 7, as a cursory glance might indicate.[4]

The problem of finding a solution of a differential equation that must pass through a given point — such as $(-1, 0)$ or $(e, 6)$ in the case of (1.3) — is called an INITIAL VALUE PROBLEM, and the point is called an INITIAL VALUE, INITIAL CONDITION, or INITIAL POINT.

Initial condition

Writing Solutions As Integrals

There is another way of expressing the solution of (1.3) when an initial point is specified as (x_0, y_0). If we use the fact that $\int_{x_0}^{x} g(t)\,dt$ is an antiderivative of $g(x)$, our solution in (1.4) may also be expressed as

$$y(x) = \int_{x_0}^{x} \frac{1}{t}\,dt + C, \tag{1.10}$$

if x_0 and x have the same sign. If we substitute $x = x_0$ into (1.10), use the initial condition $y(x_0) = y_0$, and the fact that $\int_{x_0}^{x_0} 1/t\,dt = 0$ (if $x_0 \neq 0$), we obtain the value of C as $C = y_0$. Thus (1.10) can be written as

$$y(x) = \int_{x_0}^{x} \frac{1}{t}\,dt + y_0.$$

Important point

From this example we see that the explicit solution of the initial value problem $\frac{dy}{dx} = g(x)$, $y(x_0) = y_0$, can be written in the form

$$y(x) = \int_{x_0}^{x} g(t)\,dt + y(x_0) = \int_{x_0}^{x} g(t)\,dt + y_0, \tag{1.11}$$

if $g(t)$ is bounded for t between x_0 and x. This form of the solution is particularly useful when we are unable to evaluate the integral in terms of familiar functions. We demonstrate this in the following example.

EXAMPLE 1.3 *The Error Function*

An important function, used extensively in applications in probability theory and diffusion processes, is the solution of the differential equation

$$\frac{dy}{dx} = \frac{2}{\sqrt{\pi}} e^{-x^2} \tag{1.12}$$

---

[4] A function that merely satisfies a differential equation is sometimes called an **integral** of the differential equation. A solution is an integral that is also continuous. Thus, $\ln|x| + C$ is an integral of $dy/dx = 1/x$ but is not a solution.

subject to the initial condition that

$$y(0) = 0. \tag{1.13}$$

This example will recur throughout this chapter.

Explicit solutions of (1.12) can be written as

$$y(x) = \frac{2}{\sqrt{\pi}} \int e^{-x^2}\, dx + C. \tag{1.14}$$

The usual way to evaluate the constant C so (1.13) is satisfied is to substitute $x = 0$ and $y = 0$ into the solution (1.14) and solve for C. However, the integral in (1.14) cannot be expressed in terms of familiar functions, so this usual way to evaluate C does not work. To bypass this problem we change the form of our solution to the one given in (1.11) and use the fact that $x_0 = 0$ and $y_0 = 0$ to obtain

$$y(x) = \frac{2}{\sqrt{\pi}} \int_0^x e^{-t^2}\, dt.$$

This explicit solution is called the Error Function, and it is usually denoted by $erf(x)$; that is,

$$erf(x) = \frac{2}{\sqrt{\pi}} \int_0^x e^{-t^2}\, dt. \tag{1.15}$$

We might ask how we can determine the graph of this function from its form in (1.15).[5] One way would be to construct a table of values of $(x, erf(x))$ by using a numerical method of approximating the integral for specific choices of x (see Exercise 2 on page 7). However, numerical techniques require considerable computation to plot enough points to be confident of the shape of the graph (see Exercises 3 and 4 on page 7). For that reason, in the next sections we develop methods for obtaining qualitative properties of the graph of the solution of our differential equation directly from the differential equation. ■

EXERCISES

1. Solve each of the following first order differential equations.[6] Sketch the explicit solution for three different values of the arbitrary constant C. Then find the specific value of C and the formula for $y(x)$ giving the particular explicit solution that passes through the given point P.

   (a) $dy/dx = x^3$ $\quad P = (1, 1)$

   (b) $dy/dx = x^4$ $\quad P = (1, 1)$

   (c) $dy/dx = \cos x$ $\quad P = (0, 0)$

   (d) $dy/dx = \sin x$ $\quad P = (\pi, 2)$

   (e) $dy/dx = e^{-x}$ $\quad P = (0, 1)$

   (f) $dy/dx = 1/x^2$ $\quad P = (1, 1)$

   (g) $dy/dx = 1/x$ $\quad P = (-1, 1)$

   (h) $dy/dx = 1/(1 + x^2)$ $\quad P = (1, \pi/4)$

---

[5] You might also ask how we were able to draw the graph of the function $y = \ln x$ in the previous example. This was done by a computer/calculator. So how did the computer/calculator do it? It constructed a table of numerical values. If the computer/calculator had the ability to construct functions of the sort $\int_0^x g(t)\, dt$, we could use it to graph $erf(x)$. Not many computers/calculators have this facility built in.

[6] The expression "solve the differential equation" is synonymous with "find the explicit solution of the differential equation."

(i) $dy/dx = 1/[x(1-x)]$ $P = (2, 1)$

(j) $dy/dx = \ln x$ $P = (1, 1)$

(k) $dy/dx = x^2 e^{-x}$ $P = (0, 1)$

(l) $dy/dx = e^{-x} \sin x$ $P = (0, 1)$

2. The Error Function. The purpose of this exercise is to graph the Error Function defined by (1.15), namely,

$$erf(x) = \frac{2}{\sqrt{\pi}} \int_0^x e^{-t^2}\, dt,$$

by constructing a table of values of $(x, erf(x))$.

(a) What is the value of $erf(0)$?

(b) What is the relationship between $erf(x)$ and $erf(-x)$?

(c) Use a computer/calculator program that performs numerical integration to obtain approximate values (say to 3 decimal places) for $erf(x)$ at $x = 1$, 2, and 3. Use this information, and the results from parts (a) and (b), to plot $erf(x)$ in the interval $[-3, 3]$. How confident are you that the graph you have is fairly accurate?

(d) Now repeat part (c) for $x = 0.5$, 1.5, and 2.5. Did this change the accuracy of your previous graph for $erf(x)$?

(e) Now repeat part (c) for $x = 0.25$, 0.75, 1.25, and 1.75. Did this change the accuracy of your previous graph for $erf(x)$?

(f) What do you think happens to $erf(x)$ as $x \to \infty$?

3. The Fresnel Sine Integral. Using (1.11), write down an integral that represents the solution of the initial value problem

$$\frac{dy}{dx} = \sqrt{\frac{2}{\pi}} \sin x^2, \qquad y(0) = 0.$$

This solution, known as the Fresnel Sine Integral and denoted by $S(x)$, cannot be expressed in terms of familiar functions.

(a) What is the value of $S(0)$?

(b) What is the relationship between $S(x)$ and $S(-x)$?

(c) Use a computer/calculator program that performs numerical integration to obtain approximate values (say to 3 decimal places) for $S(x)$ at $x = 2$ and 4. Use this information, and the results from parts (a) and (b), to plot $S(x)$ in the interval $[-5, 5]$. How confident are you that the graph you have is fairly accurate?

(d) Now repeat part (c) for $x = 1$, 3, and 5. Did this change the accuracy of your previous graph for $S(x)$?

(e) Now repeat part (c) for $x = 0.5$, 1.5, 2.5, 3.5, and 4.5. Did this change the accuracy of your previous graph for $S(x)$?

(f) What do you think happens to $S(x)$ as $x \to \infty$?

4. The Sine Integral. Using (1.11), write down an integral that represents the solution of the initial value problem $\frac{dy}{dx} = g(x)$, $y(0) = 0$, where

$$g(x) = \begin{cases} \sin x/x & \text{if } x \neq 0 \\ 1 & \text{if } x = 0 \end{cases}.$$

This solution, known as the Sine Integral and denoted by $Si(x)$, cannot be expressed in terms of familiar functions. Use the ideas from Exercise 3 to graph the solution of this differential equation. What do you think happens to $Si(x)$ as $x \to \infty$?

5. Find the family of solutions for each of the differential equations $\frac{dy}{dx} = e^x$ and $\frac{dy}{dx} = -e^{-x}$. Graph these two families of solutions on one plot, using the same scale for the x- and y-axes. What do you notice about the angle of intersection between these two families of solutions?[7] Could you have seen that directly from the differential equations, without solving them?

6. Write down some odd functions and find their antiderivatives.[8] What property do these antiderivatives share? Make a conjecture that starts: *The antiderivative of an odd function is always ....* Prove your conjecture.

1.2 GRAPHICAL SOLUTIONS USING CALCULUS

In the previous section we used antiderivatives to determine the behavior of solutions of $dy/dx = g(x)$ by finding the explicit solution. In this section we discover that

[7]The angle of intersection between two curves at a point is the angle between the tangent lines to the curves at that point.

[8]Odd functions have the property that $f(-x) = -f(x)$ for all values of x in the domain of f.

there is a wealth of information available about the behavior of such solutions by considering the differential equation itself without finding the explicit solution.

EXAMPLE 1.4 *The Natural Logarithm Function*

We return to the second example,

$$y' = \frac{dy}{dx} = \frac{1}{x}, \tag{1.16}$$

where $'$ indicates differentiation with respect to x. Look at the 14 particular solutions in Figure 1.2. They were sketched directly from the functions $\ln x + C$ for $x > 0$, and $\ln(-x) + C$ for $x < 0$, for different values of C. Based on Figure 1.2, we ask the following questions:

1. *Monotonicity.*[9] Where are the solutions increasing and where are they decreasing?
2. *Concavity.* Where are the solutions concave up and where are they concave down?
3. *Symmetry.* Are there any symmetries?
4. *Singularities.*[10] Is it possible to start on a solution curve where $x < 0$ and proceed along this curve and eventually arrive at positive values of x?
5. *Uniqueness.* Do any solutions intersect?

Based on the graphs in Figure 1.2, the answers to these questions seem to be:

1. *Monotonicity.* Decreasing for $x < 0$ and increasing for $x > 0$.
2. *Concavity.* Concave down for $x \neq 0$.
3. *Symmetry.* Yes — across the y-axis. However, no particular solution has this symmetry. **It is the family of solutions that has this symmetry.**
4. *Singularities.* Not for any particular solution if the y-axis is a vertical asymptote, as it appears to be.
5. *Uniqueness.* On this graph, the answer appears to be yes — near the y-axis.

Now imagine that we are unable to integrate (1.16) explicitly, and so we are unable to draw the particular solutions in Figure 1.2. How much of this information (monotonicity, concavity, symmetry, singularities, and uniqueness) can we obtain directly from the differential equation (1.16) using our knowledge of calculus?

From calculus we know that $y' > 0$ on an interval requires that y be increasing on that interval, and $y' < 0$ means that y is decreasing. We also know that $y'' > 0$ on an interval requires the function to be concave up on that interval, whereas $y'' < 0$ means the function is concave down. In Figure 1.4 we show the general shapes of solution curves for the four cases: $y' < 0$, $y'' > 0$; $y' > 0$, $y'' > 0$; $y' > 0$, $y'' < 0$; and $y' < 0$, $y'' < 0$.

---

[9]A function is monotonic on an interval if it is either increasing on the entire interval or decreasing on the entire interval.

[10]A function $f(x)$ is singular at $x = a$ if $\lim_{x \to a} f(x)$ does not exist.

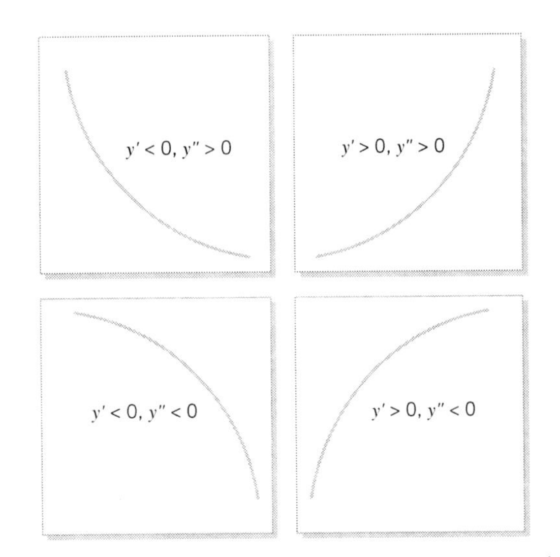

FIGURE 1.4 Possible shapes of solution curves determined by the first and second derivatives

With this information, let's return to our original questions and try to justify these answers.

1. *Monotonicity*. From (1.16) we see that the derivative of $y(x)$ is positive for $x > 0$, and so y increases there. Similar reasoning shows that y decreases when $x < 0$.

2. *Concavity*. If we differentiate (1.16) with respect to x, we have

$$y'' = \frac{d^2y}{dx^2} = -\frac{1}{x^2},$$

which is negative for $x \neq 0$. Thus, the solutions must be concave down for all $x \neq 0$.

3. *Symmetry*. To be symmetric across the y-axis means that we have no change in the family of solutions if x is replaced by $-x$ on both sides of (1.16). If in (1.16) we replace x by $-x$ we have

$$\frac{dy}{-dx} = \frac{1}{-x},$$

or $y' = 1/x$, which is exactly (1.16). Thus, the family of solutions that satisfies (1.16) is unchanged by the interchange of x with $-x$, and so must be symmetric across the y-axis.

4. *Singularities*. Because (1.16) is undefined at $x = 0$, we anticipate problems at $x = 0$.

5. *Uniqueness*. The statement that two solutions intersect means that there is a common point (x_0, y_0) through which two distinct particular solutions of (1.16), say $y_1(x)$ and $y_2(x)$, pass. Because both y_1 and y_2 are solutions of (1.16), we must have $y_1' = 1/x$ and $y_2' = 1/x$, so that $y_1' = y_2'$, or $\left(y_1 - y_2\right)' = 0$. From this we have $y_1(x) - y_2(x) = C$. The fact that $y_0 = y_1(x_0)$ and $y_0 = y_2(x_0)$ requires that $C = 0$,

so that $y_1(x) = y_2(x)$. In other words, the two curves $y_1(x)$ and $y_2(x)$ are one and the same. This means that only one solution of (1.16) can pass through any point (x_0, y_0). Another way of saying this is that a solution of the differential equation (1.16) that passes through any given point is unique. Consequently, contrary to our conjecture based on Figure 1.2, solutions do not intersect. In fact, this argument can be used on any differential equation of the form $y' = g(x)$ to show that their solutions cannot intersect (see Exercise 7 on page 11).

Caution! **Exercise care when drawing conclusions from graphical analysis about whether curves intersect.**

From the preceding analysis we see that just by using calculus we can obtain much qualitative information about solution curves without knowing the explicit solution. Let's see how we can use this to sketch the Error Function, *erf*(x), by going through the checklist we have developed.

EXAMPLE 1.5 *The Error Function*

Using the techniques of calculus, sketch the family of solutions of the differential equation

$$y' = \frac{2}{\sqrt{\pi}} e^{-x^2}. \tag{1.17}$$

1. *Monotonicity*. The derivative of y is always positive, so all solutions are increasing.
2. *Concavity*. If we differentiate (1.17) with respect to x, we find $y'' = -\frac{4}{\sqrt{\pi}} x e^{-x^2}$. From this we see that $y'' > 0$ when $x < 0$, and $y'' < 0$ when $x > 0$. Thus, all solutions are concave up when $x < 0$ and concave down when $x > 0$.
3. *Symmetry*. If we replace x with $-x$ on both sides of (1.17), the right-hand side is unchanged but the left-hand side changes sign. So the family of solutions is not symmetric across the y-axis. However, if we simultaneously replace x with $-x$ and replace y with $-y$, then we obtain (1.17) back again. So the family of solutions of (1.17) is unchanged under simultaneous interchange of x with $-x$ and y with $-y$. This means that the family of solutions is symmetric about the origin.[11]
4. *Singularities*. There are no obvious points where the derivative fails to exist.
5. *Uniqueness*. From arguments similar to those at the end of Example 1.4 on page 8, we see that solutions cannot intersect.

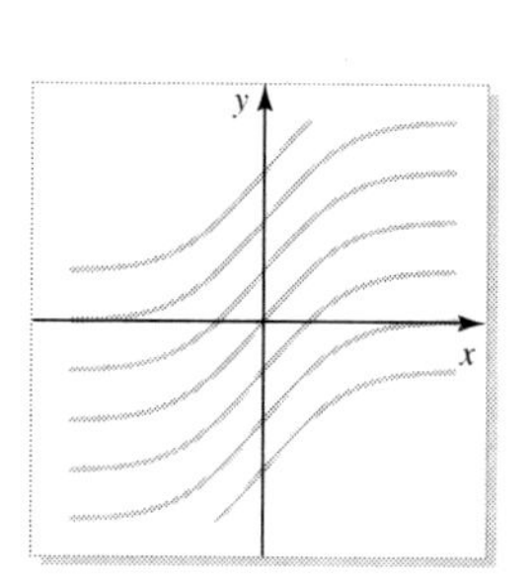

FIGURE 1.5
Solution curves of $y' = 2e^{-x^2}/\pi^{1/2}$

Based on this qualitative information, we can sketch by hand the family of solutions of (1.17), which is shown in Figure 1.5.[12] This family of solutions contains the particular solution curve that passes through the point with coordinates (0, 0) – namely, *erf*(x).

[11] Recall that a graph is symmetric about the origin if it is unchanged when rotated 180° about the origin.

[12] Throughout the text we make reference to hand-drawn solutions. Of course, they were drawn by machine.

EXERCISES

1. Use monotonicity, concavity, symmetry, singularities, and uniqueness to sketch various solution curves for each of the following first order differential equations. Then draw the particular solution curve that passes through the point P. When you have finished, compare your answers with those you found for Exercise 1, Section 1.1.

(a) $y' = x^3$ $\quad P = (1, 1)$
(b) $y' = x^4$ $\quad P = (1, 1)$
(c) $y' = \cos x$ $\quad P = (0, 0)$
(d) $y' = \sin x$ $\quad P = (\pi, 2)$
(e) $y' = e^{-x}$ $\quad P = (0, 1)$
(f) $y' = 1/x^2$ $\quad P = (1, 1)$
(g) $y' = 1/x$ $\quad P = (-1, 1)$
(h) $y' = 1/(1 + x^2)$ $\quad P = (1, \pi/4)$
(i) $y' = 1/[x(1 - x)]$ $\quad P = (2, 1)$
(j) $y' = \ln x$ $\quad P = (1, 1)$
(k) $y' = x^2 e^{-x}$ $\quad P = (0, 1)$
(l) $y' = e^{-x} \sin x$ $\quad P = (0, 1)$

2. Intuition suggests that if $g(x) \to 0$ as $x \to \infty$ then the solutions of the differential equation $y' = g(x)$ will have horizontal asymptotes. Explain why this suggestion might be plausible. Explain why this suggestion is wrong.

3. For the differential equation

$$y' = \frac{4}{x\,(x-4)},$$

find the explicit solution satisfying the initial condition (a) $y(-1) = 0$, (b) $y(1) = 0$, (c) $y(5) = 0$.

4. For what values of a and x_0 is the solution of the initial value problem

$$y' = \frac{1}{x\,(x-a)}, \qquad y(x_0) = 0,$$

valid for all $x > 0$?

5. **The Fresnel Sine Integral.** Use monotonicity, concavity, symmetry, singularities, and uniqueness to sketch various solution curves for the differential equation $y' = \sqrt{\frac{2}{\pi}} \sin x^2$. Then draw the graph of the particular solution that satisfies $y(0) = 0$. What do you think happens to $y(x)$ as $x \to \infty$? Compare your answers with the one you found for Exercise 3, Section 1.1.

6. **The Sine Integral.** Use monotonicity, concavity, symmetry, singularities, and uniqueness to sketch various solution curves for the differential equation $y' = g(x)$, where

$$g(x) = \begin{cases} \sin x/x & \text{if } x \neq 0 \\ 1 & \text{if } x = 0 \end{cases}.$$

Then draw the graph of the particular solution that satisfies $y(0) = 0$. What do you think happens to $y(x)$ as $x \to \infty$? Compare your answers with the one you found for Exercise 4, Section 1.1.

7. **The Uniqueness Theorem.** Show that if $y_1(x)$ and $y_2(x)$ are solutions of the initial value problem $y' = g(x)$, $y(x_0) = y_0$, where $g(x)$ is continuous, then $y_1(x) = y_2(x)$. How does this guarantee that different solutions of the differential equation $y' = g(x)$ cannot intersect?

8. Show that the family of antiderivatives of an even function is symmetric about the origin.[13] Under what conditions will an antiderivative of an even function be an odd function? Give some examples.

1.3 SLOPE FIELDS AND ISOCLINES

In the previous section we saw that using the techniques of calculus, we can determine much qualitative information about solutions of $y' = g(x)$ from the signs of the first and second derivatives. However, there is still more information contained

[13] Even functions have the property that $f(-x) = f(x)$ for all values of x in the domain of f.

in the differential equation, because, in addition to the signs, it also gives us the magnitude of the slope at each point on a solution curve. In this section we exploit this information.

What Are Slope Fields?

EXAMPLE 1.6 *Lines*

We start with a very simple differential equation that describes the function whose rate of change is always 1, namely,

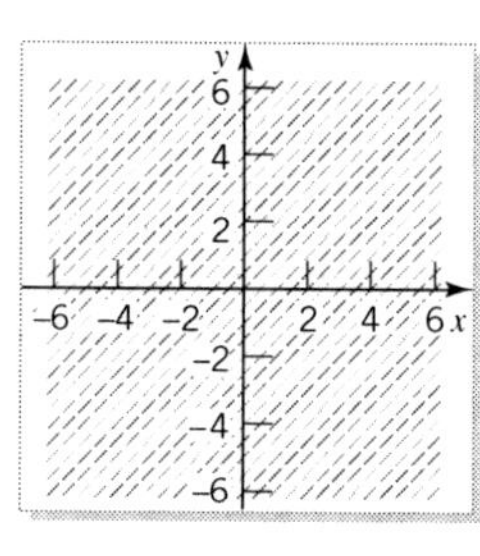

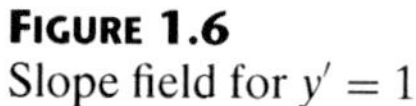

FIGURE 1.6
Slope field for $y' = 1$

$$y' = 1. \tag{1.18}$$

Let's use our knowledge of calculus to sketch some solution curves of (1.18). Because the right-hand side of (1.18) is positive (namely, 1), we know that the solutions of (1.18) are increasing everywhere. From (1.18) we also know that for all values of x and y, the solution of this differential equation has a tangent line whose slope is 1. To transfer this information to a graph we can select various coordinates (x, y) and draw short line segments with slope 1, as shown in Figure 1.6.

From calculus we know that a differentiable function may be approximated near a point on the curve by its tangent line at that point. Another way of stating this is that each tangent segment gives the slope of the solution of a differential equation at that point. Such a collection of short line segments is known as a SLOPE FIELD of the differential equation, as it gives a short segment of the tangent line to the solution curve at each selected point.[14]

Slope field

We now construct a solution curve such that the tangent lines to this curve are consistent with the slope field. If we try to draw a curve whose tangent line has the slope 1 everywhere, we will end up drawing a straight line with slope 1. In fact the solution curves of (1.18) are the family of straight lines $y = x + C$. ■

How to Sketch the Slope Field for $y' = g(x, y)$

Purpose To sketch the slope field for $y' = g(x, y)$.

Process

1. Select a rectangular window in the xy-plane in which to view the slope field.
2. Subdivide this rectangular region into a grid of equally spaced points (x, y). The number of points in the x and y directions may be different.
3. At each of these points (x, y), find the numerical value of $g(x, y)$ and draw a short line segment at (x, y) with slope $g(x, y)$.

---

[14] Slope fields are sometimes called direction fields.

Comments about Slope Fields

- All the slope fields in this book were computer generated.
- **From now on we assume you either have access to a computer/calculator program that displays slope fields or are willing to construct slope fields by hand.**

EXAMPLE 1.7 *Parabolas*

Now consider the differential equation that models the situation where the rate of change of the unknown function is equal to the value of the independent variable,

$$y' = x. \tag{1.19}$$

This is the first example we considered in Section 1.1.

Monotonicity, concavity, symmetry

Again using our knowledge of calculus, we see that (1.19) tells us that solutions $y = y(x)$ increase if $x > 0$ and decrease if $x < 0$. Moreover, because $y'' = 1$, the second derivative of y is always positive, so y is concave up everywhere. Finally, if we replace x by $-x$ in (1.19), the differential equation remains unchanged, so the family of solutions is symmetric across the y-axis.

Singularities and uniqueness

Because the right-hand side of (1.19) is defined for all values of x, there are no singularities. As shown in Exercise 7 on page 11, all solutions of differential equations of the form $y' = g(x), y(x_0) = y_0$ are unique if $g(x)$ is continuous, so distinct solutions of (1.19) do not intersect.

Slope field

If we construct the slope field as shown in Figure 1.7, we can obtain more information: we have a zero slope when $x = 0$, and the slopes of the short line segments of the slope field increase as x increases. Notice that the slope field appears to be symmetric across the y-axis.

Because the slope field for a differential equation gives the inclination of the tangent line to solutions at many points, the graph of any particular solution of this differential equation must be consistent with these tangent lines. Notice that the solution curves of (1.19) have horizontal tangents for $x = 0$, positive slopes for $x > 0$, and negative slopes for $x < 0$. Also note that these slopes become larger as x increases.

Drawing solution curves

To manually draw a solution curve on the graph of the slope field, we start at some point; for instance, where we have already drawn a short tangent line. Because the solution will be a differentiable function, its graph is well approximated by its tangent line near every point. Thus, we may proceed in the direction given by this tangent line for a short distance to the right and then see what the slope field looks like there. We then adjust the direction of our curve so it changes in a manner consistent with this slope field.

To show this for a specific case in Figure 1.7, consider the solution curve that passes through the point $(0, 1)$. As we move to the right from this point, the curve changes from being horizontal in such a manner that the slope is continually increasing. This gives the curve labeled A shown in Figure 1.8. (Note that this results in a curve that is concave up.) Figure 1.8 also shows some other hand-drawn solution curves (all of which are parabolas), each having a different y-intercept.

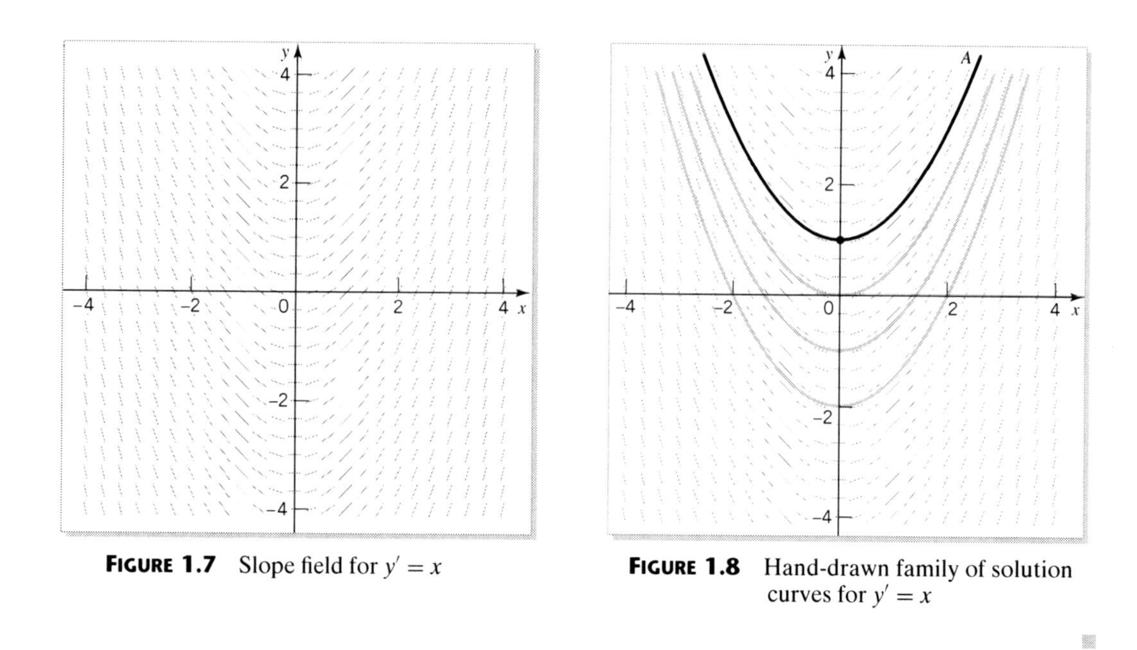

FIGURE 1.7 Slope field for $y' = x$

FIGURE 1.8 Hand-drawn family of solution curves for $y' = x$

How to Manually Sketch Solution Curves from the Slope Field for $y' = g(x, y)$

Purpose To sketch, by hand, solution curves from the slope field for

$$y' = g(x, y). \tag{1.20}$$

Process

1. Sketch the slope field for (1.20).
2. Start at some initial point and put a dot there. If this dot lies on a short line segment, you have the slope of your solution curve at that point. If not, estimate the value of the slope of the tangent line at that point by looking at nearby slopes. This gives the direction of the slope field at that point.
3. Proceed in this direction for a short distance to the right. Place a dot at the point where you finish.
4. Adjust your direction so it is consistent with the direction of the slope field in the vicinity of the point where you finished.
5. Repeat steps 3 and 4, as often as needed, joining the dots with a curve.
6. Start with a new initial point, and return to step 2.

EXAMPLE 1.8 *The Natural Logarithm Function*

Once more we return to the differential equation

$$y' = \frac{1}{x}. \tag{1.21}$$

Based on our previous analysis we know we have a concave down, decreasing shape for $x < 0$, and a concave down, increasing shape for $x > 0$. We also know the family of solutions will be symmetric across the y-axis.

Slope field We now draw the slope field for (1.21) (see Figure 1.9) and from it confirm the major properties of its solution. We see from Figure 1.9 that the solution curves consistent with this slope field are increasing for $x > 0$ and decreasing for $x < 0$ and that all the slopes above a specific x location are equal. Figure 1.10 shows a few solution curves drawn on this slope field. Note that the solution curves are concave down everywhere, and that the slope field appears to be symmetric across the y-axis. Also note that the solution curves appear to be vertical translations of each other. You should measure the vertical distances between the curves to verify this conjecture.

Vertical translation

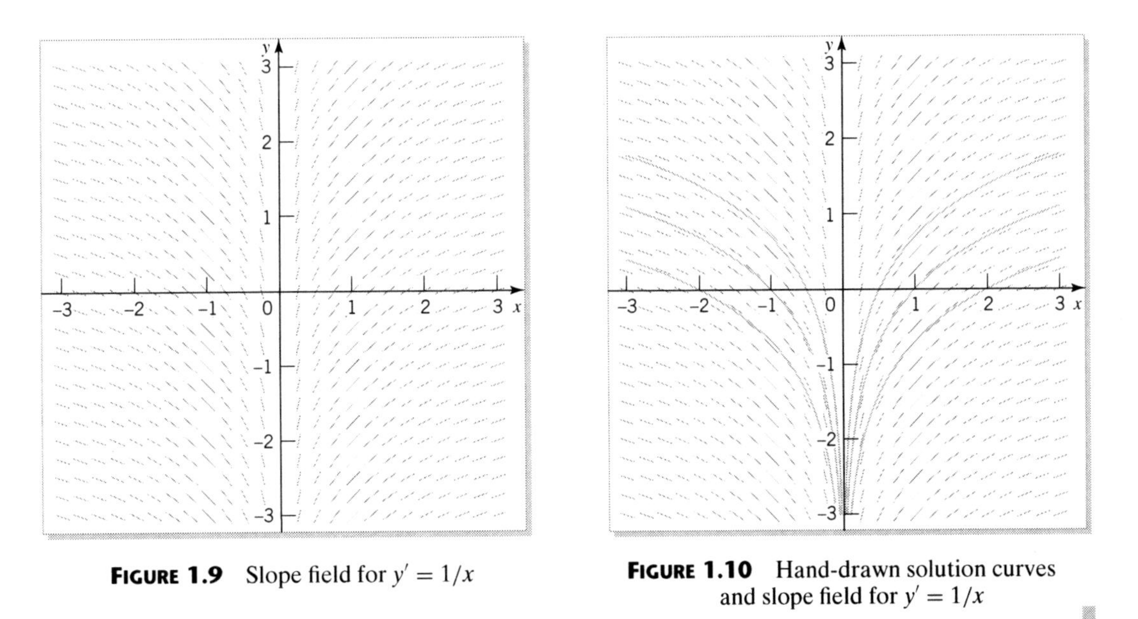

FIGURE 1.9 Slope field for $y' = 1/x$

FIGURE 1.10 Hand-drawn solution curves and slope field for $y' = 1/x$

Now that we have some solution curves on the slope field in Figure 1.10, it is apparent that **we can think of a slope field as what remains after plotting many solution curves and then erasing parts of them, leaving only some short segments here and there that look like straight lines**. In this sense, the challenge in finding solution curves from a slope field is to fill in the gaps between the short line segments of the slope field. This is a major use of slope fields — namely, to determine the graph of a solution of a differential equation whether or not an explicit solution is readily obtainable in terms of familiar functions. Slope fields are also used to check consistency with your findings concerning monotonicity and concavity.

EXAMPLE 1.9 *The Error Function*

We already have an example where we needed to draw a solution curve of an initial value problem without having an explicit solution in terms of familiar functions — namely, the initial value problem that generates the Error Function,

$$y' = \frac{2}{\sqrt{\pi}} e^{-x^2}, \qquad y(0) = 0. \tag{1.22}$$

FIGURE 1.11 Slope field for $y' = 2e^{-x^2}/\pi^{1/2}$

Because we want the solution of this equation that starts at the point $(0, 0)$, we construct the slope field that includes this point, as shown in Figure 1.11.

Monotonicity, concavity, symmetry

As expected, the slope field indicates that the solution curve that passes through $(0, 0)$ is increasing everywhere, concave up when $x < 0$ and concave down when $x > 0$. Also notice that the slope field appears to be symmetric about the origin. Figure 1.12 shows a hand-drawn solution curve for (1.22) that passes through the origin, the graph of $y = erf(x)$.

We could also use a numerical integration technique to obtain values for $erf(x)$ at different values of x from its definition, namely,

$$erf(x) = \frac{2}{\sqrt{\pi}} \int_0^x e^{-t^2}\, dt.$$

Table 1.1
Simpson's rule for *erf*(x)

x	$y(x)$
0.0	0.000
0.5	0.520
1.0	0.843
1.5	0.966
2.0	0.995

For example, we used Simpson's rule on this integral to create Table 1.1.[15] (Here we set the number of subintervals to 16 and rounded the answers to three decimal places.) Figure 1.13 shows the slope field, these numerical values, and a hand-drawn solution curve. Notice the agreement between this solution curve and a plot of these numerical values.

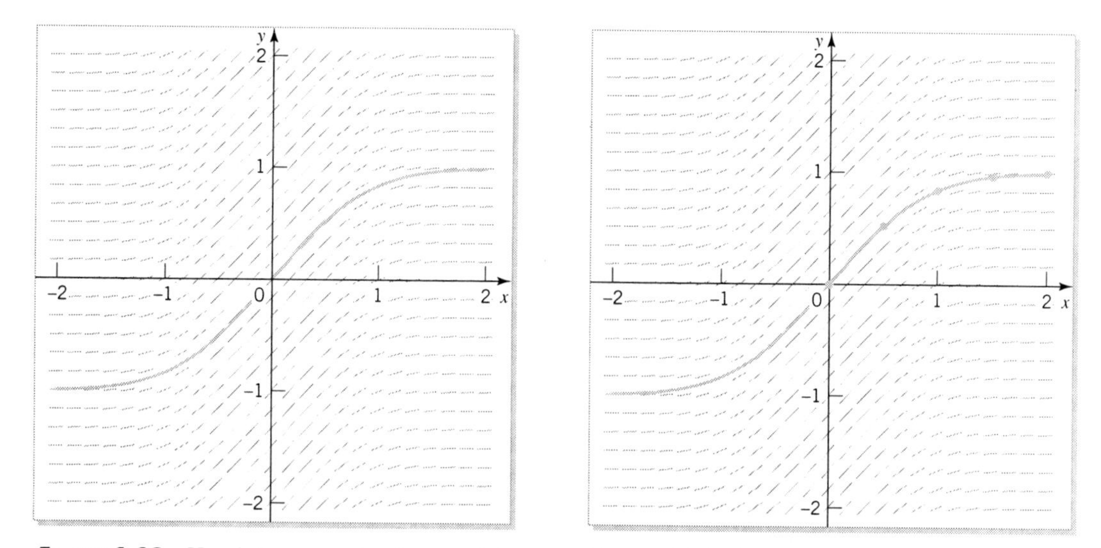

FIGURE 1.12 Hand-drawn graph of $y = erf(x)$

FIGURE 1.13 Numerical values and hand-drawn graph of $y = erf(x)$

[15] See a calculus text to remind yourself of Simpson's rule.

Caution! **Slope fields can sometimes be misleading — see Exercise 8 on page 21. We must make sure that any conclusions drawn from slope fields are confirmed by other means.** One way is to make use of the analytical and graphical techniques learned from the previous two sections, where the first and second derivatives of a function give us information about the function itself.

What Are Isoclines?

We now exploit the slope field concept in a different way, to obtain additional properties of solution curves directly from the differential equation.

EXAMPLE 1.10 *The Natural Logarithm Function*

We start by reconsidering the differential equation

$$y' = \frac{1}{x}, \tag{1.23}$$

Slope field

which has the slope field shown in Figure 1.9. Our objective is to show that this slope field is reasonable.

In calculus, a first step in plotting the graph of a function was to compute its derivative, but (1.23) supplies the derivative of all the solution curves without any further work. In calculus we then set the derivative equal to 0 to find horizontal tangents. Because the derivative in this case, $1/x$, is never equal to 0, no solution curve has a horizontal tangent.

Even though there are no points on any of our solution curves that have a horizontal tangent, we can set the derivative equal to another constant and see for what value (or values) of x the solution curves would have that constant slope. For example, because $1/x = 1$ for $x = 1$, the short line segments will have a slope of 1 at all points on the slope field where x equals 1. In general we can say that solution curves will have a slope equal to m whenever

$$\frac{1}{x} = m,$$

that is,

$$x = \frac{1}{m},$$

which is a vertical line through the point $(1/m, 0)$.

Isoclines

This vertical line is called an ISOCLINE (equal inclination) of the differential equation (1.23). All solution curves of (1.23) will have the same slope, m, as they cross the isocline at this value of x. For example, the solution curves of (1.23) will have a slope of 1 when $x = 1$, a slope of $1/2$ when $x = 2$, a slope of 2 when $x = 1/2$, a slope of -1 when $x = -1$, a slope of $-1/2$ when $x = -2$, and a slope of -2 when $x = -1/2$. We can see that Figure 1.14 is consistent with this information, which shows isoclines for $m = \pm 1/2$ and $m = \pm 1$. To make sure you understand isoclines, add the isoclines for $m = 2$ and $m = -2$ to this figure. (Is there an isocline for $m = 0$?)

We hope that with this additional information you feel very confident in drawing solution curves consistent with the slopes, the isoclines, the monotonicity, and the concavity we have determined (Figure 1.15).

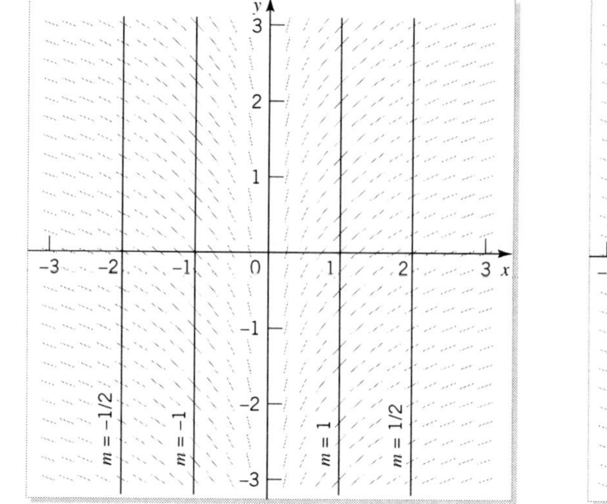

FIGURE 1.14 Isoclines ($m = \pm 1/2, \pm 1$) and slope field for $y' = 1/x$

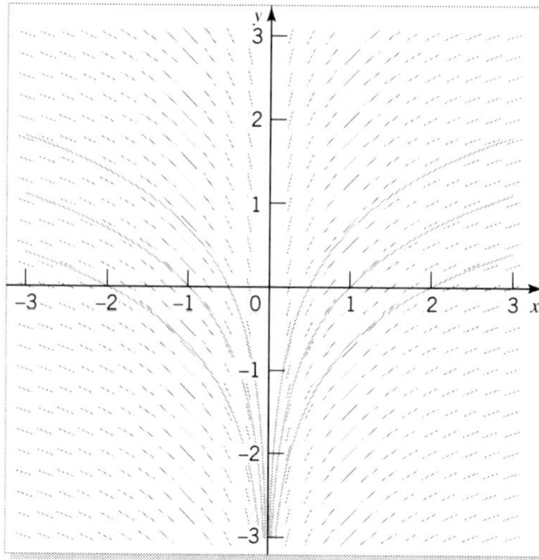

FIGURE 1.15 Hand-drawn solution curves and slope field for $y' = 1/x$

◆ *Definition 1.3:* **An ISOCLINE CORRESPONDING TO SLOPE m of the differential equation $y' = g(x, y)$ is the curve characterized by the equation $g(x, y) = m$.** ◆

How to Sketch Isoclines for $y' = g(x, y)$

Purpose To sketch isoclines for $y' = g(x, y)$.

Process

1. Set

$$g(x, y) = m, \tag{1.24}$$

where m is constant.

2. Pick several different values for m. For each m try to solve (1.24) for y in terms of x and m, or for x in terms of y and m. If this cannot be done, try to identify the curves defined implicitly by (1.24). This gives the isocline corresponding to slope m.
3. For each value of m, plot the isocline corresponding to slope m.
4. If you are constructing slope fields by hand, draw short line segments with slope m crossing the appropriate isocline.

Comments about Isoclines

- For any particular m, the isocline corresponding to slope m may consist of more than one curve.
- An isocline corresponding to slope m is also called an isocline for slope m.
- If $g(x, y)$ does not include y, isoclines are vertical lines.
- In the general case where $g(x, y)$ depends on both x and y, isoclines may not be lines. For example, if $g(x, y) = x^2 + y^2$ then the isoclines $x^2 + y^2 = m$ are circles centered at the origin with radius $\sqrt{m}$.

EXAMPLE 1.11 *The Error Function*

Now we return to the differential equation giving rise to the Error Function,

$$y' = \frac{2}{\sqrt{\pi}} e^{-x^2}.$$

Isocline The isocline corresponding to slope m is given by

$$\frac{2}{\sqrt{\pi}} e^{-x^2} = m. \tag{1.25}$$

Notice that this guarantees that there are no isoclines for slope $m \leq 0$ or for slope $m > 2/\sqrt{\pi} \approx 1.128$. (Why?) If we solve (1.25) for x we obtain $x = \pm \left\{\ln\left[2/\left(m\sqrt{\pi}\right)\right]\right\}^{1/2}$ as the equation of the isocline corresponding to slope m. The isoclines for slope 0.1, 0.3, and 0.7 are shown in Figure 1.16. Notice that in this case each isocline consists of two vertical lines.

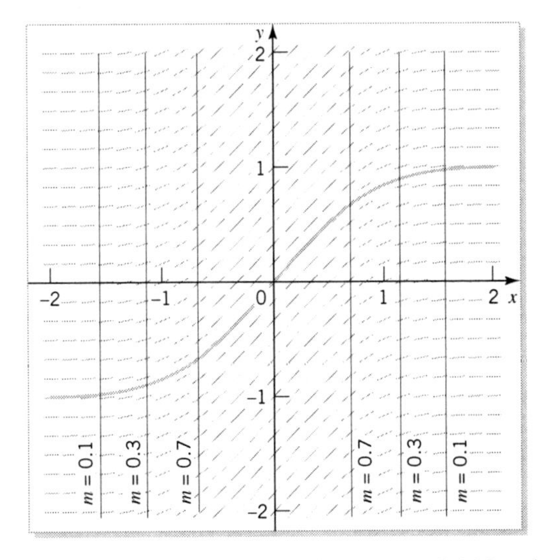

FIGURE 1.16 Isoclines ($m = 0.1, 0.3, 0.7$) and slope field for $y' = 2e^{-x^2}/\pi^{1/2}$

To summarize results found to this point, we gather together some **general observations about differential equations of the special form** $y' = g(x)$.

Summary

- All solutions are explicit.
- Once we have found one member of the family of solutions, other members of the family can be generated from this member by vertical translations.
- If $y' = g(x)$ remains unchanged after the interchange of x with $-x$, then the family of solutions is symmetric across the y-axis.
- If $y' = g(x)$ remains unchanged after the simultaneous interchange of y with $-y$ and x with $-x$, then the family of solutions is symmetric about the origin.
- For the case $y' = g(x)$, all isoclines are vertical lines; that is, parallel to the y-axis.

EXERCISES

1. Sketch the slope field for each of the following first order differential equations. In each case draw some isoclines to confirm your sketch. Use your sketch to draw various solution curves. Then draw the solution curve that passes through the point P. When you have finished, compare your answers with those you found for Exercise 1, Section 1.1, and Exercise 1, Section 1.2.

(a) $y' = x^3$ $\quad P = (1, 1)$
(b) $y' = x^4$ $\quad P = (1, 1)$
(c) $y' = \cos x$ $\quad P = (0, 0)$
(d) $y' = \sin x$ $\quad P = (\pi, 2)$
(e) $y' = e^{-x}$ $\quad P = (0, 1)$
(f) $y' = 1/x^2$ $\quad P = (1, 1)$
(g) $y' = 1/x$ $\quad P = (-1, 1)$
(h) $y' = 1/(1 + x^2)$ $\quad P = (1, \pi/4)$
(i) $y' = 1/[x(1 - x)]$ $\quad P = (2, 1)$
(j) $y' = \ln x$ $\quad P = (1, 1)$
(k) $y' = x^2e^{-x}$ $\quad P = (0, 1)$
(l) $y' = e^{-x}\sin x$ $\quad P = (0, 1)$

2. Explain why it is useful to plot isoclines corresponding to an infinite slope, even though no point on a solution curve can have a vertical tangent.

3. The Fresnel Sine Integral. Use slope fields and isoclines for the differential equation $y' = \sqrt{\frac{2}{\pi}}\sin x^2$ to draw various solution curves. Then draw the solution curve that satisfies $y(0) = 0$. What do you think happens to $y(x)$ as $x \to \infty$? Compare your answers with those you found for Exercise 3, Section 1.1, and Exercise 5, Section 1.2.

4. The Sine Integral. Consider the differential equation $y' = g(x)$, where

$$g(x) = \begin{cases} \sin x/x & \text{if } x \neq 0 \\ 1 & \text{if } x = 0 \end{cases}.$$

Use slope fields and isoclines to draw various solution curves. Then draw the solution curve that satisfies $y(0) = 0$. What do you think happens to $y(x)$ as $x \to \infty$? Compare your answers with those you found for Exercise 4, Section 1.1, and Exercise 6, Section 1.2.

5. Figure 1.17 shows one member of a family of solutions of the differential equation $y' = g(x)$, where $g(x)$ is a given function.

(a) Use this information to plot other members of the family of solutions. Do not attempt to find $y(x)$ or $g(x)$.

(b) Can every solution of the differential equation be obtained by the technique used in part (a)?

6. Figures 1.18 and 1.19 are mystery slope fields, believed to be the slope fields for two of the following differential equations:

$$y' = \frac{x^2 + 1}{x^2 - 1}, \qquad y' = \frac{x^2 - 1}{x^2 + 1},$$

$$y' = -\frac{x^2 + 1}{x^2 - 1}, \qquad y' = -\frac{x^2 - 1}{x^2 + 1}.$$

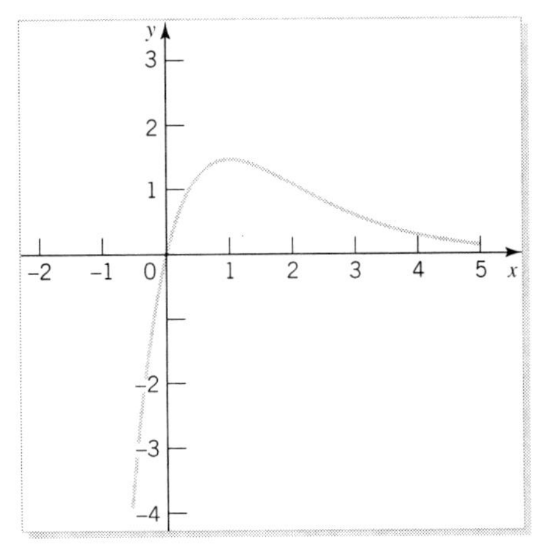

FIGURE 1.17 Graph of $y(x)$

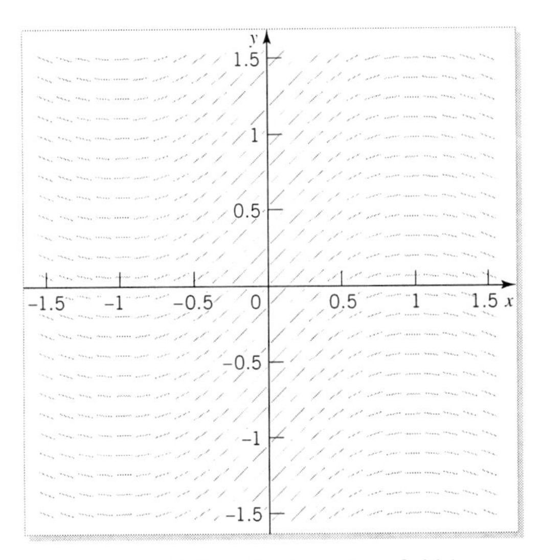

FIGURE 1.18 Mystery slope field 1

(a) Identify to which differential equation each of the mystery slope fields belongs. Confirm all the information using calculus and isoclines. Do not plot any slope fields to answer this question.

(b) Now superimpose the two mystery slope fields, perhaps by placing one on top of the other and holding both up to the light. [Another way to do this is to plot the slope fields for

$$y' = a\frac{x^2+1}{x^2-1} + b\frac{x^2-1}{x^2+1}$$

with $a = \pm 1$ and $b = 0$, and then with $a = 0$ and $b = \pm 1$.] What do you notice? Would this have made your previous analysis in part (a) easier?

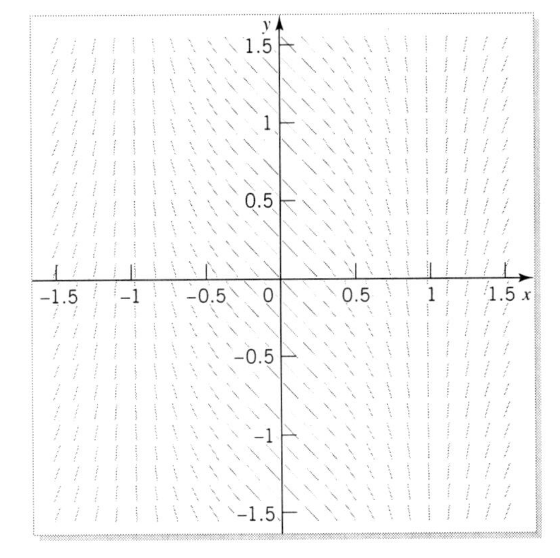

FIGURE 1.19 Mystery slope field 2

7. Consider the following four differential equations:

$$y' = x + 1, \quad y' = x - 1, \quad y' = \ln|x|, \quad y' = x^2 - 1.$$

(a) The slope field of one of the preceding equations is given in Figure 1.20. Match the correct equation with the figure, carefully stating your reasons. Do not plot any slope fields to answer this question.

(b) Briefly outline a general strategy for this matching process.

8. Use a computer/calculator program to sketch the slope field for $y' = 1 + \cos(1000x)$ in the window $-10 < x <$

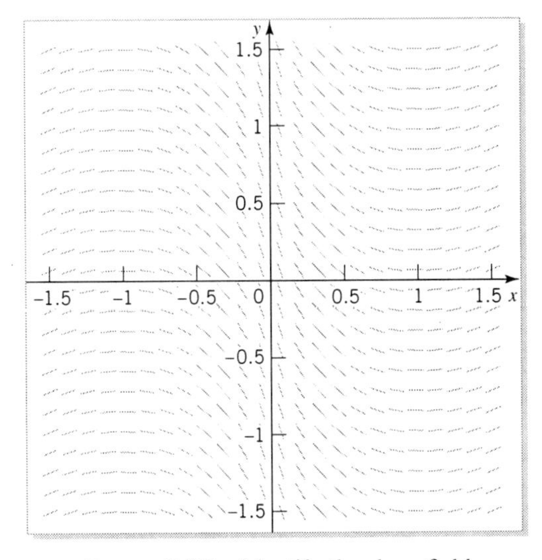

FIGURE 1.20 Identify the slope field

$10, -10 < y < 10$. Use your sketch to draw the solution curve that passes through the point $(0, 0)$. Now solve the differential equation and find the formula for the particular explicit solution that passes through the point $(0, 0)$. Plot this solution on top of your previous sketches, and comment on what you see. What lesson can be learned from this exercise?

9. If a particular object falls out of an airplane, its downward velocity after x seconds is crudely approximated by

$$y' = \frac{g}{k}\left(1 - e^{-kx}\right),$$

where $g = 9.8$ m/sec^2 and $k = 0.2$ sec^{-1}. Here $y(x)$ is the distance fallen at time x so $y(0) = 0$. If this object falls from 5000 meters above the ground, estimate how many seconds it falls before it hits the ground, by

(a) using slope fields, monotonicity, isoclines, and concavity, and

(b) finding the explicit solution.

1.4 FUNCTIONS AND POWER SERIES EXPANSIONS

So far in this chapter we have developed methods for sketching solution curves of $y' = g(x)$ that pass through specific points. For the case when we were unable to discover an antiderivative in terms of familiar functions, we still could write down the explicit solution

$$y(x) = \int_{x_0}^{x} g(t)\, dt + y_0$$

that satisfied the initial condition $y(x_0) = y_0$. To evaluate $y(x)$ at specific values of x from this expression requires the use of a numerical approximation for the definite integral.

However, there is an alternative expression for this explicit solution that takes the form of an infinite series. To obtain such an expression, we expand the integrand $g(t)$ in a Taylor series,[16] and then integrate the resulting powers of t term by term.

EXAMPLE 1.12 *The Error Function*

To illustrate this procedure, we consider the Error Function defined by

$$erf(x) = \frac{2}{\sqrt{\pi}} \int_0^x e^{-t^2}\, dt. \tag{1.26}$$

First recall that the function e^x has the Taylor series expansion[17]

$$e^x = \sum_{k=0}^{\infty} \frac{x^k}{k!} = 1 + x + \frac{x^2}{2!} + \frac{x^3}{3!} + \frac{x^4}{4!} + \frac{x^5}{5!} + \cdots, \tag{1.27}$$

[16] A summary of Taylor series results is in Appendix A.3.

[17] Remember $0! = 1$, $n! = n(n-1)(n-2)\cdots 3 \cdot 2 \cdot 1$.

valid for all x. If we replace x by $-t^2$ in (1.27), we find

$$e^{-t^2} = \sum_{k=0}^{\infty}(-1)^k \frac{t^{2k}}{k!} = 1 - t^2 + \frac{t^4}{2!} - \frac{t^6}{3!} + \frac{t^8}{4!} - \frac{t^{10}}{5!} + \cdots,$$

which, when integrated from 0 to x, gives

$$\int_0^x e^{-t^2}\,dt = \sum_{k=0}^{\infty}(-1)^k \int_0^x \frac{t^{2k}}{k!}\,dt = \sum_{k=0}^{\infty}(-1)^k \frac{x^{2k+1}}{(2k+1)k!}. \tag{1.28}$$

(Recall that the integral of a convergent power series is also convergent.) Combining (1.26) and (1.28), we see that the Error Function can be written in the form

$$\begin{aligned} erf(x) &= \frac{2}{\sqrt{\pi}} \sum_{k=0}^{\infty}(-1)^k \frac{x^{2k+1}}{(2k+1)k!} \\ &= \frac{2}{\sqrt{\pi}}\left(x - \frac{x^3}{3} + \frac{x^5}{5\cdot 2!} - \frac{x^7}{7\cdot 3!} + \frac{x^9}{9\cdot 4!} - \frac{x^{11}}{11\cdot 5!} + \cdots\right), \end{aligned} \tag{1.29}$$

valid for all x. This is an alternative representation for the Error Function.

The first term on the right-hand side of (1.29), denoted by

$$P_1(x) = \frac{2}{\sqrt{\pi}}x,$$

is the Taylor polynomial of degree one for $erf(x)$. The sum of the first two terms on the right-hand side of (1.29), denoted by

$$P_3(x) = \frac{2}{\sqrt{\pi}}\left(x - \frac{x^3}{3}\right),$$

gives the Taylor polynomial of degree three for $erf(x)$, and so on. Figure 1.21 shows the graph of the Error Function together with the Taylor polynomials of degrees one, three, five, and so on to eleven — that is, $P_1(x), P_3(x), P_5(x), \cdots, P_{11}(x)$. Notice how the graph of the Error Function gradually emerges as the curve common to these graphs. It appears that $erf(x)$ is trapped by the other curves. Also notice that near the origin, $erf(x)$ behaves like the first term in the Taylor series (1.29), namely the straight line $y = 2x/\sqrt{\pi}$.

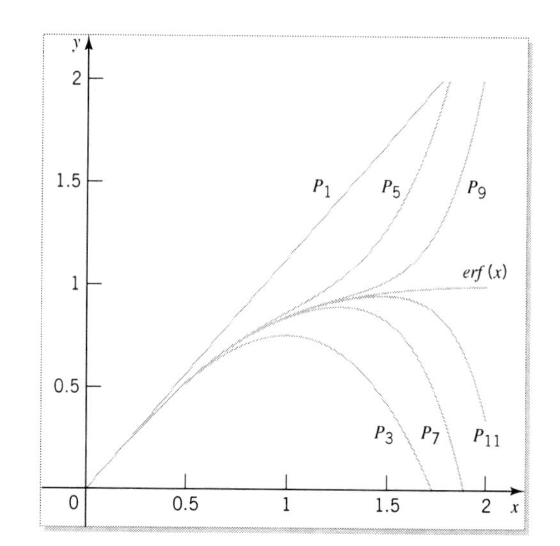

FIGURE 1.21 The Error Function trapped by its Taylor polynomials $P_1(x), P_3(x), P_5(x), \cdots, P_{11}(x)$

EXERCISES

1. **The Fresnel Sine Integral.** Calculate the power series expansion for the Fresnel Sine Integral, $S(x)$, defined as

$$S(x) = \sqrt{\frac{2}{\pi}} \int_0^x \sin t^2 \, dt.$$

Confirm that it is consistent with the answers you found for Exercise 3, Section 1.1; Exercise 5, Section 1.2; and Exercise 3, Section 1.3.

2. **The Sine Integral.** Consider the differential equation $y' = g(x)$, where

$$g(x) = \begin{cases} \sin x/x & \text{if } x \neq 0 \\ 1 & \text{if } x = 0 \end{cases}.$$

Explain why the solution of this differential equation can be written as

$$y(x) = \int_0^x \frac{\sin t}{t} \, dt,$$

paying particular attention to the lower limit of integration. Calculate the power series expansion for this function $y(x)$. Confirm that it is consistent with the answers you found for Exercise 4, Section 1.1; Exercise 6, Section 1.2; and Exercise 4, Section 1.3.

What Have We Learned?

Main Ideas

- A FIRST ORDER ORDINARY DIFFERENTIAL EQUATION is an equation that involves at most the first derivative of an unknown function. If y, the unknown function, is a function of x, then we write the first order equation as

$$\frac{dy}{dx} = g(x, y)$$

where $g(x, y)$ is a given function of x and y.

- An EXPLICIT SOLUTION of

$$\frac{dy}{dx} = g(x, y) \tag{1.30}$$

is any function $y = y(x)$ (differentiable in some interval $a < x < b$) that identically satisfies the differential equation (1.30).

- If an explicit solution contains an arbitrary constant, the infinite number of solutions it generates is called a FAMILY OF EXPLICIT SOLUTIONS. If an explicit solution contains no arbitrary constant, it is called a PARTICULAR EXPLICIT SOLUTION.
- The graph of a particular solution is called a SOLUTION CURVE of the differential equation.
- Looking for a solution of a differential equation that must pass through a given point is called an INITIAL VALUE PROBLEM, and the point is called an INITIAL VALUE, INITIAL CONDITION, or INITIAL POINT.
- There is one and only one solution of any differential equation of the form $y' = g(x)$ that passes through a given point (x_0, y_0) if $g(x)$ is continuous. See the Uniqueness Theorem on page 11.
- To sketch slope fields, see *How to Sketch the Slope Field for* $y' = g(x, y)$ on page 12.
- To hand-draw solution curves from slope fields, see *How to Manually Sketch Solution Curves from the Slope Field for* $y' = g(x, y)$ on page 14.
- An ISOCLINE CORRESPONDING TO SLOPE m of the differential equation $y' = g(x, y)$ is the curve characterized by the equation $g(x, y) = m$.
- To sketch isoclines, see *How to Sketch Isoclines for* $y' = g(x, y)$ on page 18.
- When the explicit solution is left in the form of a definite integral, it is frequently possible to put it in an alternative form. This is done by expanding the integrand using Taylor series, and then integrating the result term by term. See page 22.

Words of Caution

- Exercise care when drawing conclusions from graphical analysis about whether curves intersect.
- Make sure that any conclusions drawn from one technique are confirmed by other techniques.

CHAPTER 2

AUTONOMOUS DIFFERENTIAL EQUATIONS

Where We Are Going – and Why

In this chapter we consider first order differential equations of the form

$$\frac{dy}{dx} = g(y), \tag{2.1}$$

where the right-hand side of (2.1) depends only on the dependent variable y. Such differential equations are called autonomous. They are important for two practical reasons: they are used to model many situations, and their properties form the basis for tackling more-complicated situations, which we will discuss in Chapters 6, 9, and 10.

One of the objectives of this chapter is to show you how the techniques developed in Chapter 1 can be applied to help us fully understand the qualitative and quantitative behavior of solutions of autonomous differential equations. These different techniques are needed because a single method — including analytical solutions — can lead to misleading conclusions, as we show in several examples.

In Chapter 1 we saw that only one solution of $y' = g(x)$ could pass through a specified point (x_0, y_0) if $g(x)$ is continuous. This guaranteed that solution curves did not intersect. We need to know what conditions must be imposed on $g(y)$ so that the solution curves of (2.1) do not intersect. Thus, we need an existence and uniqueness theorem for initial value problems associated with (2.1). This is given, and its significance is very apparent when we hand-draw solution curves using information given by slope fields. Equilibrium solutions, phase line analysis, and bifurcation diagrams are new topics introduced in this chapter to help us determine properties of solution curves of $y' = g(y)$.

We give examples of how autonomous differential equations are used in developing mathematical models for population dynamics, epidemics, and drug behavior. The exercises contain many more models of other situations. We use real data sets to evaluate the parameters that occur in the differential equations. This allows us to check the predictions of these mathematical models. It is vital to test a model's predictions against real data.

2.1 AUTONOMOUS EQUATIONS

In this section we apply the techniques developed in Chapter 1 — graphical analysis and explicit solutions — to differential equations of the form (2.1) where the right-hand side depends only on y. We first consider a simple example, and then make some general observations about what we have learned. The important concept of an equilibrium solution is introduced.

A Definition and an Example

We start with a definition that governs all the differential equations considered in this chapter.

◆ *Definition 2.1:* **A first order differential equation of the form**

$$\frac{dy}{dx} = g(y), \tag{2.2}$$

where $g(y)$ is a given function of y alone, is called AUTONOMOUS. ◆

EXAMPLE 2.1

As our first example of an autonomous differential equation, we look at the counterpart of an example in Chapter 1 — $y' = 1/x$ — which in this case is

$$y' = \frac{1}{y}. \tag{2.3}$$

In other words, find a function which has the property that its derivative is the reciprocal of the original function.

Let's see what qualitative information we can discover about solutions of (2.3) if we apply the ideas of calculus.

Monotonicity and concavity

If $y > 0$, then $y' > 0$ and solution curves are increasing. If $y < 0$, then $y' < 0$ and solution curves are decreasing. To consider the concavity of the solution curves, we need y''. If we differentiate (2.3) with respect to x, we find

$$y'' = -\frac{1}{y^2}y'. \tag{2.4}$$

[Why is there a y' in (2.4)?] This differs from the situation in Chapter 1, where all second derivatives were explicit functions of x. Here we find the second derivative in terms of the first. However, we can substitute the value of y' from (2.3) into (2.4) to find

$$y'' = -\frac{1}{y^3}. \tag{2.5}$$

Even though the right-hand side of (2.5) is not an explicit function of x, it does tell us that the sign of the second derivative is determined by the sign of y. So if $y > 0$, solution curves are concave down, and if $y < 0$ they are concave up.

Slope field

Now let's consider the construction of the slope field for (2.3). As we calculate the slopes of solution curves at equally spaced points (x, y), we see that along the line $y = 0$ (the x-axis), the slopes are all infinite. Along the horizontal line $y = 1$ the slopes are all 1. Along the line $y = 2$ the slopes are all $1/2$. Along the line $y = -1$ the slopes are all -1. Along the line $y = -2$ the slopes are all $-1/2$. Just as for the situation $y' = g(x)$, we can construct a slope field, which is shown in Figure 2.1.

Isoclines

We now consider the isocline question — namely, where is the slope of the solution curve equal to m? This will be when $y' = m$; that is, when $y = 1/m$. These are horizontal lines, and Figure 2.1 is consistent with this information. (To make sure you understand this, add isoclines for $m = 1$, $m = -1$, $m = 2$, and $m = -2$ to Figure 2.1.)

Symmetry

Notice that the slope field in Figure 2.1 appears symmetric across the x-axis. How can we tell whether this is actually correct? Symmetry across the x-axis means that interchanging y with $-y$ gives the same picture. If in (2.3) we interchange y with $-y$, on both sides of the equation, we find exactly (2.3) again, so the family of solutions corresponding to it must be symmetric across the x-axis.

Singularities and uniqueness

We noted that (2.3) gave infinite slopes if $y = 0$. This means that $y = 0$ (the x-axis) is excluded from the domain of any solution curve. While it appears that no two solution curves intersect, we cannot verify this at this point.

Thus, without solving (2.3) explicitly, we have been able to use graphical techniques to obtain the general shape of the solution curves, which are shown in Figure 2.2. (Explain why this figure contains 12 particular solutions, and not 6.)

Analytical solution

We now look at the analytical approach. If we try to apply the technique of Chapter 1 to (2.3), we would write down

$$y(x) = \int \frac{1}{y}\, dx.$$

This will not give our solution for $y(x)$ because to evaluate the integral we would have to know y explicitly as a function of x. But, if we knew y explicitly as a function

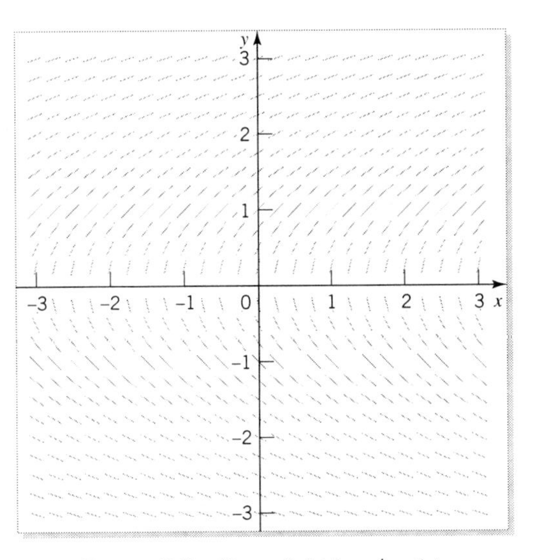

FIGURE 2.1 Slope field for $y' = 1/y$

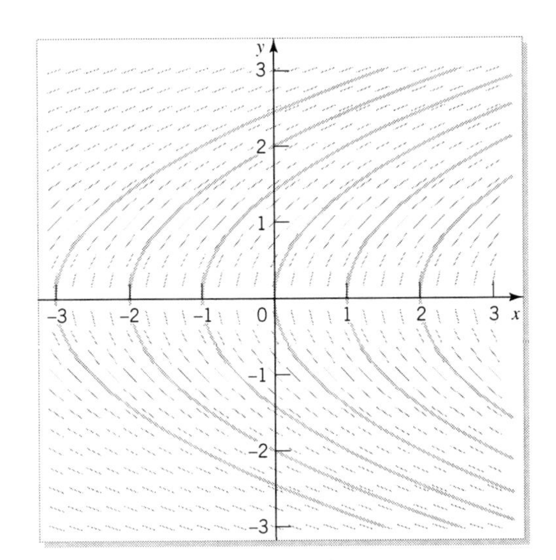

FIGURE 2.2 Hand-drawn solutions and slope field for $y' = 1/y$

of x we would already have the solution. This does not mean that there are no explicit solutions of (2.3). If both y and y' were on the same side of the equal sign we would be in a better position to integrate. Thus we multiply both sides of $y' = 1/y$ by y, giving

$$yy' = 1.$$

In this form we can integrate both sides of the equation with respect to the independent variable x to find

$$\int yy'\,dx = \int 1\,dx.$$

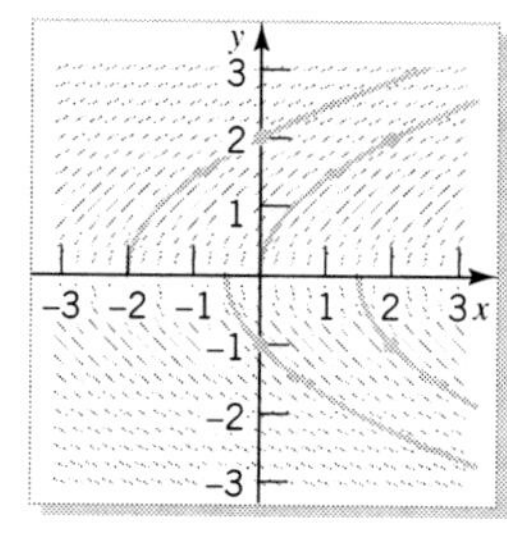

FIGURE 2.3
Solution curves and slope field for $y' = 1/y$

The left-hand side of this equation can be integrated to give

$$\int yy'\,dx = \int y\frac{dy}{dx}\,dx = \int y\,dy = \frac{1}{2}y^2 + C_1,$$

which, because the right-hand side equals $x + C_2$, gives $\frac{1}{2}y^2 + C_1 = x + C_2$, where C_1 and C_2 are arbitrary constants. If we define $C = C_2 - C_1$, then the solution of (2.3) is

$$\frac{1}{2}y^2 = x + C, \tag{2.6}$$

Explicit solution

where C is an arbitrary constant (because C_1 and C_2 are arbitrary constants). Notice that this can be expressed as an explicitly defined function for $x(y)$. We can also solve (2.6) for y, and we find the family of solutions

$$y(x) = \pm\sqrt{2\,(x + C)}. \tag{2.7}$$

Horizontal translation

It should be apparent that contrary to the situation in Chapter 1, the arbitrary constant in (2.7) does not represent a vertical translation between any two solutions, but represents a horizontal translation. Several particular solutions of this family — with initial values (0, 2), (0, −1), (2, 2), (2, −1) — are shown in Figure 2.3. (Why do these particular solutions remain in either the upper or lower half-plane?) The agreement between our hand-drawn solutions (Figure 2.2) and graphs of our explicit solutions (Figure 2.3) is obvious.

The results of the preceding example suggest some **general observations about autonomous differential equations** $y' = g(y)$.

Summary

- Solutions of an autonomous differential equation are always explicitly defined functions for $x(y)$ but are not necessarily explicitly defined functions for $y(x)$. Sometimes we may be able to solve $x = x(y)$ for $y = y(x)$.
- Isoclines of autonomous differential equations are always horizontal lines.

- If $y' = g(y)$ remains unchanged after the interchange of y with $-y$ on both sides of the differential equation, then the family of solutions is symmetric across the x-axis.
- If $y' = g(y)$ remains unchanged after the simultaneous interchange of y with $-y$ and x with $-x$ on both sides of the differential equation, then the family of solutions is symmetric about the origin.
- We use the expression ANALYTICAL SOLUTION CURVE to describe the graph of an explicit solution of a differential equation. This distinguishes it from a hand-drawn solution curve obtained directly from the slope field of the differential equation.
- If we know one solution of an autonomous differential equation — say $y(x) = f(x)$ — then $y(x) = f(x + C)$, where C is any constant, is another solution, because the graph of $f(x + C)$ is a horizontal translation of the graph of $f(x)$.

The Exponential Function and Equilibrium Solutions

EXAMPLE 2.2 *The Exponential Function*

Often, beginning calculus courses contain examples of differential equations that concern the growth or decay of some substance. These applications model situations where the rate of change of a substance is proportional to the amount of substance present;[1] that is,

$$y' = ay, \tag{2.8}$$

where y is the amount of substance, x is time, and a is a given constant, $a \neq 0$. If the application concerns growth of bacteria, then $a > 0$ (the amount of bacteria, y, is increasing), whereas for radioactive decay $a < 0$ (the amount of radioactive material, y, is decreasing). The understanding in (2.8) is that we are given a and want to find $y = y(x)$.

For the purposes of this example, we will consider all values of y, although for models dealing with growth or decay we would consider only nonnegative values of y, because y gives the amount of substance present at time x.

Monotonicity and concavity

First, let's see what the ideas of calculus tell us about the solution curves of (2.8). If $a > 0$, then y is increasing when $y > 0$ and decreasing when $y < 0$. The reverse is true if $a < 0$. To consider the concavity of the solution curves, we need y''. If we differentiate (2.8) with respect to x, we find

$$y'' = ay',$$

which, substituting for y' from (2.8), becomes

$$y'' = a^2 y. \tag{2.9}$$

[1]An alternative way of stating this is that the relative rate of change of a substance, $\frac{1}{y}y'$, is constant.

Because $a^2 > 0$, (2.9) tells us that the sign of the second derivative is determined solely by the sign of y. So if $y > 0$, the solution curves are concave up, and if $y < 0$, they are concave down.

Slope field and symmetry

We now consider the construction of the slope field for (2.8). As we calculate the slopes of solution curves at equally spaced points (x, y), we see that along the line $y = 0$ (the x-axis) the slopes are all 0. Along the horizontal line $y = 1$ the slopes are all a. Along the line $y = 2$ the slopes are all $2a$. Along the horizontal line $y = -1$ the slopes are all $-a$. Continuing in this manner we construct slope fields, which for $a = 1$ and $a = -1$ are shown in Figures 2.4 and 2.5, respectively. Notice that both slope fields in these figures appear symmetric across the x-axis. (How would you prove that this observation is correct?)

Isoclines

We now consider the isocline question — namely, where is the slope of solution curves equal to m? This will occur when $y' = m$; that is, $y = m/a$. These are horizontal lines, and Figures 2.4 and 2.5 are consistent with this information.

Thus, without solving (2.8) explicitly, we have been able to use graphical techniques to obtain the general shape of the solution curves, which are shown in Figure 2.6 for $a = 1$ and in Figure 2.7 for $a = -1$.

Analytical solution

We now look at the analytical approach. If we divide

$$y' = ay \tag{2.10}$$

by y, we have

$$\frac{1}{y}y' = a. \tag{2.11}$$

In this form we may now integrate with respect to x, and find

$$\ln|y(x)| = ax + c, \tag{2.12}$$

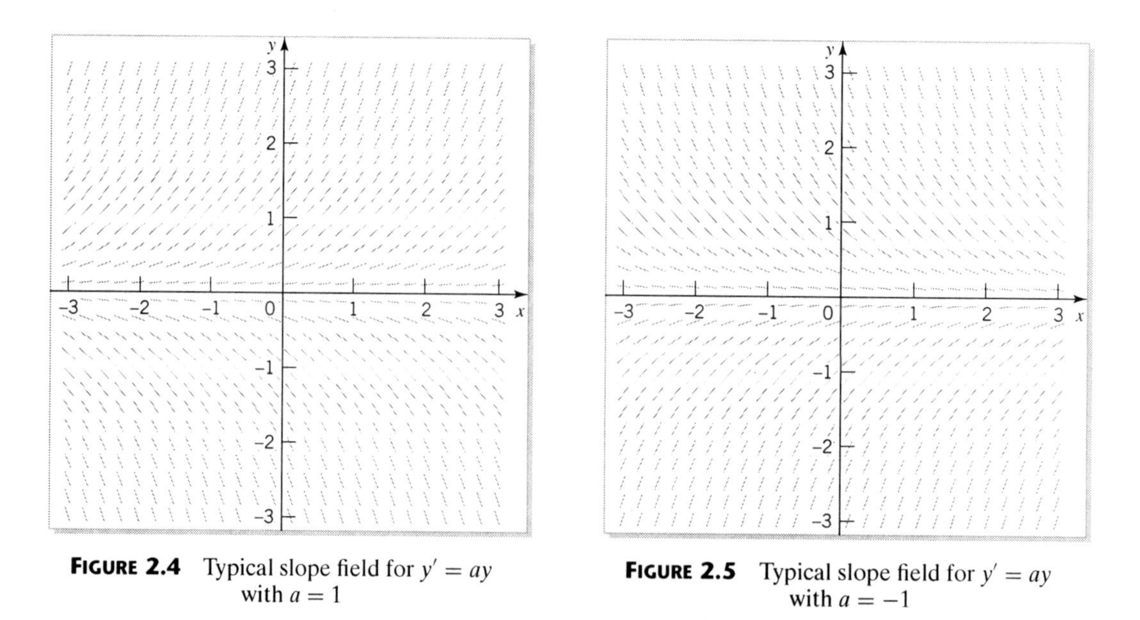

FIGURE 2.4 Typical slope field for $y' = ay$ with $a = 1$

FIGURE 2.5 Typical slope field for $y' = ay$ with $a = -1$

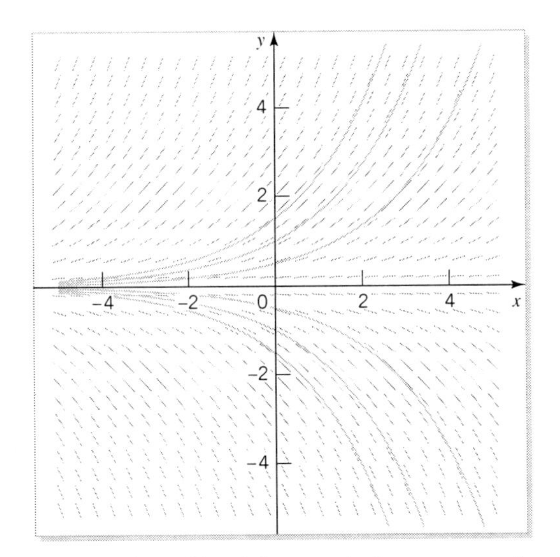

FIGURE 2.6 Hand-drawn solution curves and slope field for $y' = ay$ with $a = 1$

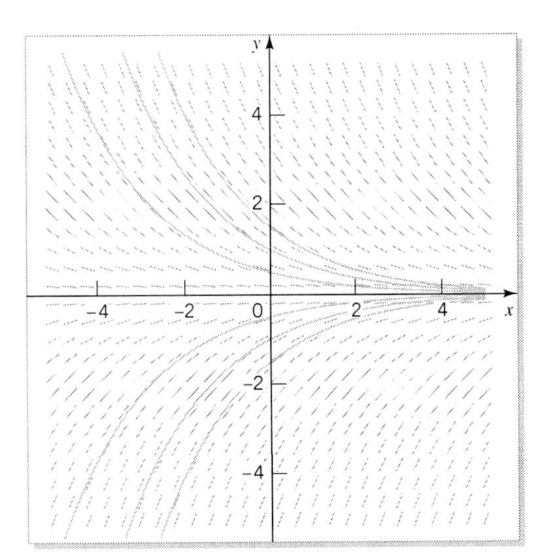

FIGURE 2.7 Hand-drawn solution curves and slope field for $y' = ay$ with $a = -1$

where c is an arbitrary constant. This gives x as an explicit function of y. To find y as an explicit function of x, we apply the exponential function to both sides of this equation to get $|y(x)| = e^{ax+c} = e^c e^{ax}$, or, equivalently, $y(x) = \pm e^c e^{ax}$. If we define $C = \pm e^c$ (which is never 0), we have the family of solutions

$$y(x) = Ce^{ax}, \qquad C \neq 0, \tag{2.13}$$

the graph of which is in good agreement with Figures 2.6 and 2.7.

At first sight this may seem acceptable. However, this analysis requires closer scrutiny. The problem is that in going from (2.10) to (2.11) we implicitly assumed that $y(x) \neq 0$. But (2.10) is clearly satisfied by

$$y(x) = 0, \tag{2.14}$$

which we lost in going from (2.10) to (2.11). This is a solution that is not contained in the solution given by (2.13), although it could be if we allow $C = 0$. Thus, we can combine the two distinct solutions (2.13) and (2.14) into the single equation

$$y(x) = Ce^{ax}, \tag{2.15}$$

where C can now take on any value, including 0.

Note that the value of C is the y-intercept because $y(0) = C$. Several members of this family, with $y(0) = 0, 0.5, -1$, and 1.5, are shown in Figure 2.8 for $a = 1$, and in Figure 2.9 for $a = -1$. The agreement between our hand-drawn solutions (Figures 2.6 and 2.7) and our analytical solution curves (Figures 2.8 and 2.9) is obvious.

In the application of this solution to population growth (where $a > 0$), we see that if $y(0) > 0$, the population grows without bound as $x \to \infty$. In the case of radioactive decay (where $a < 0$ and $y(0) > 0$), we see that the amount of radioactive substance approaches 0 as $x \to \infty$. In both cases, if there is no substance present

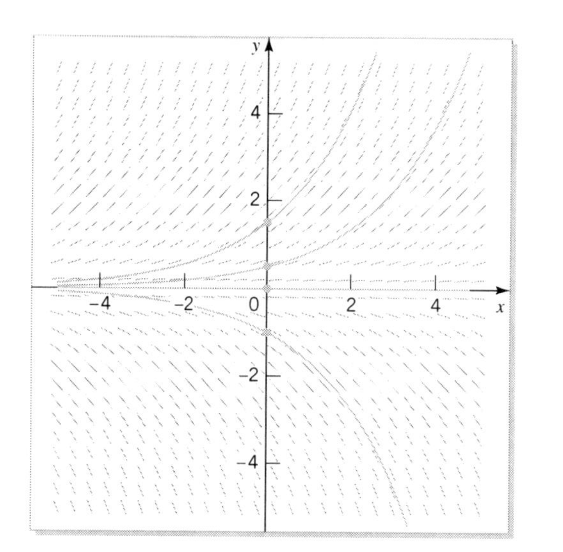

FIGURE 2.8 Four analytical solution curves and slope field for $y' = ay$ with $a = 1$

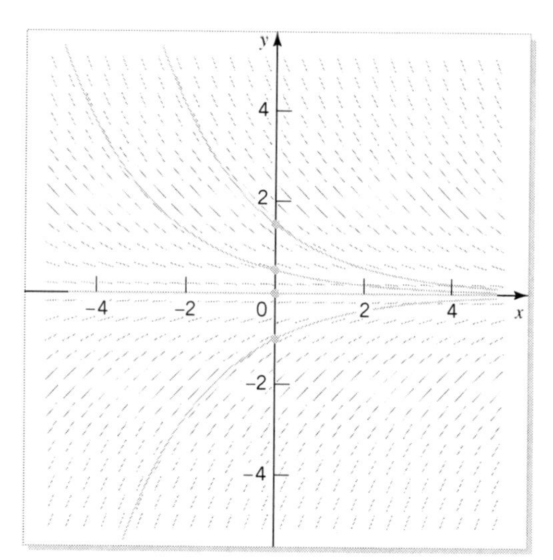

FIGURE 2.9 Four analytical solution curves and slope field for $y' = ay$ with $a = -1$

at the beginning — that is, $y(0) = 0$ — then there is never any substance present, $y(x) = 0$ for all x. ■

Equilibrium Solutions

In the preceding example we found that the constant function $y(x) = 0$ was a particular explicit solution of (2.8). This solution is an example of an equilibrium solution, defined as follows.

◆ *Definition 2.2:* **If the differential equation $y' = g(x, y)$ has a solution of the form $y(x) = $ *constant*, then this solution is called an EQUILIBRIUM SOLUTION.** ◆

Caution! **It is a common mistake to forget about the equilibrium solutions when solving a differential equation.**

Solutions near the equilibrium solution $y(x) = 0$ of $dy/dx = ay$ have different characteristics depending on whether $a < 0$ or $a > 0$. If $a < 0$ and we start on a solution near the equilibrium solution $y(x) = 0$, we move closer to $y = 0$ as x increases. Such an equilibrium solution is called a STABLE EQUILIBRIUM SOLUTION. If $a > 0$ and we start near the equilibrium solution $y(x) = 0$, we move farther from $y = 0$ as x increases. Such an equilibrium solution is called an UNSTABLE EQUILIBRIUM SOLUTION. It is possible for an equilibrium solution to appear stable from one side but unstable from the other — see Exercise 9 on page 36. Such an equilibrium solution is called SEMISTABLE.

This leads to the following definition.

◆ *Definition 2.3:* **Let the differential equation $y' = g(x, y)$ have an equilibrium solution; that is, a solution of the form $y(x) = $ *constant*. An equilibrium solution is called**

STABLE if all solutions of the differential equation that start near this equilibrium solution remain near this equilibrium solution as $x \to \infty$. An equilibrium solution is called UNSTABLE if all other solutions of the differential equation that start near this equilibrium solution move away from this equilibrium solution as $x \to \infty$. If an equilibrium solution is neither stable nor unstable, it is called SEMISTABLE. ◆

As an example of the importance of stability, consider the case $a = 1$; that is, $y' = y$. If the initial value $y(x_0)$ is positive, then $y(x) \to \infty$ as x increases. If $y(x_0) = 0$, then $y(x) = 0$ for all x. Finally, if $y(x_0) < 0$, then $y(x) \to -\infty$ as x increases. Notice that a small change in the initial condition may give a large change in the final behavior. For example, if $y(x_0) = 10^{-17}$, then $y(x) \to \infty$ as x increases. If $y(x_0) = 0$, then $y(x) = 0$ for all x. If $y(x_0) = -10^{-17}$, then $y(x) \to -\infty$ as x increases. This is called SENSITIVITY TO INITIAL CONDITIONS. Many computers and calculators cannot distinguish between 10^{-17}, 0, and -10^{-17}.

EXERCISES

1. For each of the differential equations

(a) $y' = -y^{-2}$ (b) $y' = -1 - y^2$

(c) $y' = y^2$ (d) $y' = y^{1/5}$

(e) $y' = y^{3/2}$ (f) $y' = \sqrt{1 - y^2}$

i. Use graphical methods (calculus, slope field, isoclines) to sketch some solution curves.

ii. Use analytical methods to solve the differential equation for $y(x)$.

iii. Confirm that parts i and ii are consistent.

2. Find the family of explicit solutions for each of the differential equations $y' = e^y$ and $y' = -e^{-y}$. Graph these two families of solutions on one plot, using the same scale for the x- and y-axes. What do you notice about the angle of intersection between these two families of solutions? Could you have seen that directly from the differential equations, without solving them?

3. Solve $y' = 1 - y$ subject to the initial condition

(a) $y(0) = 0$.

(b) $y(0) = 1$.

(c) $y(0) = 2$.

(d) $y(x_0) = y_0$.

4. Consider the following four differential equations:

$$y' = y + 1, \quad y' = y - 1, \quad y' = \ln|y|, \quad y' = y^2 - 1.$$

(a) The slope field of one of these equations is given in Figure 2.10. Match the correct equation with the figure, carefully stating your reasons. Do not plot any slope fields to answer this question.

(b) Briefly outline a general strategy for this matching process.

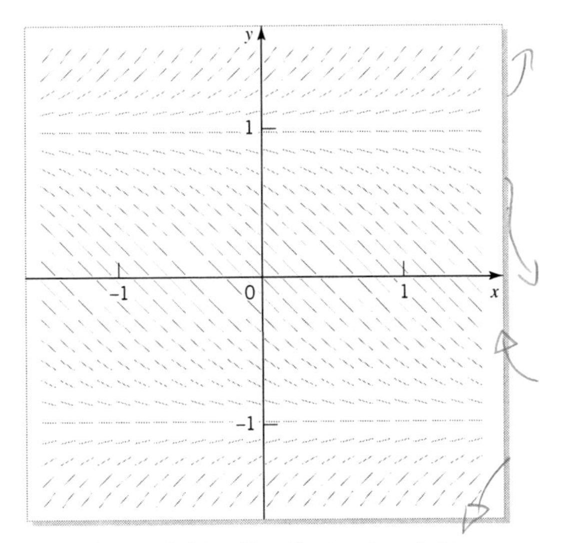

FIGURE 2.10 Identify the slope field

5. Figures 2.11 and 2.12 are mystery slope fields, believed to be the slope fields for two of the following differential equations:

$$y' = \frac{y^2+1}{y^2-1}, \quad y' = \frac{y^2-1}{y^2+1},$$

$$y' = -\frac{y^2+1}{y^2-1}, \quad y' = -\frac{y^2-1}{y^2+1}.$$

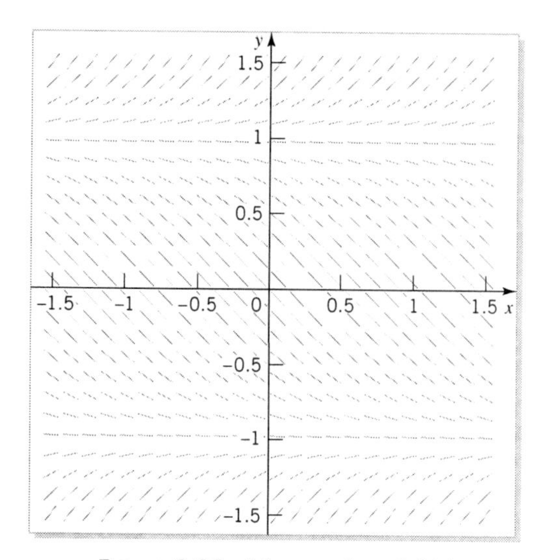

FIGURE 2.11 Mystery slope field 1

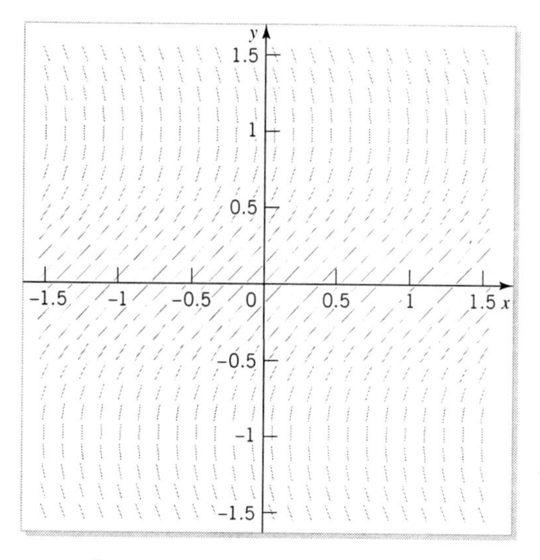

FIGURE 2.12 Mystery slope field 2

(a) Identify to which differential equation each of the mystery slope fields belongs. Confirm all the information using calculus and isoclines. Do not plot any slope fields to answer this question.

(b) Now superimpose the two mystery slope fields, perhaps by placing one on top of the other and holding both up to the light. [Another way to do this is to plot the slope fields for

$$y' = a\frac{y^2+1}{y^2-1} + b\frac{y^2-1}{y^2+1}$$

with $a = \pm 1$ and $b = 0$, and then with $a = 0$ and $b = \pm 1$.] What do you notice? Would this have made your previous analysis in part (a) easier?

6. Explain why in Figure 2.1 none of the solution curves touches the horizontal axis.

7. Can the slope field of the differential equations discussed in Chapter 1 — namely, $y' = g(x)$ — be symmetric across the x-axis if $g(x)$ is not the zero function? If your answer is yes, give the required conditions on $g(x)$ and prove it. If your answer is no, give an example where it is not symmetric.

8. Can the slope field of the autonomous differential equation $y' = g(y)$ be symmetric across the y-axis if $g(y)$ is not the zero function? If your answer is yes, give the required conditions on $g(y)$ and prove it. If your answer is no, give an example where it is not symmetric.

9. Solve the differential equation $y' = -y^2$. Show that the equilibrium solution is semistable. Explain why a semistable equilibrium solution is not stable.

10. Describe the behavior of the solutions of $y' = a + by$ if

(a) $a < 0$ and $b < 0$, $b = 0$, $b > 0$.

(b) $a = 0$ and $b < 0$, $b = 0$, $b > 0$.

(c) $a > 0$ and $b < 0$, $b = 0$, $b > 0$.

11. Concerning "sensitivity to initial conditions," explain the difference between a pendulum being at equilibrium hanging straight down and balancing vertically.

12. Give another example of a process to which sensitivity to initial conditions would apply.

2.2 SIMPLE MODELS

A significant use of mathematical models is to provide a quantitative description of some process of interest. Often we start with experimental data from which we are to make predictions. One of the important ways of making such predictions is by first developing an appropriate differential equation that models the process.

In this section we consider two examples that illustrate models that involve the differential equation $y' = ay$; the growth of a population, and the amount of a drug in the bloodstream. We will also discover how to use real data sets to estimate numerical values for parameters in a differential equation, which allows us to predict future behavior.

EXAMPLE 2.3 *The Population of Botswana (1975–1990)*

The population of Botswana from 1975 to 1990 is shown in Table 2.1 and Figure 2.13.[2] It is claimed that this data set is consistent with the model where the population growth per unit population — the relative population growth — is equal to a constant. Let y be the population in millions at time t in years measured from 1975. Because the population growth per unit population is the quantity y'/y, we have the differential equation

$$\frac{1}{y}\frac{dy}{dt} = k,$$

where k is a positive constant (the growth rate). This is the differential equation (2.8) with x and a replaced by t and k, respectively. It is common practice to use t to denote time, and so we will follow this practice when appropriate. We know that the solution of this differential equation — see (2.15) — is

Explicit solution

$$y(t) = Ce^{kt}, \tag{2.16}$$

where C is a constant.

Table 2.1 Population of Botswana

Year	Population (millions)
1975	0.755
1980	0.901
1985	1.078
1990	1.285

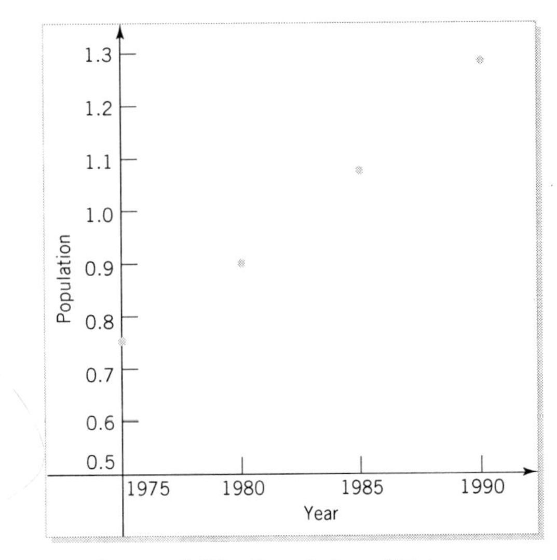

FIGURE 2.13 Population of Botswana

[2] *World Population Growth and Aging* by N. Keyfitz, University of Chicago Press, 1990, page 118.

Check model against data

How do we estimate values for C and k from the data set? If the data set was approximated by a line, then we could fit the data set by eye and estimate the slope and intercept. Unfortunately these data are not linear. However, if we take the logarithm of (2.16) we find

$$\ln y(t) = \ln C + kt.$$

Note that this corresponds to (2.12). Consequently, if we plot $\ln y(t)$ versus t, we would have a straight line with slope k and y-intercept $\ln C$, from which we can estimate both C and k.[3] This is done in Figure 2.14, from which we estimate $\ln C \approx -0.281$ (so $C \approx 0.755$) and $k \approx 0.0355$.

Figure 2.15 shows the population data as well as the function

$$y(t) = 0.755e^{0.0355t} \tag{2.17}$$

(adjusted so $t = 0$ corresponds to the year 1975), which seem to be in good agreement.

Prediction

Let's see where this model leads. It predicts that the population of Botswana is doubling about every 20 years (see Exercise 1 on page 40), compared with the world population's doubling about every 34 years. According to this model, the population of Botswana will increase to about 25 million by 2075. If we continue in this way, we find that, about 160 years from 1975, the population of Botswana will be about 255 million, which exceeds the 1990 population of the entire United States. However, the area of Botswana is about that of the state of Texas. Although this model seems reasonable for 1975–1990, it is unlikely to be valid forever.

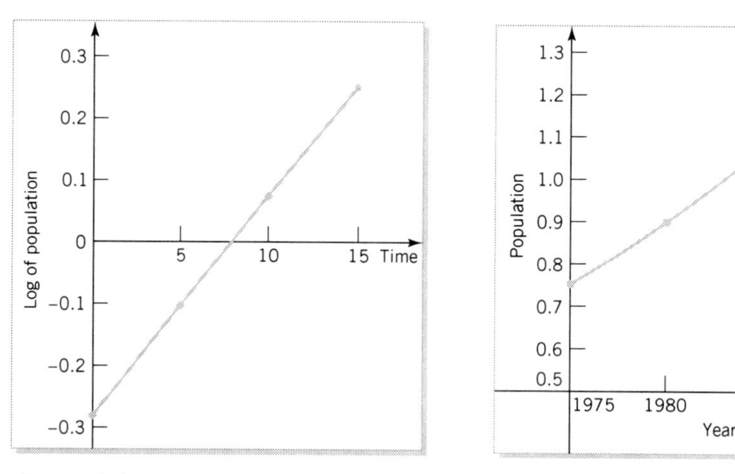

FIGURE 2.14 Logarithm of population of Botswana versus time

FIGURE 2.15 Actual and theoretical population of Botswana

---

[3]Throughout the text we will be fitting straight lines through data sets. This can be done by eye, or by using the least squares method described in Appendix A.6. We will consistently use the latter without further reference to it.

EXAMPLE 2.4 *Administering Drugs*[4]

Table 2.2 shows the concentration of theophylline, a common asthma drug, in the bloodstream as a function of time after an injection of an initial dose into a subject. Figure 2.16 shows these data plotted against time.

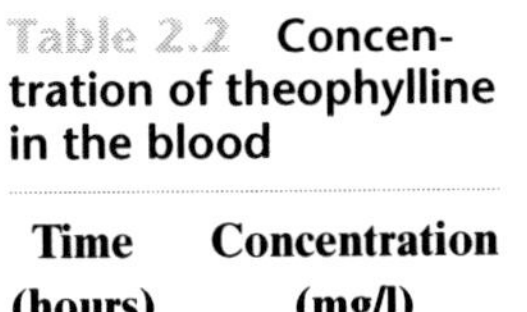

Table 2.2 Concentration of theophylline in the blood

Time (hours)	Concentration (mg/l)
0	12.0
1	10.0
3	7.0
5	5.0
7	3.5
9	2.5
11	2.0
13	1.5
15	1.0
17	0.7
19	0.5

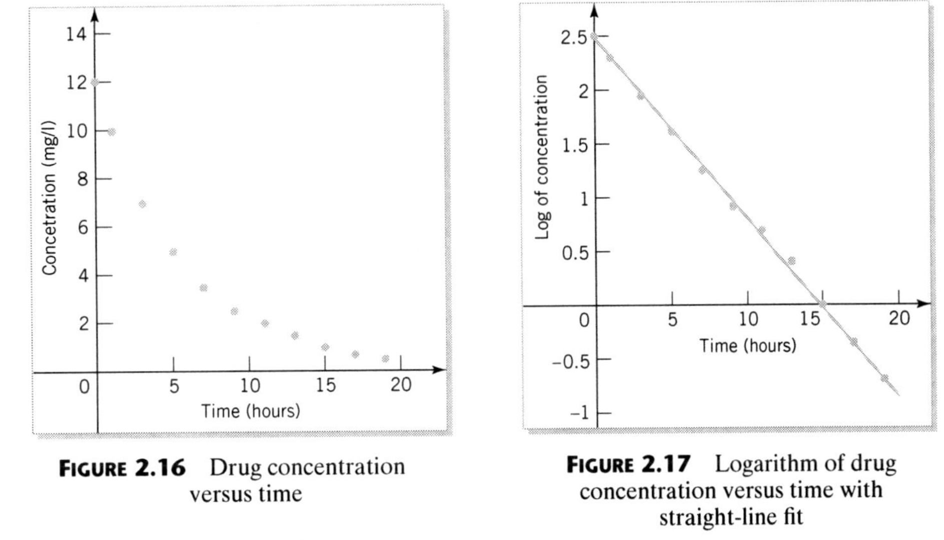

FIGURE 2.16 Drug concentration versus time

FIGURE 2.17 Logarithm of drug concentration versus time with straight-line fit

Figure 2.17 shows the logarithm of the concentration plotted against time with a straight-line fit. This suggests that the concentration $y(t)$ at time t might decay exponentially and, if so, is modeled by the differential equation

$$y' = -ky, \tag{2.18}$$

where $k > 0$, which has an explicit solution of $y(t) = Ce^{-kt}$. Figure 2.17 allows us to estimate $\ln C$ because it is the y-intercept. Thus, we have $\ln C \approx 2.485$, so $C \approx 12$. Similarly, $-k$ is the slope, so $k \approx 0.167$. This gives

$$y(t) = 12e^{-0.167t}. \tag{2.19}$$

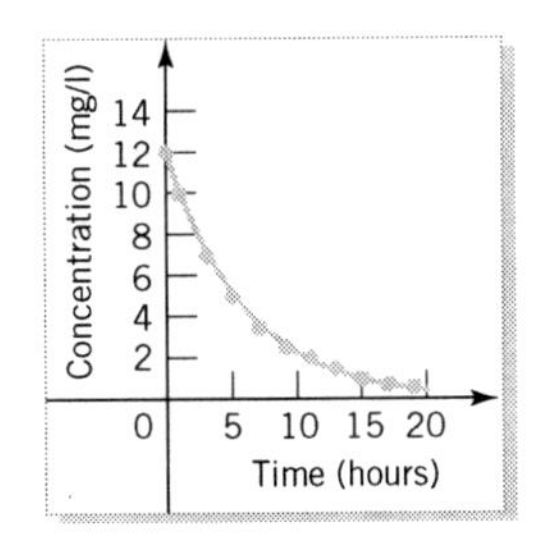

FIGURE 2.18 Concentration versus t with exponential fit

Check model against data

Figure 2.18 shows these data plotted against time with the exponential function (2.19).

[4] Based on *Applying Mathematics* by D.N. Burghes, I. Huntley, and J. McDonald, Ellis Horwood, 1982, page 125.

EXERCISES

1. **Doubling Time.** A measure of exponential growth, $y(t) = Ce^{kt}$ where $k > 0$, is the DOUBLING TIME, T_d — the time it takes for the value of y at any particular time, t_0 — that is, $y(t_0)$ — to double to $2y(t_0)$. Show that $T_d = \ln 2/k$. Notice that T_d is independent of $y(t_0)$. What does this tell you?

2. **Half-Life, Time Constant, and Settling Time.** There are various measures of exponential decay, $y(t) = Ce^{kt}$ where $k < 0$.

   (a) The HALF-LIFE, T_h — the time it takes for the value of y at any particular time, t_0 — that is, $y(t_0)$ — to halve to $y(t_0)/2$. Show that $T_h = -\ln 2/k$. Notice that it is independent of $y(t_0)$.

   (b) The TIME CONSTANT, T_c — the time where the tangent through the point $(0, y(0))$ crosses the t-axis. Show that $T_c = -1/k$. Confirm that the half-life is about 70% of the time constant. In some disciplines, the value of y at five time constants — that is, $5T_c$ — is regarded as zero. What is the value of y at five time constants?

   (c) The SETTLING TIME, T_s — the time after which the value of y never exceeds 1% of the maximum value of y. Show that, for exponential decay, $T_s = -2\ln 10/k$. Confirm that the settling time lies between four and five time constants.[5]

   (d) Figure 2.19 shows the graph of an exponentially decaying function $y = y(t)$. From the preceding definitions identify the half-life, the time constant, and the settling time for $y = y(t)$. Do this geometrically. Do not attempt to identify the function.

   (e) Example 2.4 on page 39 deals with administering the drug theophylline. Using (2.19) — namely, $y(t) = 12e^{-0.167t}$ — evaluate the half-life, the time constant, and the settling time for theophylline.

   (f) What is the settling time for $y(t) = te^{-t}$?

3. **World Population.** In 1800 the world's population was approximately 1 billion, whereas in 1900 it was 1.7 billion. If the population P at time t obeys the differential equation $\frac{dP}{dt} = kP$, estimate the world's population in the year 2000. What is the doubling time? Do you think these are good estimates if the world's population in 1990 was 5.3 billion, and was doubling every 40 years?

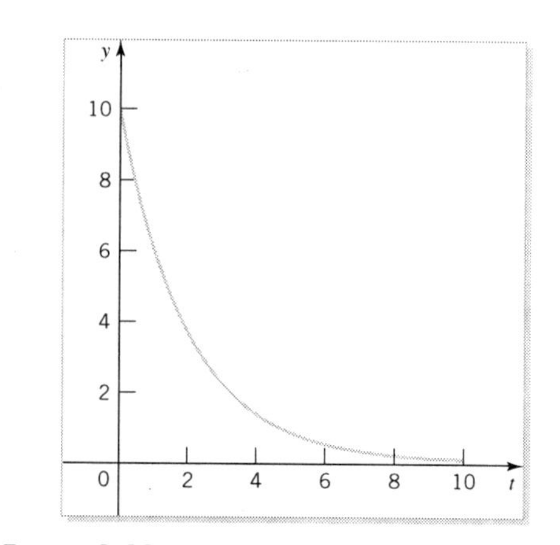

FIGURE 2.19 Exponentially decaying function

4. At time $t = 0$ a bacteria culture has N_0 bacteria. One hour later the population has grown by 25%. If the population P at time t obeys the differential equation $\frac{dP}{dt} = kP$, how long will it take the population to double?

5. After a drug dose is administered, the concentration of the drug in someone's body decreases by 50% in 10 hours. How long will it take for the concentration of the drug to reach 10% of its original value, if the concentration C at time t obeys the differential equation $\frac{dC}{dt} = kC$?

6. The land area of Botswana is 220,000 square miles (about the size of Texas). Show that based on the model characterized by (2.17), the time when the population density of Botswana will be one person per square foot will occur in about 450 years from 1975.

7. **The Population of Houston.**[6] Table 2.3 and Figure 2.20 show the population of Houston from 1850 to 1980. How well does the exponential model fit this data set? The world population in 1990 was estimated to be 5.3 billion and doubling about every 40 years. If the population of Houston and the population of the world continue to grow at these rates, when will the population of Houston exceed the population of the world? Explain this paradox.

---

[5] The half-life and the time constant apply only to exponential decay. The settling time is more general and applies to any quantity that has the property that $y \to 0$ as $t \to \infty$.

[6] "An Example of Exponential Growth" by B. Franklin, *The Physics Teacher*, October 1987, pages 436–437.

Table 2.3 The population of Houston in thousands

Year	Population	Year	Population
1850	18.632	1920	272.475
1860	35.441	1930	455.570
1870	48.986	1940	646.869
1880	71.316	1950	947.500
1890	86.224	1960	1430.394
1900	134.600	1970	1999.316
1910	185.654	1980	2905.344

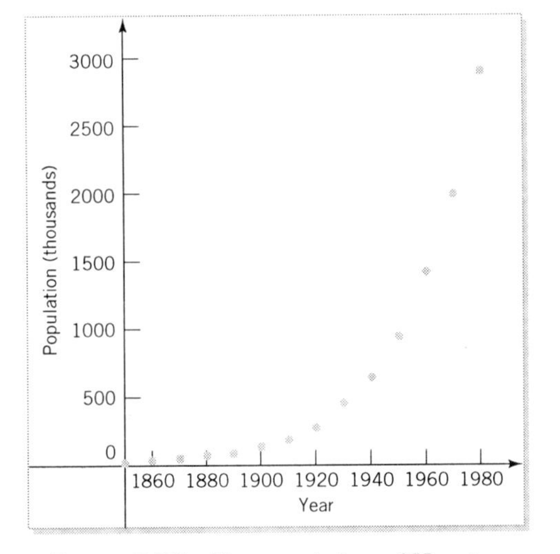

FIGURE 2.20 The population of Houston

8. **The Half-Life of ^{64} Cu.**[7] Table 2.4 and Figure 2.21 show the decay of the ^{64}Cu isotope as a function of time in hours. How well does the exponential model fit this data set? What estimate does this data set give for the half-life of ^{64}Cu? Compare your estimate to the value given in the *CRC Handbook of Chemistry and Physics* for the half-life of the ^{64}Cu isotope — 12.701 hours. Estimate the time constant and the settling time for the ^{64}Cu isotope.

9. A bullet passes through a piece of wood 0.1 m thick. It enters the wood at 200 m/sec, and it leaves at 80 m/sec. It is assumed that the velocity v of the bullet while in the wood obeys the differential equation $dv/dt = -kv^2$, where k is a given positive constant. How long does it take the bullet to pass through the wood? How thick should the wood be to bring the bullet to rest? Does this differential equation produce a realistic model?

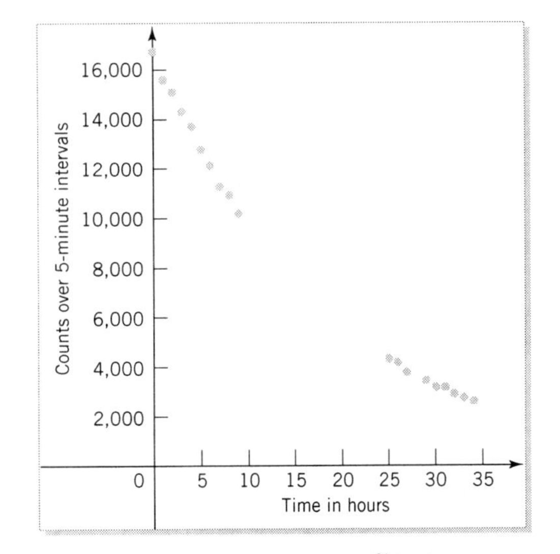

FIGURE 2.21 The decay of the ^{64}Cu isotope as a function of time in hours

Table 2.4 Data for ^{64}Cu decay

Time	Counts	Time	Counts	Time	Counts
0	16744	7	11277	30	3184
1	15596	8	10949	31	3177
2	15120	9	10174	32	2910
3	14325	25	4317	33	2766
4	13723	26	4181	34	2598
5	12788	27	3784		
6	12141	29	3454		

[7] "Data analysis in the undergraduate nuclear laboratory" by B. Curry, D. Riggins, and P. B. Spiegel, *American Journal of Physics* 63, 1995, pages 71–76. Table 1, page 73.

10. Growing Duckweed.[8] Table 2.5 shows the results of an experiment in which duckweed fronds — large compound leaves — were counted weekly. Figure 2.22 shows these data plotted by the week. How well does the exponential model fit this data set?

Table 2.5 Total number of duckweed fronds by the week

Week	Total number
0	20
1	30
2	52
3	77
4	135
5	211
6	326
7	550
8	1052

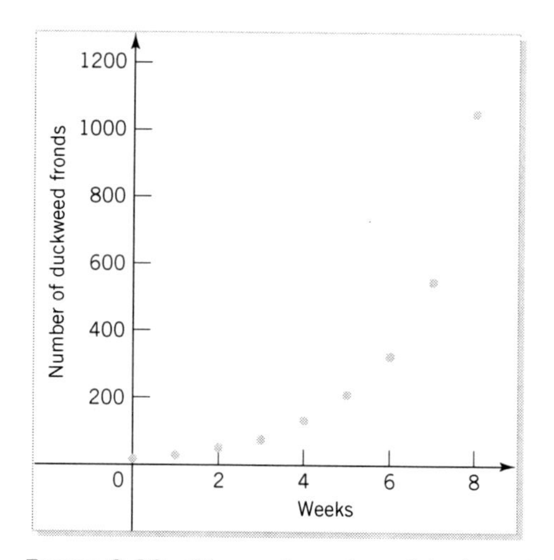

FIGURE 2.22 The total number of duckweed fronds by the week

11. From your own experience, or a recent newspaper article or magazine, develop or find a data set, of at least five data points, that exhibits exponential growth. Some possibilities include the out-of-state tuition at your local college, the weight of your new puppy, or the cost of mailing a first-class letter. Comment on how long this exponential growth could continue. Now repeat this for a data set that exhibits exponential decay.

2.3 THE LOGISTIC EQUATION

Using $y' = ay$, $a > 0$, to model the growth of any substance y always gives an exponential solution that is unbounded as $x \to \infty$. Thus, this model does not yield realistic results over long periods of time for substances whose growth is bounded. In our next example we consider the growth of sunflowers, which is clearly limited.

Graphical Analysis of the Logistic Equation

EXAMPLE 2.5 ***Growth of Sunflower Plants***[9]

Table 2.6 shows the height of sunflower plants, y, as a function of time, x. This data set is plotted in Figure 2.23. How can we model this?

---

[8] "Some accessory factors in plant growth and nutrition" by W. B. Bottomley, *Proceedings of the Royal Society B* 88, 1914, pages 237–247, as reported in *A Handbook of Small Data Sets* by D. J. Hand, F. Daly, A. D. Lunn, K. J. McConway, and E. Ostrowski, Chapman and Hall, 1994, page 119.

[9] "Growth of Sunflower Seeds" by H.S. Reed and R.H. Holland, *Proc. Nat. Acad. Sci.* 5, 1919, page 140.

Table 2.6
Height of sunflower plants

Time (days)	Height (cm)
7	17.93
14	36.36
21	67.76
28	98.10
35	131.00
42	169.50
49	205.50
56	228.30
63	247.10
70	250.50
77	253.80
84	254.50

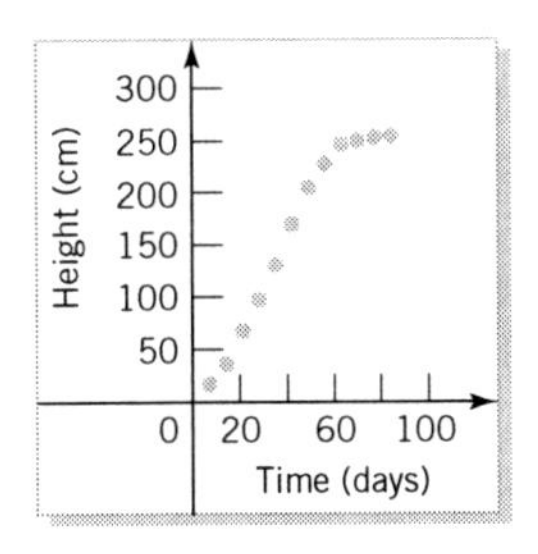

FIGURE 2.23 Height of sunflower versus time

From Figure 2.23, we see that for small values of time the height increases rapidly, but for larger values of time the increases diminish. In fact, the last two data points show very little change in height.

One way to model this growth via differential equations is to add a term to the right-hand side of $y' = ay$ that would decrease the growth rate for large values of y. This should be done in such a way that the rate of change of height per unit height — the relative growth rate — decreases as the height increases. The equation

$$\frac{1}{y}y' = a(b - y),$$

where a and b are given positive constants, describes such a situation. We can rewrite this differential equation in the form

$$y' = ay(b - y). \tag{2.20}$$

Equation (2.20) is called the LOGISTIC DIFFERENTIAL EQUATION, and it is commonly used for population models where the population's growth rate decreases with increasing population due to factors such as limited food supply, overcrowding, and disease.

We want to discover all the qualitative information we can about solutions of (2.20) from the differential equation itself. We start with the construction of the slope field. Because we are dealing with heights of sunflower plants, we consider only nonnegative values of y. As we use (2.20) to evaluate the slopes of solution curves at equally spaced points (x, y), we see that along the horizontal lines $y = 0$ (the x-axis) and $y = b$, the slopes are all zero. Along the horizontal line $y = 1$, the slopes are all $a(b - 1)$. Along the horizontal line $y = 2$, the slopes are all $2a(b - 2)$, and so on. The slope field for (2.20), with $a = 1$ and $b = 2$, is shown in Figure 2.24. Because we are concerned only with $x \geq 0$ and $y \geq 0$, we don't look for symmetries.

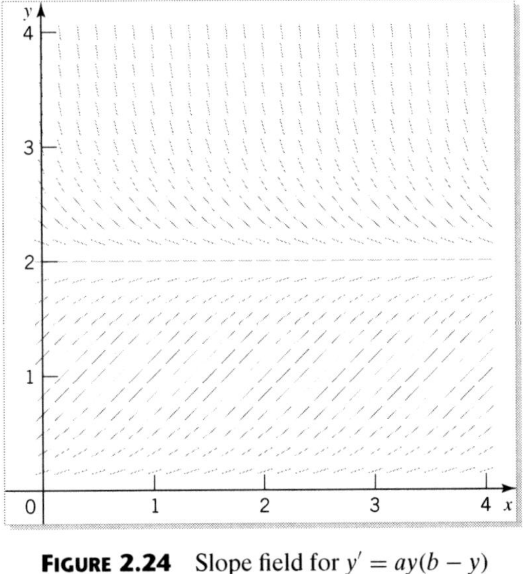

FIGURE 2.24 Slope field for $y' = ay(b - y)$ with $a = 1, b = 2$

Isoclines

The isoclines for (2.20) are given by

$$ay(b - y) = m. \tag{2.21}$$

With $m = 0$ we see that the only places where the solution curves have horizontal tangents are along the horizontal lines $y = 0$, and $y = b$. To consider other values of m we rewrite (2.21) as a quadratic equation in y,

$$y^2 - by + \frac{m}{a} = 0,$$

and use the quadratic formula to solve for our isoclines as

$$y = \frac{1}{2}\left(b \pm \sqrt{\frac{ab^2 - 4m}{a}}\right), \tag{2.22}$$

where m is the specified slope. This tells us that when $ab^2 - 4m < 0$ — that is, $m > ab^2/4$ — there are no solution curves with slope m. In other words, the only solution curves of (2.20) that may have slope m occur if m is in the interval $-\infty < m \le ab^2/4$. Equation (2.22) also tells us that the maximum slope, $m = ab^2/4$, will occur at $y = b/2$, halfway (in the y direction) between the horizontal lines $y = 0$ and $y = b$. The slope field shown in Figure 2.24 is consistent with this information.

Monotonicity

The slope field also suggests that all solutions of (2.20) are either increasing or decreasing depending on the initial value of y. This may be verified from (2.20) — namely, $y' = ay(b - y)$ — in which the derivative is positive if $0 < y < b$ and negative if $y > b$. Thus, if we start with an initial value of y_0 at $x = 0$, where $y_0 > b$, the solution will decrease to the limiting value b. Likewise, if we start with an initial value of y_0, where $0 < y_0 < b$, the solution will increase toward its limiting value b. In contrast to the exponential growth model for populations, all solutions of the logistic differential equation for $y_0 > 0$ will approach the value of b for large values of time. This limiting value of b is often called the CARRYING CAPACITY of a specific population for situations where y denotes a population.

Equilibrium solutions

There are two specific initial conditions for which the preceding analysis breaks down — namely, $y_0 = 0$ and $y_0 = b$. If $y_0 = b$, then from the differential equation we see that its rate of change is always zero (recall that $y = b$ was an isocline for slope zero). Because $y(x) = b$ is a solution of (2.20) for all values of time, it is an equilibrium solution. Note that $y(x) = 0$ is also an equilibrium solution, and it makes sense that if we have a process in which the rate of change is proportional to the number present, and we start with zero present, we will stay at zero. Likewise, as seen from the slope field, for any $y_0 > 0$ all the solutions tend toward b as time increases, so if we start at this limiting value of b, there will be no change in the population.

Inflection points

Figure 2.25 shows an enlargement of the slope field between the two equilibrium solutions, where we see that there are inflection points. To discover the exact location of the inflection points, we differentiate the original differential equation (2.20) with respect to x to obtain

$$y'' = a\left[y'(b - y) + y\left(-y'\right)\right] = a(b - 2y)y' = a^2 y(b - 2y)(b - y).$$

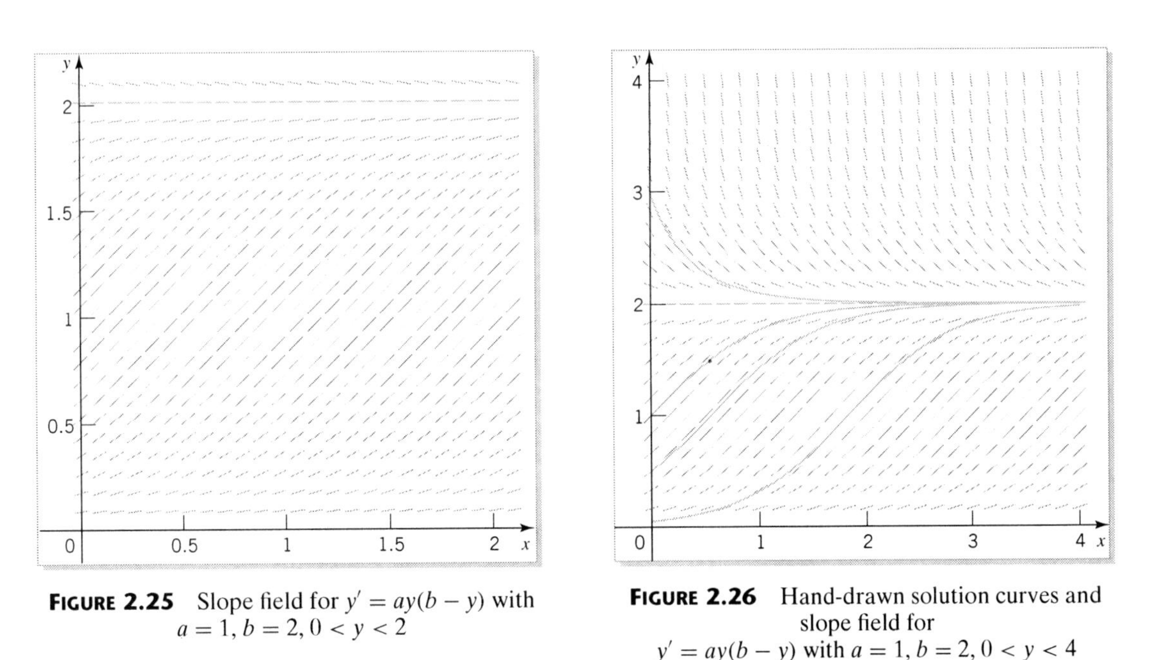

FIGURE 2.25 Slope field for $y' = ay(b - y)$ with $a = 1, b = 2, 0 < y < 2$

FIGURE 2.26 Hand-drawn solution curves and slope field for $y' = ay(b - y)$ with $a = 1, b = 2, 0 < y < 4$

There are three places where this second derivative is zero, two coinciding with the equilibrium solutions and the third when $b = 2y$. Therefore, the inflection point must occur when $y = b/2$; that is, at one-half the carrying capacity b.

Concavity

Because y is monotonically increasing for $0 < y < b$, we have just discovered that solutions of the logistic equation for $y > 0$ will have an inflection point only when $0 < y_0 < b/2$. For all other situations the solution curves are either increasing and concave down (for $b/2 < y_0 < b$) or decreasing and concave up (for $y_0 > b$). Some hand-drawn solution curves for the logistic equation are shown in Figure 2.26. Note that we have obtained many essential features of solutions of the logistic equation without obtaining an explicit solution.

Explicit Solution of the Logistic Equation

However, we can find explicit solutions of the logistic differential equation

$$y' = ay(b - y). \tag{2.23}$$

Equilibrium solutions

First, we observe that there are two equilibrium solutions; namely,

$$y(x) = 0 \text{ and } y(x) = b.$$

If we concentrate on the nonequilibrium solutions, from (2.23) we find

$$\int \frac{dy}{y(b - y)} = \int a\,dx,$$

which, by partial fractions,[10] leads to the equation

$$\int \frac{1}{b}\left(\frac{1}{y} + \frac{1}{b-y}\right) dy = \int a\, dx.$$

After integration this becomes

$$\ln |y| - \ln |b-y| = abx + c,$$

where c is the constant of integration. Using the properties of logarithmic functions, we can rewrite this as

$$\ln \left| \frac{y}{b-y} \right| = abx + c. \tag{2.24}$$

This is an explicit function of x and an implicit function of y. In this case it is possible to solve (2.24) for y to obtain an explicit solution. Taking the exponential of each side of (2.24), we have

$$\frac{y}{b-y} = Ce^{abx}, \tag{2.25}$$

where $C = \pm e^c \neq 0$. Now we multiply (2.25) by $b - y$ and solve the resulting equation for y to find

$$y(x) = \frac{bCe^{abx}}{1 + Ce^{abx}}.$$

Explicit solution This family of solutions can be rewritten in the alternative form

$$y(x) = \frac{bC}{e^{-abx} + C}. \tag{2.26}$$

Notice that one equilibrium solution, $y(x) = 0$, can be absorbed into (2.26) if we let $C = 0$, but that the other one, $y(x) = b$, is not contained in (2.26) for any finite choice of C.

If we are given the initial condition $y(0) = y_0$, then we can find the value of the constant C from (2.25); namely,

$$C = \frac{y_0}{b - y_0},$$

which can be substituted into (2.26) to give

$$y(x) = \frac{by_0}{(b - y_0)e^{-abx} + y_0}. \tag{2.27}$$

This is the explicit solution of the initial value problem associated with (2.23).

---

[10] A summary of techniques of integration is in Appendix A.1.

Let's look at this solution (2.27) a little more carefully for the physically meaningful case when $0 < y_0 < b$. We see that $y(x) \to b$ as $x \to \infty$, in complete agreement with our previous graphical analysis, which suggested b was the carrying capacity of the population.

By dividing (2.27) by b, we find

$$y(x) = \frac{y_0}{(1 - y_0/b)e^{-abx} + y_0/b}.$$

From this equation we can see that if the carrying capacity, b, is large compared with the initial population, y_0, so that y_0/b can be neglected when compared to 1, then $y \approx y_0 e^{abx}$ for small x. In other words, in this case, the initial growth of the logistic model is approximately exponential. Consequently, at the beginning of the growth of a plant or a population, it is difficult to distinguish between exponential and logistical growth.

Some analytical solution curves for the logistic equation are shown in Figure 2.27. There is good agreement with our hand-drawn solutions in Figure 2.26.

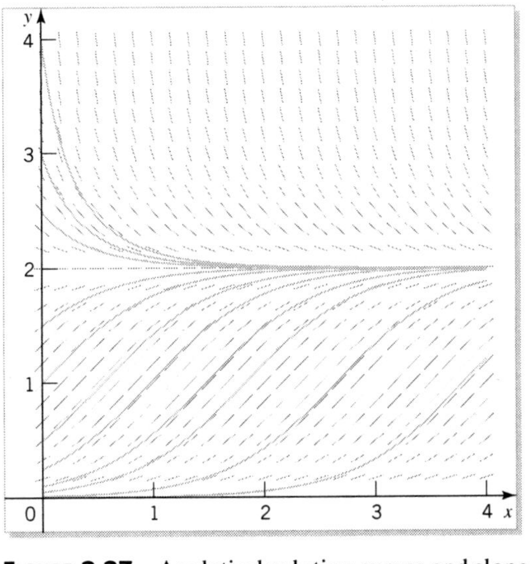

FIGURE 2.27 Analytical solution curves and slope field for $y' = ay(b - y)$ with $a = 1, b = 2$

Stability of Equilibrium Solutions

Solutions near the two equilibrium solutions in this last example — namely, $y(x) = b$ and $y(x) = 0$ — have different characteristics. If we start near the equilibrium solution $y(x) = b$, we move closer to $y = b$ as x increases, so this equilibrium solution is stable. If we start near the equilibrium solution $y(x) = 0$, we move farther from $y = 0$ as x increases, so this equilibrium solution is unstable.

Check model against data

Let us now return to the sunflower data set and see if the explicit solution (2.26) — namely,

$$y(x) = \frac{bC}{e^{-abx} + C} \tag{2.28}$$

where C is an arbitrary constant — gives a good fit. Notice that there are three unknown constants in the solution — a, b, and C. Unlike the last two examples involving exponential growth and decay, there is no way to manipulate (2.28) to construct a linear function involving these three constants. (Why?) However, from an intermediate step of the solution, (2.24) — that is,

$$\ln\left[\frac{y(x)}{b - y(x)}\right] = abx + c,$$

where $e^c = C$ — we see that if we can estimate b, the carrying capacity, we can plot $\ln[y/(b - y)]$ against x to see whether these manipulated data are approximately linear. If they are, we can then estimate ab and c.

There are two ways we can estimate b: directly, by estimating the value of y on the graph of the data where the logistic curve levels off, in which case b equals this value; or indirectly, by estimating the value of y where the curve through the data has an inflection point, in which case b equals twice this value. (Remember, for the logistic equation, the point of inflection occurs at $b/2$.) Looking at Figure 2.23 on page 43, we can try to estimate b based on these two estimates. It appears that $b \approx 260$ is reasonable from both points of view. Figure 2.28 shows a plot of $\ln[y/(260 - y)]$ against x for the data in Table 2.6, together with a straight-line fit. From this we estimate $ab \approx 0.087$, because ab is the slope of the straight line. Similarly, c is the y-intercept, so $c \approx -3$; thus, $C = e^c \approx 0.0498$. In this case, from (2.26), the height of the sunflower over time is given by

$$y(x) = \frac{(260)\,(0.0498)}{e^{-0.087x} + 0.0498} = \frac{12.94}{e^{-0.087x} + 0.0498}. \tag{2.29}$$

Figure 2.29 shows the data set plotted against time with the function (2.29).

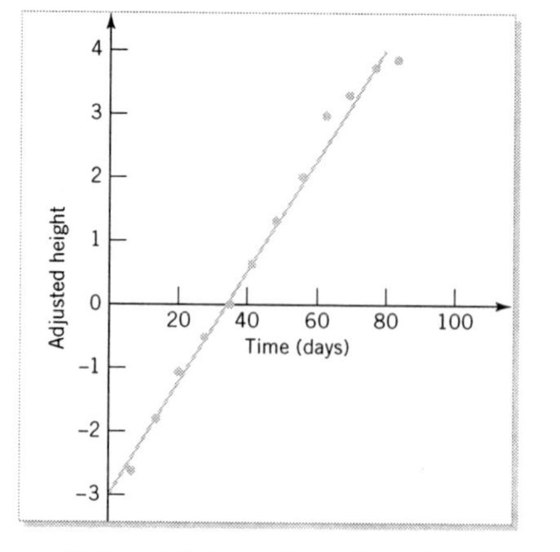

FIGURE 2.28 Adjusted height of sunflower versus time

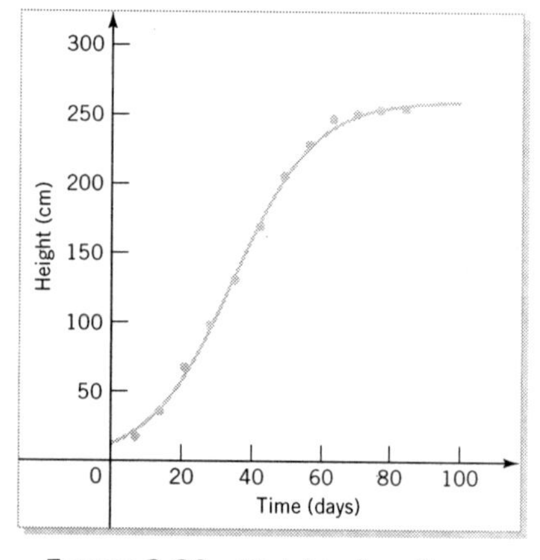

FIGURE 2.29 Height of sunflower versus time with logistic fit

EXERCISES

1. **Spread of Rumors.** The University of Arizona has 35,000 students. On the first day of the semester, a class of 35 students thought they heard their mathematics professor say that everyone would receive the grade of A for the course. The next day a carefully conducted survey of the entire student population showed that by now 700 students had heard this rumor. If the rumor spreads according to the logistic equation $y' = ay(b - y)$, where y is the number of students who have heard the rumor, at what time will 90% of the students be aware of this rumor?

2. **Technology Adoption.** Of 10,000 companies, 100 have adopted a new technology at time $t = 0$. If the number, $y(t)$, of companies that have adopted the technology at time t (units of years) satisfies the differential equation $y' = 0.0001y(10000 - y)$, find the number of companies that can be expected to adopt the technology after 5 years.

3. **The Logistic Equation.** In the logistic equation $y' = ay(b - y)$, make the change of variable $y(x) = 1/u(x)$, and show that u satisfies the autonomous equation $u' = a(1 - ub)$. Solve this equation for $u(x)$ and use this to find the solution of the logistic equation. Is this method easier than the one presented in the text?

4. **The Logistic Equation.** Show that the solution of the logistic equation $y' = ay(b - y)$ is symmetric about the point of inflection. [Hint: Translate the differential equation so that the inflection point $(x_0, b/2)$ lies at the origin.]

5. **The Logistic Equation.** Figure 2.30 shows the graphs of two functions. One of them is definitely not a solution of a logistic equation. Which one? Explain.

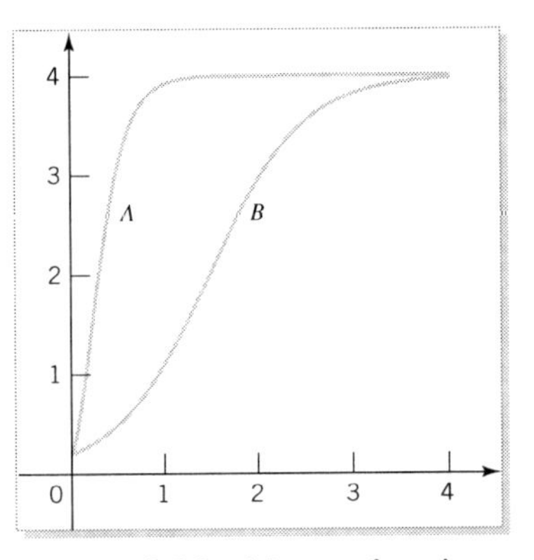

FIGURE 2.30 Mystery functions

6. **The Logistic Equation.** Solve the differential equation $y' = -y^2$ for $y(x)$. According to (2.26), the solution of $y' = ay(b - y)$ is $y(x) = bC/(e^{-abx} + C)$. Thus, with $a = 1$ and $b = 0$, the solution of $y' = -y^2$ is $y(x) = 0$. Compare this to the solution you obtained for $y' = -y^2$. Explain any discrepancies.

7. Solve $y' = \frac{1}{2}(1 - y^2)$ subject to the initial condition

   (a) $y(0) = 0$. (b) $y(0) = 1$. (c) $y(0) = -1$.

8. Describe the behavior of the solutions of $y' = ay(b - y)$ if

   (a) $a < 0$ and $b < 0, b = 0, b > 0$.

   (b) $a = 0$ and $b < 0, b = 0, b > 0$.

   (c) $a > 0$ and $b < 0, b = 0, b > 0$.

9. **Chemical Reaction.** An initial value problem that arises in the study of a chemical of concentration y (as a function of time x) that is undergoing both first and second order reactions is $y' = \alpha y - \beta y^2$, $y(0) = y_0$. The quantities α and β are constants.

   (a) Find the solution of this differential equation assuming $\alpha \neq 0$ and $\beta \neq 0$.

   (b) Find the solution of this differential equation assuming $\alpha \neq 0$ and $\beta = 0$. Is this solution contained in part (a)?

   (c) Find the solution of this differential equation assuming $\alpha = 0$ and $\beta \neq 0$. Is this solution contained in part (a)?

   (d) Find the equilibrium solutions of this differential equation. Could they be included in the results of parts (b) and (c)?

10. **Virus Infection.** Suppose there is a homogeneous group of N individuals which at time t (measured in days) consists of x individuals who are susceptible to infection — say by some virus — and y individuals who are already infected. Also assume that there is no removal from circulation by recovery, isolation, or death. (This would be the situation for the early stages of an upper respiratory infection.) This means we have $x + y = N$. If we assume that the rate of change of infected individuals is proportional to both the number of infected individuals and the number of susceptible individuals, we have the differential equation $\frac{dy}{dt} = kxy = k(N - y)y$. If we assume that at time $t = 0$, one person becomes ill and the rest are not infected, our initial condition is $y(0) = 1$.

(a) By analyzing the differential equation directly, find the number of people who are infected at the time the infection has the most rapid rate of change.

(b) Solve the initial value problem for y.

(c) Public records for epidemics record the number of new cases appearing each day. The quantity is given by $-dx/dt$ and the graph of $-dx/dt$ versus t is called the epidemic curve. Compute this quantity and graph the result. What is the maximum value of this function? Show how you could have found this maximum value without solving the differential equation.

11. The Gompertz Equation. Observations of the growth of animal tumors indicate that the size $y(t)$ of the tumor at time t may be modeled by the differential equation $y' = -ky \ln(y/b)$, where k and b are positive constants. This differential equation is sometimes called the Gompertz growth law.

(a) Construct slope fields for this equation for several values of b and k. What other slope field have you observed with the same general characteristics?

(b) Use graphical analysis to determine regions in which the solutions of the differential equation are increasing, decreasing, concave up, or concave down. Confirm that points of inflection occur along the line $y = be^{-1}$.

(c) Does this differential equation have any equilibrium solutions? If so, what are they? If not, explain why not.

(d) Solve this differential equation subject to the initial condition $y(0) = y_0, y_0 > 0$.

(e) Show that the inflection point $y = be^{-1}$ occurs when $t = \frac{1}{k} \ln\left[-\ln\left(\frac{y_0}{b}\right)\right]$. Find a relation between y_0 and b such that the graph of y versus t will have no inflection points.

(f) The solution of the logistic equation, $y' = ay(b - y)$, was given in (2.27) as $by_0/\left[(b - y_0)e^{-abt} + y_0\right]$, where $y(0) = y_0$. Show that if this solution and the solution of the Gompertz equation have the same initial condition, the same carrying capacity, and the same time for the inflection point, then $k = ab \ln\left[-\ln\left(\frac{y_0}{b}\right)\right] / \ln\left(\frac{b-y_0}{y_0}\right)$. On one figure, graph the solution of the logistic equation and the solution of the Gompertz equation for the same initial condition, the same carrying capacity, and the same time for the inflection point. Describe the similarities and differences between the two graphs. Give some general guidelines as to when one model would be more appropriate to use than the other.

12. Leaf Growth. Table 2.7 shows the total area of the leaves of a plant as a function of time.[11] This data set is sketched in Figure 2.31. It is claimed that the Gompertz equation models this situation very well. Investigate this claim.

Table 2.7 The area of the leaves of a plant as a function of time

Time (days)	Area (cm^2)
0	9.0
20	39.7
40	92.5
60	142.7
80	186.6
100	209.7
120	230.5
140	235.4

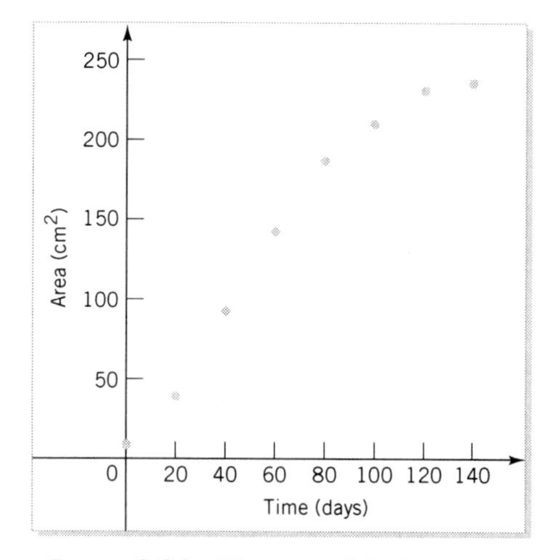

FIGURE 2.31 The area of the leaves of a plant as a function of time

13. The Population of Botswana. Using the values for C and k we obtained for the exponential model of

[11] *Growth and Diffusion Phenomena* by R. T. Banks, Springer-Verlag, 1991, page 153.

population growth of Botswana, show that the logistic model could also approximate the population of Botswana with many different carrying capacities.

14. Bacteria.[12] Table 2.8 shows the size of a bacterial colony as a function of time. Figure 2.32 shows these data plotted against time. How well does the logistic model fit this data set?

Table 2.8 **Size of bacteria colony**

Time (days)	Area (sq cm)
0	0.24
1	2.78
2	13.53
3	36.30
4	47.50
5	49.40

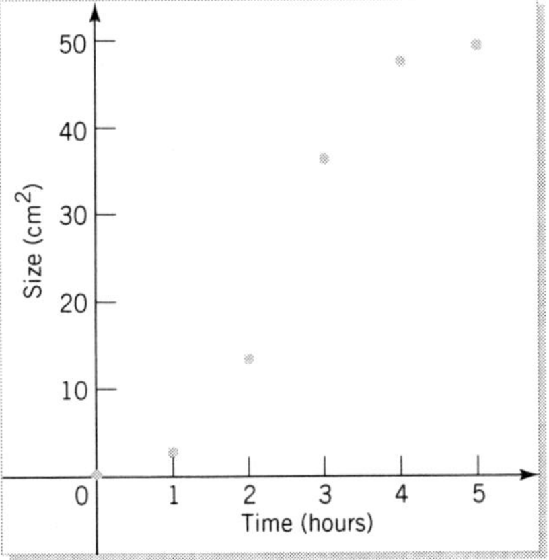

FIGURE 2.32 Size of bacteria colony versus time

15. Rise Time. Consider an increasing function, $f(t)$, that is bounded as $t \to \infty$, so that $\lim_{t\to\infty} f(t) = M$, the limiting value of $f(t)$. The RISE TIME, T_R, is that value of t after which $f(t)$ is within 1% of its limiting value M. That is,

$$M - f(t) < 0.01M \text{ for } t > T_R.$$

(a) Use the explicit solution given in (2.29) on page 48 to determine the rise time for the growth of sunflowers.

(b) Use the explicit solution you obtained in Exercise 1 to find the rise time for the spread of rumors.

(c) Use the explicit solution you obtained in Exercise 2 to find the rise time for the adaptation of technology.

(d) Use the explicit solution you obtained in Exercise 14 to find the rise time for the growth of bacteria.

16. Collared Doves.[13] During the first half of the twentieth century, the collared dove spread across Europe from east to west. This bird was very rare in Britain before 1955, so rare that it is not even included in the *1952 Checklist of the Birds of Great Britain and Ireland*. Thus, the spread of the collared dove in Britain from 1955 was of great interest to bird watchers. Consequently, there are very good records of its spread throughout Britain from 1955 to 1964, when the dove was so plentiful that little attention was then paid to it. The population of the collared dove is believed to be directly proportional to the number of locations where it was resident. Table 2.9 shows the number of locations where it was resident each year. Figure 2.33 shows these data plotted against time. How well does the logistic model fit this data set?

Table 2.9 **Number of locations of collared doves**

Year	Total
1955	1
1956	2
1957	6
1958	15
1959	29
1960	58
1961	117
1962	204
1963	342
1964	501

[12] "Growth of Bacterial Colony" by H.G. Thornton, *Ann. Appl. Biol.*, 1922, page 265.

[13] "The spread of the Collared Dove in Britain and Ireland" by R. Hudson, *British Birds* 58, No. 4, April 1965, pages 105–139.

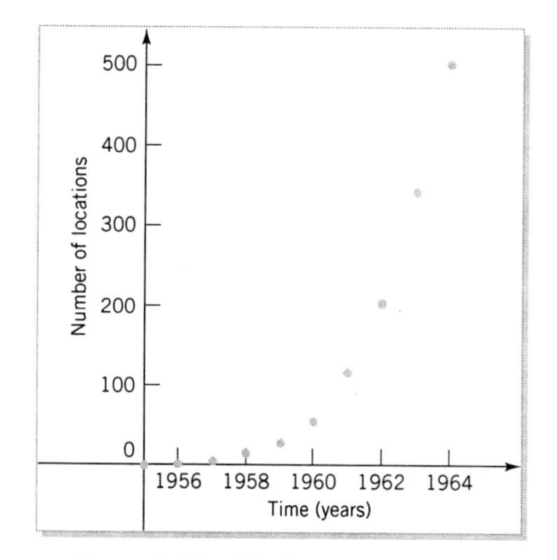

FIGURE 2.33 Number of locations of collared doves by year

17. The Bombay Plague.[14] Table 2.10 shows the number of deaths in Bombay from a plague spread by rats during the period December 1905 to July 1906. This data set is plotted in Figure 2.34. How well does the logistic model fit this data set?

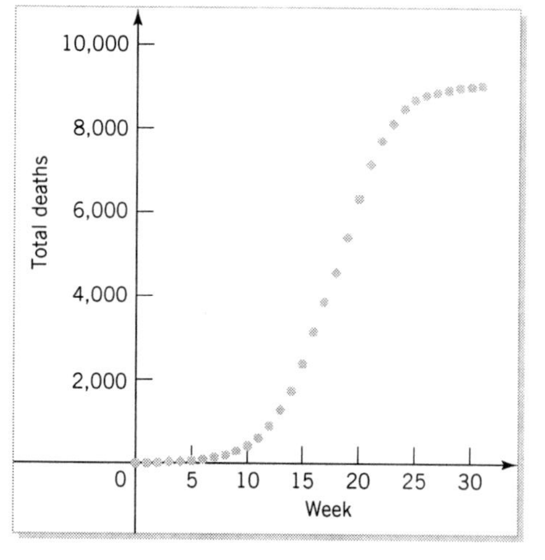

FIGURE 2.34 The number of deaths due to the Bombay plague

Table 2.10 The Bombay plague data

Week	Deaths	Week	Deaths	Week	Deaths
1	4	12	900	23	8129
2	14	13	1290	24	8480
3	29	14	1738	25	8690
4	47	15	2379	26	8803
5	68	16	3150	27	8868
6	99	17	3851	28	8920
7	150	18	4547	29	8971
8	203	19	5414	30	9010
9	300	20	6339	31	9043
10	425	21	7140		
11	608	22	7720		

18. Growing Beans.

(a) A student[15] placed beans in a styrofoam cup, covered them lightly with soil, and watered them regularly. On the tenth day sprouts emerged, and from then on he carefully measured the height (to the nearest $\frac{1}{8}$ inch) as a function of time (in days). Table 2.11 shows the approximate height of the beans. Figure 2.35 shows these data plotted against time. How well does the logistic model fit this data set?

(b) Conduct your own experiment, by growing your own beans and recording the height as a function of time. How well does the logistic model fit your data set?

Table 2.11 Beans' height

Time (days)	Height (in.)	Time (days)	Height (in.)
10	0.125	18	7.000
11	0.250	19	9.000
12	0.750	20	10.000
13	1.125	21	10.500
14	1.500	22	11.000
15	2.500	23	11.250
16	4.000	24	11.500
17	5.500	40	12.000

[14] "A contribution to the mathematical theory of epidemics" by W. O. Kermack and A. G. McKendrick, *Proc. Roy. Soc.* 115A, 1927, pages 700–721.

[15] Ian Scott, University of Arizona.

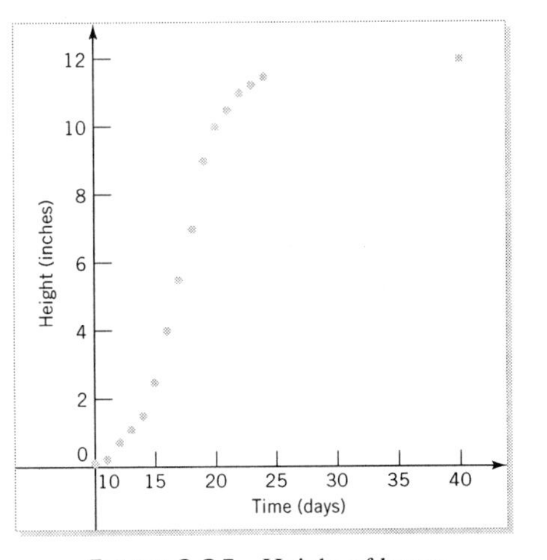

FIGURE 2.35 Height of beans

19. **Change of Scale.** Show that the logistic differential equation $y' = ay(b - y)$, where a and b are positive constants, can be transformed into

$$\frac{dY}{dX} = Y(1 - Y)$$

by the substitutions $x = X/(ab)$, $y = bY$. Solve this differential equation for $Y = Y(X)$. Now substitute for X and Y in terms of the original variables x and y. Compare your analysis to the derivation of (2.26) from (2.23). Comment on this process.

20. **Change of Scale.** In each of the following cases, imagine that you are given the solution of the first differential equation involving X and Y. Explain how you could immediately write down the solution of the second differential equation involving x and y. (Constants a, b, and k are positive.)

(a) $\frac{dY}{dX} = Y$ $\qquad \frac{dy}{dx} = ay$

(b) $\frac{dY}{dX} = X$ $\qquad \frac{dy}{dx} = ax$

(c) $\frac{dY}{dX} = -Y \ln Y$ $\qquad \frac{dy}{dx} = -ky \ln \frac{y}{b}$

(d) $\frac{d^2Y}{dX^2} \pm Y = 0$ $\qquad \frac{d^2y}{dx^2} \pm a^2 y = 0$

21. Devise your own simple experiment that will result in a graph that resembles a graph of a solution of the logistic equation. See if you can give rational arguments as to why the logistic equation is a reasonable model for your experiment. [Hint: Two examples of such experiments are the number of kernels of popcorn popped in a popper as a function of time, and the growth of mold on a moist piece of bread.]

2.4 EXISTENCE AND UNIQUENESS OF SOLUTIONS, AND WORDS OF CAUTION

In this section we consider four examples where things can go wrong. We first look at an example that illustrates how graphical analysis may not show everything about our solutions. This leads to a discussion of initial conditions and the uniqueness of solutions for an autonomous equation. Uniqueness guarantees that different solutions do not intersect. The next three examples expose different problems that can occur because we fail to use all available information.

Limitations of Graphical Analysis

EXAMPLE 2.6 *Vertical Asymptotes*

Let us look at what happens if the right-hand side of $y' = g(y)$ is a quadratic function of y, namely,

$$y' = y^2. \tag{2.30}$$

Slope field The slope field for (2.30) is shown in Figure 2.36. It is symmetric about the origin. (Why?)

Monotonicity, concavity

Equation (2.30) requires solution curves to be increasing for $y \neq 0$. Also, by differentiating (2.30), we find

$$y'' = 2yy' = 2y^3,$$

so that solution curves are concave up if $y > 0$ and concave down if $y < 0$. The slope field in Figure 2.36 is consistent with this information. Figure 2.37 shows some hand-drawn solution curves.

Equilibrium solution

Now let's find the explicit solution. We can see that $y(x) = 0$ is an equilibrium solution. (Why?) If $y(x) \neq 0$, then (2.30) can be written as $y^{-2}y' = 1$, which, after integration, yields

$$-\frac{1}{y} = x + C. \tag{2.31}$$

Explicit solution

If we solve (2.31) for y, we find the family of solutions

$$y(x) = \frac{-1}{x + C}. \tag{2.32}$$

The family of solutions (2.32) does not contain the equilibrium solution $y(x) = 0$ for any finite C.

Vertical asymptote

Notice that the denominator of (2.32) is 0 when $x + C = 0$, so that the function defined by (2.32) has a vertical asymptote at $x = -C$. For example, the solution that passes through the point $y(1) = 1$ is

$$y(x) = \frac{1}{2 - x},$$

which goes to $+\infty$ as x goes from 1 to 2. Don't forget that a solution cannot continue past its vertical asymptote. (Why?)

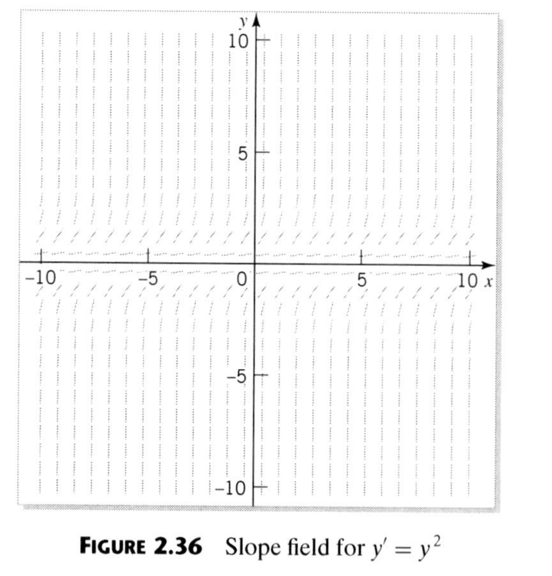

FIGURE 2.36 Slope field for $y' = y^2$

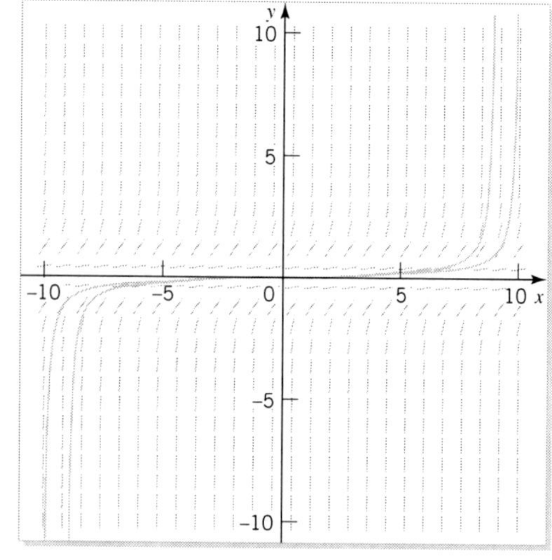

FIGURE 2.37 Hand-drawn solution curves and slope field for $y' = y^2$

Horizontal asymptote

Also notice that as $x \to \infty$, the function given by (2.32) goes to 0. For example, the solution that passes through the point $y(-1) = -1$ is

$$y(x) = -\frac{1}{x+2},$$

which goes to 0 (through negative values) as x goes from -1 to ∞.

We can evaluate the arbitrary constant in (2.32) for any initial condition. If our initial condition is $y(0) = y_0 \neq 0$, we have $C = -1/y_0$, and so

$$y(x) = \frac{-1}{x - 1/y_0}.$$

Thus, we have two cases for $y_0 \neq 0$:

(a) if $y_0 > 0$, then $y(x) \to \infty$ as $x \to 1/y_0$ from the left — a vertical asymptote;

(b) if $y_0 < 0$, then $y(x) \to 0$ (from below) as $x \to \infty$ — a horizontal asymptote.

However, if $y_0 = 0$, then we have only the equilibrium solution $y(x) = 0$.

Important point

Notice that when we drew Figure 2.37 we missed the vertical and horizontal asymptotes. The moral is that **graphical analysis may not show everything**. Solutions need not go on forever.

Figure 2.38 shows the original slope field and some particular analytical solution curves passing through the initial points $(1, 1), (-1, -1), (3, 1/3), (-3, -1/3), (5, 1/5)$, and $(-5, -1/5)$, which demonstrate these properties.

Wait! It's not over yet. Carefully compare Figure 2.38 with Figure 2.37, which shows solution curves drawn by hand from the slope fields. Something is different, but what? Do you notice that in Figure 2.37 our hand-drawn curves intersect the equilibrium solution $y(x) = 0$, whereas the analytical solution curves in Figure 2.38

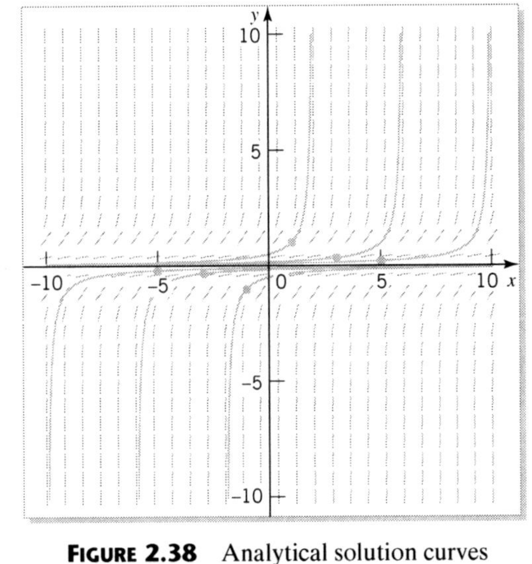

FIGURE 2.38 Analytical solution curves and slope field for $y' = y^2$

do not? Had we known before drawing our solutions by hand that these solutions cannot intersect, we would not have drawn the curves we did in Figure 2.37. ■

Caution! **Graphical analysis will not show whether a solution has a vertical asymptote at a finite value of x. Solutions need not go on forever.**

The Existence-Uniqueness Theorem

In the last example we observed that the knowledge that solution curves cannot intersect would have been very valuable when sketching solution curves. We now state an important theorem that guarantees that different solution curves cannot intersect.

▶ **Theorem 2.1: The Existence-Uniqueness Theorem.** *If $g(x, y)$ and $\partial g/\partial y$ are defined*[16] *and continuous in a finite rectangular region containing the point (x_0, y_0) in its interior, then the differential equation $y' = g(x, y)$ has a unique solution passing through the point $y(x_0) = y_0$. This solution is defined for all x for which the solution remains inside the rectangle.* ◀

Comments about the Existence-Uniqueness Theorem

- As we mentioned in Chapter 1, solving a first order differential equation $y' = g(x, y)$ subject to an initial condition $y(x_0) = y_0$ is called an INITIAL VALUE PROBLEM.
- This theorem is an existence-uniqueness theorem because it guarantees that a solution exists and that there is only one solution. This Existence-Uniqueness Theorem says that if we want to solve the initial value problem $y' = g(x, y)$, $y(x_0) = y_0$, then we are guaranteed that there is a solution curve that passes through this point (x_0, y_0), and that no other solution curve can pass through this point, provided that $g(x, y)$ and $\partial g/\partial y$ are continuous for all points near (x_0, y_0).
- This theorem gives no hints on how to find a solution. However, it guarantees that there is one to look for.
- The uniqueness part of this theorem guarantees that no matter how we find a solution, whether it be by diligence, luck, intuition, or skullduggery, it is THE solution.
- This theorem does **not** say that if $g(x, y)$ and $\partial g/\partial y$ are continuous for all x and y, then there is a unique solution valid for all x and y. There is a unique solution, but it may not be valid for **all** x and y.
- This theorem does **not** say that if either $g(x, y)$ or $\partial g/\partial y$ is discontinuous, then there is no solution. In that case, the theorem is inconclusive and tells us nothing.

---

[16] $\partial g/\partial y$ is the partial derivative of $g(x, y)$ with respect to y and is obtained by differentiating $g(x, y)$ with respect to y, treating x as a constant.

How to See Whether Unique Solutions of $y' = g(x, y)$ Exist

Purpose To see whether the initial value problem $y' = g(x, y), y(x_0) = y_0$ has a unique solution.

Process

1. Confirm that $g(x, y)$ is continuous in the vicinity of (x_0, y_0).
2. Compute $\partial g/\partial y$, which may be a function of both x and y. Confirm that this function is continuous in the vicinity of (x_0, y_0).
3. If steps 1 and 2 are confirmed, then there is a unique solution through (x_0, y_0), valid for all points near (x_0, y_0). If either step 1 or 2 fails, a solution may or may not exist, and if one exists, it may or may not be unique.

EXAMPLE 2.7 *The Vertical Asymptotes Example Revisited*

Let's see how the Existence-Uniqueness Theorem applies to the differential equation (2.30), namely,

$$y' = y^2. \tag{2.33}$$

Here $g(x, y) = y^2$, so $\partial g/\partial y = 2y$. Both $g(x, y)$ and $\partial g/\partial y$ are continuous for all x and y. The theorem thus states that there is a unique solution satisfying the initial condition $y(x_0) = y_0$, or, to put it another way, there is a unique curve that passes through the point $y(x_0) = y_0$. This means that solution curves cannot intersect, because otherwise there would be two distinct solutions with the same initial condition at the point where they intersect. Now the function $y(x) = 0$ is a solution of (2.33), so no other solutions can intersect it. If we return to the slope field for (2.33) — namely, Figure 2.36 — armed with this information, and draw in the solution $y(x) = 0$, the fact that other solutions must never intersect this solution shows us that Figure 2.37 is incorrect. Thus, when we draw solution curves by hand, we must check to see that this theorem is satisfied and, if so, make sure that any solution curve we draw does not intersect another. ■

Caution! **If the Existence-Uniqueness Theorem is satisfied, then solution curves cannot intersect.**

The Existence-Uniqueness Theorem allows us to settle a subtle point that we have avoided mentioning so far. To illustrate the point, let's return to (2.8), namely,

$$y' = ay, \tag{2.34}$$

which satisfies the conditions of the Existence-Uniqueness Theorem, so solutions do not intersect.

When we analyzed (2.34), we first noted that $y(x) = 0$ was an equilibrium solution. Then we searched for nonequilibrium solutions by dividing (2.34) by y and integrating. However, how do we know that there isn't a particular x, say x_0, for which $y(x_0) = 0$? Our division by $y(x)$ would not be valid for this x_0, and we might miss a particular solution because of this. The answer is that the equilibrium solution $y(x) = 0$ passes through the point $(x_0, 0)$, so no other solution can. Thus, we have not lost any solutions in this way.

We now look at three different examples, all of which expose different problems that can occur. The first demonstrates that without the Existence-Uniqueness Theorem, graphical analysis may not show everything. The second shows that without the graphical analysis, the solution we obtain analytically may be wrong. The third shows that not only can we get our analytical calculations wrong, but so can most of the computer algebra system software packages that give solutions analytically for differential equations.

EXAMPLE 2.8

Consider the autonomous differential equation

$$y' = (y - 1)^{2/3} \tag{2.35}$$

subject to the initial condition

$$y(1) = 1. \tag{2.36}$$

Slope field

Figure 2.39 shows the slope field for (2.35), and in Figure 2.40 we have hand-drawn a solution curve through the point $(1, 1)$.[17]

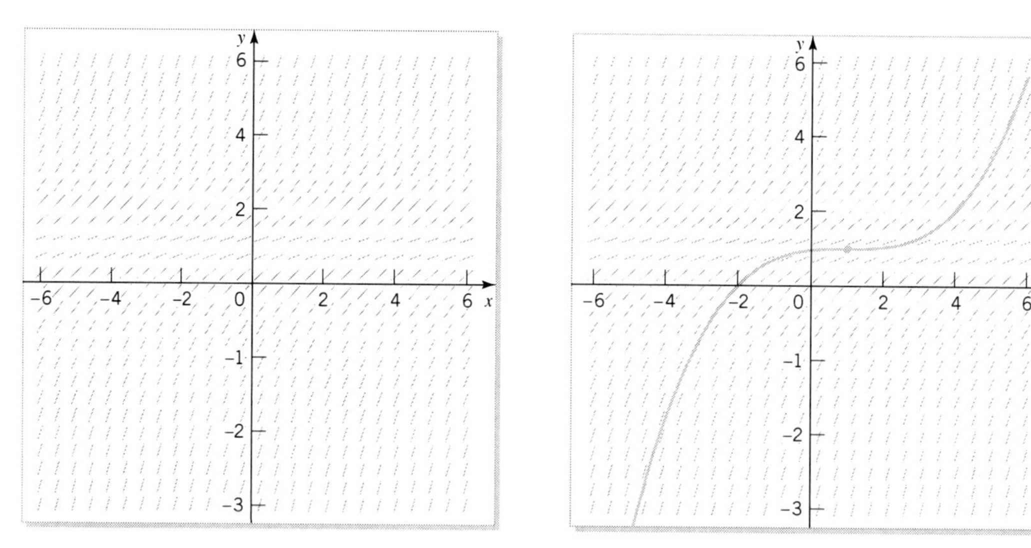

FIGURE 2.39 Slope field for $y' = (y - 1)^{2/3}$

FIGURE 2.40 Hand-drawn solution curve through $(1, 1)$ and slope field for $y' = (y - 1)^{2/3}$

[17]Some computer programs may draw the slope field only for $y > 1$. If this happens, try using $\left[(y - 1)^2\right]^{1/3}$ in place of $(y - 1)^{2/3}$.

Explicit solution

Now let's look at the analytical approach. Although the function $(y-1)^{2/3}$ is continuous everywhere, its derivative with respect to y — namely, $(2/3)\,(y-1)^{-1/3}$ — is discontinuous whenever $y=1$, which includes our initial condition. Thus, in spite of Figure 2.40, we are not guaranteed that there is only one solution through the point $(1,1)$.

We can write (2.35) in the form

$$\frac{1}{(y-1)^{2/3}}y' = 1,$$

which when integrated yields $3\,(y-1)^{1/3} = x + C$, or

$$y(x) = \frac{1}{27}\,(x+C)^3 + 1. \tag{2.37}$$

If we apply the initial condition $y(1) = 1$ to (2.37), we find $C = -1$, so that (2.37) reduces to

$$y(x) = \frac{1}{27}\,(x-1)^3 + 1. \tag{2.38}$$

Equilibrium solution

If we stop here, we would conclude that there is only one solution through $(1,1)$ and that Figure 2.40 is correct. We would be wrong, because (2.35) has an additional solution that we have overlooked, the equilibrium solution $y(x) = 1$. Furthermore, it also satisfies the initial condition (2.36). Thus,

$$y(x) = 1 \tag{2.39}$$

is another solution through $(1,1)$.

Figure 2.41 shows the slope field for (2.35), together with the two solutions (2.38) and (2.39). Notice that we missed the equilibrium solution $y(x) = 1$ when we drew Figure 2.40.

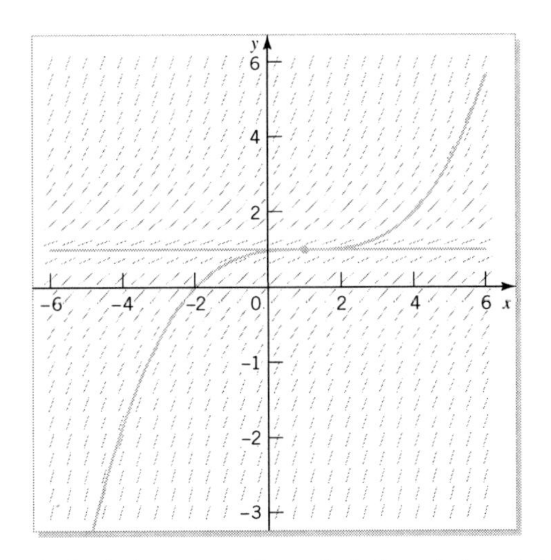

FIGURE 2.41 Analytical solution curves through $(1,1)$ and slope field for $y' = (y-1)^{2/3}$

If we look at Figure 2.41, we can actually see four solutions. The two obvious ones are (2.38) and (2.39). However, taking $y(x) = 1$ for $x \le 1$ and then taking the $y(x) = (x-1)^3/27 + 1$ branch is a third. Finally, taking $y(x) = (x-1)^3/27 + 1$ for $x \le 1$ and then taking $y(x) = 1$ is a fourth. In fact these last two are not only continuous at $x = 1$, but also differentiable there. (See Exercise 4 on page 64.) ■

Caution! **It is possible to miss nonunique solutions using graphical analysis.**

Limitations of Explicit Solutions

EXAMPLE 2.9

Now we consider an initial value problem for the autonomous differential equation

$$y' = y^{3/2}. \tag{2.40}$$

Uniqueness Both $y^{3/2}$ and its derivative with respect to y, $(3/2)\,y^{1/2}$, are continuous throughout their domains, so the Existence-Uniqueness Theorem applies for any x_0 and y_0. Equation (2.40) has the equilibrium solution $y(x) = 0$.

Calculus In the nonequilibrium solution case, if $y > 0$ the solution is increasing, and if $y < 0$ the differential equation is undefined. (Why?)

Because

$$y'' = \frac{3}{2}y^{1/2}y' = \frac{3}{2}y^{1/2}y^{3/2} = \frac{3}{2}y^2,$$

Slope field the solution is concave up for $y \neq 0$. The slope field for (2.40) is shown in Figure 2.42. Notice that nothing is drawn for $y < 0$. (Why?)

Explicit solution Now let's find the explicit solution of (2.40) in the nonequilibrium case, when $y(x) \neq 0$. We divide (2.40) by $y^{3/2}$ to find $y^{-3/2}y' = 1$, which when integrated yields $-2y^{-1/2} = x + C$, or

$$\sqrt{y} = \frac{-2}{x+C}. \tag{2.41}$$

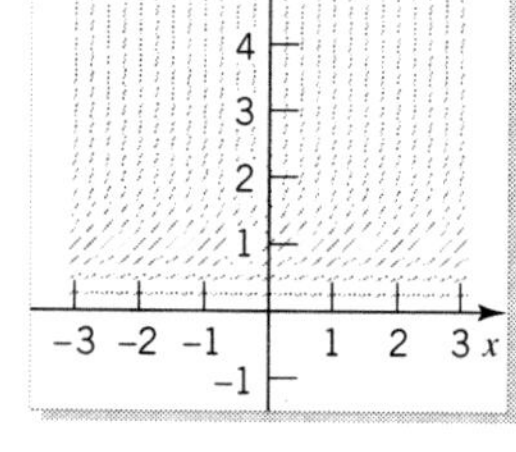

FIGURE 2.42
Slope field for $y' = y^{3/2}$

Squaring (2.41) gives

$$y(x) = \frac{4}{(x+C)^2}. \tag{2.42}$$

Notice that there is no finite value of C that allows the equilibrium solution $y(x) = 0$ to be absorbed into (2.42).

If we consider the particular case of (2.42) when $C = 0$ – namely, $y(x) = 4/x^2$ – we see that it passes through the points $(\pm 2, 1)$, for example. The slope field for (2.40), along with the function (2.42) with $C = 0$ – that is, $y(x) = 4/x^2$ – and the points $(\pm 2, 1)$, are shown in Figure 2.43.

Mistake! There is something wrong here! The right-hand part of $y(x) = 4/x^2$ and the slope field disagree. Where is our mistake? Well, we have made a very common oversight. In fact, had we merely obtained the explicit solution, without doing the graphical analysis, we wouldn't even realize that anything was wrong. Our mistake is going from (2.41) to (2.42). The right-hand side of (2.41) must be positive, which

Solution means that (2.41) is valid only for $x < -C$. Consequently, the correct way of expressing the family of solutions using (2.42) is

$$y(x) = \frac{4}{(x+C)^2}, \qquad x < -C.$$

Thus, $y = 4/x^2$ is the explicit solution of this initial value problem valid only for $x < 0$ and is plotted in Figure 2.44.

To illustrate what happens for another initial point, we note the explicit solution for the initial condition $y(2) = 1$ is $4/(x-4)^2$, valid for $x < 4$. It is shown in Figure 2.44.

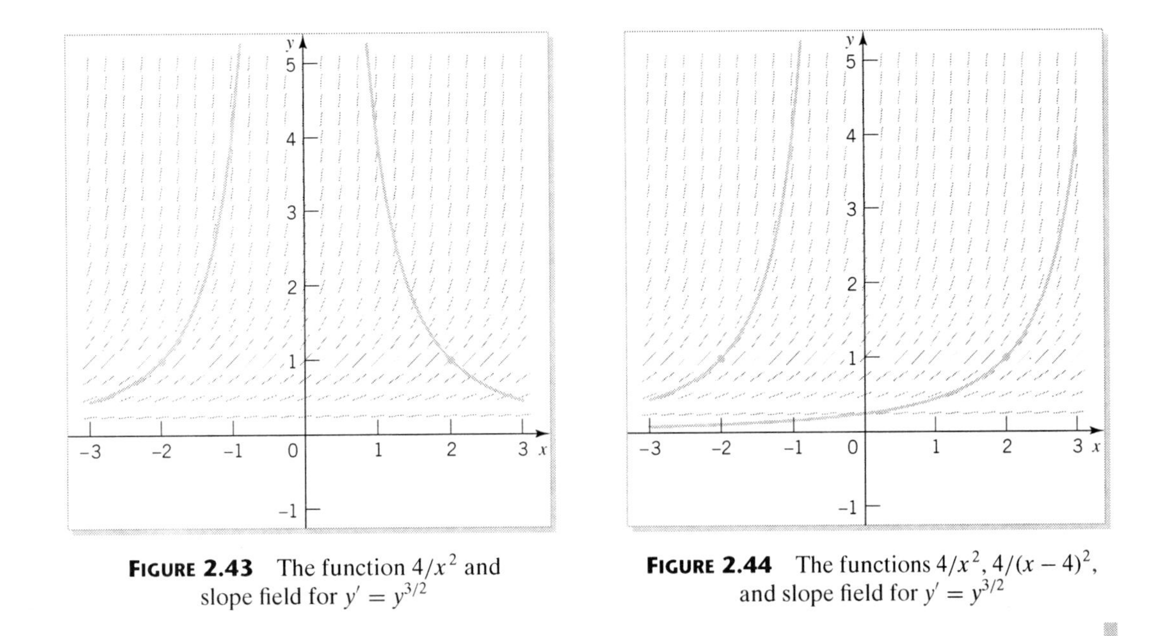

FIGURE 2.43 The function $4/x^2$ and slope field for $y' = y^{3/2}$

FIGURE 2.44 The functions $4/x^2$, $4/(x-4)^2$, and slope field for $y' = y^{3/2}$

Caution! **Be very careful when solving for $y(x)$. It is a common mistake to introduce extra functions that are not solutions.**

This last example has something in common with the solution (2.26) of the logistic equation $y' = ay(b-y)$, namely,

$$y(x) = \frac{bC}{e^{-abx} + C}.$$

In both cases there is a particular solution (in each case an equilibrium solution) that is not obtained from the family of solutions by selecting a finite value for C. Such solutions are called singular.

◆ *Definition 2.4:* **A particular solution of $y' = g(x, y)$ that cannot be obtained from the family of solutions (containing the arbitrary constant C) by selecting a finite value for C is called a SINGULAR SOLUTION.** ◆

EXAMPLE 2.10

Consider the autonomous differential equation

$$y' = \sqrt{1 - y^2} \tag{2.43}$$

subject to the initial condition

$$y(0) = 0. \tag{2.44}$$

Uniqueness

Both $\sqrt{1-y^2}$ and its derivative, $-y/\sqrt{1-y^2}$, are continuous in the vicinity of $(0, 0)$, so we don't anticipate any nonuniqueness problems.

Equilibrium solution

Equation (2.43) has the equilibrium solutions

$$y(x) = \pm 1. \tag{2.45}$$

Explicit solution

If we divide (2.43) by $\sqrt{1-y^2}$ and integrate, we have

$$\int \frac{dy}{\sqrt{1-y^2}} = \int dx,$$

which yields

$$\arcsin y = x + C. \tag{2.46}$$

If we use the initial condition (2.44) in (2.46), we find $C = 0$, so (2.46) reduces to

$$\arcsin y = x, \tag{2.47}$$

from which we get

$$y = \sin x. \tag{2.48}$$

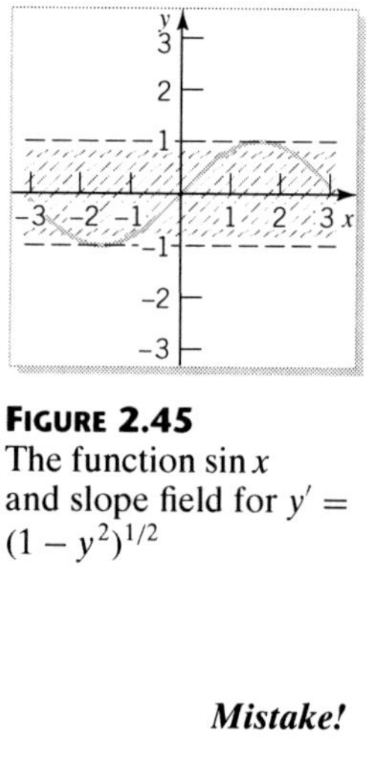

FIGURE 2.45
The function $\sin x$ and slope field for $y' = (1 - y^2)^{1/2}$

Mistake!

But wait a minute! According to (2.43), all solutions should be increasing (or have horizontal tangent lines when $y = \pm 1$), whereas the solution we have found, (2.48), oscillates. Something is wrong! This can also be seen from Figure 2.45, where the slope field for (2.43) and the solution (2.48) are sketched together.

Correct solution

We have made a similar mistake in going from (2.47) to (2.48) as we did in Example 2.9 on page 60. The range of the arcsin function is $[-\pi/2, \pi/2]$, so that (2.47) is valid only for $-\pi/2 \le x \le \pi/2$. Thus, (2.48) is valid only for $-\pi/2 \le x \le \pi/2$. The correct solution to (2.43) subject to (2.44) is

$$y = \sin x \quad \text{where} \quad -\frac{\pi}{2} \le x \le \frac{\pi}{2}. \tag{2.49}$$

This mistake is easy to make. In fact, most software packages that solve differential equations give (2.48) as the solution, which it is not.

Slope field

In Figure 2.46 we have sketched the slope field for (2.43), the solution (2.49), and the equilibrium solutions (2.45). (Why is the slope field drawn only in the region

$-1 \le y \le 1$?) At the points $(\pi/2, 1)$ and $(-\pi/2, -1)$, there appear to be nonunique solutions. (How is this consistent with the Existence-Uniqueness Theorem?)

In fact, it is possible to piece together parts of the equilibrium solutions and the function (2.49) to create a solution valid for $-\infty < x < \infty$. The function

$$y(x) = \begin{cases} -1 & \text{if } -\infty < x < -\pi/2 \\ \sin x & \text{if } -\pi/2 \le x \le \pi/2 \\ 1 & \text{if } \pi/2 < x < \infty \end{cases} \tag{2.50}$$

is a differentiable function that satisfies (2.43) subject to (2.44) for $-\infty < x < \infty$. See Figure 2.47. [You should check that (2.50) is differentiable at $x = \pm\pi/2$.]

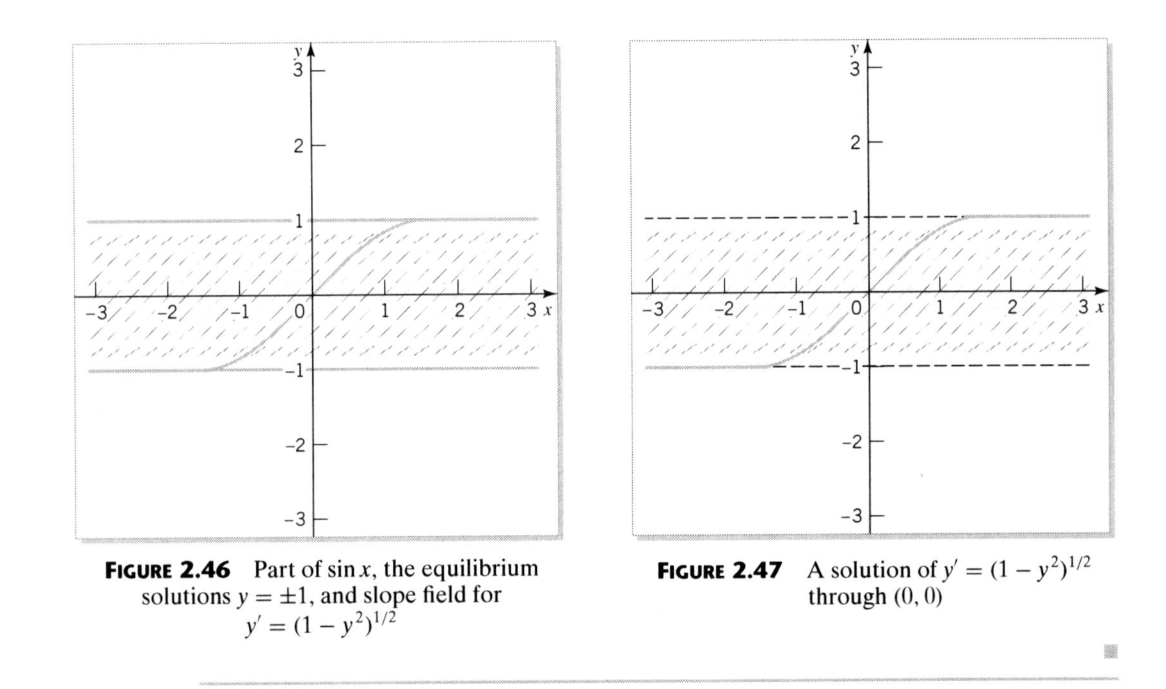

FIGURE 2.46 Part of $\sin x$, the equilibrium solutions $y = \pm 1$, and slope field for $y' = (1 - y^2)^{1/2}$

FIGURE 2.47 A solution of $y' = (1 - y^2)^{1/2}$ through $(0, 0)$

Caution! **Software packages that find solutions of differential equations analytically are not infallible. Many of the popular ones give $y(x) = \sin x$ as the solution of $y' = \sqrt{1 - y^2}$, subject to $y(0) = 0$. Slope fields are useful!**

EXERCISES

1. Does $y' = 3y - 2$ have any equilibrium solutions? For what initial condition(s) would this equilibrium solution be the unique solution of an initial value problem?

2. Consider the initial value problem $y' = y^{5/2}$, $y(0) = 1$.

(a) According to the Existence-Uniqueness Theorem, does this initial value problem have a unique solution?

(b) Solve $y' = y^{5/2}$ to find $-\frac{2}{3}y^{-3/2} = x + C$, or $y(x) = \left(-\frac{2}{3}\frac{1}{x+C}\right)^{2/3}$. Show, from this last equation, that the initial condition $y(0) = 1$ implies that $C = \frac{3}{2}$ or $C = -\frac{3}{2}$. Thus, there are two different solutions to the initial value problem, so the initial value problem does not have a unique solution.

(c) Reconcile the answers you gave to parts (a) and (b).

3. Solve the differential equation $y' = y^{1/3}$. Someone says that there are at least three different solutions passing through the point $(0, 0)$. Is that person right? Comment.

4. Show that the function defined by

$$f(x) = \begin{cases} 1 & \text{if } x \le 1 \\ \frac{1}{27}(x-1)^3 + 1 & \text{if } x > 1 \end{cases}$$

is differentiable everywhere.

5. The Logistic Equation. Consider the equilibrium solution $y = b$ of the logistic equation $y' = ay(b - y)$. Is it possible for a nearby solution to reach this solution for a finite value of x? [Hint: Use the Existence-Uniqueness Theorem.]

6. The Logistic Equation. Consider the logistic equation $y' = ay(b - y)$.

(a) If $y(0) > b$, show that the solution has a vertical asymptote to the left of $x = 0$. [Hint: Use (2.27).] What happens to this solution as you approach the asymptote from the right?

(b) If $y(0) < 0$, show that the solution has a vertical asymptote to the right of $x = 0$. What happens to this solution as you approach the asymptote from the left?

7. Poaching. One of the problems faced by national parks worldwide is the poaching of animals, some endangered. We will develop a very simple model to see what we can predict. We will assume that without poaching, the animal population obeys the logistic equation $y' = ay(b - y)$, where a and b are positive constants, and that poaching is done at a constant rate c, another positive constant. Thus, we assume that the animal population $y(x)$ at time x obeys the differential equation

$$y' = ay(b - y) - c. \tag{2.51}$$

(Why is there a minus sign in front of c?)

(a) What does your intuition suggest about the ultimate animal population if c is very large?

(b) Show that the change of scale $y = bY/2$, $x = 2X/(ab)$ allows the differential equation (2.51) to be written as

$$\frac{dY}{dX} = Y(2 - Y) - C, \tag{2.52}$$

where $C = 4c/(ab^2)$. Do not attempt to solve this equation.

(c) Find the equilibrium solutions of (2.52). There should be three cases.

i. No equilibrium solutions.

ii. One repeated equilibrium solution.

iii. Two distinct equilibrium solutions.

(d) Which equilibrium solutions are stable?

(e) Explain why case i of part (c) predicts that the animal population will die out in a finite time. Does this agree with the answer you gave to part (a)?

(f) Discuss under what circumstances cases ii and iii of part (c) predict that the populations will die out. What is needed for the population to survive? Under these circumstances, what is the ultimate size of the population?

8. Figure 2.48 shows one member of a family of solutions of the differential equation $y' = g(y)$, where $g(y)$ is a given continuous function. Use this information to plot the family of solutions. Do not attempt to find $y(x)$ or $g(y)$. How do you know that there are an infinite number of points where dg/dy is discontinuous?

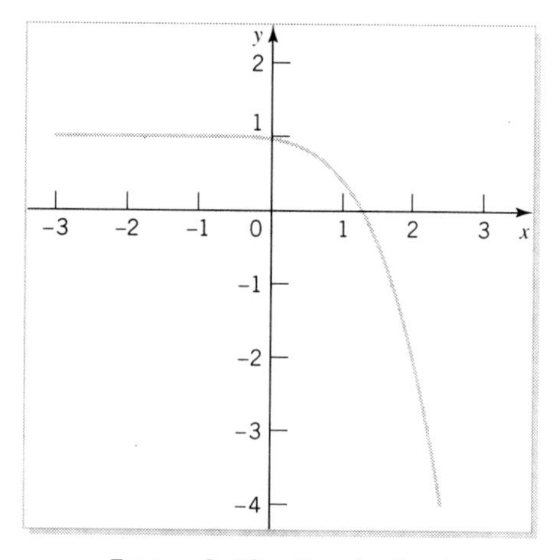

FIGURE 2.48 Graph of $y(x)$

9. Consider the differential equation $y' = g(y)$, where $g(y)$ is a given function for which dg/dy is continuous. If $y' = g(y)$ has exactly three equilibrium solutions, is it possible to write down a function $g(y)$ that satisfies the following conditions? If it is possible, write down $g(y)$ and comment on its uniqueness. If it is not possible, explain why not.

(a) The equilibrium solutions $y(x) = 1$, $y(x) = 2$, and $y(x) = 3$ are all stable.

(b) The equilibrium solutions $y(x) = 1$ and $y(x) = 2$ are stable, whereas $y(x) = 3$ is unstable.

(c) The equilibrium solution $y(x) = 1$ is stable, whereas $y(x) = 2$ and $y(x) = 3$ are both semistable.

10. The Doomsday Model. Table 2.12 shows the world population in billions from 1650 to 1990.[18] This data set is sketched in Figure 2.49. It is claimed that the differential equation

$$\frac{dP}{dt} = aP^2, \tag{2.53}$$

where a is a constant, models this situation very well. Investigate this claim. What is the long-term implication of this model?

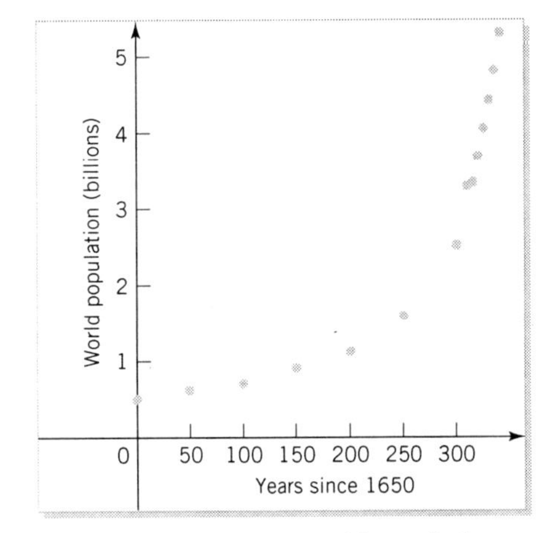

FIGURE 2.49 The world population in billions from 1650

Table 2.12 The world population in billions from 1650

Year	Population	Year	Population
0	0.510	310	3.307
50	0.625	315	3.354
100	0.710	320	3.696
150	0.910	325	4.066
200	1.130	330	4.432
250	1.600	335	4.822
300	2.525	340	5.318

11. Saving Money. When savings are compounded continuously at rate r, the principal, P, changes with time, t, according to the differential equation $dP/dt = rP$. In order to attract new business, a local bank offers to do better than this by compounding the principal of a new investor according to $dP/dt = rP^n$, where n is a constant, and $n > 1$. Discuss the pros and cons of opening a new account at this bank.

2.5 QUALITATIVE BEHAVIOR OF SOLUTIONS USING PHASE LINES

There is another way of analyzing autonomous differential equations, using what is called a PHASE LINE ANALYSIS. This method does not produce an explicit solution, but it can give the qualitative behavior of the solutions with very little calculation. The key to this technique is the Existence-Uniqueness Theorem, which guarantees that solutions cannot intersect. In this section we will demonstrate this phase line analysis by concentrating on a special case of the logistic equation.

EXAMPLE 2.11

We will perform a phase line analysis of

$$y' = y(2 - y). \tag{2.54}$$

[18] *Introduction to the Mathematics of Population* by N. Keyfitz, Addison-Wesley (1968), as reported in *Growth and Diffusion Phenomena* by R. T. Banks, Springer-Verlag, 1991, page 17.

Uniqueness

As usual we first find the equilibrium solutions of (2.54), which are $y(x) = 0$ and $y(x) = 2$. From the Existence-Uniqueness Theorem we know that no other solutions can intersect either of these two solutions. Thus, we will concentrate on the three regions created by these equilibrium solutions — namely, $y < 0$, then $0 < y < 2$, and finally $2 < y$.

1. $y < 0$. If $y < 0$, then by (2.54) $y' < 0$, so in this region y is a decreasing function of x. Thus, if $y < 0$ at any time — that is, if $y(x) < 0$ for any value of x — then, as x increases, y moves away from the equilibrium solution $y(x) = 0$.

2. $0 < y < 2$. In this case we see that (2.54) requires that $y' > 0$, so if at any value of x we have $0 < y < 2$, then the solution $y(x)$ will be an increasing function of x. It cannot intersect $y(x) = 2$.

3. $2 < y$. In this case $y' < 0$, so if at any value of x we have $2 < y$, then the solution $y(x)$ will be a decreasing function of x. It cannot intersect $y(x) = 2$.

Phase line

We can represent this graphically. We draw a horizontal straight line to represent the y values and identify the equilibrium solutions by the points 0 and 2 — called EQUILIBRIUM POINTS. Then in the three regions $y < 0$, $0 < y < 2$, and $2 < y$, we place arrows to indicate whether a solution $y(x)$ is moving away from or toward the equilibrium solution at its boundary, as x increases. For example, in case 1, if $y < 0$, then y moves away from the equilibrium solution $y = 0$; we indicate this with arrows pointing left. We do this for all three regions, and the result is Figure 2.50, called the PHASE LINE of (2.54).

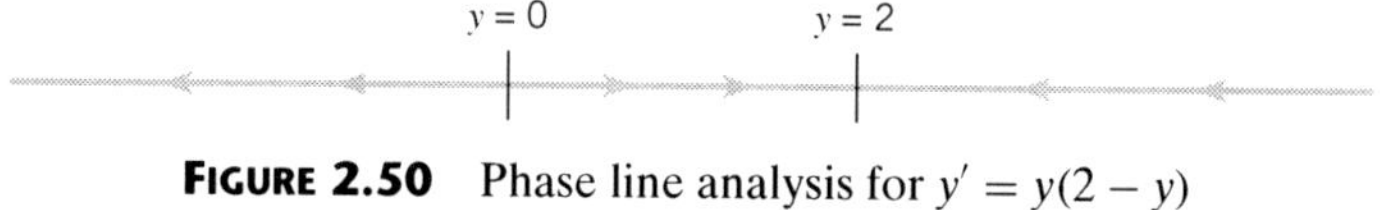

FIGURE 2.50 Phase line analysis for $y' = y(2 - y)$

Stability

That $y = 0$ is an unstable equilibrium solution is characterized by arrows on either side of the equilibrium point, each pointing away from it. Arrows pointing toward an equilibrium point indicate a stable equilibrium solution, in this case $y = 2$.

If initially $y(x_0) = y_0 > 2$, we expect the solution $y(x)$, which passes through the point (x_0, y_0), to decrease toward 2 as x increases. However, this solution $y(x)$ cannot reach 2 for any value of x, because $y(x) = 2$ is another solution, and the Existence-Uniqueness Theorem guarantees that two solutions cannot intersect.

Figure 2.51 shows typical solutions consistent with the phase line analysis. Notice that the y-axis is the phase line.

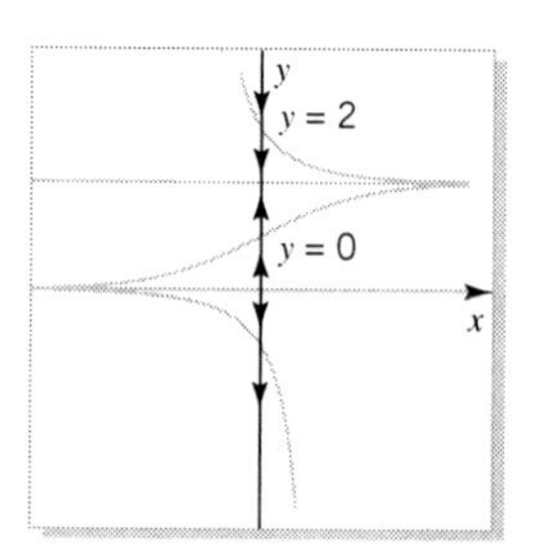

FIGURE 2.51 Solution curves of $y' = y(2 - y)$

In Exercise 6 on page 64, we saw that some solutions of the logistic equation $y' = y(2 - y)$ had vertical asymptotes. If initially $y(0) > 2$, then there is a negative value of x at which y came in from $+\infty$. If $y(0) < 0$, then there is a positive value of x at which $y \to -\infty$. Thus, if $y(0) > 2$, the solution does not go back all the way to $x = -\infty$, and if $y(0) < 0$, the solution does not go forward all the way to $x = +\infty$. However, Exercise 5 on page 64 shows that if $0 < y(0) < 2$, then the solution extends from $-\infty$ to $+\infty$. The phase line analysis, like the graphical analysis, does not show these properties. ■

We now collect some ideas about phase lines.

Summary

- A simple way to construct the phase line for $y' = g(y)$ is to graph $g(y) = y(2 - y)$ as a function of y (see Figure 2.52), making y the horizontal axis.

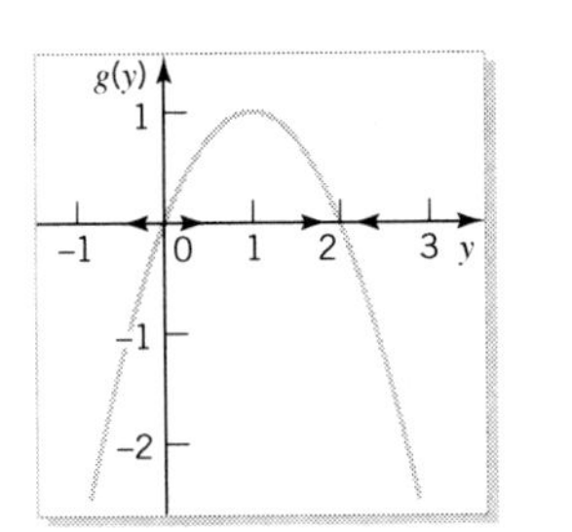

FIGURE 2.52
The function $y(2 - y)$ versus y

  (a) The equilibrium points are where the graph of $g(y)$ crosses the y-axis, so identify those points on the y-axis.
  (b) The values of y for which $g(y) > 0$ are where the function y is increasing. Identify those regions on the y-axis with arrows pointing to the right.
  (c) The values of y for which $g(y) < 0$ are where the function y is decreasing. Identify those regions on the y-axis with arrows pointing to the left.
  (d) The y-axis is now the phase line.

- Another way of thinking of the phase line is to imagine the solution curves projected onto the y-axis. In Figure 2.51, if we imagine looking down the x-axis from $x = \infty$, and then project the directions of all solution curves onto the y-axis, we could characterize the properties of these solutions with arrows along the vertical axis. Rotating this line by 90° gives the phase line in Figure 2.50.

EXAMPLE 2.12

We will perform a phase line analysis of

$$y' = y. \tag{2.55}$$

We now look at the phase line for (2.55). There is only one equilibrium point — namely, $y = 0$ — and the phase line in Figure 2.53 is easily constructed, by realizing that $g(y) = y$ is negative for $y < 0$ and positive for $y > 0$.

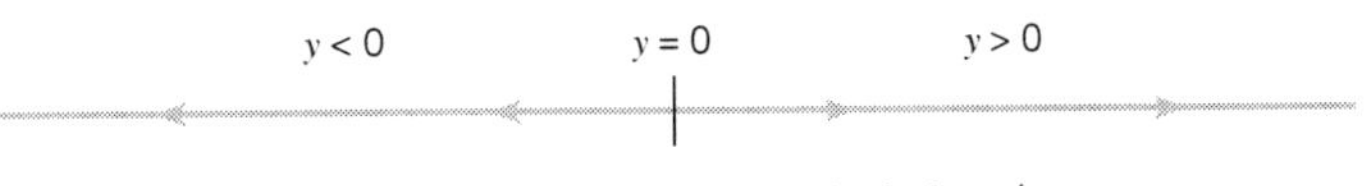

FIGURE 2.53 Phase line analysis for $y' = y$

Unstable equilibrium

Now let us analyze this phase line. From the direction of the arrows we see that $y = 0$ is an unstable equilibrium. If a solution curve starts out with $y < 0$, it will decrease as x increases and move away from the equilibrium solution $y(x) = 0$. In the same way, a solution that starts with $y > 0$ increases with x. If we compare this observation with Figure 2.8 (on page 34), we see that Figure 2.53 can be obtained by projecting Figure 2.8 onto the y-axis, inserting arrows to indicate increasing or decreasing solutions, and rotating 90° clockwise. ■

EXAMPLE 2.13

We will perform a phase line analysis of

$$y' = y^2. \tag{2.56}$$

There is one equilibrium solution — namely, $y(x) = 0$. Because $g(y) = y^2$ is a parabola opening up that touches the y-axis at $y = 0$, the phase line for (2.56) is given in Figure 2.54.

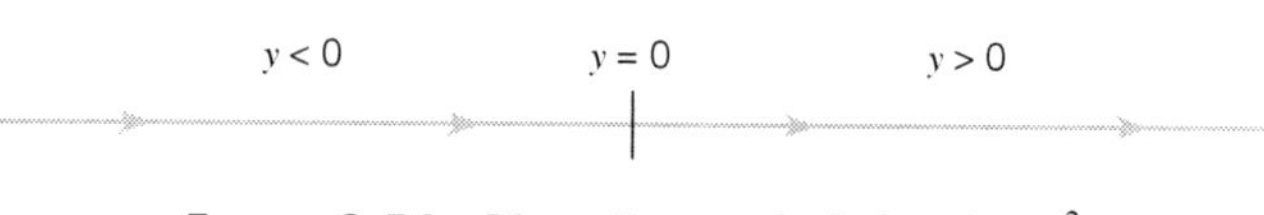

FIGURE 2.54 Phase line analysis for $y' = y^2$

Semistable equilibrium

Solutions of (2.56) that start with $y < 0$ increase toward the equilibrium solution $y = 0$ as x increases, whereas solutions that start with $y > 0$ also increase and move away from the equilibrium solution $y = 0$ as x increases. We see that the equilibrium solution $y(x) = 0$ is neither a stable nor an unstable equilibrium. It is semistable.

We should not think that the arrows in the phase line in Figure 2.54 imply that as x increases, a solution that starts with $y < 0$ passes through $y = 0$ and continues increasing. The equation $y(x) = 0$ is a solution and, because of the Existence-Uniqueness Theorem, no other solution can pass through it. So Figure 2.54 tells us that if a solution starts with $y < 0$, it stays with $y < 0$. If we compare this observation with Figure 2.38 (on page 55), we see that Figure 2.54 can be obtained by projecting Figure 2.38 onto the y-axis, inserting arrows to indicate increasing solutions, and rotating 90° clockwise.

Caution!

When analyzing the phase line, realize that if the Existence-Uniqueness Theorem is satisfied, no solutions intersect equilibrium solutions.

How to Perform a Phase Line Analysis on $y' = g(y)$

Purpose To construct and analyze the phase line of the autonomous differential equation

$$y' = g(y). \tag{2.57}$$

Process

1. Make sure that $g(y)$ satisfies the conditions of the Existence-Uniqueness Theorem on page 56. (It will if dg/dy is continuous.)
2. Solve $g(y) = 0$ to determine all the equilibrium solutions of (2.57), and mark them on the phase line. Remember these are solutions.

3. Decide on the sign of y' – that is, $g(y)$ – by sketching $g(y)$ as a function of y. Between successive equilibrium solutions, mark the phase line with an arrow pointing left if $g(y) < 0$ and right if $g(y) > 0$.
4. Identify the type of equilibrium solution (stable, unstable, semistable) according to Figure 2.55.

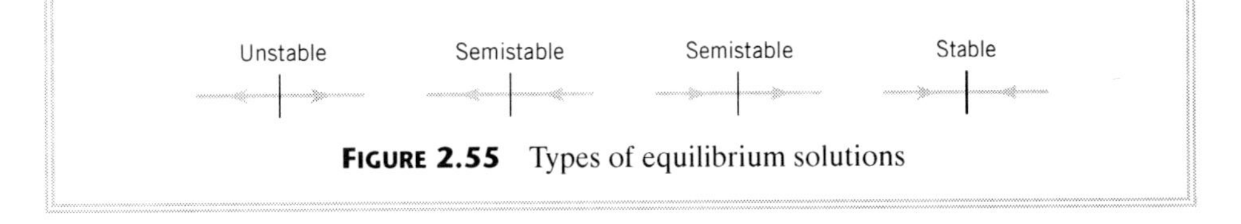

FIGURE 2.55 Types of equilibrium solutions

We now consider an example where we use the phase line to discover properties of a solution of a differential equation when the explicit solution is not available.

EXAMPLE 2.14 *An Epidemic Model*[19]

Consider the differential equation

$$y' = 5 - 0.5y - 5e^{-y}, \tag{2.58}$$

subject to $y(0) = 0.1$. This differential equation occurs when modeling epidemics, where y represents the total number of people who have contracted a disease and have been "removed" at time x. Depending on the epidemic, "removed" might mean isolated, immune, or deceased, for example.

Phase line

The differential equation (2.58) is autonomous and satisfies the conditions of the Existence-Uniqueness Theorem. Thus, we have unique solutions so we can do a phase line analysis. This means that we first plot the function $g(y) = 5 - 0.5y - 5e^{-y}$ – but what domain should we consider? After inspecting $g(y)$ we notice that $g(0) = 0$, which gives us one equilibrium solution, $y(x) = 0$. Are there any others? If we think of $g(y)$ for y large, we see that the term $-5e^{-y}$ will be negligible when compared to the other two terms; thus, for large y, $g(y) \approx 5 - 0.5y$, which is zero when $y = 10$, so we anticipate another equilibrium solution at approximately $y(x) = 10$. Using a numerical root-finding method to solve $g(y) = 0$ near $y = 10$, we find $y \approx 9.999546$, so that the second equilibrium solution is at $y(x) \approx 9.999546$.

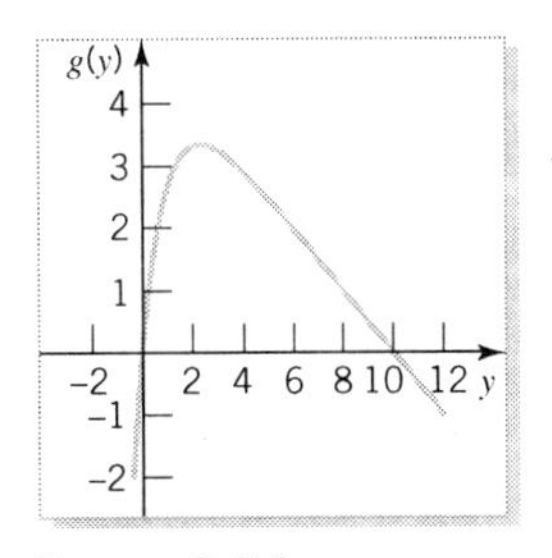

FIGURE 2.56 The graph of $g(y) = 5 - 0.5y - 5e^{-y}$

Furthermore,

$$\frac{dg}{dy} = -0.5 + 5e^{-y} = -0.5e^{-y}(e^{y} - 10), \tag{2.59}$$

Monotonicity

so $g(y)$ is increasing if $e^y - 10 < 0$ – that is, for $-\infty < y < \ln 10 \approx 2.3$ – and decreasing if $e^y - 10 > 0$ – that is, for $\ln 10 < y < \infty$. (Why?) Figure 2.56 shows the graph of $g(y)$. This is similar to the graph of a parabola, and so will have a similar

[19] "A contribution to the mathematical theory of epidemics" by W. O. Kermack and A. G. McKendrick, *Proc. Roy. Soc.* 115A, 1927, pages 700–721.

phase line to the logistic equation — that is, the equilibrium solution $y(x) = 0$ is unstable while $y(x) \approx 9.999546$ is stable. See Figure 2.57. Thus, we know that the solutions of (2.58) will have the general appearance of the logistic equation with a carrying capacity of 9.999546. However, from (2.58) and (2.59) we know that the point of inflection occurs at $y \approx 2.3$, rather than at half the carrying capacity, namely, 4.99973.

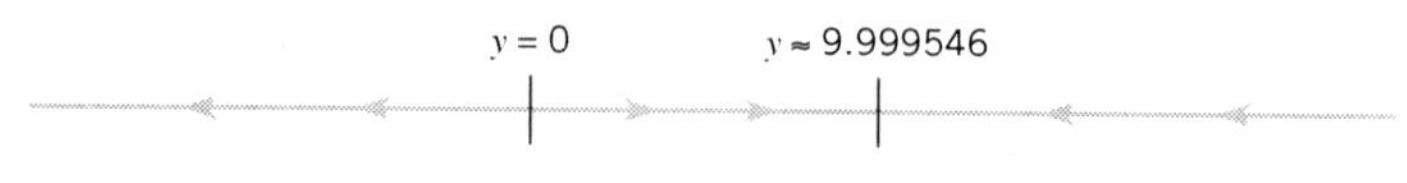

FIGURE 2.57 Phase line analysis for $y' = 5 - 0.5y - 5e^{-y}$

If we try to solve (2.58) analytically for nonequilibrium solutions, we are led to

$$\int \frac{1}{5 - 0.5y - 5e^{-y}}\, dy = x + C.$$

The integral cannot be evaluated in terms of familiar functions, so we cannot find an explicit solution for y. So how can we graph the solution through (0, 0.1) accurately?

Explicit solution

If we use the initial condition $y(0) = 0.1$, we can write the solution in the form

$$\int_{0.1}^{y} \frac{1}{5 - 0.5t - 5e^{-t}}\, dt = x.$$

This gives x as an explicit function of y for $0 < y < 9.999546$. We can use a standard numerical integration technique to evaluate the integral for different values of y. Table 2.13 was constructed in this way, using Simpson's rule. Notice that it gives x as a function of y. Figure 2.58 shows the slope field and the numerical solution of (2.58)

Table 2.13
Numerical approximation to the solution of $y' = 5 - 0.5y - 5e^{-y}$

y	x
0.1	0.000
1.0	0.637
2.0	0.962
3.0	1.264
4.0	1.587
5.0	1.959
6.0	2.410
7.0	2.988
8.0	3.801
9.0	5.189

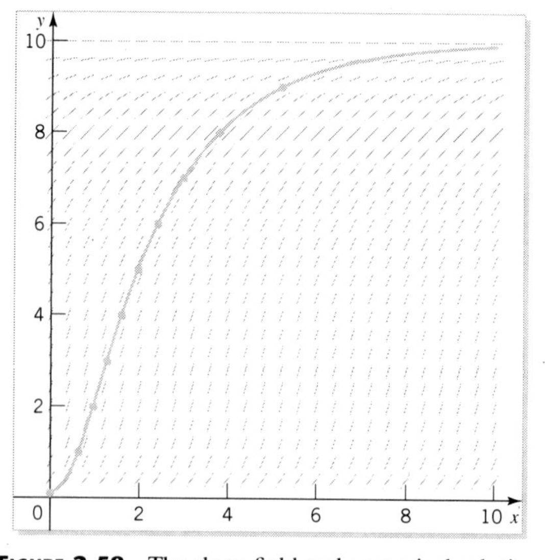

FIGURE 2.58 The slope field and numerical solution of $y' = 5 - 0.5y - 5e^{-y}$ subject to $y(0) = 0.1$

subject to $y(0) = 0.1$ from Table 2.13. This is in good agreement with our phase line analysis and the location of the inflection point.

Thus, even though we have not found an explicit solution for y, we have a good idea what the solution looks like. The phase line and the techniques of calculus provided a powerful tool for understanding this solution. This is not true just for this case; it is also true in many other situations. ■

In the previous example we used the sign of the derivative of $g(y)$ to determine the stability of the equilibrium solutions. In fact, if we look at the two points in Figure 2.56 where $g(y)$ crosses the horizontal axis ($y = 0$ and $y \approx 9.999546$), we see that $\frac{dg}{dy} > 0$ when $y = 0$ (the unstable equilibrium), and $\frac{dg}{dy} < 0$ when $y = 9.999546$ (the stable equilibrium). This observation leads to the following test to decide on the stability of an equilibrium solution.

► **Theorem 2.2: The Derivative Test for Stable or Unstable Equilibrium Solutions.** *Consider the autonomous differential equation $dy/dx = g(y)$. Let $y(x) = y_e$ be an equilibrium solution of the differential equation so that $g(y_e) = 0$. Let dg/dy be continuous in the neighborhood of y_e.*

(a) If $\frac{dg}{dy} < 0$ at $y = y_e$, then the equilibrium solution $y(x) = y_e$ is stable.

(b) If $\frac{dg}{dy} > 0$ at $y = y_e$, then the equilibrium solution $y(x) = y_e$ is unstable.

(c) If $\frac{dg}{dy} = 0$ at $y = y_e$, then the equilibrium solution $y(x) = y_e$ may be stable, unstable, or semistable. ◄

To prove this theorem, we expand $g(y)$ as a Taylor series about $y = y_e$ so we have

$$g(y) \approx g(y_e) + (y - y_e)g'(y_e),$$

where $g'(y_e)$ is the value of $\frac{dg}{dy}$ at $y = y_e$. Because $g(y_e) = 0$, we have

$$g(y) \approx (y - y_e)g'(y_e).$$

Thus, near y_e, we have $dy/dx = g(y) \approx (y - y_e)g'(y_e)$.

(a) If $g'(y_e) < 0$, then $dy/dx > 0$ if $y < y_e$ and $dy/dx < 0$ if $y > y_e$ for y near y_e. This characterizes $y(x) = y_e$ as stable.

(b) If $g'(y_e) > 0$, then $dy/dx < 0$ if $y < y_e$ and $dy/dx > 0$ if $y > y_e$ for y near y_e. This characterizes $y(x) = y_e$ as unstable.

(c) If $g'(y_e) = 0$, then from this analysis alone we cannot characterize the equilibrium solution.

This result is particularly useful if the differential equation contains parameters.

EXAMPLE 2.15

Identify the stability of the equilibrium solutions of the differential equation $dy/dx = y(a - y)$, where a is a constant.

Here $g(y) = ay - y^2$ and the equilibrium solutions are $y(x) = 0$ and $y(x) = a$. Because $g'(y) = a - 2y$, we find $g'(0) = a$, and $g'(a) = -a$. Thus, if $a < 0$, the equilibrium solution $y(x) = 0$ is stable, whereas the equilibrium solution $y(x) = a$ is unstable. If $a > 0$, the equilibrium solution $y(x) = 0$ is unstable, whereas the equilibrium solution $y(x) = a$ is stable. If $a = 0$, then $g'(0) = 0$, and the test is inconclusive. (In fact, in this case the equilibrium solution is semistable.)

The preceding theorem deals with equilibrium solutions. What can we say about nonequilibrium solutions? If we look at all the autonomous equations we have discussed in this chapter, we see that their solutions are either equilibrium solutions, or monotonic — there are no oscillating solutions. This is true in general, as the following theorem shows.

Theorem 2.3: *Nonequilibrium solutions of $y' = g(y)$ are monotonic if $g(y)$ is continuous.*

We outline the proof of this theorem. We suppose that there is a nonequilibrium solution of $y' = g(y)$ that is not monotonic. This would mean that there exist numbers x_1 and x_2 ($x_1 < x_2$) for which $y(x_1) = y(x_2)$, with $y(x) \neq y(x_1)$ for $x_1 < x < x_2$. From $y(x_1) = y(x_2)$ and $y'(x) = g(y)$, we have $y'(x_1) = y'(x_2)$. Without loss of generality, we can assume that $y'(x_1) = y'(x_2) > 0$, which means that $y(x)$ is increasing in the intervals around x_1 and x_2. Thus, for $x_1 < x$ and x near x_1, we must have $y(x_1) < y(x)$, whereas for $x < x_2$ and x near x_2, we must have $y(x) < y(x_2) = y(x_1)$. So between x_1 and x_2, say at $x = c$, $y(x)$ crosses the horizontal line joining $y(x_1)$ to $y(x_2)$. Thus, there is a point $x = c$ for which $x_1 < c < x_2$ and $y(c) = y(x_1)$. See Figure 2.59. This is a contradiction, so all nonequilibrium solutions are monotonic.

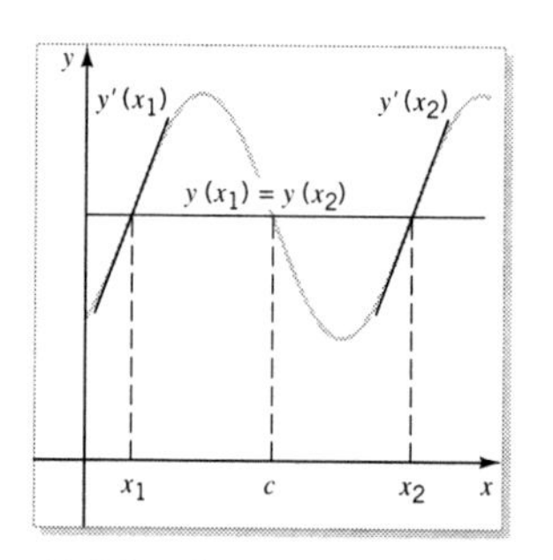

FIGURE 2.59 Geometric representation

No oscillations

An immediate consequence of this theorem is that **solutions of autonomous equations cannot oscillate**, so it is pointless trying to model periodic behavior with an autonomous equation.

EXERCISES

1. **The Gompertz Equation.** The Gompertz equation is $y' = -ky \ln(y/b)$, where k and b are positive constants. Use a phase line analysis to sketch the solution curves of this equation. What other differential equation has solution curves with some of these properties?

2. On the same graph sketch the functions $y(2 - y)$ and $\sin\left(\frac{\pi}{2}y\right)$ as functions of y for $0 \leq y \leq 2$. What does this suggest about solutions of $y' = y(2 - y)$ and $y' = \sin\left(\frac{\pi}{2}y\right)$, with $y(0) = y_0$, where $0 < y_0 < 2$? What happens if $y_0 > 2$?

3. **Passenger Pigeons.** During the nineteenth century, there was a massive population of passenger pigeons in the United States. In order to breed successfully, a large number of pigeons had to be present. Successful breeding rarely occurred when small numbers were present. In the latter part of the century, passenger pigeons were hunted more heavily than in the past, and within 30 years they were extinct. It has been suggested that the population, P, of passenger pigeons, as a function of time, t, could be modeled by the differential equation $\frac{dP}{dt} = aP(P-b)(c-P)$, where a, b, and c are positive constants, and $b < c$. Without solving this differential equation, but by sketching the family of solutions after using a phase line analysis, decide whether this is a reasonable model. If it is, what are the physical interpretations of b and c?

4. Consider the differential equation $y' = \cos y$.
   (a) Do any of the solution curves have inflection points? If so, where are they; if not, why not?
   (b) List all equilibrium solutions and use a phase line analysis to discuss the qualitative behavior of the solution of this differential equation subject to $y(0) = y_0$ for various y_0. What happens to the solution as $x \to \infty$? [Hint: It may be informative to look at the slope field and draw solutions through the following points — $(0, -2)$, $(0, 0)$, $(0, \pi/2)$.]
   (c) Find the explicit solution of the initial value problem in part (b). From this solution, can you tell what happens to the solution as $x \to \infty$?

5. Consider the differential equation

$$y' = (y-a)(y-b)(y-c). \tag{2.60}$$

   (a) Does the Existence-Uniqueness Theorem apply to this equation?
   (b) Confirm that $y(x) = a$, $y(x) = b$, and $y(x) = c$ are the equilibrium solutions of (2.60).
   (c) Draw the phase line for each of the following cases i through iii. In each case identify all the equilibrium solutions as being stable, unstable, or semistable.
      i. $a = b = c = 1$
      ii. $a = b = 1, c = -1$
      iii. $a = 1, b = 0, c = -1$
   (d) Based on the information you obtained in part (c), describe qualitatively the behavior of the solutions in each of the three cases in part (c). Are your descriptions consistent with part (a), in particular as far as intersecting solutions are concerned?
   (e) Use a computer/calculator program to draw several solution curves for each of the cases in part (c). Make sure the results you obtain here are consistent with the answers you gave to part (d).

6. **Sandhill Cranes.**[20] If a population is being poached at the constant rate c, then the logistic equation has a term $-c$ added to the right-hand side, giving $y' = ay(b-y) - c$.
   (a) Use a phase line analysis to draw typical solution curves of this differential equation. Consider a and b as fixed positive numbers, and explain what happens for different values of c. Determine if there are any equilibrium solutions, and if so, determine their stability. Compare your answer to the one you found when doing Exercise 7 on poaching on page 64.
   (b) Solve the differential equation for the case $ab = 0.87$, $b = 194600$, and $y(0) = 194600$. This corresponds to the hunting of sandhill cranes at the constant annual rate of c cranes per year. What value of c gives an equilibrium solution in this case? What happens if $c = 5000$ cranes/year? If $c = 4000$ cranes/year?

7. In each of the following differential equations, explain how the behavior of the solution varies with the value of the constant a. Are there any values of a at which this behavior changes dramatically? (Do not attempt to solve these equations.)

   (a) $y' = a - y^2$ (b) $y' = y(a - y^2)$

8. **Qualitatively Equivalent Differential Equations.** Two autonomous differential equations are **qualitatively equivalent** if they have essentially the same phase line; namely, the same number of equilibrium points, the same type of stability at each of these points, and the points in the same order. If two autonomous differential equations are qualitatively equivalent, will their solutions behave in similar ways (increasing, decreasing concave up, concave down, asymptotes)? Why? Which of the following differential equations are qualitatively equivalent? Do not attempt to solve any of them.

   (a) $y' = e^y$ (b) $y' = y$
   (c) $y' = 1 - y^2$ (d) $y' = y(1 - y^2)$
   (e) $y' = 1 - y^4$ (f) $y' = ye^y$
   (g) $y' = 1 + y^2$ (h) $y' = (1-y)(2-y)(3-y)$
   (i) $y' = \cosh y$ (j) $y' = y^3$
   (k) $y' = -y \ln y$ (l) $y' = \sin y$

[20] *Growth and Diffusion Phenomena* by R. T. Banks, Springer-Verlag, 1991, pages 123–125.

9. Consider the differential equation $y' = g(y)$ where $g(y)$ has a continuous derivative. Show that between any two equilibrium solutions, the solution curves change concavity. [Hint: Sketch $g(y)$.]

10. Give an example of an autonomous differential equation $y' = g(y)$ with an equilibrium solution $y(x) = y_e$ where $g'(y_e) = 0$, and the equilibrium solution is (a) stable, (b) unstable, (c) semistable.

11. The Mile Race. The men's world record for the mile race has been decreasing as the years pass by. Sketch a possible graph of time, T, for the world record (in seconds) versus the year, t, in which the world record was broken. Use the fact that Roger Bannister broke the 4-minute mile barrier in 1954 as a reference point. Make sure you give some thought to what happens to T as $t \to \infty$ and as $t \to -\infty$. Now write down an autonomous differential equation, relating dT/dt to a function of T that would have your graph as a possible solution. This equation should contain three arbitrary parameters. Table 2.14 shows how the mile record has dropped since 1911, and Figure 2.60 graphs this data set.[21] Does this data set have the shape of your graph? Try to estimate the three parameters in your differential equation from this data set. What does your model predict for the ultimate men's world record for the mile?

Table 2.14 The men's world record for the mile in seconds

Year	Record	Year	Record
1911	255.4	1957	237.2
1913	254.4	1958	234.5
1915	252.6	1962	234.4
1923	250.4	1964	234.1
1931	249.2	1965	233.6
1933	247.6	1966	231.3
1934	246.8	1967	231.1
1937	246.4	1975	229.4
1942	244.6	1979	229.0
1943	242.6	1980	228.8
1944	241.6	1981	227.33
1945	241.4	1985	226.31
1954	238.0	1993	224.39

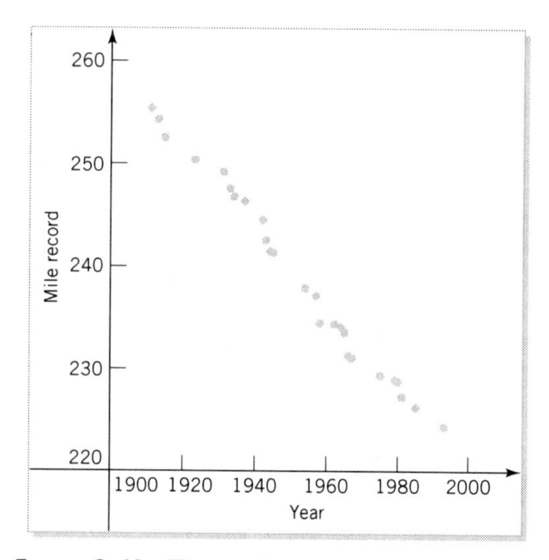

FIGURE 2.60 The men's world record for the mile

12. It is claimed that Figure 2.61 shows a solution curve of $y' = g(y)$ where $g(y)$ is continuous. Explain why this claim is false.

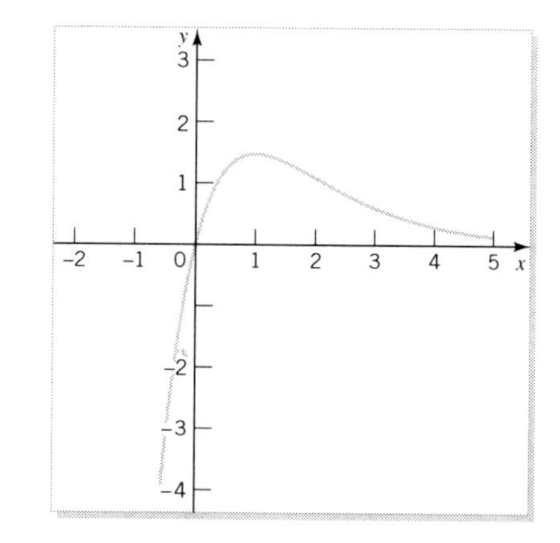

FIGURE 2.61 Is this a solution curve of $y' = g(y)$?

13. Consider the differential equation $y' = \frac{y^2-4y}{4-2y}$.

(a) Find all the equilibrium solutions of this differential equation and use the Derivative Test to determine their stability.

[21] "History of the Record for the Mile Run," *Information Please Almanac* 1995, page 957.

(b) Find the explicit solution of this differential equation subject to the initial condition $y(0) = y_0$. How does the behavior of this solution, as $x \to \infty$, depend upon y_0? Is this consistent with the answer you found in part (a)?

(c) Based only on the answers you found in parts (a) and (b), try to sketch a phase line consistent with this information. What problems do you encounter?

(d) By plotting the function $\frac{y^2-4y}{4-2y}$ versus y, explain the phase line you obtained in part (c).

2.6 BIFURCATION DIAGRAMS

The behavior of a solution of a differential equation containing a parameter may vary dramatically depending on the value of the parameter.

EXAMPLE 2.16 *The Logistic Equation*

As an example of this we consider the logistic equation $y' = y(a - y)$, where the parameter a may be any number: negative, zero, or positive.

Figure 2.62 shows typical solutions corresponding to $a = -2$, $a = 0$, and $a = 2$. As we increase a, from negative to zero to positive, the equilibrium solution $y(x) = a$ gradually rises. Notice that if $a < 0$ the equilibrium solution $y(x) = 0$ is stable, whereas the equilibrium solution $y(x) = a$ is unstable. If $a = 0$ there is only one equilibrium solution — namely, $y(x) = 0$ — and it is semistable. If $a > 0$ the equilibrium solution $y(x) = 0$ is unstable, whereas the equilibrium solution $y(x) = a$ is stable. Thus, the behavior of the solution changes at $a = 0$.

If in Figure 2.62 we add arrows along the vertical axes to produce a phase line for each of these three cases, we find Figure 2.63. Imagine we were to repeat this process for many different values of a. We have pointed out that as a increases through negative values toward 0, the two equilibrium solutions get closer together until they coincide when $a = 0$. Then as a increases further, the equilibrium solutions get farther apart. Thus, if we plot y versus a, we find Figure 2.64 — called the BIFURCATION DIAGRAM of $y' = y(a - y)$.

We now look at this bifurcation diagram to understand what it represents. We will concentrate on $a = -2$. If for a particular x the value of y is positive, then as x increases, the solution $y(x)$ approaches the equilibrium solution $y(x) = 0$. That is what the downward pointing arrow above the point $a = -2$, $y = 0$, in Figure 2.64

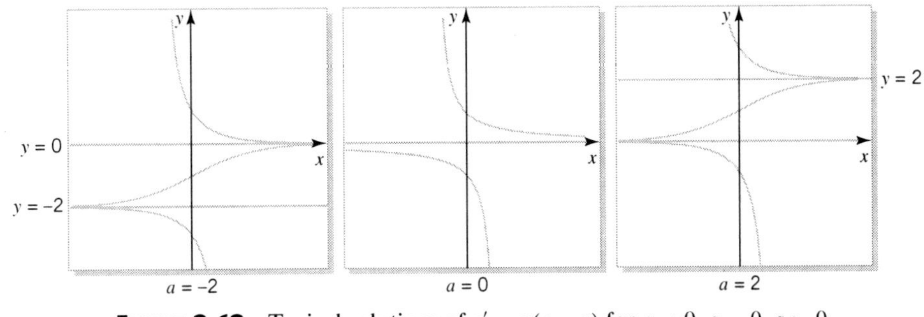

FIGURE 2.62 Typical solutions of $y' = y(a - y)$ for $a < 0$, $a = 0$, $a > 0$

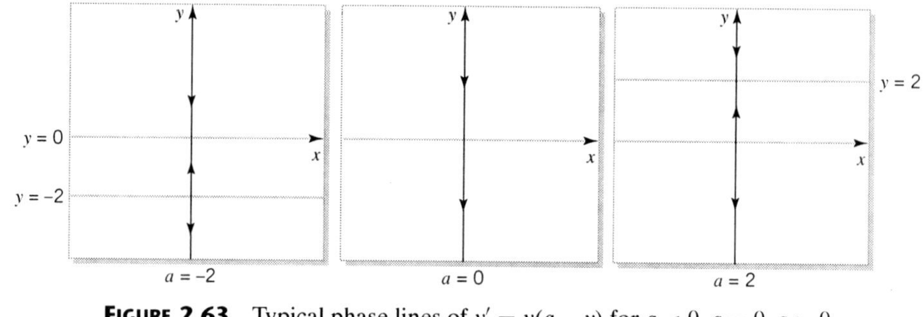

FIGURE 2.63 Typical phase lines of $y' = y(a - y)$ for $a < 0, a = 0, a > 0$

represents. In the same way, if $a = -2$ and $y < -2$, then $y \to -\infty$ as x increases. If $a = -2$ and $-2 < y < 0$, then $y(x) \to 0$ as x increases. If $y(x) = 0$ or $y(x) = -2$, then $y(x)$ remains at $y(x) = 0$ or $y(x) = -2$ as x increases (the equilibrium solutions). Thus, with $a = -2$, the equilibrium solution $y(x) = 0$ is stable and the equilibrium solution $y(x) = -2$ is unstable. However, if $a > 0$ the roles of these equilibrium solutions are reversed. If we indicate the stable equilibrium solutions by a solid line, and the nonstable equilibrium solutions by a dotted line, we can eliminate the arrows from Figure 2.64, and draw Figure 2.65 — another form of the bifurcation diagram.

Now that we have understood the bifurcation diagram of $y' = y(a - y)$, we will show an easy way to construct it. In the ay-plane we plot $y(a - y) = 0$; that is, the two lines $y = 0$ and $y = a$. This gives us the diagram shown in Figure 2.64. All we need to do is identify the equilibrium solutions as stable or unstable. Here we may use the Derivative Test for Stable or Unstable Equilibrium Solutions described on page 71.

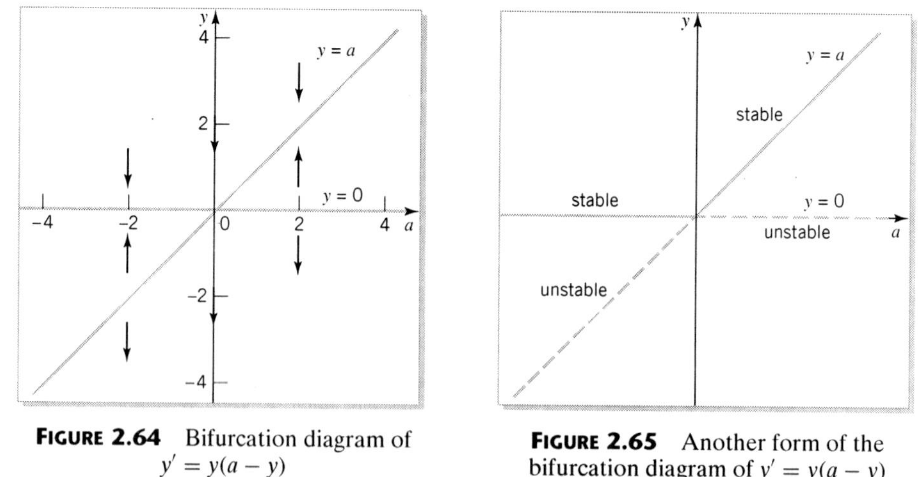

FIGURE 2.64 Bifurcation diagram of $y' = y(a - y)$

FIGURE 2.65 Another form of the bifurcation diagram of $y' = y(a - y)$

The bifurcation diagram for $y' = y(a - y^2)$ is shown in Figure 2.66 — confirm that it is correct. It is called a PITCHFORK BIFURCATION.

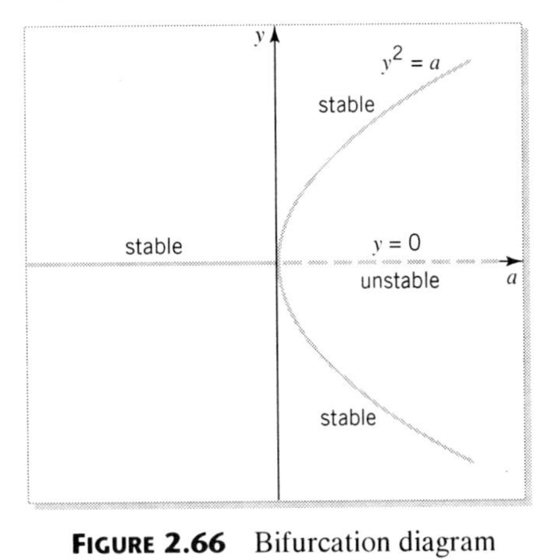

FIGURE 2.66 Bifurcation diagram of $y' = y(a - y^2)$

EXERCISES

1. Construct the bifurcation diagrams for the following differential equations.

   (a) $y' = a - y^2$ (b) $y' = y(a - y)(2a - y)$
   (c) $y' = a + y + y^2$ (d) $y' = a - y - y^2$

2. Write down an autonomous differential equation, containing a parameter a that has the property that if $a < 0$, there are two equilibrium points, one stable and the other unstable; if $a = 0$ there is one equilibrium point that is semistable; and if $a > 0$, there are no equilibrium points.

3. Create a *How to Construct a Bifurcation Diagram for* $y' = g(y, a)$ by adding statements under Purpose, Process, and Comments that summarize what is discussed in this section.

What Have We Learned?

Main Ideas

- An AUTONOMOUS first order differential equation is of the form $y' = g(y)$, where $g(y)$ is a given function of y alone.
- If $y' = g(y)$ remains unchanged after the interchange of y with $-y$ on both sides of the differential equation, then the family of solutions is symmetric across the x-axis.
- If $y' = g(y)$ remains unchanged after the simultaneous interchange of y with $-y$ and x with $-x$ on both sides of the differential equation, then the family of solutions is symmetric about the origin.

- If the differential equation $y' = g(x, y)$ has a solution of the form $y(x) = constant$, then this solution is called an EQUILIBRIUM SOLUTION.
- Let the differential equation $y' = g(x, y)$ have an equilibrium solution; that is, a solution of the form $y(x) = constant$. An equilibrium solution is called STABLE if all other solutions of the differential equation that start near this equilibrium solution remain near this equilibrium solution as $x \to \infty$. An equilibrium solution is called UNSTABLE if all other solutions of the differential equation that start near this equilibrium solution move away from this equilibrium solution as $x \to \infty$. An equilibrium solution is called SEMISTABLE if it is neither stable nor unstable.
- If $g(x, y)$ and $\partial g/\partial y$ are defined and continuous in a finite rectangular region containing the point (x_0, y_0) in its interior, then the differential equation $y' = g(x, y)$ has a unique solution passing through the point $y(x_0) = y_0$. This solution is valid for all x for which the solution remains inside the rectangle.
- A particular solution of $y' = g(x, y)$ that cannot be obtained from the family of solutions (containing the arbitrary constant C) by selecting a finite value for C is called a SINGULAR SOLUTION.
- A PHASE LINE is a line giving values of the dependent variable including the equilibrium solutions. Arrows on this line show regions where the solution is increasing or decreasing, which is useful in determining the stability of equilibrium solutions.
- A BIFURCATION DIAGRAM associated with $y' = g(y)$, which contains a parameter a, is a graph of y versus a that gives the curves where $g(y) = 0$. These are useful in determining the stability of equilibrium solutions.

How to Analyze $y' = g(y)$ Graphically

Purpose To use the techniques of calculus, isoclines, and slope fields to sketch the solutions $y = y(x)$ of

$$y' = g(y). \tag{2.61}$$

Process Use the following operations — the order is flexible.

1. Make sure a solution exists. See *How to See Whether Unique Solutions of $y' = g(x, y)$ Exist* on page 57.
2. Identify and characterize the equilibrium points. See *How to Perform a Phase Line Analysis on $y' = g(y)$* on page 68. Use the phase line to determine regions in the xy-plane where the solution curves are increasing or decreasing.
3. Determine regions in the xy-plane in which the solution curves are concave up or concave down. Differentiate (2.61) with respect to x, and then use (2.61) to eliminate y' from the result. This leads to

$$\frac{d^2y}{dx^2} = \frac{dg}{dy}\frac{dy}{dx} = \frac{dg}{dy}g = f(y).$$

(a) Consider $f(y) = 0$. These will define regions in the xy-plane where the solutions $y = y(x)$ may have inflection points (change concavity).

(b) Consider $f(y) > 0$. These will define regions in the xy-plane where solutions $y = y(x)$ are concave up.

(c) Consider $f(y) < 0$. These will define regions in the xy-plane where solutions $y = y(x)$ are concave down.

4. Determine the isoclines. Consider $g(y) = m$, where m is constant. Solve this equation for y in terms of m. These are horizontal lines, and they identify where all the solution curves have the same slope, namely, m.

5. Construct the slope field for (2.61), using a computer/calculator program.

6. Check whether the slope field has any symmetries.

7. Sketch several curves that are consistent with all the preceding information.

Comments about Graphical Analysis

- The operations 2 through 6 in the preceding technique need not be done in this order, depending on the problem. For example, we might first construct the slope field and then check where the solution is increasing.
- Operations 2 through 4 usually require some algebra in order to solve equations like $g(y) = m$ and $f(y) = 0$ for y. Even if you have great algebraic skills, solving the equations is not always possible.

Words of Caution

- It is a common mistake to forget about the equilibrium solutions.
- When analyzing the phase line, realize that if the Existence-Uniqueness Theorem is satisfied, no solutions intersect equilibrium solutions.
- Graphical analysis will not show whether a solution goes to infinity for finite x. Solutions need not go on forever.
- If the Existence-Uniqueness Theorem is satisfied, then solution curves cannot intersect.
- It is possible to miss nonunique solutions using graphical analysis.
- Be very careful when solving for $y(x)$. It is a common mistake to introduce functions that are not solutions.
- Software packages that find solutions of differential equations analytically are not infallible. They too can get the wrong answer. Slope fields are useful!

CHAPTER 3

FIRST ORDER DIFFERENTIAL EQUATIONS—QUALITATIVE AND QUANTITATIVE ASPECTS

Where We Are Going — and Why

In this chapter we examine the general first order differential equation

$$\frac{dy}{dx} = g(x, y) \tag{3.1}$$

and discover that the graphical techniques of Chapter 1 (where $g(x, y)$ is independent of y) and Chapter 2 (where $g(x, y)$ is independent of x) apply with little or no modification. This is important, because, unlike the situation in Chapters 1 and 2, in general we cannot find solutions of (3.1), and even when we are able to do so in special cases, such solutions are not always useful.

Because of this, finding numerical solutions of (3.1) is important. Thus, we introduce a simple method of obtaining a numerical solution of (3.1) — namely, Euler's method — and show how it can be improved. Sometimes numerical solutions are misleading and give incorrect conclusions. This leads to a discussion of period doubling and chaos. As an aid to determining the accurate behavior of solutions, we also give a theorem that allows us to compare the solution of a differential equation we are unable to solve explicitly with one that we can solve.

Such alternative methods are also useful when it is difficult or impossible to graph implicit solutions. In all cases our primary goal is to fully understand the solution's behavior even though we do not necessarily have an explicit solution in terms of familiar functions.

3.1 GRAPHICAL SOLUTIONS USING CALCULUS

In this section we apply the techniques developed in Chapters 1 and 2 — graphical analysis and explicit solutions — to differential equations of the form (3.1) where both variables x and y are present on the right-hand side. We first analyze a simple example and then make some general observations about what we have learned. These techniques are especially important in cases where we cannot find an analytical solution, or when its representation cannot be easily graphed. We use these ideas to show how an implicit function may be graphed by converting it to a differential equation.

Graphical Analysis of $dy/dx = g(x,y)$

EXAMPLE 3.1

Consider the differential equation

$$y' = x - y. \tag{3.2}$$

We want to sketch the solution curves of this differential equation using graphical techniques.

Notice that although $y = x$ makes the right-hand side of (3.2) zero, it makes the left-hand side 1, so $y = x$ is not a solution of (3.2). There are no equilibrium solutions.

Monotonicity

First we use the techniques of calculus to decide where the solutions are increasing and where they are decreasing. Solutions will have horizontal tangent lines when $x - y = 0$ — that is, when $y = x$. Solutions will be increasing when $y' = x - y > 0$ — that is, when $x > y$ — and decreasing when $y' < 0$ — that is, when $x < y$. So the line $y = x$ divides the xy-plane into two regions. Solutions that lie above the line $y = x$ have the property that $x < y$, so they must be decreasing. Similarly, solutions that lie below the line $y = x$ must be increasing. These regions are sketched in Figure 3.1.

Isoclines

The line $y = x$ is the curve along which the solution curves will have slope 0; in other words, $y = x$ is the isocline corresponding to slope 0. Let's look at the isoclines corresponding to slope m — that is, the curves $x - y = m$, giving $y = x - m$. These isoclines are straight lines with slope 1 and y-intercept given by $-m$. For example, the isocline $y = x - 1$ is where the solutions have slope 1. This isocline is parallel to but below the isocline $y = x$, which separates the increasing and decreasing solutions.

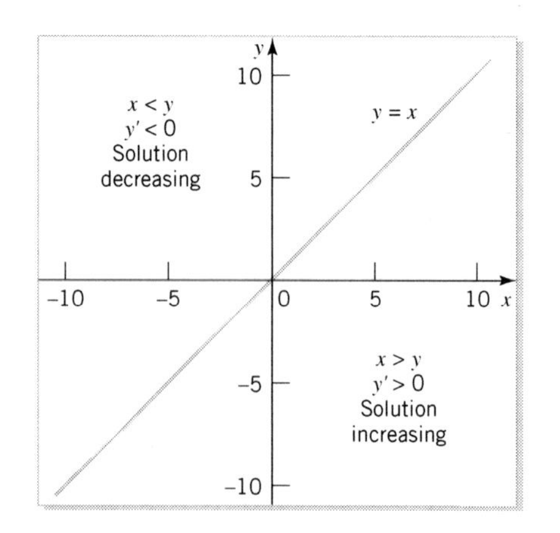

FIGURE 3.1 Regions where solutions of $y' = x - y$ are increasing and decreasing

Unlike the previous examples, these isoclines are neither vertical, as was the case for $y' = g(x)$, nor horizontal, as was the case for $y' = g(y)$. Isoclines corresponding to $m = 0$, $m = \pm 2$, and $m = \pm 4$ are shown in Figure 3.2.

Concavity

Now let's look at the concavity of the solutions of (3.2), $y' = x - y$. To do this we need y'', which we get from $y' = x - y$ by differentiating with respect to x,

$$y'' = 1 - y'.$$

If we substitute for y' from (3.2) into this last equation, we find

$$y'' = 1 - x + y.$$

Solutions will be concave up when $y'' > 0$ — that is, when $1 - x + y > 0$ — or, equivalently, when $y > x - 1$. Solutions will be concave down when $y'' < 0$ — that is, when $y < x - 1$ — so points of inflection might occur along the line $y = x - 1$, where $y'' = 0$. Thus, the line $y = x - 1$ divides the xy-plane into two regions. Solutions that lie above $y = x - 1$ have the property that $y > x - 1$, and so this will be where the solution curves are concave up. In the same way, the region $y < x - 1$ is where the solution curves will be concave down. These regions are sketched in Figure 3.3.

Slope field

If we calculate the slopes corresponding to (3.2) for several points, we obtain

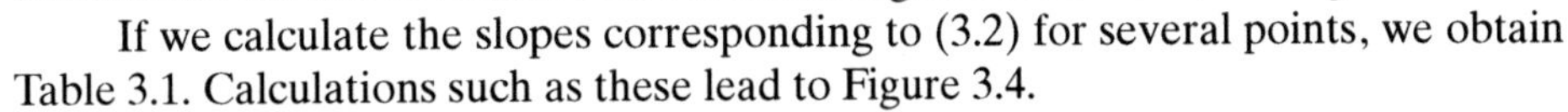

Table 3.1. Calculations such as these lead to Figure 3.4.

If we hand-draw a few solution curves on the slope field of Figure 3.4, we find Figure 3.5. These are in complete agreement with our previous information concerning monotonicity, isoclines, and concavity.

However, there are two things to notice in Figure 3.5. First, the line $y = x - 1$ appears to be a solution curve. Second, all solution curves seem to tend to the line $y = x - 1$. If we substitute $y = x - 1$ into (3.2), we find that $y = x - 1$ is indeed a solution. To confirm the suggestion that all solutions tend to $y = x - 1$ as $x \to \infty$ leads us to wonder whether we can find analytical solutions of (3.2). In Exercise 1 on page 91 you are asked to show that $y(x) = x - 1 + Ce^{-x}$ is the solution of (3.2). Notice

Table 3.1
Slopes for $y' = x - y$

x	y	$x - y$
−2	−2	0
−2	0	−2
−2	2	−4
0	−2	2
0	0	0
0	2	−2
2	−2	4
2	0	2
2	2	0

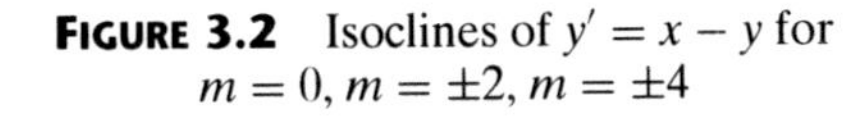

FIGURE 3.2 Isoclines of $y' = x - y$ for $m = 0$, $m = \pm 2$, $m = \pm 4$

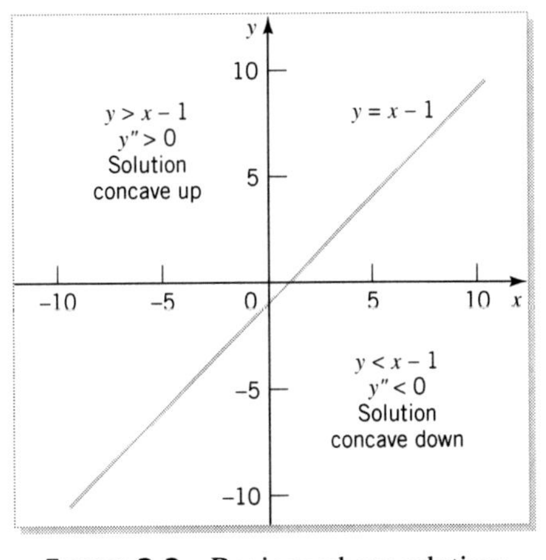

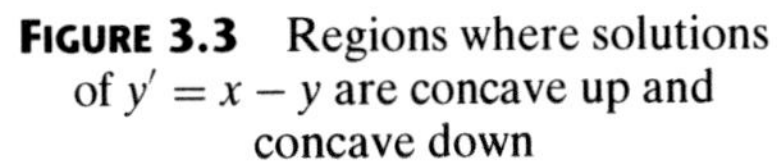

FIGURE 3.3 Regions where solutions of $y' = x - y$ are concave up and concave down

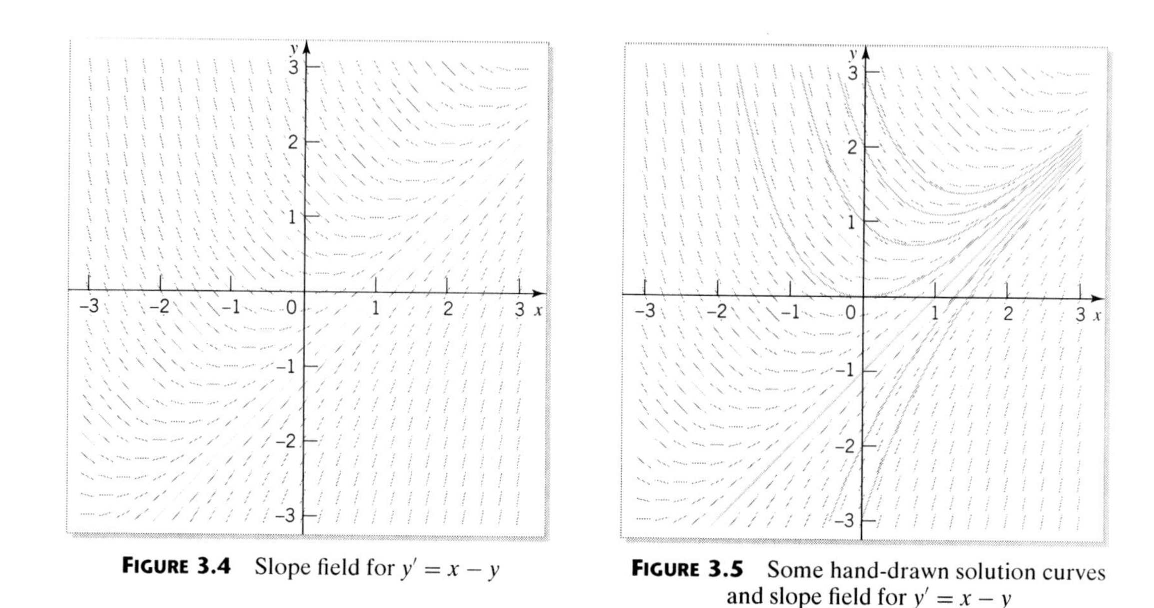

FIGURE 3.4 Slope field for $y' = x - y$

FIGURE 3.5 Some hand-drawn solution curves and slope field for $y' = x - y$

that the term involving C goes to 0 as $x \to \infty$, so, as expected, all solutions tend to $y = x - 1$ as x increases. For this reason we call $y(x) = x - 1$ a STABLE solution.[1]

Earlier we noticed that inflection points might occur along the line $y = x - 1$. They don't because $y = x - 1$ is a solution, and by the Existence-Uniqueness Theorem on page 56, no solution may intersect this solution. [Explain why the differential equation (3.2) satisfies the hypothesis of the Existence-Uniqueness Theorem.]

How to Analyze $y' = g(x, y)$ Graphically

Purpose To use the techniques of calculus, isoclines, and slope fields to sketch the solutions $y = y(x)$ of

$$y' = g(x, y). \tag{3.3}$$

Process Use the following operations — the order is flexible.

1. Construct the slope field for (3.3), using a computer/calculator program.
2. Determine regions in the xy-plane where the solution curves are increasing or decreasing.
   (a) Consider $g(x, y) = 0$. These will define regions in the xy-plane where solutions $y = y(x)$ have slope zero. These are possible locations of local maxima and minima.[2]

---

[1] This terminology is consistent with that used for stable and unstable equilibrium solutions for autonomous differential equations.

[2] Local extrema (maxima and minima) are also called relative extrema (maxima and minima).

(b) Consider $g(x, y) > 0$. These will define regions in the xy-plane where solutions $y = y(x)$ are increasing.

(c) Consider $g(x, y) < 0$. These will define regions in the xy-plane where solutions $y = y(x)$ are decreasing.

3. Determine the isoclines. Consider $g(x, y) = m$, where m is a constant. Solve this equation, obtaining the isoclines as y in terms of x and m (or as x in terms of y and m). These isoclines identify where all the solution curves have the same slope, namely, m. Note that these isoclines need not be straight lines.

4. Determine regions in the xy-plane where the solution curves are concave up or concave down. Differentiate (3.3) with respect to x (be sure to use the chain rule), and then substitute (3.3) into the result to eliminate y'. This leads to $y'' = f(x, y)$.

(a) Consider $f(x, y) = 0$. These will define regions in the xy-plane where the solutions $y = y(x)$ may change concavity and where they may have inflection points.

(b) Consider $f(x, y) > 0$. These will define regions in the xy-plane where solutions $y = y(x)$ are concave up.

(c) Consider $f(x, y) < 0$. These will define regions in the xy-plane where solutions $y = y(x)$ are concave down.

5. Sketch several curves that are consistent with all the preceding information.

Comments about Graphical Analysis

- Operations 1 through 4 need not be done in this order. For example, we might first determine regions in the xy-plane where the solution curves are increasing or decreasing, and then construct the slope field.
- Operations 2, 3, and 4 usually require some algebra in order to solve the equations $g(x, y) = m$ and $f(x, y) = 0$. Even if you have great algebraic skills, solving these equations is not always possible.

We will devote the rest of this section to these graphical techniques, applying them to various examples.

Two Examples

EXAMPLE 3.2 *Graphing Implicit Functions*

There are situations in mathematics in which having a simple expression does not mean we know much about this expression. For example, consider the simple equation

$$(x - y)^3 + (x + y)^3 = 1000. \tag{3.4}$$

There does not appear to be an easy way to obtain the graph of this equation without plotting many points, even though it is apparent that the graph will be symmetric across the x-axis [replace y by $-y$ in (3.4) and observe that the equation does not change].

One of the important uses of calculus is as an aid in drawing graphs of functions, mainly to find local extreme values as well as intervals where the functions are increasing or decreasing. Thus, we need the derivative of a function, but here we have no function. The procedure given in calculus is to assume that the equation (3.4) defines a function implicitly and then implicitly differentiate (3.4) with respect to x. Thus, we assume that y may be found as a function of x and differentiate (3.4) to obtain

$$3\,(x-y)^2\,(1-y') + 3\,(x+y)^2\,(1+y') = 0.$$

Dividing by 3 and collecting together the terms involving y' gives

$$y'\left[-(x-y)^2 + (x+y)^2\right] + \left[(x-y)^2 + (x+y)^2\right] = 0,$$

or

$$4xyy' + 2\left(x^2+y^2\right) = 0. \tag{3.5}$$

Solving (3.5) for y', we find

$$y' = -\frac{x^2+y^2}{2xy}. \tag{3.6}$$

Monotonicity This is a differential equation whose solutions contain the curve we want. By looking at the sign of this derivative, we see that the curve we seek will be decreasing if $xy > 0$ and increasing if $xy < 0$. Thus, the curve we seek will be decreasing in the first and third quadrants, and will be increasing in the second and fourth quadrants. Also, if $y \neq 0$ the curve will have vertical tangent lines along $x = 0$, the y-axis; and if $x \neq 0$ the curve will have vertical tangent lines along $y = 0$, the x-axis.

Slope field Notice from (3.6) that the slope field will be symmetric across the x-axis, the y-axis, and about the origin. This means that if we know the slope field in the first quadrant, we can obtain the slope field in the other three quadrants using this symmetry. If we calculate the slopes corresponding to (3.6) for various values of x and y in the first quadrant, we obtain Table 3.2. Plotting results of calculations such as these leads to Figure 3.6.

Table 3.2 Slopes for $y' = -(x^2+y^2)/(2xy)$

x	y	y'
3	3	−1
3	2	−13/12
3	1	−5/3
2	3	−13/12
2	2	−1
2	1	−5/4
1	3	−5/3
1	2	−5/4
1	1	−1

We noted that the only place where the graph of our original equation has vertical tangents is where $x = 0$ and $y \neq 0$, or where $x \neq 0$ and $y = 0$. Other isoclines of slope m are given by

$$-\frac{x^2+y^2}{2xy} = m.$$

If we multiply by $-2xy$ and bring all the terms to the left-hand side, we have

$$x^2 + y^2 + 2mxy = 0.$$

We can think of this as a quadratic equation in y, which when solved for y gives

$$y = -mx \pm \sqrt{m^2x^2 - x^2},$$

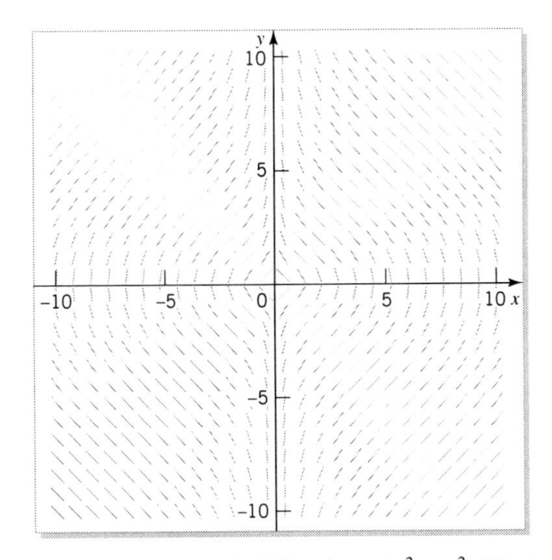

FIGURE 3.6 Slope field for $y' = -(x^2 + y^2)/(2xy)$

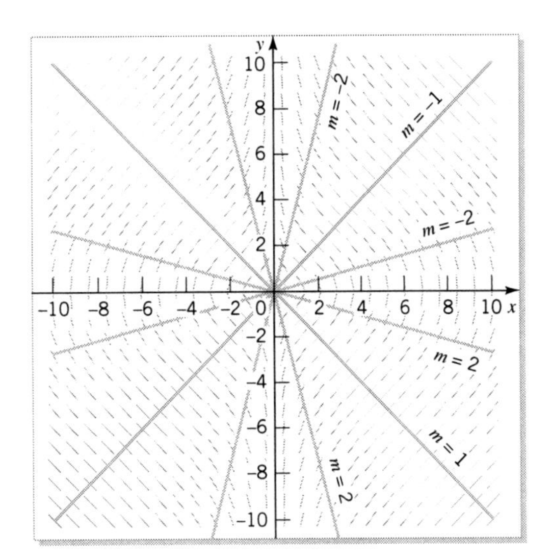

FIGURE 3.7 Straight-line isoclines and slope field for $y' = -(x^2 + y^2)/(2xy)$

or

$$y = \left(-m \pm \sqrt{m^2 - 1}\right)x. \tag{3.7}$$

Isoclines

The isoclines (3.7) are lines through the origin with slope $-m \pm \sqrt{m^2 - 1}$. Notice that these lines will exist only if $m^2 - 1 \geq 0$ — that is, if $m \leq -1$ or $m \geq 1$. In particular, $m = 0$ is excluded, so solution curves never have horizontal tangents, a fact we can observe from (3.6). In Figure 3.7 we have drawn the slope field corresponding to (3.6) with the isoclines (3.7) for $m = \pm 1$ and ± 2 superimposed.

In order to sketch the graph of (3.4), namely,

$$(x - y)^3 + (x + y)^3 = 1000, \tag{3.8}$$

from the slope field, we need a specific point on this curve. If we put $x = y$ in (3.8) we find that $x = y = 5$, and the hand-drawn curve through the point $(5, 5)$ is shown in Figure 3.8. Notice that this curve is symmetric across the x-axis, which it should be. Also notice that the hand-drawn curve is actually two particular solutions of (3.8), one through $(5, 5)$ and the other through $(5, -5)$, joined at the point $(500^{1/3}, 0) \approx (7.94, 0)$.

Concavity

We confirm that Figure 3.8 is reasonable by considering the concavity of the solution curve. We differentiate (3.6) to find

$$y'' = -\frac{1}{2x^2y^2}\left[(2x + 2yy')\,xy - \left(x^2 + y^2\right)\left(y + xy'\right)\right].$$

We collect together the terms involving y' so that

$$y'' = -\frac{1}{2x^2y^2}\left\{2x^2y - \left(x^2 + y^2\right)y + y'\left[2xy^2 - x\left(x^2 + y^2\right)\right]\right\},$$

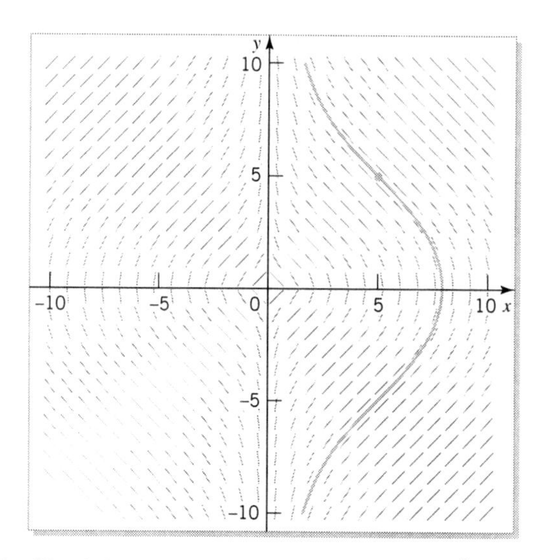

FIGURE 3.8 Hand-drawn graph of the equation $(x-y)^3 + (x+y)^3 = 1000$

or

$$y'' = -\frac{1}{2x^2y^2}\left[\left(x^2y - y^3\right) + y'\left(xy^2 - x^3\right)\right].$$

If we factor both terms in parentheses and use (3.6) to substitute for y' in this last equation, we find

$$y'' = -\frac{1}{2x^2y^2}\left[y\left(x^2 - y^2\right) - \frac{x^2+y^2}{2xy}x\left(y^2 - x^2\right)\right].$$

Factoring the common term $x^2 - y^2$ we have

$$y'' = -\frac{x^2-y^2}{2x^2y^2}\left[y + \frac{x^2+y^2}{2y}\right],$$

which can be written as

$$y'' = -\frac{x^2-y^2}{4x^2y^3}\left(3y^2 + x^2\right).$$

If we use the fact that $-\left(x^2 - y^2\right) = (y-x)(y+x)$, we may rearrange this last equation in the form

$$y'' = \frac{(y-x)(y+x)}{y}\frac{3y^2+x^2}{4x^2y^2}. \tag{3.9}$$

Looking at the right-hand side of this equation we see that solution curves are concave up if $(y-x)(y+x)/y > 0$ and are concave down if $(y-x)(y+x)/y < 0$. If we concentrate on the first quadrant — which is all we need to do because of the symmetry of the slope field — we see that the solution curve is concave up if $y > x$

and concave down if $y < x$. Points of inflection lie on the line $y = x$. This is consistent with the graph in Figure 3.8. ■

To summarize, we started with an implicit equation relating x and y, and via implicit differentiation we obtained a differential equation as an aid to obtaining the graph of this implicit equation. Thus, the original implicit equation is a solution of the differential equation; it is called an implicit solution of the differential equation. This leads to the following definition.

◆ *Definition 3.1:* **An IMPLICIT SOLUTION of a first order differential equation is any equation involving x and y that through implicit differentiation yields the original differential equation.** ◆

Comments about Implicit Solutions

- Sometimes it is possible to obtain an explicit solution from an implicit solution; sometimes it is impossible. In our last example, the implicit solution $(x - y)^3 + (x + y)^3 = 1000$ can be solved giving two explicit solutions, namely,

$$y = \sqrt{\frac{500 - x^3}{3x}} \text{ and } y = -\sqrt{\frac{500 - x^3}{3x}},$$

  both of which are needed to obtain the complete graph of our original equation. However, in general it is not possible to obtain an explicit solution from an implicit solution; for example, the implicit equation $(x - y)^6 + (x + y)^5 = 1000$ cannot be solved for y to give an explicit solution.
- Implicit solutions can be very difficult to graph.
- If you are given an implicit function and want to check that it satisfies a particular differential equation (as is the case in Exercise 6 on page 91), you should not try to find an explicit solution, but rather mimic the steps that we used to go from (3.4) to (3.6).

EXAMPLE 3.3

Sometimes important information comes by considering specific isoclines, as we now show by investigating the differential equation

$$y' = y^2 + x^2 - 1. \tag{3.10}$$

Monotonicity The solution curve of this equation will be increasing whenever the derivative is positive — namely, when $y^2 + x^2 - 1 > 0$ — and decreasing when $y^2 + x^2 - 1 < 0$. Thus, the curve $y^2 + x^2 - 1 = 0$ separates the xy-plane into regions of increasing and decreasing solutions. We know that the curve $y^2 + x^2 - 1 = 0$ is a circle centered at the origin. Because $y^2 + x^2 - 1 > 0$ outside this circle, our solution curves are increasing there, whereas inside this circle they are decreasing. These regions are shown in Figure 3.9.

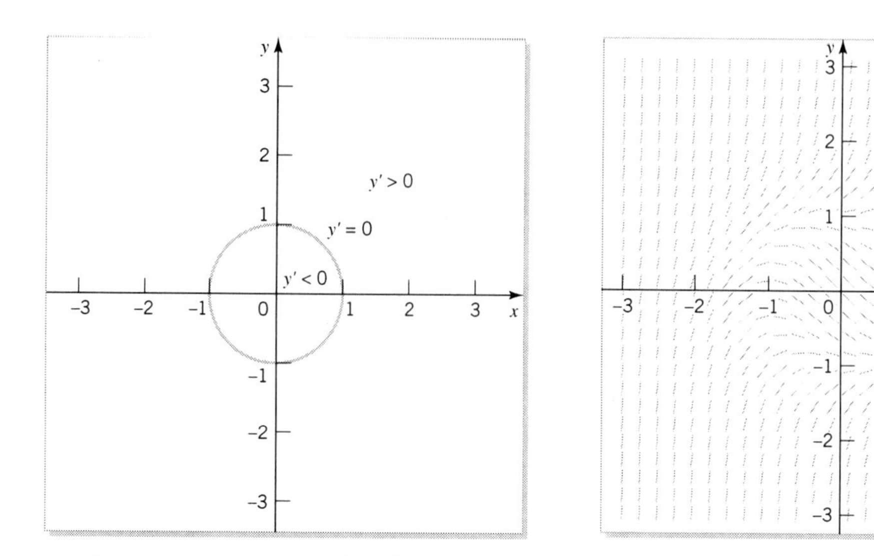

FIGURE 3.9 The circle $y^2 + x^2 - 1 = 0$

FIGURE 3.10 Slope field for $y' = y^2 + x^2 - 1$

Slope field The slope field for (3.10) is shown in Figure 3.10. This figure suggests that the slope field is symmetric about the origin, which can be confirmed by noting that replacing x with $-x$ and y with $-y$ on both sides of (3.10) leaves it unchanged. This figure also suggests that the behavior near the origin is different from that farther from the origin. No horizontal tangents appear in the figure other than those in a circle-shaped region around the origin. It appears that the maximum negative slope occurs at the origin. This is made obvious in Figure 3.11, which shows both the slope field and the circle where the slopes are zero.

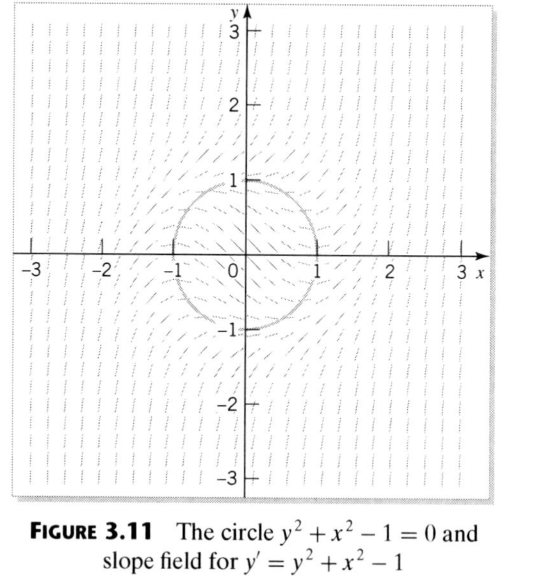

FIGURE 3.11 The circle $y^2 + x^2 - 1 = 0$ and slope field for $y' = y^2 + x^2 - 1$

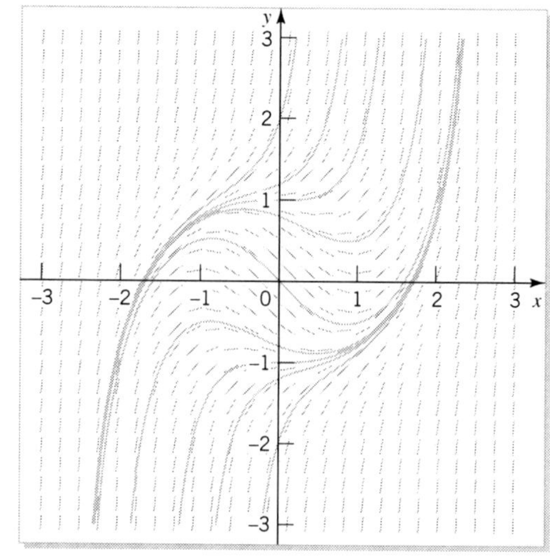

FIGURE 3.12 Hand-drawn solution curves and slope field for $y' = y^2 + x^2 - 1$

Isoclines To explore this further we note that the isoclines for slope m occur when

$$y^2 + x^2 - 1 = m, \qquad \text{or} \qquad x^2 + y^2 = m + 1. \tag{3.11}$$

This equation cannot be satisfied if $m < -1$. (Why?) If $m = -1$, then (3.11) requires that $x = 0$, $y = 0$. Thus, the smallest slope that can occur is $m = -1$, and this occurs only at the origin.

Also note that from (3.11), we have horizontal tangents ($m = 0$) only when $x^2 + y^2 = 1$. Thus, the isocline for zero slope is the circle $y^2 + x^2 = 1$ as shown in Figure 3.11. Figure 3.12 shows some hand-drawn solution curves of (3.10) consistent with these facts. ∎

EXERCISES

1. Use the following outline to find the explicit solution of $y' = x - y$.

(a) Introduce a new dependent variable $u(x)$, where $u(x) = y(x) - x$, to transform the differential equation $y' = x - y$ into the autonomous equation $u' = -1 - u$.

(b) Solve the autonomous equation to find $u(x) = -1 + Ce^{-x}$, where C is an arbitrary constant. (Note the equilibrium solution is included here if $C = 0$.)

(c) Change from the new variable u back to the old variable y by $y(x) = x + u(x)$, to find the solution of the original equation $y(x) = x - 1 + Ce^{-x}$.

2. Use the following outline to find the explicit solution of $y' = g(y + ax + b)$ where a and b are given constants, and g is a given function of $y + ax + b$.

(a) Introduce a new dependent variable $u(x)$ where $u(x) = y(x) + ax + b$ to transform the differential equation $y' = g(y + ax + b)$ into the autonomous equation $u' = g(u) + a$.

(b) Solve the autonomous equation.

(c) Use this technique to solve each of the following equations.

i. $y' = 1 + e^{x-y}$.
ii. $y' = (y - 2x)^2$.
iii. $y' = -(y - 2x)^2$.
iv. $y' = \sin(y - x)$.

3. Substitute the given function $y(x)$ into the differential equation that follows, and determine n so that $y(x)$ is an explicit solution.

(a) $y(x) = x^n$, $\quad xy' - 3y = 0$

(b) $y(x) = nx + x^2$, $\quad xy' - 2y + x = 0$

(c) $y(x) = 1 - e^{-nx}$, $\quad y' + 32y - 32 = 0$

(d) $y(x) = (x - n)^2 e^x$, $\quad (x - 2)y' - xy = 0$

(e) $y(x) = \exp(1 - ne^x)$, $\quad y' + e^x y = 0$ (Note: $\exp u = e^u$.)

4. Substitute the given function $y(x)$ into the differential equation that follows, to show that $y(x)$ is an explicit solution for any choice of the constant C.

(a) $y(x) = Cx^4 - 1$, $\quad xy' - 4y - 4 = 0$

(b) $y(x) = 1 - Ce^{-2x}$, $\quad y' + 2y - 2 = 0$

(c) $y(x) = C(x - 2)^2 e^x$, $\quad (x - 2)y' - xy = 0$

(d) $y(x) = C\exp(1 - e^x)$, $\quad y' + e^x y = 0$ (Note: $\exp u = e^u$.)

5. For what value(s) of the constant C does the graph of the solution for each part of Exercise 4 pass through the point $(0, -1)$?

6. Show that the following are implicit solutions of the given differential equations for any choice of the constant C. [Do not try to solve the first equation for y and then substitute in the differential equation, but rather mimic the process whereby (3.6) was obtained from (3.4).]

(a) $x^2 + y^2 = C$, $\quad C > 0$
$y' + x/y = 0$

(b) $\sin xy + x^3 + xy^2 = C$,
$(x\cos xy + 2xy)y' + 3x^2 + y^2 + y\cos xy = 0$

(c) $xe^{-xy} + x^2 y = C$,
$(x^2 - x^2 e^{-xy})y' - xye^{-xy} + e^{-xy} + 2xy = 0$

(d) $\ln|ye^C| + e^{x^2} = y^3$
$(-3y^2 + 1/y)y' + 2xe^{x^2} = 0$

(e) $xy + (Cy)^{-1} = 3x, \quad C \neq 0$
$(3x - 2xy)y' - y^2 + 3y = 0$

7. Sketch the slope field corresponding to $yy' = x$.

(a) What symmetry does the slope field have?

(b) What is the equation of the isocline where all solutions have zero slope?

(c) Draw isoclines for $m = 0, \pm 0.5, \pm 1$, and ± 2.

(d) Decide where the solution curves are increasing, decreasing, concave up, concave down.

(e) Draw the solution curve that passes through the point $x = 0, y = 1$.

(f) Show that $y^2 - x^2 = 1$ passes through $(0, 1)$ and satisfies $yy' = x$.

8. Sketch the slope field corresponding to $y' = x^2 + y^2$.

(a) What symmetry does the slope field have?

(b) When will the solution of this differential equation have a horizontal tangent?

(c) Draw isoclines for $m = 1/4, 1, 4$, and 9.

(d) Draw the solution curve that passes through $(-1, 1)$.

9. Use graphical methods to sketch solution curves for: (a) $y' = -2xy$, (b) $y' = \sqrt{x^2 + y^2}$.

10. Consider the differential equation $y' = 1 - 2xy$.

(a) Sketch the isoclines for slope -1, 0, and 1 in the window $1 \leq x \leq 3, 0 \leq y \leq 2$.

(b) Using only your results from part (a) sketch by hand several solution curves of this differential equation with initial conditions $x_0 = 1$, and y_0 taking on various values in $0 < y_0 < 1$. What appears to happen to these solutions as $x \to \infty$?

(c) The region between two of the isoclines you drew in part (a) is often called a "funnel." Which region fits this description, and why?

(d) The isoclines that bound the region you identified in part (c) are often called "fences." Which isoclines fit this description, and why? Which one would you describe as an "upper fence" and which one a "lower fence"?

11. Look at the slope field for $y' = e^{-x} - 2y$ in the window $-3 < x < 3, -3 < y < 3$.

(a) Does it appear that solutions of this differential equation have a local maximum? (You may want to consider the isocline that corresponds to a slope of zero.)

(b) Does it appear that all solutions of this differential equation have a horizontal asymptote? (Use the differential equation to justify your answer.)

(c) Confirm that $y(x) = e^{-x} + (y_0 - 1)e^{-2x}$ satisfies $y' = e^{-x} - 2y$ subject to $y(0) = y_0$. Explain how this solution is consistent with your answers in parts (a) and (b).

12. Consider the following eight first order differential equations:

i. $y' = x + 1$	v. $y' = y^2 - 1$
ii. $y' = x - 1$	vi. $y' = y + x$
iii. $y' = y + 1$	vii. $y' = y - x$
iv. $y' = y - 1$	viii. $y' = y - x + 1$

(a) The slope fields for three of the preceding equations are given in Figures 3.13, 3.14, and 3.15. Match an equation with each of the three figures, carefully stating the reasons for your choices. Do not plot any slope fields to answer this question.

(b) Briefly outline a general strategy for this matching process.

13. Figures 3.16 and 3.17 are mystery slope fields, believed to be the slope fields for two of the following differential equations:

$$y' = \frac{x^2 + 1}{y^2 + 1}, \qquad y' = -\frac{x^2 + 1}{y^2 + 1},$$

$$y' = \frac{y^2 + 1}{x^2 + 1}, \qquad y' = -\frac{y^2 + 1}{x^2 + 1}.$$

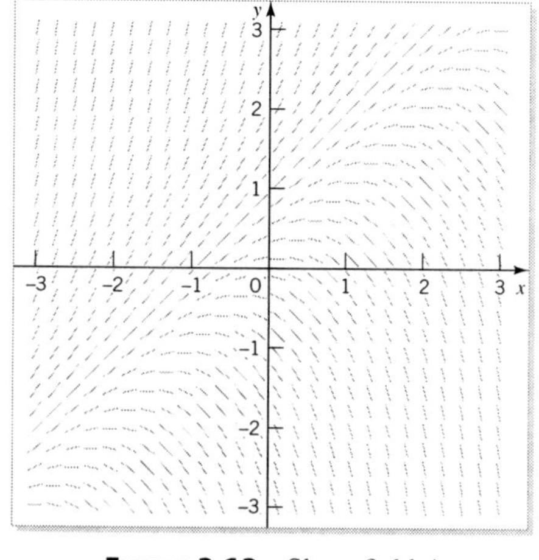

FIGURE 3.13 Slope field A

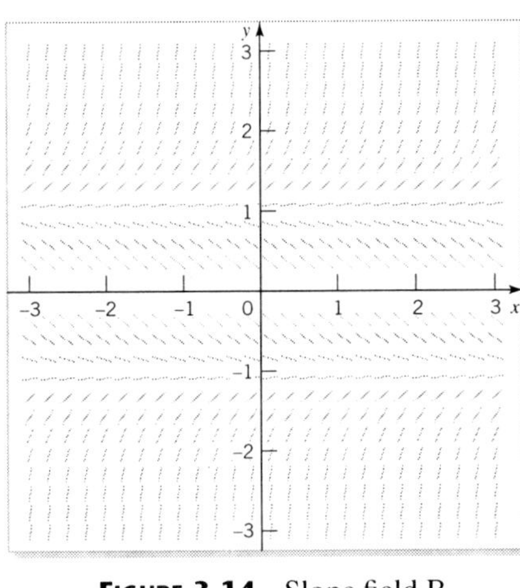

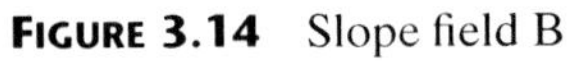

FIGURE 3.14 Slope field B

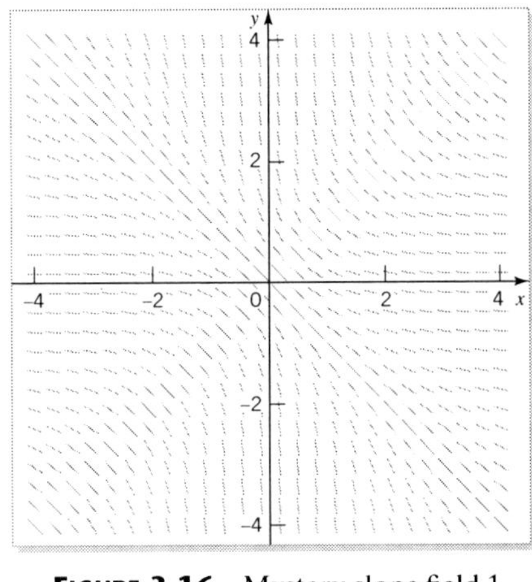

FIGURE 3.16 Mystery slope field 1

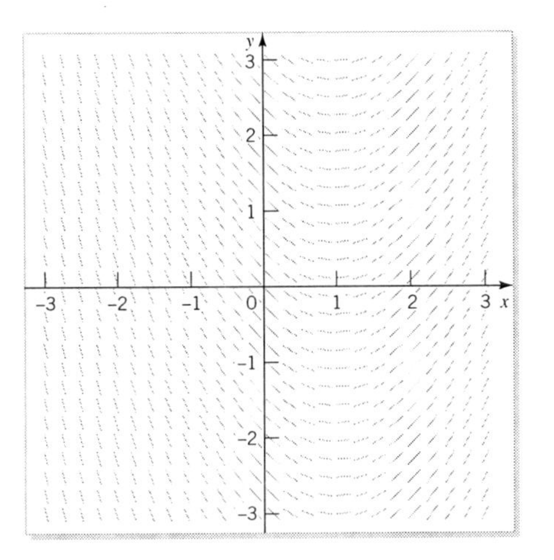

FIGURE 3.15 Slope field C

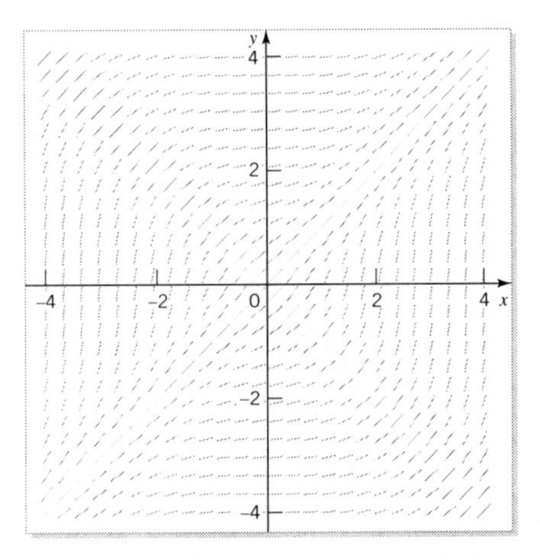

FIGURE 3.17 Mystery slope field 2

(a) Identify to which differential equation each of the mystery slope fields belongs. Confirm all the information in the slope fields using isoclines and concavity. Do not plot any slope fields to answer this question.

(b) Now superimpose the two mystery slope fields, perhaps by placing one on top of the other and holding both up to the light. [Another way to do this is to plot the slope fields for

$$y' = a\frac{x^2+1}{y^2+1} + b\frac{y^2+1}{x^2+1}$$

with $a = \pm 1$ and $b = 0$, and then $a = 0$ and $b = \pm 1$.] What do you notice? Would this have made your previous analysis in part (a) easier?

14. Someone claims that the solution of the initial value problem $y' = -0.2\left[y - 45 - 10\sin\left(\frac{\pi}{12}x\right)\right]$, $y(0) = 70$, is $y = u(x)$, where $u(x) = 45 + 10\sin\left(\frac{\pi}{12}x\right) + 25e^{-0.2x}$. Plot the function $u(x)$ and slope field for this differential equation in the window $0 \le x \le 80$, $0 \le y \le 80$. What do you conclude?

15. Consider the implicitly defined function $x^2 + y^2 = C$. Show that, for this function, $y' = -x/y$. Use slope fields to show that the curves characterized by this differential equation are circles.

16. Consider the implicitly defined function $x^2 - y^2 = C$. Show that, for this function, $y' = x/y$. Use slope fields to show that the curves characterized by this differential equation are hyperbolas.

17. Consider the implicitly defined function $(x^2 + y^2)^2 = 12(x^2 - y^2)$. Show that, for this function,

$$y' = \frac{x(6 - x^2 - y^2)}{y(6 + x^2 + y^2)}.$$

Use slope fields to sketch solution curves of this differential equation. Notice that far from the origin, the curves look like circles, whereas near the origin they look like hyperbolas. Explain why this is the case by comparing this differential equation with those in Exercises 15 and 16.

18. Use the preceding techniques to sketch the following implicit functions.

(a) $y^2 - xy + x^2 - 1 = 0$. Notice that the points $(0, \pm 1)$ lie on the curve.

(b) $y^4 - 4(x + 1)y^3 + 3 = 0$. Notice that the point $(0, 1)$ lies on the curve.

(c) $x^3 + y^3 = -8$.

(d) $x^3 + xy = 8$.

(e) $x^3 + x + y^3 + y = 4$. Notice that the point $(1, 1)$ lies on the curve.

3.2 SYMMETRY OF SLOPE FIELDS

Several examples in the previous sections made use of symmetry properties of the slope field to reduce the work in discovering the behavior of solutions of differential equations. Types of slope field symmetries encountered so far are across the x-axis, across the y-axis, or about the origin. For example, slope fields for $y' = g(x)$ could be symmetric across the y-axis or about the origin, while those for $y' = g(y)$ could be symmetric across the x-axis or about the origin. In this section we bring these ideas together in one place for general first order differential equations.

How to Test $y' = g(x, y)$ for Symmetry

Purpose To determine the symmetry of the slope field directly from the differential equation

$$y' = g(x, y). \tag{3.12}$$

Process

1. Replace x with $-x$ in (3.12). If the resulting equation is identical to (3.12) — that is, if (3.12) is unchanged — then the slope field is symmetric across the y-axis.
2. Replace y with $-y$ in (3.12). If the resulting equation is identical to (3.12), then the slope field is symmetric across the x-axis.
3. Replace x with $-x$ and y with $-y$ in (3.12). If the resulting equation is identical to (3.12), then the slope field is symmetric about the origin.

Comments about Testing for Symmetry

- This is a very useful test and is easily applied. Symmetry is one thing we should always check when analyzing a differential equation graphically. If a slope field has a symmetry, we need consider only part of the xy-plane.

- If a slope field is symmetric across any two of the x-axis, the y-axis, and the origin, it is automatically symmetric across the third (see Exercise 4 on page 96).
- Symmetry of the slope field does not guarantee that a particular solution curve has the same symmetry (see Exercise 3 on page 96).
- Symmetry of the slope field guarantees the same symmetry for the **family** of solutions. For example, consider a slope field that is symmetric across the x-axis (replacing y with $-y$ leaves the differential equation unchanged). If a particular solution curve passes through the point (x_0, y_0), then there is a solution curve (which could be the same curve) that passes through the point $(x_0, -y_0)$. This solution curve can be obtained by reflecting the original solution curve across the x-axis.
- The word that mathematicians use in place of the word "unchanged" is "invariant." Thus, for example, if the differential equation (3.12) is invariant under the interchange of x with $-x$, then the slope field is symmetric across the y-axis.

Caution! **Remember, when checking (3.12) for symmetry, make all substitutions on both sides of the equation, not just on the right-hand side of (3.12).**

Caution! **Symmetry of the slope field does not guarantee that a particular solution curve has the same symmetry (see Exercise 3 on page 96).**

EXAMPLE 3.4

Identify the symmetries of the slope field for

$$\frac{dy}{dx} = \frac{y^2}{x}. \tag{3.13}$$

y-axis symmetry

1. If we replace x with $-x$ on both sides of (3.13), we find

$$\frac{dy}{-dx} = \frac{y^2}{-x},$$

which reduces to

$$\frac{dy}{dx} = \frac{y^2}{x}.$$

Because this is identical to (3.13), the slope field is symmetric across the y-axis.

x-axis symmetry

2. If we replace y with $-y$ on both sides of (3.13), we find

$$\frac{-dy}{dx} = \frac{(-y)^2}{x},$$

which reduces to

$$\frac{dy}{dx} = \frac{-y^2}{x}.$$

Because this is not identical to (3.13), the slope field is not symmetric across the x-axis.

Origin symmetry **3.** By the second statement under *Comments about Testing for Symmetry,* and because the slope field is symmetric across the y-axis but not symmetric across the x-axis, we know that the slope field cannot be symmetric about the origin. (Why?) Nevertheless, for illustrative purposes, we will confirm this directly. If we replace x with $-x$, and y with $-y$ on both sides of (3.13), we find

$$\frac{-dy}{-dx} = \frac{(-y)^2}{-x},$$

which reduces to

$$\frac{dy}{dx} = \frac{-y^2}{x}.$$

Because this is not identical to (3.13), the slope field is not symmetric about the origin.

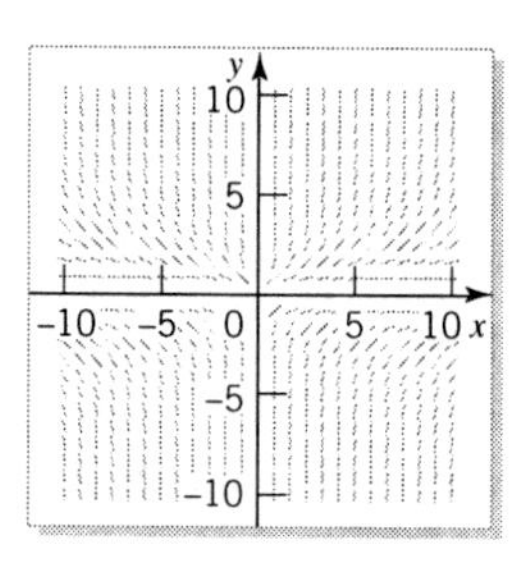

FIGURE 3.18
Slope field for $y' = y^2/x$

These observations are confirmed by Figure 3.18, which shows the slope field for (3.13).

EXERCISES

1. Without sketching the slope field for $y^3 y' = x^3$, decide what type of symmetry (if any) it possesses. Sketch the slope field to confirm your analysis.

2. Without sketching the slope field for $y' = x^2 + y^2$, decide what type of symmetry (if any) it possesses. Sketch the slope field to confirm your analysis.

3. Consider the differential equation $y' = \frac{y}{2x}$.

   (a) Show that the slope field is symmetric across the x-axis, symmetric across the y-axis, and symmetric about the origin.

   (b) Show that $y = \sqrt{x}$ $(x > 0)$ is a solution curve that passes through the point $(1, 1)$.

   (c) Show that the solution curve in part (b) has no symmetries.

4. Show that

   (a) A slope field that is symmetric across both the x-axis and the y-axis is automatically symmetric about the origin.

   (b) A slope field that is symmetric across the x-axis and about the origin is automatically symmetric across the y-axis.

   (c) A slope field that is symmetric about the origin and across the y-axis is automatically symmetric across the x-axis.

5. If the slope field for $y' = g(x, y)$ is symmetric across the y-axis, and $y(0)$ is defined, what can you say about the solution curve through the point $(0, y(0))$?

6. The slope field for $y' = g(x, y)$ — with a particular $g(x, y)$ — is shown in Figure 3.19 for the window $0 < x < 5$, $0 < y < 5$.

   (a) Construct four members of the family of solutions on this figure by hand-drawing solution curves that pass through the points $(0, 1)$, $(0, 2)$, $(0, 3)$, and $(0, 4)$.

   (b) Now suppose that you know that the slope field is symmetric across the y-axis. In which other quadrant(s) does this allow you to extend the slope field? Draw the four members of the family of solutions that are associated with those you drew in part (a).

   (c) Repeat part (b) if the slope field is symmetric across the x-axis.

   (d) Repeat part (b) if the slope field is symmetric about the origin.

   (e) Repeat part (b) if the slope field is symmetric across both the x-axis and the y-axis.

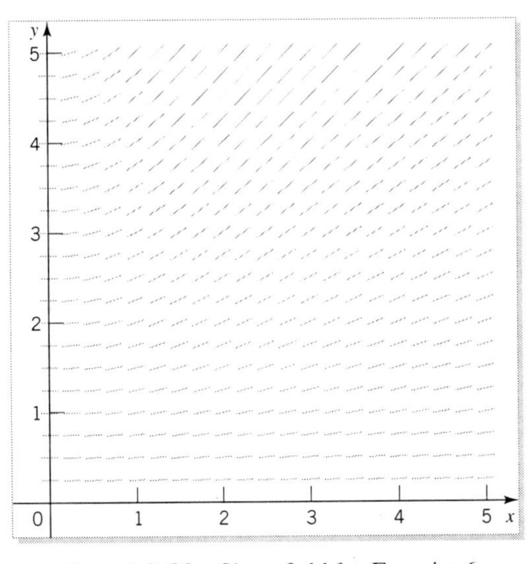

FIGURE 3.19 Slope field for Exercise 6

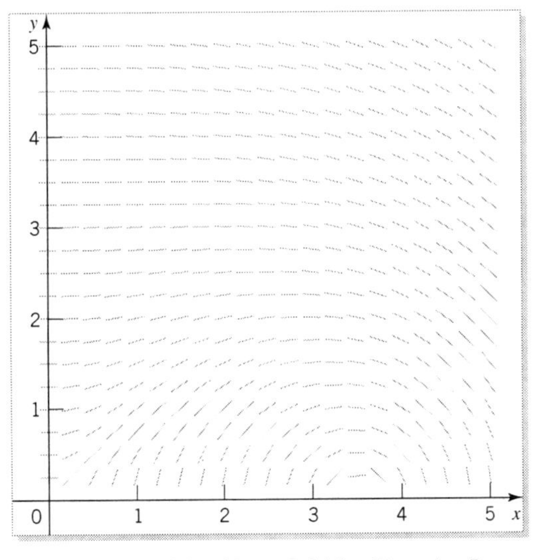

FIGURE 3.20 Slope field for Exercise 7

7. Repeat Exercise 6 for the slope field shown in Figure 3.20.

8. (a) Give an example of a differential equation whose slope field is not symmetric across the x-axis, across the y-axis, or about the origin.

(b) Give an example of a differential equation whose slope field is symmetric about the origin but is not symmetric across either the x-axis or the y-axis.

(c) Give an example of a differential equation whose slope field is symmetric across the y-axis but is not symmetric across the x-axis or about the origin.

(d) Give an example of a differential equation whose slope field is symmetric across the x-axis but is not symmetric across the y-axis or about the origin.

9. Find conditions on $f(y)$ and $g(x)$ such that the family of solutions of differential equations of the type $y' = f(y)g(x)$ will be (a) symmetric across the x-axis, (b) symmetric across the y-axis, (c) symmetric about the origin.

10. Show that differential equations of the type $y' = g(x)$ are invariant under the interchange of y with $y + C$ for every constant C. What does this tell you about the family of solutions of these differential equations?

11. Show that differential equations of the type $y' = g(y)$ are invariant under the interchange of x with $x + C$ for every constant C. What does this tell you about the family of solutions of autonomous differential equations?

12. Consider the differential equation $y' = ax + by$, where a and b are constants. What conditions must these constants satisfy if this differential equation is symmetric (a) across the x-axis, (b) across the y-axis, (c) across both the x-axis and the y-axis? Confirm your answers by constructing the slope field for the appropriate differential equation.

3.3 NUMERICAL SOLUTIONS AND CHAOS

The preceding techniques let us graphically sketch the solution curve $y = y(x)$ for the differential equation

$$y' = g(x, y) \tag{3.14}$$

subject to the condition

$$y(x_0) = y_0 \tag{3.15}$$

without knowing the analytical solution. However, there are times when we need not only a sketch of the solution, but also need accurate numerical values of $y = y(x)$.

In Chapter 1 we found that the solution of $y' = g(x)$ subject to (3.15) can be written in the form

$$y(x) = \int_{x_0}^{x} g(t)\, dt + y_0.$$

In this case, a numerical approximation for $y(x)$ can be computed for given values of x by any of the standard techniques used in the numerical evaluation of definite integrals, such as the left-hand rule, the right-hand rule, the midpoint rule, the trapezoidal rule, Simpson's rule, and so on.[3] Indeed, we used this technique to determine specific values of the Error Function in Chapter 1. In Chapter 2, similar comments applied to finding $x = x(y)$ through standard numerical integration techniques. However, these numerical integration techniques fail on more general differential equations like (3.14).

Euler's Method

We now introduce a method of approximating the solution of (3.14) subject to (3.15) that relies only on the form of the differential equation itself. This technique is known as EULER'S METHOD and is based on locally approximating a function by its tangent line.[4]

In calculus we often find it convenient to approximate the graph of a differentiable function $y(x)$ near a point $x = a$ by its tangent line at that point, namely,[5]

$$y(x) \approx y(a) + y'(a)(x - a).$$

This approximation is called a "linear approximation" or "local linearity" and is frequently written as

$$y(a + h) \approx y(a) + y'(a)h, \tag{3.16}$$

where we have replaced $x - a$ by h. The geometric interpretation of (3.16) is shown in Figure 3.21, where the exact value of y at $x = a + h$, $y(a + h)$, is approximated by $y(a) + y'(a)h$.

---

[3] You should refer to your calculus text for a discussion of the pros and cons of these methods.

[4] Euler is pronounced "oiler."

[5] This equation is based on $y'(a) \approx [y(x) - y(a)]/(x - a)$.

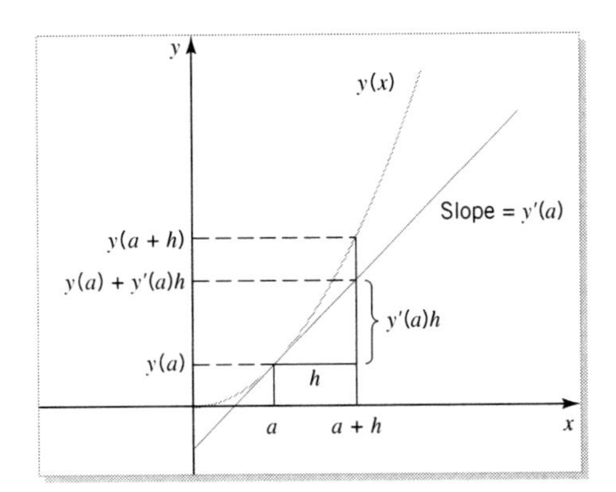

FIGURE 3.21 Geometric interpretation of $y(a+h) \approx y(a) + y'(a)h$

EXAMPLE 3.5

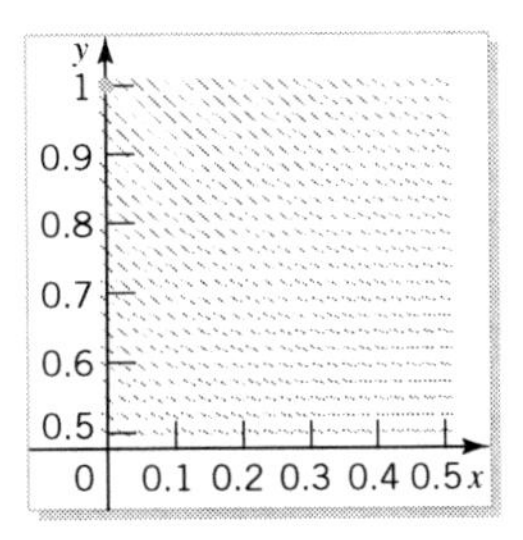

FIGURE 3.22 The initial point (0, 1) and slope field for $y' = x - y$

Suppose we start with the differential equation from Example 3.1,

$$y' = x - y, \tag{3.17}$$

along with an initial condition $y(0) = 1$, where we want to obtain a numerical value for y when $x = 0.5$ — namely, $y(0.5)$. Figure 3.22 shows the slope field for (3.17) and the initial point (0, 1), from which we could sketch a solution curve. From this hand-drawn solution curve we could estimate $y(0.5)$, but not very accurately.

We now use (3.16) to find a numerical approximation at $x = 0.5$. We start at the initial point (0, 1). We use the differential equation to discover that the slope of the solution curve at the point $x = 0$, $y = 1$, is $0 - 1 = -1$, from (3.17). We now draw a line from (0, 1) to the right with slope -1 and assume that near the point (0, 1) the graph of the actual solution and its tangent line will be close together. If we use this tangent line to approximate the solution curve for $0 \le x \le 0.1$, then at $x = 0.1$, (3.16) — with $a = 0$ and $h = 0.1$ — gives an approximation for the solution as $1 + (-1)(0.1)$, or 0.9; that is, $y(0.1) \approx 0.9$. This is recorded at the left-hand side of Table 3.3 where $x = 0.1$.

Thus, starting at the point (0, 1), we have found that the point (0.1, 0.9) is approximately on the solution curve. We now use the point (0.1, 0.9) in the same way we used (0, 1) to find an approximate value for $y(0.2)$. At the point (0.1, 0.9), we

Table 3.3 Numerical approximation to the solution of $y' = x - y$

x	y	$y' = x - y$	h	$y'h$
0.0	1.0000	$0.0 - 1.0000 = -1.0000$	0.1	$(-1.0000)(0.1) = -0.1000$
0.1	$1.0000 - 0.1000 = 0.9000$	$0.1 - 0.9000 = -0.8000$	0.1	$(-0.8000)(0.1) = -0.0800$
0.2	$0.9000 - 0.0800 = 0.8200$	$0.2 - 0.8200 = -0.6200$	0.1	$(-0.6200)(0.1) = -0.0620$
0.3	$0.8200 - 0.0620 = 0.7580$	$0.3 - 0.7580 = -0.4580$	0.1	$(-0.4580)(0.1) = -0.0458$
0.4	$0.7580 - 0.0458 = 0.7122$	$0.4 - 0.7122 = -0.3122$	0.1	$(-0.3122)(0.1) = -0.0312$
0.5	$0.7122 - 0.0312 = 0.6810$			

use the differential equation to compute the slope of the tangent line as $0.1 - 0.9$, or -0.8. We now construct this new tangent line, starting at $(0.1, 0.9)$, with slope -0.8. We use this tangent line and (3.16) to approximate the solution of the differential equation for $x = 0.2$ as $0.9 + (-0.8)(0.1)$, or $y(0.2) \approx 0.82$. A series of five steps using this pattern of moving along the tangent line, at each new point, to approximate the solution is shown in Table 3.3 (to 4 decimal places). This results in $y(0.5) \approx 0.6810$. This point and the intermediate points are shown in Figure 3.23. A better approximation for $y(0.5)$ could be obtained by making h smaller. The quantity h is called the STEP-SIZE.

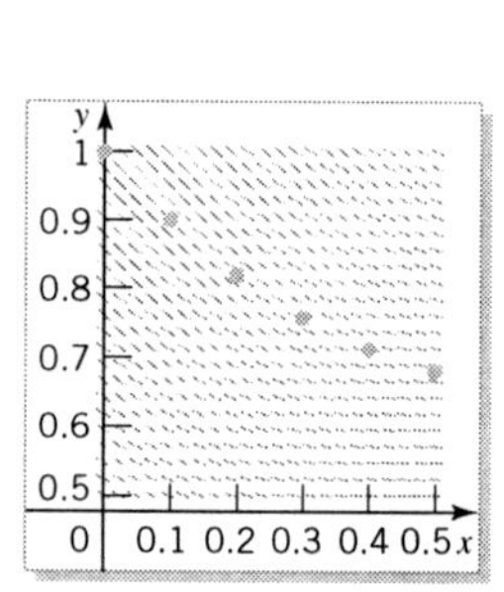

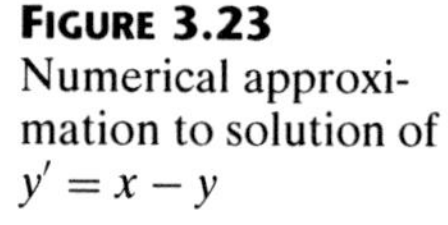
FIGURE 3.23 Numerical approximation to solution of $y' = x - y$

In Exercise 1 on page 91 we obtained the explicit solution of (3.17), namely $y(x) = x - 1 + Ce^{-x}$, which, subject to the condition $y(0) = 1$, gives the explicit solution

$$y(x) = x - 1 + 2e^{-x}. \tag{3.18}$$

Because the accuracy of Euler's method depends on the step-size h, in Table 3.4 we compare the Euler approximation with the exact solution for different step-sizes. In this table we can see how the accuracy depends on the step-size h. Figure 3.24 shows

Table 3.4 Comparison of solutions of $y' = x - y$, $y(0) = 1$

x	Euler ($h = 0.1$)	Euler ($h = 0.01$)	Euler ($h = 0.001$)	Exact
0.0	1.0000	1.0000	1.0000	1.0000
0.1	0.9000	0.9088	0.9096	0.9097
0.2	0.8200	0.8358	0.8373	0.8375
0.3	0.7580	0.7794	0.7814	0.7816
0.4	0.7122	0.7379	0.7404	0.7406
0.5	0.6810	0.7100	0.7128	0.7131

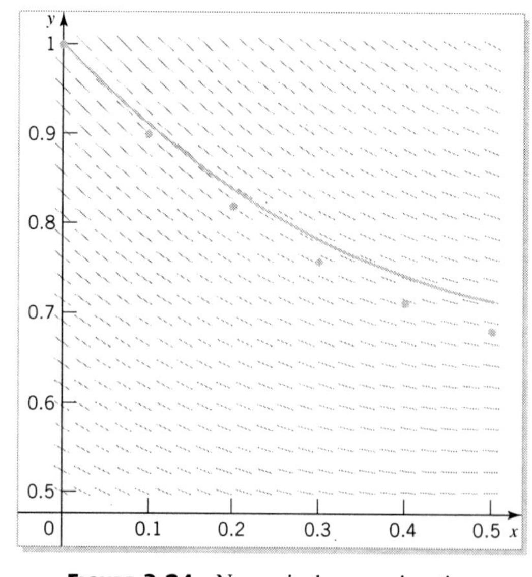

FIGURE 3.24 Numerical approximation ($h = 0.1$) and exact solution of $y' = x - y$

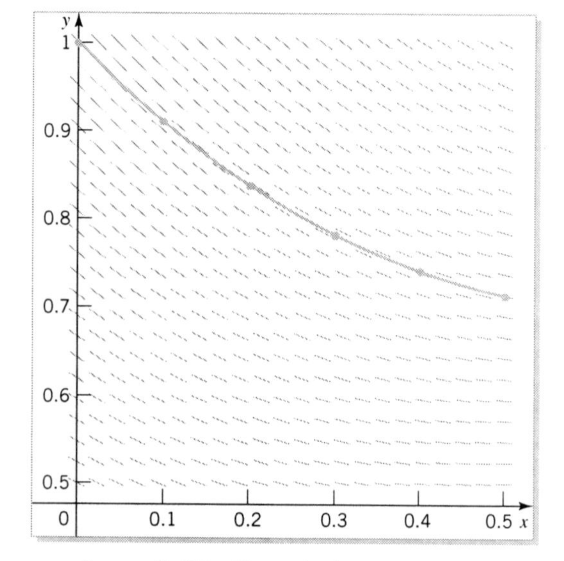

FIGURE 3.25 Numerical approximation ($h = 0.001$) and exact solution of $y' = x - y$

the numerical approximation for $h = 0.1$ and the exact solution. Figure 3.25 shows the numerical approximation for $h = 0.001$ and the exact solution. (Can you explain why, in Table 3.4, all the numerical values using Euler's method are never larger than the exact value for every value of x and all three values of h?) ■

Constructing the algorithm

We can construct an algorithm for this process — namely, that the y value at a new value of x equals the y value at the old value of x plus the slope at the old value of x times the distance between these two values of x. This is perhaps more clearly stated in terms of an equation. We do it for a general first order differential equation, $y' = g(x, y)$, starting at the point (x_0, y_0). We first note that the slope of the tangent line to the solution curve at the point (x_0, y_0) is given by $g(x_0, y_0)$. This means the equation of the tangent line at this point is

$$y - y_0 = g(x_0, y_0)(x - x_0).$$

If we move to the right along this tangent line a horizontal distance h to $x_0 + h$, we estimate the y value of the solution of the differential equation at $x_0 + h$ as

$$y(x_0 + h) \approx y(x_0) + g(x_0, y_0)h, \tag{3.19}$$

where $y(x_0) = y_0$. Continuing this process gives

$$y(x_0 + 2h) \approx y(x_0 + h) + g(x_0 + h, y_0 + g(x_0, y_0)h)h. \tag{3.20}$$

Because the initial value of y was y_0, we denote by y_1 the numerical approximation to the solution of the differential equation at $x_0 + h$, we denote by y_2 the numerical approximation at $x_0 + 2h$, and so on. This means we can rewrite (3.19) and (3.20) as

$$y_1 = y_0 + g(x_0, y_0)h$$

and

$$y_2 = y_1 + g(x_1, y_1)h, \text{ where } x_1 = x_0 + h,$$

or, after n steps,

$$y_n = y_{n-1} + g(x_{n-1}, y_{n-1})h$$

where

$$x_{n-1} = x_0 + (n - 1)\,h.$$

This way of constructing a numerical approximation to the solution of a differential equation is called Euler's method.

How to Analyze $y' = g(x, y)$ Numerically

Purpose To use Euler's method to obtain a numerical approximation to the solution of

$$y' = g(x, y), \tag{3.21}$$

starting at the point (x_0, y_0).

Process

1. Select a step-size h.
2. Compute $y_1 = y(x_0) + g(x_0, y_0)h$, where $y(x_0) = y_0$. The value of y_1 is a numerical approximation to the solution of (3.21) at $x_1 = x_0 + h$; that is, $y(x_1) \approx y_1$.
3. Use x_1 and y_1 from step 2 to compute $y_2 = y_1 + g(x_1, y_1)h$. The value of y_2 is a numerical approximation to the solution of (3.21) at $x_2 = x_1 + h = x_0 + 2h$; that is, $y(x_2) \approx y_2$.
4. Continue in this way so that after n steps the value of y_n, where $y_n = y_{n-1} + g(x_{n-1}, y_{n-1})h$, is a numerical approximation to the solution of (3.21) at $x_n = x_0 + nh$; that is, $y(x_n) \approx y_n$.

Comments about Analyzing $y' = g(x,y)$ Numerically

- The usual advice in choosing the step-size h is "the smaller the better."
- If you want to find a numerical solution over an interval $a \le x \le b$, starting at $(a, y(a))$ and ending when $x = b$, choose the step-size h so that $h = (b - a)/n$, where n is the number of steps you will take along the x-axis.
- Euler's method also works going from larger values of x to smaller. In this case h is negative.
- Euler's method is one of many ways to obtain a numerical approximation to a solution of a differential equation, or, in short, a numerical solution. Some of these ways are mentioned later in this section.
- Every numerical method has its problems.
- **From now on, we assume that either you have access to a computer/calculator program that performs Euler's method or you are willing to do these calculations by hand.**

Another View of Euler's Method

We know that the differential equation

$$y' = x - y \tag{3.22}$$

has the solution $y(x) = x - 1 + Ce^{-x}$, where C is determined by an initial condition. In order to understand what we are doing when we use Euler's method, we will look at the numerical solution of (3.22) that we obtained by starting from $y(0) = 1$ and using a step-size of $h = 0.1$. We know that the exact solution, subject to the initial condition $y(0) = 1$, is $y(x) = x - 1 + 2e^{-x}$.

Using Euler's method to approximate $y(0.1)$ gives $y(0.1) \approx y_1 = 0.9$; $y(0.2) \approx y_2 = 0.82$, and so on. These three points are shown and joined in Figure 3.26.

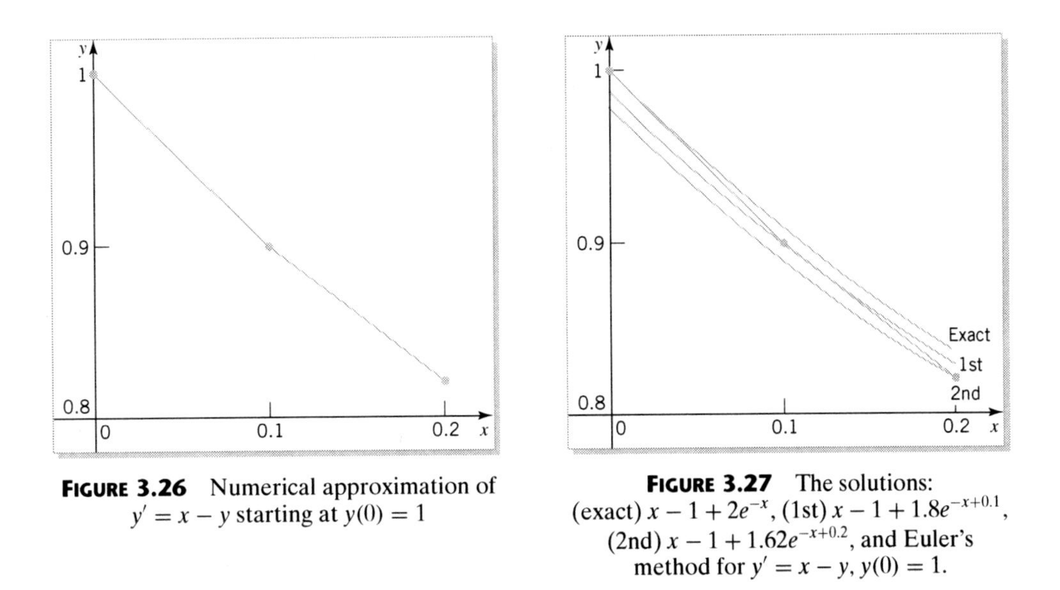

FIGURE 3.26 Numerical approximation of $y' = x - y$ starting at $y(0) = 1$

FIGURE 3.27 The solutions: (exact) $x - 1 + 2e^{-x}$, (1st) $x - 1 + 1.8e^{-x+0.1}$, (2nd) $x - 1 + 1.62e^{-x+0.2}$, and Euler's method for $y' = x - y$, $y(0) = 1$.

The exact value for $y(0.1)$ is $y(0.1) = 0.1 - 1 + 2e^{-0.1} \approx 0.9097$, and $y(0.2) = 0.2 - 1 + 2e^{-0.2} \approx 0.8375$. Thus, the point $(x_1, y_1) = (0.1, 0.9)$ does not lie on the solution curve $y(x) = x - 1 + 2e^{-x}$. On which solution curve does $(0.1, 0.9)$ lie? That is, which solution $y(x) = x - 1 + Ce^{-x}$ satisfies $y(0.1) = 0.9$? This is another initial value problem, with exact solution $y(x) = x - 1 + 1.8e^{-x+0.1}$. So in going from $x = 0$ to $x = 0.1$, our numerical answer jumped from the exact solution curve $y(x) = x - 1 + 2e^{-x}$ to the first solution curve $y(x) = x - 1 + 1.8e^{-x+0.1}$. In the same way, in going from $x = 0.1$ to $x = 0.2$, our numerical answer jumped from the first solution curve $y(x) = x - 1 + 1.8e^{-x+0.1}$ to the second solution curve $y(x) = x - 1 + 1.62e^{-x+0.2}$. This is shown in Figure 3.27.

Thus, it is important when using numerical methods that the solution we seek has the property that nearby solutions remain close to it as x increases. If not, we may get further and further away from the exact solution as we take values of x further from x_0.

Improvements on Euler's Method

We will now describe various ways to improve Euler's method. These improvements are based on the following observation. By integrating both sides of the differential equation $y' = g(x, y)$ with respect to x from x_0 to $x_0 + h$, we can write the initial value problem

$$y' = g(x, y) \qquad y(x_0) = y_0$$

in the form

$$y(x_0 + h) = y_0 + \int_{x_0}^{x_0+h} g(x, y(x))\, dx.$$

Thus, the problem of estimating $y(x_0+h)$ reduces to the problem of estimating $\int_{x_0}^{x_0+h} g(x, y(x))\, dx$. We will obtain estimates for $y(x_0+h)$ by using three of the standard ways used to estimate integrals — namely, the left-hand rule, the trapezoidal rule, and Simpson's rule.

Euler's method

If we use the left-hand rule to approximate this integral — which is equivalent to assuming that $g(x, y(x))$ takes on its value at the left-hand endpoint (x_0, y_0); that is, $g(x, y(x)) \approx m_0$, where $m_0 = g(x_0, y(x_0))$ — we find

$$\int_{x_0}^{x_0+h} g(x, y(x))\, dx \approx \int_{x_0}^{x_0+h} g(x_0, y_0)\, dx = g(x_0, y_0)\, h,$$

so that, according to this approximation,

$$y(x_0+h) \approx y_0 + m_0 h,$$

the first step in Euler's method. Here m_0 is the slope of the solution curve at the left-hand endpoint.

Heun's method

This procedure suggests that we could improve on Euler's method by using a better approximation to estimate the integral $\int_{x_0}^{x_0+h} g(x, y(x))\, dx$. One such approximation is the trapezoidal rule where we assume that $g(x, y(x))$ takes on the average of its values at the endpoints (x_0, y_0) and $(x_0+h, y(x_0+h))$ — that is, $g(x, y(x)) \approx \frac{1}{2}(m_0+m_1)$, where $m_0 = g(x_0, y_0)$ and $m_1 = g(x_0+h, y(x_0+h))$ are the slopes of the actual solution at the two endpoints. However, the evaluation of m_1 requires the value of $y(x_0+h)$, which is the quantity we seek. To circumvent this, we estimate $y(x_0+h)$ in the expression for m_1 by the Euler approximation so that $m_1 \approx k_1 = g(x_0+h, y_0+g(x_0, y_0)h)$ — that is, we approximate the slope of the solution curve through the right endpoint $(x_0+h, y(x_0+h))$ by the slope of the solution curve through the point $(x_0+h, y_0+m_0 h)$. According to this approximation,

$$y(x_0+h) \approx y_0 + \frac{1}{2}(m_0+k_1)\, h,$$

the first step in what is usually called HEUN'S METHOD or MODIFIED EULER'S METHOD. In this method, we initially use m_0 to predict the value at $y(x_0+h)$ by Euler's method and then use k_1 — the slope of the solution at the right-hand endpoint predicted by Euler's method — to correct this prediction. This is an example of a PREDICTOR-CORRECTOR METHOD for solving initial value problems numerically.

Runge-Kutta 4 method

A popular technique used to approximate integrals is Simpson's rule, which assumes that the slope of the solution curve $g(x, y(x))$ is a weighted average of the slopes at the two endpoints (x_0, y_0) and $(x_0+h, y(x_0+h))$ and the midpoint $(x_0+\frac{1}{2}h, y(x_0+\frac{1}{2}h))$ — that is, $g(x, y(x)) \approx \frac{1}{6}(m_0+4m_{1/2}+m_1)$, where $m_0 = g(x_0, y_0)$, $m_{1/2} = g(x_0+\frac{1}{2}h, y(x_0+\frac{1}{2}h))$, and $m_1 = g(x_0+h, y(x_0+h))$. (What do the slopes $m_{1/2}$ and m_1 represent?) However, the evaluation of $m_{1/2}$ and m_1 requires the values of $y(x_0+\frac{1}{2}h)$ and $y(x_0+h)$. Using ideas similar to those developed when deriving the modified Euler's method, the slopes $m_{1/2}$ and m_1 can be approximated by

$$m_{1/2} \approx \frac{1}{2}(k_1+k_2),$$

and

$$m_1 \approx k_3,$$

where

$$\begin{aligned} k_1 &= g\left(x_0 + h/2, y_0 + m_0 h/2\right) && \text{slope of solution curve through } \left(x_0 + h/2, y_0 + m_0 h/2\right), \\ k_2 &= g\left(x_0 + h/2, y_0 + k_1 h/2\right) && \text{slope of solution curve through } \left(x_0 + h/2, y_0 + k_1 h/2\right), \\ k_3 &= g\left(x_0 + h, y_0 + k_2 h\right) && \text{slope of solution curve through } \left(x_0 + h, y_0 + k_2 h\right). \end{aligned}$$

According to this approximation,

$$y(x_0 + h) \approx y_0 + \frac{1}{6}\left(m_0 + 2k_1 + 2k_2 + k_3\right) h,$$

the first step in what is usually called the RUNGE-KUTTA 4 METHOD. This too is a predictor-corrector method.

To compare these three methods, we return to the initial value problem

$$y' = x - y, \qquad y(0) = 1,$$

to see what these methods predict for $y(0.5)$.

This initial value problem has the exact solution

$$y(x) = x - 1 + 2e^{-x}$$

so $y(0.5) = -0.5 + 2e^{-0.5} \approx 0.7130613194$. The ERROR made by a particular method is the magnitude of the difference between the exact value at $x = 0.5$ — that is, $y(0.5) \approx 0.7130613194$ — and the estimated value at $x = 0.5$.

Table 3.5 shows the error made in the numerical estimates of $y(0.5)$ for the three methods, as the step-size h is halved (0.1, 0.05 , 0.025, 0.0125). Figure 3.28 shows a plot of the logarithm of the error versus the logarithm of h, for these three cases. Each set appears to lie on a line, so we have drawn the lines of best-fit — namely, $\ln(\text{error}) = 1.02344 \ln h - 1.088824$ (Euler), $\ln(\text{error}) = 2.031356 \ln h - 2.151619$ (Heun), and $\ln(\text{error}) = 4.131532 \ln h - 4.865449$ (Runge-Kutta).

Table 3.5 Errors as a function of step-size

h	Euler	Heun	Runge-Kutta
0.1000	0.0320813194	0.0010902112	0.0000005494
0.0500	0.0155874409	0.0002624159	0.0000000329
0.0250	0.0076859590	0.0000643785	0.0000000020
0.0125	0.0038166890	0.0000159440	0.0000000001

All three of these equations lead to a similar formula — namely, $\ln(\text{error}) = n \ln h + c$ — which implies that the error is proportional to h^n, where the constant

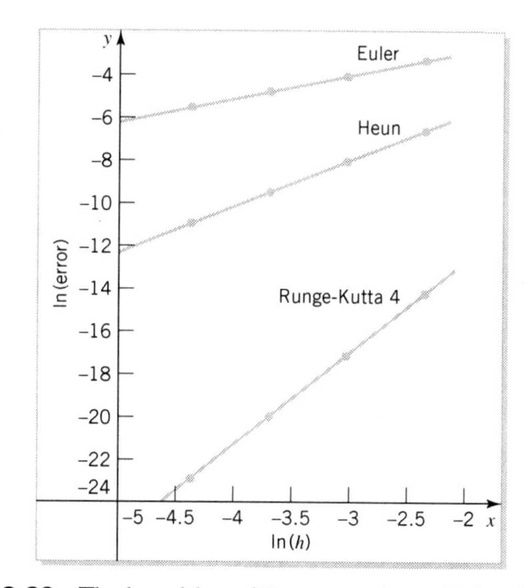

FIGURE 3.28 The logarithm of the errors due to Euler, Heun, and Runge-Kutta methods versus the logarithm of the step-size

n depends on the technique. These results — which can be proved generally[6] — suggest that

- the error due to Euler's method is proportional to h,
- the error due to Heun's method is proportional to h^2, and
- the error due to Runge-Kutta 4 is proportional to h^4.

The exponent of h is called the ORDER OF THE METHOD, so Euler, Heun, and Runge-Kutta are first, second, and fourth order methods, respectively. Thus, if we halve the step-size, the error is reduced by $\frac{1}{2}$ using Euler's method, by $\frac{1}{4}$ using Heun's method, and by $\frac{1}{16}$ using Runge-Kutta 4. This can be seen by looking at Table 3.5. Runge-Kutta 4 is the method of choice for finding numerical solutions of most ordinary differential equations.

These are three of the many different methods available to solve differential equations numerically.[7] **From now on we assume that when you are doing Examples and Exercises identified as** COMPUTER EXPERIMENTS, **you have access to a computer/calculator program that performs the Runge-Kutta 4 method.**

Period Doubling and Chaos

Because many useful differential equations do not possess explicit solutions, numerical methods are frequently used. However, care must be taken in using these

[6]See, for example, *Numerical Methods* by J.T. Mathews, Prentice Hall, 1987.

[7]See, for example, *Numerical Recipes in C* by W.H. Press, S.A. Teukolsky, W.T. Vetterling, and B.P. Flannery, Cambridge University Press, 2nd edition, 1992.

methods because they give only approximate solutions. We now provide some examples where we are misled by the numerical solutions.

We will concentrate on numerical solutions of the initial value problem

$$y' = 10y\,(1-y)\,, \qquad y(0) = 0.1.$$

This is the logistic equation, so we know that the solution that starts at $(0, 0.1)$ increases toward the equilibrium solution $y(x) = 1$.

Figures 3.29, 3.30, 3.31, and 3.32 show some numerical solutions using the Runge-Kutta 4 method with different step-sizes h. In Figure 3.29 we see that with $h = 0.27$, the numerical solution approaches $y(x) = 1$, as it should. With $h = 0.30$ and $h = 0.34$, the numerical solutions approach $y(x) \approx 0.76$ and $y(x) \approx 0.63$, respectively, with the latter oscillating initially. The numerical solution in Figure 3.30 (with $h = 0.35$) does not approach a single value but eventually oscillates between the two values 0.72 and 0.51. We say that such a numerical solution has period 2. (In the same way, those previous numerical solutions that approach a single value — namely, 1, 0.76, and 0.63 — are said to have period 1.) The numerical solution in Figure 3.31 (with $h = 0.365$) does not oscillate between two values, but eventually oscillates between four (0.39, 0.82, 0.41, 0.80), and so the numerical solution has period 4. The numerical solution in Figure 3.32 (with $h = 0.39$) is quite chaotic, showing no periodic behavior. All of these numerical solutions except the one corresponding to $h = 0.27$ are completely wrong and do not predict the correct behavior of the solution.

We can gain insight into this process if we solve the differential equation numerically for particular values of the step-size h, and then look at the last few values generated by the numerical method to see whether they eventually follow a pattern. One way to see this graphically is to start with a specific h, calculate the numerical solution for, say, 100 steps, and then plot the next 100 values corresponding to that value of h. Then increase h and repeat the process, ultimately plotting these

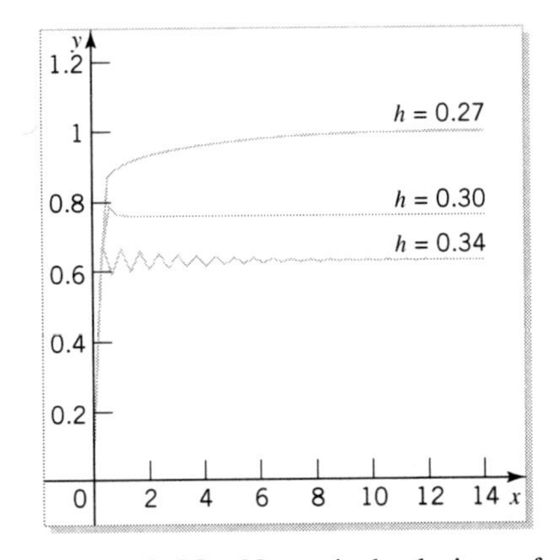

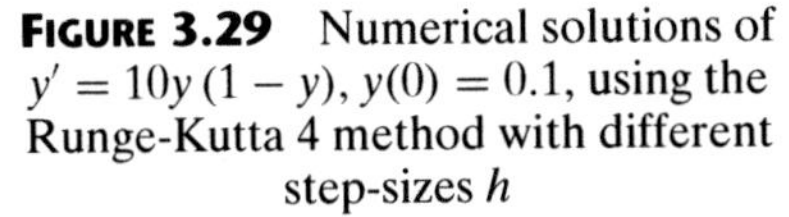
FIGURE 3.29 Numerical solutions of $y' = 10y\,(1-y)$, $y(0) = 0.1$, using the Runge-Kutta 4 method with different step-sizes h

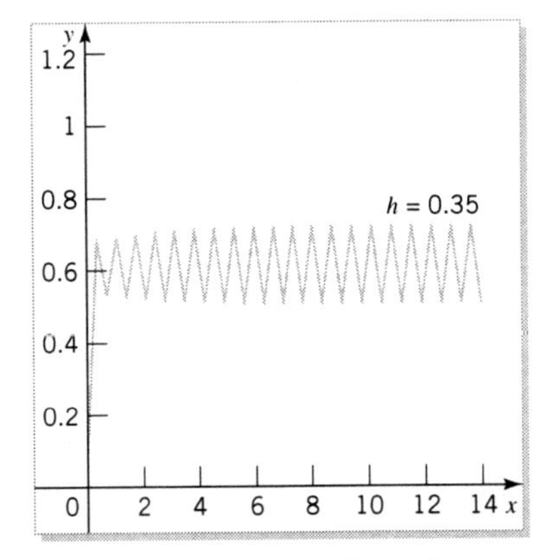

FIGURE 3.30 Numerical solution with period 2 of $y' = 10y\,(1-y)$, $y(0) = 0.1$, using the Runge-Kutta 4 method

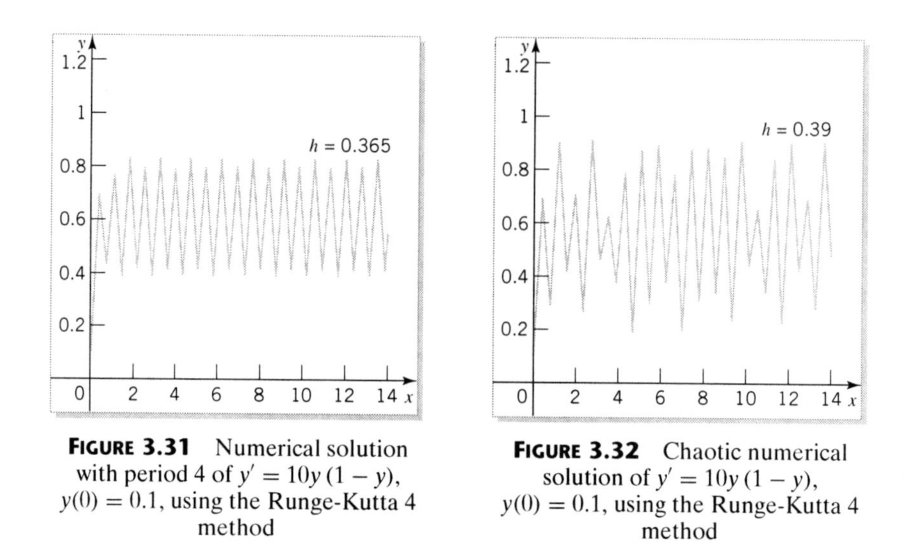

FIGURE 3.31 Numerical solution with period 4 of $y' = 10y(1-y)$, $y(0) = 0.1$, using the Runge-Kutta 4 method

FIGURE 3.32 Chaotic numerical solution of $y' = 10y(1-y)$, $y(0) = 0.1$, using the Runge-Kutta 4 method

Orbit diagram

100 values for each value of h as a function of the step-size h. Figure 3.33 shows such a plot, often called an ORBIT DIAGRAM.

Let's make sure we understand Figure 3.33. If we look at the point on the diagram that corresponds to a step-size of 0.27, we see that the last 100 numerical values fell approximately on the same number, namely, 1. This agrees with the $h = 0.27$ curve in Figure 3.29. In the same way, with $h = 0.30$, Figure 3.33 tells us that the last 100 numerical values fell approximately on the same number – namely, 0.76 – again in agreement with Figure 3.29. With $h = 0.35$, Figure 3.33 tells us that the last 100 numerical values fell approximately on two different numbers – namely, 0.72 and 0.51 – in agreement with Figure 3.30. Thus, when h is between about 0.34 and 0.36 we expect period 2 solutions. We can see the period 4 solutions ($h = 0.365$)

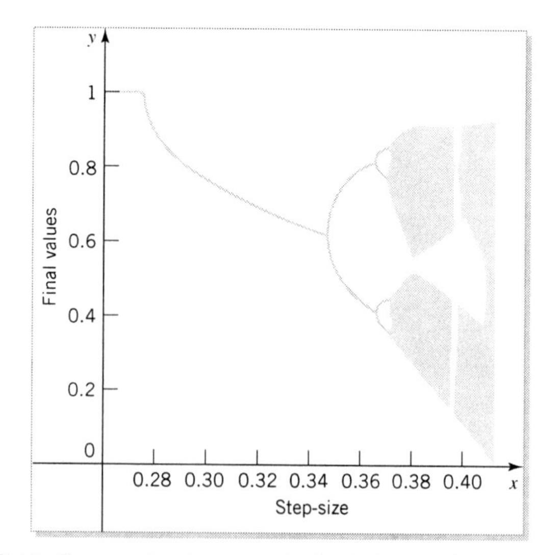

FIGURE 3.33 Orbit diagram for the numerical solution of $y' = 10y(1-y)$, $y(0) = 0.1$, using the Runge-Kutta 4 method

and the chaotic solutions (with $h = 0.39$) in Figure 3.33. Can you see any period 8 solutions? Where would you look for a period 16 solution? Figure 3.33 is sometimes described by the expression "periodic doubling is the road to chaos." Why is this an accurate description?

This property of period doubling leading to chaos is not restricted to the Runge-Kutta method. Other methods show similar properties. See Exercises 10 and 11 on pages 110–111.

This is one aspect of the subject of CHAOS, an exciting area of mathematics that is being studied very intensively at present.[8]

EXERCISES

1. Complete steps i through iv for each of the following initial value problems (a) through (f):

i. Find the exact solution (see Exercise 1, Section 1.1), and evaluate it at x_1, x_2, x_3, and x_4.

ii. Use Simpson's rule to find approximate values for $y(x_1), y(x_2), y(x_3)$, and $y(x_4)$. Each time start with the point $(x_0, y(x_0))$. Compare these results with the exact values.

iii. Use Euler's method with $h = 0.1$ to find approximate values for $y(x_1)$, $y(x_2)$, $y(x_3)$, and $y(x_4)$. Compare these results with the exact values.

iv. Repeat iii with $h = 0.05$, making the appropriate changes in $x_1, x_2, x_3, \cdots, x_8$.

(a) $y' = x^3$, $y(1) = 1$. $x_0 = 1$
$x_1 = 1.1$ $x_2 = 1.2$ $x_3 = 1.3$ $x_4 = 1.4$

(b) $y' = \cos x$, $y(0) = 0$. $x_0 = 0$
$x_1 = 0.1$ $x_2 = 0.2$ $x_3 = 0.3$ $x_4 = 0.4$

(c) $y' = e^{-x}$, $y(0) = 1$. $x_0 = 0$
$x_1 = 0.1$ $x_2 = 0.2$ $x_3 = 0.3$ $x_4 = 0.4$

(d) $y' = 1/x$, $y(-1) = 1$. $x_0 = -1$
$x_1 = -0.9$ $x_2 = -0.8$ $x_3 = -0.7$ $x_4 = -0.6$

(e) $y' = 1/(1+x^2)$, $y(1) = \pi/4$. $x_0 = 1$
$x_1 = 1.1$ $x_2 = 1.2$ $x_3 = 1.3$ $x_4 = 1.4$

(f) $y' = x^2e^{-x}$, $y(0) = 1$. $x_0 = 0$
$x_1 = 0.1$ $x_2 = 0.2$ $x_3 = 0.3$ $x_4 = 0.4$

2. Solve the following initial value problems numerically on the indicated interval using Euler's method with $h = 0.1$. Compare your value of y at the right-hand end-point of the interval with the value given by the explicit solution.

(a) $y' = 1 + y^2$ $y(0) = 0$ $0 \le x \le 0.5$
(b) $y' = e^{-y}$ $y(0) = 0$ $0 \le x \le 0.5$
(c) $y' = 2x - 3y$ $y(1) = 1$ $1 \le x \le 1.5$

3. Consider the differential equation
$y' = (y-a)(y-b)(y-c)$.

(a) Draw slope fields for each of the following cases i through iii, and sketch by hand what you think the solution curves will look like.

i. $a = b = c = 1$
ii. $a = b = 1, c = -1$
iii. $a = 1, b = 0, c = -1$

(b) Use a computer/calculator program — changing the step-size if needed — to draw solution curves for each of the cases in part (a) for various initial conditions. Make sure the results you obtain here are consistent with the answers you gave to part (a).

4. Table 3.6 shows the errors produced when evaluating the solution of a particular initial value problem using one of the Euler, Heun, or Runge-Kutta methods with different step-sizes. Which one do you think was used? Why?

Table 3.6 Errors as a function of step-size

h	Error
0.1000	0.004201
0.0500	0.001091
0.0250	0.000278
0.0125	0.000070

[8] For a popular account of this subject, see *Chaos – Making a New Science* by J. Gleick, Penguin Books, 1987. For mathematical treatments, see *An Introduction to Chaotic Dynamical Systems* by R.L. Devaney, Addison-Wesley, 1989, and *Nonlinear Dynamics and Chaos* by S.H. Strogatz, Addison-Wesley, 1994.

5. Figure 3.34 shows the slope field and a numerical solution — produced by Euler's method — for a particular initial value problem. These were generated by a computer software program. How do you know that the slope field and the numerical solution are not consistent? Which do you think is correct, the slope field or the numerical solution? What would you try to change in the software program to make them consistent?

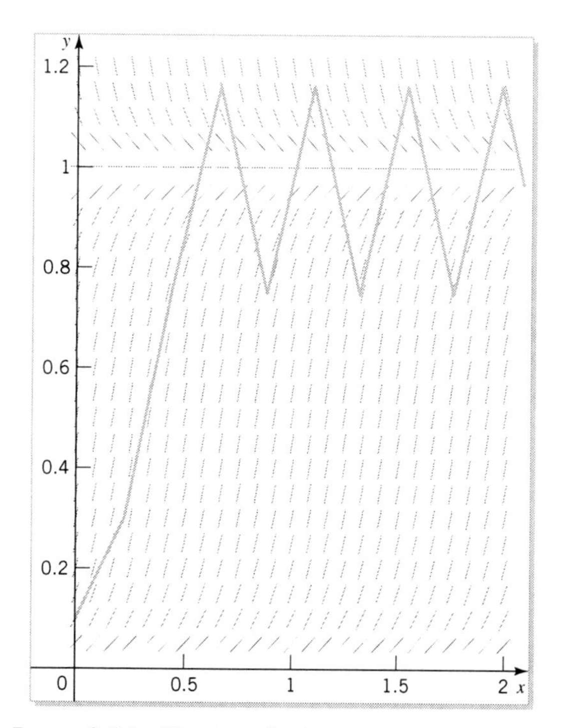

FIGURE 3.34 The slope field and a numerical solution

6. **Midpoint Method.** A numerical method sometimes used to solve the initial value problem $y' = g(x, y)$, $y(x_0) = y_0$ is the MIDPOINT METHOD. According to this method, the first step is

$$y(x_0 + h) \approx y_0 + hg\left(x_0 + \frac{1}{2}h, y_0 + \frac{1}{2}hg(x_0, y_0)\right).$$

(a) Describe this technique geometrically in terms of the slopes of solutions of the differential equation. Do you think this technique is well-named?

(b) Table 3.7 shows the errors produced when evaluating the numerical solution of a particular initial value problem using the midpoint method with different step-sizes. Based on this, estimate the order of this method.

Table 3.7 Errors as a function of step-size

h	Error
0.1000	0.000332446
0.0500	0.000083129
0.0250	0.000020784
0.0125	0.000005120

(c) If $g(x, y)$ is only a function of x — that is, $g(x, y) = g(x)$ — the midpoint method reduces to a standard method for numerically evaluating the integral $\int_{x_0}^{x_0+h} g(x)\,dx$. What is that standard method?

7. **Computer Experiment.** Consider the initial value problem $y' = y - e$, $y(0) = e - 1$.

(a) Solve this problem exactly. What is $y(1)$?

(b) Estimate $y(1)$ by Euler's method using step-sizes h of 0.1, 0.05, 0.025, and 0.0125. Plot the logarithm of the error as a function of the logarithm of h. Is this consistent with Euler's method being of first order?

(c) Estimate $y(1)$ by Heun's method using step-sizes h of 0.1, 0.05, 0.025, and 0.0125. Plot the logarithm of the error as a function of the logarithm of h. Is this consistent with Heun's method being of second order?

(d) Estimate $y(1)$ by the Runge-Kutta 4 method using step-sizes h of 0.1, 0.05, 0.025, and 0.0125. Plot the logarithm of the error as a function of the logarithm of h. Is this consistent with the Runge-Kutta 4 method being of fourth order?

8. **Computer Experiment.** Consider the initial value problem $y' = y^2$, $y(0) = 1$.

(a) Solve this problem exactly. What is $y(2)$?

(b) Estimate $y(2)$ by solving the initial value problem by any numerical method. What do you notice? What is causing the problem?

9. **Computer Experiment.** Consider the initial value problem $y' = x^2 + y^2$, $y(0) = 1$, which cannot be solved in terms of familiar functions. Estimate $y(2)$ by solving the initial value problem by different numerical methods. What do you notice? What is causing the problem?

10. Figure 3.35 shows the orbit diagram corresponding to numerical solutions of the initial value problem $y' = 10y(1 - y)$, $y(0) = 0.1$, using Euler's method. For what values of the step-size h do we find period 1 solutions?

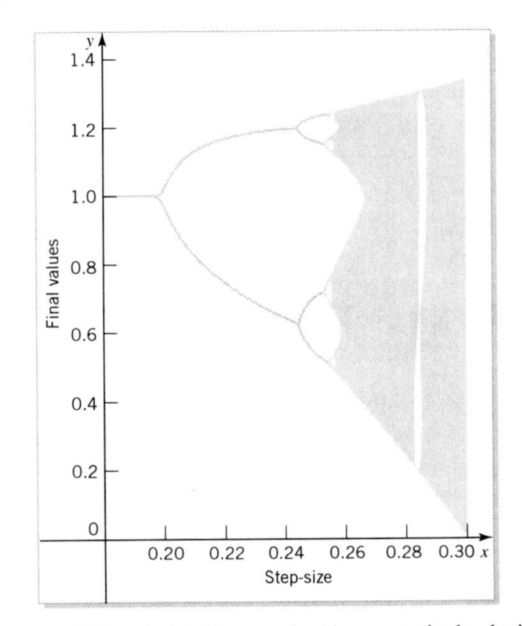

FIGURE 3.35 Orbit diagram for the numerical solution of $y' = 10y(1-y)$, $y(0) = 0.1$, using Euler's method

Period 2 solutions? Period 4 solutions? Chaos? Do all period 1 solutions have the same ultimate behavior?

11. Figure 3.36 shows the orbit diagram corresponding to numerical solutions of the initial value problem $y' = 10y(1-y)$, $y(0) = 0.1$, using Heun's method. For what values of the step-size h do we find period 1 solutions?

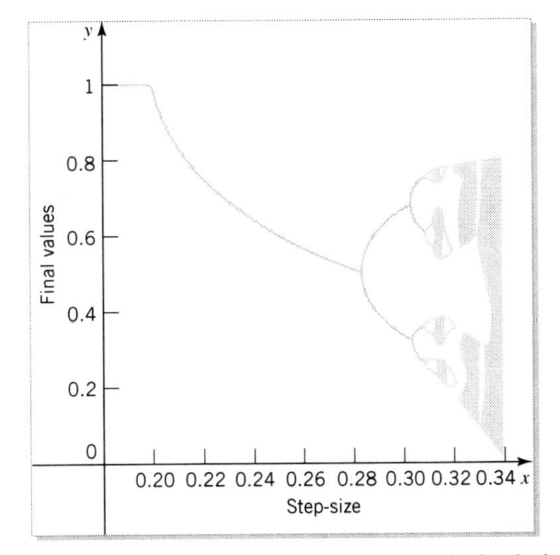

FIGURE 3.36 Orbit diagram for the numerical solution of $y' = 10y(1-y)$, $y(0) = 0.1$, using Heun's method

Period 2 solutions? Period 4 solutions? Chaos? Do all period 1 solutions have the same ultimate behavior?

12. Computer Experiment. Use Runge-Kutta 4 to graph solution curves of the initial value problem $y' = y^2 - x$, $y(x_0) = 0$, for various x_0 near -1, in the window $-2 \le x \le 4$, $-3 \le y \le 3$. You should see two dramatically different types of behavior.

(a) Can you find an initial condition that separates these two behaviors?

(b) Explain this behavior by plotting the isoclines corresponding to slopes -1, 0, and 1.

(c) Someone comments that this behavior is an example of "sensitivity to initial conditions." Explain what you think is meant by this comment, and in what sense this is true.

13. Computer Experiment. Use Runge-Kutta 4 to graph solution curves of the initial value problem $y' = -1 + 2xy$, $y(0) = y_0$, for various y_0 between 0 and 1, in the window $0 \le x \le 4$, $0 \le y \le 4$. You should see two dramatically different types of behavior.

(a) Can you find an initial condition that separates these two behaviors?

(b) Explain this behavior by plotting the isoclines corresponding to slopes -1 and 0.

(c) Someone comments that this behavior is an example of "sensitivity to initial conditions." Explain what you think is meant by this comment and in what sense this is true.

(d) Confirm that the solution of the initial value problem is

$$y(x) = \left[y_0 - \frac{\sqrt{\pi}}{2} erf(x)\right] e^{x^2}$$

by substituting $y(x)$ into the differential equation and the initial value.

(e) What happens to this solution as $x \to \infty$? [Hint: $\lim_{x\to\infty} erf(x) = 1$.] Is this consistent with the answer you found in part (a)?

14. Computer Experiment. Use Runge-Kutta 4 with $h = 0.01$ to graph solution curves of the initial value problem $y' = -2(x+2)e^{-x^2} + 4y$, $y(0) = y_0$, for various y_0 near 1, in the window $0 \le x \le 6$, $-1 \le y \le 5$. You should see two dramatically different types of behavior.

(a) Can you find an initial condition that separates these two behaviors?

(b) Someone comments that this behavior is an example of "sensitivity to initial conditions." Explain what you think is meant by this comment and in what sense this is true.

(c) Confirm that the solution of the initial value problem is $y(x) = e^{-x^2} + (y_0 - 1)\, e^{4x}$ by substituting $y(x)$ into the differential equation and the initial value.

(d) What is this solution if $y(0) = 1$? What happens to this solution as $x \to \infty$? Is this consistent with the answer you found in part (a)? What do you think has happened?

3.4 COMPARING SOLUTIONS OF DIFFERENTIAL EQUATIONS

We have seen examples where numerical solutions led to believable but incorrect results. For cases where we cannot obtain explicit solutions, we need other ways of testing the accuracy of our numerical solution. Graphical analysis is one way. Another is to use a comparison theorem, which compares the behavior of the solution of one differential equation to the solution of another.

In calculus you may have met the following theorem regarding continuous functions f and g — if not, looking at its graphical representation will illustrate its intention.

▶ **Theorem 3.1:** *If $f(t) < g(t)$ for $x_0 < t < x$, then*

$$\int_{x_0}^{x} f(t)\, dt < \int_{x_0}^{x} g(t)\, dt. \tag{3.23}$$

◀

Let's assume that $f(t) < g(t)$ for $x_0 < t < x$ and apply this theorem to two differential equations of the type discussed in Chapter 1, namely

$$y' = f(x) \tag{3.24}$$

and

$$y' = g(x) \tag{3.25}$$

subject to the same initial condition $y(x_0) = y_0$. If $y_1(x)$ is a solution of (3.24) subject to $y_1(x_0) = y_0$, and $y_2(x)$ is a solution of (3.25) subject to the same initial condition $y_2(x_0) = y_0$, then (3.23) would imply that

$$\int_{x_0}^{x} \frac{dy_1}{dt}\, dt < \int_{x_0}^{x} \frac{dy_2}{dt}\, dt, \tag{3.26}$$

that is,

$$y_1(x) - y_1(x_0) < y_2(x) - y_2(x_0).$$

Because $y_1(x_0) = y_2(x_0)$, this equation implies that $y_1(x) < y_2(x)$. Thus, we have the following result.

Theorem 3.2: *Let $f(t) < g(t)$ for $x_0 < t < x_1$. If $y_1(x)$ is a solution of the initial value problem $y' = f(x)$, $y(x_0) = y_0$, and if $y_2(x)$ is a solution of the initial value problem $y' = g(x)$, $y(x_0) = y_0$, then*

$$y_1(x) < y_2(x) \text{ for } x_0 < x < x_1. \tag{3.27}$$

This theorem says that if the solutions of (3.24) and (3.25) pass through the same point, and if the slope of the second solution always exceeds the slope of the first solution on an interval, then the second solution always exceeds the first solution on that interval. We can restate this in another way if we interpret y' as the velocity of an object — namely, if two objects start at the same time from the same place, and the velocity of the first object always exceeds the velocity of the second object, then the first object will be ahead of the second object at any later time.

A natural question to ask is, How does this generalize to a larger class of differential equations? In order to answer this question we will first show that the previous theorem can be justified in a way that lends itself to generalization. We will assume that the theorem is false, that is there is another point x^*, $x_0 < x^*$, for which the solutions intersect — that is, $y_1(x_0) = y_2(x_0)$ and $y_1(x^*) = y_2(x^*)$. We will show that this leads to a contradiction and so $y_1(x^*) = y_2(x^*)$ cannot happen.

For this we will need the Mean Value Theorem.

Theorem 3.3: Mean Value Theorem *If $h(x)$ is a continuous function for $a \le x \le b$, and is differentiable for $a < x < b$, then there is at least one value c between a and b, $a < c < b$, for which*

$$h'(c) = \frac{h(b) - h(a)}{b - a}.$$

We apply this to the function $h(x) = y_1(x) - y_2(x)$ with $a = x_0$ and $b = x^*$. Because $h(a) = y_1(x_0) - y_2(x_0) = 0$, and $h(b) = y_1(x^*) - y_2(x^*) = 0$, the Mean Value Theorem says that there is a c, where $x_0 < c < x^*$, for which $h'(c) = 0$. But, $h'(c) = y_1'(c) - y_2'(c)$, so there is a c for which $f(c) = g(c)$, which contradicts the condition $f < g$, and so the theorem is proved in a different way. However, this proof applies immediately to general first order differential equations, so we have the theorem.

Theorem 3.4: The Comparison Theorem *Let $f(x, y) < g(x, y)$ for all $x > x_0$ and $y > y_0$. If $y_1(x)$ is a solution of the initial value problem*

$$y' = f(x, y), \qquad y(x_0) = y_0, \tag{3.28}$$

and if $y_2(x)$ is a solution of the initial value problem

$$y' = g(x, y), \qquad y(x_0) = y_0, \tag{3.29}$$

then wherever $y_1(x)$ and $y_2(x)$ are both defined, we have $y_1(x) < y_2(x)$.

This theorem says that if the solutions of (3.28) and (3.29) pass through the same point, and if the slope of the second solution always exceeds the slope of the first solution, then the second solution always exceeds the first solution. Thus, if the solution of $y' = f(x, y)$, $y(x_0) = y_0$ has a vertical tangent at $x = a$, then the solution of $y' = g(x, y)$, $y(x_0) = y_0$ has a vertical tangent at $x = b$ where $x_0 < b \leq a$.

EXAMPLE 3.6

In Exercise 9 on page 110 you were asked to solve the initial value problem $y' = x^2 + y^2$, $y(0) = 1$ — which cannot be solved in terms of familiar functions — by different numerical methods, and then to estimate $y(2)$. This may have caused problems. We can use the Comparison Theorem to show that this initial value problem has a vertical asymptote between $\pi/4$ and 1. Thus, we can never reach $y(2)$ from $y(0)$.

The key to establishing this result is to realize that

$$y^2 < x^2 + y^2 < 1 + y^2 \text{ for } 0 < x < 1,$$

and that, although we cannot solve $y' = x^2 + y^2$, we can solve both $y' = y^2$ and $y' = 1 + y^2$.

First, we use the Comparison Theorem on the differential equations

$$y' = y^2, \qquad y' = x^2 + y^2,$$

subject to $y(0) = 1$. We know that the initial value problem $y' = y^2$, $y(0) = 1$, has the solution

$$y(x) = \frac{1}{1 - x},$$

which has a vertical asymptote at $x = 1$. Because

$$y^2 < x^2 + y^2 \text{ for } 0 < x$$

the Comparison Theorem guarantees that the initial value problem

$$y' = x^2 + y^2, \qquad y(0) = 1,$$

must have a vertical asymptote between 0 and 1.

Second, we use the Comparison Theorem on the differential equations

$$y' = 1 + y^2, \qquad y' = x^2 + y^2,$$

subject to $y(0) = 1$. We can show that the initial value problem $y' = 1 + y^2$, $y(0) = 1$, has the solution

$$y(x) = \tan\left(x + \frac{\pi}{4}\right).$$

(Do this.) This solution has a vertical asymptote at $x = \pi/4 \approx 0.785$.

Now, because

$$y^2 < x^2 + y^2 < 1 + y^2 \text{ for } 0 < x < 1,$$

the initial value problem $y' = x^2 + y^2$, $y(0) = 1$, must have a vertical asymptote between $\pi/4$ and 1. ■

EXERCISES

1. Explain why the initial value problem $y' = x^n + y^2$, $y(0) = 1$ where $n > 1$, has a vertical asymptote between $\pi/4$ and 1.

2. Explain why the initial value problem $y' = y^2 - x$, $y(0) = 2$, has a vertical asymptote between 0.5 and $0.5 \ln 3$. [Hint: Consider $0 < x < 1$ so that $y^2 - 1 < y^2 - x < y^2$.]

3. The Generalized Logistic Equation. The Generalized Logistic Equation is $y' = ay(B^n - y^n)$, where a, B, and n are positive constants. Show that $B^{1/2} - y^{1/2} < B - y < B^2 - y^2$ if $1 < y < B$. What does this imply about the three solutions of the general logistic equation — using the same values for a, B, and $y(0)$ — corresponding to $n = 1/2$, $n = 1$, and $n = 2$?

4. Consider the initial value problem $y' = a^2y^2 + \beta$, $y(0) = \alpha$, where, without loss of generality, a is a positive constant.

(a) Show that, if $\beta = b^2$, where b is a positive constant, then $y(x)$ is an increasing function, and

$$\arctan(ay/b) = abx + \arctan(a\alpha/b).$$

Now show that $y(x)$ has a vertical asymptote at $x = [\pi/2 - \arctan(a\alpha/b)]/(ab)$.

(b) Show that, if $\beta = 0$, then $y(x)$ is an increasing function, and $y(x) = \alpha/\left(1 - \alpha a^2 x\right)$. Now show that $y(x)$ has a vertical asymptote at $x = 1/\left(\alpha a^2\right)$.

(c) Show that, if $\beta = -b^2$, where b is a positive constant, and if $\alpha > b/a$, then $y(x)$ is an increasing function, and

$$\ln\left(\frac{ay - b}{ay + b}\right) = 2abx + \ln\left(\frac{a\alpha - b}{a\alpha + b}\right).$$

Now show that $y(x)$ has a vertical asymptote at

$$x = \frac{1}{2ab} \ln\left(\frac{a\alpha + b}{a\alpha - b}\right).$$

5. Use the results from Exercise 4 together with the Comparison Theorem to show that the initial value problem $y' = x^2 + y^2 - xy - 1$, $y(0) = \sqrt{3}$, has a vertical asymptote less than $(\ln 5)/\sqrt{3} \approx 0.929$ by the following steps.

(a) Show that $x^2 + y^2 - xy - 1 = \left(x - \frac{1}{2}y\right)^2 + \frac{3}{4}y^2 - 1$, so that $\frac{3}{4}y^2 - 1 < x^2 + y^2 - xy - 1$.

(b) Use part (c) of Exercise 4 to show that the initial value problem $y' = \frac{3}{4}y^2 - 1$, $y(0) = \sqrt{3}$, has a vertical asymptote at $x = (\ln 5)/\sqrt{3} \approx 0.929$. Explain how the Comparison Theorem guarantees that the original equation has a vertical asymptote less than $(\ln 5)/\sqrt{3} \approx 0.929$.

3.5 FINDING POWER SERIES SOLUTIONS

In Chapter 1 we used a Taylor series expansion for the exponential function to obtain an alternative form of an explicit solution when we could not find the antiderivative of $\exp(-x^2)$. We want to generalize this procedure for differential equations of the form $y' = g(x, y)$. These solutions are an alternative to numerical solutions in certain situations.

In the preceding sections we considered the differential equation

$$y' = x - y \tag{3.30}$$

subject to the initial condition

$$y(0) = 1. \tag{3.31}$$

We found that its solution is

$$y(x) = x - 1 + 2e^{-x}. \tag{3.32}$$

In this section we will obtain this result using power series. We introduce this new method on a problem that we have solved already to increase our confidence in the new method. The method we introduce hinges on the fact that (3.30) satisfies the conditions of the Existence-Uniqueness Theorem and so we know that there is a unique solution to this problem.

At the end of Chapter 1 we found that although we could not express the Error Function in terms of familiar functions, we could obtain some idea of its behavior by expressing it as a power series. We use the same idea here to find a solution $y(x)$ of (3.30) in the form of a power series

$$y(x) = \sum_{k=0}^{\infty} c_k x^k = c_0 + c_1 x + c_2 x^2 + c_3 x^3 + \cdots + c_n x^n + \cdots, \tag{3.33}$$

where $c_0, c_1, c_2, \cdots$, are constants. Once we find such a solution, then the Existence-Uniqueness Theorem guarantees that it is the solution. So the idea is simple: we assume that (3.30) has a solution of the form (3.33) and then determine the constants c_0, c_1, c_2, $\cdots$. The crucial feature is not whether we are allowed to make such an assumption about the solution being a power series, but whether the assumption allows us to find a solution.

To determine the constants $c_0, c_1, c_2, \cdots$, we substitute (3.33) and its derivative into (3.30), which gives

$$c_1 + 2c_2 x + 3c_3 x^2 + \cdots + nc_n x^{n-1} + \cdots = x - (c_0 + c_1 x + c_2 x^2 + \cdots + c_m x^m + \cdots).$$

This equation can be rewritten in the form

$$(c_1 + c_0) + (2c_2 + c_1 - 1)\,x + (3c_3 + c_2)\,x^2 + (4c_4 + c_3)\,x^3 + (5c_5 + c_4)\,x^4 + \cdots = 0.$$

We want this to be an identity in x — that is, valid for every x. Equating the coefficients of $x^0, x^1, x^2, \cdots$, to zero, we get

$$\begin{cases} c_1 + c_0 &= 0, \\ 2c_2 + c_1 - 1 &= 0, \\ 3c_3 + c_2 &= 0, \\ 4c_4 + c_3 &= 0, \\ 5c_5 + c_4 &= 0, \\ 6c_6 + c_5 &= 0, \end{cases} \tag{3.34}$$

and so on. From the first equation we see that

$$c_1 = -c_0, \tag{3.35}$$

which, when used in the second equation of (3.34), requires that

$$c_2 = \frac{1}{2}\left(1 - c_1\right) = \frac{1}{2}\left(1 + c_0\right),$$

which, when used in the third equation of (3.34), requires that

$$c_3 = -\frac{1}{3}c_2 = -\frac{1}{3 \cdot 2}\left(1 + c_0\right) = -\frac{1}{3!}\left(1 + c_0\right),$$

which, when used in the fourth equation of (3.34), requires that

$$c_4 = -\frac{1}{4}c_3 = (-1)^2\frac{1}{4!}\left(1 + c_0\right).$$

If we take more terms in (3.35), we find that

$$c_n = (-1)^{n-2}\frac{1}{n!}\left(1 + c_0\right) = (-1)^n\frac{1}{n!}\left(1 + c_0\right), \qquad n = 2, 3, 4, \cdots. \tag{3.36}$$

If we substitute (3.35) and (3.36) into (3.33), we find

$$y(x) = c_0 - c_0x + \left(1 + c_0\right)\left(\frac{1}{2}x^2 - \frac{1}{3!}x^3 + \frac{1}{4!}x^4 - \frac{1}{5!}x^5 + \cdots\right).$$

We recognize the quantity in parentheses as part of the Taylor series expansion for e^{-x} — convergent for all x — namely,

$$e^{-x} = 1 - x + \frac{1}{2}x^2 - \frac{1}{3!}x^3 + \frac{1}{4!}x^4 - \frac{1}{5!}x^5 + \cdots,$$

so we have

$$y(x) = c_0 - c_0x + \left(1 + c_0\right)\left(e^{-x} - 1 + x\right) = -1 + x + \left(1 + c_0\right)e^{-x}, \tag{3.37}$$

where c_0 is an undetermined constant. This constant is determined by our initial condition $y(0) = 1$ which gives $c_0 = 1$. When the latter is used in (3.37), we find that $y(x) = -1 + x + 2e^{-x}$ is the solution of $y' = x - y$ subject to $y(0) = 1$. Thus, (3.32) is the solution of this initial value problem.

EXERCISES

1. Solve the following differential equations by assuming that they each have a power series solution of the form (3.33). Try to express your series solution in terms of familiar functions.

(a) $y' = -y$ (b) $y' = 2y$ (c) $(1 + x)y' = 2y$

2. Solve the differential equation $y' = 1 - 2xy$ by assuming that it has a power series solution of the form (3.33). Do not try to express the solution in terms of familiar functions.

3. Solve the differential equation $y' = -2xy$ by assuming that it has a power series solution of the form (3.33). Express your series solution in terms of familiar functions. Compare your answer with the one you obtained in Exercise 9(a) on page 92.

What Have We Learned?

Main Ideas

- An IMPLICIT SOLUTION of a first order differential equation is any relationship between x and y that through implicit differentiation yields the original differential equation.

- The graphical methods from Chapters 1 and 2 apply directly to equations of the form $y' = g(x, y)$.
- Symmetry considerations are simple to apply and can be very useful.
- One method for obtaining a numerical approximation to $y' = g(x, y)$ is Euler's method. It starts at a point (x_0, y_0) and computes subsequent values of x and y using $x_n = x_0 + nh$, $y_n = y_{n-1} + g(x_{n-1}, y_{n-1})h$, where h is the step-size.
- Euler's method may be made more accurate in several ways. This chapter considers Heun's method and the Runge-Kutta method to illustrate the possibilities.
- Period doubling and chaos can occur using numerical methods if the step-size is not chosen properly.
- The Comparison Theorem may sometimes be used to verify that a numerical solution is not accurate.
- We can obtain solutions of differential equations in the form of power series. Chapter 12 contains a detailed discussion of this method.

How to Fully Analyze $y' = g(x, y)$

Purpose To summarize the major steps required to fully analyze the differential equation $y' = g(x, y)$.

Process Note that the order is flexible.

1. Check that a solution exists. See *How to See Whether Unique Solutions of $y' = g(x, y)$ Exist* on page 57.
2. Perform a graphical analysis.
   (a) Use a computer/calculator program to plot the slope field. Confirm using isoclines.
   (b) Check where solutions are increasing, decreasing, concave up, concave down, and have inflection points. See *How to Analyze $y' = g(x, y)$ Graphically* on page 84.
   (c) Check whether the slope field has any symmetries. See *How to Test $y' = g(x, y)$ for Symmetry* on page 94.
   (d) If the equation is autonomous, analyze the phase line. See *How to Perform a Phase Line Analysis on $y' = g(y)$* on page 68.
3. Perform a numerical analysis using a computer/calculator program. See *How to Analyze $y' = g(x, y)$ Numerically* on page 101.
4. Find an explicit or implicit solution. (Some methods are given in Chapters 4 and 5.)

CHAPTER 4

MODELS AND APPLICATIONS LEADING TO NEW TECHNIQUES

Where We Are Going – and Why

In this chapter we continue the discussion started in Chapter 3 by looking at a special type of first order differential equation, namely,

$$y' = f(y)g(x),$$

where the right-hand side is the product of a function of y and a function of x. Such equations are called separable. They are important for three reasons: they can be used to model various situations, they can be solved analytically (up to an integration), and they can be used for solving other types of differential equations.

In the first part of the chapter we focus on obtaining analytical solutions and developing a technique for graphing such solutions if we cannot solve explicitly for one of the variables. We introduce a new type of differential equation — one with homogeneous coefficients — that reduces to a separable one by an appropriate change of variable.

We also give examples on how to proceed from a data set to a differential equation to model the process giving rise to this data set. Other applications include models of sky diving, orthogonal trajectories, and the population of Ireland during the potato famine.

4.1 SOLVING SEPARABLE DIFFERENTIAL EQUATIONS

In this section we look at a simple model of a mixture problem that leads to a type of differential equation we have not yet discussed — a separable differential equation. We discover how to find analytical solutions of these equations and then apply the technique to a number of examples.

A Mixture Problem

The problem we consider now is to determine the concentration of solute in a container as a substance is being simultaneously added to and removed from the container at different rates. To derive a differential equation that models this process, we let y be a function that represents the amount of substance in a given container at time x, and we assume that the instantaneous rate of change of y with respect to x is given by

$$y' = \begin{matrix}\textit{rate at which substance} \\ \textit{is added to the container}\end{matrix} - \begin{matrix}\textit{rate at which substance} \\ \textit{is leaving the container}\end{matrix}. \tag{4.1}$$

Conservation equation

This equation is often called a CONSERVATION EQUATION or an EQUATION OF CONTINUITY. In setting up this equation for a particular case, it is important to pay careful attention to the units, which we demonstrate in the following example.

EXAMPLE 4.1 *Solute in a Container*

A 300-gallon container is 2/3 full of water containing 50 pounds of salt. At time $x = 0$ minutes, valves are turned on so pure water is added to the container at a rate of 3 gallons per minute. If the well-stirred mixture is drained from the container at the rate of 2 gallons per minute, how many pounds of salt are in the container when it is full (and all valves are turned off)?

We first note that more of the liquid is being added per minute than is being drained, so the liquid level is rising and the number of gallons in the container is increasing. In fact, we may use a conservation-type argument to note that the rate of change of volume V of liquid in the container (in gallons per minute) equals the rate being added (3 gallons per minute) minus the rate being drained (2 gallons per minute). Thus,

$$V' = 3 - 2 = 1 \text{ gallon per minute.}$$

Find volume

Integration gives

$$V(x) = x + 200 \text{ gallons.} \tag{4.2}$$

(Why is the constant of integration 200?)

Let y represent the number of pounds of salt in the container at time x. We want to find an equation for y' based on the conservation equation (4.1), so we need to find the rate at which salt is added and the rate at which salt is leaving, both at time x. The key is to concentrate on the units of y' at time x, which in this case are pounds of salt per minute. Thus, the rates added and leaving must also be in pounds of salt per minute at time x.

First, we consider the contribution from the rate at which salt is added. Pure water is being added to the container at 3 gallons per minute, and to obtain the units of pounds of salt per minute, we need to multiply the 3 gallons per minute by the number of pounds of salt per gallon being added, which in this case is 0 pounds of salt per gallon. Thus, $0 \times 3 = 0$ pounds of salt per minute is the rate being added at any time.

Second, we consider the contribution from the rate at which salt is leaving. The solute is leaving the container at 2 gallons per minute, and to obtain the units of pounds of salt per minute we need to multiply the 2 gallons per minute by the number of pounds of salt per gallon leaving. At time x the number of pounds of salt in the container is $y(x)$, so to obtain the number of pounds of salt per gallon at time x, we divide $y(x)$ by the number of gallons in the container at time x — namely, $V(x) = x + 200$ — giving $y/(x+200)$ pounds of salt per gallon at time x. Thus, the rate at which salt is leaving the container is $2y/(x+200)$ pounds of salt per minute.

Conservation equation

With this information we see that the conservation equation gives $y' = 0 \times 3 - 2y/(x+200)$, or

$$y' = -\frac{2y}{x+200}. \tag{4.3}$$

Initial condition

Because there are 50 pounds of salt in the container at $x = 0$, the proper initial condition is $y(0) = 50$. This differential equation is valid for $0 < x < 100$. (Why don't we consider values of time greater than 100?)

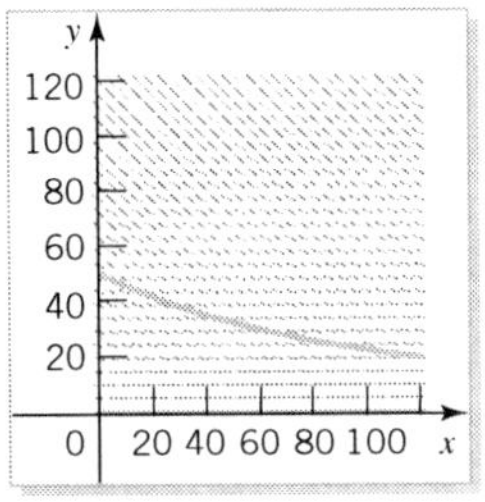

FIGURE 4.1
Solution curve and slope field for $y' = -2y/(x+200)$

If we look at the slope field for (4.3) in Figure 4.1, we observe that all solutions will be decreasing and concave up for the region of interest ($x > 0$, $y > 0$). [To fully convince yourself of these facts, consider (4.3) and the result of differentiating (4.3).] Note that the slope field ignores the condition that the container is full when $x = 100$. We have also hand-drawn the solution curve that passes through the initial point $(0, 50)$. From this curve we can estimate the value of $y(x)$ for $x = 100$ at about 25 pounds, which is an approximate answer to our original question. We can obtain a better numerical approximation by using the Runge-Kutta 4 method described in Chapter 3, and, with $h = 1$, we find $y(100) \approx 22.222$ pounds. However, to obtain an exact answer we must obtain an explicit solution.

Separate variables

Equation (4.3) is a type of differential equation that we have not seen before. However, if we rewrite it in the form

$$\frac{1}{y}y' = \frac{-2}{x+200},$$

and then integrate both sides of the equation with respect to the independent variable x, we see that

$$\int \frac{1}{y}\frac{dy}{dx}\,dx = \int \frac{-2}{x+200}\,dx,$$

or

$$\int \frac{1}{y}\,dy = \int \frac{-2}{x+200}\,dx.$$

Integrate

Evaluating these integrals, we find

$$\ln y + C_1 = -2\ln(x+200) + C_2,$$

where C_1 and C_2 are arbitrary constants. Notice we used the fact that x and y are positive. (Where did we use this fact?) This equation can be rewritten as

$$\ln y = -2\ln(x+200) + C,$$

where $C = C_2 - C_1$ is an arbitrary constant. This is the solution of (4.3), where both x and y occur implicitly. Fortunately we can solve this for y, and in doing so we find

Explicit solution

the explicit solution of (4.3), namely,

$$y(x) = \frac{e^C}{(x+200)^2}.$$

From the initial condition $y(0) = 50$, we find $e^C = 50\left(200^2\right)$, so our final form of the solution is

$$y(x) = \frac{50\left(200^2\right)}{(x+200)^2}. \tag{4.4}$$

To answer the original question about how many pounds of salt are in the container when full, we note from (4.2) that the container will be full when $x = 100$, so $y(100)$ will be the amount of salt in the container at this time. From (4.4) we have that $y(100) = 50\left(200^2\right)/300^2 = 22\frac{2}{9}$ pounds of salt, which isn't far from our initial estimate of 25 pounds and in excellent agreement with the numerical solution of 22.222.

Solving Separable Differential Equations

Equation (4.3) is an example of a special type of differential equation that may be solved using straightforward integration techniques — a SEPARABLE DIFFERENTIAL EQUATION.

◆ *Definition 4.1:* **A SEPARABLE DIFFERENTIAL EQUATION has the form**

$$y' = f(y)g(x), \tag{4.5}$$

where the right-hand side is the product of a function of y and a function of x. ◆

For example, $y' = xy + x + y + 1$ is separable, but $y' = x - y$ is not. (Why?)

Separate variables

Nonequilibrium solutions of separable equations may be obtained by separating the y and x dependence as

$$\frac{1}{f(y)}\frac{dy}{dx} = g(x) \tag{4.6}$$

Integrate

and then integrating both sides with respect to x, so that

$$\int \frac{1}{f(y)}\frac{dy}{dx}\,dx = \int g(x)\,dx,$$

or

$$\int \frac{1}{f(y)}\,dy = \int g(x)\,dx. \tag{4.7}$$

Observe that after isolating the two variables y and x, we are left with a calculus problem — namely, finding antiderivatives of both sides of an equation.

Comments about Separable Differential Equations

- If we apply the Existence-Uniqueness Theorem on page 56 to (4.5), we see that we are guaranteed a unique solution through the point (x_0, y_0) if $g(x)$ and df/dy are continuous in the vicinity of (x_0, y_0). [Strictly speaking, the Existence-Uniqueness Theorem requires the condition that $f(y)$ is continuous in the vicinity of (x_0, y_0). Why is this condition unnecessary in the current situation?]
- Separable differential equations contain the differential equations considered in Chapter 1, namely, $y' = g(x)$, and Chapter 2, $y' = g(y)$, as special cases.
- Earlier we defined equilibrium solutions as solutions for which $y(x) = c$, for all x. From (4.5), we see that equilibrium solutions of separable equations will occur at the roots $y(x) = c$ of $f(y) = 0$. Thus, to find equilibrium solutions of (4.5), we should solve $f(y) = 0$ for $y = c$.
- If we look at the prescription we have just given to solve (4.5), we realize that we have implicitly assumed that $f(y) \neq 0$ in going from (4.5) to (4.6). Thus, to ensure that we do not inadvertently discard some solutions, we suggest the following procedure:
  (a) First, find all the equilibrium solutions — that is, find all values of y that satisfy $f(y) = 0$.
  (b) Second, for $f(y) \neq 0$, perform the indicated integrations in (4.7).

How to Solve Separable Differential Equations

Purpose To find all solutions $y = y(x)$ of

$$y' = f(y)g(x) \tag{4.8}$$

for given $f(y)$ and $g(x)$.

Process

1. Check that the Existence-Uniqueness Theorem is satisfied.
2. Find the equilibrium solutions. Solve $f(y) = 0$ for y. The solutions $y = c_1$, $y = c_2, \cdots$, are the equilibrium solutions.
3. Find the nonequilibrium solutions. Rewrite (4.8) in the form

$$\frac{1}{f(y)}\frac{dy}{dx} = g(x),$$

and integrate with respect to x:

$$\int \frac{1}{f(y)}\frac{dy}{dx}\,dx = \int g(x)\,dx,$$

or

$$\int \frac{1}{f(y)}\,dy = \int g(x)\,dx.$$

Comments about Solving Separable Differential Equations

- Frequently, the nonequilibrium solutions are implicitly defined. Sometimes, implicitly defined functions can be solved for $y = y(x)$, or for $x = x(y)$. If this can be done, do so.
- When writing down antiderivatives in step 3, we need include an arbitrary constant only on one side of the equation. (Why?) It is a common mistake to forget all constants of integration.
- The solutions obtained in steps 2 and 3 are all the solutions of (4.8) if the Existence-Uniqueness Theorem is satisfied.

Caution! **A common mistake in solving separable equations is to go straight to step 3, missing any equilibrium solutions.**

More Examples

In Example 4.1 on page 120, we discussed a simple mixture problem that led to a separable differential equation. Now we look at a similar example, but the rate of draining the container is not constant.

EXAMPLE 4.2 *Solute in a Container Draining in Periodic Way*

A 300-gallon container is 2/3 full of water containing 50 pounds of salt. At time $x = 0$, valves are turned on so pure water is added to the container at a rate of 3 gallons per minute. If the well-stirred mixture is drained from the container at the rate of $2 + \sin x$ gallons per minute, how many pounds of salt are in the container when it is full (and all valves are turned off)?

The only difference between this example and Example 4.1 is the draining rate of $2 + \sin x$ gallons per minute. We can make an estimate of the answer in this case by observing that the average value of $2 + \sin x$ over any time interval of length 2π is 2, the same as the constant draining rate in Example 4.1.[1] Thus, the final answer to this example should be close to the answer of that example, namely, $22\frac{2}{9}$ pounds.

As before, we use a conservation argument and note that the rate of change of volume V of liquid in the container equals the rate being added minus the rate being drained, so

$$V' = 3 - 2 - \sin x = 1 - \sin x \text{ gallon per minute.}$$

Find volume Integration gives

$$V(x) = x + \cos x + 199. \tag{4.9}$$

(Why is the constant of integration 199?)

If y represents the number of pounds of salt in the container at time x, the concentration of salt at time x is $y/V = y/(x + \cos x + 199)$ pounds per gallon. Thus,

[1] The average value of $f(x)$ over the interval $a \le x \le b$ is $\frac{1}{b-a}\int_a^b f(x)\,dx$.

the rate at which salt is leaving the container is $(2 + \sin x)\, y/(x + \cos x + 199)$ pounds per minute. Because the rate at which salt is arriving in the container is 0 — there is no salt in the incoming pure water — the conservation equation gives

Conservation equation

$$y' = -\frac{(2 + \sin x)\, y}{x + \cos x + 199}, \tag{4.10}$$

which is valid from $x = 0$ until the container is full, which will occur at the value of x for which $x + \cos x + 199 = 300$ — that is, $x + \cos x = 101$. This equation cannot be solved exactly, but we can see that it must have a solution between 100 and 102. (Why?) With this information we use a numerical or graphical technique to estimate that the tank will be full when $x \approx 100.09$.

Initial condition

Because there are 50 pounds of salt in the container at $x = 0$, the proper initial condition is $y(0) = 50$.

Slope field

If we look at the slope field for (4.10) in Figure 4.2, we observe that all solutions will be decreasing in agreement with (4.10). We have also hand-drawn the solution curve that passes through the initial point (0, 50). From this curve we can estimate the value of $y(x)$ for $x = 100.09$ at about 22.5 pounds, which is an approximate answer to our original question. This is in agreement with our initial estimate of $22\frac{2}{9}$. However, to obtain an accurate answer we must obtain an explicit solution.

Separate variables

The differential equation (4.10) is a separable equation, and the non-equilibrium solution can be written in the form

$$\int \frac{1}{y}\, dy = -\int \frac{2 + \sin x}{x + \cos x + 199}\, dx.$$

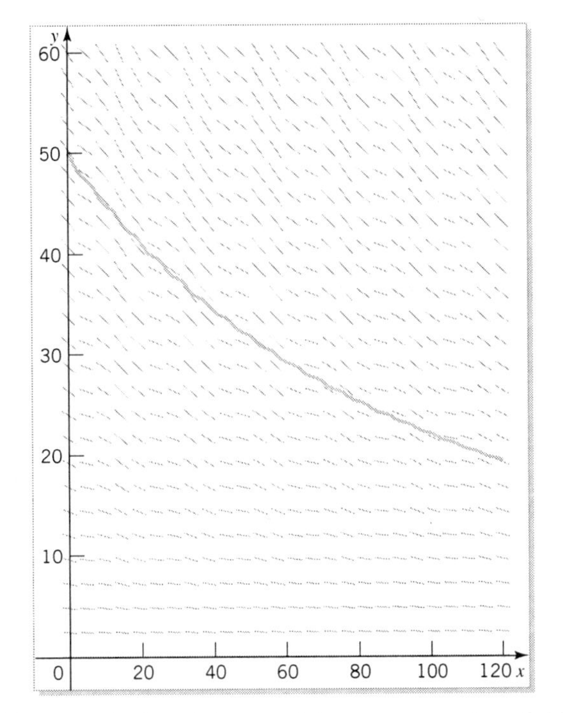

FIGURE 4.2 Hand-drawn solution curve and slope field for $y' = -(2 + \sin x)\, y/(x + \cos x + 199)$

Integrate

The integral on the left-hand side is $\ln y$, but the one on the right-hand side cannot be expressed in terms of familiar functions. However, we can use the ideas from Chapter 1 to write its solution in the form

$$\ln y = -\int_0^x \frac{2+\sin t}{t+\cos t+199}\,dt + \ln 50.$$

Explicit solution

(Why is the constant of integration $\ln 50$?) Solving for $y(x)$ gives

$$y(x) = 50\exp\left(-\int_0^x \frac{2+\sin t}{t+\cos t+199}\,dt\right).$$

Thus, the amount of salt in the tank at $x = 100.09$ is

$$y(100.09) = 50\exp\left(-\int_0^{100.09} \frac{2+\sin t}{t+\cos t+199}\,dt\right) \text{ pounds.}$$

Using Simpson's rule to evaluate this integral, we find $y(100.09) \approx 22.09$ pounds, in agreement with all our estimates. Using the Runge-Kutta 4 method (described in Chapter 3) to solve (4.10) subject to $y(0) = 50$ also gives 22.09 as the value of y when $x = 100.09$.

EXAMPLE 4.3

As our next example of separation of variables, consider the differential equation

$$y' = \frac{x^3+1}{y^3+1} \tag{4.11}$$

in the first quadrant, so $x > 0$ and $y > 0$.

Separate variables

Integrate

Writing (4.11) in the separable form $(y^3+1)\,y' = x^3+1$ and integrating yields the implicit solution $\frac{1}{4}y^4 + y = \frac{1}{4}x^4 + x + c$, or

$$y^4 - x^4 + 4(y-x) = C. \tag{4.12}$$

Slope field

We cannot solve this either for $y(x)$ or for $x(y)$. So even though we have a family of implicit solutions, we have no graph of them. However, we can construct the slope field and draw in a few solutions by hand. These are shown in Figure 4.3. This figure suggests that $y = x$ is a solution. In fact $y = x$ satisfies (4.12) when $C = 0$ and is therefore a solution of (4.11). This figure also suggests that if $y > x$, the solution will be increasing and concave up, whereas if $y < x$, the solution will be increasing and concave down. We will use calculus to see whether these suggestions are accurate.

Monotonicity

The derivative given in (4.11) is positive for $x > 0$, $y > 0$, so the solutions will be increasing, as suggested by Figure 4.3. Also note that the slopes are larger than 1 if $x > y$ and smaller than 1 if $x < y$.

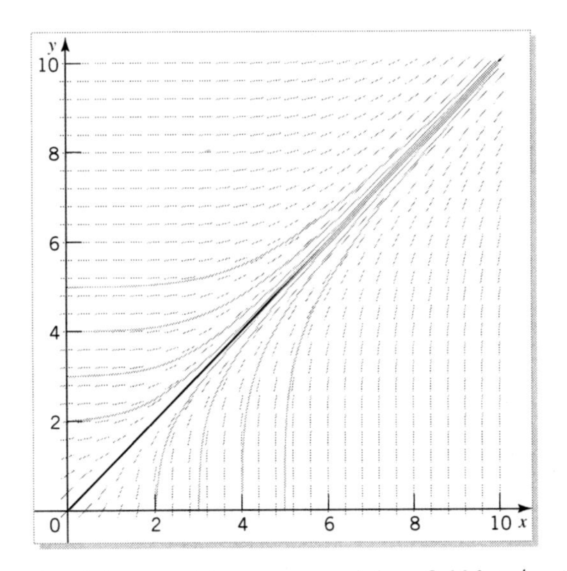

FIGURE 4.3 Some hand-drawn solution curves and slope field for $y' = (x^3+1)/(y^3+1)$

Concavity

To decide on the concavity, we differentiate (4.11) and substitute for y' from (4.11), giving, after some lengthy algebra,

$$y'' = \frac{3}{\left(y^3+1\right)^3}\left(xy^3+x+yx^3+y\right)\left(xy^3+x-yx^3-y\right). \tag{4.13}$$

We are interested only in the first quadrant where $x > 0$ and $y > 0$. Here the sign of y'' is determined entirely by the sign of the last term on the right-hand side of (4.13), namely, xy^3+x-yx^3-y, which we will now analyze. Figure 4.3 suggests that the concavity changes at $y = x$, which suggests that $y - x$ will be a factor of xy^3+x-yx^3-y. It is, because

$$xy^3+x-yx^3-y = (y-x)(xy^2+x^2y-1) = x(y-x)\left(y^2+xy-\frac{1}{x}\right). \tag{4.14}$$

So the sign of y'' is determined by the product of these three terms of which the first, x, is always positive. Figure 4.3 suggests that the only change of sign of y'' occurs at $y = x$; that is, when $y - x = 0$. This would mean that the third term, $y^2+xy-\frac{1}{x}$, does not change sign. Let's investigate this possibility. If we think of the equation

$$y^2+xy-\frac{1}{x} = 0$$

as a quadratic in y and solve it, we find

$$y = \frac{-x \pm \sqrt{x^2+4/x}}{2}.$$

Thus, (4.14) can be written as

$$xy^3 + x - yx^3 - y = x(y - x)\left(y + \frac{x - \sqrt{x^2 + 4/x}}{2}\right)\left(y + \frac{x + \sqrt{x^2 + 4/x}}{2}\right).$$

Because we are concerned with $x > 0$ and $y > 0$, the last term is always positive, but there is a possibility of y'' changing sign along the curve

$$y = -\frac{x - \sqrt{x^2 + 4/x}}{2}. \qquad (4.15)$$

We can ask where the curve (4.15) crosses the curve $y = x$, and answer that this will occur when

$$x = -\frac{x - \sqrt{x^2 + 4/x}}{2}.$$

Solving this equation gives $x = (1/2)^{1/3} \approx 0.7937$. If we look at Figure 4.3 again, we see that something unusual is happening near the origin. Figure 4.4 shows the slope field in the window $0 < x < 2, 0 < y < 2$, together with the curve given by (4.15) and some hand-drawn solution curves. The change in concavity near the origin is now obvious. Without the concavity analysis, a cursory examination of Figure 4.3 would have led us to believe that an individual solution could not change concavity.

Long-term behavior

Figure 4.3 suggests that as $x \to \infty$, $y(x) \to x$. It does. See Exercise 6 on page 133.

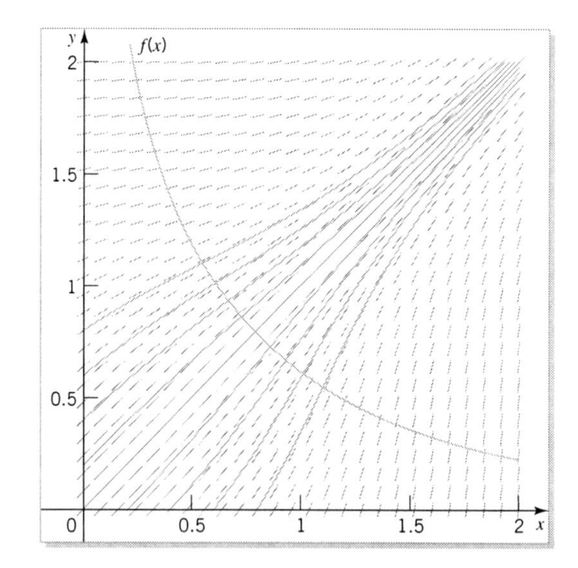

FIGURE 4.4 Some hand-drawn solution curves, the functions $f(x) = -[x - (x^2 + 4/x)^{1/2}]/2$, $y = x$, and slope field for $y' = (x^3 + 1)/(y^3 + 1)$

Sketching Solutions of Separable Equations

When we integrate separable equations, our solutions will always have the form

$$G(y) = F(x); \tag{4.16}$$

that is, solutions are defined implicitly, with the dependence on x and y separated. As we have seen, sometimes it is possible to solve (4.16) for $y = y(x)$ or for $x = x(y)$, in which case we can sketch the graph of the solution directly. However, whether or not we can solve (4.16), it is always possible to sketch the solution curve relating x and y. We now demonstrate this technique.

EXAMPLE 4.4

Solve the initial value problem

$$y' = \frac{18x^2}{-15y^2 - 6y - 4}, \qquad y(0) = 1. \tag{4.17}$$

Separate variables The differential equation in (4.17) is separable, and so we are led to

$$-\int \left(15y^2 + 6y + 4\right) dy = 18 \int x^2\, dx,$$

Integrate which yields

$$-5y^3 - 3y^2 - 4y + C = 6x^3. \tag{4.18}$$

Implicit solution If we substitute the initial condition $y(0) = 1$ into this equation, we find that $C = 12$, so (4.18) becomes

$$-5y^3 - 3y^2 - 4y + 12 = 6x^3. \tag{4.19}$$

Notice that this solution is of the form (4.16), where the left-hand side is a function of y and the right-hand side is a function of x.

To sketch the graph of the solution (4.19), we proceed as follows. We first divide the graph paper into four boxes. In the lower left-hand box we sketch the function of x — namely, $F(x) = 6x^3$ — and in the lower right-hand box we sketch the function of y — namely, $G(y) = -5y^3 - 3y^2 - 4y + 12$. In the upper right-hand box, we draw a diagonal line with slope 1 that connects the corners. In this way we create Figure 4.5. The scales used along the horizontal x- and y-axes in Figure 4.5 are independent of each other. However, it is critical in this construction that the scales used on the vertical axes for $F(x)$ and $G(y)$ are identical.

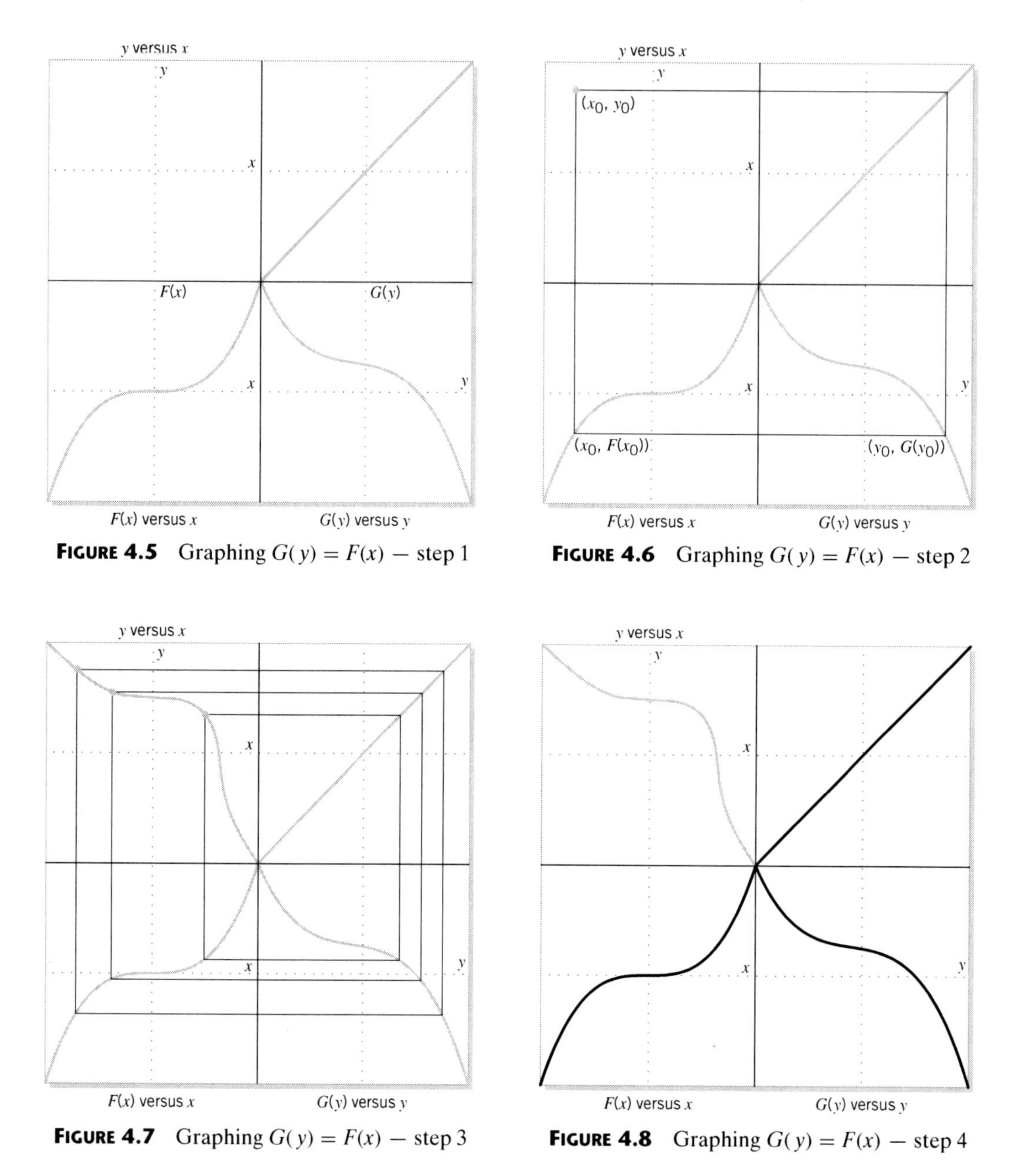

FIGURE 4.5 Graphing $G(y) = F(x)$ — step 1

FIGURE 4.6 Graphing $G(y) = F(x)$ — step 2

FIGURE 4.7 Graphing $G(y) = F(x)$ — step 3

FIGURE 4.8 Graphing $G(y) = F(x)$ — step 4

We proceed by selecting a value of x in the lower left-hand box — say, x_0 — and drawing a vertical line through the point $(x_0, F(x_0))$ into the upper left-hand box, which is where we will finally sketch the solution curve relating x and y. We now draw a horizontal line through the point $(x_0, F(x_0))$ until we cross the curve $G(y)$ in the lower right-hand box at the point $(y_0, G(y_0))$. At this point we have $F(x_0) = G(y_0)$ (remember the vertical axes have the same scale), so the point with numerical values $x = x_0$, $y = y_0$ must lie on the graph of $F(x) = G(y)$. What we need to do is plot this point in the upper left-hand box where we plot y versus x.

We have already drawn the vertical line $x = x_0$ in the upper left-hand box, so the next step is to draw a horizontal line corresponding to $y = y_0$ in the same box.

This is where the line with slope 1 comes into play. We draw a vertical line through $(y_0, G(y_0))$ until we intersect the line with slope 1 in the upper right-hand box. We then draw a horizontal line through this point of intersection into the upper left-hand box. This is the line $y = y_0$. The point where this line meets the vertical line $x = x_0$ is the point (x_0, y_0) that lies on the graph of $F(x) = G(y)$. This is demonstrated in Figure 4.6.

We repeat this process, starting with different values of x in the lower left-hand box, and then join the points in the upper left-hand box as is done in Figure 4.7. Finally, we remove the construction lines and have the graph of $F(x) = G(y)$ in the upper left-hand box, as shown in Figure 4.8. ■

How to Sketch Solutions of $G(y) = F(x)$

Purpose To use graphical techniques to sketch the solution $y = y(x)$ of

$$G(y) = F(x). \tag{4.20}$$

Process

1. Divide the graph paper into four boxes. Label the boxes (see Figure 4.9).

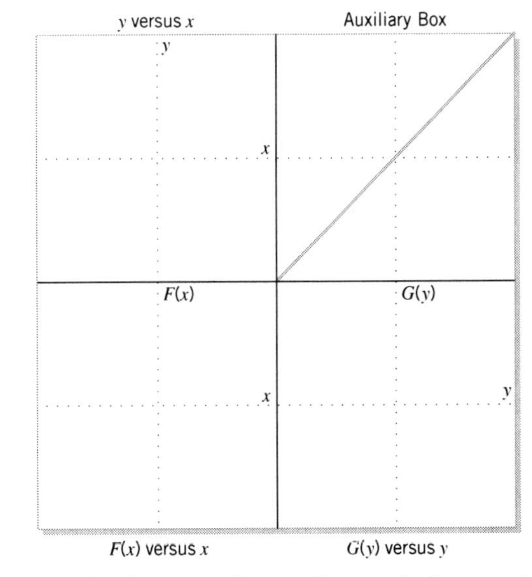

FIGURE 4.9 Graphing $G(y) = F(x)$ — the four quadrants

(a) The lower left-hand box is where we graph $F(x)$ versus x.

(b) The lower right-hand box is where we graph $G(y)$ versus y.

(c) The upper left-hand box is where we will eventually sketch $y = y(x)$.

(d) The upper right-hand box is the auxiliary box.

2. Draw the graph of the function $F(x)$ in the lower left-hand box. Draw the graph of the function $G(y)$ in the lower right-hand box, making sure that the vertical scale is the same as the vertical scale used in the lower left-hand box. Draw the diagonal with slope 1 in the auxiliary box.
3. Select any point x_0 in the lower left-hand box. Draw the vertical line $x = x_0$ into the upper left-hand box. Draw a horizontal line through the point $(x_0, F(x_0))$ until it crosses the curve $G(y)$ in the lower right-hand box at the point $(y_0, G(y_0))$. [If it crosses the curve $G(y)$ at more than one point, consider every one of these points in the next step.]
4. Draw a vertical line through $(y, G(y_0))$ until it intersects the line with slope 1 in the auxiliary box. Draw a horizontal line through this point of intersection into the upper left-hand box. The point where this line meets the vertical line $x = x_0$ is the point (x_0, y_0), which lies on the graph of $F(x) = G(y)$.
5. Repeat the last two steps for different values of x_0, obtaining as many points in the upper left-hand box as you desire. Join them. The curve is the graph of the solution $y = y(x)$ of $G(y) = F(x)$.

EXERCISES

1. Identify the following differential equations as autonomous, separable, or neither of these. (Some may fall into more than one category.) Do not attempt to solve any of these equations.

   (a) $x^2y' - y^2 = 0$
   (b) $y' - \sin(x/y) = 0$
   (c) $yy' - e^{xy} = 0$
   (d) $yy' - e^{x+y} = 0$
   (e) $y' - x + y = 0$
   (f) $y' - y\sin x = 0$
   (g) $y' + 4 + y^2 = 0$
   (h) $y' - \ln x + \ln y = 0$

2. Solve the differential equation $y' = -2xy$, and compare your answer with the one you obtained in Exercise 9(a) on page 92, and Exercise 3 on page 117.

3. Solve the differential equation $y' = x/y$, and compare your answer with the one you obtained in Exercise 7 on page 92.

4. Solve the following differential equations, finding an explicit solution wherever possible. Make sure your solutions are consistent with the corresponding slope field.

   (a) $x^2y' - y^2 = 0$
   (b) $e^{-x}y' - \sec y = 0$
   (c) $yy' - x\sin x^2 = 0$
   (d) $(4y + x^2y)y' - 2x - xy^2 = 0$
   (e) $x^2y' + 4 + y^2 = 0$
   (f) $(y^2 + 1)y' + y\tan x = 0$
   (g) $y' - e^{x+y} = 0$
   (h) $yy' - e^{x+y} = 0$
   (i) $yy' - (1 + y)\cos^2 x = 0$
   (j) $(1 + x^2)y' - 1 - y^2 = 0$

5. Solve the following initial value problems, finding an explicit solution wherever possible. Confirm your results by using graphical and numerical methods.

   (a) $xyy' - y = 0,\quad y(e) = 1$
   (b) $yy' + y^2\cos x = 0,\quad y(0) = 1/2$
   (c) $xyy' - (4 - y^2)^{1/2} = 0,\quad y(8) = 1$
   (d) $(1 - x^2)^{1/2}yy' + y^3 = 0,\quad y(1) = 1$
   (e) $(4x^2 + x - 1)yy' + (8x + 1)(y + 2) = 0,\quad y(1) = -2$
   (f) $2(2 + y)yy' + y(1 - x^2) = 0,\quad y(0) = -1$
   (g) $y\sin x + (y^2 + 1)e^{\cos x}yy' = 0,\quad y(\pi/2) = 1$
   (h) $3ye^{x^2}yy' + 1 = 0,\quad y(1) = 2$
   (i) $3y^2yy' - \sin x^2 = 0,\quad y(0) = 3$

(g) $y \sin x + (y^2 + 1)e^{\cos x} y' = 0, \quad y(\pi/2) = 1$

(h) $3ye^{x^2} y' + 1 = 0, \quad y(1) = 2$

(i) $3y^2 y' - \sin x^2 = 0, \quad y(0) = 3$

(j) $y' - 2xy/(x^2 - 1) = 0, \quad y(2) = -6$

(k) $y' - 2xy/(x^2 - 1) = 0, \quad y(0) = 2$

(l) $y' - 2xy/(x^2 - 1) = 0, \quad y(-2) = -6$

6. Show that all solutions of (4.11), namely,

$$y^4 - x^4 + 4(y - x) = C, \tag{4.21}$$

have the property that as $x \to \infty$, $y \to x$. [Hint: Use the fact that the solution is increasing (why?) and that $y^4 - x^4 = (y - x)\left(y^3 + xy^2 + x^2 y + x^3\right)$ to write (4.21) in the form $y - x = C/\left(y^3 + xy^2 + x^2 y + x^3 + 4\right)$, and then let $x \to \infty$.]

7. **Mixture Problem.** Initially a 200-gallon container is filled with pure water. At time $x = 0$ a salt concentration with 3 pounds of salt per gallon is added to the container at the rate of 4 gallons per minute, and the well-stirred mixture is drained from the container at the same rate.

   (a) Find the number of pounds of salt in the container as a function of time.

   (b) How many minutes does it take for the concentration in the container to reach 2 pounds per gallon?

   (c) What does the concentration in the container approach for large values of time? Does this agree with your intuition?

8. **Allometric Growth of Fish**. Allometric growth is the study of the relative size of different parts of an organism as a consequence of growth. One of the simplest models of allometry is one in which it is assumed that the relative growth rates of the two components $w(t)$ and $L(t)$ satisfy

$$\frac{dw}{dL} = a\frac{w}{L},$$

where a is a constant. Table 4.1 and Figure 4.10 show the results[2] of an experiment relating the weight of plaice $w(t)$ to its length $L(t)$.

   (a) Find an explicit solution of this differential equation. How well does this model fit the data set?

   (b) The following argument, based on units, is sometimes used to relate the weight $w(t)$ of a fish to its length $L(t)$. If L is a typical unit of length then L^3 is a typical unit of volume, and volume and weight are proportional. Thus, $w(t)$ is proportional to L^3, so $w(t) = bL^3(t)$ where b is a constant. Is this consistent with the result you found in part (a)?

Table 4.1 Weight of plaice as a function of length

Length (cm)	Weight (gm)	Length (cm)	Weight (gm)
23.5	124	37.5	455
24.5	146	38.5	500
25.5	155	39.5	538
26.5	174	40.5	574
27.5	190	41.5	623
28.5	213	42.5	674
29.5	236	43.5	724
30.5	259	44.5	808
31.5	284	45.5	812
32.5	308	46.5	909
33.5	332	47.5	1039
34.5	363	48.5	1124
35.5	391	49.5	1163
36.5	419		

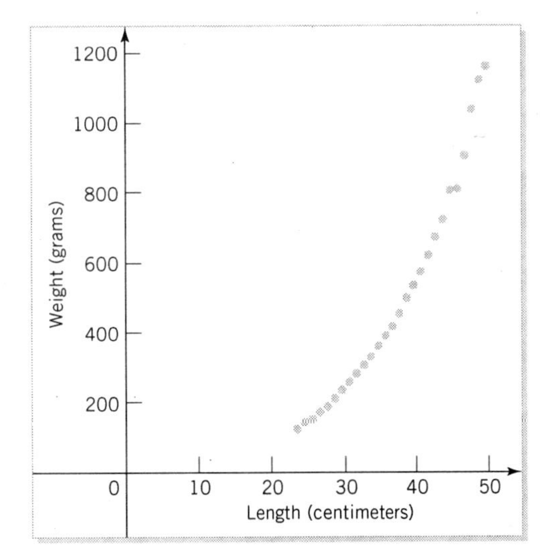

FIGURE 4.10 Weight of plaice as a function of length

[2] "On the Dynamics of Exploited Fish Populations" by R. J. H. Beverton and S. J. Holt, *Fishery Investigations*, Series II, 19, 1957, page 281.

9. The Clapeyron Equation.[3] The Clapeyron equation

$$\frac{dP}{dT} = a\frac{P}{T^2},$$

where a is a positive constant, governs how the vapor pressure P of a substance varies with its temperature T (in degrees Kelvin). Solve this differential equation, calling the constant of integration C. Explain how, by plotting $\ln P$ against $1/T$, you can decide whether a given data set obeys the Clapeyron equation, and how this allows you to estimate a and C for the data set. Table 4.2 and Figure 4.11 show the vapor pressure of toluene[4] as a function of the temperature in °K. How well does the Clapeyron equation model this data set?

Table 4.2 The vapor pressure of toluene as a function of temperature

Temperature (°K)	Pressure (mm Hg)	Temperature (°K)	Pressure (mm Hg)
268.75	5	325.05	100
279.55	10	342.65	200
291.55	20	362.65	400
304.95	40	383.75	760
313.45	60	409.65	1520

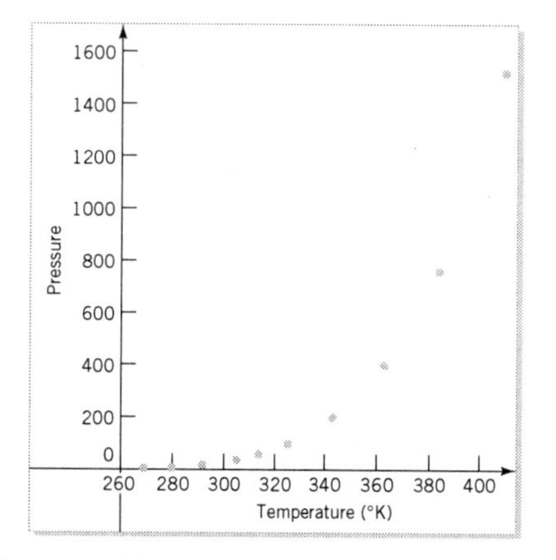

FIGURE 4.11 The vapor pressure of toluene as a function of temperature

10. Killing Mosquitoes. Nicotine sulphate is sprayed in a closed room containing mosquitoes to determine how efficiently the chemical kills the insects. Table 4.3 and Figure 4.12 show the results of this experiment for various doses of nicotine sulphate.[5] It is believed that the killing efficiency $E(x)$ of the chemical as a function of the dose x (in g/100 cc) is modeled by the differential equation $E' = ax^{-0.8}E(1 - E)$ where a is a positive constant.

Table 4.3 The efficiency of nicotine sulphate at killing mosquitoes

Dose	Number in Room	Deaths	Efficiency
0.10	47	8	0.170
0.15	53	14	0.264
0.20	55	24	0.436
0.30	52	32	0.615
0.50	46	38	0.826
0.70	54	50	0.926
0.95	52	50	0.962

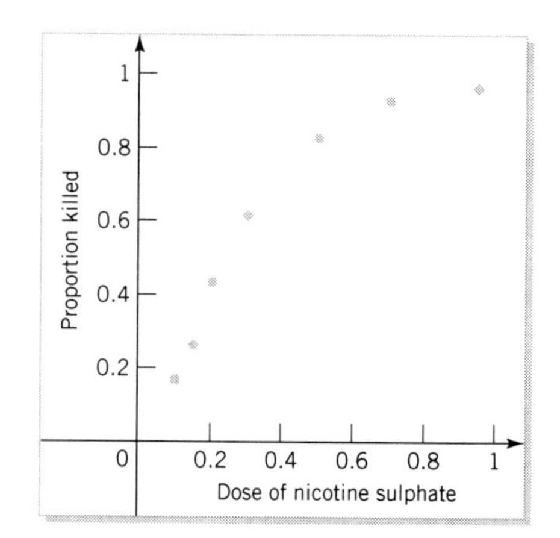

FIGURE 4.12 The efficiency of nicotine sulphate at killing mosquitoes

(a) Justify why E' might be proportional to $E(1 - E)$.

(b) Solve the differential equation for $E(x)$.

(c) Use the data to identify all the constants in $E(x)$.

(d) What dosage will destroy 50% of the mosquitoes in the room?

---

[3] We thank Herbert Yee, a student at the University of Arizona, for bringing this equation and data set to our attention. See *Elementary Principles of Chemical Processes* by R.M. Felder and R.W. Rousseau, Wiley, 1986, 2nd edition, page 230.

[4] The data set is taken from *Computational Methods in Chemical Engineering* by O.T. Hanna and O.C. Sandall, Prentice Hall, 1995, page 136.

[5] We thank Javier Osante Vazquez of the Escuela Nacional de Ciencias Basicas, Mexico, for these data.

11. Use the technique described in this section to sketch $x^2 = 1 - y^2$. Use Figure 4.13 for this purpose. It shows the functions x^2 versus x and $1 - y^2$ versus y.

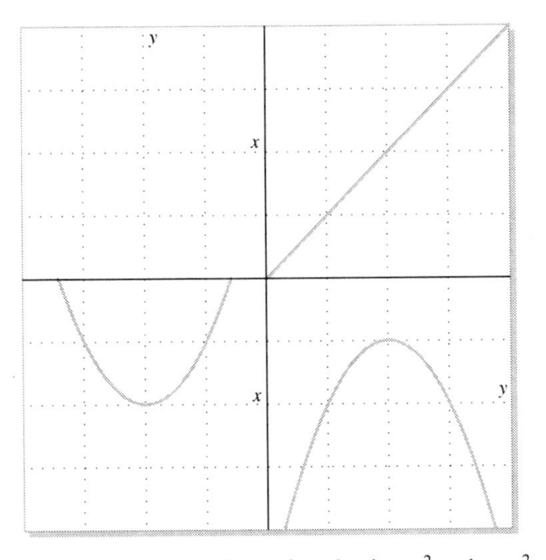

FIGURE 4.13 The figure for plotting $x^2 = 1 - y^2$

12. Use the technique described in this section to sketch the following curves.

(a) $x^2 = 10 + y^2$ (b) $x^3 + 3y^3 = 27$

13. The Lotka-Volterra Differential Equation. A differential equation that models how the number of prey y change with the number of predators x is

$$y' = \frac{y(1 - x)}{x(-1 + y)},$$

where x and y are both positive.

(a) Solve this differential equation subject to $y(1) = 3.5$, and show that

$$\frac{1}{x}e^{x-1} = \frac{2}{7}ye^{-y+3.5}. \quad (4.22)$$

(b) In the lower left-hand box of Figure 4.14 we have plotted the function of x on the left-hand side of (4.22) — namely e^{x-1}/x. In the lower right-hand box of Figure 4.14 we have plotted the function of y on the right-hand side of (4.22) — namely, $2ye^{-y+3.5}/7$. Complete Figure 4.14. What shape do you get?

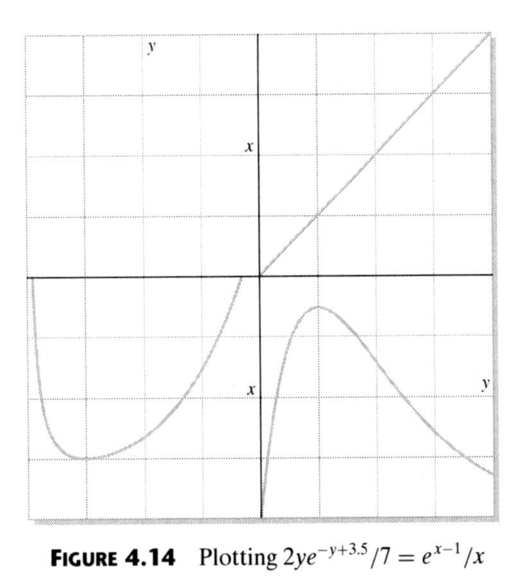

FIGURE 4.14 Plotting $2ye^{-y+3.5}/7 = e^{x-1}/x$

14. Look in the textbooks of your other courses and find an example that uses a separable differential equation. Write a report about this example that includes the following items:

(a) A brief description of background material so a classmate will understand the origin of the differential equation.

(b) How the constants in the differential equation can be evaluated and what the initial condition means.

(c) The solution of this differential equation.

(d) An interpretation of this solution, and how it answers a question posed by the original discussion.

4.2 SOLVING DIFFERENTIAL EQUATIONS WITH HOMOGENEOUS COEFFICIENTS

In this section we start with an example from optics. However, the differential equation that arises in this case is not one that we have encountered previously. We develop a technique for solving this differential equation and show how it is used in another example. This technique is used later in the text.

An Example from Optics

EXAMPLE 4.5 *The Focal Property of the Parabolic Reflector*

An important property of the parabola is that rays coming from the focus are reflected parallel to the axis. Or, equivalently, incoming parallel rays are focused at a single point after reflection. This focal property (shown in Figure 4.15) is why parabolas are used in the design of car headlights and radio telescopes. However, constructing a parabolic telescope is very expensive, and so a natural question to ask is whether any other shapes have this focal property.

To be specific, consider a ray parallel to the y-axis coming from infinity along the line $x = x_0$ (see Figure 4.16). It strikes the mirror with shape $f(x)$ at the point a with coordinates (x_0, y_0) and is then focused at the origin O. (The ray and mirror continue into the left half-plane, but are not drawn in Figure 4.16.) When a ray strikes the mirror, it is reflected so that the angle between the incoming ray and the mirror equals the angle between the reflected ray and the mirror. At the place where the ray strikes the mirror, the mirror acts as if it were a straight line with slope $f'(x_0)$. Thus, the angle between the incoming or reflected ray and the mirror is actually the angle A between the incoming or reflected ray and the straight line with slope $f'(x_0)$.

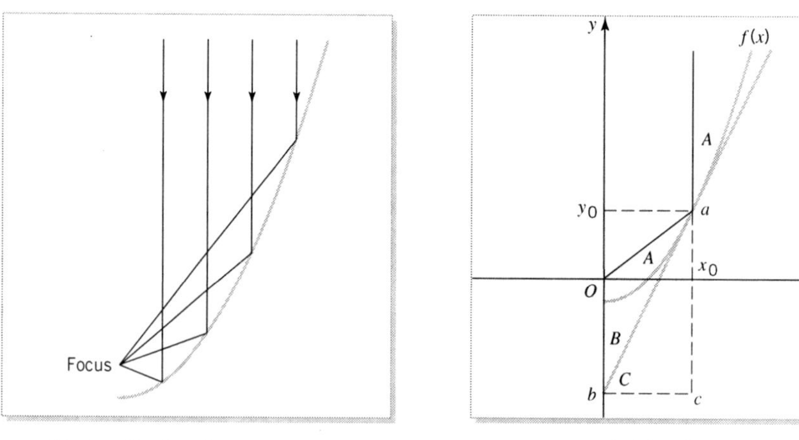

FIGURE 4.15 Focal property

FIGURE 4.16 Geometry for focal property

We want to find the differential equation that $f(x)$ must satisfy.[6] From the geometry of Figure 4.16, where A and B are angles, we see that $A = B$ and so the triangle bOa is isosceles. This means that the distance from O to b is the same as the distance from O to a, namely, $\sqrt{x_0^2 + y_0^2}$. From this, and considering the triangle abc in Figure 4.16, we see that

$$\tan C = \frac{\text{length of } ac}{\text{length of } bc} = \frac{y_0 + \sqrt{x_0^2 + y_0^2}}{x_0}.$$

Because $f'(x_0) = \tan C$, the previous equation becomes

$$f'(x_0) = \frac{y_0 + \sqrt{x_0^2 + y_0^2}}{x_0}.$$

This is the slope of the tangent line to the curve $f(x)$ at the point (x_0, y_0) that causes a line parallel to the y-axis to be reflected through the origin. Because we want this to happen for every point (x_0, y_0) – namely, (x, y) – we have the differential equation

$$y' = \frac{y + \sqrt{x^2 + y^2}}{x}, \tag{4.23}$$

where $y = f(x)$ and $x > 0$. (A similar equation is valid if $x < 0$.) If we divide numerator and denominator of the right-hand side of (4.23) by x, we find[7]

$$y' = \frac{y}{x} + \sqrt{1 + \left(\frac{y}{x}\right)^2} \text{ if } x > 0, \tag{4.24}$$

$$y' = \frac{y}{x} - \sqrt{1 + \left(\frac{y}{x}\right)^2} \text{ if } x < 0. \tag{4.25}$$

Slope field

We will concentrate on the $x > 0$ case. Before trying to solve this analytically, let's look at the slope field for (4.24) to see whether this differential equation is reasonable. Figure 4.17 shows the slope field. The slope field is consistent with (4.24) in that all curves are increasing, and isoclines for slope m are given by $y = \left[(m^2 - 1)/(2m)\right]x$, which are straight lines through the origin. (See Exercise 1 on page 144.) These are included in Figure 4.17 for the cases $m = 0.5$, 1, and 2. Figure 4.18 shows the slope field again, and a few hand-drawn solution curves for (4.24).

Change variable

Now let us turn to finding the explicit solution of (4.24) to confirm these observations. Although (4.24) is not a separable differential equation, we do notice the repeated presence of the ratio y/x on the right-hand side. Because x and y occur

---

[6] We thank our colleague Bill Mueller for suggesting this construction.

[7] Remember, $\sqrt{a^2} = a$ if $a > 0$, and $\sqrt{a^2} = -a$ if $a < 0$.

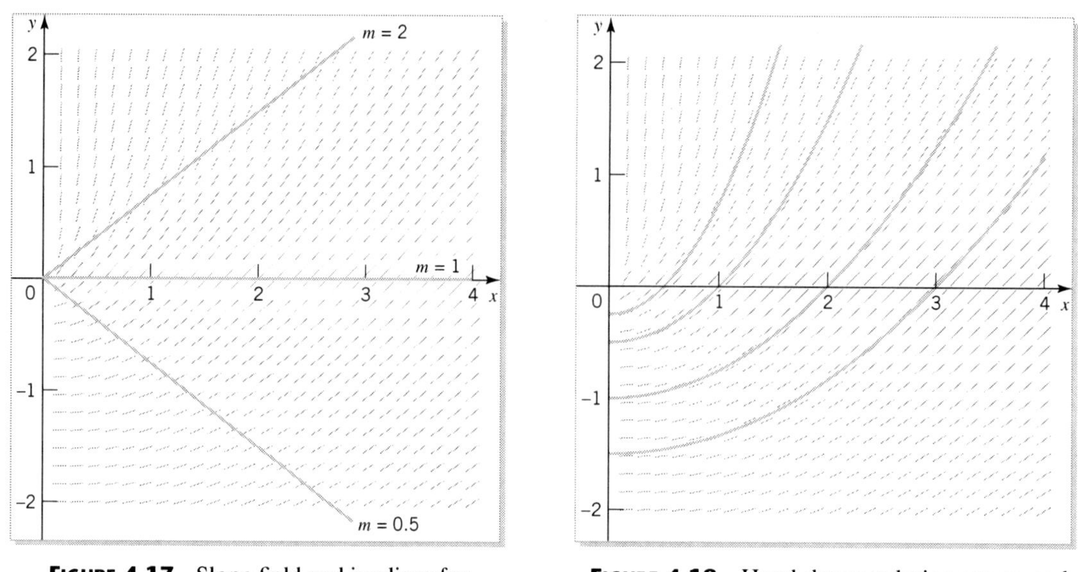

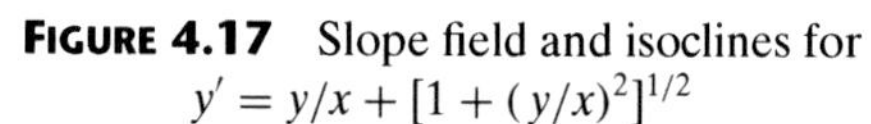

FIGURE 4.17 Slope field and isoclines for $y' = y/x + [1 + (y/x)^2]^{1/2}$

FIGURE 4.18 Hand-drawn solution curves and slope field for $y' = y/x + [1 + (y/x)^2]^{1/2}$

only in this ratio, let us try a change of the dependent variable from $y(x)$ to $z(x)$ by using this ratio; that is,

$$z = \frac{y}{x}.$$

This means that

$$y = xz,$$

and differentiating this equation gives us

$$\frac{dy}{dx} = z + x\frac{dz}{dx}.$$

Making these two substitutions in (4.24) gives the transformed differential equation

$$z + x\frac{dz}{dx} = z + \sqrt{1 + z^2},$$

or

$$x\frac{dz}{dx} = \sqrt{1 + z^2}.$$

Separate variables This equation is separable and may be rearranged as

$$\int \frac{dz}{\sqrt{1 + z^2}} = \int \frac{dx}{x}.$$

Integrate The integral on the left-hand side may not be familiar, but can be found in a table of integrals or evaluated directly (see Exercise 2 on page 144), namely,

$$\int \frac{dz}{\sqrt{1+z^2}} = \ln\left(z + \sqrt{z^2+1}\right) + C.$$

Thus, we have

$$\ln\left(z + \sqrt{z^2+1}\right) + C = \ln x.$$

By taking exponentials, we find

$$z + \sqrt{z^2+1} = cx,$$

where $c = e^{-C} > 0$. To eliminate the square root, we write this equation in the form $\sqrt{z^2+1} = cx - z$, and square, obtaining

$$z^2 + 1 = (cx - z)^2 = c^2x^2 - 2cxz + z^2,$$

or

$$1 = c^2x^2 - 2cxz.$$

Explicit solution To put this expression in terms of our original variables, we use the fact that $z = y/x$, so our solution becomes

$$1 = c^2x^2 - 2cx\left(\frac{y}{x}\right),$$

or

$$y = \frac{1}{2}cx^2 - \frac{1}{2c}.$$

As expected, this is a family of parabolas, opening up because $c > 0$. The case $x < 0$ — namely, (4.25) — is treated in the same way with the same result. (See Exercise 3 on page 144.) Thus, **the parabola is the only curve with the focal property.** ■

Solving Differential Equations with Homogeneous Coefficients

It was no accident that the differential equation (4.24) was transformed to a separable one after the change of variable $y = xz$. The following few steps show that this would happen to any equation of the form

$$y' = g\left(\frac{y}{x}\right). \tag{4.26}$$

Change variable If we make the change of variable $z = y/x$ in equation (4.26), it becomes

$$x\frac{dz}{dx} + z = g(z),$$

Separate variables

which is seen to be separable by rearranging it as

$$\frac{1}{g(z) - z}\frac{dz}{dx} = \frac{1}{x}.$$

So a first order differential equation of the type (4.26) can always be converted to a separable differential equation by a change of variable. We now formalize this discussion.

◆ *Definition 4.2:* **A DIFFERENTIAL EQUATION WITH HOMOGENEOUS COEFFICIENTS is one that may be put in the form $y' = g\left(\frac{y}{x}\right)$.** ◆

How to Solve Differential Equations with Homogeneous Coefficients

Purpose To solve first order differential equations of the form

$$y' = g\left(\frac{y}{x}\right). \tag{4.27}$$

Process

1. Make the change of variable $z = \dfrac{y}{x}$, so that

$$y = xz \tag{4.28}$$

and

$$\frac{dy}{dx} = x\frac{dz}{dx} + z. \tag{4.29}$$

2. Substitute (4.28) and (4.29) into (4.27) to eliminate all y dependence.
3. Rearrange the result so that terms involving z are separated from terms involving x.
4. Integrate this separable equation to find either an explicit or implicit solution, $z = z(x)$.
5. If the last step yields an explicit solution for $z = z(x)$, use (4.28) to find the explicit solution of (4.27) as $y = xz(x)$. If, on the other hand, there is an implicit solution relating z with x, replace z in this implicit solution with y/x to obtain an implicit solution of (4.27).

Comments about Differential Equations with Homogeneous Coefficients

- This type of differential equation is so named because functions of the form $g(y/x)$ are homogeneous of degree 0 in x and y.[8]

[8] A function $F(x, y)$ is homogeneous of degree n if $F(tx, ty) = t^n F(x, y)$.

- Sometimes it is quite difficult to recognize that a differential equation is of the type (4.27). One way to test whether a differential equation

$$y' = g(x, y) \tag{4.30}$$

is of the type (4.27) is to substitute $y = xz$ into $g(x, y)$, obtaining $g(x, xz)$. If the resulting function may be simplified to be independent of x, then (4.30) is of the type (4.27). This shows that $g(x, y)$ is homogeneous of degree 0.

- A common mistake is to omit the z on the right-hand side of (4.29).
- Sometimes the resulting separable equation is easier to integrate if we make the change of variable $v = x/y$ instead of $z = y/x$.
- The slope fields of differential equations with homogeneous coefficients, (4.27), are unchanged if the x- and y-axes are rescaled by equal amounts. (See Exercise 6 on page 144.)
- The slope fields of differential equations with homogeneous coefficients, (4.27), are always symmetric about the origin. (Why?)

EXAMPLE 4.6

We show how this technique works in practice by considering

$$y' = \frac{x^3 y}{x^4 + y^4}, \qquad (x, y) \neq (0, 0). \tag{4.31}$$

We notice this is not separable, but is it an equation with homogeneous coefficients?

We determine this by dividing the numerator and denominator by x^4, to find

$$y' = \frac{y/x}{1 + (y/x)^4}.$$

The right-hand side is a function of y/x, so it is a differential equation with homogeneous coefficients. [Alternatively, we could test whether the right-hand side of (4.31) is of the form $g(y/x)$ by substituting $y = xz$ to find

$$\frac{x^3 y}{x^4 + y^4} = \frac{x^3 xz}{x^4 + x^4 z^4} = \frac{z}{1 + z^4}.$$

Because this is independent of x, we have a differential equation with homogeneous coefficients.]

Change variable

We make the substitution $z = y/x$ in (4.31) and obtain

$$x\frac{dz}{dx} + z = \frac{z}{1 + z^4},$$

which can be rewritten as

$$x\frac{dz}{dx} = \frac{z}{1 + z^4} - z = \frac{-z^5}{1 + z^4}.$$

Separate variables This equation is separable. The equilibrium solution, $z = 0$, gives the solution $y(x) = 0$, while non-equilibrium solutions are found from

$$\frac{1+z^4}{z^5}\frac{dz}{dx} = -\frac{1}{x},$$

or

$$\left(\frac{1}{z^5}+\frac{1}{z}\right)\frac{dz}{dx} = -\frac{1}{x}.$$

Integrate Integration and combination of logarithmic terms yield the implicit solution $4\ln|zx| = z^{-4} + C$. Replacing z with y/x, we find the implicit solution of (4.31) in the form

$$4\ \ln|y| = \left(\frac{x}{y}\right)^4 + C.$$

Although this is the analytical solution of the differential equation — and some might be content to stop here — it is not very helpful in trying to understand the behavior of the solution. Thus, we turn to a graphical analysis to obtain whatever qualitative information we can.

Symmetry The first thing we notice is that (4.31) is symmetric across the x-axis and across the y-axis, and is therefore symmetric about the origin. (Check this.) Thus, we need only consider the solution in the first quadrant, $x \geq 0,\ y \geq 0$. Concentrating on this region, we see that $y' > 0$ if $x > 0$ and $y > 0$, so solutions are increasing there, while $y' = 0$ if $x = 0$ or $y = 0$. To decide on concavity we need

Monotonicity

$$y'' = \frac{1}{\left(x^4+y^4\right)^2}\left[\left(3x^2y + x^3y'\right)\left(x^4+y^4\right) - 4x^3y\left(x^3+y^3y'\right)\right].$$

We now substitute for y' from (4.31) and, after some algebraic manipulations, we find

$$y'' = \frac{x^2y^5\left(3y^4 - x^4\right)}{\left(x^4+y^4\right)^3}.$$

Concavity Thus, in the first quadrant, the solutions will be concave up when $3y^4 - x^4 > 0$ — that is, $y > x/3^{1/4}$ — and concave down when $3y^4 - x^4 < 0$ — that is, $y < x/3^{1/4}$. The straight line through the origin $y = x/3^{1/4} \approx 0.76x$ separates the concave up region from the concave down one.

Isocline Finally, we look at isoclines for slope m, which must satisfy

$$\frac{y/x}{1+(y/x)^4} = m,$$

which in theory we can solve for y/x. However, that is not so simple in practice. But, if we could solve for y/x, we would find $y/x =$ constant; that is, the isoclines are straight lines through the origin. So we can turn the problem around, and start with $y = \alpha x$ for various α and see what m's they give rise to, because the isocline $y = \alpha x$ must have an m which satisfies

$$\frac{\alpha}{1+\alpha^4} = m.$$

For example, if $\alpha = 1$ then $m = 0.5$; $\alpha = 0$, $m = 0$; $\alpha = 2$, $m = 2/9$, and so on. Thus, when a solution crosses the line $y = x$ it must have slope $m = 0.5$. The isocline where the concavity changes, $y = x/3^{1/4}$, corresponds to $m = 1/(4 \times 3^{1/4}) \approx 0.19$.

With so many straight lines, it is natural to ask whether (4.31) has any straight-line solutions $y = mx$. In this case m would have to satisfy

$$\frac{m}{1+m^4} = m,$$

which is satisfied only by $m = 0$, which gives the equilibrium solution $y = 0$.

Figure 4.19 shows the slope field in the first quadrant for (4.31) together with the information we have discovered, and some numerical solutions. All this is consistent. Figure 4.20 shows all four quadrants and some numerical solutions.

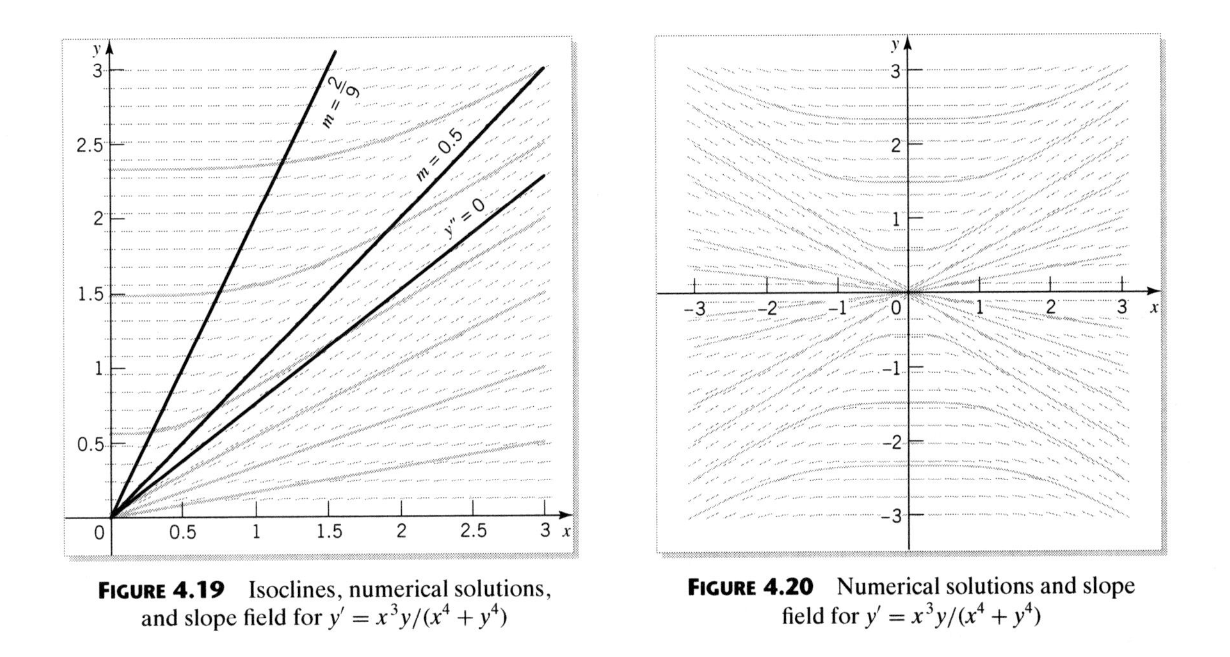

Figure 4.19 Isoclines, numerical solutions, and slope field for $y' = x^3y/(x^4 + y^4)$

Figure 4.20 Numerical solutions and slope field for $y' = x^3y/(x^4 + y^4)$

These examples suggest two more observations about differential equations with homogeneous coefficients.

More Comments about Differential Equations with Homogeneous Coefficients

- All isoclines are straight lines through the origin.
- If a straight-line solution $y = mx$ of $y' = g(y/x)$ exists, then m is a solution of $g(m) = m$.

EXERCISES

1. Show that the isoclines for slope m of (4.24), namely,

$$y' = \frac{y}{x} + \sqrt{1 + \left(\frac{y}{x}\right)^2},$$

are given by $y = \left[(m^2 - 1)/(2m)\right]x$.

2. By making the substitution $u = z + \sqrt{z^2 + 1}$ show that

$$\int \frac{dz}{\sqrt{1 + z^2}} = \ln\left(z + \sqrt{z^2 + 1}\right) + C.$$

3. Show that the solution of (4.25), namely,

$$y' = \frac{y}{x} - \sqrt{1 + \left(\frac{y}{x}\right)^2}$$

for $x < 0$, is

$$y = \frac{1}{2}cx^2 - \frac{1}{2c},$$

where c is an arbitrary positive constant.

4. Consider the two differential equations

$$y' = \frac{x + \sqrt{x^2 + y^2}}{y}, \text{ and } y' = \frac{1 + \sqrt{1 + (y/x)^2}}{y/x}.$$

Without solving these equations, would you expect their solutions to be identical? Many people would. Plot the slope fields for each of these equations, and explain why they are not identical.

5. First confirm that the following are differential equations with homogeneous coefficients, and then solve them analytically. If the solution is explicit, graph it and compare with its slope field. If the solution is implicit, compare the isoclines and any straight-line solutions with its slope field. [In part (d) assume $x > 0$.]

   (a) $(x^2 + 3y^2)y' + 2xy = 0$
   (b) $3xyy' + x^2 + y^2 = 0$
   (c) $(x^2 - xy + y^2)y' + y^2 = 0$
   (d) $xy' + \sqrt{x^2 + y^2} - y = 0$
   (e) $(xy - x^2)y' - y^2 = 0$
   (f) $[y\tan(x/y) - x]\, y' + y = 0$
   (g) $x^2y' + y^2 - xy = 0$
   (h) $3xyy' + x^2 + y^2 = 0$
   (i) $2xyy' + x^2 + y^2 = 0$
   (j) $(x + 2y)y' + x + y = 0$
   (k) $xy' - y\ln(y/x) - y = 0$

6. Select any differential equation with homogeneous coefficients.

   (a) Plot its slope field in the window $-10 < x < 10$, $-10 < y < 10$; then in the window $-1 < x < 1$, $-1 < y < 1$; and finally in the window $-0.1 < x < 0.1$, $-0.1 < y < 0.1$. Compare the three slope fields.
   (b) Make the change of scale $X = ax$, $Y = ay$, where a is a positive constant, in your differential equation to obtain a differential equation in X and Y. Compare this differential equation with your original differential equation. How does the differential equation change? Explain.
   (c) How are parts (a) and (b) related? Does this result apply to all differential equations with homogeneous coefficients?

7. A function $F(x, y)$ is homogeneous of degree n if $F(tx, ty) = t^nF(x, y)$. For example,

$$F(x, y) = \frac{x^4 + y^4}{x}$$

is homogeneous of degree 3 because

$$F(tx, ty) = \frac{(tx)^4 + (ty)^4}{tx} = t^3\frac{x^4 + y^4}{x} = t^3F(x, y).$$

Check to see if the following functions are homogeneous. If so, give the degree.

   (a) $F(x, y) = x^2 + 3xy - x^3(x + y)^{-1}$
   (b) $F(x, y) = x^2 + 3xy - x^2(x + y)$
   (c) $F(x, y) = x\sin(x/y) + x^2/y$
   (d) $F(x, y) = \sqrt{x^4 + 4x^2y^2 + x^2}$
   (e) $F(x, y) = \ln x - \ln y + e^{x/y}$

8. Consider the differential equation

$$y' = f\left(\frac{a_1x + b_1y + c_1}{a_2x + b_2y + c_2}\right),$$

where a_1, b_1, c_1, a_2, b_2, and c_2 are constants.

   (a) If $a_1b_2 - a_2b_1 \neq 0$, show that the transformation $x = u + \alpha$, $y = v + \beta$, through careful choice of the constants α and β, reduces the differential equation to the differential equation with homogeneous coefficients

$$\frac{dv}{du} = f\left(\frac{a_1u + b_1v}{a_2u + b_2v}\right).$$

   (b) If $a_1b_2 - a_2b_1 = 0$, show that the transformation $z = a_1x + b_1y$ reduces the differential equation to a separable differential equation involving z and x.

9. Using the technique described in Exercise 8, solve the following differential equations.

(a) $y' = -\dfrac{x+y-2}{x-y+4}$ (b) $y' = -\dfrac{x+y+1}{2x+2y-4}$

(c) $y' = \dfrac{x-2y+1}{2x-4y+3}$

10. Identify the following differential equations as separable equations, those with homogeneous coefficients, or none of these. (Some may fall into more than one category.) Do not attempt to solve any of these equations.

(a) $x^2y' - y^2 = 0$ (b) $y' - \sin(x/y) = 0$

(c) $yy' - xe^y/e^x = 0$ (d) $yy' - e^{x+y} = 0$

(e) $y' - 1 + x/y = 0$ (f) $y' - y\sin(x) = 0$

(g) $y' + 4 + y^2 = 0$ (h) $y' - \ln x + \ln y = 0$

4.3 MODELS: DERIVING DIFFERENTIAL EQUATIONS FROM DATA

In this section we discuss two applications of differential equations that utilize real data sets. The main purpose is to illustrate how to obtain a reasonable differential equation that models the situation by analyzing the data set numerically.[9]

EXAMPLE 4.7 *The Heating of a Probe*

Table 4.4
The heating of a temperature probe with time

Time	Temp.
0	32.78
2	33.12
4	33.37
6	33.54
8	33.68
10	33.78
12	33.85
14	33.93
16	33.98
18	34.03
20	34.05

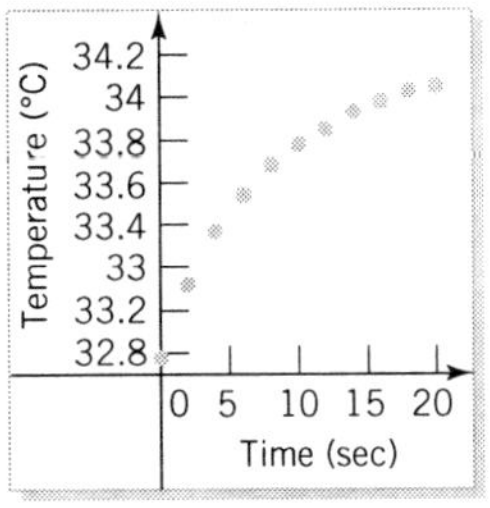

A student[10] held a temperature probe firmly between her thumb and forefinger while the temperature of the probe was recorded. Table 4.4 shows the resulting data set where the temperature is in degrees centigrade and the time is in seconds. This data set is plotted in Figure 4.21. Our goal is to determine a formula that will allow us to predict the temperature at other times.

Because the temperature is clearly changing with time, we ask, What is the relationship between the temperature of the probe T and the time t? To answer this question, we try to find a differential equation governing this process, which means we want to relate the derivative dT/dt to a function of t and T. We can obtain an approximate numerical value for dT/dt from Table 4.4 by using one of several approximate forms of the derivative. Here we list three such forms — the right-hand difference quotient[11]

$$\frac{dT}{dt} \approx \frac{\Delta_R T}{\Delta t} = \frac{T(t+h) - T(t)}{h}, \tag{4.32}$$

the left-hand difference quotient

$$\frac{dT}{dt} \approx \frac{\Delta_L T}{\Delta t} = \frac{T(t-h) - T(t)}{-h}, \tag{4.33}$$

FIGURE 4.21
Temperature of probe versus time

[9]Different disciplines view the data-modeling process in different ways. In some disciplines, the model is constructed based on established principles, and then the data set is used to test the model. In other disciplines, the data set is used to construct the model empirically, and then the model is used to predict subsequent behavior. This section deals with the latter approach.

[10]Melisa Enrico, University of Arizona.

[11]Here $T(t+h)$ is the value of T at $t+h$, not T times $t+h$.

and the average of the last two, the central difference quotient,

$$\frac{dT}{dt} \approx \frac{\Delta_C T}{\Delta t} = \frac{T(t+h) - T(t-h)}{2h}. \tag{4.34}$$

Of these, the central difference quotient is the approximation most commonly used. Table 4.5 shows the central difference quotient calculations $\Delta_C T/\Delta t$ for Table 4.4.

In Figure 4.22 we plot this numerical approximation for dT/dt against the time t. We do not see any obvious relation between these variables. In Figure 4.23 we plot the numerical approximation for dT/dt against the temperature T, and it appears that the relationship is approximately linear with a negative slope. Figure 4.24 is Figure 4.23 with the addition of the straight line of best fit, which has slope -0.14 and vertical intercept 4.78. (Note that the horizontal scale is from 33 to 34.)

Table 4.5 The numerical approximation for dT/dt

Time	Temp.	$[T(t+2) - T(t-2)]/4$
0	32.78	
2	33.12	0.1475
4	33.37	0.1050
6	33.54	0.0775
8	33.68	0.0600
10	33.78	0.0425
12	33.85	0.0375
14	33.93	0.0325
16	33.98	0.0250
18	34.03	0.0175
20	34.05	

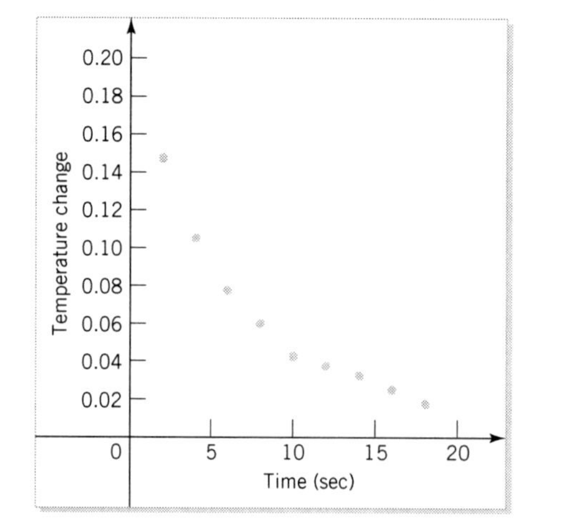

FIGURE 4.22 Numerical approximation for dT/dt versus time t

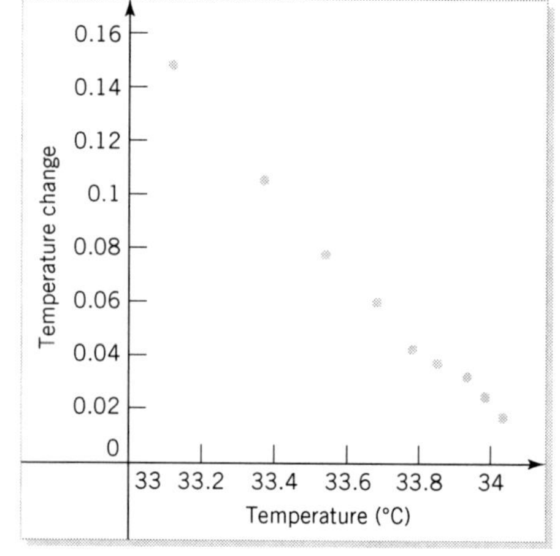

FIGURE 4.23 Numerical approximation for dT/dt versus temperature T

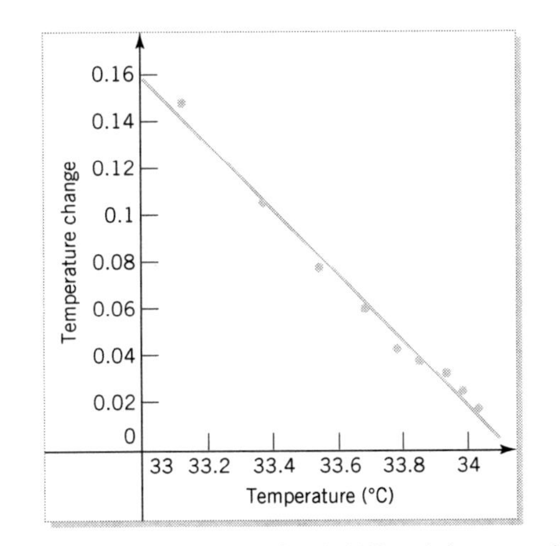

FIGURE 4.24 The line $4.78 - 0.14T$ and the numerical approximation for dT/dt versus temperature T

This suggests that the differential equation governing the temperature change might be of the form $T' = a + kT$, where a and k are constants, with $k < 0$. This differential equation has the equilibrium solution $-a/k$, which we label T_a, so $T_a = -a/k$. We can then rewrite this differential equation in terms of T_a as

Equilibrium solution

$$\frac{dT}{dt} = k(T - T_a). \tag{4.35}$$

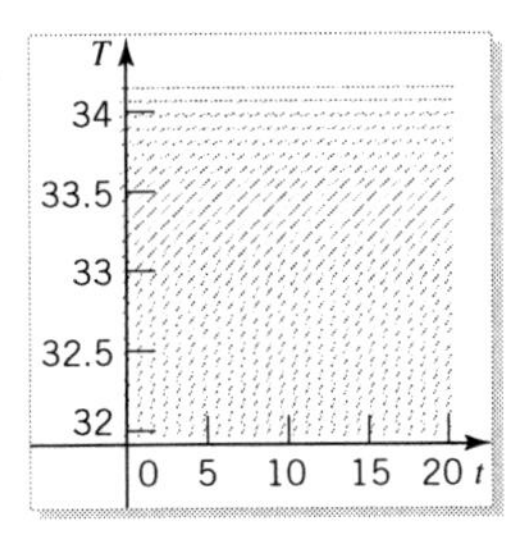

FIGURE 4.25
Slope field for $dT/dt = k(T - T_a)$

Before trying to solve (4.35), we will use graphical arguments to analyze this model. For this to be a reasonable model for the heating of the probe, we expect the solution curves to be increasing, concave down, and tend to the exterior temperature of the finger as time increases. This agrees with the slope field for (4.35), as shown in Figure 4.25 for $T_a = -a/k = 4.78/0.14 \approx 34.14$ and $k = -0.14$.

From (4.35) we see that if T, the temperature of the probe, is less than T_a, then $k(T - T_a)$ is positive (remember $k < 0$), so that T is an increasing function. Furthermore, the nearer T gets to T_a, the smaller the increase. If we take the derivative of (4.35), we find $T'' = kT' = k^2(T - T_a)$. This equation tells us that for $T < T_a$, the solution curves are concave down. All of this seems to suggest that (4.35) may be a reasonable model.

Concavity

Because (4.35) is an autonomous differential equation (the right-hand side is independent of t), we can perform a phase line analysis, and in doing so in Figure 4.26 we find that T_a is a stable equilibrium solution. This means that the temperature will tend to the quantity T_a as time increases, so that T_a is the exterior temperature of the finger. The temperature T_a is often called the ambient temperature. (Note that this analysis also implies that if $T > T_a$, then T decreases to T_a.)

Phase line

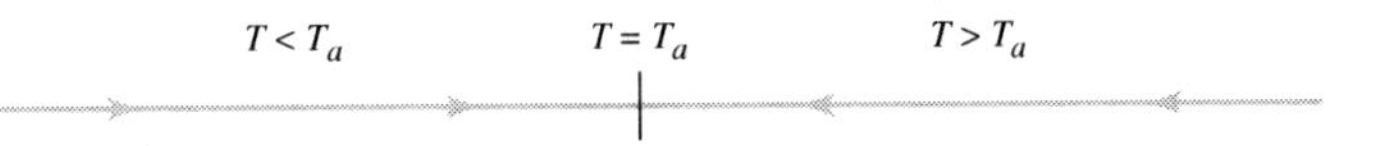

FIGURE 4.26 Phase line analysis for $dT/dt = k(T - T_a)$

We now turn to finding the explicit solution of (4.35), which has the equilibrium solution

$$T(t) = T_a. \tag{4.36}$$

Separate variables

By rewriting (4.35) in the form

$$\frac{1}{T - T_a}\frac{dT}{dt} = k$$

Integrate

and then integrating, we find $\ln|T - T_a| = kt + c$. We isolate the temperature by taking the exponential of this equation, $|T - T_a| = e^{kt+c}$, which can be written as $T - T_a = Ce^{kt}$, where the arbitrary constant C may be either positive or negative. (Why?) If we permit $C = 0$, we can absorb the equilibrium solution (4.36) into this solution. Rearranging gives $T(t) = T_a + Ce^{kt}$, where the value of C is determined by requiring the value of the temperature at $t = 0$ equal the initial temperature T_0; that is, $T_0 = T_a + C$. Thus, the final form of the solution is

Explicit solution

$$T(t) = T_a + (T_0 - T_a)e^{kt}. \tag{4.37}$$

Long-term behavior

A brief look at this equation leads us to the conclusion that for large values of time, the exponential term will be insignificant (remember that $k < 0$) and the temperature of the probe will approach the ambient temperature, in this case the exterior temperature of the finger. This agrees with the slope field for (4.35) as shown in Figure 4.27, and it also agrees with our common sense.

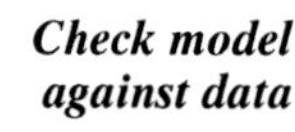

Check model against data

We now return to the original experimental data set (Table 4.4 on page 145) to see how well it compares to our mathematical model. It is clear that Figure 4.24 gives a crude estimate (based on the differential equation) of $k \approx -0.14$ and $kT_a = -a \approx 4.78$, from which we can estimate the ambient temperature as $T_a \approx 34.14$. However, we expect to find a more accurate estimate of the parameters by comparing the data with the exact solution (4.37). If we rewrite (4.37) in the form $T_a - T(t) = \left(T_a - T_0\right)e^{kt}$ and take the logarithm of both sides, we find $\ln\left[T_a - T(t)\right] = \ln\left(T_a - T_0\right) + kt$. Thus, if we use the estimate $T_a \approx 34.14$ and plot $\ln[34.14 - T(t)]$ against t and then find a straight line, its slope will determine k, and its vertical intercept will determine $\ln\left(T_a - T_0\right)$. Figure 4.28 shows this plot, together with the straight line $0.308 - 0.135t$. Thus, we estimate $k \approx -0.135$ and $T_a - T_0 \approx e^{0.308}$, which gives $T_0 \approx 32.78$. This is consistent with the experimental value of $T_0 = 32.78$ and our crude initial estimate of $k \approx -0.14$. When these estimates, together with $T_a \approx 34.14$, are substituted in (4.37), we find $T(t) = 34.14 - e^{0.308-0.135t}$. The agreement between the graph of this function and the original data set may be seen in Figure 4.29.

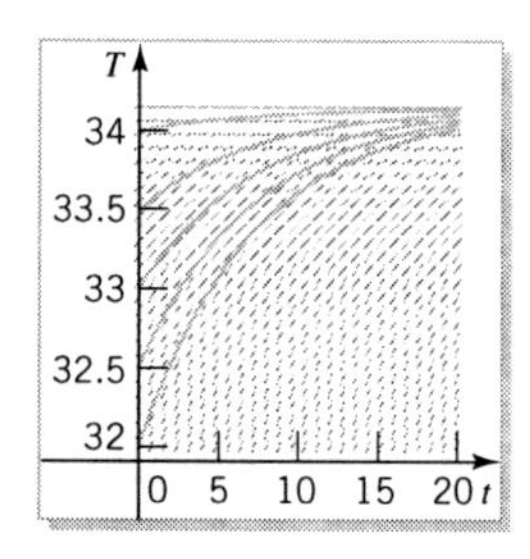

FIGURE 4.27 Solution curves and slope field for $dT/dt = k(T - T_a)$

Equation (4.35) on page 147, $T' = k(T - T_a)$, is called NEWTON'S LAW OF HEATING and is consistent with the assumption that the change in temperature of a body being heated is proportional to the difference between the temperature of the body and the temperature of the ambient medium. Equation (4.35) is also NEWTON'S LAW OF COOLING and is consistent with the assumption that the change in temperature of a body being cooled is proportional to the difference between the temperature of the body and the temperature of the ambient medium.

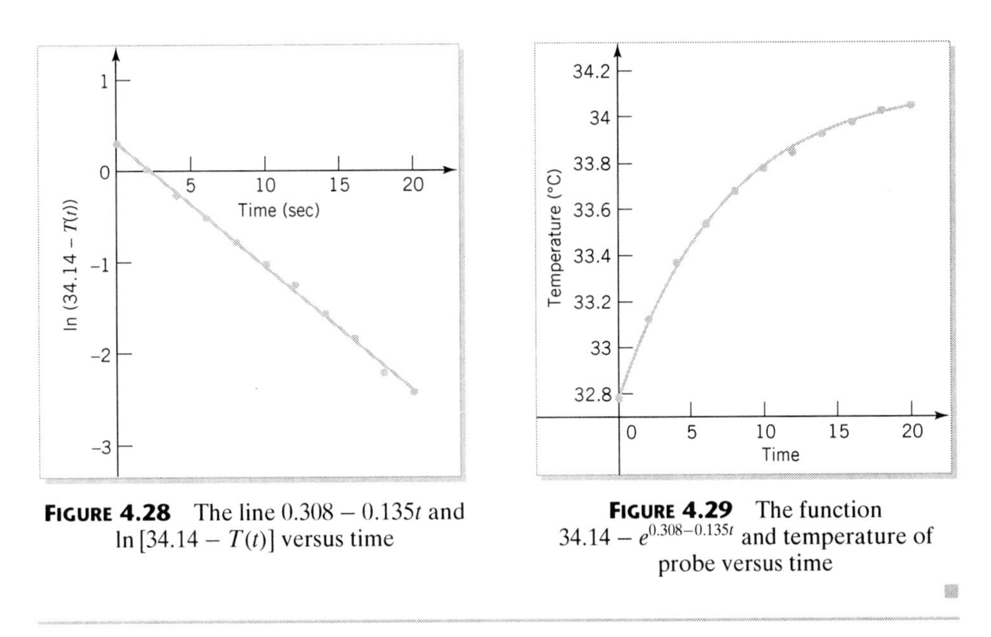

FIGURE 4.28 The line $0.308 - 0.135t$ and $\ln[34.14 - T(t)]$ versus time

FIGURE 4.29 The function $34.14 - e^{0.308-0.135t}$ and temperature of probe versus time

EXAMPLE 4.8 *The Population of Kenya*[12] *(1950–1990)*

In recent years the population of Kenya has grown rapidly, as can be seen from Table 4.6 and Figure 4.30. Can we find a differential equation that models this growth? Once we find this, its solution, subject to an appropriate initial condition, will give us a formula that can be used to predict the population at future times.

Table 4.6
Population of Kenya

Year	Population (in millions)
1950	6.265
1955	7.189
1960	8.332
1965	9.749
1970	11.498
1975	13.741
1980	16.632
1985	20.353
1990	25.130

If we look carefully at Table 4.6, we see that the 1950 population of 6.265 million doubled somewhere between 1970 and 1975 (taking between 20 and 25 years), and then doubled again by 1990 (taking between 15 and 20 years). Thus, the doubling time is not constant, but is decreasing. This population growth therefore cannot satisfy the type of simple differential equation we used to model Botswana's population growth, denoted by P – namely, $dP/dt = kP$ – because that differential equation results in a constant doubling time.

In modeling population growth, the common yardstick is the population growth per unit population – that is, the quantity $(1/P)(dP/dt)$. For example, in the case of Botswana,

$$\frac{1}{P}\frac{dP}{dt} = k. \tag{4.38}$$

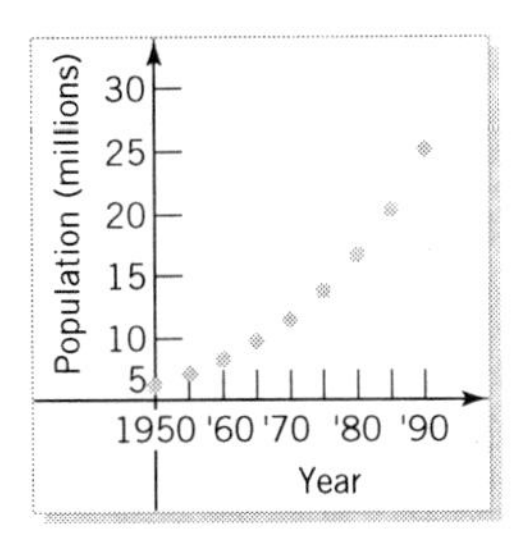

FIGURE 4.30
Population of Kenya

We can discover what information the data set for Kenya can give us about $(1/P)(dP/dt)$ by approximating this quantity by one of the numerical derivatives given on pages 145 and 146 – say, the central difference quotient – in which case

$$\frac{1}{P}\frac{dP}{dt} \approx \frac{1}{P(t)}\frac{\Delta_C P}{\Delta t} = \frac{1}{P(t)}\frac{P(t+h) - P(t-h)}{2h}.$$

Table 4.7 shows the result of these numerical calculations.

[12] *World Population Growth and Aging* by N. Keyfitz, University of Chicago Press, 1990, page 137.

Table 4.7 Estimating $(1/P)dP/dt$ for Kenya

t	$P(t)$	$[P(t+5)-P(t-5)]/(10P(t))$
1950	6.265	
1955	7.189	0.028752
1960	8.332	0.030724
1965	9.749	0.032475
1970	11.498	0.034719
1975	13.741	0.037363
1980	16.632	0.039755
1985	20.353	0.041753
1990	25.130	

In Figure 4.31 we have plotted the numerical approximation for $(1/P)(dP/dt)$ against P, and we see no obvious straight-line approximation. In Figure 4.32 we have plotted the numerical approximation for $(1/P)(dP/dt)$ against t, and we see a striking linear relationship, which we may characterize by the straight line $0.026227 + 0.000443t$.

Thus, our findings suggest that Kenya's population growth is similar to Botswana's, (4.38), except that the growth rate k is not constant but is a linearly increasing function of time, yielding the differential equation

$$\frac{1}{P}\frac{dP}{dt} = at + b,$$

Integrate

where a and b are constants ($a > 0$). This is a separable differential equation, with implicit solution

$$\ln P(t) = \frac{1}{2}at^2 + bt + C. \tag{4.39}$$

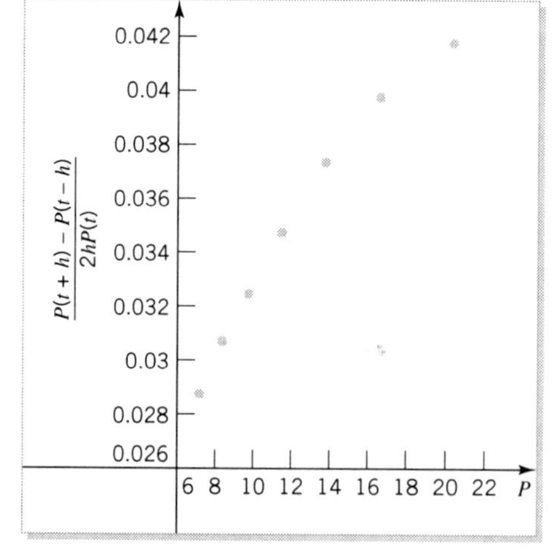

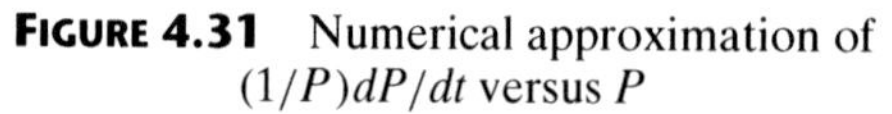

FIGURE 4.31 Numerical approximation of $(1/P)dP/dt$ versus P

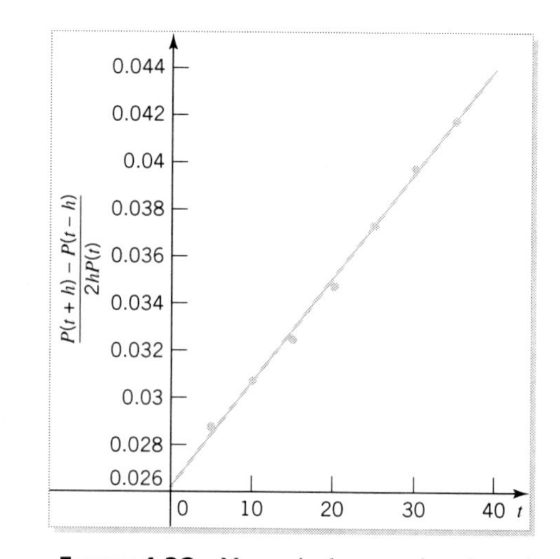

FIGURE 4.32 Numerical approximation of $(1/P)dP/dt$ versus t

If we let $t = 0$ correspond to the year 1950 and put $t = 0$ in (4.39), we find $C = \ln P(0)$, so that

$$\ln P(t) = \frac{1}{2}at^2 + bt + \ln P(0). \tag{4.40}$$

Explicit solution

From (4.40) we find

$$P(t) = P(0)e^{\frac{1}{2}at^2 + bt}. \tag{4.41}$$

This is exponential growth, but here the exponent is quadratic in t rather than being linear in t, as in Botswana's case.

Check model against data

Now let's see how this fits the actual data for Kenya. Equation (4.41) contains three unknowns, $P(0)$, a, and b, which we have to estimate. We already have the initial estimates based on Figure 4.32, of $a \approx 0.000443$ and $b \approx 0.026227$, but we should be able to find better estimates now by comparing the data with the exact solution rather than with the differential equation. (Why?) In order to obtain a linear equation, we rewrite (4.40) in the form

$$\frac{\ln P(t) - \ln P(0)}{t} = \frac{1}{2}at + b.$$

Thus, plotting $[\ln P(t) - \ln P(0)]/t$ against t should yield a straight line from which we could estimate the slope, $a/2$, and the vertical intercept, b. However, this means we must first know $P(0)$. If, from Table 4.6, we accept $P(0) = 6.265$, then Figure 4.33 shows $[\ln P(t) - \ln P(0)]/t$ plotted against t together with the straight line $y(t) = 0.026434 + 0.00203t$. This gives $a \approx 2\,(0.00203) = 0.00406$ and $b \approx 0.026434$, which are consistent with our earlier crude estimates. So we have the population of Kenya growing according to

$$P(t) = 6.265 \exp(0.000203t^2 + 0.026434t). \tag{4.42}$$

Figure 4.34 shows the population as well as the graph of the function (4.42) (adjusted so $t = 0$ corresponds to the year 1950), which seem in good agreement.

This model has the property that the population doubling time is not constant but is getting smaller as time progresses. Thus, in 1990 the population of 25 million is predicted to double in 15 years. A century later, in the year 2090, the population of Kenya is predicted to double in about 8 years. The population at that time is predicted to be about 13.5 billion, which is over twice the 1990 world population of 5.2 billion. In 2098 the population of Kenya is predicted to be 27 billion. In the year 2153 the population density of Kenya is predicted to be about one person per square foot. (The land areas of Kenya and Botswana are approximately the same, about the area of Texas.)

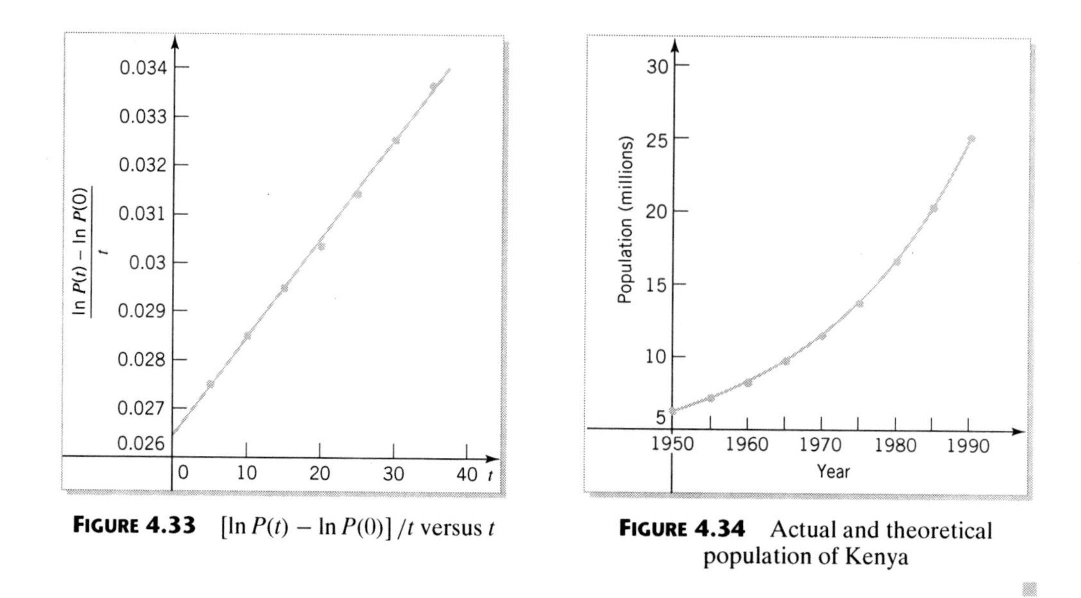

FIGURE 4.33 $[\ln P(t) - \ln P(0)]/t$ versus t

FIGURE 4.34 Actual and theoretical population of Kenya

There are a variety of reasons why we would not expect a simple population model for any country to predict accurately far into the future. Among these are the inability to predict future discoveries (for example, starting in 1970 the widespread use of "the pill," a birth control device, drastically reduced the birthrate in England); the inability to predict future diseases (for example, the AIDS epidemic has not yet had a major impact on published population figures, but will definitely change the deathrate); the inability to predict wars (for example, the Second World War raised the deathrate in several countries); and the effect of feedback, whereby the prediction of a population explosion changes the attitudes, habits, or behavior of a society, thereby lowering its birthrate (for example, China's decision to limit each couple to having only one child had a definite impact on the birthrate).

EXERCISES

1. **The Cooling of Coffee.** Some students measured the temperature of coffee to see how rapidly it cooled in a room that was 24°C. The coffee temperature in degrees centigrade, taken at 1 minute intervals, is given in Table 4.8. This data set is plotted in Figure 4.35. Let T be the temperature of the coffee in °C at time t in minutes.

Table 4.8 The temperature of coffee as a function of time

Time	Temp.	Time	Temp.
0	82.3	6	69.3
1	79.7	7	67.0
2	77.4	8	65.4
3	75.1	9	63.4
4	73.2	10	62.1
5	70.8		

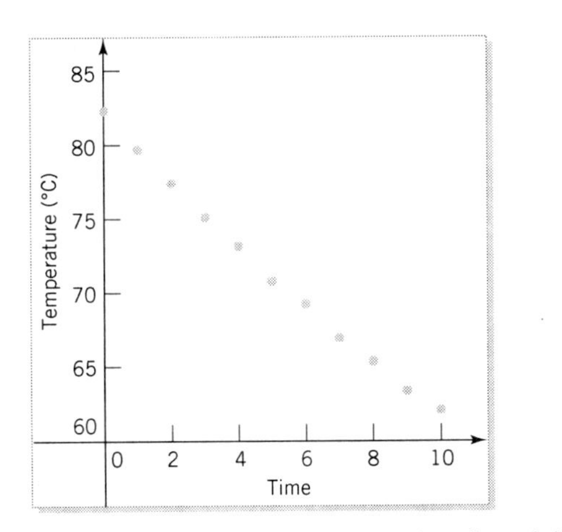

FIGURE 4.35 Temperature of coffee as a function of time

(a) Plot the approximate numerical value for dT/dt (by the central difference quotient) against the time t. This plot appears approximately linear. If this is accurate, it would mean that the cooling of coffee is governed by a differential equation of the type $dT/dt = \alpha t + \beta$, where α and β are constants. What is the sign of α? Solve this differential equation and describe what happens to T as $t \to \infty$. Is this a reasonable model for the cooling of coffee?

(b) Plot the approximate numerical value for dT/dt (by the central difference quotient) against the temperature T. This plot appears approximately linear. If this is accurate, it would mean that the cooling of coffee is governed by a differential equation of the type $dT/dt = aT + b$, where a and b are constants. What is the sign of a? Solve this differential equation and describe what happens to T as $t \to \infty$. Is this a reasonable model for the cooling of coffee?

(c) Expand the explicit solution you obtained in part (b) in a Taylor series about the origin. Explain how this Taylor series is related to the explicit solution you obtained in part (a).

2. Saving Money.[13] Five years ago Joe quit smoking. Before quitting, he smoked an average of two packs a day, which cost him about \$30 a week. For the past five years he put the \$30 that he saved into a savings account earning a fixed annual interest rate, compounded continuously. The balance of the account for the past five years is shown in Table 4.9 and Figure 4.36. Let t represent the number of years since Joe quit smoking, so that $t = 0$ represents five years ago. Let P represent the account balance after t years, and let $P' = dP/dt$.

(a) Plot the approximate numerical value of P' (by the central difference quotient) against the time t. This plot is approximately linear. If this is accurate, it would mean that the change in Joe's balance is governed by a differential equation of the type $P' = \alpha t + \beta$, where α and β are constants. From your plot, estimate the values of α and β. Using these values for α and β, solve the differential equation. Is this a reasonable model? What does this model predict the account balance will be after 40 years when Joe is about to retire?

(b) Plot the approximate numerical value of P' (by the central difference quotient) against the account balance P. This plot is approximately linear. If this is accurate, it would mean that the change in Joe's balance is governed by a differential equation of the type $P' = aP + b$, where a and b are constants. From your plot, estimate the values of a and b. Using these values for a and b, solve the differential

Table 4.9
Saving money

Year	Balance
0	0
1	1,639
2	3,451
3	5,453
4	7,620
5	10,061

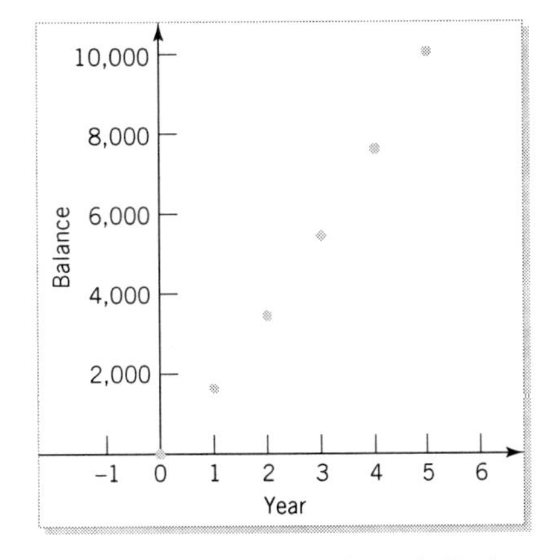

FIGURE 4.36 The balance in Joe's account for five years

[13] This exercise is based on a suggestion from Jeffrey Igo of the Henry Ford Community College, Dearborn, Michigan.

equation. Is this a reasonable model? What does this model predict the account balance will be after 40 years when Joe is about to retire?

(c) Joe's younger brother, Bill, who also smoked an average of two packs a day, decided to quit one year after Joe quit, and invest his \$30 a week in exactly the same way that Joe did. How much will Bill have in his account after one year? On Figure 4.36 plot Bill's balance over his five years, where $t = 1$ is when Bill started with \$0. Joe's older brother, Pete, quit one year before Joe did, and invested his \$30 savings in the same way that Joe did later — in fact, Joe got the idea from Pete. On Figure 4.36 plot Pete's balance over his five years, where $t = -1$ is when Pete started with \$0. Explain how, from Joe's graph, you can obtain both Bill's and Pete's graphs. Is this property characteristic of an equation of the type $P' = g(t)$ or $P' = g(P)$? Which of the differential equations in parts (a) and (b) is the better model?

(d) Expand the explicit solution you obtained in part (b) in a Taylor series about the origin. Explain how this Taylor series is related to the explicit solution you obtained in part (a).

3. At noon a cup of coffee at 180°F is placed in a room where the air temperature is 70°F. After 30 minutes the temperature of the coffee is 120°F. At this time the coffee is placed in an oven, which was preheated to 400°F, until the coffee is once more at 180°F. Assume that Newton's law of cooling (heating) applies to both time periods.

(a) Construct a separate phase line for each of these two time periods and then sketch the temperature of the coffee as a function of time for the entire time period.

(b) Find the analytical solution to this problem, and compare the graph of this solution with your sketch.

(c) At what time does the temperature of the coffee in the oven reach 180°F?

4. Estimating Parameters. If we have a data set that is to be modeled by the exponential law $y(x) = ae^{bx}$, the standard way of estimating the parameters a and b is to rewrite it in the form $\ln y = \ln a + bx$, and plot $\ln y$ against x. This idea can be extended to $y(x) = c + ae^{bx}$, by rewriting it in the form $y(x) - c = ae^{bx}$, if we can first estimate c. This exercise shows three ways to estimate c.

(a) Draw a smooth curve through the data set and select three points, (x_1, y_1), (x_2, y_2), and (x_3, y_3), on the curve where x_1 and x_2 are selected arbitrarily, but $x_3 = \frac{1}{2}(x_1 + x_2)$. Show that

$$c = \frac{y_1 y_2 - y_3^2}{y_1 + y_2 - 2y_3}.$$

(b) If $b < 0$, what happens to e^{bx} as $x \to \infty$? How does this help estimate c?

(c) This third way also applies to the case $b < 0$ but relies heavily on the data set being recorded at equal x intervals, say, h units apart. From $y(x) = c + ae^{bx}$ we have $y(x+h) = c + ae^{b(x+h)} = c + ae^{bx}e^{bh}$. If we solve $y(x) = c + ae^{bx}$ for ae^{bx}, we have $ae^{bx} = y(x) - c$, so $y(x+h) = c + [y(x) - c]\,e^{bh} = y(x)e^{bh} + c\left[1 - e^{bh}\right]$. Thus, if we plot $y(x+h)$ against $y(x)$ for all the data points, we should see a straight line. Someone says that where this line crosses the line through the origin with slope 45° will give you an estimate of c. Why is this true? [Hint: As $x \to \infty$, what happens to $y(x+h)$ and $y(x)$?] Use this technique to estimate T_a in Example 4.7 on page 145.

5. Explain how the three techniques described in Exercise 4 can be used to estimate the carrying capacity b of the logistic equation $y' = ay(b - y)$, by rewriting the solution (2.26) on page 46 — namely,

$$y(x) = \frac{bC}{e^{-abx} + C},$$

in the form

$$\frac{1}{y(x)} = \frac{1}{bC}e^{-abx} + \frac{1}{b}.$$

6. Expanding Pupil. Table 4.10 shows the result of an experiment in which 10 subjects were briefly exposed to a bright light, causing their pupils to dilate, and then the diameter of each pupil, D, in millimeters was measured every 0.5 seconds as the pupil expanded.[14]

(a) A model proposed for this phenomena is $dD/dt = aD + b$, where t is the time in seconds, and a and b are constants.

i. What is the equilibrium solution of this differential equation? Show that this differential equation can be rewritten in terms of its equilibrium solution, D_e, as $dD/dt = a\left(D - D_e\right)$. Explain why this differential equation is a reasonable model. What physical meaning can be attached to D_e? What is the sign of a?

[14]Adapted from *Quantitative Methods in Psychology* by D. Lewis, McGraw-Hill, New York, 1960.

Table 4.10 Average diameter D in millimeters of an expanding pupil at time t in seconds

t	1.0	1.5	2.0	2.5	3.0	3.5	4.0	4.5	5.0	5.5	6.0	6.5	7.0
D	4.31	4.77	5.06	5.23	5.35	5.60	5.66	5.76	5.86	5.94	5.94	6.00	6.04

ii. What other phenomena does this equation model?

iii. Solve this differential equation. How many parameters are there in this solution? Using the ideas from Exercise 4, estimate these parameters for the data set in Table 4.10. How well does this model fit the data set?

iv. What does this model predict about the ultimate diameter of the pupil?

(b) Another model proposed for this phenomena is $dA/dt = aA + b$, where $A = \pi D^2/4$ is the area of the pupil, t is the time in seconds, and a and b are constants.

i. What is the equilibrium solution of this differential equation? Show that this differential equation can be rewritten in terms of its equilibrium solution, A_e, as $dA/dt = a(A - A_e)$. Explain why this differential equation is a reasonable model. What physical meaning can be attached to A_e? What is the sign of a?

ii. What other phenomena does this equation model?

iii. Solve this differential equation. How many parameters are there in this solution? Using the ideas from Exercise 4, estimate these parameters for the data set in Table 4.10. How well does this model fit the data set?

iv. What does this model predict about the ultimate diameter of the pupil?

(c) Which of the models proposed in parts (a) and (b) do you think is more reasonable?

7. **The Length of a Sturgeon.** A biologist measured the length of a sturgeon over a 21 year period.[15] Table 4.11 shows the resulting data set where the length L is in centimeters and the time t is in years. This data set is plotted in Figure 4.37. A model proposed for this phenomena is $dL/dt = \alpha + kL$, where α and k are constants.

Table 4.11 Length of sturgeon as a function of time

Time (years)	Length (cm)	Time (years)	Length (cm)
0	21.1	11	107.6
1	32.0	12	112.7
2	42.3	13	117.7
3	51.4	14	122.2
4	60.1	15	126.5
5	68.0	16	130.9
6	75.3	17	135.3
7	82.3	18	140.2
8	89.0	19	145.0
9	95.3	20	148.6
10	101.6	21	152.0

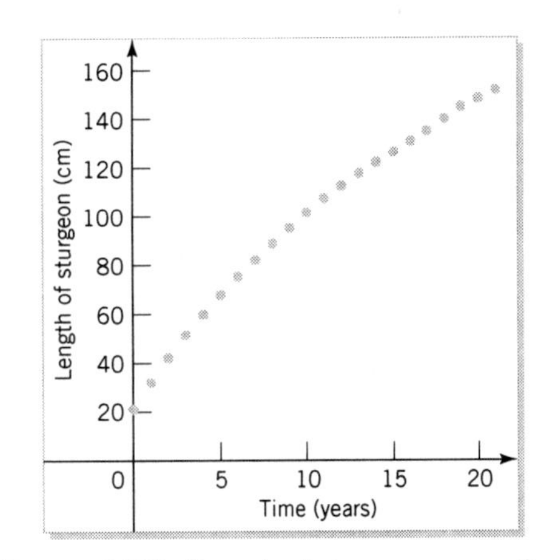

FIGURE 4.37 Length of sturgeon versus time

(a) What is the equilibrium solution of this differential equation? Show that this differential equation can be rewritten in terms of its equilibrium solution, L_e, as $dL/dt = a(L_e - L)$, where $a = \alpha/L_e$. Explain why this differential equation is a reasonable model. What physical meaning can be attached to L_e? What is the sign of a?

(b) What other phenomena does this equation model?

[15] *General System Theory* by L. von Bertalanffy, Braziller, 1968, page 177, Table 7.5.

(c) Solve this differential equation. How many parameters are there in this solution? Estimate these parameters for the data set in Table 4.11. How well does this model fit the data set?

8. **Air Drag on a Car — A Simple Experiment.**[16] Drive a car with manual transmission on a long straight horizontal road at 55 mph — assuming that speed is legal. Take the car out of gear and let it roll to a stop. Record the time at which the car is traveling at 50, 45, $\cdots$, 10, 5, and 0 mph. Table 4.12 and Figure 4.38 show the results of such an experiment. It is believed that the motion of the car is determined by the differential equation $v' = -a - bv^2$. Here a and b are positive constants, $v = dx/dt$, and x is the distance the car has traveled measured from when the car was taken out of gear. The term a accounts for the rolling friction force, and bv^2 is the air drag on the car. From the data, plot the approximate numerical value for dv/dt (by the central difference quotient) against v^2. The data set may be approximately linear. Use this to estimate a and b. Now plot the solution of the differential equation, using these estimated values for a and b with the data. How well do your data fit this model? Would you expect the constants a and b to be the same for all cars? Explain.

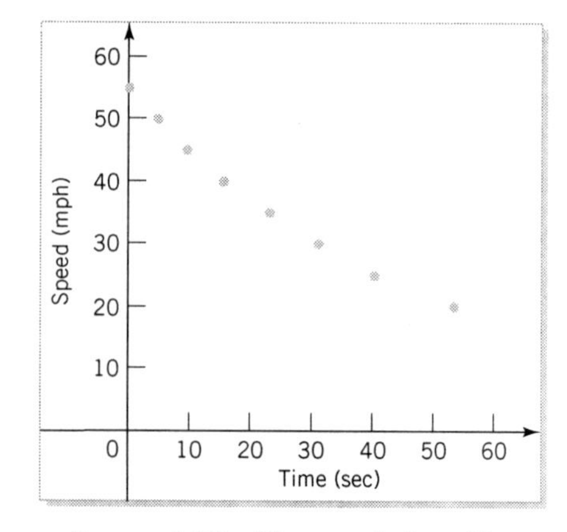

FIGURE 4.38 The speed of a rolling car as a function of time

Table 4.12 The speed of a rolling car as a function of time

Time (seconds)	Speed (mph)
0.0	55
4.7	50
9.6	45
15.6	40
23.1	35
31.1	30
40.3	25
53.3	20

9. **Other Laws of Cooling.**[17] Several other laws of cooling have been proposed in place of Newton's law of cooling, including the DULONG-PETIT LAW, $T' = k\left(T - T_a\right)^n$, where n is a constant between 1.25 and 1.6; and the NEWTON-STEFAN LAW, which can be approximated by $T' = k\left(T - T_a\right) + h\left(T - T_a\right)^2$, where k and h are negative constants.

(a) Use a phase line analysis to explain why these are reasonable alternatives to Newton's law of cooling.

(b) Solve both of these equations for $T(t)$ explicitly.

10. **AIDS.** Table 4.13 shows the total number of cases of AIDS in the United States from 1982 in three-month intervals.[18] It is possible to analyze the data set in a manner similar to the way we analyzed the population of Kenya. We can show that the same differential equation — namely, $P' = (at + b)\,P$, where here $a < 0$ and $b > 0$ are selected constants — fits the data set well. (Do not do this.) A particular solution of this differential equation and the data set is plotted in Figure 4.39 where this fit can be seen. In the case of Kenya's population, a is positive, whereas in this case a is negative. What impact will $a < 0$ have on predicting the subsequent number of cases of AIDS?

---

[16] "Determining the air drag on a car" by J. E. Farr, *The Physics Teacher*, May 1983, pages 320–321.

[17] "Newton's law of cooling — A critical assessment" by C. T. O'Sullivan, *The American Journal of Physics* 58, 1990, pages 956–960.

[18] "Backcalculation of flexible linear models of the Human Immunodeficiency Virus infection" by P. S. Rosenburg and M. H. Gail, *Applied Statistics* 40, 1991, pages 269–282, Table 1, as reported in *Growth and Diffusion Phenomena* by R. T. Banks, Springer-Verlag, 1991, page 372.

Table 4.13 Total number of cases of AIDS in the U.S.A. from 1982

Quarter since 1982	Number	Quarter since 1982	Number
0	374	13	12260
1	559	14	14785
2	759	15	17736
3	1052	16	20896
4	1426	17	24715
5	1980	18	29036
6	2693	19	33899
7	3456	20	39091
8	4313	21	45246
9	5460	22	52062
10	6829	23	59553
11	8392	24	67279
12	10118	25	75762

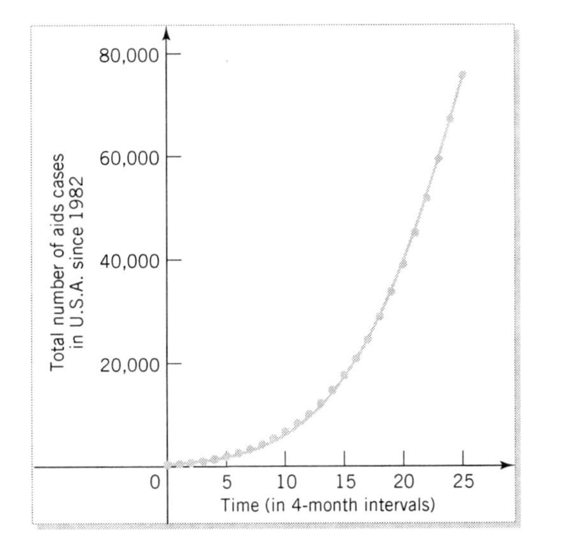

FIGURE 4.39 The total number of cases of AIDS in the U.S.A. from 1982 and a solution of $dP/dt = (at + b)P$

11. **Time of Death.** A rule of thumb used during homicide investigations is that if the temperature of the body of an average-sized person is within 1% of the room temperature, the person has been dead for at least 24 hours. Assume that at the time of death a person's body temperature was 98.6°F and that Newton's law of cooling (4.35) applies; that is, $dT/dt = k(T - T_a)$, where $k < 0$ and T_a is the room temperature. If the room temperature was a constant 72°F, what can you say about the limitations of k for an average-sized person?

12. To apply the central difference quotient (4.34), the data set must be sampled at equal time intervals. If the data set is sampled at unequal time intervals, what formula would you use in place of (4.34)?

13. Show that if $T(t)$ is a linear function of t — that is, $T(t) = a + bt$ — then

$$\frac{dT}{dt} = \frac{T(t+h) - T(t)}{h}$$

for all $h \neq 0$. What does this suggest about accuracy if you use the right-hand quotient to estimate the derivative of a linear function? A nonlinear function?

14. Show that if $T(t)$ is a quadratic function of t — that is, $T(t) = a + bt + ct^2$ — then

$$\frac{dT}{dt} = \frac{T(t+h) - T(t-h)}{2h}$$

for all $h \neq 0$. What does this suggest about accuracy if you use the central difference quotient to estimate the derivative of a quadratic function? A nonquadratic function?

15. Show that if $T(t)$ is an exponential function of t — that is, $T(t) = ce^{at}$ — then

$$\frac{1}{T(t)}\frac{dT}{dt} = \frac{\ln T(t+h) - \ln T(t)}{h}$$

for all $h \neq 0$. What does this suggest about accuracy if you use the right-hand or central difference quotient to estimate the derivative of an exponential function?

16. Show that if $T(t)$ is an exponential function of t^2 — that is, $T(t) = ce^{at+bt^2}$ — then

$$\frac{1}{T(t)}\frac{dT}{dt} = \frac{\ln T(t+h) - \ln T(t-h)}{2h}$$

for all $h \neq 0$. What does this suggest about accuracy if you use the right-hand or central difference quotient to estimate the derivative of such an exponential function?

17. Show that if $T(t) = c + ae^{bt}$, then the ratio of the central difference quotient $\frac{\Delta_c T}{\Delta t} = \frac{T(t+h)-T(t-h)}{2h}$ to the derivative $T'(t)$, can be written as

$$\frac{1}{T'(t)}\frac{\Delta_c T}{\Delta t} = \frac{e^{bh} - e^{-bh}}{2bh}.$$

If T_H is the half-life of ae^{bt} — so $bT_H = \ln\frac{1}{2}$ — and if $2h = \lambda T_H$ (that is, the data are collected at equal time intervals λT_H), show that

$$\frac{1}{T'(t)}\frac{\Delta_c T}{\Delta t} = \frac{e^{(\lambda/2)\ln 2} - e^{-(\lambda/2)\ln 2}}{\lambda \ln 2}.$$

Someone says that if a data set is modeled exactly by $T(t) = c + ae^{bt}$ and the data are collected at equal time intervals T_H (that is, at one half-life intervals), then the error in using $\frac{\Delta_c T}{\Delta t}$ to estimate T' is about 2%. Justify this statement.

4.4 MODELS: OBJECTS IN MOTION

In this section we discuss applications of differential equations used to model objects in motion and compare the models with real data sets.

EXAMPLE 4.9 *Skydiving*

In calculus you most likely saw the differential equation that modeled an object falling under the influence of gravity, obtained by using Newton's second law of motion (mass × acceleration = applied force) as

$$m\frac{dv}{dt} = mg, \tag{4.43}$$

where m is the mass of the object, v is velocity (positive velocities, or increasing distances, associated with downward movements), and g is acceleration due to gravity.

Table 4.14 and Figure 4.40 show data obtained from the United States Parachute Association that correspond to a sky diver in free fall in a stable spread-eagle position when falling from rest. It shows the sky diver's distance fallen (in feet) as a function of the time (in seconds). The third column in Table 4.14 contains the central difference quotients of the distance with respect to time, which is a numerical approximation of the velocity.

From the third column of Table 4.14 (or from the data points in Figure 4.40, which seem to lie along a straight line as you move to the right), we conclude that the velocity — the rate of change of distance x — is approaching a constant value as t increases. Thus, the rate of change of velocity with respect to time approaches zero for larger times, giving rise to what is commonly called a TERMINAL VELOCITY.

Terminal velocity

However, if we look at the slope field for (4.43) [or use isoclines; see *How to Analyze* $y' = g(x, y)$ *Graphically* on page 84], we observe that the rate of change of the velocity in this differential equation $mdv/dt = mg$ is always a nonzero constant. Thus, (4.43) does not model these data. Missing in the differential equation is a term to account for air resistance.

Table 4.14 **Distance fallen (in feet) against time (in seconds)**

Time	Distance	Difference Quotient	Time	Distance	Difference Quotient
0	0				
1	16	31.0	7	652	152.0
2	62	61.0	8	808	159.5
3	138	90.0	9	971	165.0
4	242	114.0	10	1138	169.0
5	366	131.0	11	1309	172.5
6	504	143.0	12	1483	

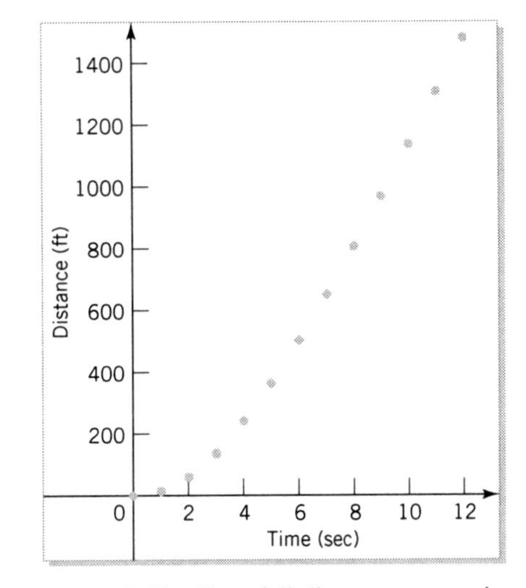

FIGURE 4.40 Free-fall distance versus time

To obtain a better model, we consider a more general situation in which the object of mass m, besides moving under the influence of an external force mg, is subjected to a frictional force, the air resistance. A widely quoted rule of thumb concerning this frictional force is that for a slow-moving object (like a ball bearing falling in honey), the resistance is proportional to the velocity, whereas for a fast-moving object (like a ball bearing falling in air), resistance is proportional to the square of the velocity. The second case is appropriate for skydiving, and so Newton's second law of motion gives

$$m\frac{dv}{dt} = mg - kv|v|, \tag{4.44}$$

where $-kv|v|$ represents this frictional force ($k > 0$). The reason for the term $v|v|$ rather than v^2 is that the air resistance always opposes the motion. If $v < 0$, a term like v^2 would be in the same direction as the motion and therefore aid it. We will concentrate on the case in which the sky diver falls downward from the aircraft — that is, $v \geq 0$. In this case, (4.44) reduces to

$$\frac{dv}{dt} = g - \frac{k}{m}v^2, \tag{4.45}$$

where $k > 0$.

Equilibrium solution

We observe that (4.45) has equilibrium solutions — namely, when $g = kv^2/m$. Because we are concerned only with positive values of v, we denote the positive equilibrium solution by V, so we have

$$V = \sqrt{\frac{mg}{k}}. \tag{4.46}$$

Because V is a quantity that can be measured, we replace m/k in (4.45) by V^2/g which results in

$$\frac{V^2}{g}\frac{dv}{dt} = V^2 - v^2. \tag{4.47}$$

(This will also simplify future algebraic calculations.)

Phase line

Because this is an autonomous differential equation — the right-hand side is independent of t — we do a phase line analysis, giving Figure 4.41. Here we see clearly that V is a stable equilibrium. Thus, for the usual situation where the sky diver's initial velocity v_0 is less than V, we see that the sky diver's velocity increases with time but never reaches the velocity V in finite time. (Why?) The velocity V is the terminal velocity.

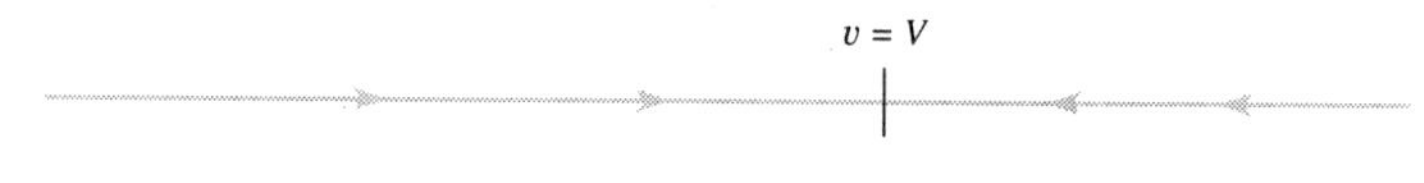

FIGURE 4.41 Phase line analysis for $(V^2/g)dv/dt = V^2 - v^2, v \geq 0$

Initial condition

The data set in Table 4.14 was obtained for a sky diver falling from rest. If we measure the time t from time $t = 0$, we have the initial condition $v(0) = 0$.

To solve (4.47) for nonequilibrium solutions, we write it in the form

$$\frac{1}{V^2 - v^2}\frac{dv}{dt} = \frac{g}{V^2}. \tag{4.48}$$

We use partial fractions on the left-hand side of (4.48) to obtain

$$\frac{1}{2V}\int\left(\frac{1}{V-v} + \frac{1}{V+v}\right)dv = \int\frac{g}{V^2}\,dt.$$

Integrate

Integration leads to

$$\frac{1}{2V}\ln\left(\frac{V+v}{V-v}\right) = \frac{gt}{V^2} + C.$$

From the initial condition, $v(0) = 0$, we have that $C = 0$, and our solution may be written as

$$\frac{V+v}{V-v} = e^{2gt/V}. \tag{4.49}$$

Because $2g/V$ is a constant, to simplify this expression we replace $2g/V$ by another constant, α, with $\alpha = 2g/V$, so (4.49) can be written as

$$\frac{V+v}{V-v} = e^{\alpha t}.$$

Explicit solution

Solving this for v, we find

$$v(t) = V\frac{e^{\alpha t}-1}{e^{\alpha t}+1}. \tag{4.50}$$

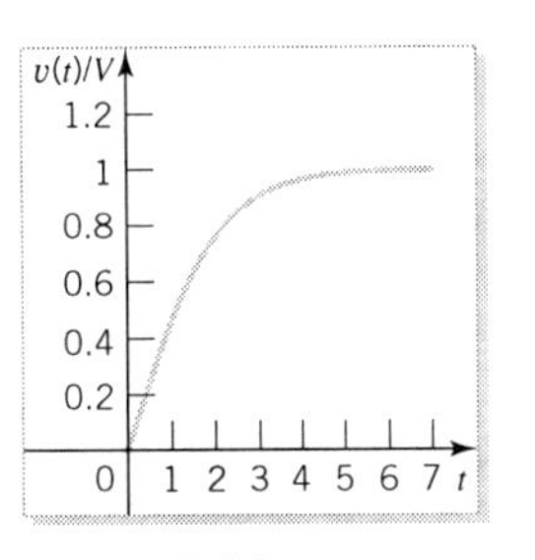

FIGURE 4.42
The velocity $(e^t - 1)/(e^t + 1)$

Notice that $v(t) \to V$ as $t \to \infty$, in complete agreement with our understanding that V is the terminal velocity. Figure 4.42 shows $v(t)/V$ from (4.50) for the case $V = 2g$ (so $\alpha = 1$).

If we denote the distance that the body has fallen from its initial position by x (positive being downward), we see that (4.50) is

$$\frac{dx}{dt} = V\frac{e^{\alpha t}-1}{e^{\alpha t}+1}. \tag{4.51}$$

By using the substitution $u = e^{\alpha t} + 1$, we can integrate (4.51), subject to the initial condition $x(0) = 0$. Doing so gives

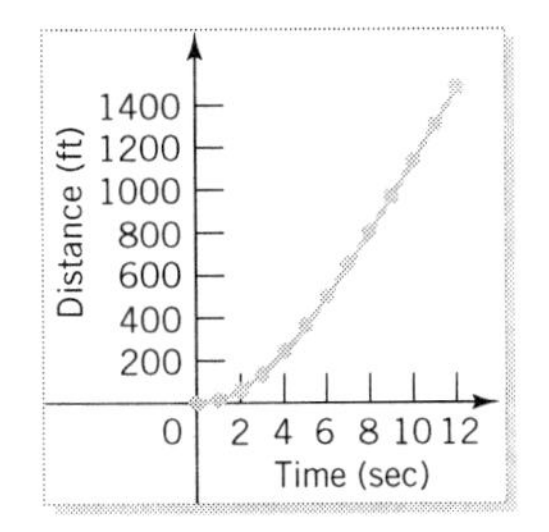

FIGURE 4.43
The data set and the theoretical function

$$x(t) = \frac{2V}{\alpha}\ln\left[\frac{1}{2}\left(e^{\alpha t}+1\right)\right] - tV \tag{4.52}$$

as the distance fallen as a function of the time t.

If we now return to the original data set, there is only one constant we need to estimate — namely, V — because we know $g \approx 32.2$ ft/sec^2. Table 4.14 suggests a terminal velocity in excess of 172.5 ft/sec. Figure 4.43 shows the data set and the function (4.52) for one such choice of V; namely, $V = 182$ ft/sec.

EXAMPLE 4.10 *Braking in an Emergency Stop*

A number of experiments have been performed that measure the distance a car travels when braking in an emergency stop as a function of the velocity of the car at the time the brakes are applied. Typical results are shown in Table 4.15 and presented graphically in Figure 4.44. Someone claims that these results are in good agreement with the model of a vehicle coming to rest under constant deceleration. Is that person right?

Table 4.15

Observed car-stopping distances

Velocity (mph)	Stopping distance (feet)
10	5.6
15	12.8
20	21.0
25	32.2
30	48.8
35	63.4
40	83.9
45	113.5
50	133.0
55	166.2
60	185.7
65	233.5

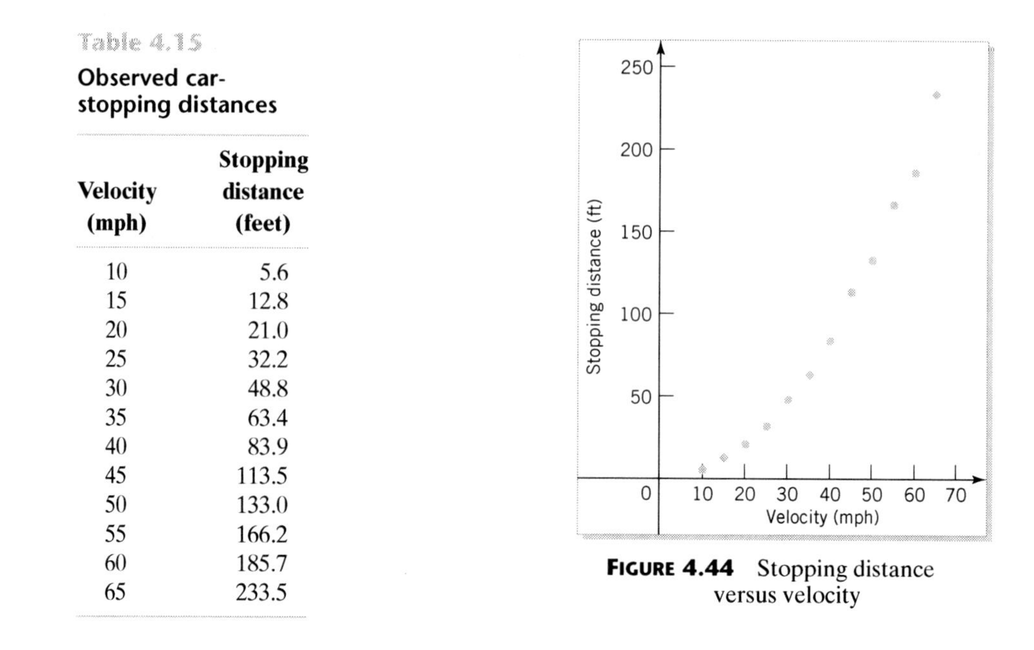

FIGURE 4.44 Stopping distance versus velocity

If we let $x(t)$ represent the distance in feet the vehicle travels from the time when the brakes are applied [say at $t = 0$, so $x(0) = 0$], then the velocity, $v(t)$, and the acceleration, $a(t)$, of the vehicle are given by

$$v(t) = \frac{dx}{dt}, \text{ and } a(t) = \frac{dv}{dt}. \tag{4.53}$$

The assumption of constant deceleration is the statement that

$$a(t) = -k, \tag{4.54}$$

where k is a positive constant. If we take the units of time as seconds, then the units of k will be ft/sec^2. In view of (4.53), we see that (4.54) can be written as the differential equation

$$\frac{dv}{dt} = -k. \tag{4.55}$$

Although we could solve this to find $v(t)$, what we seek is not a relationship between v and t, but a relationship between v and x. (Why?) If we use the chain rule, we can write

$$\frac{dv}{dt} = \frac{dv}{dx}\frac{dx}{dt} = \frac{dv}{dx}v,$$

where we have used (4.53). Thus, (4.53) can be written

$$\frac{dv}{dx}v = -k,$$

Integrate

which has the solution $v^2(x) = -2kx + C$. If we let V_0 be the velocity of the vehicle when the brakes are first applied at $x = 0$, then

$$v^2(x) = -2kx + V_0^2. \tag{4.56}$$

Explicit solution

If we let d be the distance the car travels before it stops (at $v = 0$), then $0 = -2kd + V_0^2$, so

$$d = \frac{1}{2k}V_0^2. \tag{4.57}$$

This tells us that the assumption of constant deceleration leads to a quadratic relationship between V_0, the initial velocity of the vehicle, and d, the distance it takes for the vehicle to stop. The constant k presumably has something to do with the vehicle. The only thing left is to see whether this agrees with Table 4.15.

Check model against data

There are at least two different ways we can see whether the data in Table 4.15 satisfy a quadratic equation like

$$y = hx^2, \tag{4.58}$$

where h is a constant.

1. Plot y/x on the vertical axis against x on the horizontal axis. If the data set is quadratic, the resulting graph will approximate a straight line through the origin with slope h. Figure 4.45 shows d/V_0 plotted against V_0. The line with slope 0.054 seems to be a good fit.

2. Plot $\ln y$ on the vertical axis against $\ln x$ on the horizontal axis. If the data set is quadratic, the resulting graph will approximate a straight line with slope 2 and with vertical intercept $\ln h$, because (4.58) can be written as $\ln y = \ln h + 2\ln x$. Figure 4.46 shows $\ln d$ plotted against $\ln V_0$. The line with slope 2 and intercept -2.92 seems to be a good fit. Note that $\ln 0.054 \approx -2.92$. (How does this relate to Figure 4.45?)

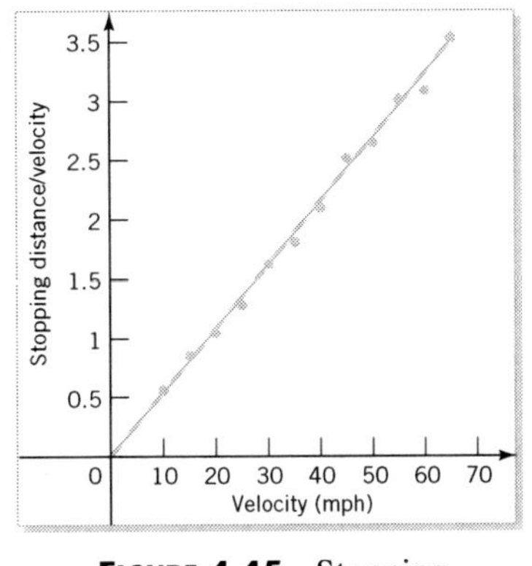

FIGURE 4.45 Stopping distance/velocity versus velocity

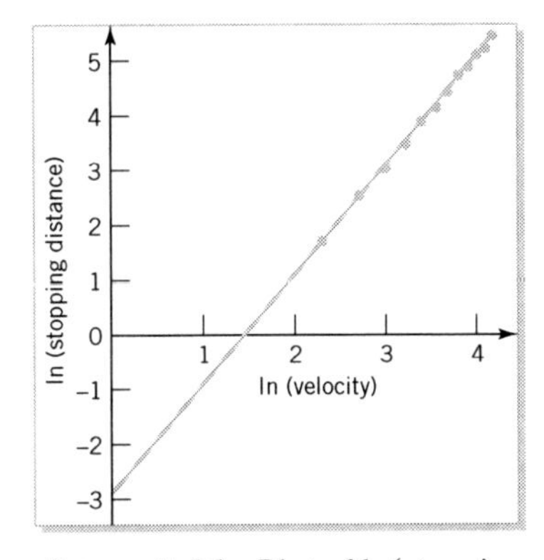

FIGURE 4.46 Plot of ln(stopping distance) versus ln(velocity)

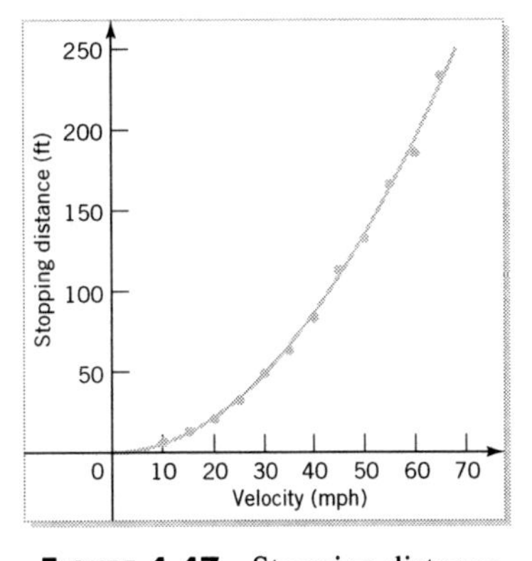

FIGURE 4.47 Stopping distance versus velocity and the quadratic function $d = 0.054v^2$

With regard to units of measure, the quantity V_0 is measured in feet per second, whereas the quantity x is measured in miles per hour, so $h \neq 1/(2k)$ but $h = (15/22)^2/(2k)$. Figure 4.47 shows the data set (Table 4.15) and the quadratic function (4.58) with $h = 0.054$.

EXERCISES

1. **Skydiving.** A sky diver falls from rest subject to the differential equation (4.48); namely,

$$\frac{1}{V^2 - v^2}\frac{dv}{dt} = \frac{g}{V^2},$$

where the terminal velocity V is 180 ft/sec. How long does it take for the sky diver to reach 90% of the terminal velocity? How far has she fallen when this happens?

2. Show that if (4.48); namely,

$$\frac{1}{V^2 - v^2}\frac{dv}{dt} = \frac{g}{V^2},$$

is solved subject to the initial conditions $v(t_0) = v_0$, $x(t_0) = x_0$, then

$$v(t) = V\frac{\beta e^{\alpha(t-t_0)} - 1}{\beta e^{\alpha(t-t_0)} + 1}$$

and

$$x(t) = x_0 + \frac{V^2}{g}\ln\left|\frac{e^{\alpha(t-t_0)} + \beta}{1+\beta}\right| - (t - t_0)V,$$

where α and β are defined by

$$\alpha = \frac{2g}{V}, \qquad \beta = \frac{V - v_0}{V + v_0}.$$

3. **Skydiving.** A sky diver falls from rest from 10,000 feet above the earth subject to the differential equation (4.48); namely,

$$\frac{1}{V^2 - v^2}\frac{dv}{dt} = \frac{g}{V^2},$$

where $g = 32.2$ ft/sec^2 and the terminal velocity V is 180 ft/sec. When he reaches 1000 feet, his parachute opens instantaneously and his terminal velocity is now 22 ft/sec.[19] Assume that the differential equation (4.48) still applies but now $V = 22$.

   (a) Construct a separate phase line for each of these two events and then sketch the velocity of the sky diver as a function of time for the entire time period.

   (b) Find the explicit solution for $v(t)$ for the first event, and compare the graph of this solution with your sketch in part (a). Use the fact that $v = dx/dt$ to find an explicit solution for $x(t)$. At what time $t = T$ does the parachute open? Use this result to determine how fast the sky diver is traveling when his parachute opens.

   (c) By using your result from part (b) and the results from Exercise 2, find the explicit solution for $v(t)$ for the second event. Compare the graph of this solution with your sketch in part (a). Also find an explicit solution for $x(t)$. Use this result to determine how fast the sky diver is traveling when he hits the ground.

4. **Reaction Time — A Simple Experiment.**[20] Assuming that a body of mass m falls under gravity according to $mx'' = mg$, where x is measured downward from the release point, show that the time t it takes for a body to fall a distance d from rest is $t = \sqrt{2d/g}$. This can be used to estimate a subject's reaction time. Take a ruler and hold it vertically so that the lower end is between the subject's thumb and index finger. Advise the subject that you are going to release the ruler at an unannounced time and the subject should catch the ruler as quickly as possible. Then, without warning, release the ruler and note how far the ruler falls before the subject catches it. Repeat a number of times. What is the subject's reaction time?

5. **Vertical Jumps in Ballet.**[21] A vertical jump performed by a ballet dancer is governed by the differential equation $mv' = -mg$. Here m is the mass of the ballet dancer,

[19] The velocity of 22 ft/sec is approximately your velocity on impact if you step off a 10-ft wall.

[20] "Ruler Physics" by R. Ehrlich, *American Journal of Physics* 62, 1994, pages 111–120.

[21] *The Physics of Dance* by K. Laws, Schirmer, 1984, pages 31–33, and Appendix A.

$g = 32.2$ ft/sec^2 is the gravitational constant, and $v' = dv/dt$, $v = dy/dt$, where y is the vertical distance measured from the ground to the center of gravity of the ballet dancer. Show that the time T that the ballet dancer is in the air during a vertical jump is related to the height H of the jump by $H = \frac{1}{8}gT^2$. When a ballet dancer performs a vertical jump, the length of time in the air is determined by the music. Consequently, the height H of the jump is determined by the tempo of the music. Show that if the music calls for vertical jumps every 1/3 second, then $H \approx 5.4$ inches for every ballet dancer, independent of height. During vertical jumps, ballet dancers are expected to point their feet. Discuss the disadvantage of being a tall ballet dancer with big feet doing vertical jumps. What happens to such ballet dancers should they choose to jump high enough to point their feet?

6. **Superterminal Velocities.**[22] In Example 4.9 on page 158, the sky diver started with a velocity smaller than the terminal velocity. Here we consider a situation where an object starts with an initial velocity greater than its terminal velocity. If an object was traveling at 30,000 m/sec and had a terminal velocity of 100 m/sec, how long would it take for the object to be traveling at 110% of its terminal velocity? How far would it have traveled in that time? What is likely to happen to such an object? Provide a realistic example of such an object.

7. In place of (4.45), $v' = g - kv^2/m$, some people have suggested that an object falling from an airplane experiences an air resistance that is linear in the velocity, namely, $v' = g - kv/m$. Find a formula that represents the terminal velocity for this equation. Also find the velocity and the distance fallen as a function of time assuming that the object falls from rest at $t = 0$ and that $x(0) = 0$. By using different values for the terminal velocity, try to match the distance as a function of time with the data set in Table 4.14. Is this model a good one?

8. **Terminal Velocity of a Styrofoam Ball.**[23] The data set in Table 4.16 and shown in Figure 4.48 was obtained by dropping a Styrofoam ball and recording its distance as a function of time. Which of the two differential equations, $mv' = mg - kv$ and $mv' = mg - kv^2$, best models the motion of the ball?

Table 4.16 Vertical distance fallen by a Styrofoam ball as a function of time

Time (seconds)	Distance (meters)	Time (seconds)	Distance (meters)
−0.132	0.000	0.400	1.270
0.000	0.075	0.500	1.730
0.100	0.260	0.600	2.230
0.200	0.525	0.700	2.770
0.300	0.870	0.800	3.350

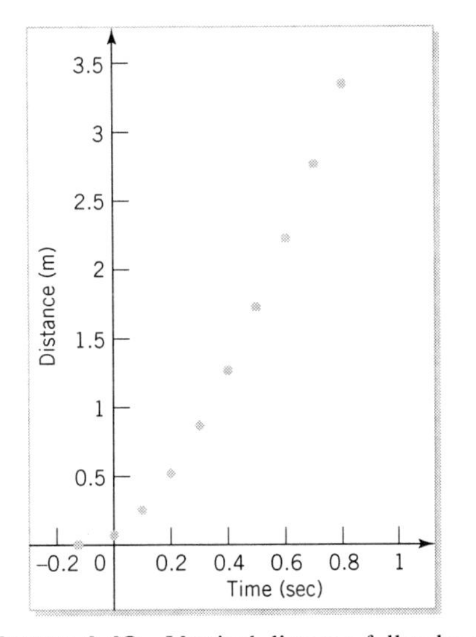

FIGURE 4.48 Vertical distance fallen by a Styrofoam ball as a function of time

9. **Just Hanging in the Air.** When slam-dunking, a basketball player seems to hang in the air at the height of his jump. This exercise gives an explanation for this phenomenon. Suppose an object of mass m is propelled vertically upwards with an initial velocity of v_0 subject to Newton's laws of motion — namely, $mv' = -mg$. Solve this equation for $v(t)$. By writing $v = dx/dt$, where x is the distance the body is above the ground, so that $x(0) = 0$, find $x(t)$. How long will it take for this object to reach its maximum height? How long will it take for the object to fall back to the ground? How long does the object spend in the top 25% of the trajectory? What percentage of the total time does the object spend in the top 25% of the trajectory?

[22] "Superterminal Velocities" by E. Zebrowski, Jr., *The Physics Teacher*, November 1989, pages 618–619.

[23] "Air Resistance Acting on a Sphere" by M. S. Greenwood, C. Hanna, and J. Milton, *The Physics Teacher*, March 1986, pages 153–159, Table V.

10. The Winchester Rifle. A student[24] shot a 120-grain hollow-point bullet from a 300 Winchester Magnum rifle. He measured the velocity, v, as a function of distance, x, and compiled the data shown in Table 4.17. This data set is plotted in Figure 4.49. We are going to construct mathematical models and analyze this data set in various ways.

(a) It is claimed that this data set looks linear, so it obeys the law $v(x) = ax + b$, where a and b are constants. Find values for a and b so this linear model gives a good fit to the data set. With these values, estimate how far the bullet travels before coming to rest. Remembering that $v = dx/dt$, estimate when the bullet comes to rest.

Table 4.17 Velocity of bullet as a function of distance

Distance (ft)	Velocity (ft/sec)
0	3290
300	2951
600	2636
900	2342
1200	2068
1500	1813

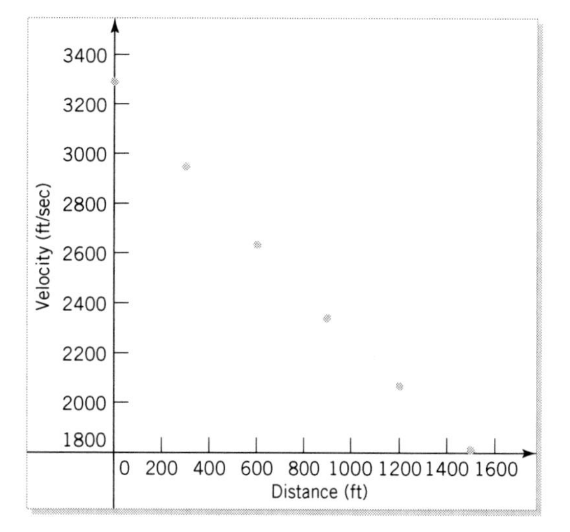

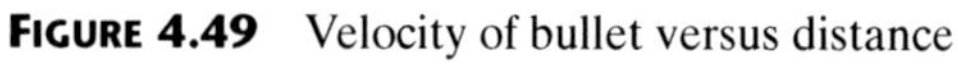
FIGURE 4.49 Velocity of bullet versus distance

(b) It is claimed that this data set looks quadratic, so it obeys the law $v(x) = ax^2 + bx + c$, where a, b, and c are constants. Find values for a , b, and c so this quadratic model gives a good fit to the data set. With these values, estimate how far the bullet travels before coming to rest. Remembering that $v = dx/dt$, estimate when the bullet comes to rest.

(c) It is claimed that the bullet, when fired horizontally, is subject only to air resistance, which is proportional to a power of the velocity; that is, $dv/dt = -kv^n$, where k and n are positive constants. Because

$$\frac{dv}{dt} = \frac{dv}{dx}\frac{dx}{dt} = \frac{dv}{dx}v,$$

we have

$$\frac{dv}{dx} = -kv^{n-1},$$

which can be written as

$$\ln\left(-\frac{dv}{dx}\right) = (n-1)\ln v + \ln k.$$

Thus, if we plot $\ln(-dv/dx)$ versus $\ln v$, we should see a straight line with slope $n - 1$ and y-intercept $\ln k$, from which we can estimate n and k. From Table 4.17, construct the central difference quotient $\Delta_C v/\Delta x$, which is an approximation to the derivative dv/dx. Now plot $\ln(\Delta_C v/\Delta x)$ versus $\ln v$. Find values for k and n so this model gives a good fit to the data set. With these values, solve the differential equation and compare the exact answer to the actual data set. Estimate how far the bullet travels before coming to rest. Remembering that $v = dx/dt$, estimate when the bullet comes to rest.

(d) If we plot $\Delta_C v/\Delta x$ versus v we see a nearly straight line. This would mean that the bullet's velocity obeys the differential equation $dv/dx = av + b$, where a and b are constants. Find values for a and b so this model gives a good fit to the data set. With these values, solve the differential equation and compare the exact answer to the actual data set. Estimate how far the bullet travels before coming to rest. Remembering that $v = dx/dt$, estimate when the bullet comes to rest. Notice that this differential equation can be recast in terms of Newton's law, namely

$$\frac{dv}{dt} = \frac{dv}{dx}v = av^2 + bv.$$

(e) If we plot $\Delta_C v/\Delta x$ versus x we see a nearly straight line. This would mean that the bullet's velocity obeys the differential equation $dv/dx = ax + b$,

[24] Richard Ziehmer, Canyon del Oro High School, Tucson, AZ.

where a and b are constants. Find values for a and b so this model gives a good fit to the data set. With these values, solve the differential equation and compare the exact answer to the actual data set. Estimate how far the bullet travels before coming to rest. Remembering that $v = dx/dt$, estimate when the bullet comes to rest. Notice that this differential equation can be recast in terms of Newton's law, namely

$$\frac{dv}{dt} = \frac{dv}{dx}v = axv + bv.$$

How is this model related to the model in part (b)?

4.5 APPLICATION: ORTHOGONAL TRAJECTORIES

The coordinate systems in two dimensions with which we are most familiar are the rectangular coordinate system and the polar coordinate system. Both have the property that setting one variable equal to a constant produces a family of curves (sometimes lines) that is perpendicular to the family obtained by setting the other variable equal to a constant.

A natural reason for the name of the rectangular coordinate system is that a rectangle is formed by any pair of vertical lines (given by $x = c_1$ and $x = c_2$) and any pair of horizontal lines (given by $y = k_1$ and $y = k_2$). It is also true in rectangular coordinate systems that lines given by $x = a$ are perpendicular to those given by $y = b$. Such lines are said to be orthogonal to each other, and the lines are examples of ORTHOGONAL TRAJECTORIES.[25] This situation can also occur for nonhorizontal and nonvertical lines or for families of curves. In this section we show how differential equations are used to construct orthogonal trajectories, and give examples of their use.

We start with two definitions.

◆ *Definition 4.3:* **A family of curves is said to be PARAMETERIZED BY λ if the family is given by a relation $f(x, y, \lambda) = 0$, where the Greek letter λ (lambda) is a parameter that may take on various values.** ◆

Thus we see that $y = 2x + \lambda$ gives a family of lines with slope 2 parameterized by λ. The equation $x^2 + y^2 = \lambda$ gives a family of circles with radius $\sqrt{\lambda}$. Notice that in the first case λ may be any real number, whereas in the second case λ may have only nonnegative values.

◆ *Definition 4.4:* **Suppose we have a family of curves given by $f(x, y, \lambda) = 0$. The ORTHOGONAL TRAJECTORIES to this family are again a family of curves that intersect those given by $f(x, y, \lambda) = 0$ at right angles.** ◆

When using this definition, we need to remember that two curves intersect at right angles — that is, are orthogonal — if their tangent lines at the point of intersection are perpendicular.

[25] From the Greek "orthogōnion" meaning "right angle."

EXAMPLE 4.11 *The Polar Coordinate System*

We first consider the family of lines through the origin parameterized by their slope, λ,

$$y = \lambda x. \tag{4.59}$$

Here λ is a parameter that gives the slope of a particular line. We know that this family of lines and the family of circles, centered at $(0, 0)$, form the polar coordinate system, and that these two families are orthogonal. But, if we didn't already know that the circles were the orthogonal trajectories to these lines, how would we find them?

We can find the equation for the orthogonal family in the following manner. If we solve (4.59) for λ, we find

$$\lambda = \frac{y}{x} \tag{4.60}$$

for $x \neq 0$. Differentiating (4.60) with respect to x, we have

$$0 = \frac{1}{x^2}\left(x\frac{dy}{dx} - y\right),$$

or

$$y' = \frac{y}{x}. \tag{4.61}$$

This is the differential equation for the family of curves (4.59). Notice it is independent of the parameter λ. It tells us that at each point (x, y) the slope of any member of the family of curves (4.59) is y/x.

How can we find the differential equation that governs the family of lines that is orthogonal to those of (4.59)? The key is to realize that if the slopes of two orthogonal curves are m_1 and m_2, then $m_1 m_2 = -1$. Thus, if we know that the slope of one curve is m_1, then the slope of the curve orthogonal to it is $-1/m_1$. Here we know that the slope of (4.59) satisfies (4.61), so the differential equation satisfied by the orthogonal trajectory can be obtained from $y' = y/x$ by replacing y' with $-1/y'$ – namely,

$$\frac{1}{-y'} = \frac{y}{x}.$$

Thus, the family of orthogonal trajectories must satisfy the differential equation

$$y' = -\frac{x}{y}.$$

Separate variables Separating the variables and integrating, we find

Integrate

$$\frac{y^2}{2} = -\frac{x^2}{2} + c, \quad \text{or } x^2 + y^2 = 2c.$$

Thus, we see that the family of circles $x^2 + y^2 = 2c$ is orthogonal to the family of straight lines $y = \lambda x$. This shown in Figure 4.50.

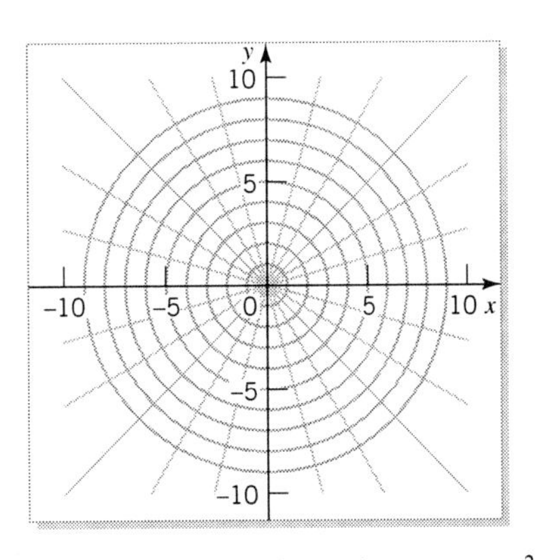

FIGURE 4.50 Orthogonal trajectories $y = \lambda x$, $y^2 + x^2 = 2c$

How to Find Orthogonal Trajectories

Purpose To construct the differential equation of the trajectories orthogonal to the family of curves

$$f(x, y, \lambda) = 0, \tag{4.62}$$

where a particular value of λ gives a particular curve.

Process

1. If possible, solve (4.62) for λ, to get

$$\lambda = g(x, y). \tag{4.63}$$

2. Differentiate (4.63) with respect to x, using the chain rule,

$$0 = \frac{\partial g}{\partial x} + \frac{\partial g}{\partial y}\frac{dy}{dx}. \tag{4.64}$$

This is the differential equation for the original family of curves where g, $\partial g/\partial x$, and $\partial g/\partial y$ are known functions.

3. To set up the differential equation for the orthogonal trajectories, replace y' with $-1/y'$ in (4.64) to get

$$0 = \frac{\partial g}{\partial x} - \frac{\partial g}{\partial y}\frac{1}{y'}. \tag{4.65}$$

Solve this differential equation for $y(x)$.

Comments about Orthogonal Trajectories

- Most people do not try to remember (4.65). It is more common to go through the technique described to reach (4.65).
- There is a different way of getting (4.64) from (4.62) without having to solve for λ. Differentiate (4.62) with respect to x to obtain

$$\frac{\partial f}{\partial x} + \frac{\partial f}{\partial y}\frac{dy}{dx} = 0. \tag{4.66}$$

  In this case, the left-hand side of (4.66) may still contain the parameter λ. We can eliminate λ from (4.66) by using (4.62). If we do this we will have an equation equivalent to (4.64).
- Although a particular curve may have no symmetry, if the original **family** of curves has a symmetry across the x-axis, the y-axis, or about the origin, then the **family** of orthogonal trajectories will inherit the same symmetry. We can see that this is the case for the slope fields because (4.64) and (4.65) will have the same symmetries. Because we are concerned with all solutions of these equations, we see that the set of all solutions will have the same symmetries.

Caution! **The differential equation for a family of curves must not contain the parameter λ. If it does, solve for λ in the original equation and use its value here.**

Two More Examples

Orthogonal trajectories occur in many situations, including fluid flow, electrostatics, and weather maps. In planar fluid flow, the fluid motion is along curves that are the orthogonal trajectories of the curves of equipotential. In electrostatics, the curves of force are the orthogonal trajectories of the curves of constant potential.

EXAMPLE 4.12

Suppose that curves of constant electrostatic potential are given by the family of ellipses

$$\frac{x^2}{2} + y^2 = \lambda, \tag{4.67}$$

and that we wish to find the curves of force associated with this potential. This family of curves is symmetric across the x-axis, the y-axis, and about the origin. (Why?) The orthogonal trajectories will have the same symmetries.

To find the family of curves orthogonal to these ellipses, we differentiate (4.67) with respect to x and obtain

$$x + 2yy' = 0.$$

The differential equation for the family of curves orthogonal to these ellipses must satisfy

$$x + 2y\frac{-1}{y'} = 0,$$

or

$$y' = \frac{2y}{x}.$$

Separate variables

This is a separable differential equation with equilibrium solution $y(x) = 0$, and nonequilibrium solution

Integrate

$$\ln|y| = 2\ln|x| + C.$$

Explicit solution

If we solve for y, we get

$$y(x) = cx^2,$$

in which we have included the equilibrium solution. Thus, we have a family of parabolas that gives the curves of force that are orthogonal to the elliptical level curves of the electrostatic potential field (see Figure 4.51).

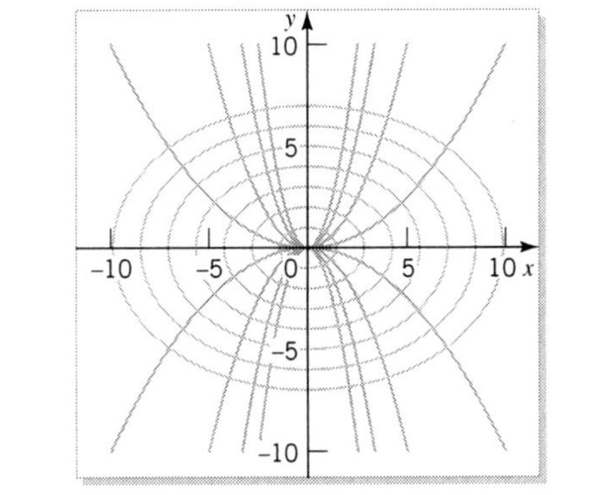

FIGURE 4.51 Orthogonal trajectories $x^2/2 + y^2 = \lambda$, $y = cx^2$

EXAMPLE 4.13

Consider the problem of finding the trajectories orthogonal to the family of curves given by

$$x^2 + y^2 = 2\lambda y. \tag{4.68}$$

These curves are "offset" circles and are shown in Figure 4.52.[26]

[26] We can see this by rewriting $x^2 + y^2 = 2\lambda y$ in the form $x^2 + (y - \lambda)^2 = \lambda^2$.

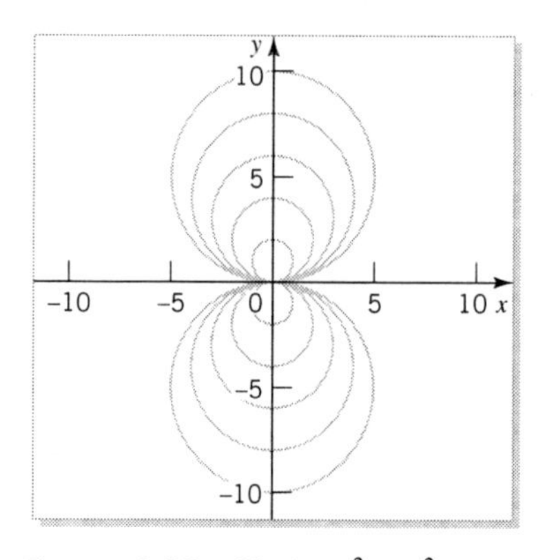

FIGURE 4.52 Circles $x^2 + y^2 = 2\lambda y$

Symmetry

This family of curves is symmetric across the x-axis, the y-axis, and about the origin. (Why?) The orthogonal trajectories will have the same symmetries. To determine the differential equation satisfied by this family of circles, we differentiate (4.68) implicitly to obtain

$$2x + 2yy' = 2\lambda y'. \tag{4.69}$$

Because we did not solve for λ before differentiating, we must now eliminate λ from (4.69) by replacing it by its value from (4.68). This gives

$$2x + 2yy' = \frac{x^2 + y^2}{y} y',$$

or, collecting terms in y',

$$\frac{x^2 - y^2}{y} y' = 2x.$$

This means that the differential equation satisfied by the trajectories orthogonal to these circles is

$$-\frac{x^2 - y^2}{y} \frac{1}{y'} = 2x,$$

which can be written as

$$y' = \frac{y^2 - x^2}{2xy} = \frac{1}{2}\left(\frac{y}{x} - \frac{x}{y}\right). \tag{4.70}$$

Change variable

This is a differential equation with homogeneous coefficients, so we make the usual change of variable $y = xz$ in (4.70), giving the transformed differential equation as

$$z + x\frac{dz}{dx} = \frac{1}{2}\left(z - \frac{1}{z}\right).$$

Separate variables This equation is separable and may be rearranged as

$$\frac{2z}{z^2+1}\frac{dz}{dx} = -\frac{1}{x},$$

Integrate so integration gives $\ln(z^2+1) = -\ln|x| + C$, or $x(z^2+1) = 2c$ where $c = \pm\frac{1}{2}e^C$.

Implicit solution Returning to the original variables and completing the square, we observe that the family of offset circles $x^2+y^2 = 2cx$, or $(x-c)^2+y^2 = c^2$, is orthogonal to the original family given by (4.68). The two families are shown in Figure 4.53.

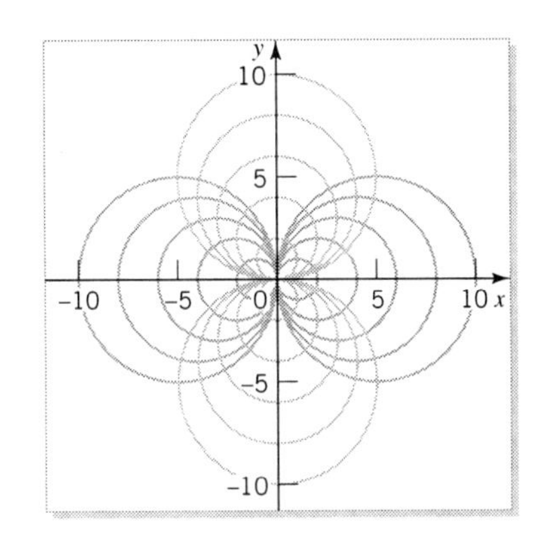

FIGURE 4.53 Orthogonal trajectories $x^2+y^2 = 2\lambda y$, $x^2+y^2 = 2cx$

EXERCISES

1. Find the orthogonal trajectories of the families of curves given by the following. In parts (g) and (k), the quantity a is a fixed constant.

   (a) $y = -2x + \lambda$
   (b) $y = \lambda x + 4$
   (c) $y^2 = \lambda x$
   (d) $y^2 = \lambda x^3$
   (e) $x^2 - \lambda^2 y^2 = 16$
   (f) $x^2 + \lambda^2 y^2 = 16$
   (g) $x^2 + ay^2 = \lambda^2$
   (h) $y = \lambda/x$
   (i) $y = \lambda e^x$
   (j) $y = \lambda e^{-x}$
   (k) $y = \lambda e^{ax}$
   (l) $y = 1 + \lambda e^{-2x}$
   (m) $y = \ln(x^3 + \lambda)$
   (n) $x^{2/3} - y^{2/3} = \lambda^{2/3}$
   (o) $e^x \cos y = \lambda$

2. Find the value of the constant a such that the family of curves $x^2 + ay^2 = \lambda^2$ is orthogonal to those of $y^2 = \lambda x^3$.

3. Use a computer/calculator program to observe graphs of $y^2 = (x+\lambda)^2$ for several values of λ. (Note that you may need to plot $y = \pm|x+\lambda|$.)

   (a) Record what you observe.
   (b) The curves you observed in part (a) are called "self-orthogonal." Give reasons why this name seems to be appropriate.
   (c) Differentiate the original equation as a first step in finding the differential equation for the orthogonal trajectories, and use the results to amplify your answer to part (b).

4. Parabolic cylinder coordinates are given by

$$x = \tfrac{1}{2}\left(u^2 - v^2\right), \qquad y = uv.$$

   (a) Eliminate u from the preceding equations, and from the result show that putting $v = \lambda$ gives parabolas in the xy-plane. Which way do these parabolas open?

(b) Find the differential equation for the family of curves orthogonal to these parabolas, and show that the solution of the differential equation also gives parabolas. Which way do these parabolas open?

5. The streamlines for irrotational, incompressible flow in the region bounded by two perpendicular lines — say, the positive x- and positive y-axes — are given by the family of hyperbolas $xy = b$. (This represents the flow inside a corner, as shown in Figure 4.54.) The orthogonal trajectories to the streamlines give lines of constant velocity potential. Show that these orthogonal trajectories are also hyperbolas.

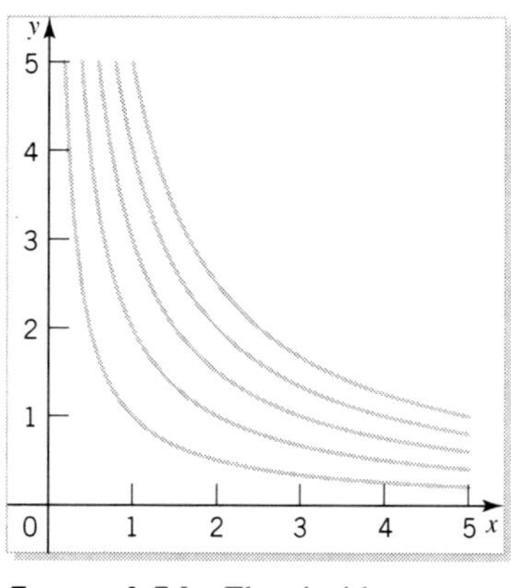

FIGURE 4.54 Flow inside a corner

6. The velocity potential for a certain flow is given by $(x+y)^2 = \lambda$. Find the streamlines for this situation — that is, the orthogonal trajectories. Describe a possible flow consistent with these streamlines.

7. On the graph of the function $r = f(\theta)$, in polar coordinates, the angle ψ_1 between the intersection of the extended radius and the tangent line is given by

$$\tan\psi_1 = \frac{r}{dr/d\theta} = \frac{f}{f'}.$$

If another graph in polar coordinates is to be orthogonal to this graph, the corresponding angle ψ_2 will be larger by adding $\pi/2$ or smaller by subtracting $\pi/2$. Thus, ψ_2 will satisfy the equation $\tan\psi_2 = \tan(\psi_1 \pm \frac{\pi}{2}) = -1/\tan\psi_1$ (justify this last step). In terms of the variables (r, θ), we have the equation for the orthogonal trajectories to the curve $r = f(\theta)$ given by

$$\tan\psi_2 = \frac{r}{dr/d\theta} = -\frac{f'}{f},$$

or

$$\frac{1}{r}\frac{dr}{d\theta} = -\frac{f}{f'}.$$

It is therefore evident that in polar coordinates, the negative reciprocal also comes into play at the start of finding orthogonal trajectories.

(a) Show that the differential equation for the orthogonal trajectories to the family of circles $r = \lambda\cos\theta$ is

$$\frac{1}{r}\frac{dr}{d\theta} = \frac{\cos\theta}{\sin\theta}.$$

(b) Integrate this equation to obtain the family $r = c\sin\theta$, and graph several members of each family.

8. Find the orthogonal trajectories to the family of cardioids given by $r = \lambda(1-\cos\theta)$.[27] Compare your results with that of Exercise 3. Is the terminology of that exercise appropriate here? Why or why not?

9. Find the trajectories orthogonal to the family of cardioids given by $r = \lambda(1-\sin\theta)$.

10. Find the orthogonal trajectories of the spirals given by $r = \lambda\theta$ (the spirals of Archimedes). Graph some of these spirals and some of the orthogonal trajectories. Were you surprised at the results? Why or why not?

11. Find the orthogonal trajectories to those of $r = \lambda/\sqrt{\theta}$.

12. The flow within a two-dimensional wedge of angle α has streamlines given by $r^{\pi/\alpha}\sin(\theta\pi/\alpha) = \lambda$. (See Figure 4.55.) Use the information from Exercise 7 to find

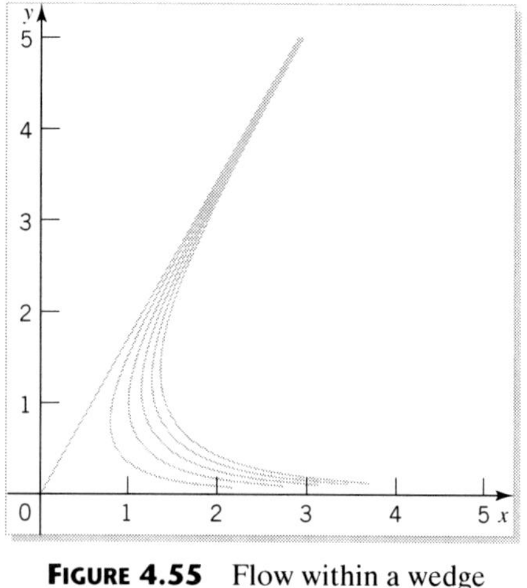

FIGURE 4.55 Flow within a wedge

[27] Cardiod is from the Greek, meaning "heart-shaped."

the associated velocity potential for this situation. Now let $\alpha = \pi/2$, and compare with the results of Exercise 5.

13. Show that the family of curves given by $x^2/(m^2 + \lambda) + y^2/(n^2 + \lambda) = 1$ (where m and n are specified constants) is self-orthogonal. [Hint: Show that the given family of curves satisfies the differential equation $(m^2 - n^2)\,y' = (xy' - y)\,(yy' + x)$, and that this equation is unchanged if y' is replaced with $-1/y'$.]

4.6 PIECING TOGETHER DIFFERENTIAL EQUATIONS

It is often too much to expect a single differential equation to model a particular situation. Circumstances change, and so must the differential equation. For example, consider investing money that is compounded continuously; that is, the rate of change of principal $P(t)$ with respect to time is proportional to the principal, $\frac{dP}{dt} = rP$, where r is a constant, the annual interest rate. At time $t = 0$, an initial amount of P_0 dollars is invested. At any time t, the solution of this initial value problem is $P(t) = P_0 e^{rt}$, so after one year we will have $P_0 e^r$ dollars. However, suppose that at the end of this first year the bank changes its interest rate from r to s. Now the differential equation is $\frac{dP}{dt} = sP$ with initial condition $P(1) = P_0 e^r$, so after the beginning of year one the principal is $P(t) = P_0 e^r e^{s(t-1)}$.

If we piece together these differential equations and corresponding solutions, we have

$$P' = \begin{cases} rP & \text{if } 0 \le t \le 1, \\ sP & \text{if } 1 < t, \end{cases} \tag{4.71}$$

subject to $P(0) = P_0$ with solution

$$P(t) = \begin{cases} P_0 e^{rt} & \text{if } 0 \le t \le 1, \\ P_0 e^r e^{s(t-1)} & \text{if } 1 < t. \end{cases} \tag{4.72}$$

As we would expect, the principal, $P(t)$, is continuous for all $t > 0$ including $t = 1$, because $\lim_{t\to 1^-} P(t) = P_0 e^r$ and $\lim_{t\to 1^+} P(t) = P_0 e^r$. However, $P(t)$ is not differentiable at $t = 1$ as can be seen from (4.72), because $\lim_{t\to 1^-} P'(t) = rP_0 e^r$ and $\lim_{t\to 1^+} P'(t) = sP_0 e^r$. This can also be seen in Figure 4.56 where we have plotted (4.72) for $r = 0.1$ and $s = 0.2$. The "corner" indicates the nondifferentiability.

However, (4.71) together with $P(0) = P_0$ are insufficient to give (4.72). They give

$$P(t) = \begin{cases} P_0 e^{rt} & \text{if } 0 \le t \le 1, \\ Ce^{s(t-1)} & \text{if } 1 < t, \end{cases} \tag{4.73}$$

where C is a constant. We need a statement about what happens at $t = 1$, such as $P(t)$ is continuous at $t = 1$, or the initial condition for $P' = sP$ is the end condition for $P' = rP$.

There are other possibilities. For example, we might decide to deposit another amount of P_0 in the bank at $t = 1$, in which case we would solve $P' = sP$ subject to $P(1) = P_0 + P_0 e^r$, which gives $P(t) = \left(P_0 + P_0 e^r\right) e^{s(t-1)}$ as the solution for $t > 1$. This gives a function $P(t)$ that is not continuous at $t = 1$, as can be seen in Figure 4.57.

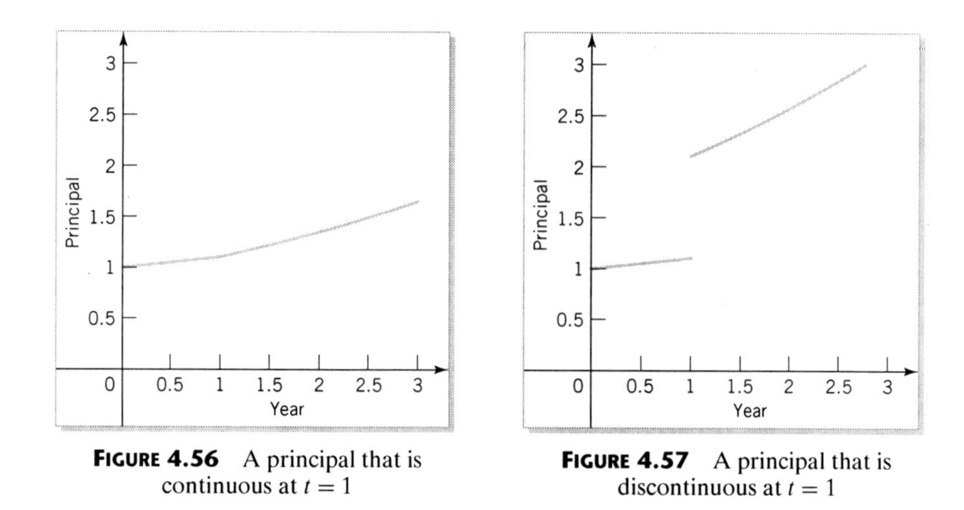

FIGURE 4.56 A principal that is continuous at $t = 1$

FIGURE 4.57 A principal that is discontinuous at $t = 1$

This discussion shows that piecing together differential equations may lead to a function $P(t)$ that is discontinuous, or is continuous but not differentiable everywhere.

EXAMPLE 4.14 *The Population of Ireland*[28] *(1780 – 1910)*

Imagine that a population is growing toward a carrying capacity, b_1, according to the logistic equation

$$\frac{dP}{dt} = a_1 P(b_1 - P),$$

when something quite catastrophic happens to the food supply at time $t = T$. As a result, the new carrying capacity, b_2, falls below the current population, so the population declines according to

$$\frac{dP}{dt} = a_2 P(b_2 - P).$$

Thus, we have

$$P' = \begin{cases} a_1 P(b_1 - P) \text{ if } 0 < t < T, \\ a_2 P(b_2 - P) \text{ if } T < t. \end{cases} \tag{4.74}$$

Now, although we expect that P (the population) will be continuous, P' will not be.

The population of Ireland from 1780 to 1910 is given in Table 4.18 and shown in Figure 4.58. We can see something quite catastrophic occurred about 1840 — the potato famines. Can this change in population be modeled by (4.74)? To consider this we need to estimate a_1, b_1, a_2, and b_2.

Table 4.18
Population of Ireland from 1780

Year	Population (millions)
0	4.0
10	4.6
20	5.2
30	5.9
40	6.7
50	7.6
60	8.2
70	6.9
80	5.8
90	5.4
100	5.2
110	4.7
120	4.5
130	4.4

[28] Based on *Mathematical Models in the Social, Management and Life Sciences* by D. N. Burghes and A. D. Wood, Ellis Horwood, 1980, page 104.

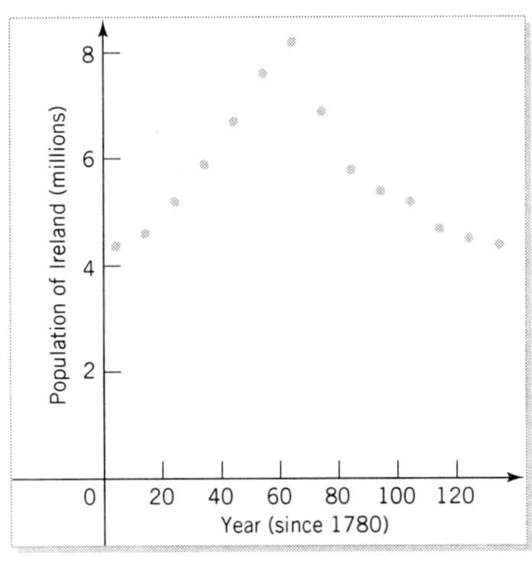

FIGURE 4.58 Population of Ireland from 1780 to 1910

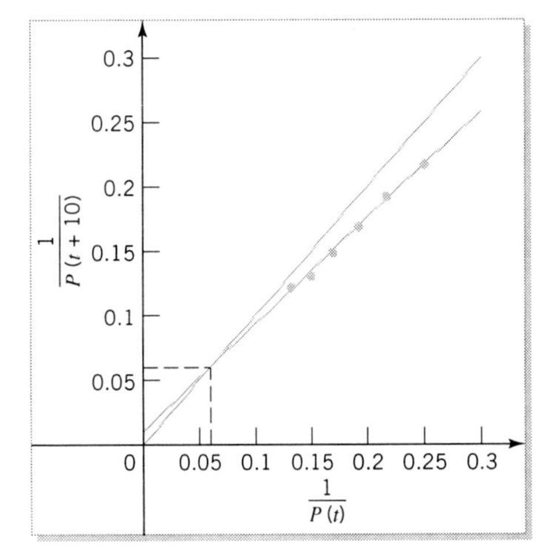

FIGURE 4.59 A plot of $1/P(t+10)$ versus $1/P(t)$ from 1780 to 1840, the line of best fit $y = 0.827584x + 0.010535$, and the line $y = x$

In Figure 4.59 we have plotted $1/P(t)$ versus $1/P(t+10)$ for the data from 1790 to 1840, and the line of best fit, $y = 0.827584x + 0.010535$. Using the idea suggested in Exercise 4 on page 154 we can estimate that $1/b_1 \approx 0.0611$, so $b_1 \approx 16.3$. With this value of b_1, we can estimate the other parameters of the logistic equation in the usual way (see page 48). We find that for the period 1790 to 1840,

$$P(t) = \frac{16.3e^{-1.13087}}{e^{-0.019293t} + e^{-1.13087}}. \tag{4.75}$$

In Figure 4.60 we have plotted $1/P(t)$ versus $1/P(t+10)$ for the data from 1840 to 1910, and the line of best fit, $y = 0.784846x + 0.052875$. In this case we estimate that $1/b_2 \approx 0.0246$, so $b_2 \approx 4.07$. With this value of b_2 we can estimate the other parameters, and we find that for the period 1840 to 1910,

$$P(t) = \frac{-4.07e^{0.618648}}{e^{-0.027547(t-60)} - e^{0.618648}}. \tag{4.76}$$

The population of Ireland from 1780 to 1910 and the functions (4.75) and (4.76) are shown in Figure 4.61.

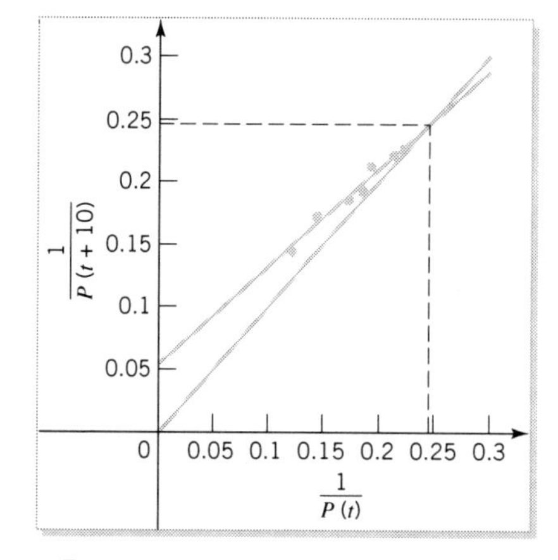

FIGURE 4.60 A plot of $1/P(t+10)$ versus $1/P(t)$ from 1840 to 1910, the line of best fit $y = 0.784846x + 0.052875$, and the line $y = x$

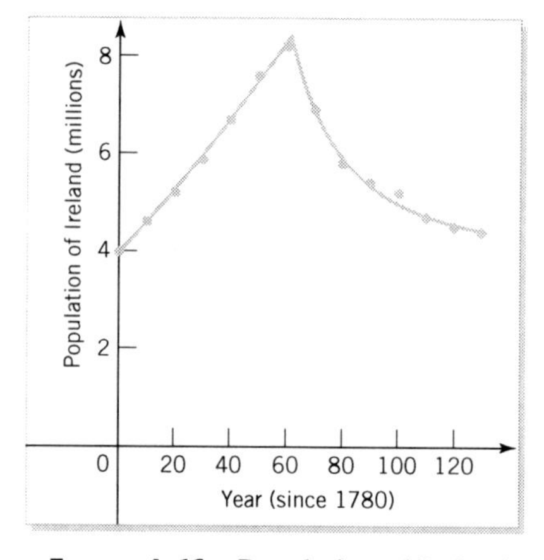

FIGURE 4.61 Population of Ireland from 1780 to 1910 modeled with two functions

EXERCISES

1. Consider the situation in which a parachutist jumps from an airplane and free falls for 6 seconds, until the parachute opens. A mathematical model for this situation has the air resistance proportional to the square of the velocity during free fall and proportional to the velocity after the parachute opens. The resulting differential equation is

$$\frac{dv}{dt} = \begin{cases} g - (k/m)v^2 & \text{if } 0 < t < 6, \\ g - (b/m)v & \text{if } 6 < t. \end{cases}$$

where $g = 32.2$, $k/m = 9.72 \times 10^{-4}$, $b/m = 1.464$, and $v(0) = 0$. Assume that the velocity is continuous at $t = 6$.

   (a) Find the velocity as a function of time, and graph your solution. Comment on its shape. Could you have predicted the shape of your solution curve from other considerations?

   (b) Compare the terminal velocity for free fall (assuming the parachute never opens) with that for this situation.

2. **Administering Drugs.**[29] It is known from experiments that theophylline has hardly any therapeutic effect if its concentration is below 5 mg/l, and that concentrations above 20 mg/l are likely to be toxic. The problem is to administer the drug to the individual in Example 2.4 on page 39 in such a way that the concentration remains between 5 and 20 mg/l. We want to do this by administering equal doses of theophylline at equal intervals of time. What advice should we give the doctor, assuming that the concentration $y(t)$ is governed by the differential equation (2.18) on page 39; namely, $y' = -ky$ with $k = 0.167$?

3. **Multiple Drug Doses I.** Consider a drug that is administered to a person and whose concentration $y(t)$ in the body satisfies $y' = -ay$. Let a dose of $y = c$ be administered every T units of time. Show that after $n+1$ doses — that is, for $nT < t < (n+1)T$ — we have $y(nT) = c\left[e^{-(n+1)aT} - 1\right] / \left(e^{-aT} - 1\right)$ and $y((n+1)T) = ce^{-aT}\left[e^{-(n+1)aT} - 1\right] / \left(e^{-aT} - 1\right)$. What happens to $y(nT)$ and $y((n+1)T)$ as $n \to \infty$? Can the drug concentration increase indefinitely?

[29] Based on *Applying Mathematics* by D.N. Burghes, I. Huntley, and J. McDonald, Ellis Horwood 1982, page 125.

4. **Multiple Drug Doses II.** If, in the previous exercise, the initial dose was $c/\left(1-e^{-aT}\right)$, and all subsequent doses of c were given at time intervals of T, what is $y(nT)$? What does this tell you about the long-term behavior of the drug?

5. **Multiple Drug Doses III.** Consider a drug that is administered to a person and whose concentration $y(t)$ in the body satisfies $y' = -a$. Let a dose of $y = c$ be administered every T units of time. Can the drug concentration increase indefinitely?

6. **Your Future.**[30] Estimate in how many years you will have graduated, found a job, married, and had your first child. You want to set aside enough money so that when your child is 19 she can afford to attend four years at your alma mater.[31] You plan to deposit a fixed sum of A dollars on her birthday every year from year 0 to year 19 inclusive. Assume that it is compounded continuously at 8% per annum and that the cost of tuition at your alma mater grows linearly at approximately the same rate it has for the past 10 years. How large must A be?

7. **Pacemaker.** Figure 4.62 represents a model of a simple pacemaker.[32] It consists of a battery of constant voltage E, a capacitor C, and a switch connected to the heart with its resistance R. Initially the switch is in position S_1, and the capacitor is charged to the value E. At time T_1 the switch is moved to S_2 and the capacitor discharges through the heart, stimulating it for time T_2. The switch continues to alternate between S_1 and S_2 at time intervals of T_1 and T_2. If V is the voltage across the heart, then V satisfies the initial value problem

$$\frac{dV}{dt} = \begin{cases} 0, & \text{where } V = 0 \text{ at the beginning of each time interval when } S_1 \text{ is closed,} \\ -\frac{1}{RC}V, & \text{where } V = E \text{ at the beginning of each time interval when } S_2 \text{ is closed.} \end{cases}$$

Solve this equation for $V(t)$ and sketch $V(t)$ from $t = 0$ until the switch is at S_1 for the fourth time.

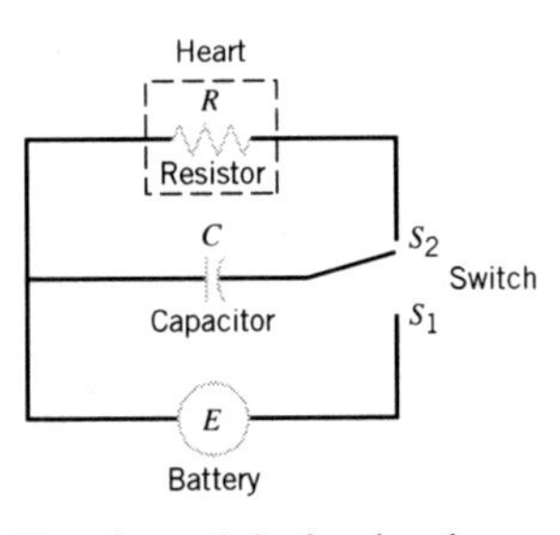

FIGURE 4.62 A model of a simple pacemaker

8. **Allometric Growth of Fiddler-Crabs.** Allometric growth is the study of the relative sizes of different parts of an organism as a consequence of growth. One of the simplest models of allometry is one in which it is assumed that the relative growth rates of the two components $y(t)$ and $x(t)$ satisfy

$$\frac{dy}{dx} = a\frac{y}{x}$$

where a is a constant. Table 4.19 and Figure 4.63 show the results of an experiment in which the relative weights of the body and the claw of fiddler-crabs were obtained.[33] The body weight is the total weight of the crab with the weight of the claw subtracted. Although it is possible to model this data set with a single allometric equation, a better model results if two allometric equations are pieced together. Construct such a model. It is believed that the onset of sexual maturity in a fiddler-crab occurs when its total weight exceeds about 1.1g. Is this consistent with your model?

9. Consider the differential equation

$$y' = \begin{cases} y & \text{if } 0 \le x \le k, \\ b - y & \text{if } k < x, \end{cases}$$

where b and k are constants. Assume that $y(x)$ is differentiable for all $x \ge 0$.

(a) Plot y' versus $y(x)$. Why is $y(k) = b/2$?

(b) What happens to $y(x)$ as $x \to \infty$? How would you interpret b if $y(x)$ is the population at time x?

(c) Solve the differential equation subject to the initial condition $y(0) = y_0$ if $0 < y_0 < b/2$. Find k in terms of b and y_0.

[30] Based on an idea of Brian Winkel, United States Military Academy.

[31] Alma mater is Latin for "nourishing mother."

[32] *Mathematics for Biomedical Applications* by S.A. Glantz, University of California Press, 1979, page 52.

[33] *Problems of Relative Growth* by J. S. Huxley, Dover, 1972, page 12, Table 1.

Table 4.19 Claw weight versus body weight of fiddler-crabs

Body Weight (mg)	Claw Weight (mg)	Body Weight (mg)	Claw Weight (mg)
57.6	5.3	680.6	271.6
80.3	9.0	743.3	319.2
109.2	13.7	872.4	417.6
156.1	25.1	983.1	460.8
199.7	38.3	1079.9	537.0
238.3	52.5	1165.5	593.8
270.0	59.0	1211.7	616.8
300.2	78.1	1291.3	670.0
355.2	104.5	1363.2	699.3
420.1	135.0	1449.1	777.8
470.1	164.9	1807.9	1009.1
535.7	195.6	2235.0	1380.0
617.9	243.0		

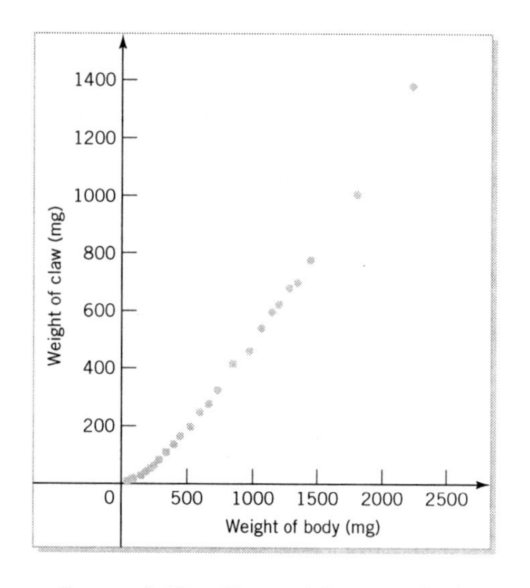

FIGURE 4.63 Claw weight versus body weight of fiddler-crabs

(d) What are the coordinates of the point of inflection of $y(x)$? Is it possible for a solution of this differential equation to have an inflection point if $y_0 > b/2$?

(e) Compare the solutions of this differential equation to the solutions of the logistic equation $y' = ay(b - y)$, by choosing a so that

i. initially both solutions grow at the same exponential rate.

ii. both solutions have the same point of inflection.

10. Consider the differential equation

$$y' = \begin{cases} a_1 y & \text{if } 0 \le x \le k, \\ a_2 (b - y) & \text{if } k < x, \end{cases}$$

where a_1, a_2, b, and k are constants. Assume that $y(x)$ is differentiable for all $x \ge 0$.

(a) Plot y' versus $y(x)$. What is $y(k)$?

(b) What happens to $y(x)$ as $x \to \infty$? How would you interpret b if $y(x)$ is the population at time x?

(c) Solve the differential equation subject to the initial condition $y(0) = y_0$ if $0 < y_0 < y(k)$. Find k in terms of a_1, a_2, b, and y_0.

(d) What are the coordinates of the point of inflection of $y(x)$? Is it possible for a solution of this differential equation to have an inflection point if $y_0 > b/2$?

(e) Compare the solutions of this differential equation to the solutions of the logistic equation $y' = ay(b - y)$, by choosing a so that

i. initially both solutions grow at the same exponential rate. Sketch typical solutions.

ii. both solutions have the same point of inflection. Sketch typical solutions.

What Have We Learned?

Main Ideas

- A first order differential equation is **separable** if it can be written in the form in which the derivative equals a product of two functions, each of which contains only one of the variables. If x and y are the variables, separable differential equations have the form

$$\frac{dy}{dx} = f(y)g(x).$$

- To solve a separable differential equation, see *How to Solve Separable Differential Equations* on page 123.
- To graph implicit equations, see *How to Sketch Solutions of* $G(y) = F(x)$ on page 131.
- If a first order differential equation can be written in the form

$$\frac{dy}{dx} = g\left(\frac{y}{x}\right),$$

  then the equation has **homogeneous coefficients**.
- To solve a differential equation with homogeneous coefficients, see *How to Solve Differential Equations with Homogeneous Coefficients* on page 140.
- Differential equations may be obtained by analyzing data sets. See page 145.
- Differential equations are found in many different applications. See the following sections for reference.
  - *Models: Objects in Motion*, page 158.
  - *Application: Orthogonal Trajectories*, page 167.
  - Exercises in Sections 4.1, 4.3, and 4.6.
- Differential equations may sometimes involve functions that are pieced together. See page 175.

CHAPTER 5

FIRST ORDER LINEAR DIFFERENTIAL EQUATIONS AND MODELS

Where We Are Going — and Why

In this chapter we consider differential equations of the form $y' + p(x)y = q(x)$ where $p(x)$ and $q(x)$ are given functions of x. Such differential equations are called first order linear. These differential equations are very important for three reasons. First, they are the most understood mathematically — all solutions are explicit, and can always be written as an integral. Second, they occur very often in models and applications. Third, certain other types of differential equations can be reduced to linear differential equations by a change of variable.

5.1 SOLVING LINEAR DIFFERENTIAL EQUATIONS

In this section we start with a variation on a preceding example involving the population of Botswana. Now emigration is assumed to occur. However, the differential equation that arises in this case is not one we have encountered previously. We develop a technique for solving this differential equation. Then by looking at the solution we are able to see a different but preferable method to obtain the same solution. We show how this method is used in other examples.

An Example that Leads to a Linear Differential Equation

EXAMPLE 5.1 ***The Population of Botswana with Emigration***

Recall in Chapter 2 we discovered that the population of Botswana has been growing exponentially with a doubling time of about 20 years. Suppose that in 1990, residents of that country worry about this growth rate and think about emigrating to other countries. It seems reasonable that any emigration for this reason would increase gradually over time, perhaps linearly. The integral of this emigration rate over the next 20 years equals the total emigration during that time period. If we hypothesize that $1/4$ of the 1990 population of 1.285 million will leave the country over the next two decades in a linear manner at the rate of at million people per year — where a is a constant and t the time in years — we have

$$\int_0^{20} at\,dt = \frac{1.285}{4},$$

from which we find $\frac{1}{2}a\,(20)^2 = 1.285/4$, so $a = 1.60625 \times 10^{-3}$. Thus, we have

$$\textit{emigration rate in millions per year} = at = 1.60625 \times 10^{-3}t.$$

If we use the growth rate for Botswana from Chapter 2 and its population in 1990 as our initial condition, we obtain the differential equation

$$\frac{dP}{dt} = kP - at \tag{5.1}$$

subject to

$$P(0) = 1.285 \text{ (million)}, \tag{5.2}$$

where $k = 0.0355$ and $a = 1.60625 \times 10^{-3}$. Here $P(t)$ is the population in millions at time t in years.

In (5.1) we have an example of a linear differential equation. The word linear is used because when (5.1) is written as

$$\frac{dP}{dt} - kP = -at,$$

the left-hand side is a linear combination of P and its derivative.[1] In terms of these variables, other linear differential equations could be expressed by

$$\frac{dP}{dt} + f(t)P = g(t),$$

where f and g are given functions of t.

Slope field

Figure 5.1 shows the slope field and the hand-drawn solution curve for (5.1) subject to (5.2).[2] It suggests that $P(20) \approx 2.2$; that is, the population after 20 years will be about 2.2 million people.

Prediction

Let us now consider other initial values of the population. If we look more carefully at Figure 5.1, particularly in the region between 0 and 1 on the vertical axis and 0 to 20 on the horizontal axis, we see that it appears possible for the population

[1] Recall that a linear combination of $r(t)$ and $s(t)$ is $C_1 r(t) + C_2 s(t)$, where C_1 and C_2 are constants.

[2] The software you are using may have trouble imitating the slope field in Figure 5.1, due to the fact that the horizontal and vertical scales are so different — in this case, 20 units horizontally and 3 units vertically. If this happens, you should rescale either the dependent or the independent variable so the new scales are approximately equal. For example, in this case you could either introduce a new dependent variable, y, where $P = y/10$ — so the new units will be 20 units horizontally and 30 units vertically — or introduce a new independent variable, x, where $t = 10x$ — so the new units will be 2 units horizontally and 3 units vertically.

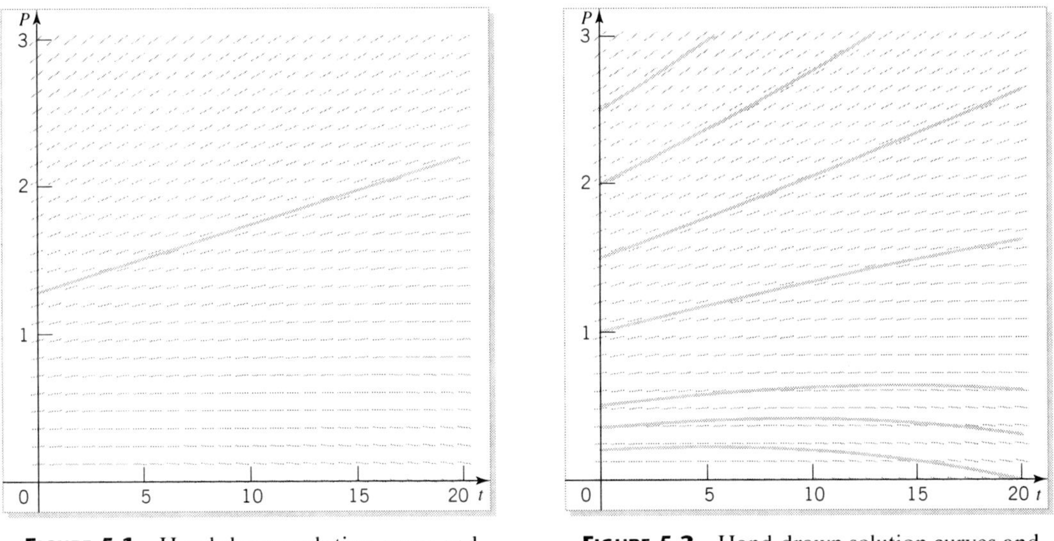

FIGURE 5.1 Hand-drawn solution curve and slope field for $dP/dt = kP - at$

FIGURE 5.2 Hand-drawn solution curves and slope field for $dP/dt = kP - at$

to die out [that is, there is a time when $P(t) = 0$] depending on the initial value $P(0)$. Figure 5.2 shows a number of hand-drawn solution curves, from which it is difficult to decide which ones will eventually die out.

Monotonicity

Isocline

We will use calculus to try to make sense of Figure 5.2. From (5.1), we see that $P(t)$ is increasing if $kP - at$ is positive and decreasing if $kP - at$ is negative. The line $P = at/k$ is where $P' = 0$, which is the isocline corresponding to horizontal tangents. So the line $P = at/k$ separates the plane into two regions: solution curves that lie above this line will be increasing, whereas below the line they will be decreasing.

If we differentiate (5.1) and use (5.1) to eliminate P' from the result, we find

$$\frac{d^2P}{dt^2} = k^2\left(P - \frac{a}{k}t - \frac{a}{k^2}\right).$$

Concavity

The line $P = at/k + a/k^2$ is where $P'' = 0$, so it separates the plane into two regions: solution curves that lie above this line will be concave up and those that lie below this line will be concave down. Notice that this line is parallel to the line $P = at/k$ but is displaced a distance a/k^2 up the vertical axis. Figure 5.3 shows these regions.

Figure 5.4 shows the line $P = at/k$ corresponding to $P' = 0$, the line $P = at/k + a/k^2$ corresponding to $P'' = 0$, and some hand-drawn solution curves to (5.1). This suggests that if a solution curve has its initial value $P(0)$ between 0 and $a/k^2 \approx 1.275$, the population will eventually die out. If $P(0) \geq a/k^2$, then the population will continue to grow.

Explicit solution

Now let's try to find an explicit solution of (5.1). Note that (5.1) is not separable nor is it an equation with homogeneous coefficients.[3] To solve (5.1) we recall that in Section 4.2 on page 138 we transformed a differential equation we could not

[3] Note also that $P(t) = at/k$ is in no sense an equilibrium solution of (5.1) because it does not satisfy the differential equation and its derivative is not zero.

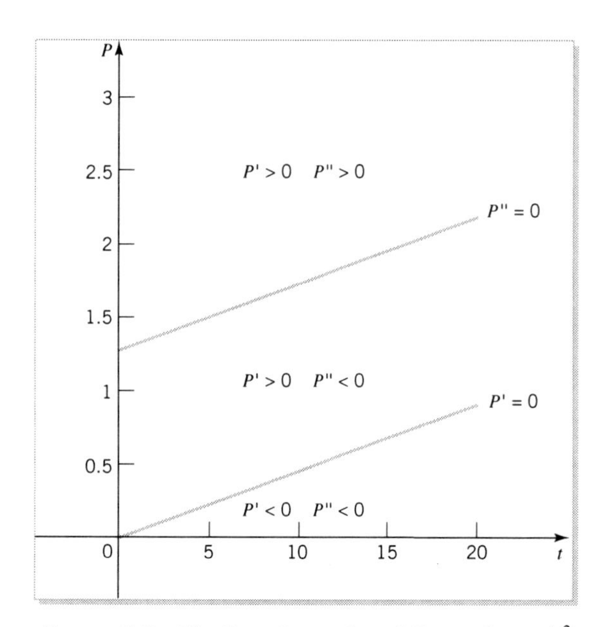

FIGURE 5.3 The lines $P = at/k$ and $P = at/k + a/k^2$

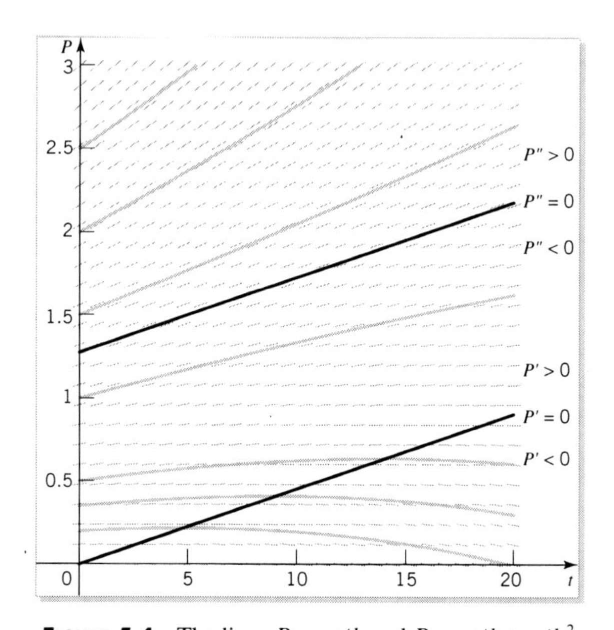

FIGURE 5.4 The lines $P = at/k$ and $P = at/k + a/k^2$, hand-drawn solution curves, and slope field for $dP/dt = kP - at$

Change variable

solve to one we could solve by making a change in the dependent variable y of the form $y(x) = xz(x)$, where x was the independent variable. We make a similar change of variable here (where P is now the dependent variable, and t the independent variable). However, because we do not know in advance what the multiplicative factor should be, we simply denote it by $u(t)$. That is, we let

$$P(t) = u(t)z(t), \tag{5.3}$$

where z is the new dependent variable and $u(t)$ is a specific function that we will choose to make the resulting differential equation solvable.

To decide how to choose $u(t)$, we differentiate (5.3) and obtain

$$\frac{dP}{dt} = u\frac{dz}{dt} + \frac{du}{dt}z, \tag{5.4}$$

and substitute (5.3) and (5.4) into (5.1), yielding

$$u\frac{dz}{dt} + \frac{du}{dt}z = kuz - at.$$

Because we want to choose a $u(t)$ that simplifies this differential equation, we put all terms involving u on the left-hand side. This yields

$$u\frac{dz}{dt} + \left(\frac{du}{dt} - ku\right) z = -at.$$

If we choose u so that the term in parentheses is zero, then we would have

$$u\frac{dz}{dt} + 0z = -at,$$

or

$$\frac{dz}{dt} = -\frac{at}{u(t)}. \tag{5.5}$$

Because $u(t)$ will be a known function of time, (5.5) may be integrated directly. The required function, $u(t)$, that lets this happen satisfies

$$\frac{du}{dt} - ku = 0,$$

or

$$\frac{1}{u}\frac{du}{dt} = k. \tag{5.6}$$

We need only one function, $u(t)$, to make this change of variable work, so we set to zero the arbitrary constant that results from integrating (5.6) and write a solution of (5.6) as

$$u(t) = e^{kt}. \tag{5.7}$$

Integrate Now we may substitute (5.7) into (5.5) and then integrate (5.5) directly, using integration by parts, to obtain

$$z(t) = \int -\frac{at}{u(t)}\,dt = \int -ate^{-kt}\,dt$$

as

$$z(t) = a\left(\frac{t}{k} + \frac{1}{k^2}\right)e^{-kt} + C. \tag{5.8}$$

Equations (5.3) and (5.7) let the solution $P(t)$ of the original problem be expressed as

$$P(t) = e^{kt}z(t),$$

which, by (5.8), can be written as

$$P(t) = a\left(\frac{t}{k} + \frac{1}{k^2}\right) + Ce^{kt}. \tag{5.9}$$

Explicit solution Because, from (5.9), $P(0) = a/k^2 + C$, then $C = P(0) - a/k^2$, and we can rewrite (5.9) as

$$P(t) = a\left(\frac{t}{k} + \frac{1}{k^2}\right) + \left(P(0) - \frac{a}{k^2}\right)e^{kt}. \tag{5.10}$$

Substituting the numerical values for a, k, and $P(0)$ — chosen to satisfy the initial condition given in (5.2) — lets us write our solution (5.10) in final form as

$$P(t) = 0.0452t + 1.275 + 0.01045e^{0.0355t}. \tag{5.11}$$

Confirmation of prediction

This means that after 20 years the population is $P(20) \approx 2.2$, which is about a 71% increase in population instead of about the 100% increase we would have without any emigration. This is in excellent agreement with the information we obtained from Figure 5.1.

Note that our solution in (5.11) is an increasing function for all positive values of t. If we look at (5.10), we see that if $P(0) - a/k^2 > 0$, the population will increase exponentially; if $P(0) - a/k^2 = 0$, the population will increase linearly; and if $P(0) - a/k^2 < 0$, the population will eventually die out. In our case $P(0) - a/k^2 = 0.01045$, which is slightly positive.

Stability

One other comment needs to be made. The solution with $P(0) - a/k^2 = 0$ [namely, $P(t) = at/k + a/k^2$] is UNSTABLE[4] because for $P(0) - a/k^2$ slightly positive or slightly negative, the associated solutions recede from the solution for $P(0) - a/k^2 = 0$ as t increases. Thus, a small change in $P(0)$ will make a large change in the eventual population, either going to exponential growth or dying out. This is consistent with Figure 5.4.

Finally, we can rewrite (5.10) in a more illuminating form, namely,

$$P(t) = P(0)e^{kt} - \frac{a}{k^2}\left(e^{kt} - 1 - kt\right).$$

In this form we may see the impact of emigration. The first term on the right-hand side corresponds to no emigration, and the second to the impact of emigration.

The technique we used to reduce the differential equation (5.1) to one that is separable will work with any first order differential equation of the form

$$y' + p(x)y = q(x), \tag{5.12}$$

where $p(x)$ and $q(x)$ are given functions of x.

◆ *Definition 5.1:* **Differential equations that have the form (5.12) are called LINEAR DIFFERENTIAL EQUATIONS. Note that there is, at most, one expression involving y or y' in each term, and it occurs to the first power.** ◆

Change variable

To show that the general linear differential equation (5.12) can always be reduced to a separable equation, we again seek a change of the dependent variable of the form

$$y(x) = u(x)z(x). \tag{5.13}$$

[4] This terminology is consistent with that used for stable and unstable equilibrium solutions for autonomous differential equations.

Here z is the new dependent variable, and we will look for a judicious choice of the function $u(x)$ so the transformed equation will be separable. Making this substitution — using the product rule to differentiate $u(x)z(x)$ — changes the form of (5.12) to

$$uz' + u'z + p(x)uz = q(x).$$

We rearrange this equation as

$$uz' + \big(u' + p(x)u\big)\, z = q(x). \tag{5.14}$$

We now choose $u(x)$ to satisfy the differential equation

$$u' + p(x)u = 0, \tag{5.15}$$

or, in separable form,

$$\frac{1}{u}u' = -p(x). \tag{5.16}$$

Choosing u in this manner reduces (5.14) to the separable differential equation $uz' = q(x)$, and by dividing this equation by u and integrating, we get

$$z(x) = \int \frac{q(x)}{u(x)}\, dx + C. \tag{5.17}$$

The function u in the denominator of this equation is found by integrating (5.16) as

$$\ln |u(x)| = \int -p(x)\, dx. \tag{5.18}$$

Again note that because we seek only one specific function, $u(x)$, we may set the arbitrary constant of integration at this step to zero and choose the positive solution of $u(x)$ from (5.18) as

$$u(x) = e^{-\int p(x)\, dx}. \tag{5.19}$$

Explicit solution

Combining (5.17) and (5.13) gives the solution, $y(x)$, of our original differential equation as

$$y(x) = u(x) \int \frac{q(x)}{u(x)}\, dx + Cu(x), \tag{5.20}$$

where $u(x)$ is given by (5.19) and C is an arbitrary constant.

Solving Linear Differential Equations

Most people find that memorizing (5.20), with $u(x)$ given by (5.19), as the solution of a general first order linear differential equation is not worth the effort. In fact, further analysis of the structure of this solution, (5.20), leads to the following simple way for its development. If we rewrite (5.20) in the form

$$\frac{1}{u(x)}y(x) = \int \frac{q(x)}{u(x)}\,dx + C$$

and differentiate, we find

$$\left(\frac{1}{u(x)}y(x)\right)' = \frac{q(x)}{u(x)}. \tag{5.21}$$

This suggests that we should be able to multiply (5.12) by $1/u(x) = e^{\int p(x)\,dx}$ and write it in the form (5.21), which can be integrated immediately to yield (5.20). This observation leads to the following three-step process, which applies to a general linear differential equation, $a_1(x)y' + a_0(x)y = f(x)$.

How to Solve Linear Differential Equations

Purpose To find $y = y(x)$ that satisfies the linear differential equation $a_1(x)y' + a_0(x)y = f(x)$, where $a_1(x)$ is not zero.

Process

1. Put the linear differential equation in the standard form of

$$y' + p(x)y = q(x) \tag{5.22}$$

by dividing both sides of the equation by $a_1(x)$. Compute the function $\mu(x) = e^{\int p(x)\,dx}$ (where the constant of integration in $\int p(x)\,dx$ is set to zero), which has the property that

$$\mu' = \mu(x)p(x). \tag{5.23}$$

2. Multiply both sides of the differential equation (5.22) by this function $\mu(x)$ to get

$$\mu(x)y' + \mu(x)p(x)y = \mu(x)q(x). \tag{5.24}$$

Notice that the term $\mu(x)p(x)$ on the left-hand side of (5.24) is equal to μ' by (5.23). If we make this replacement in (5.24), we obtain

$$\mu y' + \mu' y = \mu(x)q(x).$$

Here we see that the left-hand side of this equation may be written as the derivative of the product $\mu(x)y$. Thus

$$[\mu(x)y]' = \mu y' + \mu' y = \mu(x)y' + \mu(x)p(x)y,$$

so (5.24) is equivalent to

$$[\mu(x)y]' = \mu(x)q(x). \tag{5.25}$$

You should always check your work at this point by differentiating the left-hand side of (5.25). The resulting equation should be the same as the one obtained by multiplying the differential equation (5.22) by $\mu(x)$.

3. Integrate both sides of (5.25) to obtain

$$\mu(x)y = \int \mu(x)q(x)\,dx + C.$$

Finally, divide both sides of this last equation by the function $\mu(x) = e^{\int p(x)\,dx}$ to obtain the explicit solution of (5.22); that is,

$$y(x) = \frac{1}{\mu(x)}\int \mu(x)q(x)\,dx + \frac{C}{\mu(x)}, \tag{5.26}$$

which agrees with (5.20).

Comments about Linear Differential Equations

- The function $\mu = e^{\int p(x)\,dx}$ is called an INTEGRATING FACTOR of (5.22) because if we multiply (5.22) by this factor, we can integrate both sides of the resulting equation.
- The reason that we set the constant of integration c in $\int p(x)\,dx$ to zero in step 1 is as follows. If $\mu = e^{\int p(x)\,dx}$ is the function with $c = 0$, and $\mu_1 = e^{\int p(x)\,dx}$ is the function with $c \neq 0$, then $\mu_1 = \mu e^c$. (Why?) If in step 2 we multiply (5.22) by μ_1 we find $\mu e^c y' + \mu e^c py = \mu e^c q$, which, after cancelling the common factor of e^c, is (5.24) — the result we would have obtained had we just set $c = 0$ in the first place. For the same reason, if, in a particular example, we find $\mu = |x| = \pm x$, we need only use $\mu = x$ — the choice $\mu = -x$ is unnecessary, because it leads to the same equation after canceling the -1 each term has in common.
- Because (5.26) contains every solution of (5.22), the explicit solution in (5.26) is often called the GENERAL SOLUTION of (5.22).
- The term on the right-hand side of (5.22) — namely $q(x)$ — is frequently called the FORCING FUNCTION.
- A common mistake when going from (5.22) to (5.24) is to forget to multiply the right-hand side of (5.22) by the integrating factor. This can lead to some unexpected integrals when we reach (5.26).
- Another common mistake when going from (5.25) to (5.26) is to forget C, the constant of integration. As a result, many solutions of (5.22) will be left out.
- Solutions of first order linear differential equations are automatically explicit solutions.

- Linear differential equations are very important. They occur frequently in applications.

Two Examples

EXAMPLE 5.2

We consider the differential equation that we considered repeatedly from different points of view in Chapter 3, namely,

$$y' = x - y, \tag{5.27}$$

which we now recognize as a linear differential equation. (How?)

We apply the preceding three steps.

1. First we change the form of (5.27) to

$$y' + y = x. \tag{5.28}$$

Integrating factor

Here $p(x) = 1$ and $q(x) = x$. The integrating factor is

$$\mu(x) = e^{\int p(x)\,dx} = e^{\int 1\,dx} = e^x.$$

2. Multiply (5.28) by the integrating factor $\mu(x) = e^x$ to obtain

$$e^x y' + e^x y = xe^x.$$

Here we take advantage of the fact that the left-hand side of this equation is the derivative of the product μy to write it as

$$(e^x y)' = xe^x. \tag{5.29}$$

[You should differentiate the left-hand side of (5.29) to verify this.]

3. Integrate (5.29) to obtain

$$e^x y = \int xe^x\,dx = (x-1)\,e^x + C,$$

Explicit solution

where we used integration by parts to evaluate the integral. Now multiply by e^{-x} to obtain the explicit solution of (5.27) as

$$y(x) = x - 1 + Ce^{-x},$$

which agrees with the result we found in Chapter 3.

EXAMPLE 5.3

As another example consider the differential equation

$$xy' + 2y = 4x^2. \tag{5.30}$$

Apply the three steps previously given.

1. Divide by x to change the form of (5.30) to

$$y' + \frac{2}{x}y = 4x. \tag{5.31}$$

Integrating factor

Here $p(x) = 2/x$ and $q(x) = 4x$. The integrating factor is

$$\mu = e^{\int \frac{2}{x}\,dx} = e^{2\ln|x|} = e^{\ln x^2} = x^2.$$

2. Multiply (5.31) by $\mu = x^2$ to obtain $x^2y' + 2xy = \left(x^2\right)4x = 4x^3$, or, equivalently,

$$\left(x^2y\right)' = 4x^3. \tag{5.32}$$

3. Integrate (5.32) to obtain

$$x^2y = \int 4x^3\,dx = x^4 + C,$$

Explicit solution

and divide by x^2 to obtain the general solution of (5.30) as

$$y(x) = \frac{x^4}{x^2} + \frac{C}{x^2} = x^2 + Cx^{-2}. \tag{5.33}$$

We see that (5.33) contains two types of solutions, those in which y is defined at $x = 0$ (namely, when $C = 0$, so that $y = x^2$), and those in which y is not defined at $x = 0$ (namely, when $C \neq 0$).

Slope field

Figure 5.5 shows the slope field and various solution curves of (5.30), including $y = x^2$. We see that the slope field is symmetric about the y-axis. (How would you confirm this?)

Stability

We also notice from Figure 5.5 that as $x \to 0$ from the left, all solution curves [except $y(x) = x^2$] go to ∞ if the solution curve starts above $y = x^2$, and go to $-\infty$ if the solution curve starts below $y = x^2$. So $y(x) = x^2$ is an unstable solution as $x \to 0$ from the left.

For $x > 0$, all solution curves in Figure 5.5 appear to go to ∞ as $x \to \infty$; furthermore, they all seem to approach $y = x^2$ as $x \to \infty$. So $y(x) = x^2$ is a stable solution when $x > 0$ and $x \to \infty$.

These observations that were made by looking at Figure 5.5 can be confirmed directly from the solution (5.33). Finally, we note from (5.33) that the only solution that is valid for all x is the one where $C = 0$; namely, $y(x) = x^2$.

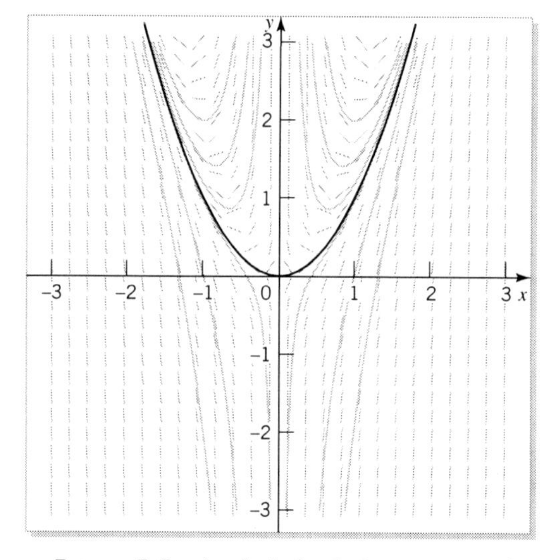

FIGURE 5.5 Analytical solution curves and slope field for $xy' + 2y = 4x^2$

Existence-Uniqueness of Solutions

If we apply the Existence-Uniqueness Theorem on page 56 to the linear differential equation $y' + p(x)y = q(x)$, we find that the conditions of that theorem are satisfied if $p(x)$ and $q(x)$ are continuous in some interval $a < x < b$. However, when these conditions are satisfied, we have the following stronger theorem, which guarantees the existence and uniqueness of solutions for **linear** differential equations.

▶ ***Theorem 5.1: The Existence-Uniqueness Theorem for First Order Linear Differential Equations.*** *Consider the initial value problem*

$$y' + p(x)y = q(x), \qquad y(x_0) = y_0. \tag{5.34}$$

If $p(x)$ and $q(x)$ are continuous for $a < x < b$, and x_0 is in this interval, then there exists a unique solution of (5.34) in $a < x < b$. ◀

Comments about the Existence-Uniqueness Theorem for First Order Linear Differential Equations

- This is a stronger theorem than our previous result. In the previous result a unique solution is guaranteed only in the vicinity of the initial point. Here a unique solution is guaranteed in the same interval that $p(x)$ and $q(x)$ are continuous, no matter how large. Thus, we need analyze only $p(x)$ and $q(x)$ to find the intervals in which a unique solution will exist.

EXERCISES

1. Solve the following differential equations. State the interval over which your solution is defined and confirm that it agrees with the prediction from the Existence-Uniqueness Theorem for First Order Linear Differential Equations. Make sure that the general solution you obtain is supported by the slope field, increasing and decreasing regions, and concavity issues.

(a) $y' + y = x$
(b) $y' - 2y = xe^{2x}$
(c) $y' + y = e^x$
(d) $y' - 6y = e^x$
(e) $y' + 3y = 3x^2e^{-3x}$
(f) $y' + y/(x \ln x) = 3x^2$
(g) $xy' + 2y = x/(x^2+2)$
(h) $y' + xy = 3x$
(i) $y' + y/(x+1) = \cos x/(x+1)$
(j) $(\cos x)\, y' + y \sin x = 3 \sin x \cos^2 x$
(k) $y' + 2xy = 2x^3$
(l) $xy' - y = x^2 \cos x$
(m) $y' - y/(2x) = 2$
(n) $y' + 2xy = 2x$
(o) $y' - 2y/x^2 = 1/x^2$
(p) $(x^2+1)y' + 4xy = x$
(q) $y' + (2 + x^{-1})y = 2e^{-2x}$
(r) $xy' + (2x+1)/(x+1)y = x - 1$
(s) $y' + 4xy/(x^2+1) = 3x$
(t) $x^4y' + 2x^3y = 1$

2. Compare the form of your solutions in Exercises 1 (a) through (e). State what is similar and what is different about the terms in your solution.

3. Identify the following differential equations as separable equations, those with homogeneous coefficients, linear equations, or none of these. (Some may fall into more than one category.) Do not attempt to solve any of these equations.

(a) $(2+y)y' - \sin x = 0$
(b) $(x^2+y^2)y' + y^2 - x^2 = 0$
(c) $2xy + y' \sin y = 0$
(d) $y \sin(x/y) + xy' = 0$
(e) $x^2y' + 3xy - \sin x = 0$
(f) $x^2y' + 3xy^2 - \sin x = 0$
(g) $y' - xy^2 + 1 = 0$
(h) $y \cos x + \ln x - y' = 0$

4. Suppose that over the next 20 years, P_0/α (that is, $1/\alpha$ of the initial population) leaves some specific country whose current rate of change of population is proportional to the population. If this number of people leaves the country at a constant rate over these 20 years, the differential equation that models this situation would be $P' = kP - \frac{1}{20\alpha}P_0$, subject to the initial condition $P(0) = P_0$. Show that if $k > 1/(20\alpha)$, all solutions of this initial value problem are increasing and concave up. (Do this using the differential equation directly and also using the explicit solution.)

5. Use the solution of the preceding exercise to compare the population of Botswana after 20 years of linear emigration to that of 20 years of constant emigration. (Assume that the total emigration is the same in the two models.)

6. Consider the following initial value problem, which models some population as a function of time: $P' = kP - at$, $P(0) = P_0$.

(a) State in words what assumptions are made in writing down this model, and give interpretations for the three constants, k, a, and P_0.

(b) Use the differential equation to discover where the solution to this initial value problem is increasing or decreasing, concave up or concave down.

(c) Solve this initial value problem and find conditions on the parameters that will guarantee that the population will (i) die out, and (ii) not die out. How much of your answer could have you determined from your results of part (b)?

7. Look at the slope field for the differential equation $(x^2+1)\, y' + xy = \sqrt{x^2+1}$. Make a conjecture about the behavior of solutions $y(x)$ as $x \to \infty$. Try to prove this conjecture by analyzing the differential equation together with its isocline for slope 0. Now obtain the explicit family of solutions of this differential equation. Look at what happens to this family as $x \to \infty$, and confirm this previous conjecture.

8. Sinking Barrel. The differential equation for a barrel sinking in a body of water under the influence of gravity and water resistance is $mv' = mg - kv$, where m is the mass, g is the gravitational constant, and kv is the frictional force.

(a) By using the slope field or the differential equation directly, determine whether there is a limiting velocity. If so, what is it? If not, why isn't there one?

(b) If the barrel has an initial velocity, v_0, at the surface of the water, find the explicit solution of the associated initial value problem. Show that this explicit solution verifies your answer in part (a).

5.2 MODELS THAT USE LINEAR EQUATIONS

In this section we look at five different applications of linear equations that show the breadth of its modeling applicability.

In Example 4.1 on page 120, we discussed a simple mixture problem that led to a separable differential equation. Now we look at a more complicated example. Recall that to derive the differential equation that describes mixture processes, we let x be a function that represents the amount of substance in a given container at time t, and we assume that the instantaneous rate of change of x with respect to t is governed by a conservation equation — also called an equation of continuity — given by

$$\frac{dx}{dt} = \textit{rate at which substance is added to the container} - \textit{rate at which substance is leaving the container}.$$

EXAMPLE 5.4 *Solute in a Container Again*

A 300-gallon container is 2/3 full of water containing 50 pounds of salt. At time $t = 0$, valves are turned on so a salt solution of concentration 1/3 pounds per gallon is added to the container at a rate of 3 gallons per minute. If the well-stirred mixture is drained from the container at the rate of 2 gallons per minute, how many pounds of salt are in the container when it is full (and all valves are turned off)?

Notice that the only difference between this example and Example 4.1 on page 120 is that the water being added is not pure, but here contains 1/3 pounds of salt per gallon. Thus, the rate at which salt is added to the container is $\frac{1}{3} \times 3 = 1$ pounds of salt per minute, instead of $0 \times 3 = 0$ pounds of salt per minute. Otherwise, we mimic the analysis used in Example 4.1.

We again note that more of the liquid is being added per minute than is being drained, so the number of gallons in the container is increasing. The rate of change of volume, V, of liquid in the container is the difference between the rate being added and the rate being drained. Thus, we have that

$$\frac{dV}{dt} = 3 - 2 = 1 \text{ gallon per minute.}$$

Find volume

Integration gives

$$V(t) = t + 200. \tag{5.35}$$

Conservation equation

If x represents the number of pounds of salt in the container at time t, the concentration of salt at time t is $x/V = x/(t + 200)$ pounds per gallon. Thus, the rate at which salt is leaving the container at time t is $2x/(t + 200)$ pounds per minute, and from the conservation equation we have

$$\frac{dx}{dt} = 1 - \frac{2x}{t + 200}, \tag{5.36}$$

which is valid for $0 < t < 100$, with initial condition $x(0) = 50$.

Slope field

If we look at the slope field for (5.36) in Figure 5.6, we observe that the slope field ignores the condition that the container is full when $t = 100$. We have hand-drawn the solution curve that passes through the initial point $(0, 50)$. From this curve we can estimate the value of $x(t)$ for $t = 100$ at about 90 pounds, which is an approximate answer to our original question. We can obtain a better numerical approximation by using the Runge-Kutta 4 method described in Chapter 3, and, with $h = 1$, we find $x(100) \approx 92.593$ pounds. However, to obtain an exact answer we must obtain an explicit solution.

Prediction

The only difference between (5.36) and the differential equation in Example 4.1 — namely, $dx/dt = -2x/(t+200)$ — is the presence of the constant 1 on the right-hand side, but that constant is enough to prevent (5.36) from being a separable differential equation. However, if we rearrange (5.36) as

$$\frac{dx}{dt} + \frac{2x}{t+200} = 1, \tag{5.37}$$

we see that it is a linear differential equation. Using our usual process, we first find the integrating factor as

Integrating factor

$$\mu = e^{\int \frac{2}{t+200}\,dt} = e^{2\ln|t+200|} = (t+200)^2.$$

If we multiply both sides of (5.37) by this integrating factor, we obtain

$$(t+200)^2\frac{dx}{dt} + 2(t+200)x = (t+200)^2,$$

or

$$\frac{d}{dt}\left[(t+200)^2 x\right] = (t+200)^2. \tag{5.38}$$

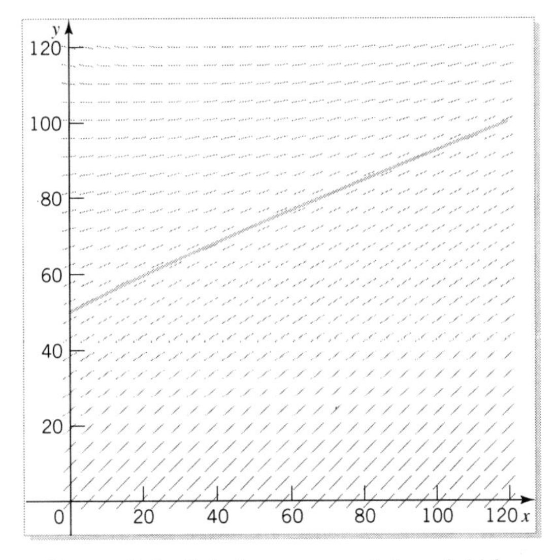

FIGURE 5.6 Solution curve and slope field for $dx/dt = 1 - 2x/(t+200)$

Integrate Integration of (5.38) yields

$$(t+200)^2 x = \frac{1}{3}(t+200)^3 + C,$$

from which we find x as

$$x(t) = \frac{1}{3}(t+200) + \frac{C}{(t+200)^2}.$$

Explicit solution The choice of C as $-50(200^2)/3$ will satisfy the initial condition $x(0) = 50$, so our final form for the solution is

$$x(t) = \frac{1}{3}(t+200) - \frac{50(200)^2}{3(t+200)^2}. \tag{5.39}$$

Confirmation of prediction To answer the original question — how many pounds of salt are in the container when it is full — we note from (5.35) that the container will be full when $t = 100$, so $x(100)$ will be the amount of salt in the container at this time. From (5.39) we have that

$$x(100) = \frac{1}{3}(300) - \frac{50(200)^2}{3(300)^2} = \frac{2500}{27} = 92\frac{16}{27} \approx 92.59259 \text{ pounds of salt,}$$

which is close to our previous estimates of 90 and 92.593 pounds. ■

Taking a closer look at (5.39), we notice that the second term approaches 0 as $t \to \infty$. Thus, as $t \to \infty$, the solution approaches the straight line $\frac{1}{3}(t+200)$. This gives rise to the following definition.

◆ *Definition 5.2:* **Consider the explicit solution of a linear differential equation. If there is a portion of this solution that does not approach zero as the independent variable approaches infinity, it is called the STEADY STATE part of the solution. For this situation, the portion of this solution that does approach zero as the independent variable approaches infinity is called the TRANSIENT PART of the solution.** ◆

Comments about the Steady State and Transient Parts of the Solution

- Often these parts of the solution are referred to as the "steady state solution" and the "transient solution," even though they may not be solutions.
- The words steady state refer to the fact that this part accurately describes the solution's behavior after a long period of time, not the fact that it is "steady."
- Frequently the long-term behavior of the solution of a differential equation is more important than the short-term behavior. Thus, we are often interested in the steady state part of the solution.

In the previous example where the general solution is $\frac{1}{3}(t+200) + C/(t+200)^2$, the transient part of the solution is $C/(t+200)^2$ and the steady state part is $\frac{1}{3}(t+200)$.

EXAMPLE 5.5 *An Electrical RL Circuit*

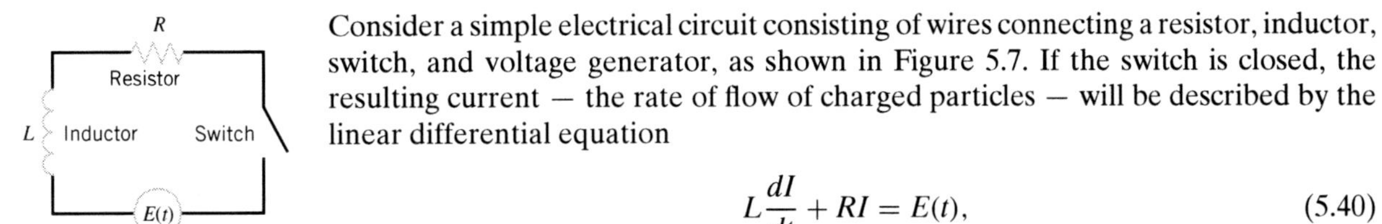

FIGURE 5.7
RL electrical circuit

Consider a simple electrical circuit consisting of wires connecting a resistor, inductor, switch, and voltage generator, as shown in Figure 5.7. If the switch is closed, the resulting current — the rate of flow of charged particles — will be described by the linear differential equation

$$L\frac{dI}{dt} + RI = E(t), \tag{5.40}$$

where L is the inductance (units of henries), R is the resistance (units of ohms), I is the current (units of amperes), and $E(t)$ is the output of the voltage generator (units of volts). Such a circuit is called an RL CIRCUIT.

Initial condition

Consider a specific circuit where the inductance is 1 henry, the resistance is 20 ohms, and the voltage generator has the form of $25e^{-20t}$, which eventually decays to zero. We consider $t = 0$ as the time when the switch is closed. A natural initial condition for this situation is that there be zero current when the switch is closed, so $I(0) = 0$. We wish to find the behavior of the current in this circuit as a function of time. To do this, we must find the explicit solution of the initial value problem

$$\frac{dI}{dt} + 20I = 25e^{-20t}, \qquad I(0) = 0. \tag{5.41}$$

Integrating factor

The integrating factor is e^{20t}. We multiply (5.41) by this integrating factor to obtain

$$e^{20t}\frac{dI}{dt} + 20e^{20t}I = 25$$

and rewrite this equation as

$$\frac{d}{dt}\left(e^{20t}I\right) = 25.$$

Finally, we integrate this to obtain

$$e^{20t}I = 25t + C$$

Explicit solution

and divide by e^{20t} to obtain the general solution $I(t) = 25te^{-20t} + Ce^{-20t}$. Choosing the arbitrary constant C as 0 lets us satisfy the initial condition of $I(0) = 0$, giving

$$I(t) = 25te^{-20t}. \tag{5.42}$$

This function is plotted in Figure 5.8. Notice that the current increases to a maximum and then decays. We could ask when the current will be zero again. The answer is never. We can show this from (5.42), where we see that the only time when $I(t) = 0$ is $t = 0$. From (5.42) we see that $I \to 0$ as $t \to \infty$. Thus, the current never reaches zero for a finite value of the time. However, from a practical point of view, there will be a time after which we are unable to detect the current.

Settling time

A useful quantity to define is the SETTLING TIME, which is the time after which a response is no larger than 1% of its maximum value. What is the settling time for the current in our example? For this we first need to find when the maximum current, I_{max},

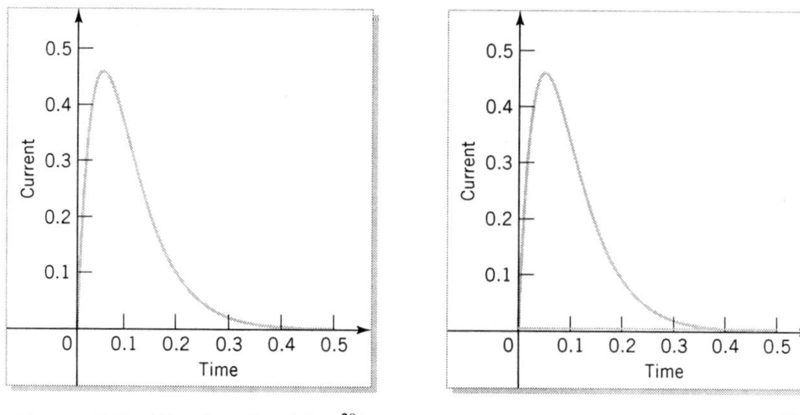

FIGURE 5.8 The function $25te^{-20t}$

FIGURE 5.9 The function $25te^{-20t}$ and the line $I_{max}/100$

occurs. This will be the time, T, when $I' = 0$. From (5.41) we see that this will occur when $20I(T) = 25e^{-20T}$, which, by (5.42), can be written $20\left(25Te^{-20T}\right) = 25e^{-20T}$, so that $T = \frac{1}{20}$. Thus, the maximum current is

$$I_{max} = I(T) = \frac{25}{20e} \approx 0.46.$$

Both of these values are consistent with Figure 5.8. The settling time, t_s, will be the time after which the current is no larger than $I_{max}/100$. Thus, the settling time satisfies

$$25t_s e^{-20t_s} = \frac{25}{20e}\frac{1}{100},$$

or

$$t_s e^{-20t_s} = \frac{1}{2000e}.$$

This equation cannot be solved using analytical techniques, but can be solved by any standard numerical root-finding method. We find two approximate solutions; namely, 0.00018 and 0.389192. The first occurs before the time when I_{max} occurs. The second is the settling time, so

$$t_s \approx 0.389192 \text{ sec.}$$

Figure 5.9 shows the current and the horizontal line $I = I_{max}/100 \approx 0.0046$.

EXAMPLE 5.6 *Fish Harvesting*

We started Section 5.1 with a mathematical model of human population growth that included emigration. Now we consider a model of the population of fish in a lake in which there are no predators but there is an abundant supply of food for the fish (for example, a fish farm). Fish are harvested at a periodic rate described by the function

$h(t) = a + b \sin 2\pi t$, where a and b are constants, $a > b$, and t is the time. Note that $h(t)$ is always positive; always oscillating between $a + b$ and $a - b$; and

$$\int_0^1 h(t)\, dt = a$$

is the number of fish harvested per year.

If we assume that fish reproduce at a rate proportional to their population, the appropriate differential equation that models this situation is

$$\frac{dP}{dt} = kP - (a + b \sin 2\pi t)\,. \tag{5.43}$$

Slope field

Here P is the fish population and $k > 0$ is the net growth rate. Figure 5.10 gives the slope field for this situation for $k = 0.5$, $a = 3$, $b = 1$, along with numerical solutions for five different initial conditions. It is apparent that in these five cases, the fish population either dies out or grows quite rapidly. There does not seem to be a simple periodic solution for this mathematical model. To investigate this further, let us find the general solution of this linear differential equation subject to $P(0) = P_0$.

We put (5.43) in the standard form

$$\frac{dP}{dt} - kP = -(a + b \sin 2\pi t) \tag{5.44}$$

Integrating factor

and compute the integrating factor as e^{-kt}. We then multiply (5.44) by this integrating factor to obtain

$$\frac{d}{dt}\left(e^{-kt}P\right) = -e^{-kt}\,(a + b \sin 2\pi t)\,. \tag{5.45}$$

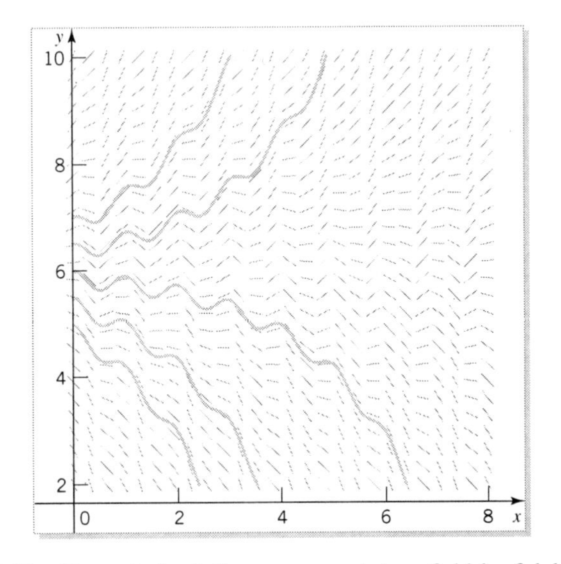

FIGURE 5.10 Numerical solution curves and slope field for fish harvesting

Integrate Finally, we integrate (5.45) — using integration by parts or a table of integrals — to obtain

$$e^{-kt}P = e^{-kt}\left(\frac{a}{k} + b\frac{k\sin 2\pi t + 2\pi\cos 2\pi t}{k^2+4\pi^2}\right) + C$$

Explicit solution and divide by e^{-kt} to obtain the general solution

$$P(t) = \left(\frac{a}{k} + b\frac{k\sin 2\pi t + 2\pi\cos 2\pi t}{k^2+4\pi^2}\right) + Ce^{kt}. \tag{5.46}$$

We use the initial condition $P(0) = P_0$ to evaluate C as

$$C = P_0 - \frac{a}{k} - \frac{2\pi b}{k^2+4\pi^2}.$$

From (5.46) we see that for any positive value of the growth rate, k, the exponential part of the solution will dominate the trigonometric part for large values of t. If $C > 0$, the solution will grow without bound, and if $C < 0$, it will become unbounded in the negative direction. The only way there will be a bounded solution is for C to equal zero. ***Stability*** This solution is unstable, because for C slightly positive or slightly negative, the associated solutions do not stay close to the solution for $C = 0$. The solution curve corresponding to the initial population, P_0, that makes $C = 0$ in (5.46) and the solution curves for $C = -0.05$ and $C = 0.05$ are shown in Figure 5.11.

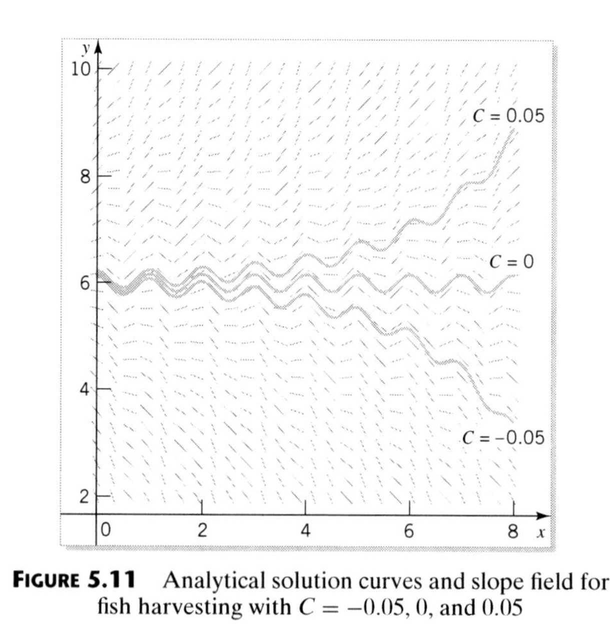

FIGURE 5.11 Analytical solution curves and slope field for fish harvesting with $C = -0.05$, 0, and 0.05

EXAMPLE 5.7 *Yam in the Oven*

Consider the situation in which a cook places a yam in an oven (at room temperature, 70°F) and simultaneously turns the oven on to 400°F. If it takes the oven 5 minutes to reach 400°F, and if it does so in a linear manner, the temperature in the oven, $T_a(t)$, is described by

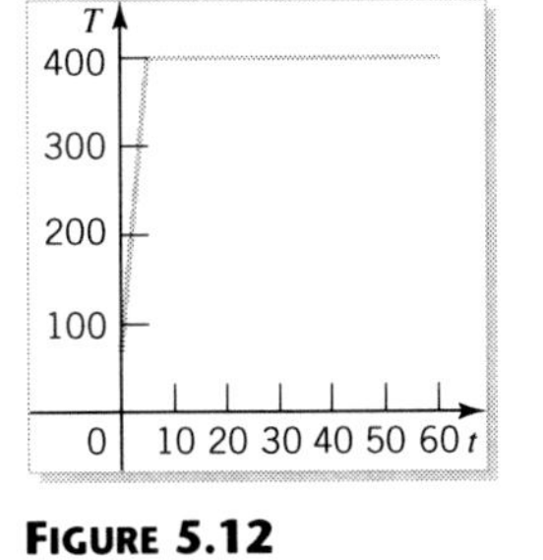

FIGURE 5.12
The function $T_a(t)$

$$T_a(t) = 70 + \frac{400 - 70}{5}t = 70 + 66t$$

for the first 5 minutes, after which the temperature remains at a constant 400°F. Thus,

$$T_a(t) = \begin{cases} 70 + 66t & \text{if } 0 \le t \le 5, \\ 400 & \text{if } 5 < t. \end{cases} \tag{5.47}$$

Figure 5.12 shows the function $T_a(t)$. It is continuous at $t = 5$. (Why?)

If the temperature of the yam is given by $T(t)$, and it obeys Newton's law of heating with $T_a(t)$ as the ambient temperature, we have

$$\frac{dT}{dt} = k\left[T - T_a(t)\right], \ k < 0. \tag{5.48}$$

Initial condition

If the yam is initially at room temperature, the appropriate initial condition is $T(0) = 70$.

Slope field

The slope field associated with (5.48) for $k = -0.04$ is shown in Figure 5.13, and it appears that regardless of the initial temperature of the yam, its temperature approaches a limiting value. Also, there seems to be an equilibrium solution near 400°F. It appears that for $t > 5$ the solution curve is increasing, but we must look at the numerical solution curve (given in Figure 5.14 for the initial condition of 70°F) to discover that this solution curve starts out concave up and then changes to concave down for larger values of t. From (5.47) and (5.48) we see that if we consider only values of t greater than 5, then $T = 400$ is an equilibrium solution. From (5.48) we

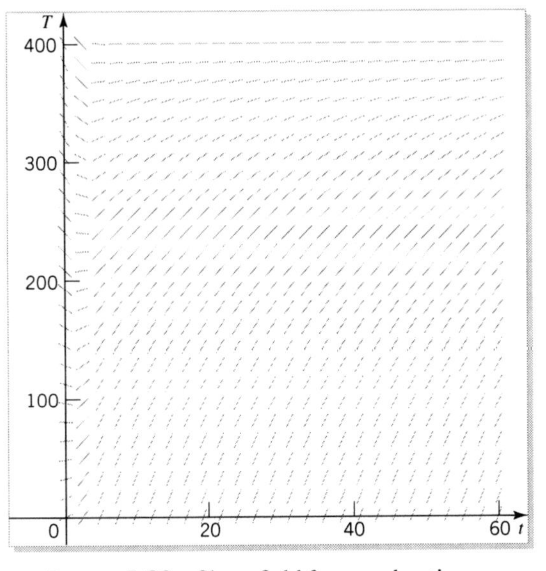

FIGURE 5.13 Slope field for yam heating up

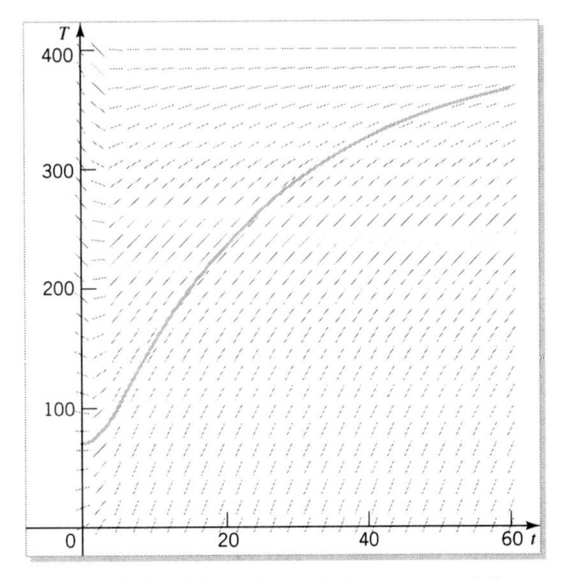

FIGURE 5.14 Numerical solution curve and slope field for yam heating up

Concavity also see that the solution is increasing for all t. But to analyze concavity in more detail, we need to differentiate (5.48) to obtain

$$\frac{d^2T}{dt^2} = k\left(\frac{dT}{dt} - \frac{dT_a(t)}{dt}\right),$$

which, by (5.47), reduces to

$$\frac{d^2T}{dt^2} = \begin{cases} k^2\,[T(t) - 70 - 66t - 66/k] & \text{if } 0 \le t < 5, \\ k^2\,[T(t) - 400] & \text{if } 5 < t. \end{cases} \tag{5.49}$$

Notice that $T_a(t)$ is not differentiable at $t = 5$, because it has a "corner" there — see Figure 5.12. Thus, d^2T/dt^2 does not exist at $t = 5$, which is why the value $t = 5$ is not included in the domain of (5.49).

Equation (5.49) shows that although initially all solutions are concave up [$k < 0$ and when $t = 0$, $T(0) = 70$], they are concave down for $t > 5$ and $T(t) < 400$. From this formulation, an inflection point can occur only if $T(t) - T_a(t) = 66/k$ at some point during the time interval $0 < t < 5$, or if d^2T/dt^2 does not exist (namely, $t = 5$). However, we cannot tell from (5.49) which will happen. We can discover this from the explicit solution.

Equation (5.48) is a linear differential equation, so we find the explicit solution by writing (5.48) as

$$\frac{dT}{dt} - kT = -kT_a(t) \tag{5.50}$$

Integrating factor and computing the integrating factor as e^{-kt}. We then multiply (5.50) by this integrating factor to obtain

$$e^{-kt}\frac{dT}{dt} - ke^{-kt}T = -kT_a(t)e^{-kt},$$

or, equivalently,

$$\frac{d}{dt}\left(e^{-kt}T\right) = -kT_a(t)e^{-kt}, \tag{5.51}$$

Integrate and integrate (5.51) to obtain

$$e^{-kt}T = -k\int T_a(t)e^{-kt}\,dt + C. \tag{5.52}$$

Writing the indefinite integral as one from 0 to t, we can choose the arbitrary constant to satisfy the initial condition $T(0) = 70$, so that $C = 70$. We then divide (5.52) by

Explicit solution the integrating factor, and write the explicit solution as

$$T(t) = -ke^{kt}\int_0^t T_a(u)e^{-ku}\,du + 70e^{kt}. \tag{5.53}$$

Equation (5.47) lets us expand (5.53) as

$$T(t) = \begin{cases} 70 + 66t + 66\left(1 - e^{kt}\right)/k & \text{if } 0 \le t \le 5, \\ 400 + \left[400 + 66(1 - e^{5k})/k\right] e^{k(t-5)} & \text{if } 5 > t. \end{cases} \tag{5.54}$$

From (5.54) we may confirm our earlier finding (as well as our common sense) that the temperature of the yam approaches 400°F for large values of time. The graph of this function coincides with the numerical solution given in Figure 5.14.

Let's return to the location of the inflection point. If we substitute (5.54) into (5.49) for $t < 5$, we find $T'' = -66ke^{kt}$, which is always positive. Thus, the inflection point occurs at exactly $t = 5$. ■

EXAMPLE 5.8 *Electromagnetic Waves*

The initial value problem

$$y' + 2xy = 1, \; y(0) = 0, \tag{5.55}$$

arises in the theory of propagation of electromagnetic waves. We want to describe the behavior of the solution.

Integrating factor This is a linear differential equation with integrating factor $\mu = e^{x^2}$, which leads to

$$\left(e^{x^2} y\right)' = e^{x^2}.$$

Explicit solution Integrating and using the initial condition gives the explicit solution

$$y(x) = e^{-x^2} \int_0^x e^{t^2}\, dt. \tag{5.56}$$

This integral cannot be evaluated in terms of familiar functions.

We have found the explicit solution of the initial value problem (5.55), and some may be satisfied with this answer. We are not. How does $y(x)$ behave? Where is the function increasing or decreasing? Where is the function concave up or concave down? What is the long-term behavior?

The first thing we notice is that, from (5.56), $y(x)$ is always positive for $x > 0$. (Why?) If in (5.55) we interchange x with $-x$ and y with $-y$, we arrive back at the same differential equation, so the slope field will be symmetric about the origin. Thus, we can use this symmetry to obtain the behavior of (5.56) for $x < 0$, from that of $x > 0$. So (5.56) gives values for $y(x)$ in the first and third quadrants. With this information, we concentrate on the first quadrant.

Symmetry

Because $y' = 1 - 2xy$, we see that the isocline for slope zero is the hyperbola $y = 1/(2x)$. Furthermore, $y' < 0$ when $y < 1/(2x)$, and $y' > 0$ when $y > 1/(2x)$. Thus, $y(x)$ will have a local maximum if it crosses the hyperbola $1/(2x)$. It will have no minima in the first quadrant, so it can have only one maximum. (Why?) Where does this maximum occur? If we let $\left(x_m, y_m\right)$ be the maximum, where $y_m = y(x_m)$, then $y_m = 1/\left(2x_m\right)$; that is, x_m must satisfy

Monotonicity

$$e^{-x_m^2} \int_0^{x_m} e^{t^2}\, dt = \frac{1}{2x_m}.$$

There is no simple way to solve this for x_m, so we resort to experimenting with different numerical values for x_m. We find (see Exercise 24 on page 211) that $x_m \approx 0.924$, in which case $y_m = 1/(2x_m) \approx 0.541$. (How do we know that there is not a second value for x_m?)

Concavity So $y(x)$ increases from (0, 0) to approximately (0.924, 0.541) and then decreases forevermore. How does $y(x)$ increase and decrease? That is a question of concavity, for which we require y'', namely,

$$y'' = -2y - 2xy' = -2y - 2x(1 - 2xy) = 2\left[y\left(2x^2 - 1\right) - x\right].$$

Thus, the only place that points of inflection can occur is where the solution crosses the curve $y = x/\left(2x^2 - 1\right)$. To find possible points of inflection, (x_i, y_i) where $y_i = y(x_i)$, we have $y_i = x_i/\left(2x_i^2 - 1\right)$; that is, x_i must satisfy

$$\frac{x_i}{(2x_i^2 - 1)} = e^{-x_i^2} \int_0^{x_i} e^{t^2}\, dt.$$

There is no simple way to solve this for x_i, so we resort to experimenting with different numerical values for x_m. We find (see Exercise 25 on page 211) $x_i \approx 1.502$, in which case $y_m = x_i/\left(2x_i^2 - 1\right) \approx 0.428$. How do we know there are no more points of inflection? There aren't any, but the justification for this is more involved (see Exercise 26 on page 211). Because $x_m < x_i$ and (x_m, y_m) is where the maximum occurs, the function $y(x)$ must be concave down before x_i and concave up forever after.

Long-term behavior To find out about the long-term behavior, we want to know what happens to $y(x)$ as $x \to \infty$, that is,

$$\lim_{x\to\infty} y(x) = \lim_{x\to\infty} \frac{\int_0^x e^{t^2}\, dt}{e^{x^2}}.$$

Because the numerator and denominator both go to ∞ as $x \to \infty$, we may apply L'Hôpital's rule, obtaining[5]

$$\lim_{x\to\infty} y(x) = \lim_{x\to\infty} \frac{\int_0^x e^{t^2}\, dt}{e^{x^2}} = \lim_{x\to\infty} \frac{e^{x^2}}{2xe^{x^2}} = \lim_{x\to\infty} \frac{1}{2x} = 0.$$

Thus, $y(x)$ approaches the x-axis (from above) as $x \to \infty$. In fact, this analysis suggests that as $x \to \infty$, the function $y(x)$ behaves like the function $\frac{1}{2x}$, which it does.

Figure 5.15 shows the function $1/(2x)$ where $y' = 0$, the function $x/\left(2x^2 - 1\right)$ where $y'' = 0$, the slope field of $y' + 2xy = 1$, and the numerical solution of (5.55) — all of which are consistent.

---

[5]One version of L'Hôpital's rule is, if $\lim_{x\to a} f(x) = \infty$, $\lim_{x\to a} g(x) = \infty$, and $\lim_{x\to a} [f'(x)/g'(x)] = L$, then $\lim_{x\to a} [f(x)/g(x)] = L$.

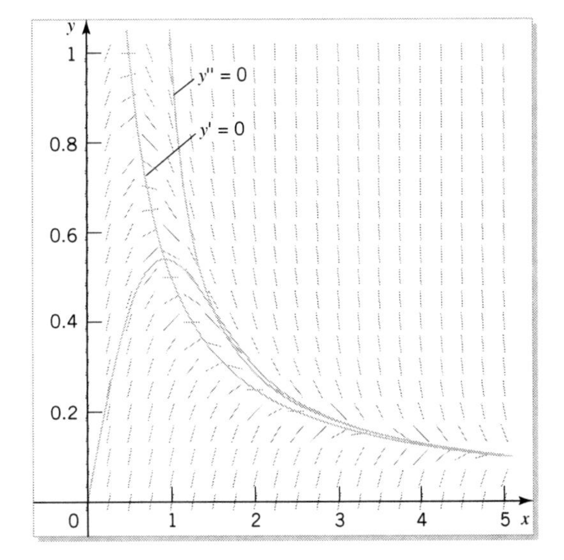

FIGURE 5.15 The slope field of $y' + 2xy = 1$ and a numerical solution

EXERCISES

1. Solve the following differential equations. (Here $y' = dy/dt$.)

   (a) $y' + 2y = 2e^{-t}$
   (b) $y' + 2y = 20e^{3t}$
   (c) $y' + 2ty = 4t$
   (d) $y' + 2y/t = 6$
   (e) $y' + 2y/t = \sin t/t^2$
   (f) $ty' + 2y = e^t$
   (g) $tP' + 3P = \ln t/t$
   (h) $tP' + 3P = 6 \sin t$
   (i) $P' + (\sin t)\, P = 4 \cos t \sin t$

2. Give the long-term behavior of each of the solutions in Exercise 1 and check that it is consistent with the corresponding slope field. Identify the transient and steady state parts of each solution.

3. Solve the following initial value problems. (Here $y' = dy/dt$.)

   (a) $y' - 3y = 6, \quad y(0) = 1$
   (b) $y' - y = \sin 2t, \quad y(0) = 0$
   (c) $y' - 7y = 14t, \quad y(0) = 0$
   (d) $y' + 2y/t = t, \quad y(1) = 1$
   (e) $y' + 2ty = t, \quad y(0) = 2$

4. **Mixture Problem.** A container is filled with 10 gallons of water containing 5 pounds of salt. A salt solution of concentration 3 pounds per gallon is pumped into the container at a rate of 2 gallons per minute, and the well-stirred mixture drains at the same rate. How much salt is in the container after 15 minutes?

5. **Mixture Problem.** Initially a 200-gallon container is filled with pure water. At time $x = 0$ a salt concentration with 3 pounds of salt per gallon is added to the container at the rate of 4 gallons per minute, and the well-stirred mixture is drained from the container at the rate of 5 gallons per minute.

   (a) Find the number of pounds of salt in the container as a function of time. For how many minutes is this solution valid?

   (b) How many minutes does it take for the concentration in the container to reach 2 pounds per gallon?

6. **Mixture Problem.** A conference room with volume 2000 cubic meters contains air with 0.002% carbon monoxide. At time $t = 0$, the ventilation system starts blowing in air that contains carbon monoxide amounting to $2 + \sin(t/5)$ (percent by volume with t measured in minutes). If the ventilation system inputs (and extracts) air at a rate of 0.2 cubic meters per minute, how long before the air in the room contains 0.015% carbon monoxide?

7. **Newton's Law of Cooling.** Consider the use of Newton's law of cooling to model the effect of temperature oscillations outside a building on the temperature within. Suppose you are leaving for four days and wonder if you can shut off your heating system during your absence. (There are plants inside that cannot tolerate temperatures below 40°F.) Assume that the outside temperature varies sinusoidally from a mean of 45°F, with a 10°F oscillation up and down. If when you leave in the morning the building is 70°F, and the outside temperature is 45°F, are your plants safe? (The surface area of the building, type of construction, insulation, and heat energy of the house are taken into account if you use the value $k = -0.2$ in Newton's law of cooling, when t is measured in hours, and T in degrees Fahrenheit.)

8. **RL Circuit.** If the simple RL circuit described by (5.40) has $L = 1$ henry and $R = 60$ ohms, and the voltage source is a 12-volt battery, find the explicit solution that describes what happens following the closure of the switch with $I(0) = 0$. What is the steady state part of the solution in this case? Are there any equilibrium solutions to this problem? Please explain.

9. **RL Circuit.** Let the circuit of Exercise 8 have a voltage source equal to $6\sin 2t$. If $I(0) = 0$, find the explicit solution to this initial value problem. What is the steady state part of the solution in this case? Are there any equilibrium solutions to this problem? Please explain.

10. Table 5.1 records the voltage, V, as a function of time for a charging capacitor, and Figure 5.16 is a graph of this voltage as a function of time.

Table 5.1 Voltage versus time

Time	Voltage	$\Delta_C V/\Delta t$
0	0.00	
2	1.95	0.813
4	3.25	0.550
6	4.15	0.385
8	4.79	
10	5.19	
12	5.45	
14	5.62	
16	5.75	
18	5.83	
20	5.89	
22	5.92	
24	5.95	
26	5.97	

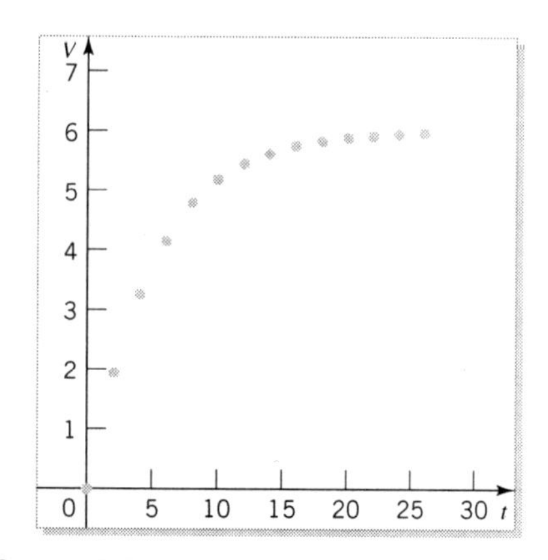

FIGURE 5.16 Graph of voltage versus time

(a) How would you describe what happens to the voltage as time increases?

(b) The third column of Table 5.1 shows the central difference approximation for the first and third, the second and fourth, and the third and fifth time periods. Complete the rest of this column, and give your observation about the change of voltage as time increases.

(c) The rate of increase of the voltage with time appears to decrease as the voltage approaches an apparent upper limit of 6, suggesting that the rate of change of voltage with time may be proportional to $(6 - V)$. To test this hypothesis, plot the central difference, $\Delta_C V/\Delta t$, from the third column in the table on the vertical axis, and $(6 - V)$ on the horizontal axis (subtract the second column of the table from 6). Describe the behavior of the resulting graph.

(d) Assuming that your answer to part (c) is that the graph is close to a straight line, estimate the slope m of this line. Because this slope gives the proportionality constant between the rate of increase of voltage and $(6 - V)$, we can conclude that the differential equation that models this process is

$$\frac{dV}{dt} = m(6 - V). \qquad (5.57)$$

(e) Figure 5.17 shows the slope field for (5.57) with a positive value for m. Can you predict the long-term behavior of the voltage? How does this depend on the initial voltage (that is, when $t = 0$)?

(f) Is there an equilibrium solution? If so, what is it? If not, why not?

(g) Solve (5.57) subject to the initial condition $V(0) = 0$, and graph your result. Compare this graph with

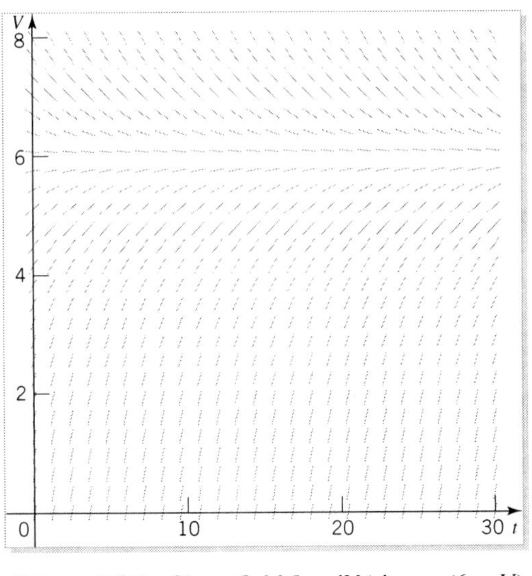

FIGURE 5.17 Slope field for $dV/dt = m(6 - V)$ for $m > 0$

Figure 5.16, describing any similarities and differences.

11. RC Circuit. An RC circuit consists of a resistor, capacitor, switch, and voltage generator, as shown in Figure 5.18. The differential equation that models this circuit is

$$R\frac{dq}{dt} + \frac{1}{C}q = E(t), \tag{5.58}$$

where q is the charge on the capacitor. (The units of C are farads, called capacitance, and the units of charge are coulombs.) Find the explicit solution of (5.58) when the resistance is 10 ohms, the capacitance is $1/100$ farads, the initial charge on the capacitor is 5 coulombs, and the voltage generator is a 12-volt battery. What is the steady state part of the solution in this case? Are there any equilibrium solutions to this problem? Please explain.

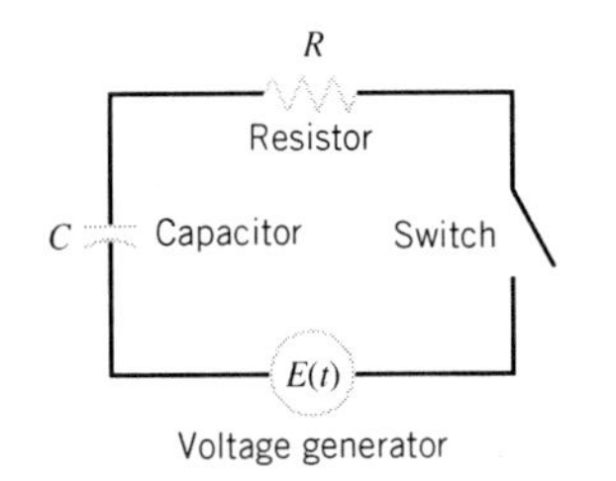

FIGURE 5.18 RC electrical circuit

12. RC Circuit. Repeat Exercise 11 if the voltage generator now has the form $E(t) = 12 \sin 4t$.

13. Solve the equation $y' + y = f(t)$, where

(a) $f(t) = \begin{cases} 2, & 0 \le t < 1 \\ 1, & 1 \le t \end{cases}, \quad y(0) = 0.$

(b) $f(t) = \begin{cases} 2, & 0 \le t < 1 \\ 0, & 1 \le t \end{cases}, \quad y(0) = 0.$

(c) $f(t) = \begin{cases} 5, & 0 \le t < 10 \\ 1, & 10 \le t \end{cases}, \quad y(0) = 6.$

(d) $f(t) = \begin{cases} e^{-t}, & 0 \le t < 2 \\ e^{-2}, & 2 \le t \end{cases}, \quad y(0) = y_0.$

14. Pumpkin Pie. A pumpkin pie recipe says to place the ingredients in a preheated oven at 425°F for 15 minutes, then turn the thermostat to 350°F and continue baking for 45 minutes. Assume that the temperature of the oven changes instantaneously when we change the thermostat from 425°F to 350°F. Use Newton's law of heating; namely, $T' = k[T - T_a(t)]$, where $T_a(t)$ is

$$T_a(t) = \begin{cases} 425 & \text{if } 0 \le t < 15, \\ 350 & \text{if } 15 < t. \end{cases}$$

If the initial temperature of the uncooked pie is 70°F, find the temperature, T, of the pie at time t. Compare the graph of this solution to the graph of a similar problem with $T_a(t) = 350$ for all the time that the pie is in the oven. From this comparison, explain the advantage of using a hotter oven for the first 15 minutes.

15. Traffic Flow. A differential equation that arises in the study of traffic flow is

$$\frac{dx}{dt} = \frac{1}{2}V + \frac{x}{2t},$$

where V is the maximum velocity of the car in traffic flow and x is the directed distance of a car from a traffic light (t is time). If the car starts from rest, then from the differential equation we have that $x = -x_0$ at $t = x_0/V$.

(a) Find the solution of this initial value problem, and show that V is in fact the maximum velocity for the car.

(b) Find the time it takes the car to reach the traffic light, and compare it with the time it would take if the car were going at its maximum speed the entire distance from its starting place to the traffic light. Does your answer surprise you?

16. Cardiovascular System. A simple model of the cardiovascular system represents arteries as a reservoir between the heart and the arterioles (smaller arteries). The output from the heart is the input to the reservoir, and the output from the reservoir is what flows into the arterioles. If we consider $P(t)$ as the pressure in the reservoir, R as the resistance to flow into the arterioles, and $f(t)$ as the output from the heart, the appropriate differential equation is

$$C\frac{dP}{dt} + \frac{1}{R}P = f(t),$$

where C is the compliance of the reservoir. [The units of $f(t)$ are volume per unit time.]

(a) Consider a case in which the interest is in the long-term behavior and the phase shift between the output from the heart and the input from the arterioles. If $f(t)$ is represented by $a + b\sin\omega t$ — where a, b, and ω are positive constants — find the effect of C and R on this phase shift. (You can get a feeling of what is happening by observing the slope field and some numerical solutions before verifying your findings analytically.)

(b) Now consider the case in which the output from the heart is represented by the following periodic function with period $2\pi/\omega$:

$$f(t) = \begin{cases} a\sin\omega t & 0 < t < \pi/\omega, \\ 0 & \pi/\omega < t < 2\pi/\omega. \end{cases}$$

Is there a nonzero long-term behavior now? [The suggestion for part (a) also applies here.]

17. Filling a Container. Consider an empty cylindrical container whose radius is 5 feet and height is 10 feet. At time $t = 0$, a tap is turned on, letting water enter the container at the constant rate of 3 cubic feet per minute. An exit valve is opened, and it lets water leave the container at a rate proportional to the square root of the depth of water in the container (let the proportionality constant be k). Solve the appropriate differential equation that describes this process. Can you determine values of k so the container will never be filled? If so, what are they? If not, why not?

18. Filling a Container. Repeat Exercise 17 with the water leaving the container at a rate equal to $k \times$ *depth*.

19. Filling a Container. Consider an empty conical container whose radius at the top is 5 feet and height is 10 feet. At time $t = 0$, a tap is turned on, letting water enter the container at the constant rate of 3 cubic feet per minute. An exit valve is opened, and it lets water leave the container at a rate proportional to the depth of water in the container (let the proportionality constant be k). Set up the appropriate differential equation to describe this process. Without finding an explicit solution of this differential equation, can you determine values of k so the container will never be filled? If so, what are they? If not, why not?

20. Filling a Container. Repeat Exercise 18 after adding the effect of loss due to evaporation, which is proportional to the surface area (let the proportionality constant be m). Does this problem have an equilibrium solution?

21. Filling a Container. Consider an open cylindrical container of height 10 feet and radius $8/\sqrt{\pi}$ feet that is initially half full of pure water. At $t = 0$, two valves are turned on. One lets in a saline solution of 3 lb/ft^3 at a rate of 4 ft^3/min, and the other lets out the well-stirred mixture at a rate equal to $0.2 \times$ *depth* ft^3/min. How long before the concentration of the solution is 1/2 of the input concentration? Does this happen before the container overflows or empties?

22. Glucose Tolerance Test. One of the standard glucose tolerance tests infuses glucose continuously into the bloodstream at the known rate of G mg/min. Blood concentrations are measured at subsequent time intervals until the long-term behavior is reached. It is assumed that the glucose is used by the body according to the differential equation $\frac{dx'}{dt} = -kx + G$, where x is the concentration of glucose, t is time, G is the infusion rate of glucose, and k is the "turnover rate" (what is checked to see if the body is processing glucose normally).

(a) Explain the relationship between this turnover rate and the equilibrium solution of the differential equation.

(b) Data from a typical glucose tolerance test are given in Table 5.2. Note that the measurements stopped before equilibrium was reached. We will now find the turnover rate by two methods. First, use the data from the table to obtain difference quotients, and plot the numerical derivative versus the concentration. If you now find the best straight line through these points, you will note that its vertical intercept is G and its slope is the turnover rate. Second, show that the explicit solution of the differential equation has the form $x(t) = ae^{-kt} + (85 - a)$, and find values of a and k

Table 5.2 Data from glucose tolerance test

Time (minutes)	$x(t)$ (mg/dl)	Time (minutes)	$x(t)$ (mg/dl)
0	85.0	30	127.3
10	105.4	40	131.9
20	120.1	50	134.4

that best fit the data. The value of k will be the turnover rate.

23. Suppose that we input a dye of concentration C at a flow rate of r into a body that removes this substance at the known rate R. Then, the concentration of the substance, x, is modeled by the differential equation $x' = -kx + Cr - R$, where k is the "reaction rate." Use the equilibrium solution to obtain an expression for the reaction rate in terms of the limiting solution.

24. In order to find a numerical solution of

$$e^{-x_m^2} \int_0^{x_m} e^{t^2}\, dt = \frac{1}{2x_m},$$

show that the change of variable, $t = x_m u$, allows this condition to be written in the form

$$\int_0^1 x_m^2 e^{(u^2-1)x_m^2}\, du = \frac{1}{2}.$$

Now, using a numerical integration package, experiment with different choices of x_m until this condition is satisfied.

25. In order to find a numerical solution of

$$\frac{x_i}{2x_i^2 - 1} = e^{-x_i^2} \int_0^{x_i} e^{t^2}\, dt,$$

show that the change of variable, $t = x_i u$, allows this condition to be written in the form

$$\int_0^1 (2x_i^2 - 1)\, e^{(u^2-1)x_i^2}\, du = 1.$$

Now, using a numerical integration package, experiment with different choices of x_i until this condition is satisfied.

26. Consider the function

$$f(x) = \frac{xe^{x^2}}{2x^2 - 1} - \int_0^x e^{t^2}\, dt.$$

(a) Show that $f(x_i) = 0$, where x_i is defined in Exercise 25.

(b) Show that $f'(x) < 0$ for $x > x_i$.

(c) Show that $f(x) < 0$ for $x > x_i$.

(d) How does this show that x_i defined in Exercise 25 is unique?

27. Assume that the initial value problem (5.55) — namely $y' + 2xy = 1$, $y(0) = 0$ — has a power series solution of the form $y(x) = \sum_{k=0}^{\infty} c_k x^k$. Show that

$$y(x) = \sum_{k=0}^{\infty} \frac{(-1)^k 2^k}{1 \cdot 3 \cdots (2k+1)} x^{2k+1}.$$

Plot the first few terms of this series to see how well it agrees with the numerical solution shown in Figure 5.15 on page 207.

28. Look in the textbooks for your other courses and find an example that uses a linear differential equation. Write a report about this example that includes the following items:

(a) A brief description of background material so a classmate will understand the origin of the differential equation.

(b) How the constants in the differential equation can be evaluated, and the meaning of the initial condition.

(c) The details of the solution of this differential equation.

(d) An interpretation of this solution, and how it answers a question posed by the original discussion.

5.3 MODELS THAT USE BERNOULLI'S EQUATION

In this section we look at an example involving the weight of fish as a function of time. The differential equation that arises in this case is not one we have encountered previously. However, a simple substitution converts this differential equation into a linear differential equation. This permits us to develop a technique for solving the original differential equation. We then show how it is used in other examples.

Bernoulli's Equation and Its Solution

In Exercise 7 on page 155 we found that the length $L(t)$ of a fish is modeled by the equation

$$\frac{dL}{dt} = a\left(L_e - L\right), \tag{5.59}$$

where a and L_e are constants. In Exercise 8 on page 133 we found that the weight $w(t)$ of a fish is related to its length $L(t)$ by $w(t) = bL^3(t)$, where b is a constant. From these two results we can find a differential equation that governs the weight of a fish as a function of time by substituting $L(t) = cw^{1/3}(t)$, where $c = b^{-1/3}$ is a constant, into the differential equation for $L(t)$. This gives

$$\frac{c}{3}w^{-2/3}\frac{dw}{dt} = a(L_e - cw^{1/3}),$$

or

$$\frac{dw}{dt} = Hw^{2/3} - kw = w^{2/3}\left(H - kw^{1/3}\right), \tag{5.60}$$

where $H = 3aL_e/c$ and $k = 3a/c$ are constants.

We will return to (5.60) shortly, but first we look carefully at what we have just done mathematically. If we rewrite (5.59) in the form

$$\frac{dL}{dt} + aL = aL_e,$$

Change variable

we notice that this is a linear differential equation. In going from (5.59) to (5.60) we changed the dependent variable from the length $L(t)$ of the fish to its weight $w(t)$, where $L(t) = cw^{1/3}$. We can ask the more general question: What happens to the general linear differential equation

$$u' + p(x)u = q(x)$$

if we make the substitution $u(x) = y^m(x)$? Substituting this into the linear differential equation gives $my^{m-1}y' + py^m = q$, or

$$y' + \frac{1}{m}p(x)y = \frac{1}{m}q(x)y^n,$$

where $n = 1 - m$. If we absorb the constant $1/m$ into $p(x)$ and $q(x)$, this equation belongs to a class of differential equations known as BERNOULLI'S DIFFERENTIAL EQUATIONS.

◆ *Definition 5.3:* **A BERNOULLI DIFFERENTIAL EQUATION has the form $y' + p(x)y = q(x)y^n$, where n is a specified constant.** ◆

However, the way we found this equation leads us directly to its solution. To solve a Bernoulli equation in $y(x)$, we first convert it to a linear differential equation in $u(x)$, by making the substitution $u(x) = y^{1-n}$, and then solve the linear differential equation for $u(x)$.

How to Solve Bernoulli's Differential Equation

Purpose To solve Bernoulli's differential equation

$$y' + p(x)y = q(x)y^n \tag{5.61}$$

for $y = y(x)$.

Process

1. Write (5.61) in the form

$$y^{-n}y' + p(x)y^{1-n} = q(x). \tag{5.62}$$

2. Use the substitution $u = y^{1-n}$ to find

$$u' = (1-n)\, y^{-n}y',$$

or

$$y^{-n}y' = \frac{1}{1-n}u'.$$

3. Substituting this and $u = y^{1-n}$ into (5.62) gives

$$\frac{1}{1-n}u' + p(x)u = q(x).$$

4. Solve this first order linear equation for $u = u(x)$, and substitute the result in $u = y^{1-n}$.

Before returning to the example of the weight of the fish, we demonstrate this method on a typical differential equation.

EXAMPLE 5.9

Consider the differential equation

$$y' + \frac{2x}{x^2+1}y = \frac{1}{(x^2+1)^2 y^2}. \tag{5.63}$$

This is a Bernoulli equation with $n = -2$, which we write as

$$y^2y' + \frac{2x}{x^2+1}y^3 = \frac{1}{(x^2+1)^2}.$$

Change variable We make the substitution $u = y^{1-(-2)} = y^3$, to find $u' = 3y^2y'$, and so the differential equation becomes

$$\frac{1}{3}u' + \frac{2x}{x^2+1}u = \frac{1}{\left(x^2+1\right)^2},$$

which is the linear differential equation

$$u' + \frac{6x}{x^2+1}u = \frac{3}{\left(x^2+1\right)^2}.$$

Integrating factor This has the integrating factor $\mu = \left(x^2+1\right)^3$, so

$$\left[\left(x^2+1\right)^3 u\right]' = 3\left(x^2+1\right),$$

and thus, $\left(x^2+1\right)^3 u = x^3 + 3x + C$, or

$$u(x) = \frac{x^3+3x+C}{\left(x^2+1\right)^3}.$$

Explicit solution Finally, from $u = y^3$, we have

$$y(x) = \frac{\left(x^3+3x+C\right)^{1/3}}{x^2+1}. \tag{5.64}$$

While (5.64) is the explicit solution, we should always try to understand the properties of the solution to make sure we fully understand its behavior.

From (5.63) we notice that if there exists an x_0 for which $y(x_0) = 0$, then y' at x_0 will be infinite, so that the curve $y = y(x)$ will approach the point $y(x_0) = 0$ (the x-axis) vertically. We can see whether there exists an x_0 for which $y(x_0) = 0$ by scrutinizing the numerator of the explicit solution (5.64). Because the function $x^3 + 3x + C$ is an increasing function of x for any given C (why?) and this function goes from $-\infty$ to ∞, the function $\left(x^3+3x+C\right)^{1/3}$ has exactly one real root.[6] Thus, for any choice of C (that is, for any solution curve), there is exactly one point x_0 for which $y(x_0) = 0$. So we know that every solution curve will have exactly one place where it approaches the x-axis vertically.

Several other things are evident at this point.

Isocline

- From (5.63) we see that the isocline for zero slope is given by $y = 1/[2x(x^2+1)]^{1/3}$. Thus, solution curves will have local extrema at places that are at larger and larger y values as x approaches the origin from the right, and at smaller and smaller y values as x approaches positive infinity.

Symmetry

- Also from (5.63) we see that the slope field has symmetry about the origin.

[6] The root is $\left(a+\sqrt{1+a^2}\right)^{1/3} + \left(a-\sqrt{1+a^2}\right)^{1/3}$ where $a = -C/2$.

- From (5.64) we see that the solution curves are bounded for all finite values of x. The y-intercept is given by $C^{1/3}$.

Slope field

Figure 5.19 gives the slope field and several solution curves for this situation. In this figure there are 14 solution curves, not 7, because the derivative is undefined on the x-axis. Note the agreement with the preceding points.

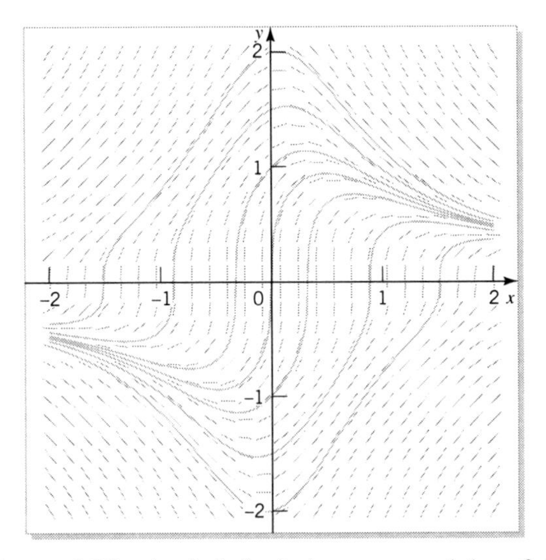

FIGURE 5.19 Analytical solution curves and slope field for $y' + 2xy/(x^2+1) = 1/[(x^2+1)^2 y^2]$

The Weight of a Fish

EXAMPLE 5.10 *Weight of Fish*

We now return to equation (5.60), namely,

$$\frac{dw}{dt} = Hw^{2/3} - kw = w^{2/3}\left(H - kw^{1/3}\right). \tag{5.65}$$

This particular differential equation goes under various names, perhaps the most common being the VON BERTALANFFY EQUATION.[7]

Slope field

The slope field and some solution curves for this equation are shown in Figure 5.20 for $H = 1$, $k = 0.6$. Note the similarities with the slope field associated with the logistic equation — namely, an apparent equilibrium solution with solutions increasing below this equilibrium and decreasing above. The equilibrium solutions occur when $w = 0$ and $w = (H/k)^3$. The equilibrium solution $w = 0$ is of no interest

Equilibrium solution

[7] "On the Dynamics of Exploited Fish Populations" by R.J.H. Beverton and S.J. Holt, *Fishery Invest. Series 2* 19, 1957, pages 1–533.

in this example, and $w = (H/k)^3$ corresponds to the limiting weight of the fish. For the case shown in Figure 5.20 — namely, $H = 1, k = 0.6$ — this limiting weight occurs when $w = (5/3)^3 \approx 4.63$, as shown in Figure 5.20.

Concavity We also note from Figure 5.20 that changes in concavity seem to occur in the lower left corner where we have small values of w. We can obtain more information about this from the differential equation. If we differentiate (5.65), we obtain $w'' = \left(\frac{2}{3}Hw^{-1/3} - k\right) w'$, from which it is clear that changes in concavity occur when

$$w = \left(\frac{2H}{3k}\right)^3;$$

that is, between $w = 0$ and the equilibrium solution at $w = (H/k)^3$. (Note that the places where $dw/dt = 0$ give equilibrium solutions and are not considered when looking for inflection points.) For the case shown in Figure 5.20 — namely, $H = 1$, $k = 0.6$ — this occurs when $w = (10/9)^3 \approx 1.37$, in agreement with Figure 5.20. Note that the times at which inflection points occur depend on the initial values.

Let us now seek an explicit solution of (5.65) for the case shown in Figure 5.20, where $H = 1, k = 0.6$:

$$\frac{dw}{dt} = w^{2/3} - 0.6w. \tag{5.66}$$

Although this equation is separable (and, in fact, autonomous), the technique to allow simple integration is obscure. (See Exercise 3 on page 220.) We will solve this differential equation by treating it as a Bernoulli equation,

$$\frac{dw}{dt} + 0.6w = w^{2/3},$$

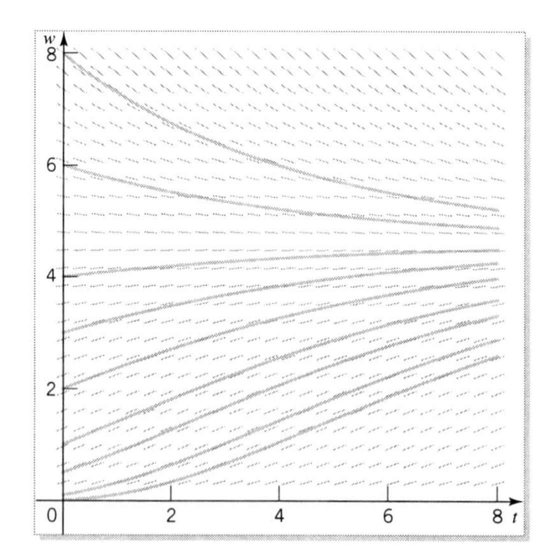

FIGURE 5.20 Analytic solution curves and slope field for $dw/dt = Hw^{2/3} - kw$

Change variable

with $n = 2/3$. Thus, we substitute $u = w^{1-n} = w^{-1/3}$ into the preceding equation after writing it as

$$w^{-2/3}\frac{dw}{dt} + 0.6w^{-1/3} = 1.$$

This gives the linear differential equation $-3u' + 0.6u = 1$, or $u' - \frac{1}{5}u = -\frac{1}{3}$, with solution $u(t) = \frac{5}{3}\left(1 - Ce^{-0.2t}\right)$. In terms of our original variable, we have

Explicit solution

$$w(t) = \left[\frac{5}{3}\left(1 - Ce^{-0.2t}\right)\right]^3, \tag{5.67}$$

where

$$C = 1 - 0.6w(0)^{1/3} \tag{5.68}$$

and $w(0)$ is the initial weight of the fish. From (5.67) we see that the equilibrium solution is $(5/3)^3$, in agreement with our earlier observation.

The explicit solution (5.67) can now be used to calculate the times at which inflection points occur — namely, the times at which $w(t) = (10/9)^3$. From (5.67) this is equivalent to asking for the times at which

$$\left(\frac{10}{9}\right)^3 = \left[\frac{5}{3}\left(1 - Ce^{-0.2t}\right)\right]^3.$$

Taking the cube root of both sides of this equation and solving for t gives $t = 5\ln(3C)$, which, in view of (5.68), can be written as

$$t = 5\ln\left\{3\left[1 - 0.6w(0)^{1/3}\right]\right\} = 5\ln\left[3 - 1.8w(0)^{1/3}\right].$$

As expected, this time depends on the initial value $w(0)$. To guarantee that $t > 0$ in this last expression, we must have $3 - 1.8w(0)^{1/3} > 1$, which implies that $w(0) < (10/9)^3$. Thus, only solution curves corresponding to initial weights less than $(10/9)^3$ will have an inflection point.

The solution of (5.65) for other choices of H and k is possible by the same technique, the solution being

$$w(t) = \left[\frac{H}{k}\left(1 - Ce^{-kt/3}\right)\right]^3. \tag{5.69}$$

Figure 5.21 shows the graph of (5.69) for $H = 4.049$, $k = 0.285$, along with data on the growth of North Sea plaice.[8] The data used in Figure 5.21 are given in Table 5.3.

Table 5.3 The weight of North Sea plaice as a function of time

Time (years)	Weight (grams)
1.250	8.330
2.000	33.330
2.125	62.500
3.000	83.330
3.250	108.330
4.625	216.667
5.625	295.830
6.625	366.670
7.625	491.670

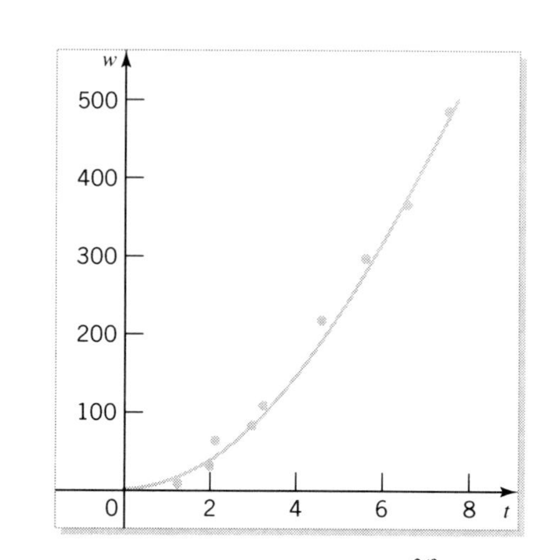

FIGURE 5.21 Solution of $dw/dt = Hw^{2/3} - kw$ and data

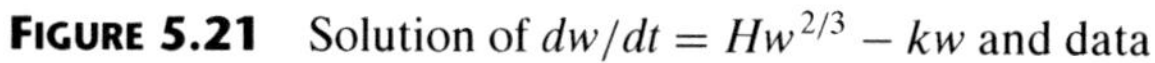

Animal Population Growth

The logistic equation was used in Chapter 2 to model human populations. In the next example, we apply it to a population of animals in nature that is confined to a specific region. A convenient form for writing the logistic model is

$$\frac{dP}{dt} = kP\left(1 - \frac{P}{b}\right),$$

where P is the number of animals, t is time, k is an effective rate of reproduction, and b is the carrying capacity. We have already seen that $P = b$ is a stable equilibrium solution. Thus, for any positive initial condition, all solutions will approach the value b as t becomes large. So by this model, b will be the number of animals in this region when this system is in equilibrium; that is, it is the carrying capacity of the region.

[8] "Modelling Exploited Marine Fish Stocks" by J. Beyer and P. Sparre, *Applications of Ecological Modelling in Environmental Management*, Part A, 1983, page 505.

EXAMPLE 5.11 *Animal Population Growth*

Often the carrying capacity of a region is dependent on rainfall or other seasonal events. Such an effect can be incorporated into this model by letting the carrying capacity have an oscillatory component. Thus, we consider the differential equation

$$\frac{dP}{dt} = kP\left(1 - \frac{P}{b + c\sin\omega t}\right),$$

where b and ω are positive constants and $b > c$, usually by a large amount. To simplify subsequent calculations, we will consider the special case where $k = 1, b = 7, c = 1$, and $\omega = 6$, so the differential equation becomes

$$\frac{dP}{dt} = P\left(1 - \frac{P}{7 + \sin 6t}\right). \tag{5.70}$$

Slope field

The slope field for this situation is shown in Figure 5.22 along with numerical solution curves for several different initial populations. Note that regardless of the initial condition, the population soon settles into a periodic pattern. Equation (5.70) is a Bernoulli equation,

$$\frac{dP}{dt} - P = -\frac{1}{7 + \sin 6t}P^2, \tag{5.71}$$

with $n = 2$, so we write it as

$$\frac{1}{P^2}\frac{dP}{dt} - \frac{1}{P} = -\frac{1}{7 + \sin 6t},$$

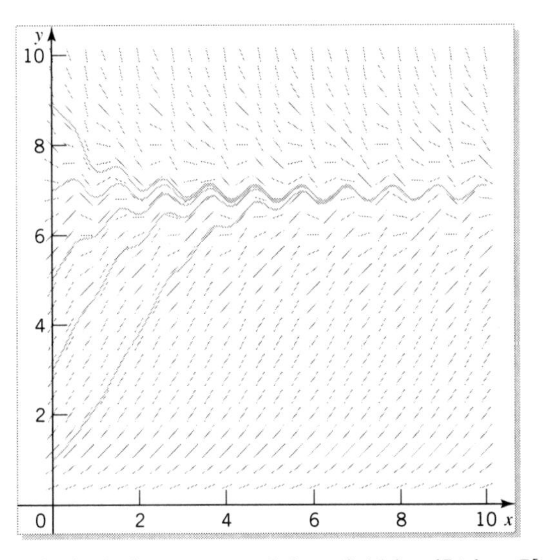

FIGURE 5.22 Numerical solution curves and slope field for $dP/dt = P[1 - P/(7 + \sin 6t)]$

Change variable and make the substitution $u = P^{1-n} = P^{-1}$, to find

$$\frac{du}{dt} + u = \frac{1}{7 + \sin 6t}. \tag{5.72}$$

Integrate Integration of (5.72) gives

$$u(t) = e^{-t} \int_0^t \frac{e^s}{7 + \sin 6s}\, ds + C,$$

Explicit solution so the explicit solution is

$$P(t) = \frac{1}{u(t)} = e^t \left(\int_0^t \frac{e^s}{7 + \sin 6s}\, ds + C \right)^{-1},$$

where $C = 1/P(0)$.

Graphical analysis Even though we cannot find a simple expression for this integral, we can discover some facts about the long-term behavior of this solution by using graphical analysis. We first note from (5.70) that $P'(t) = 0$ when $P(t) = 7 + \sin 6t$, and that $P(t)$ is an increasing function if $P(t) < 7 + \sin 6t$ and a decreasing function if $P(t) > 7 + \sin 6t$. Second, if we compute P'' and then let $P' = 0$ in the result, we find

$$P'' = \frac{6P^2}{(7 + \sin 6t)^2} \cos 6t.$$

Thus, if $P' = 0$ we have $P'' < 0$ when $\cos 6t < 0$, and $P'' > 0$ when $\cos 6t > 0$. Putting these facts together tells us that the solutions of (5.70) will have local minima ($P' = 0$, $P'' > 0$) when they cross the curve $7 + \sin 6t$ between a minimum and a maximum of $7 + \sin 6t$ (where $\cos 6t > 0$), and local maxima when they cross the curve $7 + \sin 6t$ between a maximum and minimum of $7 + \sin 6t$ (where $\cos 6t < 0$). Thus, once a solution curve enters the region $6 < y < 8$, it is trapped there, and oscillates. This oscillating behavior is indicated in Figure 5.22.

EXERCISES

1. Solve the following Bernoulli differential equations.

(a) $yy' + xy^2 - x = 0$
(b) $y' + y/x - 1/(x^3y^3) = 0$
(c) $xy' - (1 + x)y - y^2 = 0$
(d) $x^2y' + y^2 - xy = 0$
(e) $y' + 2y/x + x^9y^5 = 0$
(f) $y' + y - 2x^2y^2 = 0$
(g) $y' + \sqrt{xy} - (2/3)\sqrt{x/y} = 0$
(h) $2yy' + y^2 \sin x - \sin x = 0$
(i) $y' - y + xe^{-2x}y^3 = 0$
(j) $xy' - y/(2\ln x) - y^2 = 0$

2. Examine the slope field for each equation in Exercise 1 and look for isoclines for horizontal and vertical tangents. Does your explicit solution in Exercise 1 agree with these isoclines?

3. Show that the nonequilibrium solutions of the differential equation $w' = Hw^m - kw$ can be obtained by using the fact that

$$\frac{1}{Hw^m - kw} = \frac{1}{w\,(Hw^{m-1} - k)} = \frac{1}{k}\left(\frac{Hw^{m-2}}{Hw^{m-1} - k} - \frac{1}{w}\right).$$

Use this approach to obtain (5.67) as the solution of (5.66).

4. If $n > m > 0$, find two nonnegative equilibrium solutions of the differential equation modeling the growth of an individual fish, $w' = Hw^m - kw^n$, and show that any solution curve will be increasing between the two. The constants H and k are both positive.

(a) Show that for any size of fish at birth, the limiting value of its weight is $(H/k)^{1/(n-m)}$.

(b) Using this model, find the weight corresponding to the greatest growth rate.

(c) What would happen if you use the initial condition $w(0) = 0$?

5. Consider a Bernoulli equation of the form $y' + y = m^{n-1}y^n$, where m and n are positive constants.

(a) For the case $n = 2$, find all the equilibrium solutions and determine their stability.

(b) Repeat part (a) for the case $n = 3$.

(c) Repeat part (a) for the case $n = 5$.

(d) Can you find values of m and n such that an equilibrium solution of this equation given by $y = 0$ will be unstable? If so, what are they? If not, explain fully.

6. Consider the case in which the $p(x)$ and $q(x)$ in the Bernoulli equation $y' + p(x)y = q(x)y^n$ are bounded for all finite values of x.

(a) Find conditions on n and q such that all solutions of the Bernoulli equation will approach the x-axis vertically.

(b) What condition on n will guarantee that 0 will be an equilibrium solution? Is it possible for this equilibrium solution ever to be stable? If so, give these conditions. If not, explain fully.

7. Solve the initial value problem

$$\frac{dP}{dt} = kP\left(1 - \frac{P}{b}\right), \qquad P(0) = P_0$$

for the case when $k = k(t)$ and $b = b(t)$. Notice that (5.70) is a special case of this differential equation. Suggest an application of this differential equation.

8. Solve the initial value problem

$$\frac{dP}{dt} = kP\left(1 - \frac{P}{b}\right), \qquad P(0) = P_0,$$

where k is a constant and $b = b(t)$ satisfies the logistic equation

$$\frac{db}{dt} = ab\left(1 - \frac{b}{B}\right), \qquad b(0) = b_0,$$

where a and B are constants. Suggest an application of this differential equation. What do you expect to happen to P as $t \to \infty$?

9. Identify the following differential equations as separable equations, those with homogeneous coefficients, linear equations, Bernoulli equations, or none of these. (Some may fall into more than one category; see Table 5.4.) Do not attempt to solve any of these equations.

(a) $(2 - y)y' - x^2 - 1 = 0$
(b) $(x + y)y' + y - x = 0$
(c) $2xy + (x^2 + \cos y)\, y' = 0$
(d) $y \sin x + 1/y - y' = 0$
(e) $3x(y^2 + 1) + y(x^2 + 1)y' = 0$
(f) $xy' + 3y - x^2 = 0$
(g) $x^3 + y^3 - xy^2y' = 0$
(h) $(1 - x^2y)\, y' - xy^2 + 1 = 0$
(i) $y \sin x + e^x - y' = 0$
(j) $xy^2 - y + y' = 0$

10. Population Growth. In Chapter 2 we considered the logistic equation as a model of population growth. Here we derive a differential equation that includes the effect of births, deaths, immigration, emigration, and crowding. If we were to write

$$\textit{rate of change of population} = \textit{rate of additions to population} - \textit{rate of subtractions from population}$$

and let y be the population at time t, we would have $y' = (By + I) - (Dy + E + cy^2) = a + by - cy^2$, where

Table 5.4 Solving first order differential equations

Differential Equation	Response		Type
$y' = f(y)g(x)$?	Yes	$\longrightarrow$	Separable
$y' = g(y/x)$?	Yes	$\longrightarrow$	Homogeneous coefficients
$y' + p(x)y = q(x)$?	Yes	$\longrightarrow$	Linear
$y' + p(x)y = q(x)y^n$?	Yes	$\longrightarrow$	Bernoulli

$b = B - D =$ birthrate − deathrate (per unit population), $a = I - E =$ immigration − emigration (per unit time), and c accounts for competition or inhibition of large populations. Note that we could think of the right-hand side of the preceding equation as the first three terms in a Taylor series of a general function that describes the population growth.

(a) Look at the slope field for this differential equation and explain the effect on its solutions by independently changing a, b, and c. Pay particular attention to the existence (or nonexistence) of equilibrium solutions and their behavior as you change the parameters.

(b) Solve the equation for the case $a = c = 0$.

(c) Solve the equation for the case $b = c = 0$.

(d) Solve the equation for the case $c = 0$.

(e) Solve the equation for the case $a = 0$.

What Have We Learned?

Main Ideas

How to Identify and Solve $y' = g(x, y)$

Purpose To summarize the major steps required to identify and solve the differential equation $y' = g(x, y)$.

Process

1. If $y' = f(y)g(x)$, then the equation is **separable**. See *How to Solve Separable Differential Equations* on page 123.
2. If $y' = g(y/x)$, then the equation has **homogeneous coefficients**. See *How to Solve Differential Equations with Homogeneous Coefficients* on page 140.
3. If $a_1(x)y' + a_0(x)y = f(x)$, then we have a **linear** equation. See *How to Solve Linear Differential Equations* on page 190.
4. If $y' + p(x)y = q(x)y^n$, then we have a **Bernoulli** equation. See *How to Solve Bernoulli's Differential Equation* on page 213.

Comments about $y' = g(x,y)$

- Table 5.4 summarizes the types of first order differential equations we have covered.

CHAPTER 6

Interplay Between First Order Systems and Second Order Equations

Where We Are Going — and Why

In this chapter we extend the techniques we have developed for first order differential equations to more than one first order equation; namely, to systems of first order equations. We show how the existence-uniqueness theorems for such systems are natural extensions of our previous existence-uniqueness theorems, and that our previous numerical methods are easily extended to obtain numerical solutions of systems of first order equations.

Systems of first order equations are important because of the many situations that can be modeled by them. They are also important because of their relationship to second order differential equations. This relationship is exploited to obtain existence-uniqueness theorems and numerical solutions of second order differential equations.

Second order equations are important in their own right. First, they are used to model spring-mass systems, simple pendulums, and simple electrical circuits, which leads to the important notion of overdamped, critically damped, and underdamped motion. Second, we can obtain an explicit solution of one special class — linear with constant coefficients.

The ability to move backward and forward between systems and second order equations, taking advantage of the power of each, is the focus of this chapter. For example, we show that for second order differential equations, with the introduction of a phase plane, many of our graphical techniques carry over from the previous chapters. We describe how a curve in the phase plane can be constructed from a numerical, graphical, or analytical point of view.

The rest of this book is based on a thorough understanding of this chapter.

6.1 Simple Models

In this section we consider two examples of relationships that lead to a set of first order differential equations, called a **system** of first order differential equations. We show how our previous graphical techniques apply to this situation and how explicit solutions may be obtained by combining them into a single second order differential equation. We will use this last technique again in Chapter 9.

EXAMPLE 6.1 *Denise and Chad's Relationship*

A situation that you may have observed or experienced is the romantic attraction of two individuals ill-suited to each other. Here we try to model this situation for two such individuals, Denise and her boyfriend Chad. We observed that Denise's affection for Chad increases when her affection is reciprocated. However, Chad's affection for Denise increases when his affection is not reciprocated. If x represents Denise's affection and y represents Chad's, then

$$\frac{dx}{dt} = ay, \qquad \frac{dy}{dt} = -bx,$$

where a and b are positive constants, is a model consistent with this information. Here x and y are functions of the time, t. Positive values of affection represent a liking for the other individual, and negative values express dislike. We will analyze their stormy relationship for the case when $a = 1$ and $b = 2$; that is,

$$x' = y, \qquad y' = -2x, \tag{6.1}$$

where $x' = dx/dt$ and $y' = dy/dt$.

Equation (6.1) contains a coupled set of first order differential equations, in which a solution consists of x and y given as functions of t. However, we currently have no means of solving (6.1). We also have one more variable than usual when we draw a slope field. However, because x and y are both functions of t, we could, in principle, solve for t in terms of one of the dependent variables — say, x — and substitute the result into the expression for y. This would eliminate the variable t and give an equation relating x and y. For such situations the graph of y versus x is called an ORBIT, and the xy-plane is called the PHASE PLANE associated with (6.1). We proceed to analyze (6.1) in this light.

Orbit
Phase plane

Because the right-hand sides of the two equations in (6.1) do not involve the variable t explicitly, we may combine these two equations into one. If we treat y as a function of x, and x as a function of t, using the chain rule gives

$$\frac{dy}{dx} = \left(\frac{dy}{dt}\right)\left(\frac{dt}{dx}\right) = \frac{y'}{x'} = -\frac{2x}{y},$$

or

$$\frac{dy}{dx} = -2\frac{x}{y}, \tag{6.2}$$

a first order differential equation that is separable.

Slope field

Notice that the slope field for (6.2) has positive slopes in the second and fourth quadrants and negative slopes in the first and third quadrants. The slope field for (6.2) is shown in Figure 6.1.

Even though we do not know the explicit time dependence of x and y, we can see what happens as time increases by looking at (6.1). For example, in the first

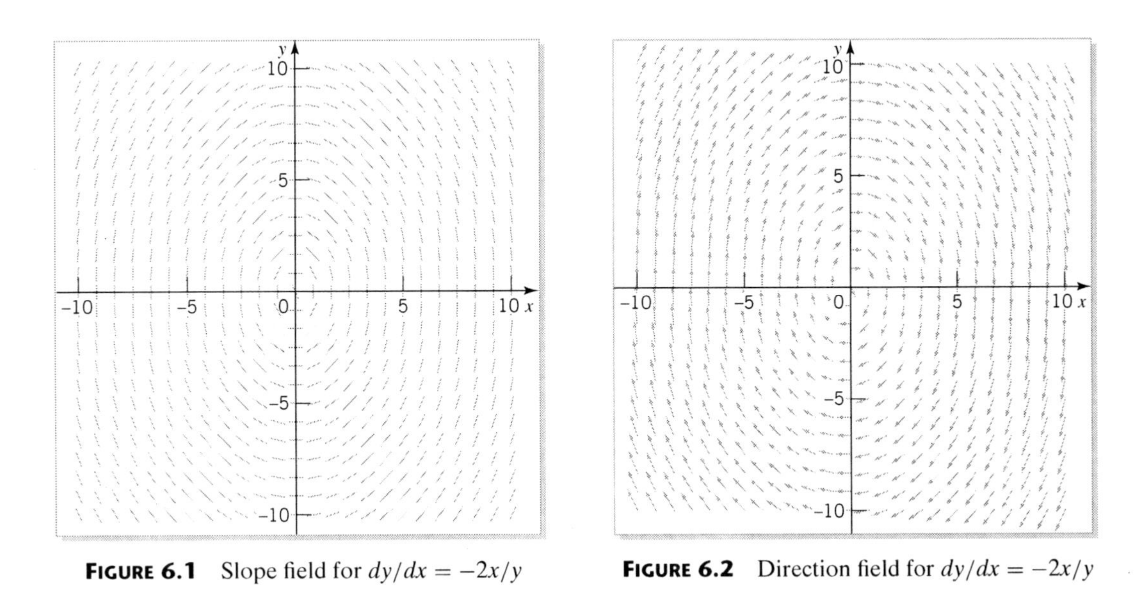

FIGURE 6.1 Slope field for $dy/dx = -2x/y$

FIGURE 6.2 Direction field for $dy/dx = -2x/y$

quadrant, where x and y are both positive, x will be an increasing function of t, and y will be a decreasing function. This means that time, as a parameter, is proceeding in a clockwise manner in the first quadrant. We analyze the other three quadrants in a similar way and find that this clockwise rotation around the origin persists. We could indicate this on the slope field by adding arrows to the slopes to indicate the direction the orbits follow as time increases, creating a field of vectors. Such a slope field is known as a DIRECTION FIELD and is shown in Figure 6.2, where the arrows indicate the direction of travel. The length of each vector has been scaled so that those vectors with small values of (x', y') are short, and those with large values are long.

Direction field

The shape of the orbits in the phase plane can be obtained by solving the separable differential equation in (6.2), and we find

$$y^2 + 2x^2 = C, \tag{6.3}$$

Orbits

where C is a constant of integration. Thus, the orbits are ellipses.

If we superimpose three of the elliptical orbits (6.3) — the ones that pass through the points $(0, 2.5)$, $(0, 5)$, and $(0, 10)$ — on Figure 6.2, we find Figure 6.3. Time, as a parameter, is proceeding clockwise around the ellipse, allowing us to tell the story of this relationship as time progresses from a fixed starting point.

For example, consider the inner orbit in Figure 6.3. We will start at the point A (where Chad has the maximum affection for Denise, and Denise has no affection for Chad), proceed clockwise to the point B (where Chad has no affection for Denise, and Denise has the maximum affection for Chad), then to C (where Chad has the least affection for Denise, and Denise has no affection for Chad), then to D (where Chad has no affection for Denise, and Denise has the least affection for Chad), and then return to A, where the cycle of affection repeats itself — forever.

In going from A to B, both x and y are positive, so Denise and Chad like each other. As x increases, y decreases, so as Denise (x) becomes more enamored with Chad (y), Chad's affection for Denise decreases.

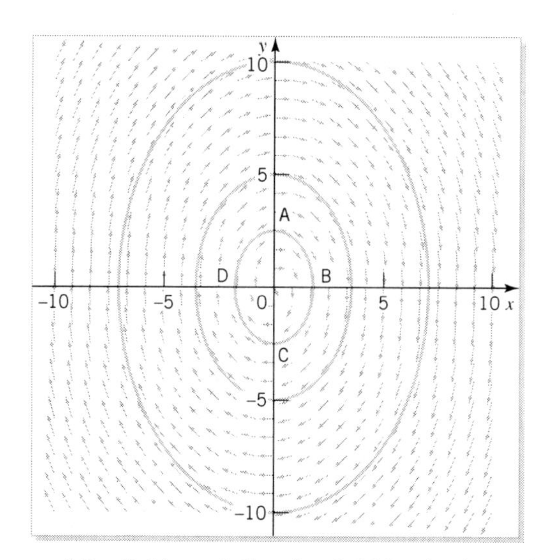

FIGURE 6.3 Orbits and direction field for $dy/dx = -2x/y$

In going from B to C, x is positive but y is negative, so Denise has affection for Chad, but that affection is not reciprocated. As x decreases, y decreases, so as Denise becomes less enamored with Chad, Chad's affection for Denise decreases further.

In going from C to D, both x and y are negative, so Denise and Chad dislike each other. As x decreases, y increases, so as Denise becomes less enamored with Chad, Chad's affection for Denise increases.

Finally, in going from D to A, x is negative but y is positive, so Denise dislikes Chad, whereas Chad likes Denise. As x increases, y increases, so as Denise becomes more enamored with Chad, Chad's affection for Denise increases further.

We can also determine from Figure 6.3 that there is no time when both are ecstatic over each other. That would occur if they each experienced maximum affection at the same time. They don't.

The previous analysis gives a very good explanation of Denise and Chad's affection for each other, but gives no information as to the time dependence. For example, as time progresses, do they spend equal time in each quadrant (which would mean that they would simultaneously like each other for only one-quarter of the time, the time they are both in the first quadrant), or do they spend more time in the first quadrant than in all the others combined (which would mean that they would simultaneously like each other for over half the time)? The answer to these questions requires an analysis in which the time is explicitly involved, so we proceed to determine the time behavior of x and y; that is, the explicit dependence of x and y on t. This cannot be determined from (6.2). To determine the time behavior we return to the original system of equations (6.1). If we solve the first equation in (6.1) for y as $y = x'$ and substitute this expression into the second equation, we obtain $y' = x'' = -2x$, or

$$x'' + 2x = 0. \tag{6.4}$$

This is a SECOND ORDER LINEAR DIFFERENTIAL EQUATION. It is second order because it contains the second derivative as its highest derivative. It is linear because it is a linear combination of x'', x', and x with coefficients 1, 0, and 2, respectively. A precise definition will be given in the next section, and Section 6.3 is devoted to solving such equations.

Explicit solution

In Section 6.3 we will show that the solution of (6.4) is

$$x(t) = C_1 \cos \sqrt{2}t + C_2 \sin \sqrt{2}t, \tag{6.5}$$

where C_1 and C_2 are arbitrary constants. [Verify that (6.5) satisfies (6.4).] If we substitute (6.5) into the first equation in (6.1), we see that

$$y(t) = \sqrt{2}\left(-C_1 \sin \sqrt{2}t + C_2 \cos \sqrt{2}t\right)$$

gives the rest of the solution of our original system of equations.

Initial conditions

If we start counting time when y is very attracted to x and x is feeling neutral toward y, we might impose the initial conditions $x(0) = 0$, $y(0) = 2.5$.[1] This requires that $C_1 = 0$ and $C_2 = 2.5/\sqrt{2}$, so for this case our solution is

$$x(t) = \frac{2.5}{\sqrt{2}} \sin \sqrt{2}t, \tag{6.6}$$

$$y(t) = 2.5 \cos \sqrt{2}t. \tag{6.7}$$

It is evident from (6.6) and (6.7) that, because $2.5/\sqrt{2} < 2.5$, the amplitude of Denise's affection is less than the amplitude of Chad's affection. So Chad will have the greater mood swing.

Figure 6.4 gives graphs of the solutions (6.6) and (6.7), which demonstrate that y has the greater mood swing. We can see that between $t = 0$ and $t = t_1$, x and y have affection for each other, although y is becoming less attracted to x while x is

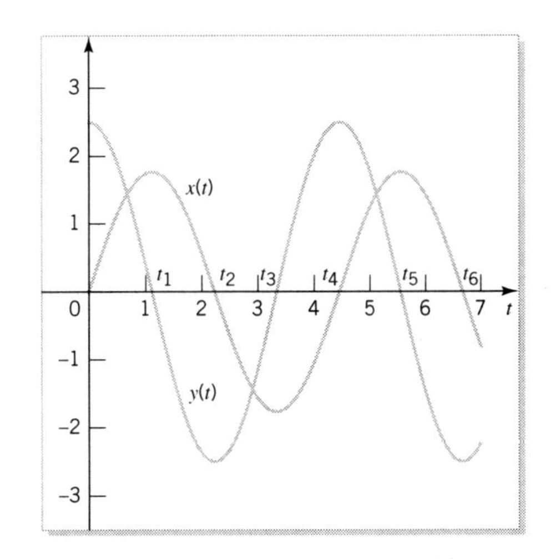

FIGURE 6.4 The solutions $x(t) = 2.5 \cdot 2^{-1/2} \sin 2^{1/2}t$, $y(t) = 2.5 \cos 2^{1/2}t$

[1]Henceforth, instead of referring to Denise and Chad by name, we will use x and y to identify them, although, in fact, x represents Denise's affection for Chad, and y represents Chad's affection for Denise.

becoming more attracted to y. This corresponds to the first quadrant in Figure 6.3. From $t = t_1$ to $t = t_2$, x has affection for y, but that affection is not reciprocated. In fact, the least attraction that y has for x occurs just as x is neutral toward y – that is, when $x = 0$ at $t = t_2$. At this stage y is more attracted to x, but x is less and less attracted to y. This corresponds to the second quadrant in Figure 6.3. Between $t = t_2$ and $t = t_3$, x and y dislike each other. This corresponds to the third quadrant in Figure 6.3. The next time interval when x and y are both positive is between $t = t_4$ and $t = t_5$, when they are both attracted to each other again. The emotional cycle from $t = 0$ to $t = t_4$ is repeated forever. Because $t_1 = t_2 - t_1 = t_3 - t_2 = t_4 - t_3$, equal times are spent in each quadrant, and so for only a quarter of the time – while in the first quadrant of Figure 6.3 – do Denise and Chad simultaneously like each other.

Notice that, from (6.1), the initial conditions $x(0) = 0$, $y(0) = 2.5$, could be expressed entirely in terms of x as $x(0) = 0, x'(0) = 2.5$. Thus, we could have solved this problem by considering (6.4) subject to the initial conditions $x(0) = 0, x'(0) = 2.5$. These initial conditions could then be used in (6.4) to find (6.6). ▪

EXAMPLE 6.2 *Parental Interference*

Suppose that the differential equation that models Denise's feelings is of the form

$$x' = ay - cx,$$

where c is a positive constant. Here we see that her change in affection for Chad is diminished by a term that is proportional to her affection. What could this mean? Well, if her parents were worried that she would become so enamored with Chad that she would leave home, then the more affection she had for Chad, the more pressure the parents might exert to have them break up.

Let's assume that the differential equation that models Chad's feelings remains the same, so we have

$$x' = ay - cx, \qquad y' = -bx.$$

We will consider the special case where $a = 1, b = 2, c = 0.4$, so that

$$x' = y - 0.4x, \qquad y' = -2x. \tag{6.8}$$

Phase plane

Combining these equations as we did in obtaining (6.2) from (6.1), we find for the phase plane in this case,

$$\frac{dy}{dx} = -\frac{2x}{y - 0.4x}. \tag{6.9}$$

Direction field

The direction field for (6.9) is shown in Figure 6.5. (Because $dy/dt = -2x$, dy/dt is negative in the first quadrant, so the direction of travel will be clockwise, as indicated by the direction of the arrows.) Note that horizontal tangents occur only along the vertical axis, while the isocline for vertical tangents is given by the line $y = 0.4x$. The slope is positive along the x-axis, so this suggests that all solution curves will spiral toward the origin. This is even clearer in Figure 6.6, where we have drawn the numerical solutions for three different initial conditions – namely, $x_0 = 0$, $y_0 = 2.5$; $x_0 = 0$, $y_0 = 5$; and $x_0 = 0$, $y_0 = 10$. This suggests that all solution

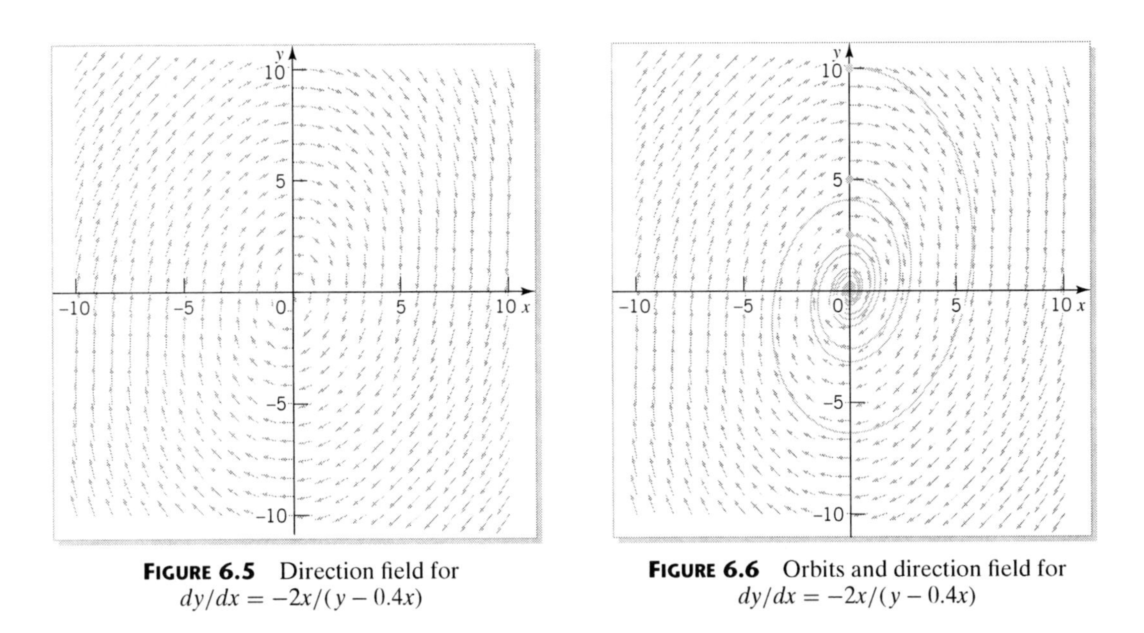

FIGURE 6.5 Direction field for $dy/dx = -2x/(y - 0.4x)$

FIGURE 6.6 Orbits and direction field for $dy/dx = -2x/(y - 0.4x)$

curves eventually end at the origin $x = 0$, $y = 0$, which means that eventually Denise and Chad will have no affection for each other.

Another way to see the effect of $-0.4x$ is by contrasting the direction fields and orbits of Figures 6.2 and 6.3 with those of Figures 6.5 and 6.6.

Equation (6.9) is a differential equation with homogeneous coefficients (see Section 4.2), which may be solved by a change in the dependent variable (see Exercise 5 on page 231). If we do that, we obtain the equations for the orbits drawn in Figure 6.6.

We can find the solutions for x and y as functions of t — that is, $x(t)$ and $y(t)$ — by substituting the first equation in (6.8), $y = x' + 0.4x$, into the second, $y' = -2x$, giving $x'' + 0.4x' = -2x$, or

$$x'' + 0.4x' + 2x = 0. \tag{6.10}$$

This is another second order linear differential equation, and in Section 6.3 we will show that the solution of (6.10) is

$$x(t) = e^{-0.2t}\left(C_1 \cos 1.4t + C_2 \sin 1.4t\right), \tag{6.11}$$

where C_1 and C_2 are arbitrary constants. [Verify that (6.11) satisfies (6.10).] If we substitute this expression for $x(t)$ into the first equation in (6.8), $y = x' + 0.4x$, we find

$$y(t) = e^{-0.2t}\left[\left(0.2C_1 + 1.4C_2\right)\cos 1.4t + \left(-1.4C_1 + 0.2C_2\right)\sin 1.4t\right]. \tag{6.12}$$

We see from (6.11) and (6.12) that as $t \to \infty$, both x and y go to zero. This confirms the earlier suggestion from the phase plane analysis that all solution curves eventually end at the origin. So eventually, neither Denise nor Chad will have any affection for each other. (Let us hope they didn't get married.)

Explicit solution

If we use the same initial conditions in (6.11) and (6.12) as we used in (6.6) and (6.7) – namely, $x(0) = 0$, $y(0) = 2.5$ – we find

$$x(t) = \frac{25}{14} e^{-0.2t} \sin 1.4t, \tag{6.13}$$

$$y(t) = 2.5e^{-0.2t} \left(\cos 1.4t + \frac{1}{7} \sin 1.4t \right). \tag{6.14}$$

Figure 6.7 shows (6.13) and (6.14). Because both amplitudes decrease with time, Denise and Chad have less affection for each other, until their affection ultimately dies out.[2] Comparing Figures 6.4 and 6.7 shows the impact of the parental interference parameter $c = 0.4$.

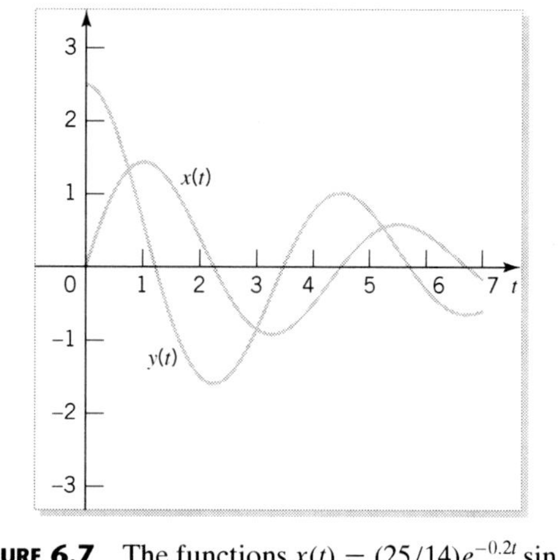

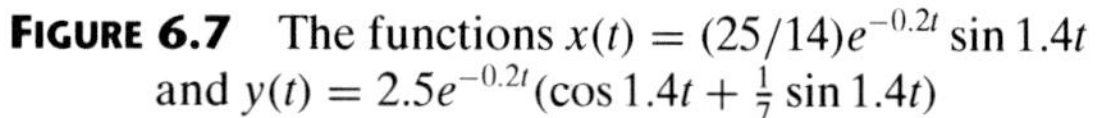

FIGURE 6.7 The functions $x(t) = (25/14)e^{-0.2t} \sin 1.4t$ and $y(t) = 2.5e^{-0.2t}(\cos 1.4t + \frac{1}{7} \sin 1.4t)$

EXERCISES

1. **Denise and Chad.** Show that for the Denise-Chad relationship – namely, (6.1) – the ratio of their emotional ranges is independent of Chad's initial emotion, provided that initially Denise had no emotion for Chad.

2. This exercise uses the following terminology: You "like" someone when you have affection for him or her. You "dislike" someone when you have negative feelings for that person. A couple is "happy" when they both like each other. A couple is "unhappy" when they both dislike each other. An individual is "happy" if he or she is liked. An individual is "unhappy" if he or she is disliked. Assume that the units of time in Figure 6.4 are weeks.
   (a) When were Denise and Chad happy as individuals? As a couple?
   (b) When were Denise and Chad unhappy as individuals? As a couple?

[2]Because this is in the best interests of both Denise and Chad, some might say that parents know best.

(c) Assuming this relationship lasted the seven weeks shown in this figure, estimate the proportion of the time Denise and Chad were happy as a couple.

(d) During this seven-week period, were Denise and Chad happy as a couple for longer periods of time than they were unhappy as a couple? Explain your reasoning.

(e) If you were a parent who was happy if either Denise was happy or Chad was unhappy, what proportion of the seven weeks were you happy? Explain.

(f) At what point during the seven weeks would you liked to have double-dated with Denise and Chad? Explain.

3. Figure 6.8 shows the slope field corresponding to $x' = -x + y$, $y' = -x - y$. Add arrows to the slope field so that it represents the direction field of this system of equations.

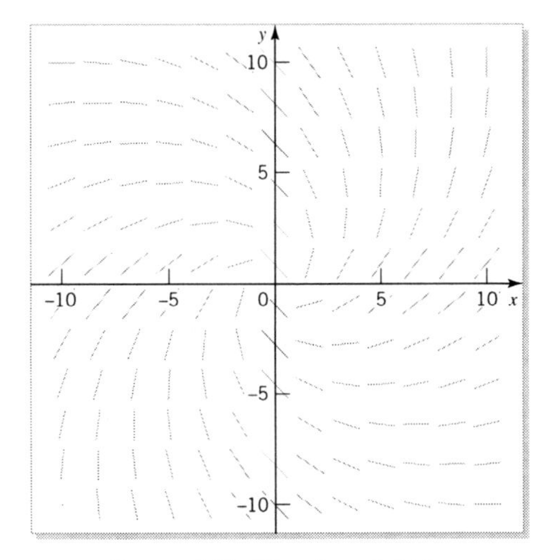

FIGURE 6.8 The slope field for $dy/dx = (-x - y)/(-x + y)$

4. **Denise and Chad — The Saga Continues.** Denise and Chad seek therapy and learn to alter their pattern of response to each other. Chad learns to accept Denise's affections, and now his own affection begins to decrease only if she becomes overly affectionate. His affection is modeled by the differential equation $y' = -0.2\,(x - 2)$. Denise learns to control her emotions. Her affection increases only when Chad becomes very affectionate, and her affection is modeled by $x' = 0.8\,(y - 2)$.

(a) Find the differential equation of the resulting orbits in the phase plane. Solve this differential equation, and comment on the long-term behavior of their romance if $x(0) = 0$, $y(0) = 2$.

(b) Who has the greater emotional range? Explain your reasoning.

5. Equation (6.9) is of the form $dy/dx = g(y/x)$, which was discussed in Section 4.2. Use the change of variables suggested in that section to find the solution curves of (6.9) — that is, find the equation for the orbits in the phase plane.

6.2 HOW FIRST ORDER SYSTEMS AND SECOND ORDER EQUATIONS ARE RELATED

In the previous section we saw examples of **systems of first order differential equations**, such as

$$\begin{cases} x' = y - 0.4x, \\ y' = -2x, \end{cases} \tag{6.15}$$

that had two dependent variables — namely, $x(t)$ and $y(t)$ — with initial conditions for $x(t_0)$, and $y(t_0)$.

In that section we also saw examples of **second order differential equations**, such as

$$x'' = -0.4x' - 2x, \tag{6.16}$$

with initial conditions for $x(t_0)$, and $x'(t_0)$.

These are each special cases of more general types of differential equations that we now bring together in one place. We first define a system of first order differential equations, and summarize its properties, and then follow the same pattern for second order differential equations.

First Order Systems

◆ *Definition 6.1:* **The set of differential equations**

$$\begin{cases} x' = P(t, x, y), \\ y' = Q(t, x, y), \end{cases} \tag{6.17}$$

where P and Q are given functions of t, x, and y, is called a SYSTEM OF FIRST ORDER DIFFERENTIAL EQUATIONS. Typical initial conditions are $x(t_0) = x_0$, $y(t_0) = y_0$. ◆

Comments about Systems of First Order Differential Equations

- The GENERAL SOLUTION of (6.17) contains every solution of (6.17).
- The problem of solving the system (6.17) subject to specified initial conditions is called an INITIAL VALUE PROBLEM.
- **Finding solutions.** Very few systems can be solved explicitly. In the same way that classifying first order equations was helpful in understanding the behavior of solutions, so too, classifying a system is helpful.
- **Finding numerical solutions.** In some cases finding numerical solutions is the only option that will give us quantitative information about a solution. As in the case of first order equations, if possible we should make sure a numerical solution is consistent with any qualitative information we have.
- **Existence and uniqueness.** Before trying to find a solution — particularly a numerical one — we should make sure such a solution exists. Existence and uniqueness theorems are therefore essential.

We now expand on these last three comments in turn.[3]

Comments about Finding Solutions and Classifying a System

Autonomous

- **Autonomous.** If P and Q are independent of t, that is,

$$\begin{cases} x' = P(x, y), \\ y' = Q(x, y), \end{cases}$$

the system is called AUTONOMOUS. For example, $x' = xy$, $y' = x - y^3$, is an autonomous system. It is rare to find the general solution of an autonomous system.

[3] These definitions, comments, and results can be extended to systems with more than two differential equations.

However, by eliminating t and working in the xy-plane, qualitative information can frequently be obtained. The xy-plane is called the PHASE PLANE, and the graph of y versus x is called an ORBIT. The graph of a collection of representative orbits is called a PHASE PORTRAIT. Autonomous systems will occur repeatedly in this book.

Phase portrait

Linear

- **Linear.** If P and Q are linear in x and y, that is,

$$\begin{cases} x' = a(t)x + b(t)y + e(t), \\ y' = c(t)x + d(t)y + f(t), \end{cases}$$

where a, b, c, d, e, and f are given functions of t, the system is called LINEAR. For example, $x' = tx + y + \sin t$, $y' = x - e^t y + t^2$, is a linear system. (A system that is not linear is called NONLINEAR.) If $e(t) = f(t) = 0$, that is,

$$\begin{cases} x' = a(t)x + b(t)y, \\ y' = c(t)x + d(t)y, \end{cases}$$

Homogeneous

the linear system is called HOMOGENEOUS — for example, $x' = tx + y$, $y' = x - e^t y$ is a homogeneous linear system — otherwise NONHOMOGENEOUS.[4] It is rare to find the general solution of a linear system that is not autonomous.

Constant coefficients

- **Constant coefficients.** A linear homogeneous autonomous system is of the form

$$\begin{cases} x' = ax + by, \\ y' = cx + dy, \end{cases}$$

where a, b, c, and d are constants. Such a system is commonly called a homogeneous system with CONSTANT COEFFICIENTS. Thus, (6.15) is an example of a linear homogeneous, autonomous system of first order differential equations. Linear systems with constant coefficients are among the very few systems that can be solved explicitly. A major part of Chapter 9 will be devoted to this topic.

Comments about Finding Numerical Solutions

Any of the numerical methods — such as Euler's or the Runge-Kutta 4 method — for first order differential equations that we saw in Chapter 3 can be extended to first order systems. For example, for (6.17), given the initial conditions $x(t_0) = x_0$, $y(t_0) = y_0$, and step-size h, we can estimate $x(t_0 + h)$ and $y(t_0 + h)$ by Euler's method as follows:

Euler's method

$$\begin{cases} x(t_0 + h) \approx x(t_0) + hx'\left(t_0, x(t_0), y(t_0)\right) = x_0 + hP\left(t_0, x_0, y_0\right), \\ y(t_0 + h) \approx y(t_0) + hy'\left(t_0, x(t_0), y(t_0)\right) = y_0 + hQ\left(t_0, x_0, y_0\right), \end{cases} \tag{6.18}$$

and in general,

$$\begin{cases} x(t_n) \approx x(t_{n-1}) + hx'\left(t_{n-1}, x(t_{n-1}), y(t_{n-1})\right) = x(t_{n-1}) + hP\left(t_{n-1}, x(t_{n-1}), y(t_{n-1})\right), \\ y(t_n) \approx y(t_{n-1}) + hy'\left(t_{n-1}, x(t_{n-1}), y(t_{n-1})\right) = y(t_{n-1}) + hQ\left(t_{n-1}, x(t_{n-1}), y(t_{n-1})\right), \end{cases} \tag{6.19}$$

[4] Notice that each side of the two equations contains a homogeneous function of x and y of degree one.

where $t_j = t_0 + jh$. Similar generalizations apply to all the numerical methods described in Chapter 3.

From now on we assume that you have access to a computer/calculator program that performs the Runge-Kutta 4 method on systems of differential equations.

If we introduce vector notation, and define[5]

$$\mathbf{X}(t) = \begin{bmatrix} x(t) \\ y(t) \end{bmatrix} \qquad \mathbf{G}(t, \mathbf{X}) = \begin{bmatrix} P(t, x, y) \\ Q(t, x, y) \end{bmatrix},$$

then the initial condition is

$$\mathbf{X}(t_0) = \begin{bmatrix} x(t_0) \\ y(t_0) \end{bmatrix} = \begin{bmatrix} x_0 \\ y_0 \end{bmatrix} = \mathbf{X}_0.$$

Vector form In this notation, (6.18) and (6.19) can be written as

$$\mathbf{X}(t_0 + h) \approx \mathbf{X}(t_0) + h\mathbf{X}'(t_0) = \mathbf{X}_0 + h\mathbf{G}\left(t_0, \mathbf{X}_0\right)$$

and

$$\mathbf{X}(t_n) \approx \mathbf{X}(t_{n-1}) + h\mathbf{G}\left(t_{n-1}, \mathbf{X}(t_{n-1})\right),$$

where

$$\mathbf{X}' = \frac{d\mathbf{X}}{dt} = \frac{d}{dt}\begin{bmatrix} x(t) \\ y(t) \end{bmatrix} = \begin{bmatrix} x' \\ y' \end{bmatrix}.$$

Comments about Existence and Uniqueness

Using vector notation, we can write (6.17) as

$$\frac{d}{dt}\begin{bmatrix} x(t) \\ y(t) \end{bmatrix} = \begin{bmatrix} P(t, x, y) \\ Q(t, x, y) \end{bmatrix},$$

or

$$\mathbf{X}'(t) = \mathbf{G}(t, \mathbf{X})$$

with initial condition $\mathbf{X}(t_0) = \mathbf{X}_0$. In this form we can see the formal analogy between this system and the first order differential equation

$$x'(t) = g(t, x)$$

with its initial condition $x(t_0) = x_0$. This analogy is more than superficial, and we now exploit this analogy to motivate two existence-uniqueness theorems.

---

[5]Appendix A.5 contains a summary of the properties of vectors and matrices.

First, the Existence-Uniqueness Theorem for $x'(t) = g(t, x)$ — Theorem 2.1 of Section 2.4 — which requires that g and $\partial g/\partial x$ be continuous in the vicinity of the initial point, can be extended to systems as follows.[6]

► Theorem 6.1: The Existence-Uniqueness Theorem for Systems *If* $\mathbf{G}(t, \mathbf{X})$ *and the partial derivatives* $\partial \mathbf{G}(t, \mathbf{X})/\partial x$ *and* $\partial \mathbf{G}(t, \mathbf{X})/\partial y$ *are defined and continuous in a region containing* $(t_0, \mathbf{X}_0)$*, then the system of differential equations*

$$\mathbf{X}'(t) = \mathbf{G}(t, \mathbf{X}) \tag{6.20}$$

has a unique solution passing through the point $\mathbf{X}(t_0) = \mathbf{X}_0$. *The solution is valid for all* t *for which the solution remains inside the region.* ◄

Second, the Existence-Uniqueness Theorem for linear differential equations — Theorem 5.1 of Section 5.1 — can also be extended to systems as follows.

► Theorem 6.2: The Existence-Uniqueness Theorem for Linear Systems *Consider the linear system of first order equations,*

$$\begin{cases} x' = a(t)x + b(t)y + e(t), \\ y' = c(t)x + d(t)y + f(t), \end{cases} \tag{6.21}$$

subject to the initial conditions $x(t_0) = x_0$, $y(t_0) = y_0$. *If* $a(t)$, $b(t)$, $c(t)$, $d(t)$, $e(t)$, *and* $f(t)$ *are continuous for* $\alpha < t < \beta$, *and* t_0 *is in this interval, then there exists a unique solution of (6.21) valid for* $\alpha < t < \beta$. ◄

Second Order Equations

We now follow the same pattern to discuss second order differential equations and show the interplay with first order systems.

◆ Definition 6.2: **The differential equation**

$$x'' = R\left(t, x, x'\right),$$

where R is a given function of t, x, and x', is called a SECOND ORDER DIFFERENTIAL EQUATION. Typical initial conditions would be $x(t_0) = x_0, x'(t_0) = x_0^*$. ◆

Comments about Second Order Differential Equations

- The GENERAL SOLUTION of $x'' = R\left(t, x, x'\right)$ contains every solution of $x'' = R\left(t, x, x'\right)$.
- The problem of solving the differential equation $x'' = R(t, x, x')$, subject to specified initial conditions $x(t_0) = x_0$, $x'(t_0) = x_0^*$, is called an INITIAL VALUE PROBLEM.

---

[6]The proof of a more general result than this can be found in *Ordinary Differential Equations* by E. L. Ince, Dover, 1956, page 71.

- The problem of solving the differential equation $x'' = R(t, x, x')$, subject to specified boundary conditions $x(t_0) = x_0$, $x(t_1) = x_1$, is called a BOUNDARY VALUE PROBLEM. Section 8.6 gives a brief introduction to these problems, and how they differ from initial value problems.
- In general, solving a second order differential equation requires two integrations, and so there will be two constants of integration, one for each integration. To evaluate these constants of integration, two conditions are needed. For initial value problems, the conditions are $x(t_0) = x_0$, $x'(t_0) = x_0^*$; for boundary value problems, the conditions are $x(t_0) = x_0$, $x(t_1) = x_1$.
- **Finding solutions.** Very few second order differential equations can be solved explicitly. In the same way that classifying first order equations was helpful in understanding the behavior of solutions, so too, classifying second order differential equations is helpful.
- **Finding numerical solutions.** In some cases, finding numerical solutions is the only option that will give us quantitative information about a solution. As in the case of first order equations, if possible we should make sure a numerical solution is consistent with any qualitative information we have.
- **Existence and uniqueness.** Before trying to find a solution — particularly a numerical one — we should make sure such a solution exists. Existence and uniqueness theorems are therefore essential.

We now expand on these last three comments in turn.

Comments about Finding Solutions and Classifying a Second Order Differential Equation

Autonomous

- **Autonomous.** If R is independent of t, that is,

$$x'' = R\left(x, x'\right),$$

the equation is called AUTONOMOUS. For example, $x'' = \left(x'\right)^2 \sin x$ is an autonomous second order differential equation. It is rare to find the general solution of an autonomous second order differential equation.

Linear

- **Linear.** If R is linear in x and x', that is,[7]

$$x'' = -q(t)x - p(t)x' + f(t),$$

where q, p, and f are given functions of t, the equation is called LINEAR. For example, $x'' = t^2x - tx' + \sin t$ is a linear second order differential equation. (A differential equation that is not linear is called NONLINEAR.) A second order linear differential equation is usually written as

$$x'' + p(t)x' + q(t)x = f(t),$$

[7]The reason for the minus signs on the right-hand side of this equation is that traditionally such an equation is written in the form $x'' + p(t)x' + q(t)x = f(t)$.

and, just like a first order linear differential equation, its solutions, if they can be found, are explicit. A second order nonlinear differential equation, just like a first order nonlinear differential equation, usually has implicit solutions.

If $f(t) = 0$, that is,

$$x'' + p(t)x' + q(t)x = 0,$$

Homogeneous

the linear equation is called HOMOGENEOUS, otherwise NONHOMOGENEOUS. It is rare to find the general solution of a linear second order differential equation.

Constant coefficients

Constant coefficients. A linear homogeneous autonomous second order equation is of the form

$$x'' + px' + qx = 0,$$

where p and q are constants. Such equations are commonly called homogeneous second order linear equations with CONSTANT COEFFICIENTS. Thus, (6.16) is an example of a linear homogeneous autonomous second order differential equation; that is, with constant coefficients. These equations are among the very few classes of second order equations that can be solved explicitly, and the next section is devoted to this topic.

Comments about Finding Numerical Solutions

It is always possible to rewrite a second order equation as a system of first order equations by introducing the variable $y = x'$, in which case $x'' = y'$ and we have the system[8]

$$\begin{cases} x' = y, \\ y' = R(t, x, y). \end{cases} \tag{6.22}$$

This ability to switch between second order equations and first order systems is very useful, as we found in the previous section. It means that **anything we know about systems can be applied to second order equations**. This interplay between systems and second order equations is very important and powerful. For example, by treating a second order equation as the system (6.22), we can generalize all the numerical methods described in Chapter 3 to the second order case by equations such as (6.18) and (6.19) with $P = y$ and $Q = R$.

From now on we assume that you have access to a computer/calculator program that performs the Runge-Kutta 4 method on second order differential equations.

Comments about Existence and Uniqueness

Also, if we apply the Existence-Uniqueness Theorem for Systems — Theorem 6.1 — to the system (6.22), we immediately have the following result.

---

[8]The introduction of the variable $y = x'$ has distinct advantages when x measures distance as a function of time. In this case, y is the velocity and y' the acceleration, and solving the system (6.22) leads to the simultaneous determination of distance and velocity. However, this is not the only way to convert a second order equation to a system of first order equations. See Exercises 6, 7, and 8.

▶ **Theorem 6.3: The Existence-Uniqueness Theorem for Second Order Differential Equations** *If $R(t, x, x')$ and the partial derivatives $\partial R(t, x, x')/\partial x$ and $\partial R(t, x, x')/\partial x'$ are defined and continuous in a region containing (t_0, x_0, x_0^*), then the second order differential equation*

$$x'' = R\left(t, x, x'\right)$$

has a unique solution passing through the point $x(t_0) = x_0$, $x'(t_0) = x_0^*$. The solution is valid for all t for which the solution remains inside the region. ◀

In the same way, a second order linear equation can be written as the linear system

$$\begin{cases} x' = y, \\ y' = -q(t)x - p(t)y + f(t). \end{cases}$$

If we apply the Existence-Uniqueness Theorem for Linear Systems — Theorem 6.2 — to this system, we immediately find the following theorem.

▶ **Theorem 6.4: The Existence-Uniqueness Theorem for Second Order Linear Differential Equations** *Consider the second order linear equation*

$$x'' + p(t)x' + q(t)x = f(t) \tag{6.23}$$

subject to the initial conditions $x(t_0) = x_0$, $x'(t_0) = x_0^*$. If $q(t)$, $p(t)$, and $f(t)$ are continuous for $\alpha < t < \beta$, and t_0 is in this interval, then there exists a unique solution of (6.23) valid for $\alpha < t < \beta$. ◀

It is very important to realize that **systems of first order equations on the one hand and second order differential equations on the other give us two different ways of looking at the same problem.** In the next three sections we will look at this interplay for specific situations.

EXERCISES

1. Classify the following systems of equations using the terms autonomous, nonautonomous, linear, nonlinear, homogeneous, nonhomogeneous, and constant coefficients. Do not attempt to solve any of these equations.

(a) $x' = y + \sin t$, $y' = x$

(b) $x' = y + \sin t$, $y' = x^2$

(c) $x' = y \cos t$, $y' = 6x$

(d) $x' = y$, $y' = x - x^3$

(e) $x' = 2\cos x - y$, $y' = x + y$

(f) $x' = x - y$, $y' = x + y$

2. Classify the following second order differential equations using the terms autonomous, nonautonomous, linear, nonlinear, homogeneous, nonhomogeneous, and constant coefficients. Do not attempt to solve any of these equations.

(a) $x'' = x' + \sin t$

(b) $x'' = \left(x'\right)^2 + t$

(c) $x'' = -1 + e^{-x}\left(x'\right)^2$

(d) $x'' + g(x)x' + x = 0$ (Liénard's equation)

(e) $x'' = x - x^3$ (Duffing's equation)

(f) $x'' - x + x^3 = \cos t$ (forced Duffing's equation)

(g) $x'' + a^2\left(1 - x^2\right)x' + x = 0$ (van der Pol's equation)

(h) $t^2x'' + tx' + \left(t^2 - n^2\right)x = 0$ (Bessel's equation of order n)

3. Show that the system (6.22) is autonomous/linear/homogeneous/constant coefficients if $x'' = R(t, x, x')$ is autonomous/linear/homogeneous/constant coefficients.

4. Show that all autonomous second order differential equations $x'' = R(x, x')$ give rise to the first order equation $\frac{dy}{dx} = \frac{R(x,y)}{y}$ by introducing $y = x'$ and $y' = \frac{dy}{dx}x'$. This first order equation gives the orbits of $x'' = R(x, x')$ in the phase plane.

5. Show that all second order differential equations of the form $x'' = R(t, x')$ give rise to the first order equation $\frac{dy}{dt} = R(t, y)$ by introducing $y = x'$.

6. Show that the differential equation $a_2(t)x'' + a_1(t)x' + a_0(t)x = 0$ $(a_2 \neq 0)$ can be written as the system

$$\begin{cases} x' = yx, \\ y' = -a_0/a_2 - a_1 y/a_2 - y^2, \end{cases}$$

by defining $y = x'/x$.

7. Convert van der Pol's equation $x'' + a^2(1 - x^2)x' + x = 0$ into

(a) the system of equations

$$\begin{cases} x' = y, \\ y' = -a^2(1 - x^2)y - x, \end{cases}$$

by defining $y = x'$.

(b) the system of equations

$$\begin{cases} x' = y - a^2\left(x - \frac{1}{3}x^3\right), \\ y' = -x, \end{cases}$$

by defining $y = x' + a^2\left(x - \frac{1}{3}x^3\right)$.

8. Convert Liénard's equation $x'' + g(x)x' + x = 0$ into the system of equations

$$\begin{cases} x' = y - \int_0^x g(u)\,du, \\ y' = -x, \end{cases}$$

by defining $y = x' + \int_0^x g(u)\,du$.

9. **Skydiving.** The equation that models the skydiving example from Section 4.4 is $x'' = g - \frac{k}{m}(x')^2$. By introducing the variable $y = x'$, write this equation as the system

$$\begin{cases} x' = y, \\ y' = g - \frac{k}{m}y^2. \end{cases}$$

Notice that in this form, the second equation is independent of x and is a first order autonomous differential equation for y. Thus, we could find $y(t)$ and substitute the result into the first equation to find $x(t)$. Compare this to the way we solved this equation in Section 4.4.

10. Show that, if $x_1(t)$ and $x_2(t)$ are both solutions of the differential equation $x'' + 2x = 0$, then so is $C_1x_1(t) + C_2x_2(t)$, where C_1 and C_2 are arbitrary constants.

11. Show that for the differential equation $a_2(t)x'' + a_1(t)x' + a_0(t)x = 0$, if $x_1(t)$ and $x_2(t)$ are solutions, then so is $C_1x_1(t) + C_2x_2(t)$, where C_1 and C_2 are arbitrary constants.

12. Show that, by eliminating all y dependence, the system of equations $x' = y$, $y' = -\sin x$ gives rise to the second order differential equation $x'' + \sin x = 0$. Explain why this is not a linear differential equation.

13. Show that any system of first order linear differential equations of the form

$$\begin{cases} x' = a(t)x + b(t)y + f(t), \\ y' = c(t)x + d(t)y + g(t), \end{cases} \tag{6.24}$$

gives rise to a single second order linear differential equation for x as a function of t ($x' = dx/dt$, $y' = dy/dt$).

14. For the following cases, use the techniques from Exercise 13 to find a second order linear differential equation obtained by eliminating one of the dependent variables ($x' = dx/dt$, $y' = dy/dt$).

(a) $x' = x - y$, $y' = -x + 2y$

(b) $x' = 3x - 2y$, $y' = 2x - 2y$

(c) $x' = 5x - 2y + 3$, $y' = 6x - 2y$

(d) $x' = -2x + y$, $y' = 2x - y - t$

(e) $x' = x - y$, $y' = -3x + 2y + \cos t$

(f) $x' = -4x - 6y + 6e^{2t}$, $y' = x + y + 3e^{2t}$

15. The method of elimination used in the previous exercise will also work for first order linear differential equations that are not in the form given by (6.24). Find a second order differential equation by eliminating one of the dependent variables in the following systems of differential equations ($x' = dx/dt$, $y' = dy/dt$).

(a) $x' + y' + y = 0$, $3x' + 2y' + x = 7$

(b) $x' + y' + 2x = -e^t$, $2x' - 2y' - y = 0$

(c) $x' - 3y' - 4x = e^{2t}$, $2x' - 2y' - y = 8$

(d) $x' + 2x + y' = e^{-t}$, $2tx' + 3y' + y = t - 3$

(e) $tx' + x + y' = \cos t$, $2x' - y' - y = -2\sin t$

(f) $x' + 2tx + y' = \sin t$, $2tx' + 2y' + 5y = 0$

16. Computer Experiment — Denise, Chad, and Parental Opinion.[9] Chad's affection for Denise increases when she feels negatively disposed toward him. Denise's affection for Chad increases not only in proportion to his affection for her as before, but also includes an additive term reflecting her parents' opinion. Their opinion is positive as long as Chad's affection for her does not exceed a certain limit, either positive or negative.

(a) Let x measure Denise's affection for Chad, and y measure Chad's affection for Denise. Explain why $x' = by + cx\left(1 - y^2\right)$, $y' = -ax$, where a, b, and c are positive numbers, is a possible model for their relationship.

(b) Use a computer to study their relationship in the xy-plane for $a = b = 1$ and various values of c. What happens to their relationship in the long term? Is their relationship stable?

17. Figure 6.9 shows the direction field corresponding to $x' = P(x, y)$, $y' = Q(x, y)$, where $P(x, y)$ and $Q(x, y)$ are given functions. From it, sketch the direction field for $x' = -P(x, y)$, $y' = -Q(x, y)$. Explain why this is sometimes called "time-reversal."

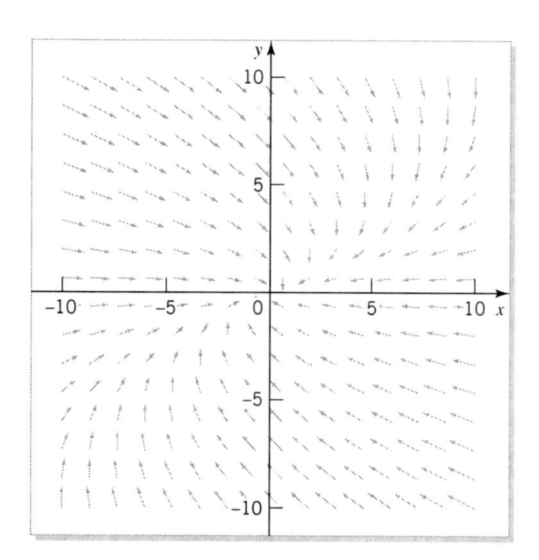

FIGURE 6.9 The direction field for $dx/dt = P(x, y)$, $dy/dt = Q(x, y)$

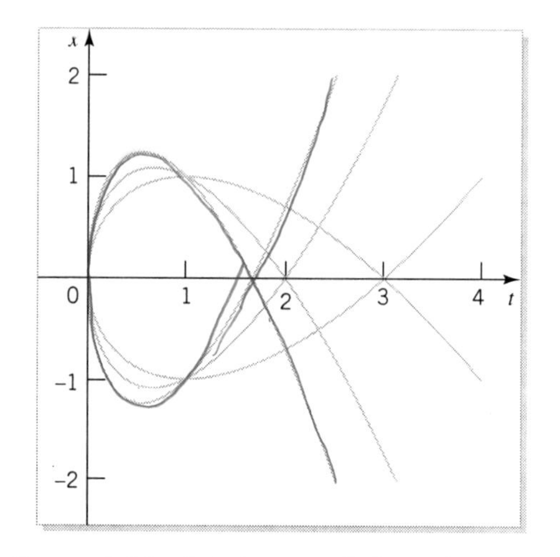

FIGURE 6.10 Solutions of $4t^2x'' - 4tx' + 3x = 0$

18. Figure 6.10 shows various solutions of $4t^2x'' - 4tx' + 3x = 0$.

(a) How many solutions are shown in the figure?

(b) From the graphs you can see that the solutions have a maximum when $x > 0$ and a minimum if $x < 0$. Explain how the differential equation can be used to show that it is not possible for a solution to have a minimum if $x > 0$, nor is it possible for a solution to have a maximum if $x < 0$.

(c) Identify the solution corresponding to the initial conditions $x(1) = 1$, $x'(1) = 0$. From the graph of this solution estimate the value of $t = t_1$ ($t_1 > 0$) for which $x(t_1) = 0$. From the graph estimate $x'(t_1)$.

19. Two people are looking at the differential equation $x'' + f(t)x = 0$, where $f(t)$ is a given continuous function, subject to $x(0) = 0$. The first person observes that the trivial solution $x(t) = 0$ satisfies these conditions, and so by the Existence-Uniqueness Theorem, the function $x(t) = 0$ is the only such solution for any $f(t)$. The second person observes that, if $f(t) = 1$, the function $x(t) = \sin t$ satisfies the differential equation subject to $x(0) = 0$. Reconcile this contradiction.

[9] This exercise is based on an idea of Alan Newell, University of Warwick.

20. The graphs shown in Figure 6.11 show two solutions of one of the differential equations $x' + p(t)x = q(t)$ or $x'' + p(t)x = q(t)$, where $p(t)$ and $q(t)$ are continuous. Explain which differential equation applies.

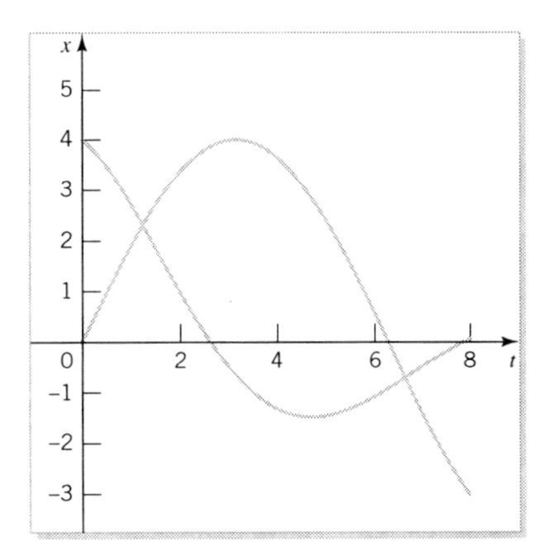

FIGURE 6.11 The mystery functions

6.3 SECOND ORDER LINEAR DIFFERENTIAL EQUATIONS WITH CONSTANT COEFFICIENTS

In this section we concentrate on the differential equation

$$ax'' + bx' + cx = 0 \tag{6.25}$$

where a, b, and c are given constants, with $a \neq 0$. By dividing by a we see that this is the standard form of a homogeneous second order linear differential equation with constant coefficients. We devote this section to such equations for three reasons. First, they can be solved explicitly. Second, they are used frequently to model different physical phenomena. Third, as we show in a later section, they can be interpreted in terms of a system of equations stressing that second order differential equations, on the one hand, and systems of first order equations, on the other, give us two different ways of looking at the same problem.

The following is the main result.

Theorem 6.5: *The solutions of $ax'' + bx' + cx = 0$, where a, b, and c are given constants, with $a \neq 0$, depend on the roots of the quadratic equation $ar^2 + br + c = 0$ in the following way. (C_1 and C_2 are arbitrary constants.)*

1. *If the roots are real and distinct — say, r_1 and r_2 — then $x(t) = C_1e^{r_1t} + C_2e^{r_2t}$.*
2. *If the roots are equal (a double, or repeated, root) — say, r — then $x(t) = (C_1 + C_2t)\, e^{rt}$.*
3. *If the roots are complex conjugates of each other — say, $\alpha \pm i\beta$ — then $x(t) = e^{\alpha t}\,(C_1 \cos \beta t + C_2 \sin \beta t)$.*

Proof If we were to solve (6.25) with $a = 0$ — namely, $bx' + cx = 0$ — we would first divide this equation by x and then integrate. This suggests that to solve (6.25) with $a \neq 0$, we divide (6.25) by x, obtaining

$$a\frac{x''}{x} + b\frac{x'}{x} + c = 0. \tag{6.26}$$

We notice that the coefficients of a and b are related by

$$\left(\frac{x'}{x}\right)' = \frac{x''}{x} - \left(\frac{x'}{x}\right)^2,$$

Change variable

which prompts us to introduce a new dependent variable, $r(t)$, where $x'/x = r$, so that $x''/x = r' + r^2$, while

$$x(t) = C\exp\left(\int r\,dt\right). \tag{6.27}$$

If we substitute for x'/x and x''/x in (6.26), we find $a\left(r' + r^2\right) + br + c = 0$, or

$$ar' = -\left(ar^2 + br + c\right). \tag{6.28}$$

This is an autonomous differential equation for $r(t)$. If we can solve (6.28) for $r(t)$, and then substitute the result into (6.27), we will have found the solution of (6.25). The differential equation (6.28) will have three different types of solutions depending on whether the quadratic $ar^2 + br + c = 0$ has real distinct roots (two equilibrium solutions), a double root (one equilibrium solution), or complex roots (no equilibrium solutions). In all cases, the nonequilibrium solutions are obtained from

$$\int \frac{a}{ar^2 + br + c}\,dr = -\int dt, \tag{6.29}$$

but we expect that the integral on the left-hand side will give three different formulas, depending on whether $ar^2 + br + c = 0$ has real distinct roots, a double root, or complex roots. We look at each of these three cases in turn.

Two distinct real roots

First, if the quadratic $ar^2 + br + c = 0$ has two distinct real roots r_1 and r_2 — so that $r_{1,2} = \left(-b \pm \sqrt{b^2 - 4ac}\right)/(2a)$ and $ar^2 + br + c = a\left(r - r_1\right)\left(r - r_2\right)$ — we find, by partial fractions, that the integral on the left-hand side of (6.29) can be written

$$\int \frac{a}{ar^2 + br + c}\,dr = \int \frac{1}{\left(r - r_1\right)\left(r - r_2\right)}\,dr = \frac{1}{r_1 - r_2}\ln\left|\frac{r - r_1}{r - r_2}\right|.$$

Thus, (6.29) yields

$$\frac{1}{r_1 - r_2}\ln\left|\frac{r - r_1}{r - r_2}\right| = -t + c_1,$$

where c_1 is an arbitrary constant. This equation can be rewritten in the form

$$\ln\left|\frac{r - r_1}{r - r_2}\right| = -\left(r_1 - r_2\right)t + c, \tag{6.30}$$

where $c = c_1(r_1 - r_2)$. We want to solve (6.30) for r and substitute the result in (6.27) to find $x(t)$. We solve for r by first taking the exponential of (6.30), obtaining

$$\frac{r - r_1}{r - r_2} = Ae^{-(r_1 - r_2)t},$$

where $A = \pm e^c$, and then solving this equation for r, finding

$$r(t) = \frac{r_1 - r_2 Ae^{-(r_1-r_2)t}}{1 - Ae^{-(r_1-r_2)t}} = \frac{r_1 e^{r_1 t} - r_2 Ae^{r_2 t}}{e^{r_1 t} - Ae^{r_2 t}}.$$

We substitute this into (6.27), which requires

$$\int r\,dt = \int \frac{r_1 e^{r_1 t} - r_2 Ae^{r_2 t}}{e^{r_1 t} - Ae^{r_2 t}}\,dt = \ln\left|e^{r_1 t} - Ae^{r_2 t}\right|,$$

Explicit solution

so

$$x(t) = C\exp\left(\int r\,dt\right) = C\exp\left(\ln\left|e^{r_1 t} - Ae^{r_2 t}\right|\right) = C_1 e^{r_1 t} + C_2 e^{r_2 t}, \tag{6.31}$$

where $C_1 = \pm C$ and $C_2 = \mp CA$. Here C_1 and C_2 are arbitrary constants. This is the solution of (6.25) for the case when the quadratic $ar^2 + br + c = 0$ has two distinct real roots, r_1 and r_2.

Double root

Second, if the quadratic $ar^2 + br + c = 0$ has a double root $r_1 = r_2$ — so that $r_{1,2} = -b/(2a)$ and $ar^2 + br + c = a(r - r_1)^2$ — we find, by the substitution $u = r - r_1$, that the integral on the left-hand side of (6.29) can be written

$$\int \frac{a}{ar^2 + br + c}\,dr = \int \frac{1}{(r - r_1)^2}\,dr = -\frac{1}{r - r_1}.$$

Using this in (6.29) yields

$$-\frac{1}{r - r_1} = -t + c_1,$$

where c_1 is an arbitrary constant, which can be solved for r as

$$r(t) = r_1 + \frac{1}{t - c_1}.$$

We substitute this into (6.27), which requires

$$\int r\,dt = \int \left(r_1 + \frac{1}{t - c_1}\right)dt = r_1 t + \ln|t - c_1|,$$

Explicit solution

so

$$x(t) = C\exp\left(\int r\,dt\right) = C\exp\left(r_1 t + \ln|t - c_1|\right) = (C_1 + C_2 t)\,e^{r_1 t}, \tag{6.32}$$

where $C_1 = \mp Cc_1$ and $C_2 = \pm C$. Here C_1 and C_2 are arbitrary constants. This is the solution of (6.25) for the case when the quadratic $ar^2 + br + c = 0$ has a double root.

Complex roots

Third, if the quadratic $ar^2 + br + c = 0$ has complex roots $r_1 = \alpha + i\beta$, $r_2 = \alpha - i\beta$ — so that $r_{1,2} = \left(-b \pm i\sqrt{4ac - b^2}\right) / (2a)$ and $ar^2 + br + c = a(r - \alpha - i\beta)(r - \alpha + i\beta) = a\left[(r-\alpha)^2 + \beta^2\right]$ — we find, by the substitution $r = \alpha + \beta \tan u$, that the integral on the left-hand side of (6.29) can be written

$$\int \frac{a}{ar^2 + br + c}\, dr = \int \frac{1}{(r-\alpha)^2 + \beta^2}\, dr = \frac{1}{\beta} \arctan\left(\frac{r-\alpha}{\beta}\right).$$

Using this in (6.29) yields

$$\frac{1}{\beta} \arctan\left(\frac{r-\alpha}{\beta}\right) = -t + c_1,$$

where c_1 is an arbitrary constant, which can be solved for r as $r(t) = \alpha + \beta \tan(-\beta t + A)$, where $A = \beta c_1$. We substitute this into (6.27), which requires

$$\int r\, dt = \int [\alpha + \beta \tan(-\beta t + A)]\, dt = \alpha t + \ln|\cos(-\beta t + A)|,$$

so

$$x(t) = C \exp\left(\int r\, dt\right) = C \exp[\alpha t + \ln|\cos(-\beta t + A)|] = \pm C e^{\alpha t} \cos(-\beta t + A).$$

Explicit solution

Using the trigonometric identity $\cos(-\beta t + A) = \cos \beta t \cos A + \sin \beta t \sin A$, we can rewrite the solution as

$$x(t) = e^{\alpha t}\left(C_1 \cos \beta t + C_2 \sin \beta t\right),$$

where $C_1 = \pm C \cos A$ and $C_2 = \pm C \sin A$. Here C_1 and C_2 are arbitrary constants. This is the solution of (6.25) for the case when the quadratic $ar^2 + br + c = 0$ has complex conjugate roots $\alpha \pm i\beta$. ◀

How to Solve Homogeneous Second Order Linear Differential Equations with Constant Coefficients

Purpose To find the general solution of the second order linear differential equation with constant coefficients,

$$ax'' + bx' + cx = 0, \tag{6.33}$$

where a, b, and c are constants, and $a \neq 0$.

Process

1. Write the associated quadratic equation $ar^2 + br + c = 0$. This equation is central to finding solutions of $ax'' + bx' + cx = 0$. It is called the CHARACTERISTIC EQUATION, because the behavior of the solutions of $ax'' + bx' + cx = 0$ are characterized by the roots of this quadratic equation.
2. Solve the characteristic equation for the two solutions r_1 and r_2. There are three possibilities.
   (a) r_1 and r_2 are real and distinct.
   (b) $r_1 = r_2$ (there is a double root).
   (c) r_1 and r_2 are complex conjugates of each other, written as $\alpha \pm i\beta$.
3. Write the explicit solution corresponding to the possibilities (a), (b), and (c) of step 2 (C_1 and C_2 are arbitrary constants) as
   (a) $x(t) = C_1 e^{r_1 t} + C_2 e^{r_2 t}$.
   (b) $x(t) = \left(C_1 + C_2 t\right) e^{r_1 t}$.
   (c) $x(t) = e^{\alpha t}\left(C_1 \cos \beta t + C_2 \sin \beta t\right)$.

Comments about Second Order Linear Differential Equations with Constant Coefficients

- Sometimes it is useful to rewrite the general solution in case (c) — namely, $x(t) = e^{\alpha t}\left(C_1 \cos \beta t + C_2 \sin \beta t\right)$ — in the form $x(t) = Ae^{\alpha t} \cos(\beta t + \phi)$, where $A = \sqrt{C_1^2 + C_2^2}$, and ϕ — the phase angle — is defined by $\cos \phi = C_1/\sqrt{C_1^2 + C_2^2}$ and $\sin \phi = -C_2/\sqrt{C_1^2 + C_2^2}$. (See Exercise 6 on page 249.) In this form we see that the solution has amplitude A and phase angle ϕ. The function $\cos(\beta t + \phi)$ is periodic with period $2\pi/\beta$.[10] For example, one solution of $x'' + 2x' + 5x = 0$ is $x(t) = e^{-t}(3 \cos 2t + 4 \sin 2t)$, which can be rewritten as $x(t) = 5e^{-t} \cos(2t + \phi)$, which has amplitude 5 and phase angle $\phi \approx 149°$. The function $\cos(2t + \phi)$ has period π.
- Sometimes it is useful to rewrite the general solution in case (c) — namely, $x(t) = e^{\alpha t}\left(C_1 \cos \beta t + C_2 \sin \beta t\right)$ — in terms of complex numbers — namely, $x(t) = Ke^{(\alpha + i\beta)t} + \widetilde{K}e^{(\alpha - i\beta)t}$, where $K = \frac{1}{2}\left(C_1 - iC_2\right)$ and $\widetilde{K}$ is the complex conjugate of K.[11] (See Exercise 7 on page 249.)
- If $r_1 = -r_2$, then the solution $x(t) = C_1 e^{r_1 t} + C_2 e^{-r_1 t}$ can be written as $x(t) = K_1 \sinh r_1 t + K_2 \cosh r_1 t$, where $K_1 = \frac{1}{2}\left(C_1 - C_2\right)$ and $K_2 = \frac{1}{2}\left(C_1 + C_2\right)$. (See Exercise 8 on page 249.)

[10]A function $f(t)$ is periodic of period P if $f(t + P) = f(t)$ for every t for which $f(t)$ is defined, and where P is the smallest positive number with this property.

[11]Appendix A.4 contains a discussion of complex numbers.

- A quick way of determining the characteristic equation associated with a second order linear differential equation with constant coefficients uses operator notation. With operator notation, the capital letter D denotes differentiation. Thus, we may write $x'(t) = dx/dt = Dx$, where the D stands for differentiation with respect to t. With this notation, the usual algebraic operations of differentiation are slightly contracted. Thus, the standard rule for differentiating a sum would be written as

$$D[f(t) + g(t)] = Df(t) + Dg(t),$$

so our basic second order linear differential equation becomes

$$aD^2x + bDx + cx = (aD^2 + bD + c)x = 0.$$

The associated characteristic equation is $ar^2 + br + c = 0$, so we may determine the characteristic equation by simply replacing the operator D in the differential equation with the parameter r and setting the result to zero.

We now apply this process to three examples, one for each of the three possible cases.

EXAMPLE 6.3

Find the explicit solution of the following differential equations.

(a) $x'' + 2x' - 3x = 0$ (6.34)

(b) $x'' + 2x' + x = 0$ (6.35)

(c) $x'' + 2x' + 5x = 0$ (6.36)

The procedure for each case is to find values of r that solve the resulting characteristic equation.

(a) The characteristic equation for (6.34) is $r^2 + 2r - 3 = 0$, which may be factored as $(r - 1)(r + 3) = 0$. The roots of this characteristic equation are 1 and -3, so the explicit solution of (6.34) is

$$x(t) = C_1 e^t + C_2 e^{-3t}.$$

(b) The characteristic equation for (6.34) is $r^2 + 2r + 1 = 0$, which may be factored as $(r + 1)(r + 1) = 0$. Because we have a double root of $r = -1$ of this characteristic equation, we write the explicit solution of (6.34) as

$$x(t) = \left(C_1 + C_2 t\right) e^{-t}.$$

(c) The characteristic equation for (6.34) is $r^2 + 2r + 5 = 0$, which has the complex roots $r_1 = -1 - 2i$, $r_2 = -1 + 2i$. We may write the explicit solution as

$$x(t) = e^{-t}\left(C_1 \cos 2t + C_2 \sin 2t\right)$$

Because we know all the possible solutions of (6.33), we will look at them a little more closely.

- If r_1 and r_2 are real and distinct, the solution is $x(t) = C_1e^{r_1t} + C_2e^{r_2t}$. This contains exponential growth ($r_1 > 0$ and $r_2 > 0$), exponential decay ($r_1 < 0$ and $r_2 < 0$), and exponential growth and decay (r_1 and r_2 have opposite signs), as well as the possibility of either of r_1 or r_2 being zero, giving a constant plus an exponential term.
- If r_1 and r_2 are real but equal, the solution is $x(t) = C_1e^{r_1t} + C_2te^{r_1t}$. This contains exponential growth ($r_1 > 0$), exponential decay ($r_1 < 0$), or a term linear in time ($r_1 = 0$).
- If r_1 and r_2 are complex conjugates of each other, written as $\alpha \pm i\beta$, the solution is $x(t) = e^{\alpha t}\left(C_1 \cos \beta t + C_2 \sin \beta t\right)$. This contains exponential growing oscillations ($\alpha > 0$), exponential decaying oscillations ($\alpha < 0$), and pure oscillations ($\alpha = 0$).

Thus, **combinations of exponential growth or decay, linear growth or decay, and oscillations are the only possible types of solutions for second order linear differential equations with constant coefficients**, (6.33).

The next example will be revisited in different ways in the next two sections.

EXAMPLE 6.4

Solve the initial value problem

$$x'' + ax' + \frac{9}{4}x = 0, \; x(0) = x_0, \; x'(0) = 0,$$

when (i) $a = 0$, (ii) $a = 1$, (iii) $a = 3$, and (iv) $a = 5$. In each case, describe the behavior of the solution.

The characteristic equation is $r^2 + ar + \frac{9}{4} = 0$, with roots $r = \frac{1}{2}(-a \pm \sqrt{a^2 - 9})$.

(i) With $a = 0$, we find $r = 0 \pm \frac{3}{2}i$, with solution $x(t) = C_1 \cos \frac{3}{2}t + C_2 \sin \frac{3}{2}t$. To apply the initial conditions, we need $x'(t) = -\frac{3}{2}C_1 \sin \frac{3}{2}t + \frac{3}{2}C_2 \cos \frac{3}{2}t$. Thus, $x(0) = C_1$ and $x'(0) = \frac{3}{2}C_2$, so the initial conditions demand that $C_1 = x_0$ and $C_2 = 0$, which leads to the solution $x(t) = x_0 \cos \frac{3}{2}t$. This solution oscillates forever with amplitude x_0. This is frequently called **simple harmonic** or **undamped motion** and is representative of a pair of complex-conjugate roots with zero real part.

Undamped motion

(ii) With $a = 1$, we find $r = -\frac{1}{2} \pm \sqrt{2}i$, with solution $x(t) = e^{-\frac{1}{2}t}(C_1 \cos \sqrt{2}t + C_2 \sin \sqrt{2}t)$. Applying the initial conditions leads to the solution $x(t) = x_0e^{-\frac{1}{2}t}(\cos \sqrt{2}t + \frac{1}{2\sqrt{2}} \sin \sqrt{2}t)$. This solution oscillates forever with a decaying amplitude that decreases to 0 as $t \to \infty$. This is frequently called **underdamped** or **subcritical motion** and is representative of a pair of complex-conjugate roots with negative real part.

Underdamped motion

(iii) With $a = 3$, we find the repeated root $r = -\frac{3}{2}$, with solution $x(t) = e^{-\frac{3}{2}t}\left(C_1 + C_2t\right)$. Applying the initial conditions leads to the solution $x(t) = x_0e^{-\frac{3}{2}t}\left(1 + \frac{3}{2}t\right)$. This solution does not oscillate — in fact, in this example it does not even cross the positive t-axis once, let alone an infinite number of times. It approaches $x = 0$ as $t \to \infty$. This is frequently called **critically damped** or **critical motion** and is representative of repeated negative real roots.

Critically damped motion

(iv) With $a = 5$, we find $r = -\frac{1}{2}$ and $r = -\frac{9}{2}$, with solution $x(t) = C_1 e^{-\frac{1}{2}t} + C_2 e^{-\frac{9}{2}t}$. Applying the initial conditions leads to the solution $x(t) = x_0(\frac{9}{8}e^{-\frac{1}{2}t} - \frac{1}{8}e^{-\frac{9}{2}t})$. This solution does not oscillate — in fact, in this example it does not cross the positive t-axis once. It approaches $x = 0$ as $t \to \infty$. This is frequently called **overdamped** or **supercritical motion** and is representative of a pair of nonrepeated negative real roots.

Overdamped motion

If in (6.33) we set $a_2 = 1, a_1 = a$, and $a_0 = \frac{9}{4}$, and then gradually increase a from $a = 0$ to values where $a < 3$, through $a = 3$, until $a > 3$, we progress from undamped to underdamped to critically damped to overdamped motion. For example, if we consider the initial values, $x(0) = 1$ and $x'(0) = 0$, then Figure 6.12 shows the motion with $a = 0$ (undamped), $a = 0.5, 1, 1.5, 2, 2.5$ (underdamped), $a = 3$ (critically damped, the dotted curve), and $a = 4, 5, 6$ (overdamped).

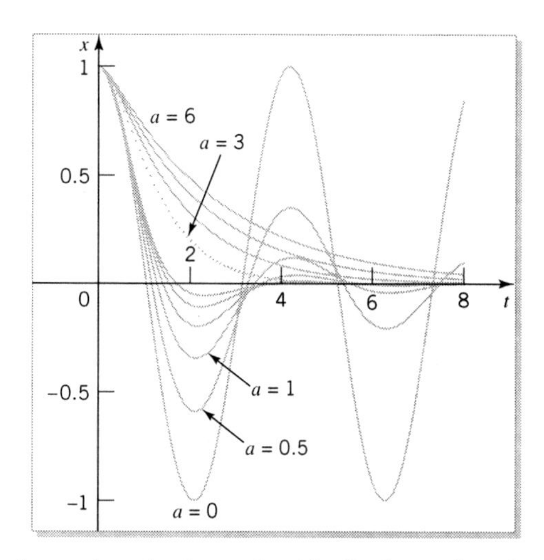

FIGURE 6.12 Undamped, underdamped, critically damped, and overdamped motion

EXERCISES

1. Find the explicit solutions of the following second order linear differential equations with constant coefficients.

(a) $x'' - x' - 6x = 0$
(b) $x'' - x' - 2x = 0$
(c) $x'' - 2x' + x = 0$
(d) $x'' - 2x' - 8x = 0$
(e) $x'' + 2x' - 8x = 0$
(f) $x'' - 6x' + 9x = 0$
(g) $x'' - x' - 12x = 0$
(h) $x'' - 2x' + 10x = 0$
(i) $x'' - 5x' + 6x = 0$
(j) $x'' - 6x' + 10x = 0$
(k) $x'' + 16x = 0$
(l) $x'' - 9x = 0$
(m) $6x'' + x' - x = 0$
(n) $9x'' - 6x' + x = 0$
(o) $x'' + 2x' + 4x = 0$
(p) $x'' + 4x' + 2x = 0$

2. Solve $x'' + 2ax' + 225x = 0$ for each of the following cases.

(a) $a = 0$ (b) $a = 9$ (c) $a = 15$ (d) $a = 25$

3. Solve the following initial value problems.

(a) $(D^2 + D - 2)x = 0$ $x(0) = 0$ $x'(0) = 3$
(b) $(D^2 + 2D - 8)x = 0$ $x(0) = 0$ $x'(0) = 4$
(c) $x'' + 6x' + 9x = 0$ $x(0) = 2$ $x'(0) = 0$
(d) $(12D^2 + D - 1)x = 0$ $x(0) = 4$ $x'(0) = 0$
(e) $x'' + 3x' = 0$ $x(0) = 4$ $x'(0) = 3$
(f) $x'' - 2\pi x' + 2\pi^2 x = 0$ $x(0) = 0$ $x'(0) = -2\pi$
(g) $x'' + 10x' + 100x = 0$ $x(0) = 15$ $x'(0) = 4$
(h) $x'' + 10x' + 100x = 0$ $x(1) = 0$ $x'(1) = 0$

4. Show that the solution of (6.4) on page 226 — namely $x'' + 2x = 0$ — is $x(t) = C_1 \cos \sqrt{2}t + C_2 \sin \sqrt{2}t$.

5. Show that the solution of (6.10) on page 229 — namely $x'' + 0.4x' + 2x = 0$ — is $x(t) = e^{-0.2t}(C_1 \cos 1.4t + C_2 \sin 1.4t)$.

6. Show that $C_1 \cos \beta t + C_2 \sin \beta t = A \cos(\beta t + \phi)$, where $A = \sqrt{C_1^2 + C_2^2}$ and ϕ is defined by $\cos \phi = C_1/\sqrt{C_1^2 + C_2^2}$ and $\sin \phi = -C_2/\sqrt{C_1^2 + C_2^2}$, by going through the following steps.

(a) Write $C_1 \cos \beta t + C_2 \sin \beta t = A(\alpha_1 \cos \beta t + \alpha_2 \sin \beta t)$, where $A = \sqrt{C_1^2 + C_2^2}$, $\alpha_1 = C_1/\sqrt{C_1^2 + C_2^2}$, and $\alpha_2 = C_2/\sqrt{C_1^2 + C_2^2}$.

(b) Show that $-1 \le \alpha_1 \le 1$. Explain why this means that an angle, ϕ, can be defined for which $\cos \phi = \alpha_1$.

(c) Solve $\cos \phi = C_1/\sqrt{C_1^2 + C_2^2}$ for $C_2/\sqrt{C_1^2 + C_2^2}$, obtaining $C_2^2/(C_1^2 + C_2^2) = \sin^2 \phi$, so that $\alpha_2 = \pm \sin \phi$. We use the convention $\alpha_2 = -\sin \phi$.

(d) Substitute $\alpha_1 = \cos \phi$, $\alpha_2 = -\sin \phi$ into part (a) and use the trigonometric identity $\cos \phi \cos \beta t - \sin \phi \sin \beta t = \cos(\beta t + \phi)$ to complete the proof.

7. Show that $e^{\alpha t}(C_1 \cos \beta t + C_2 \sin \beta t) = Ke^{(\alpha + i\beta)t} + \widetilde{K}e^{(\alpha - i\beta)t}$, where $K = \frac{1}{2}(C_1 - iC_2)$ and $\widetilde{K}$ is the complex conjugate of K, by using the identities $\cos \beta t = \frac{1}{2}(e^{i\beta t} + e^{-i\beta t})$, $\sin \beta t = \frac{1}{2i}(e^{i\beta t} - e^{-i\beta t})$.

8. Show that $C_1 e^{rt} + C_2 e^{-rt} = K_1 \sinh rt + K_2 \cosh rt$, where $K_1 = \frac{1}{2}(C_1 - C_2)$ and $K_2 = \frac{1}{2}(C_1 + C_2)$.

9. Find the amplitude, period, and phase angle for the motion described by

(a) $x'' + 9x = 0$, $x(0) = -2$, $x'(0) = -6$.
(b) $x'' + \pi^2 x = 0$, $x(0) = 1$, $x'(0) = \pi\sqrt{3}$.

10. Find the second order differential equation with real constant coefficients that has the given general solution.

(a) $x(t) = C_1 e^{-t} + C_2 e^{3t}$
(b) $x(t) = C_1 e^{-2t} \cos 3t + C_2 e^{-2t} \sin 3t$
(c) $x(t) = C_1 e^{at} \cos bt + C_2 e^{at} \sin bt$
(d) $x(t) = C_1 e^{7t} + C_2 t e^{7t}$
(e) $x(t) = A \sin(3t + \phi)$

11. Is it possible to choose the initial conditions $x(0) = x_0$, $x'(0) = x_0^*$ so that a solution of $x'' + ax' + bx = 0$, where a and b are constants, will cross the t-axis exactly no times, exactly once, exactly twice, exactly three times, ..., and so on? If so, do so; if not, explain why it is not possible.

12. Show that substituting $x = e^{rt}$ into $ax'' + bx' + cx = 0$ requires that r satisfy the characteristic equation $ar^2 + br + c = 0$.

13. Denise, Chad, and Parental Interference. The system of differential equations $x' = y - cx$, $y' = -2x$, with initial conditions $x(0) = 0$, $y(0) = 2.5$, is a model for Denise and Chad with parental interference. Is it possible to choose the constant c so that Denise and Chad eventually have no feelings for each other, but they do so in such a way that they like each other to the end?

14. Sea Battles. It is the year 1805. You are the commander of a fleet of warships about to do battle with an enemy. Your ships and those of your enemy are equal in quality. The only difference is in numbers: you have 37 ships, and your opponent has 12. You engage the enemy in battle. Who will win the battle? How many of the winner's ships will survive? How long will the battle last? We will try to answer these questions by looking at a simple model, in which we assume that the rate of change of the number of ships is proportional to the number of enemy ships. If x is the number of ships in your fleet and y the number of ships in the enemy's fleet — so x and y are nonnegative — then appropriate differential equations are $x' = -ay$, $y' = -bx$, where a and b measure the effectiveness of the two fleets and are both positive. If the fleets are equally matched in quality, then $a = b$, so $x' = -ay$, $y' = -ax$.

(a) Justify the negative signs on the right-hand side of this system.

(b) What are the initial conditions for this system? Who do you think will win the battle? How many ships will be left on the loser's side when the battle is over?

(c) Find the differential equation for the orbits in the phase plane, and solve them. Interpret the orbits in terms of the progress of the battle. Indicate where the battle starts and ends in the phase plane. From the phase plane determine who will win the battle, and how many of the winner's ships will survive. Can you tell how long the battle will last?

(d) Convert the system of differential equations to a second order differential equation for x. Solve for $x(t)$ and then use $x' = -ay$ to find $y(t)$. From these solutions determine who will win the battle, and how many of the winner's ships survive. Now can you tell how long the battle will last?

6.4 MODELING PHYSICAL SITUATIONS

In the previous section we looked at examples of initial value problems for second order differential equations with constant coefficients. We now discuss how these equations are used to model three physical situations — spring-mass systems, simple linearized pendulums, and series RLC electrical circuits.

Spring-Mass Systems

Suppose we have a mass attached to a vertical spring, as shown in Figure 6.13. If the mass is not moving, we say it is in equilibrium. In this situation the force on the spring due to the hanging mass is exactly balanced by the tension in the spring. Now consider a situation in which the mass is extended a distance, x_0, beyond its equilibrium position, $x = 0$ (the position at which the mass will hang without undergoing oscillations), and released. We take positive values of the displacement, x, to correspond to positions where the mass is below the equilibrium position.

If the spring obeys Hooke's law (where the restoring force is proportional to the distance, x), then Newton's law of motion in the absence of resistance gives

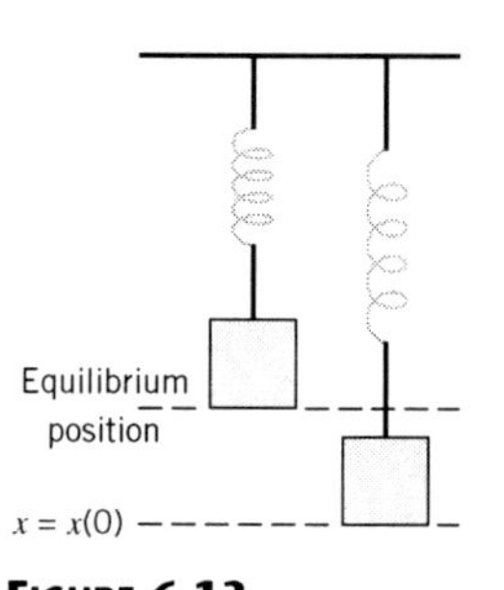

FIGURE 6.13
Mass on a spring

$$mx'' = -kx, \tag{6.37}$$

where $k > 0$ is the proportionality constant and $m > 0$ is the mass. The negative sign on the right-hand side of (6.37) indicates that the restoring force is in opposition to the acceleration.

If a friction force also acts on the spring-mass system, and if this friction force is proportional to the velocity and dampens the motion, then to include the friction force, we must add the term $-bx'$, where b is a positive constant, to the right-hand side of (6.37). This gives $mx'' = -kx - bx'$, which we can rewrite in the form

$$mx'' + bx' + kx = 0. \tag{6.38}$$

This is a homogeneous second order differential equation with constant coefficients. Here m, b, and k are positive constants.

We return to Example 6.4 and Figure 6.12 to describe the behavior of the corresponding spring-mass system (with $m = 1, b = a, k = \frac{9}{4}$). In this case, a represents the friction — the larger a is, the larger the frictional force is. So we consider the impact of a simple friction force on a spring-mass system, where we gradually increase the magnitude of this force. One way to picture this is to think of the motion of a spring-mass in a medium in which the density is gradually increased — starting with a vacuum, then successively to air, water, oil, and molasses.

The initial conditions $x(0) = x_0$ and $x'(0) = 0$ represent the mass being initially displaced x_0 units (downward if $x_0 > 0$) and then released from rest. For all values of the friction, we intuitively expect the mass to rise initially, which it does according to Figure 6.12. To fix ideas, we will consider $x_0 = 1$.

Undamped motion

(i) When $a = 0$, there is no friction, and the mass rises and falls periodically, oscillating about the equilibrium position $x = 0$. The spring is stretched the most when it is one unit below the equilibrium position, and is stretched the least (compressed the most) when it is one unit above.

Underdamped motion

(ii) When $a = 1$, there is a modest amount of friction, and the mass rises and falls about its equilibrium position at a slower rate than in the frictionless case. The amplitude decreases with time, gradually approaching the equilibrium position. As a increases toward 3, the friction increases, and the oscillations become smaller.

Critically damped motion

(iii) When $a = 3$, all oscillations cease. In this example the mass never passes through the equilibrium position, but rises slowly from its initial position. It takes an infinite time to reach $x = 0$. This is the smallest value of a for which oscillations cease.

Overdamped motion

(iv) When $a = 5$, there are no oscillations. In this example the mass never passes through the equilibrium position, but rises on average slower than the critically damped case. (See Exercise 6 on page 254.) Again it takes an infinite time to reach $x = 0$.

Simple Linearized Pendulum

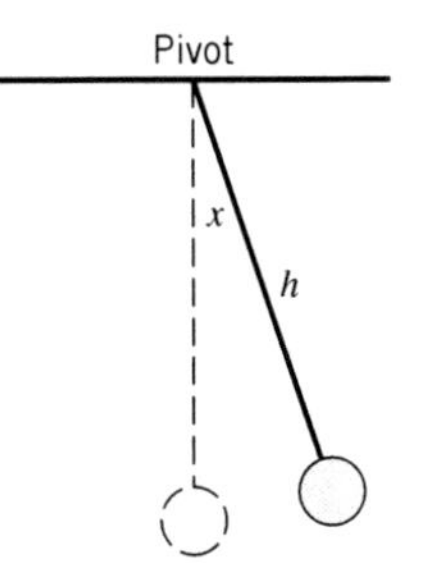

FIGURE 6.14
Simple pendulum

A simple pendulum consists of a light rigid rod of length h, hinged at one end, with a mass m attached to the other. When the pendulum is at rest in a vertical position with the weight directly below the hinge, we have the configuration from which measurements are made. If we consider the simple pendulum in motion, the position of the pendulum is characterized by the angle x (in radians) between the rod and the rest position. See Figure 6.14, in which positive angles are measured to the right, and $x' > 0$ represents counterclockwise motion. (All motion of the simple pendulum is assumed to take place in the plane of the page.)

If the weight of the rod is negligible, the hinge is frictionless, and there is no air resistance, the differential equation governing the pendulum in motion is[12]

$$mx'' = -m\lambda^2 \sin x, \tag{6.39}$$

where $\lambda = \sqrt{g/h}$. If a friction force also acts on the pendulum, and if this friction force is proportional to the velocity and dampens the motion, then to include the friction force, we must add the term $-bx'$, where b is a positive constant, to the

[12]The derivation of this equation can be found in *Physics* by R. Resnick, D. Halliday, and K.S. Krane, volume 1, 4th edition, Wiley, 1992, page 323.

right-hand side of (6.39). This gives $mx'' = -m\lambda^2 \sin x - bx'$, which we can rewrite in the form

$$mx'' + bx' + m\lambda^2 \sin x = 0. \tag{6.40}$$

This is a homogeneous second order nonlinear differential equation and cannot be solved in terms of familiar functions. It is the differential equation that models the motion of a simple pendulum.

If x is small — that is, the pendulum's displacement is always near the rest position — we can approximate $\sin x$ by the first term in its Taylor series, $\sin x \approx x$, in which case (6.40) becomes

$$mx'' + bx' + m\lambda^2 x = 0. \tag{6.41}$$

This is a homogeneous second order differential equation with constant coefficients. Here m, b, and λ^2 are positive constants. It is the differential equation that models the motion of a simple linearized pendulum.

We return to Example 6.4 and Figure 6.12 to describe the behavior of the corresponding simple linearized pendulum (with $m = 1, b = a, \lambda^2 = \frac{9}{4}$). In this case, a represents the friction — the larger a is, the larger the frictional force is. The initial conditions $x(0) = 1$ and $x'(0) = 0$ represent the mass being initially displaced to the right through 1 radian and then released from rest.[13] For all values of the friction, we intuitively expect the mass to swing to the left initially, which it does according to Figure 6.12.

Undamped motion

(i) When $a = 0$, there is no friction, and the pendulum swings backward and forward periodically, oscillating about the rest position $x = 0$. Its extreme angles are one radian to the left and right of the rest position.

Underdamped motion

(ii) When $a = 1$, there is a modest amount of friction, and the pendulum swings backward and forward about its rest position at a slower rate than in the frictionless case. The amplitude decreases with time, gradually approaching the rest position. As a increases toward 3, the friction increases, and the oscillations become smaller.

Critically damped motion

(iii) When $a = 3$, all oscillations cease. In this example the pendulum never passes through the vertical position, but moves slowly from its initial position. It takes an infinite time to reach $x = 0$. This is the smallest value of a for which oscillations cease.

Overdamped motion

(iv) When $a = 5$, there are no oscillations. In this example the pendulum never passes through the vertical position, but moves on average slower than the critically damped case. Again it takes an infinite time to reach $x = 0$.

Series RLC Electrical Circuit

Consider an electrical circuit in which wires connect a capacitor, resistor, inductor, and switch in series, as shown in Figure 6.15. We denote the current in the circuit as

[13] Because one radian is approximately 57.3 degrees, this violates the assumption that x is small. In spite of this, the qualitative motion described here is accurate.

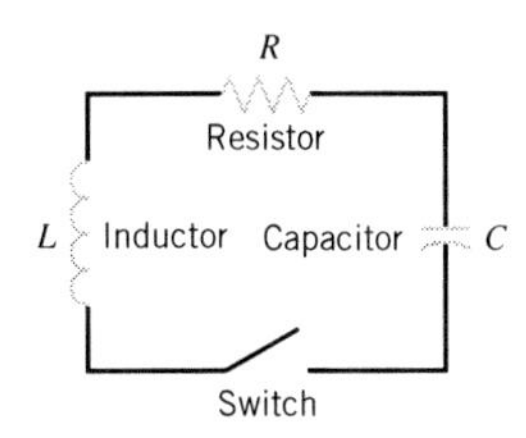

FIGURE 6.15
Series RLC circuit

I (units of amperes) and the charge on the capacitor as x (units of coulombs). The voltage drops across each of these three elements are given by LI', RI, and x/C, respectively, where the inductance is given by L (units of henries), the resistance by R (units of ohms), and the capacitance by C (units of farads). We may use one of Kirchhoff's laws to equate this applied voltage to the sum of the three voltage drops across the three circuit elements. This gives

$$Lx'' + Rx' + \frac{1}{C}x = 0, \tag{6.42}$$

where we have also made use of the fact that the current in an electrical circuit is the rate of change of charge ($I = x'$).

Notice that the differential equation for this simple electrical circuit has the same form as that of a damped spring-mass system if we replace L with the mass, m; R with the damping factor, b; and $1/C$ with the spring constant, k. In fact, because it is easier to change the resistance in a circuit than to change the value of b in an experiment involving the oscillations of a mass on a spring, people use experiments with circuits to model damped motion of a spring-mass system.

We return to Example 6.4 and Figure 6.12 to describe the behavior of the corresponding series RLC electrical circuit (with $L = 1$, $R = a$, $\frac{1}{C} = \frac{9}{4}$). In this case, a represents the resistance — larger values of a correspond to greater resistance. The initial conditions $x(0) = 1$ and $x'(0) = 0$ represent an initial charge of one unit, and no initial current.

Undamped motion (i) When $a = 0$, there is no resistance, and the charge oscillates about the zero charge $x = 0$. Its extreme values are ± 1 units of charge.

Underdamped motion (ii) When $a = 1$, there is a modest amount of resistance, and the charge oscillates about the zero charge at a slower rate than the no-resistance case. The amplitude of the charge decreases with time, gradually approaching zero. As a increases toward 3, the resistance increases, and the oscillations becomes smaller.

Critically damped motion (iii) When $a = 3$, all oscillations cease. In this example the charge never becomes negative but slowly approaches $x = 0$, taking an infinite time to reach it. This is the smallest value of a for which oscillations cease.

Overdamped motion (iv) When $a = 5$, there are no oscillations. In this example the charge never becomes negative but approaches zero charge on average slower than the critically damped case. Again it takes an infinite time to reach $x = 0$.

EXERCISES

1. Solve the following initial value problems associated with the differential equation for a damped spring-mass system $mx'' + bx' + kx = 0$.

(a) $m = 1 \quad b = 1/8 \quad k = 1 \quad x(0) = 0 \quad x'(0) = 1/2$

(b) $m = 1 \quad b = 8 \quad k = 16 \quad x(0) = 0 \quad x'(0) = -3$

(c) $m = 64 \quad b = 16 \quad k = 17 \quad x(0) = 1 \quad x'(0) = 0$

(d) $m = 9 \quad b = 6 \quad k = 37 \quad x(0) = 1 \quad x'(0) = 1$

(e) $m = 9 \quad b = 6 \quad k = 37 \quad x(0) = 0 \quad x'(0) = 0$

2. Solve the following initial value problems associated with the differential equation for a series RLC circuit $Lx'' + Rx' + \frac{1}{C}x = 0$.

(a) $L = 1 \quad R = 1/8 \quad C = 1 \quad x(0) = 6 \quad x'(0) = 0$

(b) $L = 1 \quad R = 1/8 \quad C = 1 \quad x(0) = 0 \quad x'(0) = 6$

(c) $L = 9 \quad R = 6 \quad C = 1/37 \quad x(0) = 6 \quad x'(0) = 0$

(d) $L = 9 \quad R = 6 \quad C = 1/37 \quad x(0) = 0 \quad x'(0) = 6$

3. For what values of b are the motions that are governed by $4x'' + bx' + 9x = 0$ (a) overdamped, (b) underdamped, (c) critically damped?

4. Show that

$$f(t) = \frac{e^{r_2 t} - e^{r_1 t}}{r_2 - r_1}$$

satisfies $mx'' + bx' + kx = 0$ if r_1 and r_2 satisfy $mr^2 + br + k = 0$. Set $r_2 = r_1 + \varepsilon$ in the formula for $f(t)$. Show that, for the overdamped case, $\lim_{\varepsilon \to 0} f(t) = te^{r_1 t}$. (Note that you can use L'Hôpital's rule, or the Taylor series expansion $e^z = 1 + z + z^2/2! + \cdots$.)

5. Show that if $m > 0$, $b > 0$, and $k > 0$, the general solution of $mx'' + bx' + kx = 0$ approaches 0 as $t \to \infty$.

6. For $a > 0$, find the general solution of $x'' + ax' + \frac{9}{4}x = 0$, subject to $x(0) = 1$, $x'(0) = 0$ for critically damped and overdamped motion. Show that, at any time t, the overdamped solution is always greater than the critically damped. What does this mean in terms of the motion of a spring-mass? Show that, at any time t, the overdamped solution corresponding to $a = a_1$ is always greater than the overdamped solution corresponding to $a = a_2$ if $a_1 > a_2$. What does this mean in terms of the motion of a spring-mass?

7. Show that $x = A\cos(\omega t + \phi)$ satisfies $x'' + \omega^2 x = 0$ for any choice of A and ϕ. Explain how this function fits with the Existence-Uniqueness Theorem for Second Order Linear Differential Equations on page 238.

8. Give the solution of $x'' + 16x = 0$, $x(0) = 2\sqrt{2}$, $x'(0) = 8\sqrt{2}$ in the form of Exercise 7.

9. If the total energy of the damped spring-mass system described by $mx'' + bx' + kx = 0$ is defined as $E = \frac{1}{2}m(x')^2 + \frac{1}{2}kx^2$, show that E satisfies $E' = -b(x')^2$ and is therefore a decreasing function of time.

10. Floating Object. An object in liquid is kept afloat when the upward force due to buoyancy exceeds the downward pull of gravity. If such an object is displaced slightly from its equilibrium position and released, the subsequent motion is described by $x'' + ax = 0$, where a is the ratio of the gravitational constant to the displacement of the body in equilibrium. x is positive in the downward direction.

(a) An oil drum in the form of a right circular cylinder is in equilibrium when it floats upright and half submerged in a lake. If the cylinder is 4 feet long and the gravitational constant is taken as 32 ft/sec^2, the governing differential equation is $x'' + 16x = 0$, where x is measured in feet and t in seconds. Find the resulting motion if the cylinder is further submerged so that it extends 1 foot above the water, and then is released from rest.

(b) If the initial conditions in part (a) are changed to give the cylinder an initial velocity of 2 ft/sec at its equilibrium position, find the resulting motion.

11. Hooke's Law. Hooke's law can be verified by conducting the following experiment. Suspend a spring of length L vertically from one end. Attach a mass m to the other end of the spring. Measure X, the vertical distance from the point of suspension. Repeat this for various masses. If the extension of the spring, $x = X - L$, satisfies the equation $x = \frac{g}{k}m$, where k is a positive constant, then the spring obeys Hooke's law. Under these circumstances, the constant k is called the spring constant. In general, k will differ from spring to spring. Tables 6.1 and 6.2, and Figures 6.16 and 6.17, show the results of two such experiments using different springs.[14] Show that these springs obey Hooke's law.

Table 6.1 Extension of spring as a function of mass

Mass (kg)	Extension (m)	Mass (kg)	Extension (m)
0.00	0.000	0.06	0.343
0.01	0.055	0.07	0.400
0.02	0.113	0.08	0.457
0.03	0.170	0.09	0.514
0.04	0.228	0.10	0.575
0.05	0.286		

Table 6.2 Extension of spring as a function of mass

Mass (kg)	Extension (m)	Mass (kg)	Extension (m)
0.000	0.000	0.030	0.090
0.005	0.015	0.035	0.104
0.010	0.030	0.040	0.117
0.015	0.045	0.045	0.132
0.020	0.060	0.050	0.146
0.025	0.075		

[14] These experiments were conducted by David Harman and Mark Zerella while students at the University of Arizona.

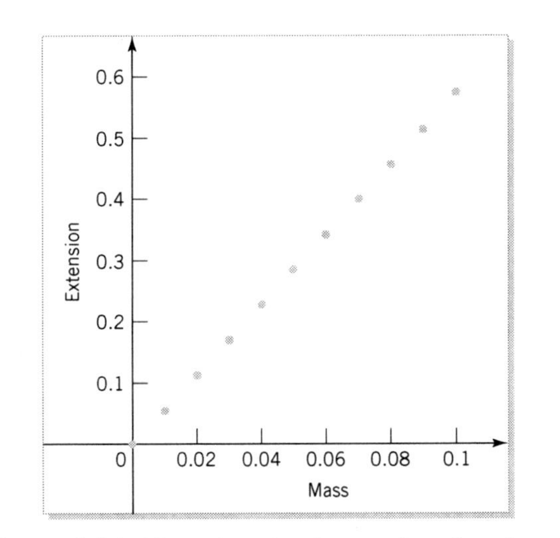

FIGURE 6.16 Extension of spring as a function of mass

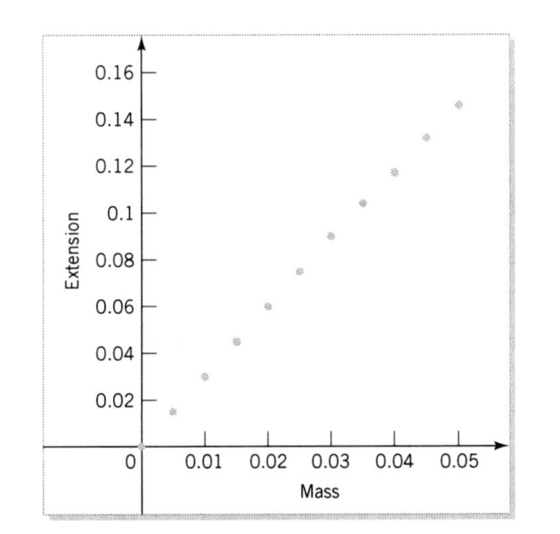

FIGURE 6.17 Extension of spring as a function of mass

12. **Simple Linear Pendulum.** The motion of an undamped linearized pendulum of length h is modeled by the differential equation $x'' + \lambda^2 x = 0$ where $\lambda^2 = g/h$. Solve this equation subject to the initial conditions $x(0) = x_0$, $x'(0) = v_0$. If the length h of the pendulum changes, how does this affect the amplitude and the period of the pendulum? Figure 6.18 shows the motion of two linearized pendulums of different lengths. What was the initial angle of the shorter pendulum? Initially was it displaced to the left or right of vertical? Which way did the pendulum first move, to the left or right? Estimate the length of the longer pendulum in terms of the length of the shorter pendulum.

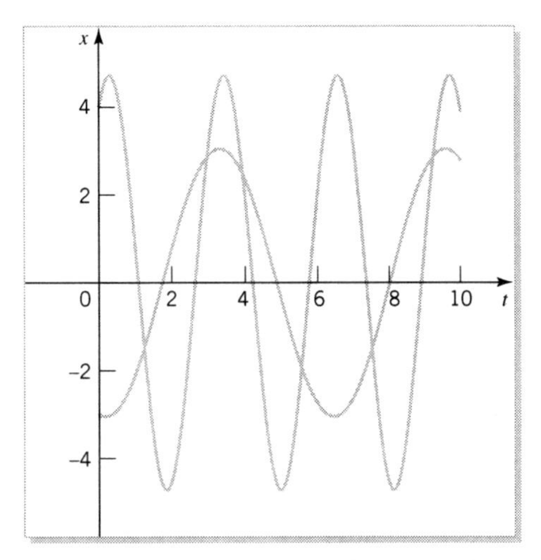

FIGURE 6.18 The motion of two linearized pendulums of different lengths

13. I have a grandfather clock that is running slow. Assuming it is modeled by a frictionless linear pendulum, should I shorten or lengthen the pendulum to attempt to have it run correctly?

14. The linearized pendulum is modeled by the differential equation $x'' + \lambda^2 x = 0$ where $\lambda^2 = g/h$. If you make a 1% error in measuring λ^2 (through inaccurate estimation of g or h), can this lead ultimately to a large difference between the model's prediction and the real situation?[15] For example, consider $\lambda^2 = 1$ as the accurate measurement and $\lambda^2 = 1.01$ as the inaccurate estimate. What do the two solutions — subject to $x(0) = x_0$, $x'(0) = 0$ — predict when $t = 300\pi$?

[15] *Applied Mathematics: An Introduction* by H. Pollard, Addison-Wesley, 1972, page 5.

15. Show that the underdamped solution of $mx'' + bx' + kx = 0$ — subject to $x(0) = x_0$, $x'(0) = v_0$ — can be written in the form $x(t) = Ae^{\alpha t} \cos(\beta t + \phi)$, where $\alpha = -b/(2m)$, $\beta = \sqrt{4km - b^2}/(2m)$, $A = \sqrt{x_0^2 + \left(\frac{v_0 - \alpha x_0}{\beta}\right)^2}$, $\phi = \arctan\left(-\frac{v_0 - \alpha x_0}{\beta x_0}\right)$.

(a) Use a computer/calculator program to plot the function $x(t) = Ae^{\alpha t} \cos(\beta t + \phi)$, with $A = \pm 5$ and $\alpha = -0.4$, for different values of β and ϕ in the intervals $0 \le t \le 10$, $-5 \le x \le 5$ — for example, $\beta = \pm 5$ and $\phi = 0, \pm 2$. Figure 6.19 shows these curves. You should notice that the plotted functions seem to be constrained by two invisible curves — one positive, the other negative. These curves are called the ENVELOPES of $x(t)$. What are the equations of these envelopes?

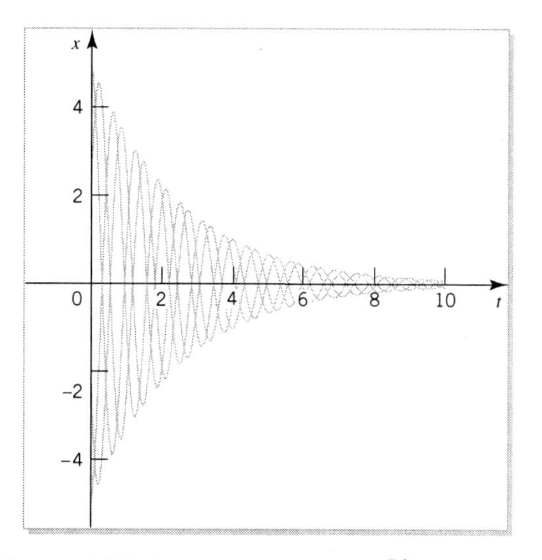

FIGURE 6.19 The curves $x(t) = \pm 5e^{-0.4t} \cos(\beta t + \phi)$

(b) Show by calculus that the curves $\pm Ae^{\alpha t}$ and the curve $x(t) = Ae^{\alpha t} \cos(\beta t + \phi)$ are tangent to each other where they touch.

(c) Show by calculus that the curves $\pm A \cos(\arctan(\alpha/\beta))e^{\alpha t}$ pass through the local extrema of the curve $x(t) = Ae^{\alpha t} \cos(\beta t + \phi)$.

(d) Figure 6.20 shows the curves referred to in parts (b) and (c). Which is which?

16. The linearized differential equation describing the oscillations of a simple pendulum including a damping term is $x'' + 2\alpha x' + \lambda^2 x = 0$, where $\lambda^2 = g/h$ and α are constants.

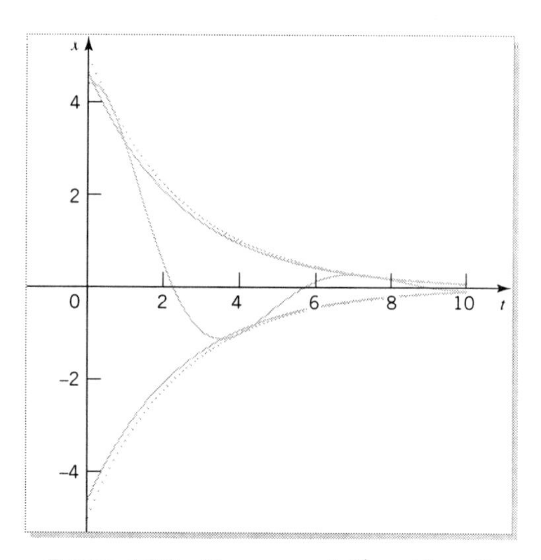

FIGURE 6.20 The curves $Ae^{\alpha t} \cos(\beta t + \phi)$, $\pm Ae^{\alpha t} \cos(\arctan(\alpha/\beta))$, and $\pm Ae^{\alpha t}$

(a) Consider a simple pendulum with $\lambda^2 = 4$.

i. For what values of α will the resulting motion of a pendulum with $x(0) = 0.5$ and $x'(0) = 0$ be underdamped?

ii. For what values of α will the pendulum in part i always stay on the same side of the vertical line on which it was released?

(b) If $\alpha = 1/10$, $\lambda^2 = 4.01$, and the pendulum is set in motion with initial conditions $x(0) = 0$, $x'(0) = 1$, find the maximum value of $|x|$ for the resulting motion.

(c) If the initial conditions of part (b) are changed to $x(0) = 0$, $x'(0) = \beta$, for what value of β will the maximum value of $|x|$ be 0.2 radian?

17. Consider the underdamped solution of the differential equation $x'' + 2\alpha x' + \lambda^2 x = 0$ subject to the initial conditions $x(0) = x_0$, $x'(0) = 0$.

(a) Show that this initial value problem has the solution $x(t) = x_0 e^{-\alpha t} \left(\cos \omega t + \frac{\alpha}{\omega} \sin \omega t\right)$, where $\omega = \sqrt{\lambda^2 - \alpha^2}$.

(b) Show that the local maxima and minima of this function occur at times $t = \left(n + \frac{1}{2}\right)\pi/\omega$, where $n = 0, \pm 1, \pm 2, \cdots$, in which case $x(t) = \pm x_0 e^{-\alpha t}$.

(c) Consider the following simple experiment. A mass is hung from a vertical spring, displaced vertically from its equilibrium position, and then released from rest. The amplitude of the mass (the maximum vertical displacement) is measured as a function of time. Based on your answer to part (b), what do you expect the graph of this amplitude versus time to look like?

(d) Either perform the experiment described in part (c) or use the experimental data in Table 6.3, shown in Figure 6.21, to confirm the answer you gave to part (c).[16] Estimate x_0 and α by plotting $\ln x(t)$ versus t.

Table 6.3 The amplitude at various times for underdamped motion

Time (sec)	Amplitude (cm)	Time (sec)	Amplitude (cm)
0.0	10.0	99.1	6.3
16.7	9.6	116.2	5.7
33.0	9.0	132.1	5.5
49.5	8.4	149.3	5.1
66.1	7.7	165.8	4.8
82.6	7.1		

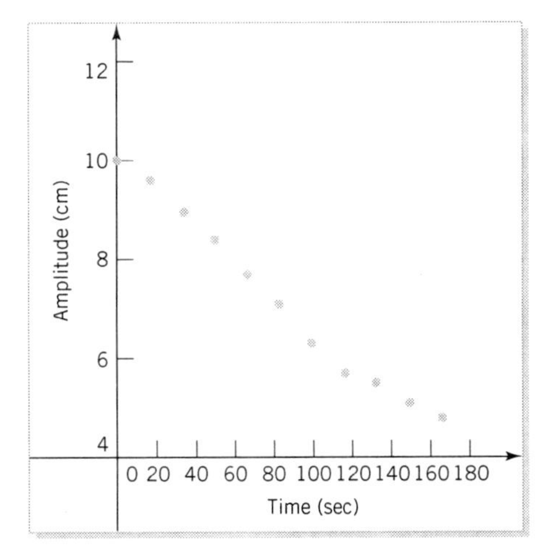

FIGURE 6.21 The amplitude at various times for underdamped motion

(e) Describe a simple experiment using a linearized pendulum that would exhibit a similar behavior to the spring-mass experiment in part (c). Perform this experiment.

18. Consider the overdamped solution of the differential equation $x'' + 2\alpha x' + \lambda^2 x = 0$, which is $x(t) = C_1 e^{r_1 t} + C_2 e^{r_2 t}$, where $r_1 = -\alpha - \sqrt{\alpha^2 - \lambda^2}$ and $r_2 = -\alpha + \sqrt{\alpha^2 - \lambda^2}$.

(a) The quantity r_1 is negative. Is r_2 positive or negative? Explain.

(b) If initially, $x(0) = x_0$ and $x'(0) = 0$, evaluate C_1 and C_2. What are the signs of C_1 and C_2?

(c) Show that as time progresses the term $C_2 e^{r_2 t}$ dominates the term $C_1 e^{r_1 t}$ if $C_2 \neq 0$, and so the solution behaves like $C_2 e^{r_2 t}$ for large values of t.

(d) Consider the following experiment. A mass is hung from a vertical spring in a beaker of glycerin. The mass is displaced vertically below its equilibrium position, and then released from rest. After a short time the position of the mass is measured as a function of time. Based on your answer to part (b), can C_2 be zero? Based on your answer to part (c), what do you expect the graph of this distance versus time to look like?

(e) Use the experimental data in Table 6.4, shown in Figure 6.22, to confirm the answer you gave to part (c).[17] Estimate C_2 and r_2 by plotting $\ln x(t)$ versus t.

Table 6.4 The distance at various times for overdamped motion

Time (sec)	Distance (cm)	Time (sec)	Distance (cm)
0.920	5.182	1.858	1.636
0.989	4.827	1.928	1.582
1.058	4.364	1.984	1.364
1.121	3.982	2.054	1.309
1.191	3.682	2.117	1.145
1.254	3.491	2.192	1.091
1.317	3.055	2.255	1.036
1.386	2.864	2.318	0.982
1.449	2.591	2.387	0.927
1.518	2.427	2.457	0.845
1.581	2.236	2.520	0.791
1.650	2.045	2.589	0.764
1.713	1.909	2.652	0.736
1.789	1.773		

[16] "Spreadsheets in the Physics Laboratory" by M. E. Krieger and J. H. Stith, *The Physics Teacher*, September 1990, pages 378–384, Table II.

[17] This data set is a subset of that shown in "Using videotapes to study damped harmonic motion and to measure terminal speeds" by M. Stautberg Greenwood, F. Fazio, M. Russotto, and A. Wilkosz, *The American Journal of Physics* 54, 1986, pages 897–904, Figure 4.

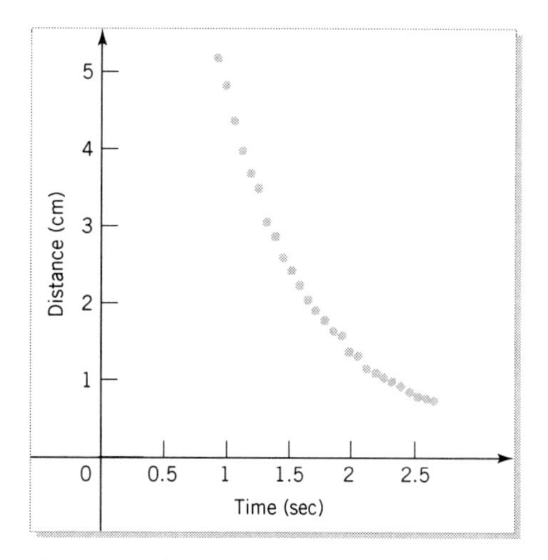

FIGURE 6.22 The distance at various times for overdamped motion

(f) Describe a simple experiment using a linearized pendulum that would exhibit a similar behavior to the spring-mass experiment in part (c). Perform this experiment.

19. A Two-Body Problem.[18] A linear spring with spring constant k has two masses, m and M, attached at either end. The system lies in a line on a horizontal surface in equilibrium. All motion takes place along this line and is assumed frictionless. The two masses are now moved to initial positions along the line and given initial velocities along the line. If $x(t)$ is the distance traveled by the mass m from its equilibrium position, and $y(t)$ is the corresponding position for M, then the motion of this system is governed by $mx'' = k(y - x)$, $My'' = -k(y - x)$.

(a) Introduce the new variables $X = (mx + My)/(m + M)$ and $Y = y - x$, and show that these differential equations may be written as $X'' = 0$, and $Y'' = -\omega^2 Y$, where $\omega^2 = k(m + M)/(mM)$. Solve these equations for X and Y. What is the physical interpretation of X?

(b) It is claimed that this system can vibrate in unison with an identical spring that is fixed at one end and has a mass $mM/(m + M)$ at the other end. Comment on this claim.

(c) Write down the solutions for $x(t)$ and $y(t)$.

20. A Falling Spring.[19] A linear spring with spring constant k has two masses, m and M, attached at either end. The end with mass m is held so the spring hangs vertically. When it has reached its equilibrium position, the mass M is a distance d below m. The spring is now allowed to fall under gravity. If $x(t)$ is the distance traveled by m and $y(t)$ is the distance traveled by M, then, neglecting the mass of the spring, the motion of this system is governed by $mx'' = mg + \frac{k}{\ell}(y - x - \ell)$, $My'' = Mg - \frac{k}{\ell}(y - x - \ell)$, where $\ell = d/(1 + Mg/k)$.

(a) Introduce the new variables $X = (mx + My)/(m + M)$ and $Y = y - x - \ell$, and show that these differential equations may be written as $X'' = gt$, and $Y'' = -\omega^2 Y$, where $\omega^2 = k(m + M)/(mM\ell)$. Solve these equations for X and Y. What is the physical interpretation of X?

(b) Write down the solutions for $x(t)$ and $y(t)$.

21. Show that the solution of the initial value problem, $(1/4)V'' + (1/R)V' + 5V = 0$, $V(0) = 1$, $V'(0) = 0$, for $R > 1/\sqrt{5}$, is given by $V(t) = e^{-2t/R}\left(\cos\beta t + \frac{2}{\beta R}\sin\beta t\right)$, where $\beta = 2\sqrt{5 - 1/R^2}$.

22. Parallel RLC Circuits. Consider the situation in which a resistor, inductor, and capacitor are connected in parallel, as shown in Figure 6.23, called a parallel RLC circuit. The voltage V in this circuit is governed by the differential equation $CV'' + \frac{1}{R}V' + \frac{1}{L}V = 0$. Find V if $L = 1/5$ henries, $C = 1/4$ farads, $R = 1/3$ ohms, $V(0) = 1$, and $V'(0) = 0$. Find the value of t at which V reaches 1% of its maximum value; that is, find the settling time.

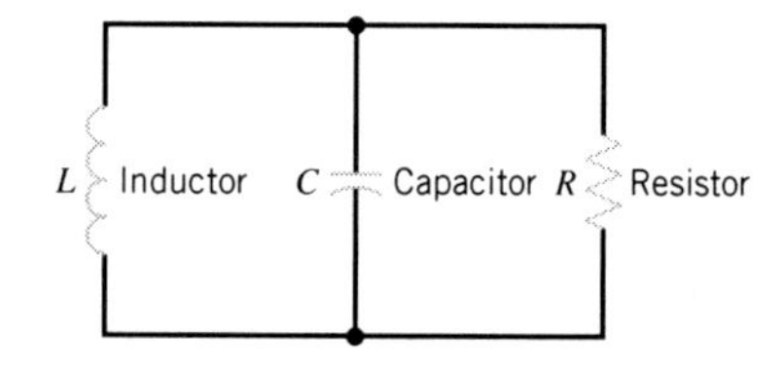

FIGURE 6.23 Parallel RLC circuit

23. For what values of the resistance, R, is the voltage in a parallel RLC circuit that is governed by $(1/9)V'' + (1/R)V' + (1/4)V = 0$ (a) overdamped, (b) underdamped, (c) critically damped?

24. Plot the solutions of $(1/4)V'' + (1/R)V' + 5V = 0$, $V(0) = 1$, $V'(0) = 0$, for $R = 0.1, 0.5, 1.0, 1.5$, and 2.0

---

[18] "A simple example for the two body problem" by E. Maor, *The Physics Teacher*, February 1973, pages 104–105.

[19] "Oscillations of a falling spring" by P. Glaster, *Physics Education* 20, 1993, pages 329–331.

on the same graph. Explain what you see and why it is the way it is.

25. Find the settling time for the voltage that satisfies $(1/4)V'' + V' + 5V = 0$, $V(0) = 0$, $V'(0) = 1$.

26. Look in the textbooks of your other courses and find an example that uses a second order linear differential equation with constant coefficients. Write a report about this example that includes the following items.

(a) A brief description of background material so a classmate will understand the origin of the differential equation.

(b) How the constants in the differential equation can be evaluated and what the initial condition means.

(c) The solution of this differential equation.

(d) An interpretation of this solution, and how it answers a question posed by the original discussion.

6.5 INTERPRETING THE PHASE PLANE

In this section we will show how the solutions of the previous section can be interpreted in terms of a phase plane. This allows us to visualize solutions for many different initial conditions. It will also give us a powerful qualitative method to understand the solutions of more general differential equations even when we are unable to solve these equations in terms of familiar functions.

We return to Example 6.4 on page 000 which dealt with

$$x'' + ax' + \frac{9}{4}x = 0,$$

where (i) $a = 0$, (ii) $a = 1$, (iii) $a = 3$, and (iv) $a = 5$.

If we introduce the variable $y = x'$, then we can write this second order equation as a system of first order equations:

$$x' = y, \qquad y' = -ay - \frac{9}{4}x.$$

Phase plane

Thus, orbits in the phase plane are solutions of

$$\frac{dy}{dx} = \frac{-ay - \frac{9}{4}x}{y}. \tag{6.43}$$

We first consider the case $a = 0$ — namely,

$$\frac{dy}{dx} = -\frac{9x}{4y}.$$

Orbits

This is the Denise and Chad relationship — equation (6.2) on page 224 — with $\frac{9}{4}$ in place of 2, so that analysis will not be repeated here. The orbits are the clockwise ellipses $4y^2 + 9x^2 = C$. (Show this.) Figure 6.24 shows the direction field and some of these ellipses with different initial conditions. As time increases, each orbit returns to its starting position, giving a closed clockwise curve that is representative of periodic solutions.

We now consider the case when $a > 0$. The differential equation (6.43) is one with homogeneous coefficients, and we can solve it to find the orbits (see Exercise 11 on page 265), but the implicit solutions are so complicated they are useless.

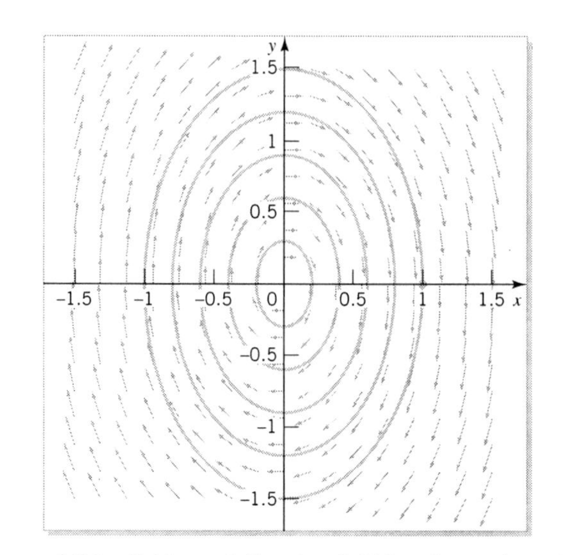

FIGURE 6.24 Orbits and direction field for $x' = y$, $y' = -9x/4$

Straight-line orbits Nevertheless, because it is homogeneous, it may have straight-line orbits that will be useful because other orbits cannot cross them. Straight-line orbits are given by $y = mx$, where, from (6.43), $m^2 + am + \frac{9}{4} = 0$. This quadratic has no real roots if $a = 1$, a repeated root of $m = -\frac{3}{2}$ if $a = 3$, and distinct real roots of $m = -\frac{1}{2}$ and $m = -\frac{9}{2}$ if $a = 5$, corresponding to zero, one, and two straight-line solutions.

Monotonicity If we apply graphical analysis to the differential equation, we find that isoclines for slope zero occur along the line $y = -\frac{9}{4a}x$. If we concentrate on solutions for which $y > 0$, they will be increasing to the left of the line $y = -\frac{9}{4a}x$ and decreasing to the right, and viceversa if $y < 0$. The orbits will have vertical tangents when $y = 0$.

Concavity To investigate concavity, we calculate the second derivative and find

$$\frac{d^2y}{dx^2} = -\frac{9}{4y^3}\left(y^2 + axy + \frac{9}{4}x^2\right).$$

This will change sign where $y = 0$ and where $y^2 + axy + \frac{9}{4}x^2$ changes sign. If $a = 1$, this polynomial has no real factors; if $a = 3$, then $y^2 + axy + \frac{9}{4}x^2 = \left(y + \frac{3}{2}x\right)^2$; and if $a = 5$, then $y^2 + axy + \frac{9}{4}x^2 = \left(y + \frac{1}{2}x\right)\left(y + \frac{9}{2}x\right)$. Thus, regions of concavity are determined by $y = 0$ and the straight-line solutions found earlier.

We can draw some general conclusions by rewriting the differential equation (6.43) in the more illuminating form

$$\frac{dy}{dx} = -\frac{9x}{4y} - a. \tag{6.44}$$

One way of interpreting this is that the slope of the orbit with $a \neq 0$ is a less than $-\frac{9x}{4y}$, which is the slope of the orbit with $a = 0$ — symbolically,

$$\left(\frac{dy}{dx}\right)_{a\neq 0} = \left(\frac{dy}{dx}\right)_{a=0} - a.$$

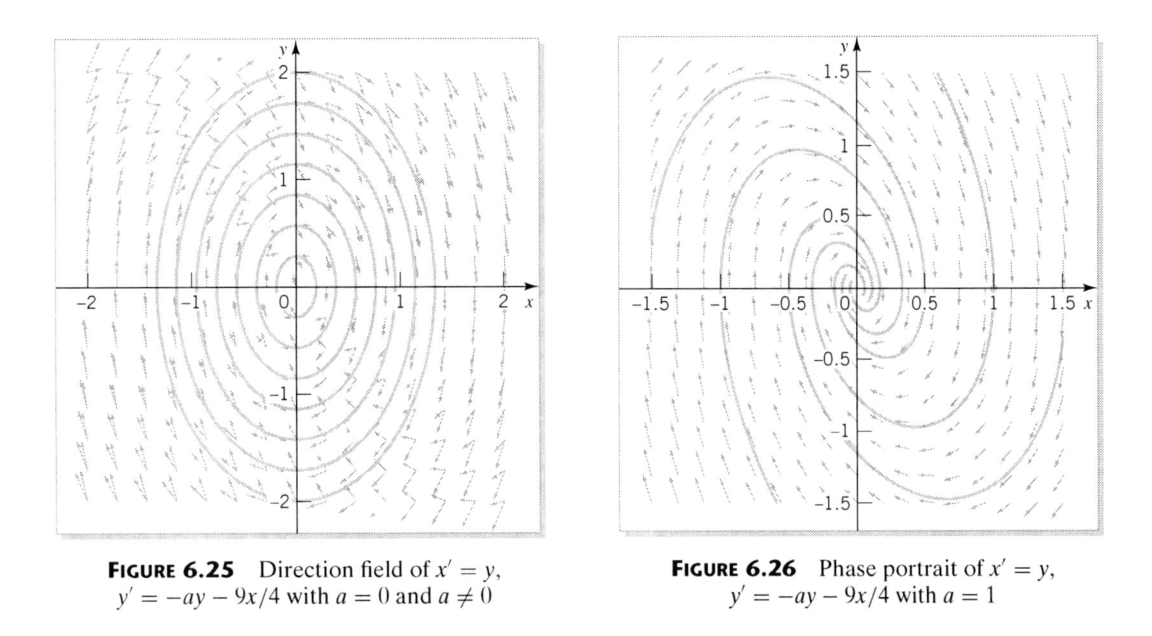

FIGURE 6.25 Direction field of $x' = y$, $y' = -ay - 9x/4$ with $a = 0$ and $a \neq 0$

FIGURE 6.26 Phase portrait of $x' = y$, $y' = -ay - 9x/4$ with $a = 1$

Direction field

Figure 6.25 attempts to demonstrate this by showing the direction fields with $a = 0$ and with $a \neq 0$. We have included some orbits (ellipses) corresponding to the $a = 0$ case to make it easier to distinguish between the $a = 0$ and $a \neq 0$ direction fields. There is a simple way to see this — imagine you are standing at the start of two vectors (arrows) facing in the general direction of the arrows. The left arrow corresponds to the $a = 0$ case, the right to the $a \neq 0$ case. Thus, the direction of travel is clockwise, and, for $a > 0$, all orbits are being attracted to $(0, 0)$.

Figure 6.26 is a phase portrait for $a = 1$. All orbits spiral around the point $(0, 0)$ and eventually end at $(0, 0)$. This is consistent with our graphical analysis.

Now imagine increasing the value of a. If we reach a value of a for which a particular orbit ends at $(0, 0)$ without spiraling, then no other orbits can cross it and so spiraling will cease for all orbits. We have already seen that when $a = 3$ there is a straight-line orbit of $y = -\frac{3}{2}x$, and if $a > 3$ there are two straight-line orbits. Thus, if $a \geq 3$, no orbits will spiral. Figures 6.27 and 6.28 are phase portraits corresponding to $a = 3$ and $a = 5$, respectively.

All this we obtained without using the explicit solutions from the previous section. Now let's interpret these phase portraits in terms of the three physical situations discussed there — spring-mass systems, simple (linearized) pendulums, and series RLC electrical circuits.

Spring-Mass Systems

Undamped motion

We start with Figure 6.24, where $a = 0$. Let's interpret one of these orbits in terms of the motion of the spring-mass — say, the outermost orbit, which has the equation $4y^2 + 9x^2 = 9$. This ellipse crosses the x-axis at $x = \pm 1$, and the y-axis at $y = \pm 1.5$. We will start at the point $(1, 0)$ in the phase plane, and proceed clockwise once around the ellipse until we return to $(1, 0)$. The point $(1, 0)$ corresponds to a vertical displacement of 1 below the equilibrium position with a velocity of zero. This is when the mass is at rest at its lowest position.

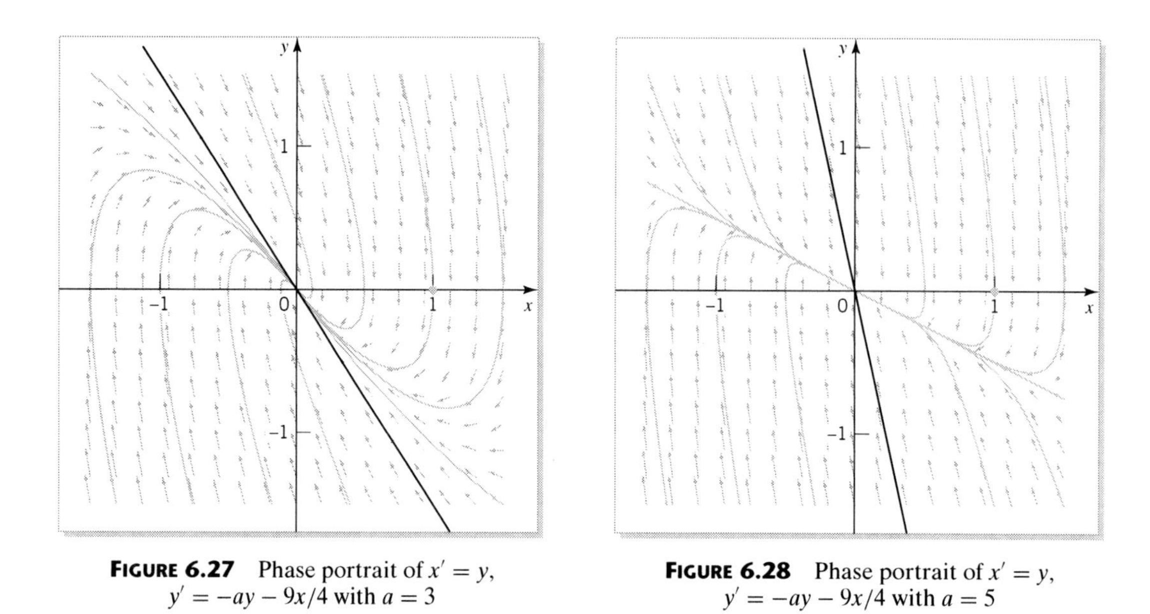

FIGURE 6.27 Phase portrait of $x' = y$, $y' = -ay - 9x/4$ with $a = 3$

FIGURE 6.28 Phase portrait of $x' = y$, $y' = -ay - 9x/4$ with $a = 5$

The part of the orbit in the fourth quadrant corresponds to the mass rising toward the equilibrium position, which it reaches at $(0, -1.5)$, when the velocity is -1.5. The negative sign indicates that the mass is moving in the direction of decreasing x values — upward.

In the third quadrant of the phase plane the mass continues to rise (above the equilibrium position) until it reaches its extreme highest position at a height of -1. This is the point $(-1, 0)$ in the phase plane, and at this stage the mass is as far above the equilibrium position as it was below.

The second quadrant characterizes the mass dropping back to the equilibrium position $(0, 1.5)$ in the phase plane, and the first quadrant is where the mass continues downward to its original starting position. Following the ellipse around for a second time repeats the process. In other words, the ellipse corresponds to a mass rising and falling forever.

Different ellipses correspond to different initial conditions. Smaller x-intercepts correspond to smaller amplitudes, and smaller y-intercepts correspond to smaller velocities. The velocity will have its extreme values when $x = 0$, and the displacement will have its extreme values when $y = 0$. In other words, the fastest the mass will move is when it is passing its equilibrium position $x = 0$, and the mass will have zero velocity when it is a maximum distance from its equilibrium position.

Underdamped motion

We now interpret Figure 6.26, where $a = 1$. Again we look at the orbit that starts at $(1, 0)$ in the phase plane, which is when the mass is at rest at its lowest position.

The part of the orbit in the fourth quadrant corresponds to the mass rising toward the equilibrium position, which it reaches at about $(0, -0.75)$, when the velocity is about -0.75, compared to -1.5 for the undamped case. In the third quadrant of the phase plane the mass continues to rise (above the equilibrium position) until it reaches its extreme highest position at a height of about -0.4. This is the point $(-0.4, 0)$ in the phase plane, and at this stage the mass is not as far above the equilibrium position as it was below. The second quadrant characterizes the mass dropping back to the equilibrium position $(0, 0.25)$, and the first quadrant is where the mass continues downward to the position $(0.1, 0)$, not as far down as its original starting position. Following the spiral around for a second time repeats the process. In other words, the spiral corresponds to a mass rising and falling forever with a decreasing amplitude. Different spirals correspond to different initial conditions.

Critically damped motion

We now consider Figure 6.27, where $a = 3$. Again we look at the orbit that starts at $(1, 0)$ in the phase plane, which is when the mass is at rest at its lowest position. This orbit is entirely in the fourth quadrant and corresponds to the mass rising toward the equilibrium position, which it reaches in infinite time.

Overdamped motion

We now consider Figure 6.28, where $a = 5$. Again we look at the orbit that starts at $(1, 0)$ in the phase plane, which is when the mass is at rest at its lowest position. This orbit is entirely in the fourth quadrant and corresponds to the mass rising toward the equilibrium position, which it reaches in infinite time.

Simple Linearized Pendulum

Undamped motion

We start with Figure 6.24, where $a = 0$. Let's interpret one of these orbits in terms of the motion of the linearized pendulum — say, the outermost orbit, which has the equation $4y^2 + 9x^2 = 9$. This ellipse crosses the x-axis at $x = \pm 1$, and the y-axis at $y = \pm 1.5$. We will start at the point $(1, 0)$ in the phase plane, and proceed clockwise once around the ellipse until we return to $(1, 0)$. The point $(1, 0)$ corresponds to an angle of one radian with an angular velocity of zero. This is when the pendulum is at rest at its extreme right-hand position.

The part of the orbit in the fourth quadrant corresponds to the pendulum swinging back toward the vertical position, which it reaches at $(0, -1.5)$, where the velocity is -1.5.

In the third quadrant of the phase plane the pendulum is moving to the left, until it reaches its extreme left-hand position at an angle of -1. This is the point $(-1, 0)$ in the phase plane, and at this stage the pendulum is as far to the left of the equilibrium position as it was to the right.

The second quadrant characterizes the pendulum swinging back to the vertical position $(0, 1.5)$, and the first quadrant is where the pendulum swings past vertical back to its original starting angle on the right-hand side. Following the ellipse around for a second time repeats the process. In other words, the ellipse corresponds to a pendulum swinging backward and forward forever.

Different ellipses correspond to different initial conditions. Smaller x-intercepts correspond to smaller amplitudes, and smaller y-intercepts correspond to smaller velocities. The velocity will have its extreme values when $x = 0$ and x will have its extreme values when $y = 0$. In other words, the fastest the pendulum will move is when it is passing its equilibrium position $x = 0$, and the pendulum will have zero velocity when it is a maximum distance from its equilibrium position.

These orbits predict that no matter how large the initial velocity, the motion is always bounded and oscillatory. This is clearly unrealistic, because if we have a sufficiently large initial velocity we would expect the pendulum to go over the top and, because there is no friction, continue going over the top forever. In this case the motion would be unbounded. This reinforces our earlier statement that for the model to be realistic, x must be restricted to small values. Later in this chapter we will discuss the general nonlinear pendulum, which does not have this restriction.

Underdamped motion

We now interpret Figure 6.26, where $a = 1$. Again we look at the orbit that starts at $(1, 0)$ in the phase plane, which is when the pendulum is at rest at its extreme right-hand position.

The part of the orbit in the fourth quadrant corresponds to the pendulum swinging back toward the vertical position, which it reaches at about $(0, -0.75)$, when the velocity is about -0.75, compared to -1.5 for the undamped case. In the third quadrant of the phase plane the pendulum is moving to the left, until it reaches its extreme left-hand position at an angle of about -0.4. This is the point $(-0.4, 0)$ in the phase plane, and at this stage the pendulum is not as far to the left of the equilibrium position as it was to the right. The second quadrant characterizes the pendulum swinging back to the vertical position — $(0, 0.25)$ in the phase plane — and the first quadrant is where the pendulum swings past vertical back to $(0.1, 0)$, not as far as its original starting position. Following the spiral around for a second time repeats the process. In other words, the spiral corresponds to a pendulum swinging backward and forward forever with a decreasing amplitude. Different spirals correspond to different initial conditions.

Critically damped motion

We now consider Figure 6.27, where $a = 3$. Again we look at the orbit that starts at $(1, 0)$, which is when the pendulum is at rest at its extreme right-hand position. This orbit is entirely in the fourth quadrant and corresponds to the pendulum swinging toward the equilibrium position, which it reaches in infinite time.

Overdamped motion

We now consider Figure 6.28, where $a = 5$. Again we look at the orbit that starts at $(1, 0)$, which is when the pendulum is at rest at its extreme right-hand position. This orbit is entirely in the fourth quadrant and corresponds to the pendulum swinging toward the equilibrium position, which it reaches in infinite time.

Series RLC Electrical Circuit

The charge x in the RLC electrical circuit of Figure 6.15 on page 253 consisting of a capacitor C, resistor R, and inductor L connected in series, is modeled by the differential equation $Lx'' + Rx' + \frac{1}{C}x = 0$, where x' is the current. This differential equation has the same form as that of

- A damped spring-mass system if we replace L with the mass, m; R with the damping factor, b; and $1/C$ with the spring constant, k. In this case x represents the vertical displacement and x' is the velocity.
- A damped linearized pendulum of length h if we replace L with the mass, m; R with the damping factor, b; and $1/C$ with the quantity mg/h. In this case x represents the angular displacement and x' is the angular velocity.

Thus, the two previous interpretations of the phase plane can be recast in the terminology of a series RLC circuit.

EXERCISES

1. Consider a spring-mass system that is governed by $x'' + 4x = 0$, where motion is started by stretching the spring 2 units from its equilibrium position and releasing it from rest.

(a) Write the corresponding differential equation for orbits in the phase plane (in terms of x and $y = x'$).

(b) Give the appropriate initial conditions for the statement of this problem both for the original second order differential equation and for the orbits.

(c) Find the equation of the orbit associated with this initial value problem.

(d) Find an explicit solution for the displacement as a function of time.

2. Solve the following initial value problems for $x(t)$, and find the equation of the corresponding orbit in the phase plane in terms of x and $y = x'$.

(a) $x'' + 16x = 0 \quad x(0) = 3 \quad x'(0) = 12$

(b) $9x'' + x = 0 \quad x(0) = -5 \quad x'(0) = 1/3$

(c) $4x'' + 9x = 0 \quad x(0) = 0 \quad x'(0) = 0$

(d) $x'' + 100x = 0 \quad x(0) = 0 \quad x'(0) = 0.2$

3. If we multiply the differential equation $mx'' + kx = 0$ by x', we should observe that each term can be integrated separately. We integrate $mx'x'' + kxx' = 0$ to obtain $\frac{1}{2}m\left(x'\right)^2 + \frac{1}{2}kx^2 = E$, where E is a constant. The first term on the left-hand side of this equation is the kinetic energy of the object; the second term represents the potential energy. For example, with the spring-mass system, $kx^2/2$ represents the energy stored in the spring. This equation is a statement of conservation of energy, $E =$ constant.

(a) Write down the conservation of energy equation for the system in Exercise 4, and evaluate the constant E.

(b) Show that the kinetic energy of the system is at a maximum when the potential energy is at a minimum. Where does this happen in the phase plane?

4. Consider a spring-mass system that is governed by $16x'' + x = 0$, subject to an initial displacement of -2 and an initial velocity of 3. What are the amplitude and period of the resulting motion? Find the equation of the orbit in the phase plane corresponding to these initial conditions.

5. Show that each orbit of $mx'' + bx' + kx = 0$ has a horizontal tangent in the phase plane.

6. The linearized differential equation for a damped pendulum is similar to those of spring-mass systems and electrical circuits. Make a table giving the equivalent terms in these models.

7. Quadratic Drag. If the atmospheric density changes with altitude in an exponential way, then the differential equation modeling an object falling vertically is $x'' = -g + ce^{-ax}\left(x'\right)^2$, where x is the height of the object above the ground, and a and c are positive constants. What sort of differential equation is this? By introducing the variable $y = x'$, write this equation as the system $x' = y$, $y' = -g + ce^{-ax}y^2$. What sort of system is this? Show that the orbits in the phase plane satisfy the Bernoulli equation

$$\frac{dy}{dx} = -\frac{g}{y} + ce^{-ax}y.$$

If the object falls from rest from height h, show that the Bernoulli equation has solution $y^2(x) = 2ge^{-\alpha(x)} \int_x^h e^{\alpha(u)}\,du$, where $\alpha(x) = (2c/a)e^{-ax}$.

8. A seconds pendulum clock is one that ticks every second at the end of each swing of the pendulum, so it has a period of 2 seconds. We want to construct such a frictionless clock. We decide to use the linearized model for the pendulum. How long should we make the pendulum if $g = 32.2$ feet/sec^2? What has this to do with the size of grandfather clocks?

9. Find the general solution of the differential equation that models the series RLC circuit; namely, $Lx'' + Rx' + \frac{1}{C}x = 0$, for the case when $R^2 = 4L/C$.

(a) Evaluate the arbitrary constants in your general solution using the initial conditions corresponding to the discharging of a charged capacitor.

(b) If we have $R = 4$, $L = 1$, and $C = 1/4$, find the value of t such that the solution in part (a) reaches 1% of its maximum value; that is, find the settling time.

10. Write out the interpretation of the differential equation $x'' + ax' + \frac{9}{4}x = 0$ for a series RLC circuit with $a = 0, 1, 3,$ and 5, in a manner similar to the interpretation of this equation for the spring-mass and linearized pendulum starting on page 250.

11. Solve

$$\frac{dy}{dx} = \frac{-ay - \frac{9}{4}x}{y}$$

to find the implicit equations of the orbits in the phase plane when (i) $a = 1$, (ii) $a = 3$, and (iii) $a = 5$.

12. If an object is thrown upward, its height x above the ground is modeled by the differential equation $mx'' + bx' = -mg$. For the purposes of this exercise we assume that units have been chosen so that $m = g = 1$.

(a) Consider the case of no resistance ($b = 0$) so that $x'' = -1$. By introducing the variable $y = x'$, write this equation as the system $x' = y$, $y' = -1$. What sort of system is this? Solve this system for the orbits in the phase plane. Draw a phase portrait for this system and interpret the orbits in terms of the motion of the object.

(b) Consider the case $b = 1$ so that $x'' + x' = -1$. By introducing the variable $y = x'$, write this equation as the system $x' = y$, $y' = -1 - y$. What sort of system is this? Solve the differential equation for the orbits in the phase plane. Draw a phase portrait for this system and interpret the orbits in terms of the motion of the object.

6.6 How Explicit Solutions Are Related to Orbits

In this section we will show how orbits and solutions of second order differential equations are related.

Example 6.5 *Denise and Chad's Relationship*

We return to the first example discussed in this chapter, where we found that

$$x(t) = \frac{2.5}{\sqrt{2}} \sin \sqrt{2}t, \qquad y(t) = 2.5 \cos \sqrt{2}t,$$

were the solutions of

$$x' = y, \qquad y' = -2x, \qquad x(0) = 0, \quad y(0) = 2.5.$$

Now that we have explicit functions to represent the affections of Denise and Chad, we can add time to the orbit of y versus x in Figure 6.3. First we construct a table of values of x and y as functions of t (Table 6.5). The variable t plays the role of a parameter, and the pairs of (x, y) values from Table 6.5 can be plotted. Figure 6.29 shows these pairs of points added to the graph from Figure 6.3.

However, we do not need to compute a table before obtaining the graph of Figure 6.3. It is possible to do this from the graphs of $x(t)$ and $y(t)$ directly. In Figure 6.30 we have plotted y versus t in the upper right-hand box and x versus t in the lower right-hand box. The upper left-hand box is where we will plot the orbit in the phase plane, and it is labeled y versus x. In the lower left-hand box we have drawn a line at a 45° angle.

Table 6.5 Values of x and y for Denise and Chad

t	x	y
0.0	0.00	2.50
0.5	1.15	1.90
1.0	1.75	0.39
1.5	1.51	−1.31
2.0	0.54	−2.38
2.5	−0.68	−2.31
3.0	−1.58	−1.13
3.5	−1.72	0.59
4.0	−1.04	2.03
4.5	0.14	2.49

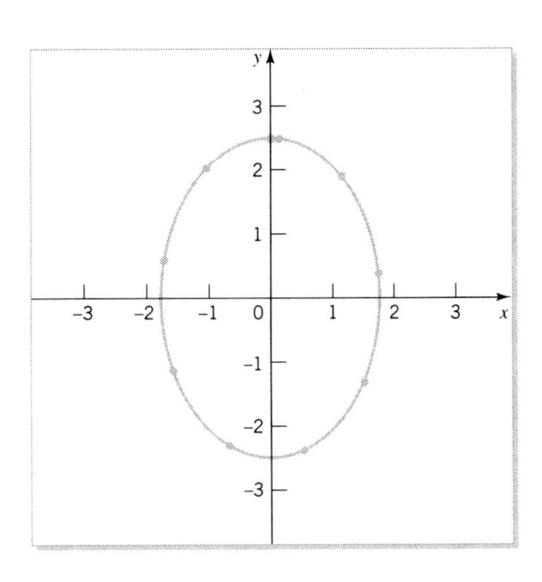

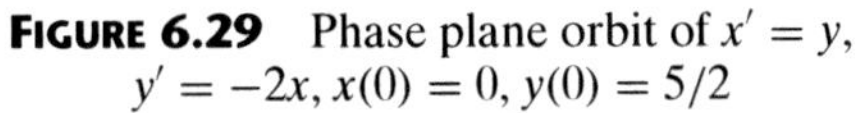
FIGURE 6.29 Phase plane orbit of $x' = y$, $y' = -2x$, $x(0) = 0$, $y(0) = 5/2$

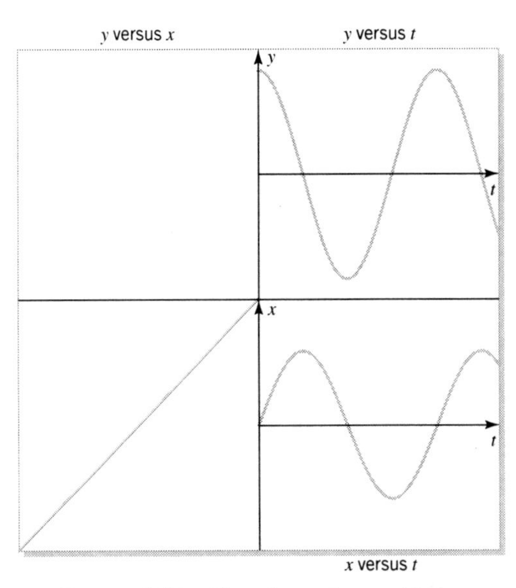

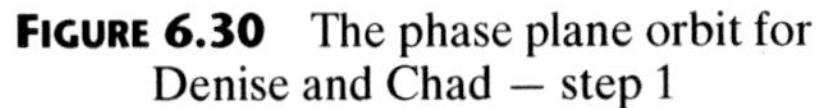
FIGURE 6.30 The phase plane orbit for Denise and Chad — step 1

How to Relate the Explicit Solution to the Orbit

Purpose To show how solution curves $x = x(t)$, $y = y(t)$ are related to the phase plane orbit.

Process

1. Select any point on the t-axis in the y versus t box; say, $t = a$.
2. Draw a vertical line through $t = a$ so that it crosses both the curve $y = y(t)$ in the y versus t box and the curve $x = x(t)$ in the x versus t box. This identifies the values of $y(a)$ and $x(a)$. We want to transfer these values to the y versus x box.
3. Draw a horizontal line through $y(a)$ into the y versus x box. $y(a)$ lies on this line.
4. Draw a horizontal line through $x(a)$ into the lower left box until it intersects the 45° line. Then draw a vertical line through this point of intersection into the y versus x box. Where this line intersects the horizontal line from step 3 gives the point $(x(a), y(a))$ in the y versus x box.
5. Repeat the preceding steps for different values of a, and then join the points in the y versus x box. Notice that the direction of the orbit is determined by the order of the points plotted for increasing t.

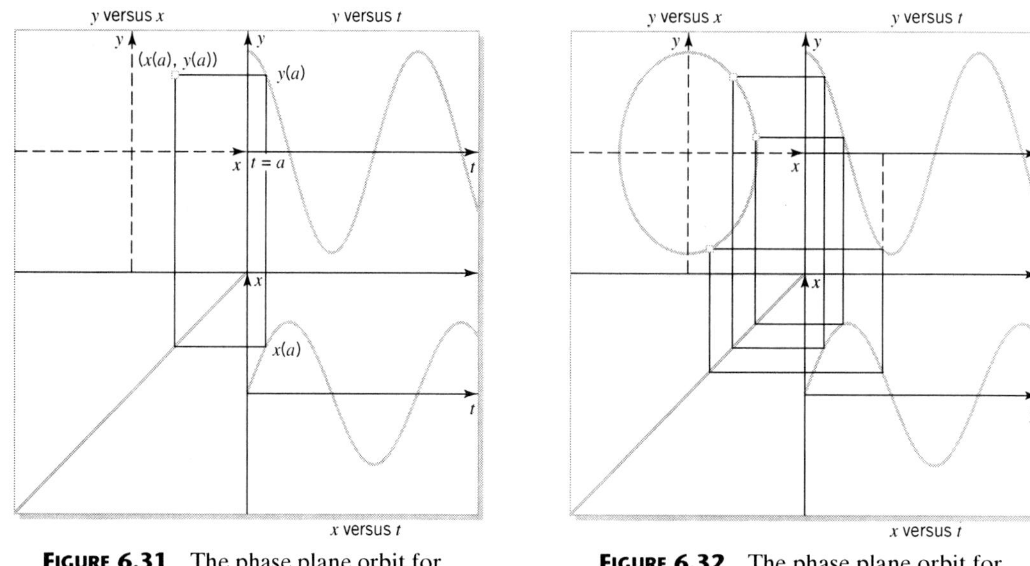

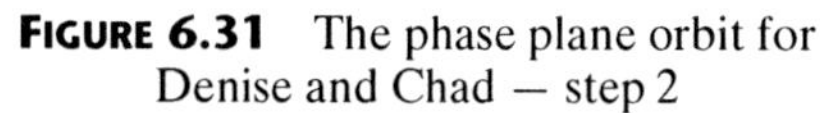

FIGURE 6.31 The phase plane orbit for Denise and Chad — step 2

FIGURE 6.32 The phase plane orbit for Denise and Chad — step 3

This process is not as complicated as it sounds. Look at Figure 6.31 and reread the steps. Some more lines and a completed orbit are shown in Figure 6.32.

Comments on Relating the Explicit Solution and the Orbit

- The orbit obtained by this graphical method is exactly the same as the curve we would obtain by eliminating t from the explicit solutions $x = x(t)$, $y = y(t)$.
- This graphical construction method can also be used to obtain the phase plane for nonautonomous systems where explicit solutions for orbits are rarely possible.

We have an additional way of looking at the phase plane. If we consider time as an axis perpendicular to the xy-plane, then the three-dimensional graph of the affections of Denise and Chad as a function of time — namely, $x(t) = \frac{2.5}{\sqrt{2}} \sin \sqrt{2}t$, $y(t) = 2.5 \cos \sqrt{2}t$ — would give a helix. If we then take the projection into the xy-plane of this curve, we obtain the phase plane. This is shown in Figure 6.33.

This view gives us a way to picture the Existence-Uniqueness Theorem for Second Order Linear Differential Equations in terms of nonintersecting curves. Imagine a three-dimensional space with coordinates (t, x, x'). Projection of a curve in this space down the t-axis gives the phase plane. Consider the plane corresponding to $t = t_0$. A point in this plane represents the initial condition $\left(x(t_0), x'(t_0)\right)$. Now imagine the curve $x = x(t)$ progressing in time in this three-dimensional space. It will start from $\left(x(t_0), x'(t_0)\right)$ and follow $x = x(t)$ and $x' = x'(t)$. The theorem says no other curve in this space can pass through the point $\left(x(t_0), x'(t_0)\right)$, so no other solution can intersect the solution at $\left(x(t_0), x'(t_0)\right)$.

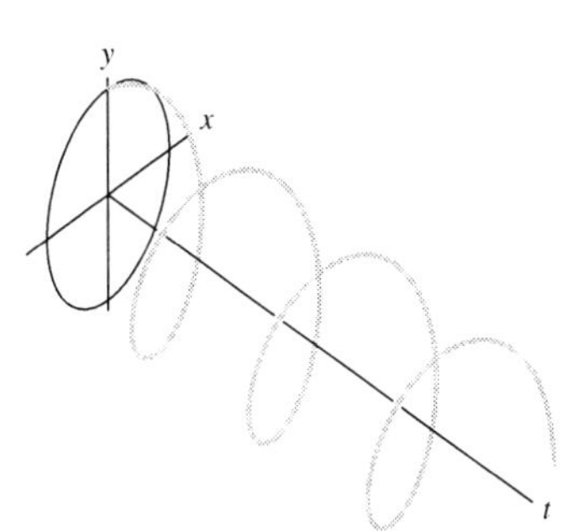

FIGURE 6.33 Helix in three-dimensional space (x, y, t)

EXAMPLE 6.6 *Parental Interference*

Finally, we look at the last example in Section 6.1, in which we had the initial value problem

$$y' = -2x, \quad x' = y - 0.4x, \quad x(0) = 0, \quad y(0) = 2.5, \tag{6.45}$$

with solution

$$x(t) = \frac{25}{14}e^{-0.2t}\sin 1.4t, \tag{6.46}$$

$$y(t) = 2.5e^{-0.2t}\left(\cos 1.4t - \frac{1}{7}\sin 1.4t\right). \tag{6.47}$$

Figure 6.34 shows a graph of (6.46) and (6.47) as functions of t in the two right-hand boxes and shows the 45° line in the lower left-hand box. The graph of the y versus x orbit in the phase plane may now be constructed by the graphical method given in the previous example. This is done in Figure 6.35.

Notice that all three graphs in Figure 6.35 suggest that as time increases, both the x and y values approach zero as a limit. In fact, from the general solution of this problem as given in (6.11) and (6.12), we see that both x and y approach zero as t approaches infinity, regardless of the choice of the arbitrary constants C_1 and C_2. Thus, in the phase plane, all orbits will spiral in toward the origin. Notice that the point $x = 0$, $y = 0$, satisfies the two differential equations in (6.45) and is therefore classified as an EQUILIBRIUM POINT (which no other orbit can intersect).

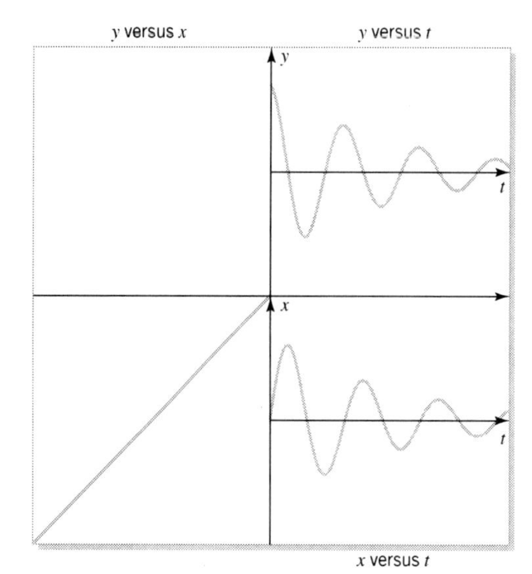

FIGURE 6.34 The phase plane orbit for parental interference — step 1

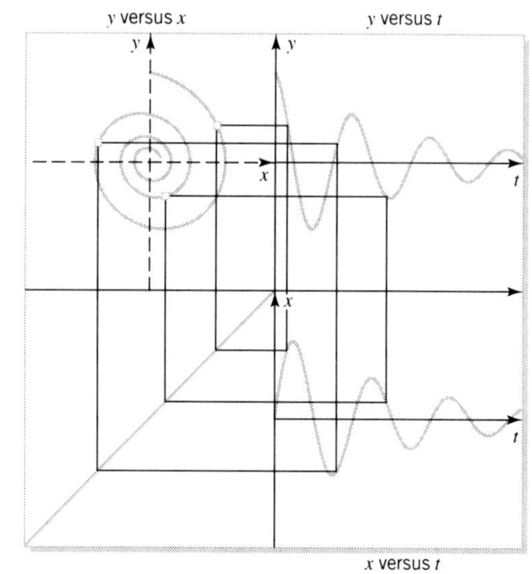

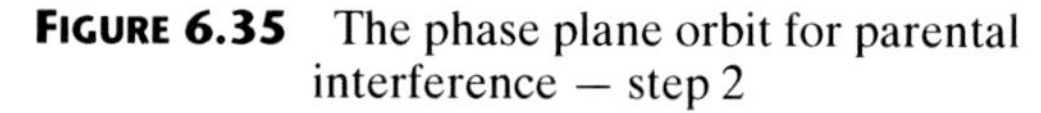

FIGURE 6.35 The phase plane orbit for parental interference — step 2

◆ Definition 6.3: **The system of autonomous differential equations**

$$\begin{cases} x' = P(x,y), \\ y' = Q(x,y), \end{cases}$$

has an equilibrium point at (x_0, y_0) if

$$\begin{cases} P(x_0,y_0) = 0, \\ Q(x_0,y_0) = 0. \end{cases}$$

◆

Equilibrium points of a system of differential equations play the same important role as that of the equilibrium solutions of a first order equation. We will return to them in Chapter 9.

EXERCISES

1. Which systems of equations in Exercise 14 on page 239 have equilibrium points? Find the equilibrium points for those systems that have them, and explain why the others don't.

2. You are given a system of differential equations, $x' = P(x,y)$, $y' = Q(x,y)$. The graphs of the explicit solution of this system — namely, $x = x(t)$ and $y = y(t)$ — are shown in Figure 6.36.

   (a) Use these graphs to sketch the associated orbit.

   (b) Do you think that $x = y = 0$ is an equilibrium point in the phase plane? Explain your reasoning. How can you tell this directly from the graphs of

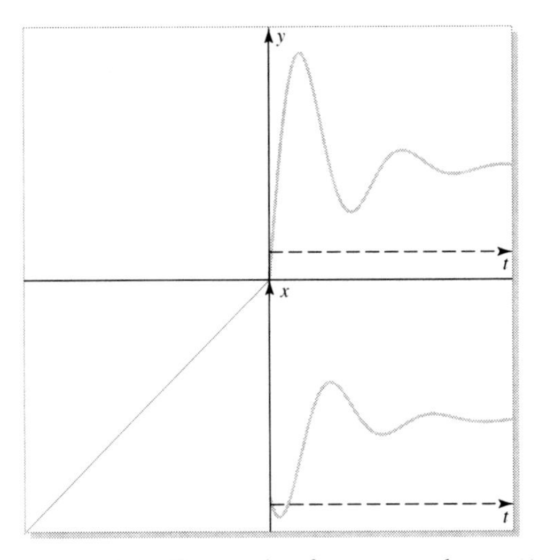

FIGURE 6.36 The graphs of $x = x(t)$ and $y = y(t)$

$x = x(t)$ and $y = y(t)$ (without constructing the phase plane)?

(c) Are there any other possible equilibrium points?

3. Give the equilibrium points for the following systems of differential equations.

(a) $x' + 2y = 6$
$2y' - 2x + 7y = 9$

(b) $x' + 3x - 2y = 0$
$3y' - 7x + 4y = 0$

(c) $x' + x - 3y = 2$
$y' - 2x + 6y = -4$

4. The differential equation for the orbits in the phase plane in Exercise 3(b) has homogeneous coefficients. Solve this differential equation to obtain an explicit solution for the orbits.

5. Exercise 14 on page 239 contains a number of systems of first order linear differential equations. The instructions in that exercise call for finding a second order differential equation by eliminating one of the dependent variables.

(a) For the following systems of differential equations, use that technique to find the associated second order differential equation and its general solution.

(b) Use this solution to solve for the other dependent variable, and choose the arbitrary constants to satisfy the given initial conditions.

(c) Use the method of this section to obtain the orbit in the phase plane from the solution of this initial value problem.

i. $x' = 2x - y \quad x(0) = 2$
$y' = -x + 2y \quad y(0) = 0$

ii. $x' = 2x - 2y \quad x(0) = 0$
$y' = 3x - 2y \quad y(0) = 2$

iii. $x' = 3x - 2y \quad x(0) = 2$
$y' = 2x - 2y \quad y(0) = 1$

iv. $x' = 5x - 2y \quad x(0) = 0$
$y' = 4x - 2y \quad y(0) = 6$

6. Exercise 15 on page 239, contains a number of systems of first order linear differential equations. The instructions in that exercise call for finding a second order differential equation by eliminating one of the dependent variables. These equations were not in the standard form $x' = P(x, y)$, $y' = Q(x, y)$, but could be put in that form by appropriate operations on the two differential equations. Consider, for example, Exercise 15(a),

$$\begin{cases} x' + y' + y = 0, \\ 3x' + 2y' + x = 7. \end{cases} \tag{6.48}$$

Multiplying the top equation by -2 and adding the result to the bottom equation results in a differential equation containing x', but not y'. In a similar manner, substituting the value of x' from the top equation into the bottom one gives an equation involving y', but not x'. These two new equations have the form

$$x' = P(x, y), \qquad y' = Q(x, y). \tag{6.49}$$

(a) For each of the following systems of equations i – iii, obtain the equivalent system of differential equations in the form (6.49).

(b) Find the associated second order differential equation for the resulting system and its general solution.

(c) Use this solution to solve for the other dependent variable, and choose the arbitrary constants to satisfy the initial conditions $x(0) = 0$, $y(0) = 1$.

(d) Use the method of this section to obtain the orbit from the solution of this initial value problem. Note any equilibrium points.

i. The system in (6.48).

ii. The system

$$\begin{cases} x' + y' = 0 \\ 2x' - 2y' - y = 0 \end{cases}.$$

iii. The system

$$\begin{cases} x' + y' - 3y = 0 \\ x' - 3x - 6y = 0 \end{cases}.$$

6.7 THE MOTION OF A NONLINEAR PENDULUM

As we have seen, if the weight of the rod is negligible, the hinge (pivot) is frictionless, and there is no air resistance, the differential equation governing the simple pendulum in motion is $mx'' = -\frac{mg}{h} \sin x$, or

$$x'' + \lambda^2 \sin x = 0, \tag{6.50}$$

where

$$\lambda = \sqrt{\frac{g}{h}}. \tag{6.51}$$

Equation (6.50) is a second order nonlinear differential equation and cannot be solved in terms of familiar functions. However, the Existence-Uniqueness Theorem guarantees that, for any initial condition, a solution curve exists and is unique. For example, the functions $x(t) = n\pi$ ($n =$ any integer) satisfy (6.50) given the initial conditions $x(t_0) = n\pi, x'(t_0) = 0$. The functions $x(t) = n\pi$ are thus solutions, and no other solution can have the property that $x(t_0) = n\pi, x'(t_0) = 0$. Physically these are solutions that correspond to the pendulum hanging at rest, with the mass either directly below the hinge ($x = 0, \pm 2\pi, \cdots$) or directly above the hinge ($x = \pm\pi, \pm 3\pi, \cdots$). By viewing the pendulum we are not able to tell the difference between the solutions $x(t) = 0$ and $x(t) = 2\pi$, unless we know that someone had first rotated the pendulum through 2π and then let it hang there. Similar comments apply to the other constant solutions.

Although we are unable to solve the nonlinear differential equation for the frictionless pendulum in terms of familiar functions, we can use it to obtain useful information by following steps similar to the linearized case. We introduce the angular velocity, $y = x'$, and write (6.50) as the system of equations

$$x' = y, \qquad y' = -\lambda^2 \sin x. \tag{6.52}$$

Notice that our constant solutions $x(t) = n\pi$ satisfy (6.52) identically because $\sin n\pi = 0$ and $x' = y = 0$. These constant solutions are the equilibrium points for (6.52).

Phase plane

The orbits in the phase plane will satisfy

$$\frac{dy}{dx} = -\lambda^2 \frac{\sin x}{y},$$

Direction field

and Figure 6.37 shows the corresponding direction field for $\lambda = 3$.

This differential equation may be integrated to yield

$$y^2(t) - 2\lambda^2 \cos x(t) = c, \tag{6.53}$$

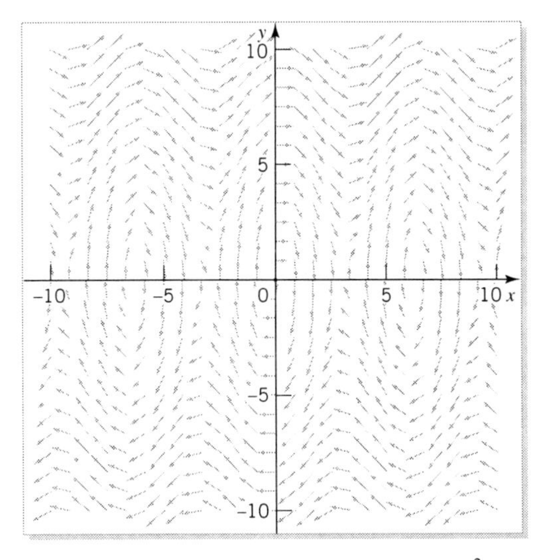

FIGURE 6.37 Direction field for $dy/dx = -(\lambda^2 \sin x)/y$

where c is an arbitrary constant. It is helpful to use the trigonometric identity $\cos x = 1 - 2\sin^2 \frac{x}{2}$ to rewrite (6.53) in the form $y^2(t) + 4\lambda^2 \sin^2 \frac{x(t)}{2} = C$, where $C = c + 2\lambda^2$. With the initial conditions $x(0) = x_0$, $y(0) = y_0$, this orbit equation becomes

$$y^2(t) + 4\lambda^2 \sin^2 \frac{x(t)}{2} = y_0^2 + 4\lambda^2 \sin^2 \frac{x_0}{2}. \tag{6.54}$$

Orbits If we consider the pendulum starting at $x(0) = 0$ with prescribed initial velocity $y(0) = y_0$, we have the orbit given by $y^2(t) + 4\lambda^2 \sin^2 \frac{x(t)}{2} = y_0^2$. Solving this for $y(t)$ gives

$$y(t) = \pm 2\lambda \sqrt{\mu^2 - \sin^2 \frac{x(t)}{2}}, \tag{6.55}$$

where $\mu = \frac{y_0}{2\lambda}$. Here $+$ applies to the case in which x is increasing [see (6.52)], and $-$ applies when x is decreasing. For $y(t)$ to be real, x must satisfy

$$\mu^2 - \sin^2 \frac{x(t)}{2} \geq 0. \tag{6.56}$$

This gives rise to four types of orbits depending on the magnitude of μ.

1. If $\mu^2 > 1$, then (6.56) is always satisfied in the sense that $\sin^2 x/2 \leq 1$ for all x. In this case the orbits (6.55) will be defined for every x, and the velocity $y(t)$ can never be zero. The condition $\mu^2 > 1$ is equivalent to $y_0 > 2\lambda$ or $y_0 < -2\lambda$. So for large initial velocities, the orbit is defined for all x. Physically this corresponds to an initial velocity large enough to propel the pendulum over the top. Because we are dealing with a frictionless pendulum, this will result in the mass circling the pivot forever. Examples of these orbits are shown in Figure 6.38.
2. If $\mu^2 = 1$, then (6.56) is satisfied for every x. However, (6.55) requires that the velocity is zero whenever $\sin x/2 = \pm 1$; that is, at $x = \pm\pi, \pm 3\pi, \pm 5\pi, \cdots =$

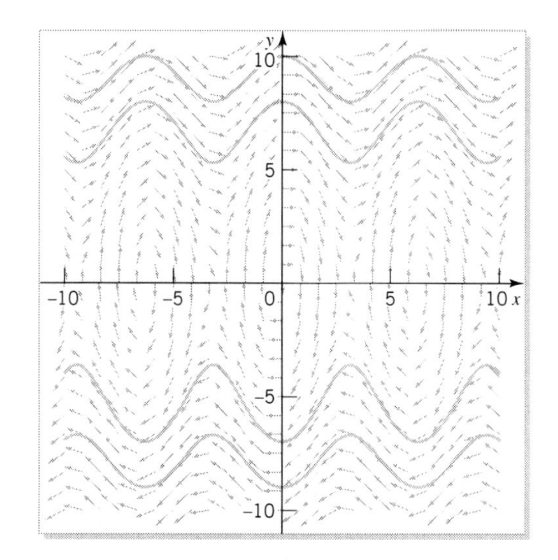

FIGURE 6.38 Orbits ($\mu^2 > 1$) and direction field for $dy/dx = -(\lambda^2 \sin x)/y$

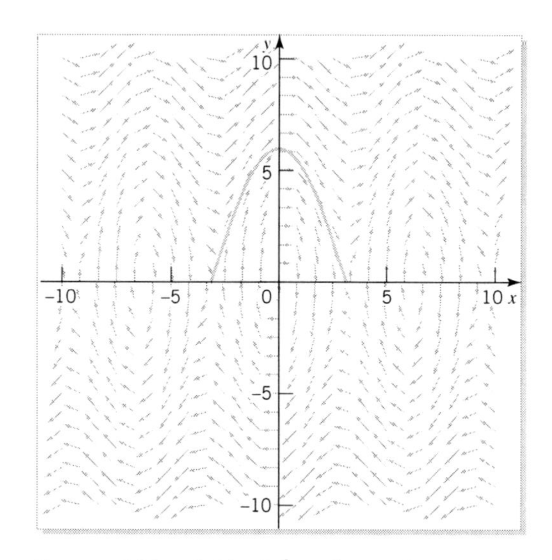

FIGURE 6.39 Orbits ($\mu^2 = 1$) and direction field for $dy/dx = -(\lambda^2 \sin x)/y$

$(2n+1)\pi$, where n is any integer. So if the orbit that starts with $x(0) = 0$ continues for all x, it would have the property that there is a time, t_0, for which $x(t_0) = \pi$ and $y(t_0) = 0$. However, we have already pointed out that the only solution for which $x(t_0) = \pi$ and $y(t_0) = 0$ is the one for which $x(t) = \pi$ for all time. Consequently, our orbit never reaches this point in finite time, so the orbit does not continue for all x but approaches the upper limit of π as t approaches ∞. Physically this corresponds to an initial velocity that is exactly right for the pendulum to come to rest in a vertical position with the mass balancing above the hinge. However, it will take the pendulum an infinite time to reach this position. An example of this orbit is shown in Figure 6.39.

3. If $0 < \mu^2 < 1$, then (6.56) will be satisfied for some values of x. For example, if $\mu^2 = 1/2$, then the only values of x for which $\sin^2 x/2 \leq 1/2$ are $(4n-1)\pi/2 \leq x \leq (4n+1)\pi/2$, where n is any integer. For the initial condition $x(0) = 0$ to be included in this set, we must have $n = 0$, so this implies that $-\pi/2 \leq x \leq \pi/2$ for $\mu^2 = 1/2$. The condition $0 < \mu^2 < 1$ is equivalent to $0 < y_0 < 2\lambda$ or $-2\lambda < y_0 < 0$. So for moderate initial velocities, the orbit is defined for these values of x. Physically this corresponds to a pendulum swinging backward and forward, and for small x it corresponds to the linearized motion described earlier. The velocity will be zero when x has its extreme values. Examples of these orbits are shown in Figure 6.40.

4. If $\mu^2 = 0$, then $\sin^2 x/2 \leq 0$, which is true only when $\sin x/2 = 0$ — that is, when $x = 2n\pi$. This corresponds to $y_0 = 0$, meaning that for zero initial velocity, we have only discrete values of x. For our initial condition $x(0) = 0$, we must have $n = 0$, so that $x(t) = 0$. This corresponds to the pendulum hanging vertically for all time.

To summarize our previous discussion, we combine all the cases in a single picture of the phase plane as shown in Figure 6.41 for $\lambda = 3$. Also shown are orbits

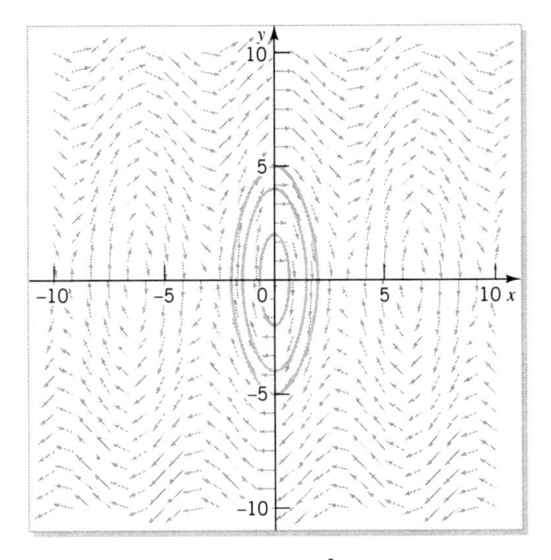

FIGURE 6.40 Orbits ($0 < \mu^2 < 1$) and direction field for $dy/dx = -(\lambda^2 \sin x)/y$

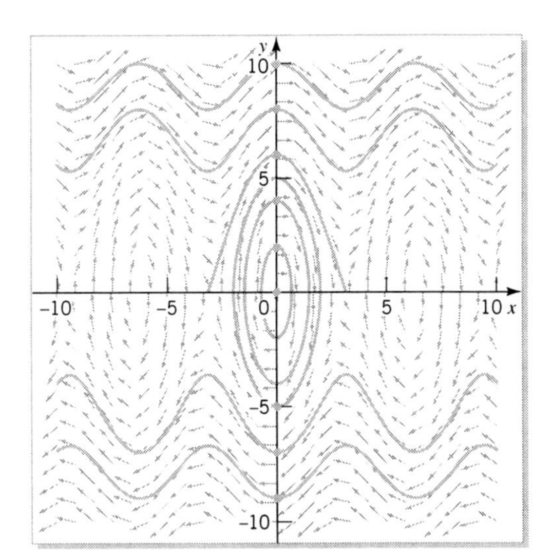

FIGURE 6.41 Orbits and direction field for $dy/dx = -(\lambda^2 \sin x)/y$

with $x(0) = 0$ and $y(0) = 10, 8, -7$, and -9 (which go on forever without oscillating); $y(0) = 6$ (which takes an infinite time to reach the point $x = \pi$); $y(0) = 4, 2$, and -5 (which oscillate forever); and $y(0) = 0$, where the pendulum is stationary for all time. If we consider the orbit through $x(0) = 0$, $y(0) = 10$, the direction of travel of the pendulum starts from left to right. The orbit for $x < 0$ corresponds to the motion of the pendulum for times prior to $t = 0$. In the same way, the motion of the pendulum corresponding to the orbit through $x(0) = 0$, $y(0) = -9$, starts from right to left. The orbit for $x > 0$ corresponds to the motion of the pendulum for times prior to $t = 0$.

It should be noted that while we cannot solve $x'' + \lambda^2 \sin x = 0$ in terms of familiar functions, we have discovered a great deal about its solutions by analyzing its orbits.

We now want to see the impact on these phase portraits of a linear friction term in the differential equation, namely,

$$x'' + bx' + \lambda^2 \sin x = 0, \tag{6.57}$$

where b is a positive constant. In this case the orbits in the phase plane satisfy

$$\frac{dy}{dx} = \frac{-\lambda^2 \sin x - by}{y}.$$

The first point to make is that the equilibrium points are unchanged, because they will still be given by $y = 0$ and the roots of $\sin x = 0$. The second point to make depends on rewriting this last differential equation in the more illuminating form

$$\frac{dy}{dx} = \frac{-\lambda^2 \sin x}{y} - b. \tag{6.58}$$

One way to interpret this is that the slope of the orbit corresponding to a pendulum with friction is b less than $-\lambda^2 \sin x/y$, which is the slope of the orbit corresponding to a pendulum without friction — symbolically,

$$\left(\frac{dy}{dx}\right)_{friction} = \left(\frac{dy}{dx}\right)_{frictionless} - b.$$

Thus, one way of constructing the phase portrait for the friction case from the frictionless case is to realize that the friction orbits must cross the frictionless orbits in such a way that the slope is reduced by an amount b. In Figure 6.42 we construct a number of orbits consistent with this principle. This is a phase portrait for the nonlinear pendulum with friction.

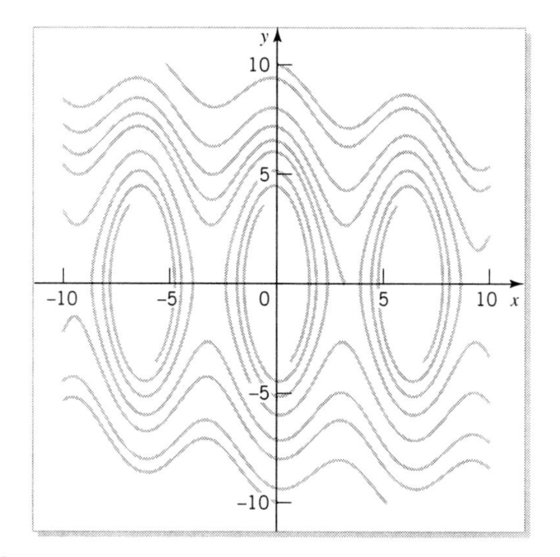

FIGURE 6.42 Phase portrait of a nonlinearized simple pendulum with friction

EXERCISES

1. On Figure 6.41, sketch the orbits corresponding to the following physical situations.

   (a) A pendulum hangs so that the mass is directly below the hinge for all time. How many such orbits are there?

   (b) A pendulum is carefully suspended so that the mass is directly above the hinge for all time. How many such orbits are there?

2. Some orbits shown in Figure 6.42 spiral in to the equilibrium point $(0, 0)$. If b in (6.57) were made larger, would you expect there to be a situation where the orbits still approach the origin but do not do so as a spiral?

3. By analyzing the orbit in the phase plane for the nonlinear pendulum $x'' + \sin x = 0$ with initial conditions $x(0) = 0, x'(0) = \beta$, answer the following questions.

   (a) For what values of β will the angle x be bounded for all times?

   (b) For what value of β will x be unbounded in the limit as $t \to \infty$?

(c) If the initial conditions are changed to $x(0) = \alpha$, $x'(0) = 0$, can you choose α so that in the resulting motion x becomes unbounded? Explain fully.

4. If the pendulum's pivot is subjected to a constant torque of $a > 0$, then the corresponding differential equation is $x'' + \sin x = a$, if there is no damping.

(a) Show that the equation for the linearized pendulum is $x'' + x = a$, and that this can be solved by making the substitution $u = x - a$. Describe the motion of the linearized pendulum that is initially hanging at rest.

(b) Use the computer to plot the numerical solutions of $x'' + \sin x = a$ for different values of a corresponding to the pendulum initially hanging at rest. For what value of a is there a sudden change in behavior? What does this correspond to physically?

(c) For the nonlinearized pendulum $x'' + \sin x = a$, use a phase plane analysis to find an equation similar to (6.53) for the orbits. Use these orbits to explain the source of the numerical value for a that you found in part (b).

5. Recall that the Taylor series for $\sin x$ (about the origin) is

$$\sin x = x - \frac{1}{3!}x^3 + \frac{1}{5!}x^5 - \cdots.$$

For oscillations where x is small, we might approximate $\sin x$ by $x - x^3/3!$ and get the differential equation for a simple pendulum,

$$x'' + \lambda^2\left(x - \frac{1}{3!}x^3\right) = 0, \qquad (6.59)$$

in place of the linearized pendulum (6.50).

(a) Write (6.59) in terms of the phase plane variables (x and $y = x'$), and compare the direction field of your result with the one we obtained for the linearized pendulum, Figure 6.24. Explain fully any similarities and differences.

(b) Find the equation for the orbits, and compare it with the equation for the orbits for $x'' + \lambda^2 x = 0$, given by $y^2 + \lambda^2 x^2 = C$.

(c) Write (6.59) in terms of the phase plane variables (x and $y = x'$) and compare the direction field of your result with the one we obtained for the nonlinear pendulum, Figure 6.41. Explain fully any similarities and differences.

(d) Compare the equation for the orbits for $x'' + \lambda^2 \sin x = 0$, given by (6.53), with the equation for the orbits from part (b).

6. In Chapter 4 we found that air resistance is sometimes proportional to the square of the velocity, rather than being linear in the velocity. Two students agree with this, but argue whether the appropriate equation for a simple pendulum should be $x'' + b\left(x'\right)^2 + \lambda^2 \sin x = 0$, or $x'' + bx'\left|x'\right| + \lambda^2 \sin x = 0$, where $b > 0$. You are asked your opinion. Use a phase plane argument similar to the one in this section to decide which student is correct.

7. Computer Experiment. The purpose of this exercise is to obtain qualitative information about the solutions of three differential equations.

(a) Use the computer to plot the numerical solutions of $x'' + 4x = 0$ subject to $x(0) = x_0$, $x'(0) = 0$ for $x_0 = 0.5$, 1, and 2. What happens to the period of the solution as you change the amplitude?

(b) Use the computer to plot the numerical solutions of $x'' + 4\sin x = 0$ subject to $x(0) = x_0$, $x'(0) = 0$ for $x_0 = 0.5$, 1, and 2. What happens to the period of the solution as you change the amplitude?

(c) Use the computer to plot the numerical solutions of $x'' + 4x^3 = 0$ subject to $x(0) = x_0$, $x'(0) = 0$ for $x_0 = 0.5$, 1, and 2. What happens to the period of the solution as you change the amplitude?

What Have We Learned?

Main Ideas

- A single second order differential equation may be obtained from a system of two first order differential equations.
- An autonomous system of two first order linear differential equations results from an appropriate substitution in any second order linear differential equation of the form $a_2 x'' + a_1 x' + a_0 x = 0$, where a_2, a_1, and a_0 are constants. This allows

our graphical analysis from earlier chapters to be used in analyzing properties of solutions of these second order equations.

- The characteristic equation may be used to find explicit solutions of all second order linear differential equations with constant coefficients. See *How to Solve Homogeneous Second Order Linear Differential Equations with Constant Coefficients* on page 244.
- If all the coefficients in a second order linear differential equation are continuous (and the one multiplying the second derivative is never zero), then all initial value problems associated with this equation will have a unique solution. See the Existence-Uniqueness Theorem for Second Order Linear Differential Equations on page 238.
- The explicit solution of a system of differential equations can be related to orbits using a geometrical technique. See *How to Relate the Explicit Solution to the Orbit* on page 267.
- Spring-mass systems, pendulums, and RLC electrical circuits are among the models that use second order linear differential equations with constant coefficients.
- For large amplitudes, the nonlinear pendulum behaves in a much different manner from the linear pendulum.

CHAPTER 7

SECOND ORDER LINEAR DIFFERENTIAL EQUATIONS WITH FORCING FUNCTIONS

Where We Are Going — and Why

In the previous chapter we developed solutions of homogeneous second order linear differential equations with constant coefficients. Because many important applications of differential equations deal with nonhomogeneous linear differential equations — those with forcing functions — we consider that situation in this chapter.

We will discover a straightforward technique that always works for a very restricted, but important, class of forcing functions. We introduce a technique — the Principle of Linear Superposition — that allows us to construct new solutions from old ones. Mathematical models of forced spring-mass systems and electrical circuits are among the applications considered.

7.1 THE GENERAL SOLUTION

Many important applications of differential equations involve solving the nonhomogeneous linear differential equation

$$a_2(t)x'' + a_1(t)x' + a_0(t)x = f(t), \tag{7.1}$$

where $a_2(t)$, $a_1(t)$, $a_0(t)$, and $f(t)$ are given functions of t. In order to discover a general method for solving (7.1), we will look at a simple example.

EXAMPLE 7.1

Solve

$$x'' + x' = e^t. \tag{7.2}$$

At first we may not see how to solve this differential equation. We notice that, if the equation were homogeneous — namely, $x'' + x' = 0$ — we could immediately write down its solution as $x(t) = C_1 + C_2 e^{-t}$ corresponding to the characteristic equation $r^2 + r = r(r+1) = 0$. But this does not seem helpful in trying to solve (7.2).

However, if we look at (7.2) long enough, we may see that the left-hand side is $(x' + x)'$. Because of this, we can integrate (7.2) with respect to t to find

$$x' + x = e^t + C_1,$$

Integrating factor

where C_1 is an arbitrary constant. We recognize this as a first order linear differential equation (see Chapter 5), which we multiply by its integrating factor, $\mu = e^t$, to find

$$(e^t x)' = e^{2t} + C_1 e^t.$$

This can be integrated to give

$$e^t x = \frac{1}{2} e^{2t} + C_1 e^t + C_2,$$

Explicit solution

where C_2 is an arbitrary constant. Thus, the solution of (7.2) is

$$x(t) = \frac{1}{2} e^t + C_1 + C_2 e^{-t}. \tag{7.3}$$

Let's look at the structure of this solution. It consists of two parts: $C_1 + C_2 e^{-t}$ and $e^t/2$. The presence of $C_1 + C_2 e^{-t}$ is not a coincidence — it is the general solution of the homogeneous equation $x'' + x' = 0$ associated with $x'' + x' = e^t$. We introduce the notation $x_h(t) = C_1 + C_2 e^{-t}$, where the subscript h indicates that x_h is the solution of the associated homogeneous equation; that is, $x_h'' + x_h' = 0$. If we introduce the notation $x_p(t) = \frac{1}{2} e^t$, the solution (7.3) of (7.2) can be written as $x(t) = x_h(t) + x_p(t)$, and we could ask what differential equation is satisfied by $x_p(t)$. To find out, we substitute $x(t) = x_h(t) + x_p(t)$ into the left-hand side of (7.2), obtaining

$$x'' + x' = (x_h + x_p)'' + (x_h + x_p)' = (x_h'' + x_h') + (x_p'' + x_p').$$

However, the first term on the right-hand side is zero, and (7.2) states that $x'' + x' = e^t$, so

$$x_p'' + x_p' = e^t.$$

Thus, the function $x_p(t) = \frac{1}{2} e^t$ satisfies the nonhomogeneous equation (7.2). However, $x_p(t)$ is not the general solution of (7.2); $x(t) = x_h(t) + x_p(t)$ is. The solution $x_p(t)$ is a particular solution of (7.2), which is why we added the subscript p to the dependent variable x. (Recall that any function that satisfies a differential equation is called a particular solution.) Putting this together, we see that the general solution $x(t) = x_h(t) + x_p(t)$ is the sum of $x_p(t)$, a particular solution of the nonhomogeneous equation, and $x_h(t)$, the general solution of the associated homogeneous differential equation. ■

We want to see whether a similar result holds for the general equation (7.1). Imagine we have found a particular solution $x = x_p(t)$ of (7.1), so that

$$a_2(t)x_p'' + a_1(t)x_p' + a_0(t)x_p = f(t). \tag{7.4}$$

Now the general solution $x = x(t)$ that we seek must satisfy

$$a_2(t)x'' + a_1(t)x' + a_0(t)x = f(t). \tag{7.5}$$

If we subtract (7.4) from (7.5) we find

$$a_2(t)(x'' - x_p'') + a_1(t)(x' - x_p') + a_0(t)(x - x_p) = 0,$$

or

$$a_2(t)(x - x_p)'' + a_1(t)(x - x_p)' + a_0(t)(x - x_p) = 0.$$

Thus, the quantity

$$x_h(t) = x(t) - x_p(t) \tag{7.6}$$

satisfies the associated homogeneous equation

$$a_2(t)x_h'' + a_1(t)x_h' + a_0(t)x_h = 0.$$

General solution

If we rewrite (7.6) in the form

$$x(t) = x_p(t) + x_h(t),$$

we see that the general solution of (7.5), $x(t)$, is again the sum of $x_p(t)$, a particular solution of the nonhomogeneous equation, and $x_h(t)$, the general solution of the associated homogeneous differential equation.

How to Solve Nonhomogeneous Linear Differential Equations

Purpose To find the general solution $x = x(t)$ of

$$a_2(t)x'' + a_1(t)x' + a_0(t)x = f(t). \tag{7.7}$$

Process

1. Find the general solution $x = x_h(t)$ of the associated homogeneous equation

$$a_2(t)x'' + a_1(t)x' + a_0(t)x = 0. \tag{7.8}$$

At present we have only one technique for solving (7.8), when the coefficients are all constants — see *How to Solve Homogeneous Second Order Linear Differential Equations with Constant Coefficients* on page 244. In

Chapter 8 we will introduce other techniques that will apply to specific cases where the coefficients are not constants.

2. Find any solution $x = x_p(t)$ of the nonhomogeneous equation (7.7).

3. The general solution of (7.7) is

$$x(t) = x_h(t) + x_p(t). \tag{7.9}$$

Comments about Nonhomogeneous Linear Differential Equations

- Chapter 6 contained techniques for solving homogeneous equations, so finding $x_h(t)$ will depend on applying those methods successfully.
- Any solution $x = x_p(t)$ of (7.7) is called a particular solution of (7.7). The differential equation (7.7) has many particular solutions. To use (7.9), all we need is one particular solution.
- It does not matter how we find a particular solution. We might find it by guesswork, luck, or insight, or perhaps by using a systematic approach.
- At present we have no general method for finding a particular solution. The next section will be devoted to this topic, as will Sections 8.2 and 8.3.
- A particular solution $x = x_p(t)$ of (7.7) cannot be a solution of the associated homogeneous equation. (Why?)
- The general solution $x = x_h(t)$ of the associated homogeneous equation (7.8) is called the COMPLEMENTARY FUNCTION of (7.7).

EXERCISES

1. In each of the following, check that the given $x_p(t)$ is a particular solution of the differential equation, and then find the general solution of the differential equation.

   (a) $x'' - x = t^2$ $\quad x_p(t) = -2 - t^2$
   (b) $x'' + 3x' + 2x = 20e^{3t}$ $\quad x_p(t) = e^{3t}$
   (c) $x'' + 3x' + 2x = e^{-t}$ $\quad x_p(t) = te^{-t}$
   (d) $x'' + 2x' + 5x = 4e^{-t}\cos 2t$ $\quad x_p(t) = te^{-t}\sin 2t$
   (e) $x'' - 4x' + 4x = 12te^{2t}$ $\quad x_p(t) = 2t^3e^{2t}$
   (f) $x'' + x' - 2x = 4\sin 2t$
   $x_p(t) = -(\cos 2t + 3\sin 2t)/5$

2. Consider the differential equation $x'' = t + 1$.

   (a) Solve the differential equation $x'' = t + 1$ by integrating twice with respect to t.
   (b) Find constants a, b, c, and d, so that $at^3 + bt^2 + ct + d$ is a particular solution of the differential equation $x'' = t + 1$. Use this information to find the general solution of $x'' = t + 1$.
   (c) Reconcile the solutions obtained in parts (a) and (b).

3. Show that the two functions t and $t + 2$ are both particular solutions of $x'' + x' = 1$.

   (a) Find the general solution of $x'' + x' = 1$ using t as the particular solution.
   (b) Find the general solution of $x'' + x' = 1$ using $t + 2$ as the particular solution.
   (c) Reconcile the solutions obtained in parts (a) and (b).

4. On the same graph, plot e^t, the forcing function in Example 7.1, and the explicit solution from (7.3) with $C_1 = 1$ and $C_2 = -1/2$. Use the window $0 < t < 5$, $0 < x < 100$. What do you observe? Often the forcing function is called the input to the differential equation

and the explicit solution the output. Comment on the long-term behavior (as $t \to \infty$) of the ratio of the output to the input. Is this consistent with your earlier observation?

5. Repeat Exercise 4 for parts (b), (c), and (e) of Exercise 1 using $C_1 = C_2 = 1$ in each general solution. Also use the following windows: (b) $0 < t < 3$, $0 < x < 900$; (c) $0 < t < 6$, $0 < x < 2$; (e) $0 < t < 3$, $0 < x < 4000$.

7.2 FINDING SOLUTIONS BY THE METHOD OF UNDETERMINED COEFFICIENTS

In this section we introduce a method — called the method of undetermined coefficients — that is used to find particular solutions for a very restricted, but important, class of nonhomogeneous second order linear differential equations. This class requires that a_2, a_1, and a_0 are constants in $a_2x'' + a_1x' + a_0x = f(t)$, and that the forcing function, $f(t)$, contains terms involving only

- exponential functions,
- polynomials,
- sine or cosine functions, or
- products and sums of these three types of functions.

If we were to write down a number of randomly generated nonhomogeneous second order linear differential equations, very few, if any, would fall into this restricted class. In spite of this, many differential equations used in applications and models do fall into this class, as we will see in the next section. In addition, in Chapter 8 we will introduce a technique for finding particular solutions — called variation of parameters — that applies to all nonhomogeneous second order linear differential equations. However, that method is considerably more work-intensive than the method of undetermined coefficients. When we deal with variation of parameters, we will redo one of the examples from this section to illustrate why the method of undetermined coefficients — when it applies — is usually preferable to variation of parameters. In Chapter 11 we introduce a third technique, which uses Laplace transforms. That technique also applies to nonhomogeneous second order linear differential equations with constant coefficients, and is particularly useful if the forcing function is discontinuous or periodic.

Motivation

To introduce the method of undetermined coefficients, we return to the previous example.

EXAMPLE 7.2

Solve $x'' + x' = e^t$.

Our problem is to find any particular solution, $x_p(t)$. A perfectly valid way of doing this is to guess. What we want is a function $x_p(t)$ that when differentiated twice — to find $x_p'(t)$ and $x_p''(t)$ — and when substituted into the left-hand side of

$x'' + x' = e^t$ — to find $x_p'' + x_p'$ — reduces to e^t, the right-hand side. A moment's thought, together with the knowledge that $(e^t)' = e^t$, leads us to guess Ae^t, where A is a constant to be determined. So let's use $x_p(t) = Ae^t$ as a trial solution. In this case, $x_p'' + x_p' = Ae^t + Ae^t = 2Ae^t$. Thus, if we choose A so that $2Ae^t = e^t$, then $x_p(t) = Ae^t$ will be a particular solution. This gives $x_p(t) = \frac{1}{2}e^t$ as a particular solution.

Trial solution

To complete the solution, we solve the associated homogeneous equation $x'' + x' = 0$. We noted previously that it has the characteristic equation $r^2 + r = 0$, roots $r_1 = 0$ and $r_2 = -1$, and solution $x_h(t) = C_1 + C_2e^{-t}$. This gives the general solution $x(t) = x_h(t) + x_p(t) = C_1 + C_2e^{-t} + \frac{1}{2}e^t$, in agreement with (7.3).

General solution

We try this same approach on another example.

EXAMPLE 7.3

Solve $x'' + x' = e^{-t}$.

Again let's look for a particular solution. The guess $x_p(t) = Ae^{-t}$ seems appropriate. In this case, $x_p'' + x_p' = Ae^{-t} - Ae^{-t} = 0$. But no choice of A will make $0 = e^{-t}$. So here Ae^{-t} is a poor guess. (How is this related to the fifth comment on page 282?)

In order to understand why the first guess was a good one — and the second guess a poor one — let's look at the general nonhomogeneous differential equation

$$a_2x'' + a_1x' + a_0x = e^{ct}, \tag{7.10}$$

where $a_2 \neq 0$, a_1, a_0, and c are given constants. As our trial solution, we guess $x_p(t) = Ae^{ct}$, where A is a constant to be determined. In this case,

$$a_2x_p'' + a_1x_p' + a_0x_p = a_2c^2Ae^{ct} + a_1cAe^{ct} + a_0Ae^{ct} = A\left(a_2c^2 + a_1c + a_0\right)e^{ct}.$$

Thus, if we can choose A so that $A\left(a_2c^2 + a_1c + a_0\right)e^{ct} = e^{ct}$, then $x_p(t) = Ae^{ct}$ will be a particular solution. To find A we must solve $A\left(a_2c^2 + a_1c + a_0\right) = 1$ for A, which gives $A = 1/\left(a_2c^2 + a_1c + a_0\right)$, provided $a_2c^2 + a_1c + a_0 \neq 0$. Thus, if $a_2c^2 + a_1c + a_0 \neq 0$, the trial solution $x_p(t) = Ae^{ct}$ will be a suitable guess for a particular solution of (7.10). This is what happened when we solved $x'' + x' = e^t$.

Trial solution

All seems well, except according to this analysis, this technique will fail if $a_2c^2 + a_1c + a_0 = 0$. Could this ever happen? The equation $a_2c^2 + a_1c + a_0 = 0$ looks familiar. Of course! It is the characteristic equation of the associated homogeneous differential equation $a_2x'' + a_1x' + a_0x = 0$. Thus, if the constant c in the forcing function e^{ct} satisfies the characteristic equation, then we will not be able to determine A, and the guess $x_p(t) = Ae^{ct}$ will fail. We can reword this statement in the following way: **If the trial solution Ae^{ct} is a solution of the associated homogeneous equation, then this trial solution will fail.**

This is what happened when we tried to solve $x'' + x' = e^{-t}$. The associated homogeneous equation is $x'' + x' = 0$ with characteristic equation $r^2 + r = 0$, roots $r_1 = 0$ and $r_2 = -1$, and solution $x_h(t) = C_1 + C_2e^{-t}$. The guess Ae^{-t} will fail because it is a solution of the homogeneous equation [in other words, it is contained in $x_h(t) = C_1 + C_2e^{-t}$ with $C_1 = 0$ and $C_2 = A$].

Is there a guess we could make in the case $a_2c^2 + a_1c + a_0 = 0$? In order to suggest what we should do, let's look at the solution of $x'' + x' = e^{-t}$, which we can obtain using the same technique that we used to solve (7.2). In this case we find $x(t) = C_1 + C_2e^{-t} - te^{-t}$. (See Exercise 2 on page 292). Notice that the particular solution has the form Ate^{-t} instead of Ae^{-t}. This suggests that if $a_2c^2 + a_1c + a_0 = 0$, we try multiplying the trial solution that did not work, Ae^{ct}, by t, and using this as a new trial solution. So if c satisfies $a_2c^2 + a_1c + a_0 = 0$, we try $x_p(t) = Ate^{ct}$. In this case,

$$a_2x_p'' + a_1x_p' + a_0x_p = a_2\left(2cAe^{ct} + Ac^2te^{ct}\right) + a_1\left(Ae^{ct} + Acte^{ct}\right) + a_0Ate^{ct},$$

or

$$a_2x_p'' + a_1x_p' + a_0x_p = A\left(a_2c^2 + a_1c + a_0\right)te^{ct} + A\left(2ca_2 + a_1\right)e^{ct}. \tag{7.11}$$

The coefficient of te^{ct} is zero (why?), so we find $a_2x_p'' + a_1x_p' + a_0x_p = A\left(2ca_2 + a_1\right)e^{ct}$. Thus, if we can choose A so that $A\left(2ca_2 + a_1\right)e^{ct} = e^{ct}$, then $x_p(t) = Ate^{ct}$ will be a particular solution. To find A we must solve $A\left(2ca_2 + a_1\right) = 1$ for A, which gives $A = 1/\left(2ca_2 + a_1\right)$, provided $2ca_2 + a_1 \neq 0$. Thus, if $a_2c^2 + a_1c + a_0 = 0$ and $2ca_2 + a_1 \neq 0$, the trial solution $x_p(t) = Ate^{ct}$ will be a suitable guess for a particular solution of (7.10).

EXAMPLE 7.4

Solve $x'' + x' = e^{-t}$.

The associated homogeneous equation is $x'' + x' = 0$ with solution $x_h(t) = C_1 + C_2e^{-t}$. As we have seen, our first guess Ae^{-t} will fail because it is contained in $x_h(t) = C_1 + C_2e^{-t}$ (with $C_1 = 0$ and $C_2 = A$). Thus, we multiply our original guess, Ae^{-t}, by t, giving a new trial solution, $x_p(t) = Ate^{-t}$.

Trial solution

In this case, $x_p'' + x_p' = \left(-2Ae^{-t} + Ate^{-t}\right) + \left(Ae^{-t} - Ate^{-t}\right) = -Ae^{-t}$. To make this equal to e^{-t}, we want $-A = 1$, so $A = -1$. This gives a particular solution, $x_p(t) = -te^{-t}$, and the general solution,

General solution

$$x(t) = x_h(t) + x_p(t) = C_1 + C_2e^{-t} - te^{-t},$$

in agreement with the previous result. ■

Is it possible for this technique to fail? Yes, from (7.11) we see that we cannot find A if both $a_2c^2 + a_1c + a_0 = 0$ **and** $2ca_2 + a_1 = 0$. Could this ever happen? Yes, these are exactly the conditions for c to be a double root of the characteristic equation $a_2c^2 + a_1c + a_0 = 0$. In other words, if the solution of the homogeneous equation is $x_h(t) = (C_1 + C_2t)e^{ct}$, then the guess Ate^{ct} will fail. So, **if the trial solution Ate^{ct} is a solution of the homogeneous equation, then this trial solution will fail.**

For example, consider $x'' + 2x' + x = e^{-t}$. Here the solution of the homogeneous equation is $x_h(t) = \left(C_1 + C_2t\right)e^{-t}$. We know the trial solution Ae^{-t} will fail because it is contained in x_h (with $C_1 = A$ and $C_2 = 0$). If we multiply our trial solution by t and try the new trial solution, $x_p(t) = Ate^{-t}$, then we would find $x_p'' + 2x_p' + x_p = \left(-2Ae^{-t} + Ate^{-t}\right) + 2\left(Ae^{-t} - Ate^{-t}\right) + Ate^{-t} = 0$, and no choice of A will make Ate^{-t} a particular solution.

Is there a guess we could make in the case $a_2c^2 + a_1c + a_0 = 0$ and $2ca_2 + a_1 = 0$? We try what worked before — that is, we multiply the guess that did not work, Ate^{ct}, by t — and use this as a new guess. So if c satisfies $a_2c^2 + a_1c + a_0 = 0$ and $2ca_2 + a_1 = 0$, we try $x_p(t) = At^2e^{ct}$. In this case,

$$a_2x_p'' + a_1x_p' + a_0x_p = a_2\left(2Ae^{ct} + 4Acte^{ct} + Ac^2t^2e^{ct}\right) + a_1\left(2Ate^{ct} + Act^2e^{ct}\right) + a_0At^2e^{ct},$$

or

$$a_2x_p'' + a_1x_p' + a_0x_p = A\left(a_2c^2 + a_1c + a_0\right)t^2e^{ct} + 2A\left(2ca_2 + a_1\right)te^{ct} + 2Aa_2e^{ct}.$$

The coefficients of t^2e^{ct} and te^{ct} are both zero in this case, so we find $a_2x_p'' + a_1x_p' + a_0x_p = 2Aa_2e^{ct}$. Thus, if we can choose A so that $2Aa_2e^{ct} = e^{ct}$, then $x_p(t) = At^2e^{ct}$ will be a particular solution. To find A we must solve $2Aa_2 = 1$ for A, which gives $A = 1/\left(2a_2\right)$, because $a_2 \neq 0$. Thus, if $a_2c^2 + a_1c + a_0 = 0$ and $2ca_2 + a_1 = 0$, the trial solution $x_p(t) = At^2e^{ct}$ will be a suitable guess for a particular solution of (7.10).

EXAMPLE 7.5

Solve $x'' + 2x' + x = e^{-t}$.

The associated homogeneous equation is $x'' + 2x' + x = 0$ with characteristic equation $r^2 + 2r + 1 = 0$, repeated roots $r = -1$, and solution $x_h(t) = \left(C_1 + C_2t\right)e^{-t}$. For a particular solution, we first guess Ae^{-t}, but immediately realize that this guess will fail because Ae^{-t} is a solution of the homogeneous equation [it is contained in $x_h(t) = \left(C_1 + C_2t\right)e^{-t}$ with $C_1 = A$ and $C_2 = 0$]. Thus, we multiply our guess, Ae^{-t}, by t, giving a new guess, Ate^{-t}. This too will fail because Ate^{-t} is a solution of the homogeneous equation [it is contained in $x_h(t) = \left(C_1 + C_2t\right)e^{-t}$ with $C_1 = 0$ and $C_2 = A$]. Thus, we multiply our latest guess, Ate^{-t}, by t, giving a new trial solution, $x_p(t) = At^2e^{-t}$. This guess is not contained in $x_h(t) = \left(C_1 + C_2t\right)e^{-t}$, so we use it.

Trial solution

In this case, $x_p'' + 2x_p' + x_p = \left(2Ae^{-t} - 4Ate^{-t} + At^2e^{-t}\right) + 2\left(2Ate^{-t} - At^2e^{-t}\right) + At^2e^{-t} = 2Ae^{-t}$. To make this equal to e^{-t}, we want $2A = 1$, so $A = \frac{1}{2}$. This gives a particular solution, $x_p(t) = \frac{1}{2}t^2e^{-t}$, and the general solution,

General solution

$$x(t) = x_h(t) + x_p(t) = (C_1 + C_2t)e^{-t} + \frac{1}{2}t^2e^{-t}.$$

This trial solution method — known as the METHOD OF UNDETERMINED COEFFICIENTS — can be used if the forcing function is

- an exponential function ae^{ct}.
- a polynomial of degree n, so that $a_2x'' + a_1x' + a_0x = b_0 + b_1t + b_2t^2 + \cdots + b_nt^n$, where $b_0, b_1, b_2, \cdots, b_n$ are given constants.

- of the form $a \sin wt + b \cos wt$, so that $a_2 x'' + a_1 x' + a_0 x = a \sin wt + b \cos wt$, where a, b, and w are given constants.
- a **product** of any of the three previous special types — namely, e^{ct}, $b_0 + b_1 t + b_2 t^2 + \cdots + b_n t^n$, and $a \sin wt + b \cos wt$.

In each case, we choose the trial solution with the same structure as the forcing function. However, if the trial solution corresponding to a specific term is contained in the solution of the homogeneous equation, $x_h(t)$, we keep multiplying that trial solution by t until the trial solution no longer has any terms that occur in $x_h(t)$. This process does come to an end!

We have summarized this information in Table 7.1. **If any part of the suggested trial solution in Table 7.1 satisfies the associated homogeneous differential equation, then we use t times the usual function indicated on the right-hand side of Table 7.1.**

Table 7.1 **Particular solutions of $a_2 x'' + a_1 x' + a_0 x = f(t)$**

$f(t)$	**Trial Solution**
ae^{ct}	Ae^{ct}
$b_0 + b_1 t + b_2 t^2 + \cdots + b_n t^n$	$B_0 + B_1 t + B_2 t^2 + \cdots + B_n t^n$
$a \sin wt + b \cos wt$	$A \sin wt + B \cos wt$
$e^{ct}(a \sin wt + b \cos wt)$	$e^{ct}(A \sin wt + B \cos wt)$
$e^{ct} \sum_{k=0}^{k=n} b_k t^k$	$e^{ct} \sum_{k=0}^{k=n} B_k t^k$
$\sum_{k=0}^{k=n} b_k t^k (a \sin wt + b \cos wt)$	$\sum_{k=0}^{k=n} A_k t^k \sin wt + \sum_{k=0}^{k=n} B_k t^k \cos wt$
$e^{ct} \sum_{k=0}^{k=n} b_k t^k (a \sin wt + b \cos wt)$	$e^{ct}(\sum_{k=0}^{k=n} A_k t^k \sin wt + \sum_{k=0}^{k=n} B_k t^k \cos wt)$

Five Examples

We demonstrate this method with a number of examples.

EXAMPLE 7.6

Solve $x'' + 16x = 16t^2$.

The solution of the associated homogeneous equation is $x_h(t) = C_1 \sin 4t + C_2 \cos 4t$. We now look for a particular solution. We recognize that the forcing function, t^2, is a polynomial of degree 2, so we make a similar guess for our trial solution; namely, $x_p(t) = A + Bt + Ct^2$, where A, B, and C are constants to be determined. This trial solution is not contained in $x_h(t)$, so we use it. Differentiating gives $x_p' = B + 2Ct$, and $x_p'' = 2C$. Substituting these expressions into $x'' + 16x = 16t^2$ gives $2C + 16\left(A + Bt + Ct^2\right) = 16t^2$, which we rewrite in the form

Trial solution

$$(2C + 16A) + 16Bt + 16Ct^2 = 16t^2.$$

Because we want this to be an identity (that is, be true for all values of t), we may equate the coefficients of the powers of t to obtain the set of algebraic equations

$2C + 16A = 0$, $16B = 0$, $16C = 16$, with solution $A = -1/8$, $B = 0$, $C = 1$. Thus, a particular solution is $x_p(t) = -1/8 + t^2$, and the general solution is

General solution

$$x(t) = C_1 \sin 4t + C_2 \cos 4t - \frac{1}{8} + t^2.$$

EXAMPLE 7.7

Solve $x'' + 5x' + 6x = 52 \sin 2t$.

The solution of the associated homogeneous equation is $x_h(t) = C_1 e^{-3t} + C_2 e^{-2t}$. We now look for a particular solution. Because the forcing function is of the form $a \sin 2t + b \cos 2t$ (with $a = 52$ and $b = 0$), we choose the trial solution of the same form, $x_p(t) = A \sin 2t + B \cos 2t$, where A and B are constants to be determined. This trial solution is not contained in $x_h(t)$ so we use it. Differentiating gives $x_p' = 2A \cos 2t - 2B \sin 2t$, and $x_p'' = -4A \sin 2t - 4B \cos 2t$. Substituting these expressions into $x'' + 5x' + 6x = 52 \sin 2t$ gives

Trial solution

$$(2A - 10B) \sin 2t + (10A + 2B) \cos 2t = 52 \sin 2t.$$

We equate the coefficients of $\sin 2t$ and $\cos 2t$ to obtain $2A - 10B = 52$, $10A + 2B = 0$, with solution $A = 1$, $B = -5$. Thus, a particular solution is $x_p(t) = \sin 2t - 5 \cos 2t$, and the general solution is

General solution

$$x(t) = C_1 e^{-3t} + C_2 e^{-2t} + \sin 2t - 5 \cos 2t.$$

EXAMPLE 7.8

Solve the initial value problem $x'' + 16x = 16 \sin 4t$, $x(0) = 0$, $x'(0) = 0$.

Before applying the initial conditions, we must first find the general solution of the differential equation.

The solution of the associated homogeneous equation is $x_h(t) = C_1 \sin 4t + C_2 \cos 4t$. We now look for a particular solution. The forcing function, $16 \sin 4t$, suggests that we make the guess $A \sin 4t + B \cos 4t$ for our trial solution (where A and B are constants to be determined). However, this trial solution is contained in $x_h(t)$ (with $A = C_1$ and $B = C_2$), so we multiply our guess by t to find a new trial solution; namely, $x_p(t) = t\,(A \sin 4t + B \cos 4t)$. This guess is not contained in x_h, so we use it. Differentiating gives $x_p' = (A \sin 4t + B \cos 4t) + t\,(4A \cos 4t - 4B \sin 4t)$, and $x_p'' = 2\,(4A \cos 4t - 4B \sin 4t) + t\,(-16A \sin 4t - 16B \cos 4t)$. Substituting these expressions into $x'' + 16x = 16 \sin 4t$ gives

Trial solution

$$[2\,(4A \cos 4t - 4B \sin 4t) + t\,(-16A \sin 4t - 16B \cos 4t)] + 16t\,(A \sin 4t + B \cos 4t) \\ = 16 \sin 4t.$$

We rewrite this in the form

$$8A \cos 4t - 8B \sin 4t + (-16A + 16A)\, t \sin 4t + (-16B + 16B)\, t \cos 4t = 16 \sin 4t.$$

Notice that the terms involving $t \sin 4t$ and $t \cos 4t$ vanish, and we are left with $8A = 0$, $-8B = 16$, so $A = 0$ and $B = -2$. Thus, a particular solution is $x_p(t) = -2t \cos 4t$, and the general solution is

General solution

$$x(t) = C_1 \sin 4t + C_2 \cos 4t - 2t \cos 4t.$$

Initial conditions

In order to apply the initial conditions $x(0) = 0$, $x'(0) = 0$, we require $x'(t)$, which, from the general solution, is $x'(t) = 4C_1 \cos 4t - 4C_2 \sin 4t - 2 \cos 4t + 8t \sin 4t$. So $x(0) = C_2 = 0$, and $x'(0) = 4C_1 - 2 = 0$, which gives $C_1 = \frac{1}{2}$ and $C_2 = 0$. Thus, the solution of the initial value problem is

$$x(t) = \frac{1}{2} \sin 4t - 2t \cos 4t.$$

EXAMPLE 7.9

Solve $x'' + 4x' - 5x = 32te^{-t}$.

The complementary function is $x_h(t) = C_1 e^{-5t} + C_2 e^{t}$. We now look for a particular solution. The forcing function, $32te^{-t}$, is the product of a polynomial of degree one, $32t$, and an exponential function, e^{-t}. This suggests that we make a guess that is the product of the corresponding guesses. The guess for the polynomial of degree one is $A + Bt$. The guess for the exponential function is Ce^{-t}. So the guess for the forcing function, $32te^{-t}$, is $(A + Bt)\,Ce^{-t}$. Although we could use this, we should notice that it can be rewritten as $(AC + BCt)\,e^{-t}$, so we would expect to determine the two quantities AC and BC rather than the three quantities A, B, and C. Consequently, all we need for our trial solution is $x_p(t) = (A + Bt)\,e^{-t}$, where we have absorbed C into A and B.

Trial solution

This trial solution, $x_p(t) = (A + Bt)\,e^{-t}$, is not contained in $x_h(t)$, so we use it. Differentiating gives $x_p' = Be^{-t} - (A + Bt)\,e^{-t} = [(B - A) - Bt]\,e^{-t}$, and $x_p'' = -Be^{-t} - [(B - A) - Bt]\,e^{-t} = [(A - 2B) + Bt]\,e^{-t}$. Substituting these expressions into $x'' + 4x' - 5x = 32te^{-t}$ gives

$$[(A - 2B) + Bt]\,e^{-t} + 4\,[(B - A) - Bt]\,e^{-t} - 5\,(A + Bt)\,e^{-t} = 32te^{-t}.$$

We rewrite this in the form

$$[(2B - 8A) - 8Bt]\,e^{-t} = 32te^{-t},$$

so we have $2B - 8A = 0$ and $-8B = 32$, giving $A = -1$ and $B = -4$. Thus, a particular solution is $x_p(t) = (-1 - 4t)\,e^{-t}$ and the general solution is

General solution

$$x(t) = C_1 e^{-5t} + C_2 e^{t} + (-1 - 4t)\,e^{-t}.$$

We now consider a final example.

EXAMPLE 7.10

Solve $x'' + 4x' - 5x = 36te^t$.

The complementary function is again $x_h(t) = C_1e^{-5t} + C_2e^t$. Following the pattern from the previous example, we consider the guess $(A + Bt)\,e^t$. However, this trial solution is contained in $x_h(t)$ (with $C_1 = 0$, $C_2 = A$, and $B = 0$), so we multiply it by t to find $x_p(t) = t\,(A + Bt)\,e^t = \left(At + Bt^2\right)e^t$. This trial solution is not contained in $x_h(t)$, so we use it. Differentiating gives $x_p' = (A + 2Bt)\,e^t + \left(At + Bt^2\right)e^t = \left[A + (A + 2B)\,t + Bt^2\right]e^t$, and $x_p'' = [(A + 2B) + 2Bt]\,e^t + \left[A + (A + 2B)\,t + Bt^2\right]e^t = \left[(2A + 2B) + (A + 4B)\,t + Bt^2\right]e^t$. Substituting these expressions into $x'' + 4x' - 5x = 36te^t$ gives

Trial solution

$$\left[(2A + 2B) + (A + 4B)\,t + Bt^2\right]e^t + 4\left[A + (A + 2B)\,t + Bt^2\right]e^t + \\ -5\left(At + Bt^2\right)e^t = 36te^t.$$

We rewrite this in the form

$$[(6A + 2B) + 12Bt]\,te^t = 36te^t,$$

so we have $6A + 2B = 0$ and $12B = 36$, giving $A = -1$ and $B = 3$. Thus, a particular solution is $x_p(t) = \left(-t + 3t^2\right)e^t$, and the general solution is

General solution

$$x(t) = C_1e^{-5t} + C_2e^t + \left(-t + 3t^2\right)e^t.$$

The Principle of Linear Superposition

The previous analysis allows us to deal with forcing functions that are products of exponentials, polynomials, and the sine or cosine function. What do we do if we have sums instead of products? For example, how would we solve $x'' + 16x = 16t^2 + 16\sin 4t$?

If we look back at the previous examples, we see that the solution of $x'' + 16x = 16t^2$ is $x(t) = C_1\sin 4t + C_2\cos 4t - 1/8 + t^2$, and the solution of $x'' + 16x = 16\sin 4t$ is $x(t) = K_1\sin 4t + K_2\cos 4t - 2t\cos 4t$. (Here C_1, C_2, K_1, and K_2 are arbitrary constants.) Based on this, we might guess that the solution of $x'' + 16x = 16t^2 + 16\sin 4t$ is $x(t) = C_1\sin 4t + C_2\cos 4t - 1/8 + t^2 + K_1\sin 4t + K_2\cos 4t - 2t\cos 4t$, which we would normally write as $x(t) = C_1\sin 4t + C_2\cos 4t - 1/8 + t^2 - 2t\cos 4t$, by absorbing K_1 into C_1 and K_2 into C_2. We now show why this guess works in this case.

Suppose we have solutions of a differential equation with two different forcing functions, $f_1(t)$ and $f_2(t)$, but the same associated homogeneous equation — that is, $x_1(t)$ satisfies $a_2(t)x_1'' + a_1(t)x_1' + a_0(t)x_1 = f_1(t)$, and $x_2(t)$ satisfies $a_2(t)x_2'' + a_1(t)x_2' + a_0(t)x_2 = f_2(t)$. Adding these two equations and using properties of derivatives gives

$$a_2(t)\left(x_1 + x_2\right)'' + a_1(t)\left(x_1 + x_2\right)' + a_0(t)\left(x_1 + x_2\right) = f_1(t) + f_2(t).$$

Thus, the sum of these two solutions, $x(t) = x_1(t) + x_2(t)$, satisfies the linear differential equation $a_2(t)x'' + a_1(t)x' + a_0(t)x = f_1(t) + f_2(t)$.

A related property of linear differential equations is that if $x_1(t)$ satisfies $a_2(t)x_1'' + a_1(t)x_1' + a_0(t)x_1 = f(t)$, then $x(t) = ax_1(t)$, where a is a constant, satisfies the linear differential equation $a_2(t)x'' + a_1(t)x' + a_0(t)x = af(t)$. (See Exercise 17 on page 293).

These two results can be combined, and form the PRINCIPLE OF LINEAR SUPERPOSITION. (See Exercise 18 on page 293.)

▶ **Theorem 7.1: Principle of Linear Superposition** *If $x_1(t)$ and $x_2(t)$ are solutions of $a_2(t)x'' + a_1(t)x' + a_0(t)x = f_1(t)$ and $a_2(t)x'' + a_1(t)x' + a_0(t)x = f_2(t)$, respectively, then $x(t) = ax_1(t) + bx_2(t)$ is a solution of $a_2(t)x'' + a_1(t)x' + a_0(t)x = af_1(t) + bf_2(t)$, where a and b are specified constants.* ◀

Comments about Linear Superposition

- It is important that the left-hand sides of the linear differential equations satisfied by $x_1(t)$ and $x_2(t)$ be exactly the same — namely, $a_2(t)x'' + a_1(t)x' + a_0(t)x$.
- To solve $a_2(t)x'' + a_1(t)x' + a_0(t)x = af_1(t) + bf_2(t)$, first solve $a_2(t)x'' + a_1(t)x' + a_0(t)x = f_1(t)$ for $x_1(t)$, and then solve $a_2(t)x'' + a_1(t)x' + a_0(t)x = f_2(t)$ for $x_2(t)$. The solution of the original equation is $ax_1(t) + bx_2(t)$.
- The principle of linear superposition is valid for $a_2(t)$, $a_1(t)$, $a_0(t)$ being functions of t, not just for the case of constant coefficients.
- The principle of linear superposition is not generally true for nonlinear differential equations. See Exercise 21 on page 294.

As an example of how we might use the principle of linear superposition, consider the following.

EXAMPLE 7.11

Solve $x'' + x' = e^t + e^{-t}$.

We can replace this problem by two problems: solve $x'' + x' = e^t$ and $x'' + x' = e^{-t}$, and then add their solutions. We found that the general solution of $x'' + x' = e^t$ was $x(t) = C_1 + C_2e^{-t} + \frac{1}{2}e^t$, and the general solution of $x'' + x' = e^{-t}$ was $x(t) = C_1 + C_2e^{-t} - te^{-t}$. Thus, we may use the principle of linear superposition to write down the general solution of $x'' + x' = e^t + e^{-t}$, as $x(t) = C_1 + C_2e^{-t} + \frac{1}{2}e^t - te^{-t}$.

General solution

■

There is an important point to make about this example. You might be tempted to look at the forcing function, $e^t + e^{-t}$, and try a guess of $Ae^t + Be^{-t}$ for the trial solution. This guess will not work because it will miss the term $-te^{-t}$ in the general solution. You might argue that because the guess $Ae^t + Be^{-t}$ is contained in the complementary function $x_h(t) = C_1 + C_2e^{-t}$ (true, $A = 0, B = C_2, C_1 = 0$), the guess should be multiplied by t, giving a new guess, $Ate^t + Bte^{-t}$. False; this guess will not work because it will miss the term e^t in the general solution. The correct trial guess is $Ae^t + Bte^{-t}$. Because of the principle of linear superposition, **when forcing functions consist of functions from different rows in Table 7.1 added together, we must make an appropriate trial guess for each of those functions separately.**

Important point

Comments about the Method of Undetermined Coefficients

- The general solution of a nonhomogeneous second order linear differential equation may be written as the sum of a particular solution and the complementary function (the general solution of the associated homogeneous differential equation).
- We can choose the appropriate form for a particular solution of a linear differential equation with constant coefficients if the forcing function is in the form of an exponential, polynomial, or sinusoidal function, or their products. These forms are shown in Table 7.1 on page 287. In this table the constants $b_0, b_1, b_2, \cdots, b_n$, a, b, c, and w occur in the forcing function for the differential equation, and the constants that appear in the trial solution are to be determined so the differential equation is satisfied.
- If the forcing function is the sum of terms from different rows in Table 7.1, treat each of these terms separately, and use the principle of linear superposition to write down the explicit solution.
- If the trial solution satisfies the associated homogeneous differential equation, then use t times the usual function on the right-hand side of Table 7.1. If, after multiplying by t, part of the resulting trial solution appears in the complementary function, then multiply the latest trial solution by t to create a new trial solution.
- If the forcing function is not listed in Table 7.1, this method fails. For example, consider the linear differential equation with constant coefficients, $x'' + x = \cot t$, which has the general solution $x(t) = C_1 \sin t + C_2 \cos t + \sin t \ln |\csc t - \cot t|$. (See Exercise 15 on page 293.) There are no simple extensions of the trial solutions in Table 7.1 to develop an appropriate trial solution for this example.
- This method cannot be generally applied to an equation that has nonconstant coefficients. For example, consider $tx'' = 1$, with the solution $x(t) = C_1 + C_2 t + t \ln |t|$. (See Exercise 16 on page 293.) Again, there is no obvious way of selecting a form for the trial solution, as we could for the forcing functions in Table 7.1.

EXERCISES

1. Solve the following differential equations and compare your answers with those you obtained for Exercise 1 on page 282.

(a) $x'' - x = t^2$ (b) $x'' + 3x' + 2x = 20e^{3t}$
(c) $x'' + 3x' + 2x = e^{-t}$ (d) $x'' + 2x' + 5x = 4e^{-t}\cos 2t$
(e) $x'' - 4x' + 4x = 12te^{2t}$ (f) $x'' + x' - 2x = 4\sin 2t$

2. Solve $x'' + x' = e^{-t}$ by the following steps. First integrate the equation with respect to t. Then solve the resulting first order linear differential equation.

3. Which of the following differential equations have forcing functions that occur in Table 7.1? Find the general solution of those that do.

(a) $x'' + x' - 6x = 28e^{4t}$ (b) $x'' + x' - 6x = 3/t$
(c) $x'' + x' - 6x = 8e^{2t}$ (d) $x'' + x' - 6x = 8\exp(t^2)$
(e) $x'' + x' - 6x = \sin t^2$ (f) $x'' + x' - 6x = te^{2t}$

4. Create a *How to Find a Particular Solution for Second Order Linear Differential Equations with Constant Coefficients*. Add statements under the headings Purpose, Process, and Comments that summarize what you discovered in this section.

5. Find the general solution of the following differential equations.
(a) $x'' + x' - 6x = t$
(b) $x'' + x' - 6x = e^{3t}$
(c) $x'' + x' - 6x = \sin t$
(d) $x'' + 3x' - 4x = t^2 + 1$
(e) $x'' + x' = 6e^{-t}$
(f) $x'' + 4x = \cos 3t$
(g) $x'' + 4x = 10 \sin 3t$
(h) $x'' + 4x = 6 \cos 3t + 20 \sin 3t$
(i) $x'' + 4x = \sin 2t$
(j) $x'' + 4x' + 4x = \cos 2t$
(k) $4x'' - 12x' + 9x = 24te^{3t/2}$
(l) $4x'' - 12x' + 9x = (t + 1)\, e^t$

6. Use your results from Exercise 5 and the Principle of Linear Superposition to write down the general solution of the following differential equations.
(a) $x'' + x' - 6x = 3t + 2e^{3t}$
(b) $x'' + x' - 6x = 5t - 2e^{3t}$
(c) $x'' + 4x = \sin 3t + 2 \sin 2t$
(d) $x'' + 4x = 5 \sin 2t - 5 \sin 3t$

7. Solve the following initial value problems.
(a) $x'' + x' - 12x = 8e^{3t}$
$x(0) = 0 \quad x'(0) = 1$
(b) $x'' + 6x' + 9x = e^{3t}$
$x(0) = 0 \quad x'(0) = 6$
(c) $x'' - 5x' + 6x = 12te^{-t} - 7e^{-t}$
$x(0) = 0 \quad x'(0) = 0$
(d) $x'' + 4x = 8 \sin 2t + 8 \cos 2t$
$x(\pi) = 2\pi \quad x'(\pi) = 2\pi$

8. Consider the three functions $x_1(t)$, $x_2(t)$, and $x_3(t)$ in Figure 7.1. Only one of these could be a solution of the initial value problem $a_2(t)x'' + a_1(t)x' + a_0(t)x = f(t)$, $x(0) = x'(0) = 0$. Which one, and why?

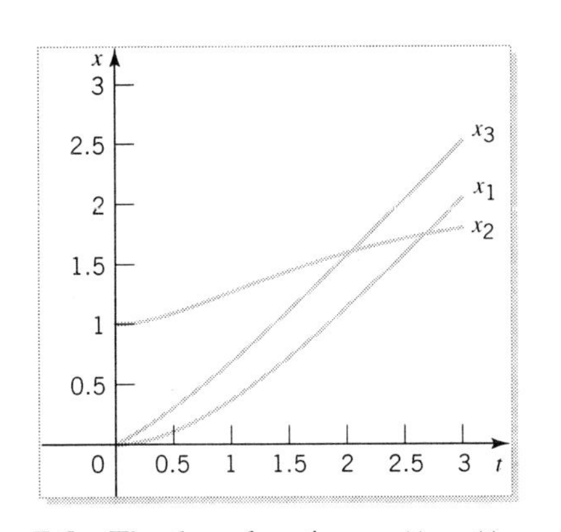

FIGURE 7.1 The three functions $x_1(t)$, $x_2(t)$, and $x_3(t)$

9. Show that if $\omega \neq a$, a particular solution of $x'' + \omega^2 x = a \cos at + b \sin at$ is

$$x_p(t) = \frac{a}{\omega^2 - a^2} \cos at + \frac{b}{\omega^2 - a^2} \sin at.$$

10. Solve $x'' = t^2 + 1$ by
(a) Using the method of undetermined coefficients.
(b) Integrating the differential equation twice.

11. Solve $x'' - x' = t^2$ by
(a) Using the method of undetermined coefficients.
(b) Using the change of variable $y = x'$ and solving the resulting first order equation for y, and then integrating $x' = y$ to find x.

12. Someone has said that for differential equations of the form $x'' + a^2 x = \cos \omega t$, we can find a particular solution from the form $x_p(t) = B \cos \omega t$ instead of from the two expressions given by Table 7.1. Explain why this could be true.

13. Someone has said that for differential equations of the form $x'' + a^2 x = \sin \omega t$, we can find a particular solution from the form $x_p(t) = A \sin \omega t$ instead of from the two expressions given by Table 7.1. Explain why this could be true.

14. Show that the general solution of $x'' + \omega^2 x = a \cos \omega t$ is $x(t) = C_1 \cos \omega t + C_2 \sin \omega t + \frac{a}{2\omega} t \sin \omega t$.

15. Show that $x_p(t) = \sin t \ln |\csc t - \cot t|$ satisfies the differential equation $x'' + x = \cot t$, and find its general solution.

16. Solve the differential equation $tx'' = 1$ by writing it in the form $x'' = t^{-1}$ and integrating twice.

17. Show that if $x_1(t)$ satisfies the differential equation $a_2(t)x'' + a_1(t)x' + a_0(t)x = f(t)$, then $ax_1(t)$ satisfies $a_2(t)x'' + a_1(t)x' + a_0(t)x = af(t)$, where a is a constant.

18. Show that if $x_1(t)$ and $x_2(t)$ are solutions of $a_2(t)x'' + a_1(t)x' + a_0(t)x = f_1(t)$ and $a_2(t)x'' + a_1(t)x' + a_0(t)x = f_2(t)$, respectively, then $ax_1(t) + bx_2(t)$ satisfies $a_2(t)x'' + a_1(t)x' + a_0(t)x = af_1(t) + bf_2(t)$, where a and b are specified constants.

19. If $x_1(t)$ is the solution of the initial value problem $a_2(t)x_1'' + a_1(t)x_1' + a_0(t)x_1 = f_1(t)$, $x_1(t_0) = h_1$, $x_1'(t_0) = k_1$, and $x_2(t)$ is the solution of the initial value problem $a_2(t)x_2'' + a_1(t)x_2' + a_0(t)x_2 = f_2(t)$, $x_2(t_0) = h_2$, $x_2'(t_0) = k_2$, what initial value problem has the solution $x_1(t) + x_2(t)$?

20. In Figure 7.2, $x_1(t)$ is the solution of $a_2(t)x'' + a_1(t)x' + a_0(t)x = 1$, $x(0) = x'(0) = 0$, and $x_2(t)$ is the solution of $a_2(t)x'' + a_1(t)x' + a_0(t)x = e^x$, $x(0) = x'(0) = 0$. Sketch the solution of $a_2(t)x'' + a_1(t)x' + a_0(t)x = 1 + e^x$, $x(0) = x'(0) = 0$.

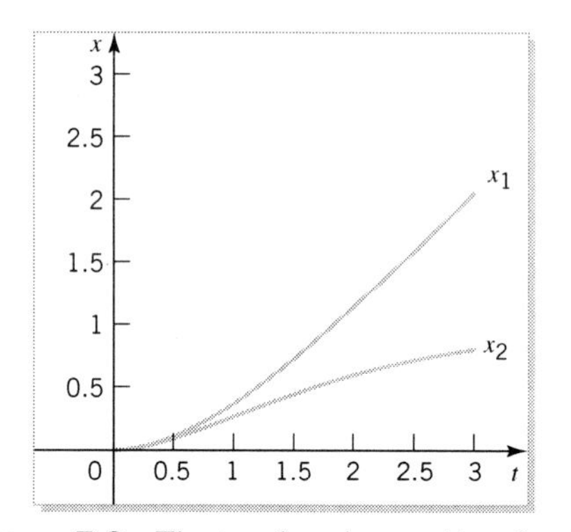

FIGURE 7.2 The two functions $x_1(t)$ and $x_2(t)$

21. Show that e^t and e^{-t} are solutions of $xx'' - 2(x')^2 + x^2 = 0$. Show that the function $x(t) = ae^t + be^{-t}$ is not a solution of $xx'' - 2(x')^2 + x^2 = 0$.

22. Show that the method of undetermined coefficients also applies to first order linear differential equations with constant coefficients by completing the following steps:

(a) Show that the general solution of $x' + kx = 0$ is $x_h(t) = Ce^{-kt}$. Show that this is the same solution you would have obtained from the characteristic equation $r + k = 0$.

(b) Show that if $x_p(t)$ is a particular solution of $x' + kx = f(t)$, then $x(t) = x_h(t) + x_p(t)$ satisfies $x' + kx = f(t)$.

(c) Use an integrating factor to solve $x' + kx = ae^{ct}$. From your solution, note that if $f(t) = ae^{ct}$, then $x_p(t) = Ae^{ct}$, where $A = a/(c + k)$ if $c + k \neq 0$. (This gives entry 1 in Table 7.1.)

(d) Show that if in part (c) we have $c + k = 0$, integration gives $x_p(t) = Ate^{ct}$, where $A = a$. (This gives the reason for multiplying the original trial solution by t if $c = -k$, as noted in the fourth point under *Comments about the Method of Undetermined Coefficients*.)

(e) Use an integrating factor to solve $x' + kx = t$. From your solution, note that if $f(t) = t$, then $x_p(t) = B_0 + B_1 t$, where $B_0 = 1/k$ and $B_1 = -1/k^2$. (This gives an example of entry 2 in Table 7.1 for $n = 1$.)

(f) Use an integrating factor to solve $x' + kx = \sin \omega t$. From your solution, note that if $f(t) = \sin \omega t$, then $x_p(t) = A \sin \omega t + B \cos \omega t$, where $A = k/(\omega^2 + k^2)$, $B = -1/(\omega^2 + k^2)$. (This gives an example of entry 3 in Table 7.1.)

(g) Create a *How to Solve Linear First Order Differential Equations with Constant Coefficients* by adding statements under Purpose, Process, and Comments.

(h) Use your *How to* ... statement in part (g) to solve Exercises 1(a) – (e) in Section 5.1, and Exercises 1(a) – (b) and 2(a) – (c) in Section 5.2.

7.3 APPLICATIONS AND MODELS

We will now apply what we have learned in this chapter to various models.

Series RLC Electrical Circuit

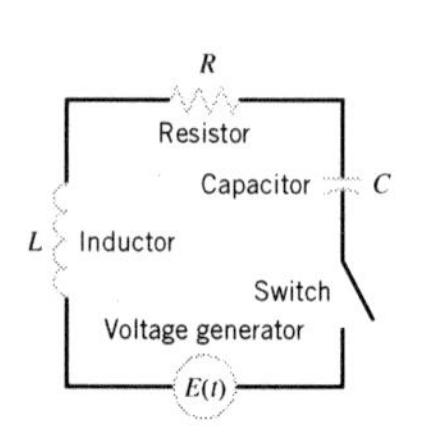

FIGURE 7.3
Series RLC circuit

Consider an electrical circuit in which wires connect a resistor R, capacitor C, inductor L, and switch in series with a voltage generator, as shown is Figure 7.3. If we let the applied voltage be $E(t)$, then the charge, x, in the circuit is modeled by the differential equation

$$Lx'' + Rx' + \frac{1}{C}x = E(t). \tag{7.12}$$

Here x' is the current.

EXAMPLE 7.12

Consider a series RLC circuit with $L = 1$, $R = 5$, $C = 1/6$, and $E(t) = 24e^{-t}$, so that (7.12) is of the form

$$x'' + 5x' + 6x = 24e^{-t}. \tag{7.13}$$

The forcing function in (7.13) models the case in which the voltage applied to the circuit decreases with time in an exponential manner from 24 to 0. If we close the switch when the charge on the capacitor is 4 coulombs and there is no current, we have the initial conditions $x(0) = 4$, $x'(0) = 0$. We want to find $x(t)$.

Following the technique described in the previous section, we first find the solution of the associated homogeneous differential equation given by $x'' + 5x' + 6x = 0$. The characteristic equation is $r^2 + 5r + 6 = 0$, with roots $r = -3$ and $r = -2$. This gives $x_h(t) = C_1e^{-3t} + C_2e^{-2t}$.

Trial solution

From Table 7.1 we have the proper form of the trial solution of (7.13) as $x_p(t) = Ae^{-t}$, with A to be determined so as to satisfy the differential equation (7.13). This trial solution is not contained in $x_h(t)$, so we substitute this expression into (7.13) and obtain $Ae^{-t} - 5Ae^{-t} + 6Ae^{-t} = 24e^{-t}$. Solving for A gives $A = 12$, and our particular solution is $x_p(t) = 12e^{-t}$. Thus, the general solution of our original problem may be written as

General solution

$$x(t) = x_h(t) + x_p(t) = C_1e^{-3t} + C_2e^{-2t} + 12e^{-t}. \tag{7.14}$$

Because we have the initial values $x(0) = 4$, $x'(0) = 0$, we can evaluate the arbitrary constants in this expression (making sure we include the particular solution). Applying these initial conditions to our general solution in (7.14) and its derivative gives

Initial conditions

$$C_1 + C_2 + 12 = 4,$$
$$-3C_1 - 2C_2 - 12 = 0.$$

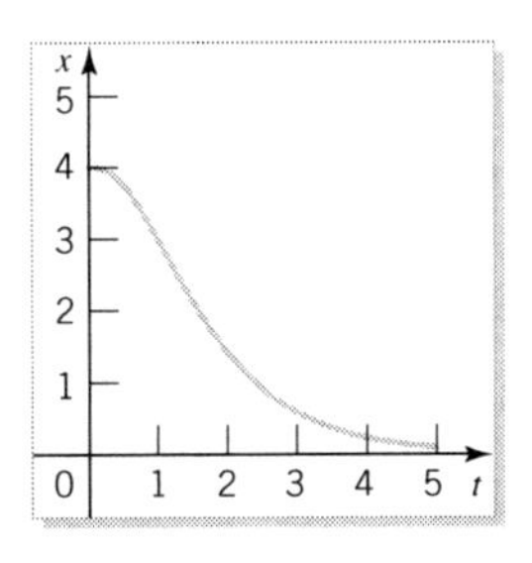

FIGURE 7.4
Graph of $4e^{-3t} - 12e^{-2t} + 12e^{-t}$

Solving this set of algebraic equations gives $C_1 = 4$, $C_2 = -12$, and our solution to this initial value problem is

$$x(t) = 4e^{-3t} - 12e^{-2t} + 12e^{-t}.$$

The graph of this solution is shown in Figure 7.4, in which the decay of the initial charge to 0 is very apparent. ■

EXAMPLE 7.13

Suppose that the applied voltage, in our last example, decayed more rapidly — namely, as $24e^{-2t}$. The appropriate differential equation then becomes

$$x'' + 5x' + 6x = 24e^{-2t}, \tag{7.15}$$

subject to $x(0) = 4$ and $x'(0) = 0$.

Our usual trial solution for this situation would be Ae^{-2t}. However, the solution of the associated homogeneous differential equation, $x'' + 5x' + 6x = 0$, is $x_h(t) = C_1e^{-3t} + C_2e^{-2t}$, and this contains Ae^{-2t} when $C_1 = 0$ and $C_2 = A$. So we multiply this trial solution by t to obtain the new trial solution, $x_p(t) = Ate^{-2t}$. This trial solution is not contained in $x_h(t)$, so we substitute this expression into (7.15), giving

Trial solution

$$A(-4e^{-2t} + 4te^{-2t}) + 5A(e^{-2t} - 2te^{-2t}) + 6Ate^{-2t} = 24e^{-2t}.$$

We solve this equation for A, finding $A = 24$. This allows us to write the general solution of (7.15) as

General solution

$$x(t) = C_1e^{-3t} + C_2e^{-2t} + 24te^{-2t}.$$

Initial conditions

From the initial conditions, $x(0) = 4, x'(0) = 0$, we find $C_1 = 16$ and $C_2 = -12$, so in this case the solution is

$$x(t) = 16e^{-3t} - 12e^{-2t} + 24te^{-2t}.$$

The graph of this solution is shown in Figure 7.5 (solid line), in which the decay of the charge to 0 is again apparent and more rapid than in the previous example (dotted line).

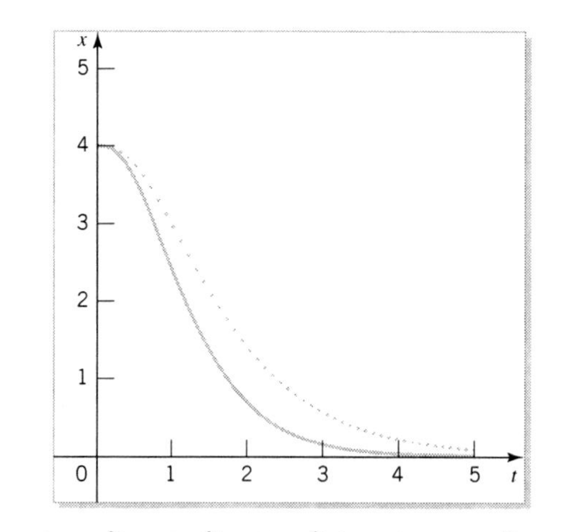

FIGURE 7.5 Graphs of $16e^{-3t} - 12e^{-2t} + 24te^{-2t}$ (solid) and $4e^{-3t} - 12e^{-2t} + 12e^{-t}$ (dotted)

EXAMPLE 7.14

In this example, we consider the case in which the applied voltage in the preceding circuit has both a component that decreases exponentially and one that is sinusoidal, giving the differential equation

$$x'' + 5x' + 6x = 24e^{-3t} + 52\sin 2t, \tag{7.16}$$

subject to $x(0) = 4, x'(0) = 0$.

In the previous section we discovered that the principle of linear superposition allowed us to find particular solutions for each term in the forcing function and then add the two solutions. From our previous examples we have $x_h(t) = C_1e^{-3t} + C_2e^{-2t}$. Using Table 7.1, we choose trial solutions as A_1te^{-3t} and $A\sin 2t + B\cos 2t$. We could substitute each of these two expressions into (7.16) separately, or substitute them in as a sum, as we do now. [Be sure to note that we considered each of the terms on the right-hand side of (7.16) separately, and only the part of the trial solution that satisfies the associated homogeneous differential equation was multiplied by t.]

Trial solution

Setting $x_p(t) = A_1te^{-3t} + A\sin 2t + B\cos 2t$ and differentiating gives $x_p' = A_1(-3t+1)e^{-3t} + 2A\cos 2t - 2B\sin 2t$, $x_p'' = A_1(9t-6)e^{-3t} - 4A\sin 2t - 4B\cos 2t$. Substituting these three expressions into (7.16) gives

$$-A_1e^{-3t} + (2A - 10B)\sin 2t + (10A + 2B)\cos 2t = 24e^{-3t} + 52\sin 2t.$$

Because we want this to be an identity, we equate coefficients of like terms to obtain the set of algebraic equations,

$$\begin{aligned} -A_1 &= 24, \\ 2A - 10B &= 52, \\ 10A + 2B &= 0, \end{aligned}$$

General solution

with solution $A = 1$, $B = -5$, $A_1 = -24$. This gives the general solution of (7.16) as

$$x(t) = C_1e^{-3t} + C_2e^{-2t} - 24te^{-3t} + \sin 2t - 5\cos 2t.$$

Initial conditions

Using the initial conditions $x(0) = 4$, $x'(0) = 0$, gives $C_1 = -40$ and $C_2 = 49$, and the solution of (7.16) becomes

$$x(t) = -40e^{-3t} + 49e^{-2t} - 24te^{-3t} + \sin 2t - 5\cos 2t. \tag{7.17}$$

From the form of (7.17) we see that the first three terms contain exponential functions with negative arguments. Thus, as time increases they will approach 0, whereas the two periodic terms $\sin 2t - 5\cos 2t$ persist. As we noted when dealing with first order linear equations in Chapter 5, the terms that persist are called the **steady state** part of the solution, and the terms that approach zero are called the **transient part** of the solution. The graphs of $x(t)$ and the function $\sin 2t - 5\cos 2t$ are shown in Figure 7.6. Notice that after about one complete cycle, the motion appears periodic and indistinguishable from $\sin 2t - 5\cos 2t$. Thus, the solution approaches the steady state $\sin 2t - 5\cos 2t$. Because $\sin 2t - 5\cos 2t = \sqrt{26}\cos(2t + \phi)$, where $\tan\phi = -1/5$ (see Exercise 6 on page 249), this is a sinusoidal function with amplitude $\sqrt{26}$ and period π. Notice that the steady state output has the same period as the "steady state" input, but different amplitude.

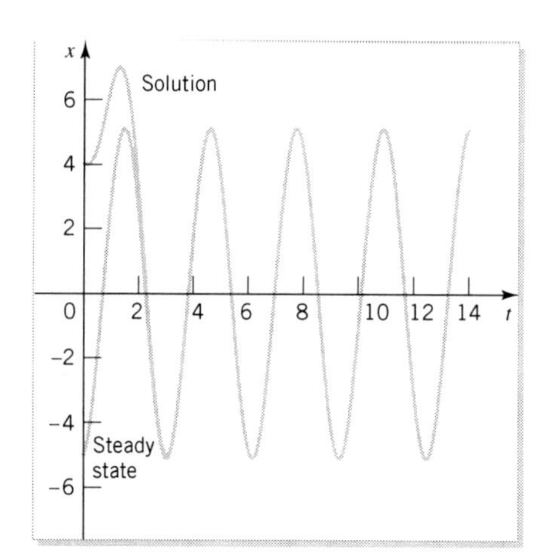

FIGURE 7.6 Graphs of $-40e^{-3t} + 49e^{-2t} - 24te^{-3t} + \sin 2t - 5\cos 2t$, and the steady state solution $\sin 2t - 5\cos 2t$

Tuning Forks

EXAMPLE 7.15 *Tuning Forks*

Our next example concerns tuning forks, devices often shaped like the letter U with a stem. When struck against a solid object, a tuning fork emits a sound with a specific frequency.

We consider an example in which we have two ideal tuning forks, whose natural frequencies are close together — say, $5/\pi$ and $4.5/\pi$. Suppose that one fork is struck so it emits a signal of the form $19\cos 9t$. If both vibrating forks are then held firmly against a table, one model governing the vibration of the second fork is the differential equation

$$x'' + 100x = 19\cos 9t, \tag{7.18}$$

where x is the displacement of one end of the fork. The initial conditions for this second tuning fork are

$$x(0) = x'(0) = 0, \tag{7.19}$$

signifying no motion in this second tuning fork at the moment the two forks are held against the table.

The characteristic equation of the associated homogeneous differential equation, $x'' + 100x = 0$, is $r^2 + 100 = 0$, with solutions of $\pm 10i$. Thus, the solution of the associated homogeneous equation is $x_h(t) = C_1 \sin 10t + C_2 \cos 10t$.

Trial solution

We obtain the proper form for a particular solution of our original differential equation (7.18) using Table 7.1. Thus, we substitute $x_p(t) = A\sin 9t + B\cos 9t$ into (7.18) to obtain

$$(-81 + 100)A\sin 9t + (-81 + 100)B\cos 9t = 19\cos 9t.$$

General solution

Solving for A and B gives $A = 0$ and $B = 1$, and the general solution is

$$x(t) = C_1 \sin 10t + C_2 \cos 10t + \cos 9t.$$

Initial conditions

If we now evaluate the two arbitrary constants by imposing the initial conditions in (7.19), we find that $C_1 = 0$, and $C_2 = -1$, giving our solution as

$$x(t) = \cos 9t - \cos 10t. \tag{7.20}$$

The graph of this explicit solution is shown in Figure 7.7, where the motion appears to have a rapid oscillation within a slower oscillation.

A more convenient form to analyze this behavior may be obtained if we use the trigonometric identity

$$\cos\alpha - \cos\beta = 2\sin\left(\frac{\beta+\alpha}{2}\right)\sin\left(\frac{\beta-\alpha}{2}\right) \tag{7.21}$$

to transform our explicit solution in (7.20) to the form

$$x(t) = 2\sin\frac{19t}{2}\sin\frac{t}{2}.$$

Beats

This form of the solution suggests that the motion is composed of a slow oscillation ($\sin t/2$) with a rapidly time-varying amplitude ($2\sin 19t/2$). This time-varying amplitude causes the sound emitted by the tuning fork to change intensity with time. This phenomenon — where there is an envelope to the oscillating function — is usually called BEATS or AMPLITUDE MODULATION.

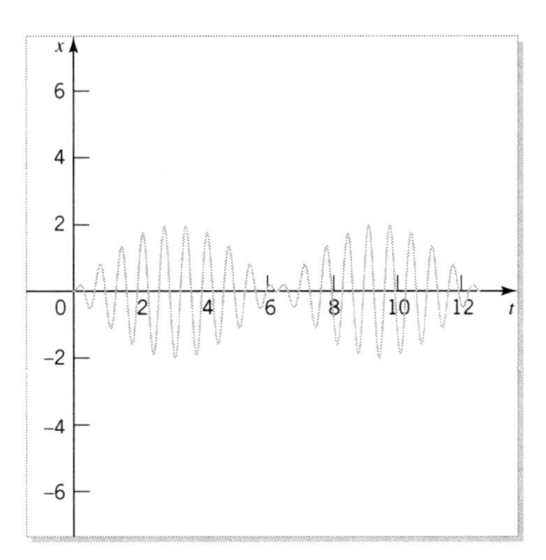

FIGURE 7.7 Graph of $\cos 9t - \cos 10t$

■

Spring-Mass Systems

EXAMPLE 7.16 *Vibration of a Linear Spring Under the Influence of a Periodic Forcing Function*

In Chapter 6 we considered the oscillatory motion of a linear spring. Here we consider the situation in which that spring is subject to a periodic external force of the form $q \cos \omega t$, where q and ω are constants. This gives the differential equation $mx'' + kx = q \cos \omega t$, where m, k, q, and ω are constants. In this example we take $m = 1$, $k = 100$, and $q = 19$ and leave ω as an unspecified parameter. This gives

$$x'' + 100x = 19 \cos \omega t, \tag{7.22}$$

and our task is to consider the effect of changes in ω on the motion of the spring-mass system. [Notice the similarity of (7.22) to the equation governing the motion of the tuning fork.]

Trial solution

We know from our previous example that $x_h(t) = C_1 \sin 10t + C_2 \cos 10t$ is the solution of the associated homogeneous differential equation. The proper form for the particular solution of (7.22) is $x_p(t) = A \sin \omega t + B \cos \omega t$, if $\omega \neq 10$. Substituting this expression into (7.22) and solving the resulting equation for A and B gives $A = 0$ and $B = 19/(-\omega^2 + 100)$. This gives our particular solution as $x_p(t) = 19/(100 - \omega^2) \cos \omega t$, and the general solution of (7.22) as

General solution

$$x(t) = C_1 \sin 10t + C_2 \cos 10t + \frac{19}{100 - \omega^2} \cos \omega t. \tag{7.23}$$

Initial conditions

Suppose that at $t = 0$, the mass on this spring is at rest at its equilibrium position and subject to the forcing function $19 \cos \omega t$. The initial conditions appropriate for this situation are $x(0) = 0$, $x'(0) = 0$. Imposing these initial conditions on our solution in (7.23) gives $C_1 = 0$, $C_2 = -19/(100 - \omega^2)$, so our solution becomes

$$x(t) = \frac{19}{100 - \omega^2} (\cos \omega t - \cos 10t). \tag{7.24}$$

Figures 7.8, 7.9, 7.10, and 7.11 show the graphs of this solution for $\omega = 9.5$, 9.7, 9.9, and 9.99, respectively. Figure 7.12 shows all four cases (using a condensed vertical scale). It is clear that as ω approaches 10, something is happening to the graphs. To understand this effect, we use the trigonometric identity given in (7.21) to change the form of our answer in (7.24) to

$$x(t) = \frac{38}{100 - \omega^2} \sin [(10 + \omega)t/2] \sin [(10 - \omega)t/2]. \tag{7.25}$$

As in the previous example, we can consider the motion as sinusoidal, $\sin [(10 - \omega)t/2]$, with a time-varying amplitude of $38 \sin [(10 + \omega)t/2] / (100 - \omega^2)$. Notice that the closer ω is to 10, the larger this amplitude is, and the longer the period of $\sin [(10 - \omega)t/2]$ is. In fact, this amplitude is unbounded as ω approaches 10, although setting $\omega = 10$ in (7.25) gives an indeterminate form of 0/0. Of course, for this application of the spring-mass system, the displacements cannot get very large, because the spring would break at some point.

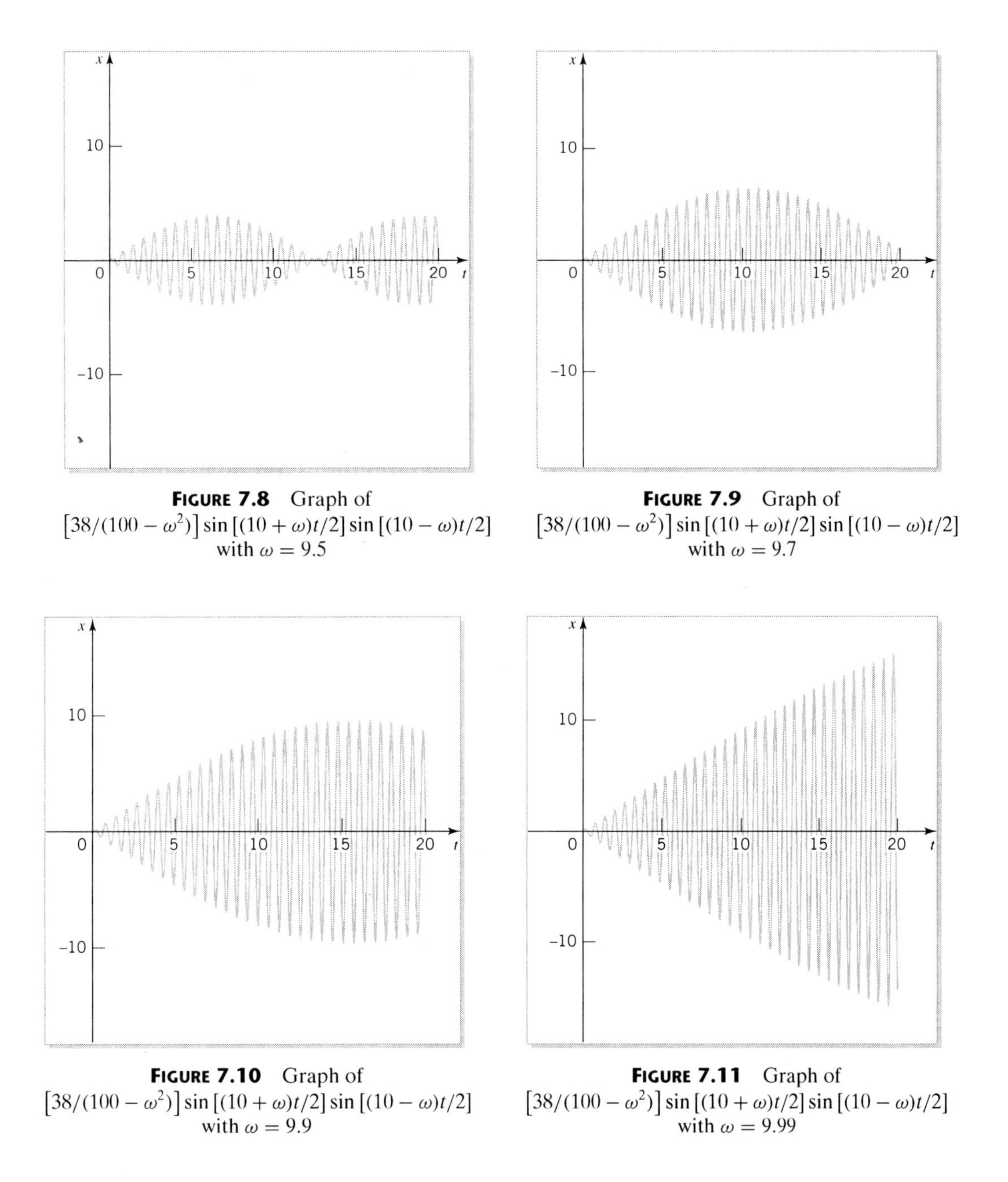

FIGURE 7.8 Graph of
$[38/(100-\omega^2)]\sin[(10+\omega)t/2]\sin[(10-\omega)t/2]$
with $\omega = 9.5$

FIGURE 7.9 Graph of
$[38/(100-\omega^2)]\sin[(10+\omega)t/2]\sin[(10-\omega)t/2]$
with $\omega = 9.7$

FIGURE 7.10 Graph of
$[38/(100-\omega^2)]\sin[(10+\omega)t/2]\sin[(10-\omega)t/2]$
with $\omega = 9.9$

FIGURE 7.11 Graph of
$[38/(100-\omega^2)]\sin[(10+\omega)t/2]\sin[(10-\omega)t/2]$
with $\omega = 9.99$

Resonance The situation in which the forcing frequency equals the natural frequency of a system (in this case when $\omega = 10$) is called RESONANCE.[1] The particular solution,

---

[1]Some mathematics and physics undergraduate texts give the Tacoma Narrows bridge failure of 1940 as an example of resonance at work. Engineers who have studied this phenomenon over the past 50 years are in substantial disagreement with this explanation. See "Resonance, Tacoma Narrows bridge failure, and undergraduate physics textbooks" by K. Y. Billah and R. H. Scanlan, *The American Journal of Physics* 59, 1991, pages 118–124.

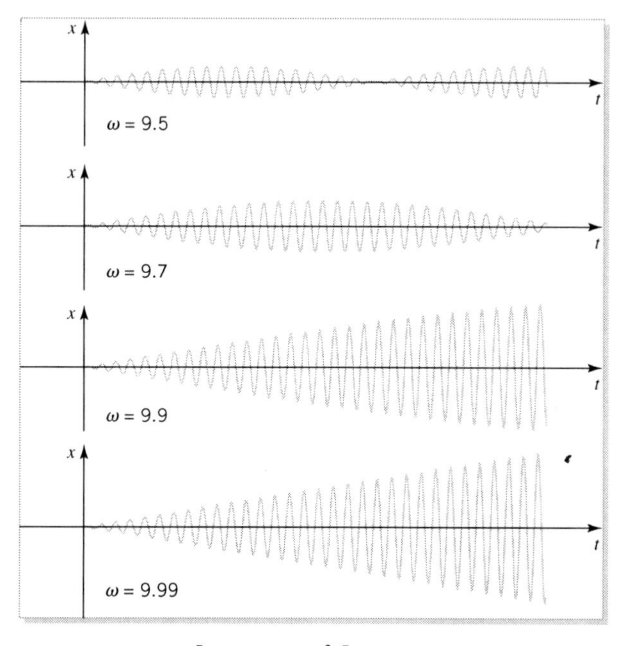

FIGURE 7.12 Graph of $[38/(100-\omega^2)]\sin[(10+\omega)t/2]\sin[(10-\omega)t/2]$ with $\omega = 9.5, 9.7, 9.9$, and 9.99

$x_p(t) = 19/(100-\omega^2)\cos\omega t$, is not valid for $\omega = 10$ (why?), so we return to the differential equation

$$x'' + 100x = 19\cos 10t \tag{7.26}$$

Trial solution and now try a particular solution of the form $x_p(t) = At\sin 10t + Bt\cos 10t$. The t is present in this particular solution because the normal trial solution, $A\sin 10t + B\cos 10t$, is a solution of the associated homogeneous differential equation.

Substituting $x_p(t)$ into (7.26) gives $20A\cos 10t - 20B\sin 10t = 19\cos 10t$, so we need $B = 0$ and $A = 19/20$. This gives our particular solution as $(19/20)t\sin 10t$, and

General solution the general solution of (7.26) as

$$x(t) = C_1\sin 10t + C_2\cos 10t + \frac{19}{20}t\sin 10t.$$

Imposing our previous initial conditions on this solution gives $C_1 = C_2 = 0$, so the final form of our solution is

$$x(t) = \frac{19}{20}t\sin 10t. \tag{7.27}$$

Here the solution will oscillate between t and $-t$ and will be unbounded as $t \to \infty$. Its graph is shown in Figure 7.13. Notice the strong similarity between this figure and Figure 7.11.

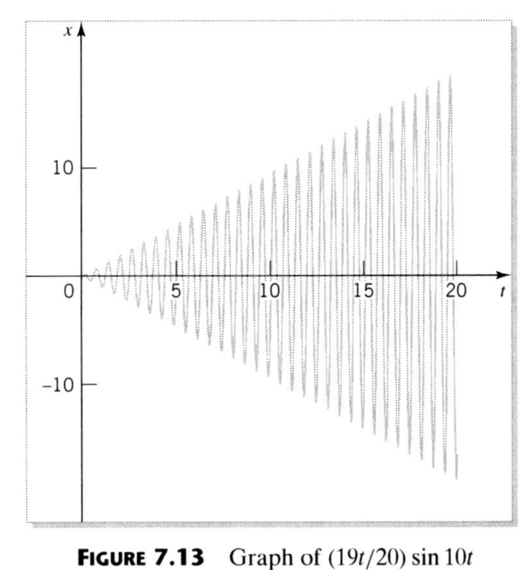

FIGURE 7.13 Graph of $(19t/20)\sin 10t$

EXAMPLE 7.17 *Forced Vibration of a Damped Spring-Mass System*

In Chapter 6 we discussed the damped motion of a spring-mass system. Here we consider that same system, but now subjected to an oscillatory force, resulting in the differential equation

$$mx'' + bx' + kx = q\cos\omega t,$$

where m, b, k, q, and ω are positive constants. For illustrative purposes, we use the values of $b/m = 2$, $k/m = 17$, and $q/m = 1$, giving

$$x'' + 2x' + 17x = \cos\omega t. \tag{7.28}$$

We know that our solution will consist of two parts. The first part is the solution of the homogeneous differential equation associated with (7.28). This leads to the characteristic equation $r^2 + 2r + 17 = 0$, with solution $r = -1 \pm 4i$, so $x_h(t) = C_1e^{-t}\sin 4t + C_2e^{-t}\cos 4t$.

Trial solution

The particular solution will have the form $x_p(t) = A\sin\omega t + B\cos\omega t$. Substituting this form for x_p into (7.28) gives

$$-\omega^2(A\sin\omega t + B\cos\omega t) + 2\omega(A\cos\omega t - B\sin\omega t) + \\ + 17(A\sin\omega t + B\cos\omega t) = \cos\omega t.$$

We equate coefficients of like terms in this equation to obtain

$$\begin{aligned} -\omega^2A - 2\omega B + 17A &= 0, \\ -\omega^2B + 2\omega A + 17B &= 1. \end{aligned}$$

From the first equation we have that $B = (\omega^2 - 17)A/(-2\omega)$. Using this fact in the second equation yields $[2\omega + (17 - \omega^2)^2/(2\omega)]A = 1$, giving A and B as

$$A = 2\omega/[4\omega^2 + (17 - \omega^2)^2],$$
$$B = (17 - \omega^2)/[4\omega^2 + (17 - \omega^2)^2]. \tag{7.29}$$

General solution

Thus, our particular solution is given by $x_p(t) = A \sin \omega t + B \cos \omega t$, and our general solution of (7.28) is

$$x(t) = C_1 e^{-t} \sin 4t + C_2 e^{-t} \cos 4t + A \sin \omega t + B \cos \omega t, \tag{7.30}$$

where C_1 and C_2 are arbitrary constants. A and B are given in (7.29). The values of C_1 and C_2 are specified once initial conditions are given, but regardless of their values, the first two terms in (7.30) become very small as t increases. Because this is so, these terms constitute the transient solution, and the remaining terms that do not decay give the steady state solution.

Steady state

We now change the form of this steady state solution — denoted by $x_{ss}(t)$ — by using the result that (see Exercise 6 on page 249)

$$x_{ss}(t) = A \sin \omega t + B \cos \omega t = \sqrt{A^2 + B^2} \cos (\omega t + \phi), \tag{7.31}$$

where

$$\cos \phi = \frac{B}{\sqrt{A^2 + B^2}}, \qquad \sin \phi = \frac{-A}{\sqrt{A^2 + B^2}}.$$

Notice that the amplitude of our steady state solution, $\sqrt{A^2 + B^2}$, depends on the value of the forcing frequency and, using (7.29), is given by

$$\sqrt{A^2 + B^2} = \frac{1}{\sqrt{4\omega^2 + (17 - \omega^2)^2}}. \tag{7.32}$$

The graph of this amplitude as a function of the forcing frequency, ω, is given in Figure 7.14, and shows that the amplitude has a maximum value. To find this maximum value we take the derivative of the amplitude with respect to ω and set the result to zero. Performing this calculation shows $\omega = \sqrt{15}$ is the forcing frequency that gives the maximum value of the steady state amplitude (see Exercise 9 on page 309).

An interesting property of this steady state solution is the fact that the value of the forcing frequency that maximizes the amplitude of the steady state velocity ($\omega\sqrt{A^2 + B^2}$) is $\omega = \sqrt{17}$ (see Figure 7.15). But this is simply the frequency for the same mass and spring constant but with no damping. This means that the frequency that maximizes the steady state velocity of a forced, damped spring-mass system is independent of the damping coefficient. (See Exercise 9 on page 309.)

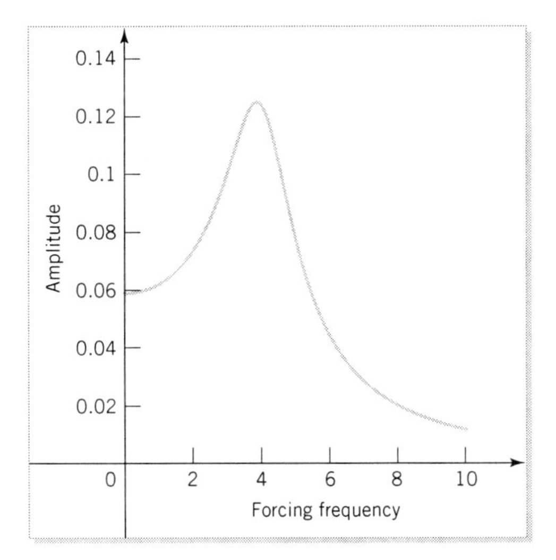

FIGURE 7.14 Amplitude of the steady state solution versus the forcing frequency

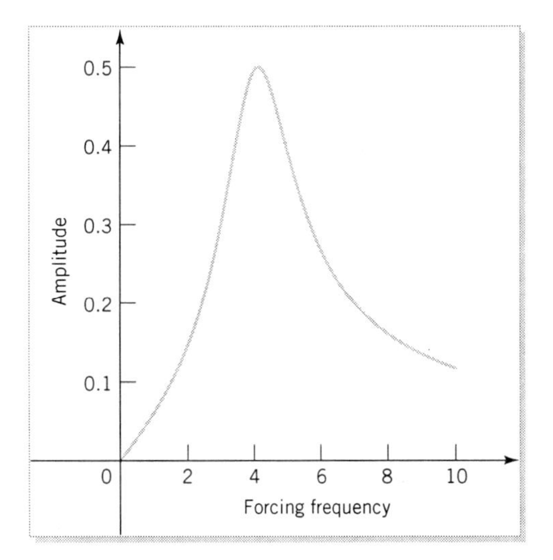

FIGURE 7.15 Amplitude of the velocity of the steady state solution versus the forcing frequency

EXAMPLE 7.18 *Spring-Mass System with Coulomb Friction*

If we place a spring-mass system on a horizontal surface, the major damping force is not necessarily proportional to the velocity. One model of this situation is to have a damping force take on a constant value that is always in opposition to the direction of motion. The differential equation for this model is

$$mx'' + kx = -Qm\, signum\,(x'), \tag{7.33}$$

where the value of the constant Q depends on the roughness of the contact between the mass and the surface, and where

$$signum\,(x) = \begin{cases} 1 \text{ if } x > 0, \\ 0 \text{ if } x = 0, \\ -1 \text{ if } x < 0. \end{cases}$$

Thus, the signum function in (7.33) has the value of 1 if $x' > 0$ and -1 if $x' < 0$, so (7.33) could be written in the form

$$mx'' + kx = \begin{cases} -Qm & \text{when the mass is moving right,} \\ 0 & \text{when } x' = 0, \\ Qm & \text{when the mass is moving left.} \end{cases}$$

This is an example of a second order differential equation being pieced together.

We will look at the special case where $m = 1$, $k = 1$, and $Q = 1$, subject to the initial conditions $x(0) = 5.5$, $x'(0) = 0$, so that

$$x'' + x = \begin{cases} -1 & \text{when the mass is moving right,} \\ 0 & \text{when } x' = 0, \\ 1 & \text{when the mass is moving left.} \end{cases}$$

General solution The solution of this differential equation for the three different situations is

$$x(t) = \begin{cases} C_1 \cos t + C_2 \sin t - 1 & \text{when the mass is moving right,} \\ C_3 \cos t + C_4 \sin t & \text{when } x' = 0, \\ C_5 \cos t + C_6 \sin t + 1 & \text{when the mass is moving left,} \end{cases} \tag{7.34}$$

where $C_1, C_2, \cdots, C_6$ are arbitrary constants, and where these solution must be pieced together so that x' is continuous. (Why?)

Because the motion is starting from rest 5.5 units to the right, we use the second equation in (7.34) to find $x(t) = 5.5 \cos t$. From this we have $x'(t) = -5.5 \sin t$, and for positive values of t near zero, $x'(t) < 0$, so the mass initially moves left — as intuition would tell us. Thus, we use the third equation in (7.34) until the first positive time, t_1, when the velocity is zero — $x'(t_1) = 0$. Then we use the second equation to decide whether the mass is trying to move left or right — intuition says it should move right. If our intuition is correct, we then use the first equation in (7.34) until the second positive time, t_2, when the velocity is zero — $x'(t_2) = 0$. Then we use the second equation in (7.34) and then, presumably, the third until the third positive time, t_3, that the velocity is zero — $x'(t_3) = 0$. And so on. Each time we do this, the values of the arbitrary constants will change. Here are the details of this process.

Initial conditions Using the initial conditions $x(0) = 5.5$, $x'(0) = 0$, in the third equation in (7.34), we find $x(t) = 4.5 \cos t + 1$ for $0 < t < t_1$. To find t_1, we calculate $x'(t)$ and find the first positive time that $x' = 0$, which is $t_1 = \pi$. At this time the mass is at its far left-hand position, $x(t_1) = 4.5 \cos \pi + 1 = -3.5$.

Now we use the initial conditions $x(t_1) = -3.5$, $x'(t_1) = 0$, in the middle equation in (7.34) to find $x(t) = 3.5 \cos t$. Because $x'(t) = -3.5 \sin t$ for values of $t > t_1$ near t_1, we find $x'(t) > 0$, and so we use the first equation in (7.34) for $t_1 < t < t_2$; that is, $x(t) = 2.5 \cos t - 1$. To find t_2, we calculate $x'(t)$ and find the second positive time that $x' = 0$, which gives $t_2 = 2\pi$. At this time the mass is at its far right-hand position, $x(t_2) = 2.5 \cos 2\pi - 1 = 1.5$. And so on.

We could continue with this process; however, it looks as though the mass is oscillating toward the origin. Nevertheless we should try to confirm our analysis. How can we do this? We notice that this differential equation is autonomous, and so we may write it as a system of first order differential equations and do a phase plane analysis. To do this we define $y = x'$ and obtain the system

$$x' = y, \qquad y' = -x - signum(y), \tag{7.35}$$

Phase plane so its equation in the phase plane is

$$\frac{dy}{dx} = \frac{-x - signum(y)}{y}. \tag{7.36}$$

The most obvious implication of (7.36) is that for positive values of x and y, the slope field will have only negative slopes, going toward vertical tangents for $y = 0$. This is clear from the direction field as seen in Figure 7.16. From (7.35) we have that in this first quadrant, x is an increasing function of t, and y is a decreasing function. We can also observe, both from (7.35) and the direction field, that the mass is at rest (zero velocity) when the displacement, x, has its extreme values, but the maximum velocity does not occur when $x = 0$. Figure 7.17 shows the numerically computed orbit in the phase plane. Notice the excellent agreement with $x(t_1) = -3.5$ when $x'(t_1) = 0$, and $x(t_2) = 1.5$ when $x'(t_2) = 0$. It also appears that $x(t_3) = -0.5$ when $x'(t_3) = 0$, which we can confirm by continuing with the next step in the previous calculation. (Do this.) Also, the orbits in the upper half plane look like semi-circles, as do those in the lower half plane. (Are they?)

Direction field

Orbits

But, wait a minute! Why did the orbit in Figure 7.17 stop at $x(t_3) = -0.5$ when $x'(t_3) = 0$? Figure 7.18 shows some more numerically computed orbits, and they all seem to halt between -1 and 1. What have we missed?

Mystery!

To try to pinpoint the problem, let's apply the general initial conditions $x(0) = x_0 > 0$, $x'(0) = 0$, to (7.34). In this case we find

$$x(t) = \begin{cases} (x_0 + 1)\cos t - 1 & \text{when the mass is moving right,} \\ x_0 \cos t & \text{when } x' = 0, \\ (x_0 - 1)\cos t + 1 & \text{when the mass is moving left.} \end{cases}$$

Because $x_0 > 0$, we expect the mass to move left, so from the third equation we should have $x'(t) < 0$; that is, $-(x_0 - 1)\sin t < 0$, for small positive values of t. Because $\sin t > 0$ for these values of t, this means that the mass will move to the left if x_0 satisfies $-(x_0 - 1) < 0$; that is, $x_0 > 1$. In other words, whenever the mass reaches a time when $x_0 < 1$, the third solution is not applicable. What about the first solution? In this case, $x'(t) = -(x_0 + 1)\sin t$, so for $x'(t)$ to be positive for small positive values of t we must have $-(x_0 + 1) > 0$; that is, $x_0 < -1$. But we assumed

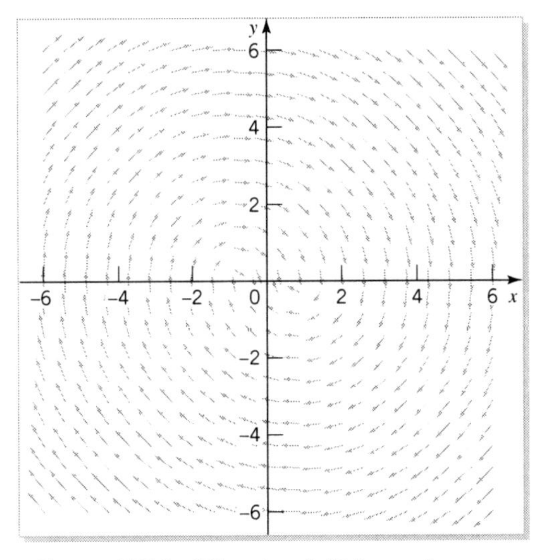

FIGURE 7.16 Direction field for spring-mass system with Coulomb friction

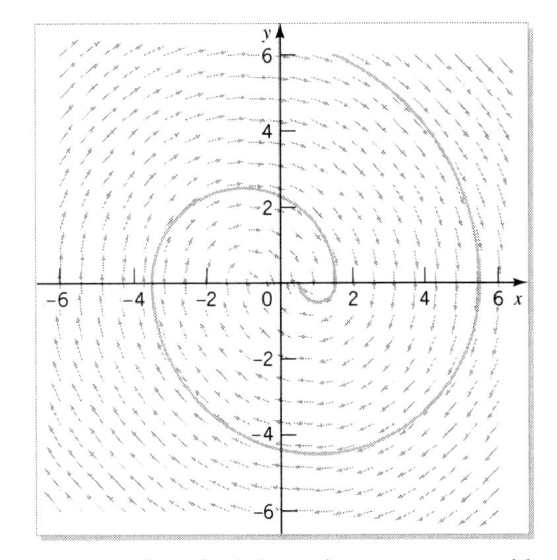

FIGURE 7.17 Orbit for spring-mass system with Coulomb friction

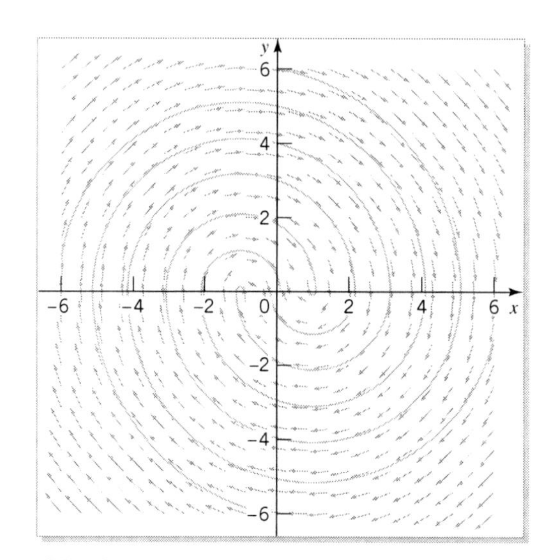

FIGURE 7.18 Orbits for spring-mass system with Coulomb friction

that $x_0 > 0$, so the first equation is not applicable. Similar remarks apply to $x_0 < 0$. In other word, if as the motion progresses we find that $-1 < x_0 < 1$, then the motion will halt. Thus, if initially we displace the mass a distance less than one from the equilibrium position, the friction is large enough to prevent the spring from moving the mass. If initially the mass is displaced a distance larger than one, it oscillates with decreasing amplitude, coming to rest between -1 and 1 when the differential equation no longer applies. That is the answer to the mystery.

Mystery solved

In fact, the differential equation (7.33) has no solution if $x(0) = x_0$, $x'(0) = 0$, and $|x_0| \leq 1$. To correct this situation, we could model what we have just discovered by

$$mx'' + kx = \begin{cases} -Qm & \text{when the mass is moving right,} \\ 0 & \text{when } x' = 0 \text{ if } |x| > Qm/k, \\ Qm & \text{when the mass is moving left} \end{cases}$$

and then demand that when the Coulomb force is greater than the restoring force,

$$mx'' = 0, \text{ when } x' = 0 \text{ if } |x| \leq Qm/k.$$

This second equation tells us that if $|x_0| \leq Qm/k$ and if there is a time T at which $x(T) = x_0$ and $x'(T) = 0$, then $x(t) = x_0$ for $t > T$. ∎

EXERCISES

1. Find $x(t)$ for the electrical circuit described by (7.12) having $L = 16, R = 8, C = 1/10, E(t) = \sin \omega t, x(0) = 0$, and $x'(0) = 0$, for given ω.

2. What value of ω will maximize the amplitude of the steady state response in Exercise 1? (In practice, a circuit is "tuned" by varying C until the steady state amplitude is maximized for a fixed value of ω.)

3. Consider the solution given by (7.24) for $\omega \neq 10$, and take the limit as $\omega \to 10$. (Use either Taylor series or L'Hôpital's rule.) Compare your answer with that given by (7.27).

4. Consider the spring-mass governed by $x'' + 16x = 15 \sin t$, $x(0) = 0, x'(0) = \sqrt{3}$.

(a) Find x as a function of time.

(b) What is the period of the solution in part (a)?

5. Consider the spring-mass governed by $x'' + 2x' + 17x = 2 \sin \omega t$, $x(0) = 0, x'(0) = 0$.

(a) Find x as a function of time.

(b) Identify the steady state and transient parts of the solution in part (a).

6. Consider the initial value problem $x'' + 2\alpha x' + (\omega^2 + \alpha^2)x = \sin \omega t$, $x(0) = 0, x'(0) = 0$.

(a) Find x as a function of time.

(b) Identify the steady state and transient parts of the solution in part (a).

(c) Determine the amplitude of the steady state portion of the solution, and, considering α as fixed, find the value of ω that maximizes this amplitude. [Hint: Use calculus.]

(d) Find the value of ω in Exercise 5 that maximizes the amplitude of the steady state solution. (When the frequency of the periodic forcing function is chosen to maximize the amplitude of the steady state response, this damped system is often said to be in resonance.)

7. Find the value of ω so that the system governed by $x'' + 6x' + 22x = \cos \omega t$ is in resonance as defined in Exercise 6.

8. Solve (7.33) for initial conditions $x(0) = 0, x'(0) = 10$, $b/m = 2, k/m = 10$, and $Q = 2$ for values of t between 0 and when it changes direction the second time. [Hint: Initially the sign of x' is positive, so solve (7.33) for x and determine the value of t for which $x' = 0$. Then, for that value of t — say, t_s — use $x(t_s)$ as the initial displacement and $x'(t_s) = 0$ as new initial conditions for (7.33), with x' now negative for times slightly greater than t_s.]

9. Show that the maximum value of the amplitude of the steady state displacement in Example 7.17 on page 303 [$\sqrt{A^2 + B^2}$ in (7.32)] is obtained when $\omega = \sqrt{15}$. Show that the maximum value of the amplitude of the steady state velocity in Example 7.17 $\left(\omega\sqrt{A^2 + B^2}\right)$ is obtained when $\omega = \sqrt{17}$.

10. Suppose that the motion of a mass is governed by the differential equation $mx'' + b\left(x'\right)^{\omega} = 0$; that is, the damping force is proportional to the velocity to a power. If the mass is subject to the initial conditions $x(0) = 0$ and $x'(0) = 10$, will the mass come to rest in a finite time? If so, how far will it travel and what will be its travel time? If it does not come to rest, explain why. You may want to consider five different cases for ω: $0 < \omega < 1, \omega = 1, 1 < \omega < 2, \omega = 2$, and $\omega > 2$.

11. If an object is thrown vertically upward, a possible model for its motion is the differential equation $mx'' + bx' = -mg$, where x is the object's height above the ground. For the purposes of this exercise we assume that units have been chosen so that $m = g = 1$.

(a) Consider the case of no resistance ($b = 0$) so that $x'' = -1$. Solve this differential equation in two different ways: (i) by integrating it twice, and (ii) by treating it as a nonhomogeneous second order equation with constant coefficients. If initially an object is thrown upward from ground level with a velocity of 1, calculate how long it will take to reach its maximum height, and how long it will take to return to ground level from that maximum height. Are these two times the same? Graph $x(t)$ versus t to confirm your conclusions.

(b) Consider the case $b = 1$ so that $x'' + x' = -1$. Solve this differential equation in two different ways: (i) by integrating it twice, and (ii) by treating it as a nonhomogeneous second order equation with constant coefficients. If initially an object is thrown upward from ground level with a velocity of 1, calculate how long it will take to reach its maximum height, and how long it will take to return to ground level from that maximum height. Are these two times the same? Graph $x(t)$ versus t to confirm your conclusions.

12. The Running Lizard. Experiments have been performed where a lizard is encouraged to run as fast as possible, starting from rest.[2] The distance run is then measured as a function of time. The data gathered are given in Table 7.2 and shown in Figure 7.19.

Table 7.2 **Distance traveled by accelerating lizard as a function of time**

Time (sec)	Distance (meters)	Time (sec)	Distance (meters)
0.000	0.000	0.336	0.760
0.044	0.040	0.416	1.020
0.076	0.100	0.500	1.280
0.104	0.160	0.576	1.520
0.150	0.275	0.661	1.780
0.252	0.520	0.750	2.050

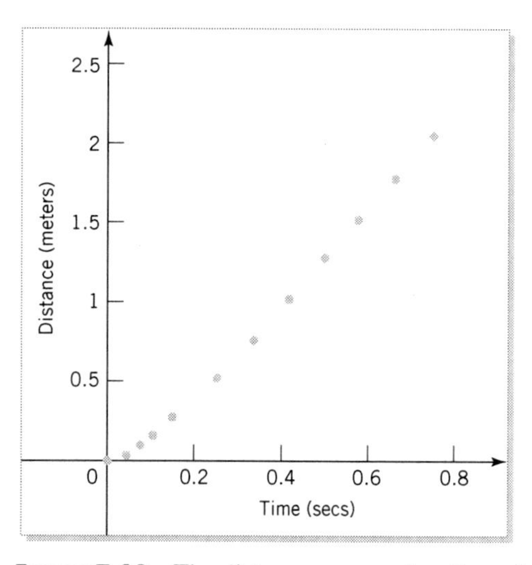

FIGURE 7.19 The distance run as a function of time by an accelerating lizard

(a) A proposed model that governs the acceleration of a lizard is

$$x'' = a - bx', \tag{7.37}$$

where a and b are positive constants and x is the distance traveled in time t. Explain what happens to the acceleration as t increases. Is this a reasonable model?

(b) Integrate (7.37) using the fact that at $t = 0$ the lizard starts from rest from $x = 0$, to find

$$x(t) = \frac{a}{b}\left[t + \frac{1}{b}\left(1 - e^{-bt}\right)\right]. \tag{7.38}$$

(c) What happens to x' as $t \to \infty$? Is this consistent with (7.37)? Explain what this means physically, and show how this can be used, in conjunction with Figure 7.19, to estimate a/b.

(d) Use one of the points from Table 7.2 — say, $x(0.661) = 1.780$ — and your estimate for a/b from part (c) to estimate the value of b. Using these estimates for a and b, plot (7.38) on Figure 7.19. Figure 7.20 is one possibility. How good is the differential equation (7.37) at modeling an accelerating lizard?

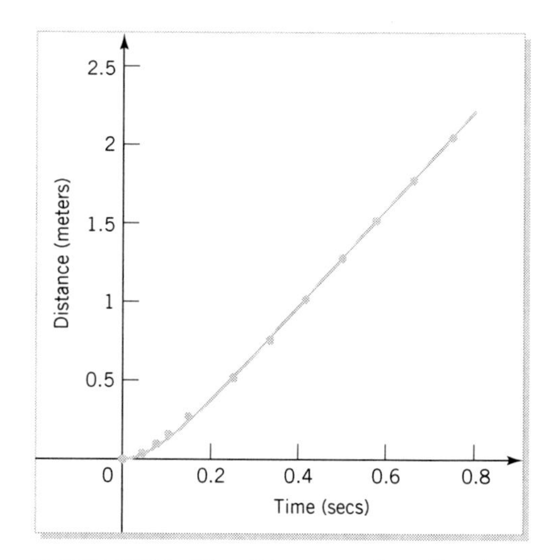

FIGURE 7.20 The distance run by a lizard and the function $x(t) = a[t + (1 - e^{-bt}/b)]/b$

(e) Table 7.3 is a data set corresponding to the times achieved every 10 meters by Carl Lewis in the 100 m final of the World Championship in Rome in 1987.[3] This data set is shown in Figure 7.21. Does

[2] "Effects of body size and slope on acceleration of a lizard (*Stellio Stellio*)" by R. B. Huey and P. E. Hertz, *J. Exp. Biol.* 110, 1984, pages 113–123.

[3] "Mathematical Models of Running" by W. G. Pritchard, *SIAM Review* 35, 1993, pages 359–379.

Table 7.3 Distance traveled by Carl Lewis as a function of time

Time (sec)	Distance (meters)	Time (sec)	Distance (meters)
0.00	0	6.50	60
1.94	10	7.36	70
2.96	20	8.22	80
3.91	30	9.07	90
4.78	40	9.93	100
5.64	50		

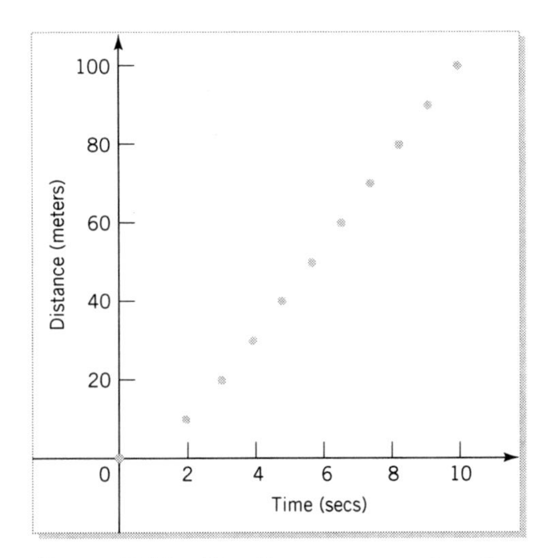

FIGURE 7.21 The distance run as a function of time by Carl Lewis

the model (7.37) also apply to world-class sprinters? According to this model, does Carl Lewis attain his maximum speed while running the 100 m race?

13. Denise and Craig.[4] Assume that after Denise and Chad break up, Chad is replaced by Craig in Denise's affection, but their affections follow much the same pattern as Denise and Chad's did. Denise and Craig take an engineering class together where they are assigned as lab partners. They have to complete a new lab every $2\pi/c$ weeks, where c is a positive constant. Suppose that the impact on Denise's affection for Craig from the experience of working together is to add the term $\sin ct$ to the rate of change of her affection at time t. For example, during the latter part of each lab, the tension involved for Denise in trying to cope with Craig's tardiness reduces her affection for him. Their emotions are modeled by

$$\begin{cases} x' = y + \sin ct, \\ y' = -4x. \end{cases}$$

(a) Show that Denise's affection for Craig satisfies the differential equation $x'' + 4x = c\cos ct$.

(b) Study this solution for the initial emotions of $x(0) = 0$, $y(0) = 0$, which means that if they were not forced to work with each other, they would have no feelings for each other. How does this relationship change as c approaches 2 and then exceeds it?

14. Cantilever Beam. If a horizontal cantilever beam of length L is subjected to a concentrated load P at its free end then the vertical displacement, $x(t)$, at the point t measured from the fixed end is governed by the differential equation $EIx'' = -P(L-t) - \frac{w}{2}(L-t)^2$, where w is the distributed load of the beam and EI is the flexural rigidity.

(a) Explain why the initial conditions are $x(0) = x'(0) = 0$, and why the differential equation is valid only for $0 \leq t \leq L$.

(b) From the differential equation determine the concavity of the beam. Does this agree with your intuition?

(c) Solve this initial value problem by integrating the differential equation twice.

(d) Solve this initial value problem by using the method of undetermined coefficients.

(e) Explain why your solution has a minimum deflection at $t = L$, and that its value there is $x = -PL^3/(3EI) - wL^4/(8EI)$.

(f) If you were to measure this deflection for various loads, P, and then plot x against P, what would you expect to see? How can you use this to estimate $-L^3/(3EI)$ and $-wL^4/(8EI)$?

(g) Table 7.4 and Figure 7.22 show the results of an experiment where the deflection at the end of a cantilevered beam was measured (in inches) as a function of the load P (in lbs).[5] How well does this model fit the data set? What are your estimates for $-L^3/(3EI)$ and $-wL^4/(8EI)$?

[4] This exercise is based on an idea of our colleague Sam Evens.

[5] This experimental data was obtained by Dale Williams while a student at the University of Arizona.

Table 7.4 The deflection of a beam as a function of the applied load

Load (lbs)	Deflection (inches)	Load (lbs)	Deflection (inches)
0.06393	0.015	0.44092	0.087
0.08818	0.019	0.66138	0.131
0.11023	0.024	0.88184	0.166
0.22046	0.043	1.10230	0.211
0.33069	0.066	2.20460	0.407

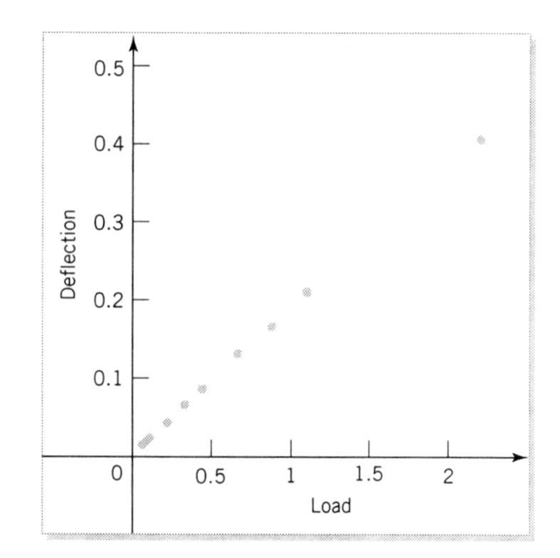

FIGURE 7.22 The deflection of a beam as a function of the applied load

What Have We Learned?

Main Ideas

How to Solve Nonhomogeneous Second Order Linear Differential Equations

Purpose To find the general solution of the nonhomogeneous linear differential equation

$$a_2x'' + a_1x' + a_0x = f(t), \tag{7.39}$$

where the coefficients a_2, a_1, and a_0 may be functions of t.

Process

1. Find the general solution of the associated homogeneous differential equation. See *How to Solve Homogeneous Linear Differential Equations with Constant Coefficients* on page 244.
2. Find a particular solution of (7.39). See Section 7.2, *Method of Undetermined Coefficients*.
3. Add the solutions found in steps 1 and 2 to obtain the general solution of (7.39).

Comments about Nonhomogeneous Second Order Linear Differential Equations

- An explicit solution of (7.39) is always possible if a_2, a_1, and a_0 are constants.
- If $f(t)$ is of the form given in Table 7.1 on page 287, use the method of undetermined coefficients to find a particular solution.
- If $f(t)$ is not of the form given in Table 7.1, use the method of variation of parameters or reduction of order (see Chapter 8) to find a particular solution.
- Be sure to include both the particular solution and the general solution of the associated homogeneous differential equation when evaluating the arbitrary constants in an initial value problem.

CHAPTER 8

SECOND ORDER LINEAR DIFFERENTIAL EQUATIONS — QUALITATIVE AND QUANTITATIVE ASPECTS

Where We Are Going — and Why

In this chapter we look at the second order linear differential equation

$$a_2(t)x'' + a_1(t)x' + a_0(t)x = 0, \tag{8.1}$$

where $a_2(t)$, $a_1(t)$, and $a_0(t)$ are continuous functions of t, and $a_2(t) \neq 0$ for $a < t < b$. We will develop several techniques for finding the general solution for special cases of (12.1).

We give an Oscillation Theorem that allows us to determine qualitative properties of solutions, such as oscillations and boundedness, even if we cannot find these solutions explicitly. The concepts of linear independence and linear dependence are introduced, which allow us to identify certain explicit solutions as general solutions. We develop additional methods for finding particular solutions of nonhomogeneous linear differential equations and discover that these methods, as well as those methods from Chapter 7, also apply to higher order linear differential equations. Explicit and numerical solutions of boundary value problems are also discussed in this chapter.

8.1 QUALITATIVE BEHAVIOR OF SOLUTIONS

We begin this chapter with a discussion of the qualitative nature of explicit solutions of the general second order linear differential equation given in (12.1). In Chapter 6 we introduced an existence-uniqueness theorem dealing with such differential equations, namely,

▶ **Theorem 8.1: The Existence-Uniqueness Theorem for Second Order Linear Differential Equations** *Consider the second order linear differential equation*

$$a_2(t)x'' + a_1(t)x' + a_0(t)x = 0, \tag{8.2}$$

subject to the initial conditions $x(t_0) = x_0, x'(t_0) = x_0^*$. *If* $a_2(t)$, $a_1(t)$, *and* $a_0(t)$ *are continuous, with* $a_2(t) \neq 0$ *for* $a < t < b$, *and* t_0 *is contained in this interval, then for the interval* $a < t < b$ *there exists a unique solution of (8.2) with a continuous first derivative.* ◀

We have seen that if a_2, a_1, and a_0 are all constants, (8.2) is autonomous and we can always analyze the qualitative behavior of its solutions by means of the phase plane. Also, for this constant coefficient case, we can always find the explicit solution of (8.2).

However, if a_2, a_1, and a_0 are not all constants, the phase plane analysis fails because the associated system of first order differential equations is not autonomous. Furthermore, if a_2, a_1, and a_0 are not all constants, obtaining a solution of (8.2) in terms of familiar functions is the exception rather than the rule. For example, the innocent-looking differential equation

$$x'' - tx = 0, \tag{8.3}$$

which is known as Airy's equation and is important in aerodynamics, cannot be solved in terms of familiar functions.

In subsequent sections we will look at various techniques for solving special cases of (8.2). However, situations exist where we can predict the qualitative behavior of solutions of (8.2) whether or not we can find an explicit solution.

The Oscillation Theorem

We introduce these ideas by looking at the behavior of solutions of

$$x'' + ax = 0, \tag{8.4}$$

where a is constant. If $a < 0$, we may put $a = -h^2$, and we know the general solution of (8.4) is $x(t) = C_1e^{ht} + C_2e^{-ht}$, whereas if $a > 0$, we may put $a = k^2$, and the general solution is $x(t) = C_1 \cos kt + C_2 \sin kt$.

There are some major behavioral differences between these two solutions. In the case $a < 0$, where $x(t) = C_1e^{ht} + C_2e^{-ht}$, any nontrivial solution[1] can be zero, at most once [at $t = \ln(-C_2/C_1)/(2h)$ if $-C_2/C_1 > 0$, and nowhere if $-C_2/C_1 \leq 0$]. In the case $a > 0$, where $x(t) = C_1 \cos kt + C_2 \sin kt$, any nontrivial solution can be zero an infinite number of times, because all solutions oscillate about zero.[2] Also in this case, if we take any two solutions that are not proportional to each other — say $\cos kt$

---

[1]Recall that $x(t) = 0$ is the TRIVIAL solution of (8.2). A NONTRIVIAL solution is a solution that is not identically zero. Unless indicated otherwise, we will assume the word solution refers to a nontrivial solution.

[2]When a function $f(t)$ is zero at $t = t_0$, mathematicians frequently say $f(t)$ **vanishes** at $t = t_0$. So a function that is zero an infinite number of times is said to vanish an infinite number of times.

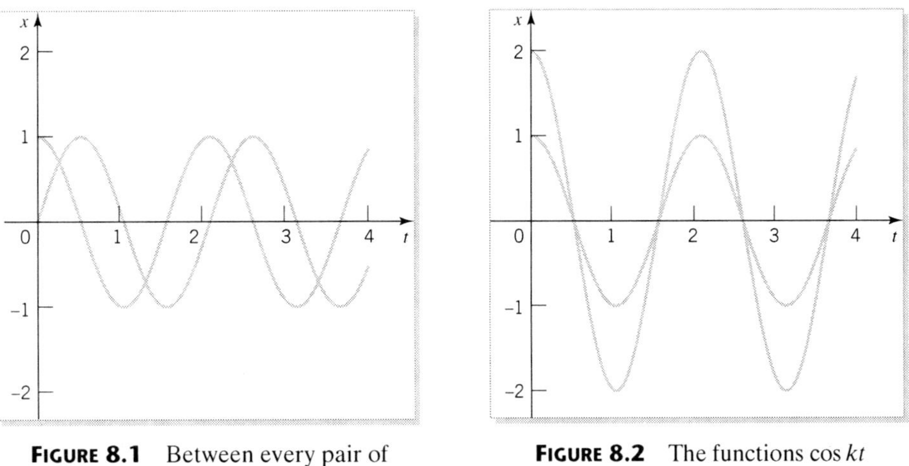

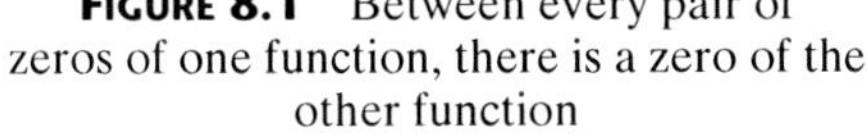
FIGURE 8.1 Between every pair of zeros of one function, there is a zero of the other function

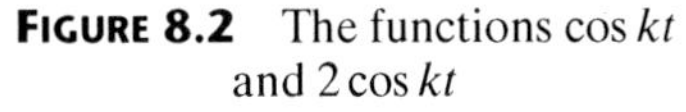
FIGURE 8.2 The functions $\cos kt$ and $2\cos kt$

and $\sin kt$ — then between every pair of zeros of one solution, there is a zero of the other solution. See Figure 8.1. (This isn't true if we select, say, $\cos kt$ and $2\cos kt$ as our solutions, which is why we restricted the observation to solutions that are **not** proportional to each other. See Figure 8.2.)

These ideas can be generalized to the case where a is not constant by the following results, which are proved and extended in Appendix A.7.

Theorem 8.2: The Oscillation Theorem *Consider the differential equation*

$$x'' + Q(t)x = 0, \tag{8.5}$$

where $Q(t)$ is continuous for $t \geq t_0$, and where $x = x_1(t)$ is a nontrivial solution of (8.5).

(a) If $Q(t) \leq 0$ for $t \geq t_0$, then $x_1(t)$ has, at most, one zero for $t > t_0$. Thus, the maximum number of times that $x_1(t)$ can be zero to the right of t_0 is once.

(b) If $Q(t) \geq k^2 > 0$ (k a real constant) for $t \geq t_0$, then $x_1(t)$ has an infinite number of zeros for $t \geq t_0$. Furthermore, if $Q'(t)$ is continuous and $Q(t)$ is monotonic for $t \geq t_0$, then $x_1(t)$ is bounded.[3]

(c) If $x = x_2(t)$ is another nontrivial solution of (8.5) and there is no constant c for which[4] *$x_1(t) = cx_2(t)$, then between two consecutive zeros of $x_1(t)$ there is exactly one zero of $x_2(t)$, and between two consecutive zeros of $x_2(t)$ there is exactly one zero of $x_1(t)$. So if $x_1(t)$ has an infinite number of zeros, so does $x_2(t)$; if $x_1(t)$ has a finite number of zeros, so does $x_2(t)$.*

Comments about the Oscillation Theorem

- Nontrivial functions that have an infinite number of zeros are said to OSCILLATE.

---

[3] A function $f(t)$ is bounded if there exists constants A and B for which $A < f(t) < B$ for all t in the domain of $f(t)$. A function that is bounded cannot approach infinity.

[4] Another way of saying this is that $x_1(t)$ and $x_2(t)$ are not proportional to each other.

- Zeros of two functions are called INTERLACED if between any two zeros of one function, there is a zero of the other function, and vice versa.
- It can be shown that a nontrivial solution of (8.5) can have only a finite number of zeros on a closed interval. Thus, the zeros of a solution are isolated.
- We might suspect that part (b) of the theorem could be replaced by "If $Q(t) > 0$ for $t \geq t_0$, then $x_1(t)$ has an infinite number of zeros for $t \geq t_0$," but it cannot. For example, the differential equation $x'' + (a^2/t^2)x = 0$ has oscillating solutions if $a^2 > 1/4$, and nonoscillating solutions if $a^2 \leq 1/4$. (See Exercise 7 on page 322.)

We now look at three examples to see how the Oscillation Theorem can be used.

EXAMPLE 8.1

What can be said about the solutions of

$$x'' - \frac{2}{t^2}x = 0 \tag{8.6}$$

for $t > 0$?

We use part (a) of the Oscillation Theorem. Because $Q(t) = -2/t^2 \leq 0$ for $t \geq t_0$, where t_0 is any positive constant, each solution of (8.6) can have, at most, one zero to the right of t_0, no matter how small t_0. Thus, the maximum number of times that a nontrivial solution of (8.6) can cross the positive t-axis is once. Numerical solutions for $x'(1) = 1$ and $x(1) = -2, -1.5, -1.0, -0.5$, and 0, are shown in Figure 8.3.

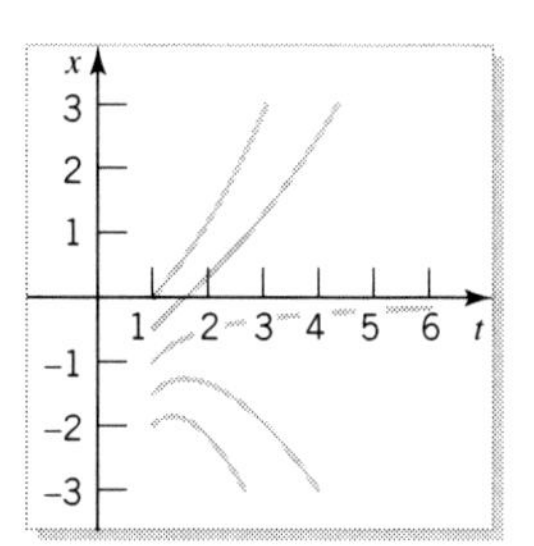

FIGURE 8.3 Numerical solutions of $x'' - 2x/t^2 = 0$

EXAMPLE 8.2

What can be said about the solutions of

$$x'' + \left(1 + \frac{1}{4t^2}\right)x = 0 \tag{8.7}$$

for $t > 0$?

We use part (b) of the Oscillation Theorem. Because $Q(t) = 1 + 1/(4t^2) \geq 1 > 0$, every nontrivial solution of (8.7) oscillates. Furthermore, because $Q'(t) = -8/t^3$ is negative for $t > 0$, part (b) also implies that every solution is bounded. Numerical solutions, for $x(1) = 0, x'(1) = 1$, for $x(1) = 0, x'(1) = 0.25$, and for $x(1) = 0.5$, $x'(1) = 0$, are shown in Figure 8.4. One of these solutions is proportional to another. (Which one?)

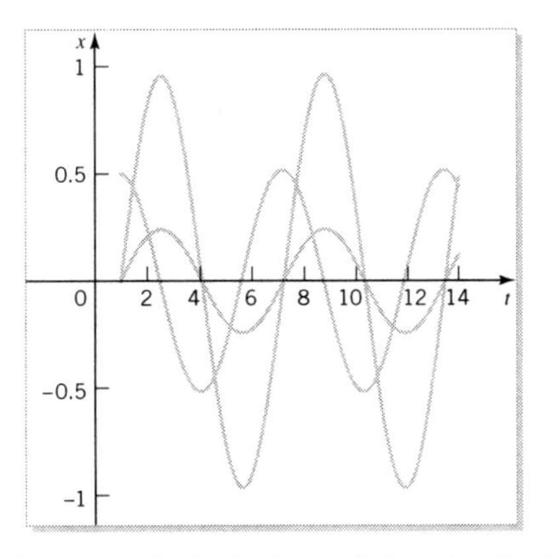

FIGURE 8.4 Numerical solutions of $x'' + (1 + 1/(4t^2))x = 0$

EXAMPLE 8.3 *Airy's Equation*

What can be said about the solutions of Airy's equation (8.3); that is,

$$x'' - tx = 0? \tag{8.8}$$

For $t > 0$ we use part (a) of the Oscillation Theorem. Because $Q(t) = -t \leq 0$, every nontrivial solution of (8.8) has at most one zero for $t > 0$.

For $t < 0$ we make the change of variables $t = -t$, which transforms (8.8) into $x'' + tx = 0$. Now consider $t \leq -t_0$ for any $t_0 > 0$, which means that $t \geq t_0$. Because $Q(t) = t \geq t_0 > 0$, part (b) of the Oscillation Theorem shows that every nontrivial solution of (8.8) oscillates for $t \leq -t_0$. These oscillations are bounded.

Thus, every nontrivial solution of (8.8) oscillates for $t < 0$ and has, at most, one zero for $t > 0$. Numerical solutions for $x(-10) = 0, x'(-10) = 1$, and for $x(-10) = 1$, $x'(-10) = 0$, are shown in Figure 8.5.

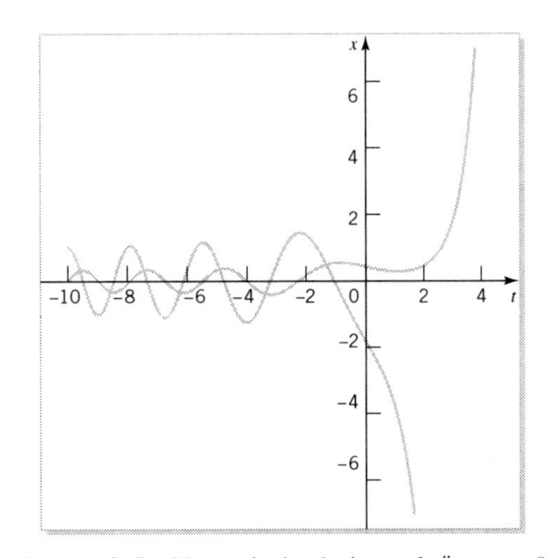

FIGURE 8.5 Numerical solutions of $x'' - tx = 0$

The Relation Theorem

The Oscillation Theorem gives us a powerful technique to determine qualitative information about solutions of differential equations of the form $x'' + Q(t)x = 0$. However, we often encounter differential equations of the form

$$x'' + p(t)x' + q(t)x = 0. \tag{8.9}$$

Thus, a natural question to ask is how we can relate the solutions of (8.9) to the solutions of $x'' + Q(t)x = 0$.

Change variable

In the past, we have found the substitution

$$x(t) = u(t)z(t) \tag{8.10}$$

very useful, so let us try it again here, where the object is to select u in a clever way to make the differential equation for z easier to solve. Substituting (8.10) into (8.9) and collecting together the terms in z, z', and z'' leads to

$$x'' + px' + qx = uz'' + \left(2u' + pu\right) z' + \left(u'' + pu' + qu\right) z = 0. \tag{8.11}$$

This equation will be of the form $z'' + Q(t)z = 0$ if we can choose $u(t)$ so that the z' term is eliminated. Thus, we choose $u(t)$ to satisfy $2u' + pu = 0$, or

$$\frac{u'}{u} = -\frac{1}{2}p,$$

which means

$$u(t) = e^{-\frac{1}{2}\int p(t)\,dt}. \tag{8.12}$$

If we substitute (8.12) and its derivatives into (8.11), we have

$$u(t)\left[z'' + \left(q - \frac{1}{2}p' - \frac{1}{4}p^2\right) z\right] = 0.$$

Thus, because from (8.12), $u(t) > 0$, we have established the following result.

► **Theorem 8.3: The Relation Theorem** *The solutions of the differential equation $x'' + p(t)x' + q(t)x = 0$ are related to the solutions of $z'' + Q(t)z = 0$, where*

$$Q(t) = q(t) - \frac{1}{2}p'(t) - \frac{1}{4}p^2(t), \tag{8.13}$$

by

$$x(t) = e^{-\frac{1}{2}\int p(t)\,dt} z(t). \tag{8.14}$$

◄

Comments about the Relation Theorem

- Because the coefficient of $z(t)$ in (8.14) is always positive, $x(t)$ and $z(t)$ have the same zeros. Thus, the number of zeros of $x(t)$ and $z(t)$ and their distances apart are the same.

- The results of the Oscillation Theorem now may be applied to the differential equation (8.9) by using $Q(t)$ defined by (8.13).
- The results of part (b) of the Oscillation Theorem concerning boundedness can also be extended to $x'' + p(t)x' + q(t)x = 0$ if $p(t) \geq 0$. In this case, the function

$$P(t) = e^{-\frac{1}{2}\int p(t)\,dt}$$

is a positive nonincreasing function of t, so $P(t)$ is bounded. Thus, by (8.14), $x(t)$ is bounded if and only if $z(t)$ is bounded.
- If $p(t)$ and $q(t)$ are continuous, then the differential equation has only a finite number of zeros in a finite interval.

We now look at two examples to see how the Oscillation Theorem and Relation Theorem can be used.

EXAMPLE 8.4

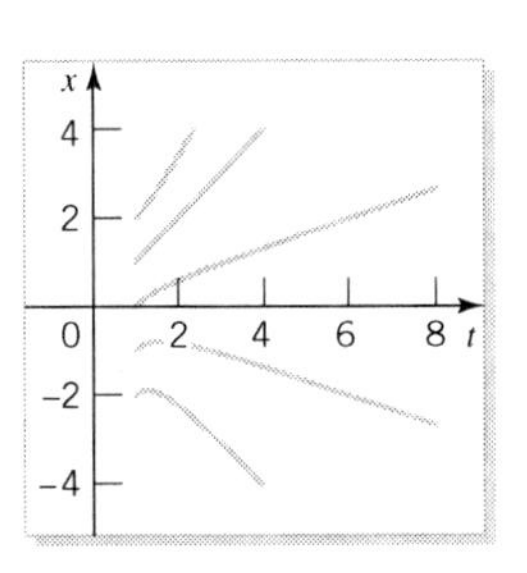

FIGURE 8.6 Numerical solutions of $t^2x'' + 2tx' - 2x = 0$

What can be said about the solutions of

$$t^2x'' + 2tx' - 2x = 0 \tag{8.15}$$

for $t > 0$?

Here $p(t) = 2/t$ and $q(t) = -2/t^2$, so, by (8.13), we have

$$Q(t) = q(t) - \frac{1}{2}p'(t) - \frac{1}{4}p^2(t) = -\frac{2}{t^2} - \frac{1}{2}\left(-\frac{2}{t^2}\right) - \frac{1}{4}\left(\frac{2}{t}\right)^2 = -\frac{2}{t^2}.$$

Because $Q(t) \leq 0$, the nontrivial solutions of (8.15) have, at most, one zero for $t > 0$.

Straight-line solution

Numerical solutions for $x'(1) = 1$ and $x(1) = -2, -1, 0, 1,$ and 2 are shown in Figure 8.6. But, wait a minute. It looks as though one of these solutions is a straight line through the origin with a slope of 1, in which case it would have the equation $x = t$. If we substitute $x = t$ into the left-hand side of (8.15), we find $t^2x'' + 2tx' - 2x = 2t - 2t = 0$. Thus, $x(t) = t$ is a solution of (8.15). In the next section we will show how this information allows us to find the general solution of (8.15). ■

EXAMPLE 8.5

We are told that the differential equation

$$x'' + 2tx' + 2x = 0 \tag{8.16}$$

has a solution $x(t) = e^{-t^2}$. Is it possible for any nontrivial solution of (8.16) to oscillate?

We first check that $x(t) = e^{-t^2}$ satisfies $x'' + 2tx' + 2x = 0$. It does. We now use part (c) of the Oscillation Theorem. It is not possible for any solution of (8.16) to oscillate, because e^{-t^2} has no zeros. If another solution were to oscillate, then between every zero of that other solution, the solution e^{-t^2} would have to equal zero. ■

EXERCISES

1. Using the results of this section, what can be said about the nontrivial solutions of the differential equation $x'' + (1 + e^t)x = 0$? Use a computer program to plot some numerical solutions of this differential equation. Do these numerical solutions agree with your predictions?

2. **Bessel's Equation.** Show that nontrivial solutions of $x'' + \frac{1}{t}x' + x = 0$ oscillate for $t > 0$. (This equation is known as BESSEL'S EQUATION OF ORDER ZERO.) Are these solutions bounded? Use a computer program to plot some numerical solutions of this differential equation. Do these numerical solutions agree with your predictions?

3. **Bessel's Equation.** Show that nontrivial solutions of $x'' + \frac{1}{t}x' + \left(1 - \frac{1}{4t^2}\right)x = 0$ oscillate for $t > 0$. (This equation is known as BESSEL'S EQUATION OF ORDER 1/2.) Are these solutions bounded? By making the substitution (8.10), find the general solution of this differential equation, and confirm that its solutions oscillate.

4. **Bessel's Equation.** Show that if μ is a constant, all nontrivial solutions of $x'' + \frac{1}{t}x' + \left(1 - \frac{\mu^2}{t^2}\right)x = 0$ oscillate for $t \geq a > 0$. (This equation is known as BESSEL'S EQUATION OF ORDER μ.)

5. **Hermite's Equation.** If n is a positive integer, it can be shown that the differential equation $x'' - 2tx' + 2nx = 0$ has a solution that is a polynomial of degree n. Using this fact, show that no nontrivial solution of this differential equation can oscillate. (This equation is known as HERMITE'S EQUATION OF ORDER n.) Use a computer program to plot some numerical solutions of this differential equation. Do these numerical solutions agree with your predictions?

6. **Laguerre's Equation.** If n is a positive integer, it can be shown that the differential equation $x'' + \left(\frac{1}{t} - 1\right)x' + \frac{n}{t}x = 0$ has a solution that is a polynomial of degree n. Using this fact, show that no nontrivial solution of this differential equation can oscillate. (This equation is known as LAGUERRE'S EQUATION OF ORDER n.) Use a computer program to plot some numerical solutions of this differential equation. Do these numerical solutions agree with your predictions?

7. Show that for $t > 0$, the differential equation $x'' + \frac{a^2}{t^2}x = 0$ is satisfied by

   (a) $x(t) = C_1 t^{r_1} + C_2 t^{r_2}$, where $r_1 = \left(1 + \sqrt{1 - 4a^2}\right)/2$, $r_2 = \left(1 - \sqrt{1 - 4a^2}\right)/2$, if $a^2 < 1/4$.

   (b) $x(t) = C_1\sqrt{t} + C_2\sqrt{t}\ln t$, if $a^2 = 1/4$.

   (c) $x(t) = C_1\sqrt{t}\cos(\omega \ln t) + C_2\sqrt{t}\sin(\omega \ln t)$, where $\omega = \left(\sqrt{4a^2 - 1}\right)/2$, if $a^2 > 1/4$.

   (d) How many zeros (for $t > 0$) can each of the solutions in parts (a), (b), and (c) have?

8. Show that if a function $f(t)$ oscillates and $f'(t)$ is continuous, then the function $f'(t)$ oscillates. Use this result to show that if the solutions of $x'' + p(t)x' + x = 0$ oscillate, then the solutions of $z'' + p(t)z' + [1 + p'(t)]z = 0$ also oscillate. [Hint: Differentiate the first equation, and set $z = x'$.] By setting $p(t) = 1/t$, show that the fact that solutions of Bessel's equation of order one oscillate is a consequence of the oscillation of solutions of Bessel's equation of order zero.

9. Figure 8.7 is the graph of a solution of one of the following second order linear differential equations, $x'' + \left(1 + e^{-t}\right)x = 0$, $x'' - \left(1 + e^{-t}\right)x = 0$. Choose the differential equation that could give rise to this figure, and give reasons for your choice. [Hint: Remember the Oscillation Theorem.]

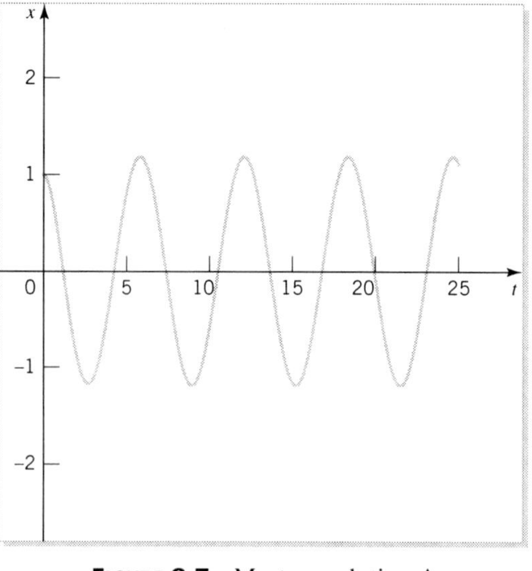

FIGURE 8.7 Mystery solution A

10. Show that the two functions $x_1(t) = \sin t$ and $x_2(t) = t$ both satisfy the linear differential equation

$(t\cos t - \sin t)x'' + (t\sin t)x' - (\sin t)x = 0$. According to the Oscillation Theorem and Relation Theorem, if x_1 has an infinite number of zeros, so does x_2. However, $\sin t$ has an infinite number of zeros, but t does not. Reconcile this apparent contradiction.

11. Show that $x_1(t) = t^2 - t$ and $x_2(t) = t$ both satisfy the differential equation $t^2x'' - 2tx' + 2x = 0$. According to the Oscillation Theorem and Relation Theorem, between consecutive zeros of x_1 there is exactly one zero of x_2. However, $t^2 - t$ has consecutive zeros of 0 and 1, whereas t has the zero of 0 that is not between 0 and 1. Reconcile this apparent contradiction.

12. Prove or disprove the following conjecture: If $Q(t)$ is continuous for $t \geq t_0$, and if $Q(t) \geq k^2 > 0$ (k a constant) for $t \geq t_0$, then between every zero of a nontrivial solution of $x'' + k^2x = 0$, there exists one and only one zero of any nontrivial solution of $x'' + Q(t)x = 0$.

13. Use the Relation Theorem to solve $x'' + 2f(t)x' + \left[c + f'(t) + f^2(t)\right]x = 0$, where c is a constant and $f(t)$ is an arbitrary function of t. Solve

(a) $x'' + 2tx' + \left(1 + t^2\right)x = 0$.

(b) $t^2x'' - 2tx' + \left(t^2 + 2\right)x = 0$.

(c) $tx'' + 2x' + 4tx = 0$.

14. Repeated Roots. Show that if $a_2(t)$, $a_1(t)$, and $a_0(t)$ are continuous in the vicinity of $t = t_0$, and if $x(t)$ is a nontrivial solution of the initial value problem $a_2(t)x'' + a_1(t)x' + a_0(t)x = 0, x(t_0) = 0, x'(t_0) = 0$, then $a_2(t_0) = 0$. [Hint: Use the Existence-Uniqueness Theorem.] Explain how this implies that a solution of $a_2(t)x'' + a_1(t)x' + a_0(t)x = 0$ cannot have a repeated root at $t = t_0$ if $a_2(t_0) \neq 0$. [Remember, a repeated root at $t = t_0$ has the property that $x(t_0) = 0$ and $x'(t_0) = 0$.]

(a) Can the function $x(t) = t^2$ be a solution of $x'' + p(t)x' + q(t)x = 0$, if p and q are continuous?

(b) **Legendre's Differential Equation.** The equation $(1 - t^2)x'' - 2tx' + m(m + 1)x = 0$, where m is a constant, is called LEGENDRE'S DIFFERENTIAL EQUATION. If m is a positive integer, it can be shown that this differential equation has m roots in the interval $-1 < t < 1$. Can any of these roots be repeated roots?

15. Explain why nontrivial solutions of the differential equation $x'' = f(t, x)$, where $f(t, x) > 0$, can never oscillate.

16. The Aging Spring. A differential equation that is used to model the motion of a mass, m, attached to a vertical spring whose spring "constant" is aging is

$$x'' + ke^{-at}x = 0, \tag{8.17}$$

where a and k are positive constants.

(a) If the spring is "young" ($t \approx 0$) or "strong" ($a \approx 0$), what will solutions look like?

(b) If the spring is "old" ($t \to \infty$) or "weak" ($a \to \infty$), what will solutions look like?

(c) Based on your observations in parts (a) and (b), do you expect all nontrivial solutions of (8.17) to oscillate?

(d) Use a computer program to plot various numerical solutions of (8.17). Are the results consistent with the conclusions you came to in parts (a) through (c)?

(e) Explain why the Oscillation Theorem does not apply to (8.17).

(f) By making the change of variable $u = 2\sqrt{k}e^{-at/2}/a$, show that the differential equation can be rewritten as

$$u^2\frac{d^2x}{du^2} + u\frac{dx}{du} + u^2x = 0.$$

This is Bessel's equation of order zero. Its nontrivial solutions oscillate as a function of u — see Exercise 2. Someone comments that this implies that all nontrivial solutions of (8.17) oscillate. What is wrong with this comment? [Hint: If $0 \leq t < \infty$, then $0 < u \leq 2\sqrt{k}/a$.]

(g) If (8.17) were replaced by $x'' + \left(b + ke^{-at}\right)x = 0$, where b is a positive constant no matter how small, how would this affect the results in parts (a) through (e)?

8.2 FINDING SOLUTIONS BY REDUCTION OF ORDER

Sometimes it is possible to find a particular nontrivial solution of $a_2(t)x'' + a_1(t)x' + a_0(t)x = 0$ by guessing, by luck, by inspection, by experimenting, or by insight. For example, in Example 8.4 on page 321, we found that $x(t) = t$ was a solution of $t^2x'' +$

$2tx' - 2x = 0$ by looking at numerical solutions. How can we use this information to find the general solution of $t^2x'' + 2tx' - 2x = 0$? In other words, if we know one (nontrivial) solution — say, $x = x_1(t)$ — of $a_2(t)x'' + a_1(t)x' + a_0(t)x = 0$, how can we find the general solution? The answer is to use the well-tried change of dependent variable, $x(t) = x_1(t)z(t)$.

Before looking at this method in general, let's look at an example to see how the technique works in a special case.

EXAMPLE 8.6

Solve

$$t^2x'' + 2tx' - 2x = 0 \tag{8.18}$$

by using the fact that $x(t) = t$ is a solution.

We first note that if we substitute $x(t) = t$ into the left-hand side of (8.18), it reduces to zero, so $x(t) = t$ is a solution of (8.18). Thus, $x_1(t) = t$, and we try the change of variable,

Change variable

$$x(t) = x_1(t)z(t) = tz(t). \tag{8.19}$$

Differentiating (8.19) twice, and substituting the results into (8.18), gives $t^2\left(tz'' + 2z'\right) + 2t\left(tz' + z\right) - 2tz = 0$, or $tz'' + 4z' = 0$. Notice that the term involving the dependent variable — z — explicitly, is missing. Thus, we can reduce this second order equation to a first order equation if we define $v = z'$. The second order equation then becomes

First order equation

$$tv' + 4v = 0, \tag{8.20}$$

which is a separable first order differential equation for $v(t)$. Notice we have reduced the order of the differential equation we are trying to solve from two to one. Solving for $v(t)$ gives $v(t) = C_1t^{-4}$, where C_1 is an arbitrary constant. Thus, $z' = v(t) = C_1t^{-4}$, which, when integrated once more, gives

$$z(t) = -\frac{1}{3}C_1t^{-3} + C_2, \tag{8.21}$$

General solution

where C_2 is an arbitrary constant. If we substitute this into (8.19), we find

$$x(t) = t\left(-\frac{1}{3}C_1t^{-3} + C_2\right) = -\frac{1}{3}C_1\frac{1}{t^2} + C_2t$$

as the solution of (8.18). This is in agreement with our findings from Example 8.4, where we discovered that nontrivial solutions of (8.18) could have no more than one zero for $t > 0$. ■

Now let's return to the general situation,

$$a_2(t)x'' + a_1(t)x' + a_0(t)x = 0, \tag{8.22}$$

where we assume that we know one solution — say, $x = x_1(t)$ — so that

$$a_2x_1'' + a_1x_1' + a_0x_1 = 0. \tag{8.23}$$

Change variable We try to find the general solution by using the change of dependent variable,

$$x(t) = x_1(t)z(t). \tag{8.24}$$

Substituting (8.24) into (8.22), we find

$$a_2\left(x_1z'' + 2x_1'z' + x_1''z\right) + a_1\left(x_1z' + x_1'z\right) + a_0x_1z = 0.$$

In this last equation we collect together like terms in z, z', and z'' to find

$$a_2x_1z'' + \left(2a_2x_1' + a_1x_1\right)z' + \left(a_2x_1'' + a_1x_1' + a_0x_1\right)z = 0.$$

Because x_1 is a known solution of (8.22) — that is, (8.23) is true — the last equation reduces to

$$a_2x_1z'' + \left(2a_2x_1' + a_1x_1\right)z' = 0.$$

First order equation If we define $v = z'$, then the last equation is

$$a_2x_1v' + \left(2a_2x_1' + a_1x_1\right)v = 0,$$

which is the counterpart of (8.20). Remember that in this equation, $a_2(t)$, $a_1(t)$, and $x_1(t)$ are all known functions. This is a first order separable equation with solution

$$z'(t) = v(t) = C_1\exp\left(-\int\frac{2a_2x_1' + a_1x_1}{a_2x_1}\,dt\right) = C_1\exp\left[-\int\left(2\frac{x_1'}{x_1} + \frac{a_1}{a_2}\right)dt\right].$$

Because

$$\exp\left(-\int 2\frac{x_1'}{x_1}\,dt\right) = \exp\left(-2\int\frac{dx_1}{x_1}\right) = \exp\left(-2\ln|x_1|\right) = \frac{1}{x_1^2},$$

we find

$$z'(t) = v(t) = C_1\frac{1}{x_1^2}\exp\left(-\int\frac{a_1}{a_2}\,dt\right).$$

Integrating once more gives

$$z(t) = C_1\int\left[\frac{1}{x_1^2}\exp\left(-\int\frac{a_1}{a_2}\,dt\right)\right]dt + C_2.$$

This is the counterpart of (8.21). Finally, in view of (8.24), we have the solution for

General solution the general case,

$$x(t) = C_1x_1\int\left[\frac{1}{x_1^2}\exp\left(-\int\frac{a_1}{a_2}\,dt\right)\right]dt + C_2x_1. \tag{8.25}$$

How to Reduce the Order

Purpose To solve

$$a_2(t)x'' + a_1(t)x' + a_0(t)x = 0 \tag{8.26}$$

if we already have one (nontrivial) solution, $x = x_1(t)$.

Process

1. Check that $x = x_1(t)$ is a solution of (8.26).
2. Introduce the new dependent variable, $z(t)$, by

$$x(t) = x_1(t)z(t). \tag{8.27}$$

3. Substitute (8.27) into (8.26) and collect together like terms in z, z', and z''. If everything has been done correctly, the coefficient of z should be zero.
4. Put $v = z'$ in the equation obtained in the last step. This should yield a first order separable equation, which we solve for $v = v(t)$.
5. Solve for $z = z(t)$ by integrating $z' = v(t)$, where $v(t)$ has been obtained from step 4.
6. Substitute this $z(t)$ into (8.27) to get the final solution.

Comments about Reduction of Order

- In steps 4 or 5 we may be unable to obtain familiar functions from the indicated integrations.
- We should always get two constants of integration in the final solution. It is a common mistake to lose one or both.
- The reduction of order technique always gives the general solution, as we will show in Section 8.4.
- The general solution should always be a linear combination of the original known solution, $x_1(t)$, and another solution.

In fact, this technique can also be used on nonhomogeneous differential equations, if we can find any nontrivial solution of the associated homogeneous equation. We illustrate this technique with two examples, one that has a forcing function not covered by Table 7.1 on page 287, and another that contains a differential equation that does not have constant coefficients. For the general case, see Exercise 4 on page 330.

EXAMPLE 8.7

Consider the differential equation

$$x'' + x = \cot t, \tag{8.28}$$

where we first find the characteristic equation of the associated homogeneous differential equation as $r^2 + 1 = 0$. The roots of this characteristic equation are i and $-i$, so the corresponding solutions of the associated homogeneous differential equation are $\cos t$ and $\sin t$. We now let one of these functions (say, $\sin t$) be included in the definition of a new dependent variable, $z(t)$, as

Change variable

$$x(t) = \sin t\, z(t). \tag{8.29}$$

Taking derivatives of (8.29) and substituting the results into (8.28) gives

$$\sin t\, z''(t) + 2\cos t\, z'(t) = \cot t.$$

Substituting $v(t) = z'(t)$ and dividing by $\sin t$ gives

$$v'(t) + 2\cot t\, v(t) = \frac{\cot t}{\sin t}, \tag{8.30}$$

First order equation

which is a linear differential equation in standard form. The integrating factor is

$$\exp\left(\int 2\cot t\, dt\right) = \exp\left(2\int \frac{\cos t}{\sin t}\, dt\right) = \exp\left(2\ln|\sin t|\right) = \sin^2 t,$$

so we multiply (8.30) by this factor and collect terms on the left-hand side of the result to obtain $\left[\sin^2 t\, v(t)\right]' = \sin t \cot t = \cos t$. Integration gives $\sin^2 t\, v(t) = \sin t$, where, because we need only one particular solution, we set the arbitrary constant to zero. This gives $v(t) = \csc t$, and because $z'(t) = v(t)$, we use a table of integrals to integrate once more, to obtain

$$z(t) = \int \csc t\, dt = \ln|\csc t - \cot t|.$$

Again we omitted the constant of integration. Thus, our particular solution is $\sin t \ln|\csc t - \cot t|$, and the general solution of (8.28) is

General solution

$$x(t) = C_1 \sin t + C_2 \cos t + \sin t \ln|\csc t - \cot t|. \tag{8.31}$$

We are familiar with the behavior of the first two functions in this solution, but we are unfamiliar with the behavior of the particular solution, $\sin t \ln|\csc t - \cot t|$. To analyze its behavior, we notice that neither $\csc t$ nor $\cot t$ is defined at $t = n\pi$, $n = 0$, ± 1, ± 2, $\cdots$. This is not surprising, because the forcing function in the differential equation (8.28) — namely, $\cot t$ — is not defined at these values of t, either. Thus, from the start we know that whatever the behavior of the solutions of (8.28), its regions of validity would be $n\pi < t < (n+1)\pi$ for $n = 0, \pm 1, \pm 2, \cdots$. The appropriate interval is determined by the value, t, used in the initial conditions.

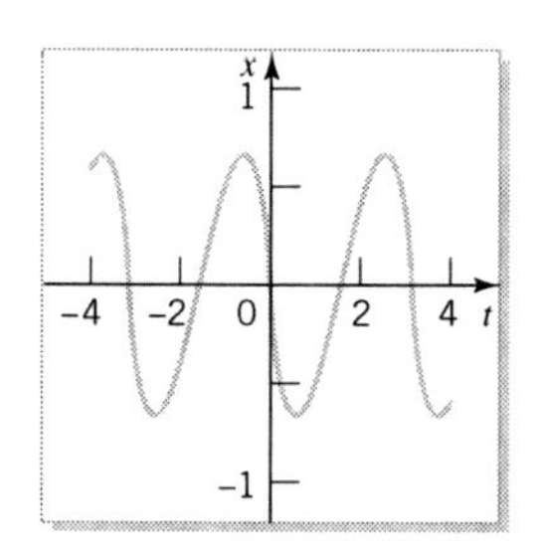

FIGURE 8.8
Graph of
$\sin t \ln |\csc t - \cot t|$

Figure 8.8 gives the graph of $\sin t \ln |\csc t - \cot t|$, where the function appears to be continuous everywhere, even though it is not defined at $t = n\pi$, $n = 0$, ± 1, ± 2, $\cdots$, and so cannot be continuous there. (What does this suggest about $\lim_{t \to n\pi} \sin t \ln |\csc t - \cot t|$? Can you prove it?)

How to Find a Particular Solution Using Reduction of Order

Purpose To find a particular solution of the second order linear differential equation,

$$a_2(t)x'' + a_1(t)x' + a_0(t)x = f(t), \tag{8.32}$$

using one solution of the associated homogeneous differential equation.

Process

1. Find a nontrivial solution, $x_1(t)$, of the associated homogeneous differential equation $a_2(t)x'' + a_1(t)x' + a_0(t)x = 0$.
2. Assume a particular solution of the form

$$x_p(t) = x_1(t)z(t), \tag{8.33}$$

and substitute (8.33) into (8.32). This will always lead to a first order differential equation in $v(t) = z'(t)$.
3. Solve the resulting first order equation for $v(t)$, and then integrate the result to find $z(t)$.
4. The final form of the particular solution is obtained by substituting the expression for $z(t)$ into (8.33).

EXAMPLE 8.8

Solve

$$t^2x'' - tx' + x = t^2 \ln t. \tag{8.34}$$

We first need a solution of the associated homogeneous differential equation, $t^2x'' - tx' + x = 0$. By inspection we see that $x_1(t) = t$ is a solution. We define a new dependent variable, $z(t)$, by

Change variable

$$x(t) = tz(t).$$

Differentiating this expression and using the results in (8.34) gives

$$t^3z''(t) + 2t^2z'(t) - t^2z'(t) = t^2 \ln t.$$

First order equation

Letting $v(t) = z'(t)$ and putting the result in the standard form of a first order linear differential equation gives

$$v' + \frac{1}{t}v = \frac{\ln t}{t}. \tag{8.35}$$

Finding the integrating factor as t and then multiplying (8.35) by this integrating factor gives $[tv(t)]' = \ln t$. Integration, by parts or by using a table of integrals, gives $tv(t) = t\ln t - t + C_1$, so $v(t) = \ln t - 1 + \frac{1}{t}C_1$, and one more integration gives

$$z(t) = \int v(t)\,dt = (t\ln t - t) - t + C_1 \ln t + C_2.$$

General solution

This gives a particular solution as $t^2 \ln t - 2t^2$, and the general solution as

$$x(t) = C_1 t \ln t + C_2 t + t^2 \ln t - 2t^2. \tag{8.36}$$

Notice that this technique gave not only a particular solution but also the general solution. This will happen if we include the arbitrary constants at the two steps in the method that involve integration (see Exercises 6 and 7 on page 330).

Figure 8.9 shows (8.36) subject to $x(1) = 0$ and $x'(1) = -2, -1, 0, 1, 2$, and Figure 8.10 has $x(1) = -4, -2, 0, 2, 4, 5$, and $x'(1) = 0$. In the latter figure it seems that all solutions that start with $x'(1) = 0$ pass through a common point. (Is this true?)

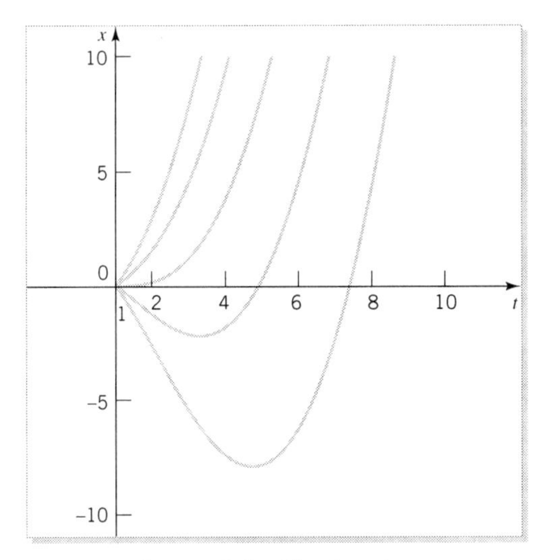

FIGURE 8.9 The function $x(t) = C_1 t\ln t + C_2 t + t^2 \ln t - 2t^2$, with $x(1) = 0$ and $x'(1) = -2, -1, 0, 1, 2$

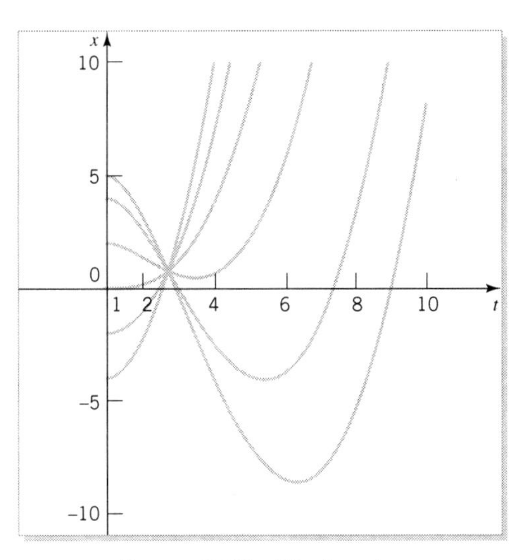

FIGURE 8.10 The function $x(t) = C_1 t\ln t + C_2 t + t^2 \ln t - 2t^2$, with $x(1) = -4, -2, 0, 2, 4, 5$, and $x'(1) = 0$

EXERCISES

1. In each of the following, check that $x_1(t)$ is a particular solution of the differential equation, and then find the general solution. Confirm that your answer is consistent with the Oscillation Theorem and Relation Theorem. If these theorems do not apply, explain fully.

(a) $x'' - 4x' + 4x = 0$, $\quad x_1(t) = e^{2t}$
(b) $t^2x'' + tx' - 4x = 0$, $\quad x_1(t) = t^2$
(c) $t^2x'' - tx' + x = 0$, $\quad x_1(t) = t$
(d) $tx'' - 2(t+1)x' + (t+2)x = 0$, $\quad x_1(t) = e^t$
(e) $t^2x'' - 2tx' + (t^2+2)x = 0$, $\quad x_1(t) = t \sin t$
(f) $t^2x'' - 3tx' + 4x = 0$, $\quad x_1(t) = t^2$
(g) $t^2x'' - t(t+2)x' + (t+2)x = 0$, $\quad x_1(t) = t$
(h) $4t^2x'' + 4tx' + (4t^2 - 1)x = 0$, $\quad x_1(t) = \sin t/\sqrt{t}\ (t > 0)$

2. Consider the differential equation $ax'' + bx' + cx = 0$, where a, b, and c are constants, and $b^2 - 4ac = 0$. Show that $x = e^{rt}$, where $r = -b/(2a)$, is a solution of this differential equation, and then find its general solution by reduction of order.

3. Find the general solution of the following differential equations.

(a) $x'' + 2x' + x = 2t^{-2}e^{-t}$, $\quad t > 0$
(b) $x'' + 6x' + 9x = t^{-1}e^{-3t}$, $\quad t > 0$
(c) $x'' + x = \sec t$, $\quad -\pi/2 < t < \pi/2$
(d) $x'' + x = \tan t$, $\quad -\pi/2 < t < \pi/2$

4. Consider the nonhomogeneous second order linear differential equation,

$$a_2(t)x'' + a_1(t)x' + a_0(t)x = f(t), \qquad (8.37)$$

where $u(t)$ is a solution of the corresponding homogeneous differential equation. Show that the substitution $x = u(t)z(t)$ gives a particular solution of (8.37) as

$$x(t) = \int \left[\frac{1}{u^2(t)\alpha(t)} \int \frac{u(t)\alpha(t)f(t)}{a_2(t)}\, dt \right] dt,$$

where[5]

$$\alpha(t) = \exp\left(\int \frac{a_1(t)}{a_2(t)}\, dt \right).$$

5. Use the method of Exercise 4 to find particular solutions for Exercise 3.

6. Consider the equation $t^3x'' + tx' - x = e^{1/t}$.

(a) Show that $x = t$ is a solution of the associated homogeneous differential equation.
(b) Show that the differential equation satisfied by $z(t)$, with the change of variable $z = u(t)t$, is $t^4z'' + (2t^3 + t^2)z' = e^{1/t}$.
(c) Solve this equation and obtain the general solution of the original differential equation as $x(t) = C_1t + C_2te^{1/t} + (1-t)e^{1/t}$.

7. Consider the equation $(t^2 + t)x'' + (2 - t^2)x' - (2+t)x = (t+1)^2$.

(a) Show that e^t is a solution of the associated homogeneous differential equation.
(b) Obtain the general solution of the original differential equation.

8.3 FINDING SOLUTIONS BY VARIATION OF PARAMETERS

The method of undetermined coefficients of Section 7.2 is guaranteed to work only when the forcing function for linear differential equations with constant coefficients has the forms listed in Table 7.1 on page 287. However, the method given in the previous section — reduction of order — requires only that the differential equation be linear, not that it have constant coefficients. Also, the forcing function is not limited to the forms of Table 7.1 for this method to work. Although the method of reduction of order is more general than the one given in Section 7.2, it may have drawbacks. Any one of the two successive integrations required by this method may not be expressible in terms of familiar functions, in which case the particular solution

[5] Remember, $\exp(a) = e^a$.

will be left as a sequence of integrals (see Exercise 4 on page 336). Thus, we proceed to develop another method that does not have this limitation.

In Example 8.7 on page 327 we had information that we did not exploit in trying to find a particular solution of

$$a_2(t)x'' + a_1(t)x' + a_0(t)x = f(t); \tag{8.38}$$

namely, we used only one of the two solutions, $x_1(t)$ and $x_2(t)$, of the associated homogeneous differential equation

$$a_2(t)x'' + a_1(t)x' + a_0(t)x = 0.$$

How might we use both? Rather than consider either $x_1(t)z(t)$ or $x_2(t)z(t)$ and try to select $z(t)$ in an appropriate way, we could consider

Trial solution

$$x_p(t) = x_1(t)z_1(t) + x_2(t)z_2(t) \tag{8.39}$$

and seek functions $z_1(t)$ and $z_2(t)$ such that $x_p(t)$ satisfies (8.38). However, we should note that we have only one condition on $x_p(t)$, whereas we are allowed to select two functions, $z_1(t)$ and $z_2(t)$. This means that we will have some flexibility in selecting $z_1(t)$ and $z_2(t)$. One choice could be to set $z_2(t) = 0$ and then determine $z_1(t)$, which is the reduction of order method we have just dealt with in Section 8.2. However, there is an alternative way to exploit this flexibility.

Because we will substitute the expression in (8.39) into the differential equation (8.38), we first calculate derivatives of this x_p. Doing so gives

$$x_p' = x_1'(t)z_1(t) + x_1(t)z_1'(t) + x_2'(t)z_2(t) + x_2(t)z_2'(t),$$

which we can write as

$$x_p' = x_1'(t)z_1(t) + x_2'(t)z_1(t) + \left[x_1(t)z_1'(t) + x_2(t)z_2'(t)\right]. \tag{8.40}$$

From this we find

$$x_p'' = x_1''(t)z_1(t) + x_1'(t)z_1'(t) + x_2''(t)z_2(t) + x_2'(t)z_2'(t) + \left[x_1(t)z_1'(t) + x_2(t)z_2'(t)\right]'. \tag{8.41}$$

If we now substitute the expressions in (8.39), (8.40), and (8.41) into the original differential equation (8.38), we obtain

$$\begin{aligned} a_2(t)\left\{x_1'(t)z_1'(t) + x_2'(t)z_2'(t) + \left[x_1(t)z_1'(t) + x_2(t)z_2'(t)\right]'\right\} + \\ + a_1(t)\left[x_1(t)z_1'(t) + x_2(t)z_2'(t)\right] = f(t). \end{aligned} \tag{8.42}$$

This is a single equation from which we expect to obtain the two functions $z_1(t)$ and $z_2(t)$. We can exploit the flexibility mentioned earlier and greatly simplify (8.42) if we require that the coefficient of $a_1(t)$ be zero; that is,

$$x_1(t)z_1'(t) + x_2(t)z_2'(t) = 0,$$

in which case (8.42) reduces to

$$a_2(t)\left[x_1'(t)z_1'(t)+x_2'(t)z_2'(t)\right]=f(t).$$

Thus, we have two equations in the two unknowns, $z_1'(t)$ and $z_2'(t)$, which we write as the system

$$\begin{cases} x_1(t)z_1'(t)+x_2(t)z_2'(t) = 0, \\ x_1'(t)z_1'(t)+x_2'(t)z_2'(t) = f(t)/a_2(t). \end{cases} \tag{8.43}$$

The system in (8.43) has a unique solution for $z_1'(t)$ and $z_2'(t)$ if the determinant of the coefficients,

$$W\left[x_1(t),x_2(t)\right]=\begin{vmatrix} x_1(t) & x_2(t) \\ x_1'(t) & x_2'(t) \end{vmatrix},$$

is never zero. It could be zero; for example, if $x_1(t)$ and $x_2(t)$ are proportional (see Exercise 1 on page 336). In the next section we will discuss when the object $W\left[x_1(t),x_2(t)\right]$ — called the Wronskian of $x_1(t)$ and $x_2(t)$ — is not zero. For the time being, let's assume that $W\left[x_1(t),x_2(t)\right]\neq 0$, so we solve for $z_1'(t)$ and $z_2'(t)$ as

Wronskian

$$z_1'(t)=-\frac{f(t)x_2(t)}{a_2(t)W\left[x_1(t),x_2(t)\right]},$$

$$z_2'(t)=\frac{f(t)x_1(t)}{a_2(t)W\left[x_1(t),x_2(t)\right]}.$$

Particular solution

To obtain the final form of the particular solution, we integrate these two equations and substitute the result into the original form of x_p from (8.39). This gives

$$x_p(t)=-x_1(t)\int\frac{f(t)x_2(t)}{a_2(t)W\left[x_1(t),x_2(t)\right]}\,dt+x_2(t)\int\frac{f(t)x_1(t)}{a_2(t)W\left[x_1(t),x_2(t)\right]}\,dt. \tag{8.44}$$

Most people do not find it useful to memorize the form of the particular solution in (8.44), but instead mimic the process leading to (8.44). We will now work an example using this procedure, called variation of parameters.

EXAMPLE 8.9

Find a particular solution of

$$x''-3x'+2x=-\frac{e^{2t}}{e^t+1}. \tag{8.45}$$

Our first step is to solve the associated homogeneous differential equation by factoring its characteristic equation, $r^2-3r+2=(r-2)(r-1)=0$. This gives the two solutions of the associated homogeneous differential equation as $x_1(t) = e^{2t}$ and $x_2(t) = e^t$. We calculate the Wronskian of these two solutions,

Wronskian

$$W\left[e^{2t}, e^t\right] = \begin{vmatrix} e^{2t} & e^t \\ \left(e^{2t}\right)' & \left(e^t\right)' \end{vmatrix} = \begin{vmatrix} e^{2t} & e^t \\ 2e^{2t} & e^t \end{vmatrix} = -e^{3t},$$

which is not zero.

Trial solution

The second step is to write the form for a particular solution as

$$x_p(t) = e^{2t} z_1(t) + e^t z_2(t), \tag{8.46}$$

along with the two equations that $z_1'(t)$ and $z_2'(t)$ must satisfy, namely,

$$\begin{aligned} e^{2t} z_1' + e^t z_2' &= 0, \\ 2e^{2t} z_1' + e^t z_2' &= -e^{2t} / \left(e^t + 1\right). \end{aligned} \tag{8.47}$$

The last step is to solve this system of equations for $z_1'(t)$ and $z_2'(t)$, integrate, and substitute the result back into (8.46). We may multiply the top equation in (8.47) by 2 and subtract the bottom equation from the result to obtain

$$e^t z_2' = \frac{e^{2t}}{e^t + 1},$$

or

$$z_2' = \frac{e^t}{e^t + 1}. \tag{8.48}$$

Using this expression in the top equation of (8.47), we obtain

$$e^{2t} z_1' = -\frac{e^{2t}}{e^t + 1},$$

or

$$z_1' = -\frac{1}{e^t + 1} = -\frac{e^{-t}}{1 + e^{-t}}. \tag{8.49}$$

Integrating the expressions in (8.48) and (8.49) gives $z_2 = \ln(e^t + 1)$ and $z_1(t) = \ln(1 + e^{-t})$, and our particular solution is obtained from (8.46) as

$$x_p(t) = e^{2t} \ln(1 + e^{-t}) + e^t \ln(e^t + 1).$$

General solution

The general solution of (8.45) is then given by

$$x(t) = C_1 e^{2t} + C_2 e^t + e^{2t} \ln(1 + e^{-t}) + e^t \ln(e^t + 1).$$

Notice that because the functions $z_1(t)$ and $z_2(t)$ multiply $x_1(t)$ and $x_2(t)$, the inclusion of the constants of integration when finding $z_1(t)$ and $z_2(t)$ will not change the form of the general solution. Including them will give us the general solution directly. ■

How to Find a Particular Solution Using Variation of Parameters

Purpose To find a particular solution of the second order linear differential equation

$$a_2(t)x'' + a_1(t)x' + a_0(t)x = f(t) \tag{8.50}$$

using two linearly independent solutions of the associated homogeneous differential equation.

Process

1. Solve the associated homogeneous differential equation — namely, $a_2(t)x'' + a_1(t)x' + a_0(t)x = 0$ — and obtain two solutions, say, $x_1(t)$ and $x_2(t)$, for which the Wronskian

$$W\left[x_1(t), x_2(t)\right] = \begin{vmatrix} x_1(t) & x_2(t) \\ x_1'(t) & x_2'(t) \end{vmatrix}$$

is nonzero.

2. Assume a particular solution of the form

$$x_p(t) = x_1(t)z_1(t) + x_2(t)z_2(t), \tag{8.51}$$

and write down the system of equations satisfied by $z_1'(t)$ and $z_2'(t)$,

$$\begin{aligned} x_1(t)z_1'(t) + x_2(t)z_2'(t) &= 0, \\ x_1'(t)z_1'(t) + x_2'(t)z_2'(t) &= f(t)/a_2(t). \end{aligned} \tag{8.52}$$

3. Solve (8.52) for $z_1'(t)$ and $z_2'(t)$, and integrate to obtain $z_1(t)$ and $z_2(t)$. (We may set the arbitrary constants to zero.) The final form of the particular solution is obtained by substituting these values of $z_1(t)$ and $z_2(t)$ into (8.51).

Comments about Variation of Parameters

- Including the arbitrary constants when integrating the expressions for $z_1'(t)$ and $z_2'(t)$ will result in the expression for x_p also being the general solution of (8.50).
- If we are solving an initial value problem and cannot evaluate the integrals for $z_1(t)$ and $z_2(t)$ in terms of familiar functions, it is often convenient to write the integrals with the independent variable as the upper limit and the value of t giving the initial condition as the lower limit. See Exercise 5 on page 336.
- A common mistake in writing down (8.52) is to forget to divide $f(t)$ by $a_2(t)$ on the right-hand side of the bottom equation.

In the method of undetermined coefficients — Section 7.2 — we stated that we would redo one of the examples from that section to illustrate why the method of undetermined coefficients — when it applies — is usually preferable to variations of parameters. We will redo Example 7.6 on page 287, $x'' + 16x = 16t^2$; there we solved it in a few lines, finding $x(t) = C_1 \sin 4t + C_2 \cos 4t - 1/8 + t^2$.

EXAMPLE 8.10

Solve $x'' + 16x = 16t^2$.

The solution of the associated homogeneous equation, $x'' + 16x = 0$, is $x_h(t) = C_1x_1(t) + C_2x_2(t)$, where $x_1(t) = \sin 4t$ and $x_2(t) = \cos 4t$. The Wronskian, $W[x_1(t), x_2(t)]$, is not zero.

Trial solution

We assume a particular solution of the form

$$x_p(t) = x_1(t)z_1(t) + x_2(t)z_2(t) = z_1(t) \sin 4t + z_2(t) \cos 4t,$$

and write down the system of equations satisfied by $z_1'(t)$ and $z_2'(t)$,

$$\begin{aligned} z_1'(t) \sin 4t + z_2'(t) \cos 4t &= 0, \\ 4z_1'(t) \cos 4t - 4z_2'(t) \sin 4t &= 16t^2. \end{aligned}$$

We solve this system for $z_1'(t)$ and $z_2'(t)$ by multiplying the first equation by $4 \sin 4t$ and the second by $\cos 4t$ and adding the result, noting that $\sin^2 4t + \cos^2 4t = 1$, to find

$$z_1'(t) = 4t^2 \cos 4t.$$

We substitute this into the first equation and have

$$z_2'(t) = -4t^2 \sin 4t.$$

Each of these last two equations can be integrated by parts (twice), giving

$$\begin{aligned} z_1(t) &= \int 4t^2 \cos 4t \, dt = t^2 \sin 4t + \frac{1}{2}t \cos 4t - \frac{1}{8} \sin 4t, \\ z_2(t) &= \int -4t^2 \sin 4t \, dt = t^2 \cos 4t - \frac{1}{2}t \sin 4t - \frac{1}{8} \cos 4t, \end{aligned}$$

where we have set the constants of integration to zero. Substituting these values of $z_1(t)$ and $z_2(t)$ into $x_p(t)$ gives

Particular solution

$$x_p(t) = \left(t^2 \sin 4t + \frac{1}{2}t \cos 4t - \frac{1}{8} \sin 4t\right) \sin 4t + \left(t^2 \cos 4t - \frac{1}{2}t \sin 4t - \frac{1}{8} \cos 4t\right) \cos 4t,$$

which simplifies to $x_p(t) = t^2 - 1/8$, in agreement with Example 7.6. ■

EXERCISES

1. Show that if $x_1(t) = cx_2(t)$, where c is a constant, then $W[x_1(t), x_2(t)] = 0$.

2. Consider the equation $(t^2 + t)x'' + (2 - t^2)x' - (2 + t)x = (t + 1)^2$.
   (a) Show that e^t is a solution of the associated homogeneous differential equation.
   (b) Find a second solution of this homogeneous equation by making a change of variable, $x = z(t)e^t$.
   (c) Use the method of variation of parameters to find a particular solution.
   (d) Write down the general solution of the original differential equation and compare your answer with the one from Exercise 7 on page 330.

3. Find the general solution of the following differential equations. Compare your answers with the ones you obtained for Exercise 3 on page 330.
   (a) $x'' + 2x' + x = 2t^{-2}e^{-t}$, $\quad t > 0$
   (b) $x'' + 6x' + 9x = t^{-1}e^{-3t}$, $\quad t > 0$
   (c) $x'' + x = \sec t$, $\quad -\pi/2 < t < \pi/2$
   (d) $x'' + x = \tan t$, $\quad -\pi/2 < t < \pi/2$.
   [Hint: Use $\sin^2 t + \cos^2 t = 1$.]

4. Use reduction of order with the solution e^t of $x'' - x = 0$, to show that the solution of the initial value problem, $x'' - x = 1/t, x(1) = A, x'(1) = B$, may be written in the form
$$x(t) = e^t \int_1^t e^{-2x}\left(\int_1^x \frac{e^s}{s}\,ds\right)dx - \frac{B}{2}e^{2-t} + \left(\frac{A}{e} + \frac{B}{2}\right)e^t.$$

5. Show that using the method of variation of parameters gives the solution of the initial value problem of Exercise 4 as
$$x(t) = \frac{1}{2}\left[(A + B)e^{t-1} + (A - B)e^{1-t} + \right.$$
$$\left. + e^t \int_1^t \frac{e^{-s}}{s}\,ds - \int_1^t \frac{e^s}{s}\,ds\right].$$

6. Solve $x'' - x = 6e^{2t}$ by (a) the method of undetermined coefficients, (b) reduction of order, and (c) variation of parameters.

8.4 THE IMPORTANCE OF LINEAR INDEPENDENCE AND DEPENDENCE

A common feature of the general solutions for the specific cases of

$$a_2(t)x'' + a_1(t)x' + a_0(t)x = 0, \tag{8.53}$$

which we have obtained in Chapters 6 and 7, is that they consist of the sum of two terms, each being an arbitrary constant multiplying a particular solution. We might wonder if all solutions of (8.53) have this same structure. That is, given two solutions, $x_1(t)$ and $x_2(t)$, of (8.53), is it possible to give the general solution by simply forming a linear combination of the two solutions; namely, $C_1x_1(t) + C_2x_2(t)$?

For example, consider the differential equation $x'' - 4x = 0$, which has the general solution $x(t) = C_1e^{2t} + C_2e^{-2t}$. Thus, $e^{2t}, 3e^{2t}, 5e^{2t}, 2e^{-2t}$, and $7e^{-2t}$ are all particular solutions (along with many others). If we select two of these particular solutions — say, $x_1(t) = e^{2t}$ and $x_2(t) = 3e^{2t}$ — is their linear combination, $C_1e^{2t} + C_23e^{2t}$, the general solution of our differential equation? Using a little algebra gives us $C_1e^{2t} + C_23e^{2t} = (C_1 + C_23)\,e^{2t}$, so this is not equivalent to our previously obtained general solution. Thus, we must take more care when choosing our two particular solutions.

Linear Independence and Dependence

In this section we will introduce the topic of linear independence of functions, which we will use to be assured that the linear combination we form with two particular solutions will indeed be the general solution of our second order linear differential equation.

The words **independence** and **dependence** in mathematics have meanings very similar to their common usage. If one function may be written as a constant times a second function, we could write

$$f(t) = cg(t),$$

where c is a nonzero constant. Here we see that to find the value of f at some point t, we compute $g(t)$ and then multiply the result by the constant c. For this reason, we say that $f(t)$ depends on $g(t)$. Equivalently, if we divide by c we find

$$g(t) = \frac{1}{c}f(t)$$

and say that $g(t)$ depends on $f(t)$. An alternative way of expressing this relationship is as follows: Two functions, f and g, are linearly dependent on an interval $a < t < b$, in the common domain of f and g, if there exist two constants, b_1 and b_2 (not both zero), such that

$$b_1 f(t) + b_2 g(t) = 0 \tag{8.54}$$

for all t in this interval. However, in differential equations, what we really need is a method for making sure we do **not** have dependent functions. Thus, we look at (8.54) and give the following definition.

◆ *Definition 8.1:* **Two functions $f(t)$ and $g(t)$ are LINEARLY INDEPENDENT if the only way for**

$$b_1 f(t) + b_2 g(t) = 0$$

to be true for $a < t < b$ in some common domain of f and g, is to have $b_1 = b_2 = 0$. If two functions are not linearly independent, they are LINEARLY DEPENDENT. ◆

Comments about Linear Independence and Linear Dependence

- If either f or g is identically zero, f and g are linearly dependent. All we need to do is choose the coefficient of the identically zero function to be nonzero and the other coefficient to be zero.
- If one function is a constant multiple of another, the two functions are linearly dependent. We show this by noting that if $f = Cg$, then $f - Cg = 0$, and we have the form of (8.54) with $b_1 = 1$ and $b_2 = -C$. Thus, the functions e^{2t} and $3e^{2t}$ from our earlier discussion are linearly dependent.
- The two functions $\sin \beta t$ and $\cos \beta t$ are linearly independent for all values of t if $\beta \neq 0$. To show this we start with the equation

$$b_1 \sin \beta t + b_2 \cos \beta t = 0 \tag{8.55}$$

and see if we can find nonzero values of b_1 and b_2 such that the equation is true for all t. Because (8.55) must be true for all values of t, it must also be true for the two specific values of 0 and $\pi/(2\beta)$. Choosing $t = 0$ gives $b_1 \sin 0 + b_2 \cos 0 = 0$, so $b_2 = 0$. Choosing $t = \pi/(2\beta)$ gives $b_1 = 0$, and, because both b_1 and b_2 must be zero, $\sin \beta t$ and $\cos \beta t$ are linearly independent functions for any nonzero β and all values of t.

- If two functions are linearly independent for all intervals in the common domain of the functions, we often omit mention of the domain. Thus, we simply say that $\sin \beta t$ and $\cos \beta t$ are linearly independent for $\beta \neq 0$.
- If $r_1 \neq r_2$, then $e^{r_1 t}$ and $e^{r_2 t}$ are linearly independent for all values of t. To show this we need to determine if there exist constants, not both zero, such that

$$b_1 e^{r_1 t} + b_2 e^{r_2 t} = 0 \tag{8.56}$$

for all t in some interval. In other words, we want (8.56) to be an identity. If (8.56) is true for t in some interval, we may differentiate this equation and obtain

$$r_1 b_1 e^{r_1 t} + r_2 b_2 e^{r_2 t} = 0. \tag{8.57}$$

Now we look upon (8.56) and (8.57) as two equations in the two unknowns, b_1 and b_2, for which we seek nonzero values that make these equations true. Solving these two equations, we find $b_1(r_2 - r_1)e^{r_1 t} = 0$ and $b_2(r_2 - r_1)e^{r_2 t} = 0$. Now, because $(r_2 - r_1)e^{r_1 t}e^{r_2 t} \neq 0$, then both b_1 and b_2 are zero, and the functions $e^{r_1 t}$ and $e^{r_2 t}$ are linearly independent for all values of t.

Two Theorems

We can use an argument similar to the last case, involving $e^{r_1 t}$ and $e^{r_2 t}$, to develop a test for linear independence of any two functions — say, f and g — that are differentiable on the interval $a < t < b$. If we form the equation $b_1 f(t) + b_2 g(t) = 0$ and also write down its derivative, $b_1 f'(t) + b_2 g'(t) = 0$, we have two equations in the two unknowns b_1 and b_2. Solving gives $b_1 [f(t)g'(t) - f'(t)g(t)] = 0$ and $b_2 [f(t)g'(t) - f'(t)g(t)] = 0$. If the quantity in brackets in these latter two expressions is not zero, then b_1 and b_2 must equal zero, so f and g are linearly independent functions. This quantity occurs often in considerations of linear independence and dependence, and given the name WRONSKIAN.

◆ *Definition 8.2:* **The Wronskian of two functions f and g, labeled $W[f,g]$, is defined as**

$$W[f,g] = f(t)g'(t) - f'(t)g(t).$$

◆

We now summarize this discussion with the following theorem.

► **Theorem 8.4: The Wronskian Theorem** *If the Wronskian of two functions is not identically zero on some nonzero interval, then the two functions are linearly independent.* ◄

Comments about the Wronskian Theorem

- Notice that the Wronskian is given by the value of the determinant

$$\begin{vmatrix} f(t) & g(t) \\ f'(t) & g'(t) \end{vmatrix}.$$

- Note that this theorem does **not** say that if the Wronskian of two functions is zero, then the two functions are linearly dependent. The theorem tells us nothing about linear independence or dependence if the Wronskian is identically zero. For example, the Wronskian of the two functions t^3 and $|t^3|$ is identically zero for all values of t. However, these functions are linearly independent on any interval $a < t < b$ that contains the origin, and are linearly dependent on any interval that does not. (Why?)
- This Wronskian is the same object that occurred in Section 8.3.

We can use this theorem to prove a result that puts the method of solving differential equations that we developed in the previous two sections on a solid footing. It also foreshadows what we will need for linear differential equations of order greater than two.

▶ *Theorem 8.5: The General Solution Theorem* *Consider the general second order linear differential equation*

$$a_2(t)x'' + a_1(t)x' + a_0(t)x = 0, \tag{8.58}$$

where $a_2(t)$, $a_1(t)$, and $a_0(t)$ are continuous, with $a_2(t) \neq 0$ for $a < t < b$. If $x_1(t)$ and $x_2(t)$ are two solutions of (8.58), and the functions x_1 and x_2 are linearly independent for $a < t < b$, then all solutions of (8.58) may be written as

$$x(t) = C_1x_1(t) + C_2x_2(t). \tag{8.59}$$

◀

Proof Choose general initial values for x and x' at $t = t_0$ (where $a < t_0 < b$) as

$$x(t_0) = x_0, \; x'(t_0) = x_0^*. \tag{8.60}$$

The coefficients in (8.58) are such that the hypothesis of the Existence-Uniqueness Theorem for Second Order Linear Differential Equations on page 238 is satisfied. Thus, we can use the conclusion of that theorem, which states that there is a unique solution of (8.58) that satisfies the initial conditions in (8.60). If we impose the initial conditions of (8.60) on our solution in (8.59), we obtain two simultaneous algebraic equations:

$$C_1x_1(t_0) + C_2x_2(t_0) = x_0,$$
$$C_1x_1'(t_0) + C_2x_2'(t_0) = x_0^*.$$

Eliminating first C_2 and then C_1 from these equations gives

$$C_1\left[x_1(t_0)x_2'(t_0) - x_1'(t_0)x_2(t_0)\right] = x_0x_2'(t_0) - x_0^*x_2(t_0),$$
$$C_2\left[x_1(t_0)x_2'(t_0) - x_1'(t_0)x_2(t_0)\right] = x_0^*x_1(t_0) - x_0x_1'(t_0).$$

From here we see that if the term $\left[x_1(t_0)x_2'(t_0) - x_1'(t_0)x_2(t_0)\right] \neq 0$, then we obtain a unique solution for C_1 and C_2. However, this expression is simply the Wronskian of x_1 and x_2, which, because $x_1(t)$ and $x_2(t)$ are both solutions of (8.58), is never zero if x_1 and x_2 are linearly independent (see Exercise 5 on page 342). Thus, C_1 and C_2 are uniquely determined, and we have obtained the unique solution of our initial value problem. Because t_0, x_0, and x_0^* were chosen arbitrarily, we see that all solutions of (8.58) may be written as a linear combination of two linearly independent solutions.

Comments about the General Solution Theorem

- Our task will be to find linearly independent solutions. This form of our solution, (8.59), where C_1 and C_2 are arbitrary constants, is the general solution of (8.58) because all solutions may be written in this form. In fact, this theorem is the justification for our prior use of the expression "general solution."
- The significance of this theorem is that no matter how we obtain $x_1(t)$ and $x_2(t)$, if they are linearly independent and satisfy (8.58), then the general solution is (8.59).
- For the variation of parameters method of the previous section to work, we needed two solutions, $x_1(t)$ and $x_2(t)$, for which $W\left[x_1(t), x_2(t)\right] \neq 0$. Thus, we needed two linearly independent solutions.
- If $x_1(t)$ and $x_2(t)$ are linearly dependent and not zero functions, then they will have the same zeros.
- The Wronskian of the two solutions of $a_2x'' + a_1x' + a_0x = 0$ is either always zero or never zero. (See Exercise 5 on page 342.)
- We could restate part (c) of the Oscillation Theorem in terms of linearly independent solutions. Also, part (c) of the Oscillation Theorem tells us that if $x_1(t)$ and $x_2(t)$ have the same zeros, then $x_1(t)$ and $x_2(t)$ are linearly dependent. (How does it tell us that?)

EXAMPLE 8.11

We found earlier that $e^{\alpha t}$ and $te^{\alpha t}$ were both solutions of the second order differential equation with constant coefficients

$$a_2x'' + a_1x' + a_0x = 0 \tag{8.61}$$

Wronskian

for the case $a_1^2 - 4a_2a_0 = 0$. Taking the Wronskian of these two solutions gives

$$W[e^{\alpha t}, te^{\alpha t}] = \begin{vmatrix} e^{\alpha t} & te^{\alpha t} \\ \alpha e^{\alpha t} & (1+\alpha t)\,e^{\alpha t} \end{vmatrix} = e^{2\alpha t}.$$

Linearly independent

Because this Wronskian is not zero, we may conclude that these two functions are linearly independent and that the general solution of (8.61) may be written as $x(t) = C_1 e^{\alpha t} + C_2 t e^{\alpha t}$.

EXAMPLE 8.12 *Reduction of Order*

We now show that in the reduction of order technique of Section 8.2, the two solutions $x_1(t)$ and

$$x_2(t) = x_1(t)z(t),$$

where

$$z(t) = \int \left[\frac{1}{x_1^2} \exp\left(-\int \frac{a_1}{a_2}\, dt\right)\right] dt$$

Wronskian

[see (8.25)], are linearly independent. The Wronskian, $W[x_1(t), x_2(t)]$, in this case becomes

$$W[x_1(t), x_2(t)] = W[x_1(t), x_1(t)z(t)] = \begin{vmatrix} x_1 & x_1 z \\ x_1' & x_1 z' + x_1' z \end{vmatrix} = x_1^2 z'.$$

But, from

$$z' = \frac{1}{x_1^2} \exp\left(-\int \frac{a_1}{a_2}\, dt\right),$$

we have

$$W[x_1(t), x_2(t)] = \exp\left(-\int \frac{a_1}{a_2}\, dt\right).$$

Linearly independent

Because the exponential function can never be zero, the Wronskian is never zero, proving that $x_1(t)$ and $x_2(t)$ are linearly independent.

EXERCISES

1. Show that the following sets of functions are linearly independent for all values of t by computing their Wronskian.

(a) t, t^2
(b) $t, 4+t$
(c) $1, e^t$
(d) t, t^{-1}
(e) $\sinh 2t, \cosh 2t$
(f) e^{2t}, e^{-3t}
(g) e^{-t}, te^{-t}
(h) $\sin t, \sin^2 t$
(i) $e^{\alpha t}\cos\beta t, e^{\alpha t}\sin\beta t,\ \beta \neq 0$
(j) $t^a, t^a \ln t$
(k) $t^a\cos(b\ln t), t^a\sin(b\ln t),\ b \neq 0$

2. Let $f(t)$ be an odd function and $g(t)$ be an even function. Neither is the zero function.

(a) Prove that f and g are linearly independent on any interval that includes the origin as an interior point.

(b) Draw graphs of typical functions for f and g, and explain part (a) from the graph.

(c) Draw the graph of a situation in which f and g are linearly dependent for $t > 0$, and linearly independent for $t < 0$.

3. Determine all values of the constant, k, that make the following sets of functions linearly independent.

(a) $t, k+t$ (b) $t, 1+kt$
(c) $1+2t, k+t$ (d) $1+2t, 1+kt$
(e) $3, 2+kt$ (f) $3t, kt$
(g) $t^2, (t+k)^2$ (h) $3e^{2t}, 6e^{(k-2)t}$
(i) $\sin^2 kt + \cos^2 kt, 6$ (j) $k \sin t \cos t, 9 \sin 2t$
(k) e^{kt}, te^{kt} (l) $\sin kt, \cos kt$

4. Sketch graphs of the functions in each part of Exercise 3, and discuss the graphical interpretation of the various situations that lead to linear independence.

5. The purpose of this exercise is to show that two solutions of $a_2(t)x'' + a_1(t)x' + a_0(t)x = 0$ are linearly independent if and only if their Wronskian is nonzero.

(a) Show that if x_1 and x_2 both satisfy the differential equation $a_2x'' + a_1x' + a_0x = 0$, then $a_2\left(x_1x_2'' - x_2x_1''\right) + a_1\left(x_1x_2' - x_2x_1'\right) = 0$.

(b) From the Wronskian $W[x_1, x_2] = x_1'x_2 - x_1x_2'$, show that W satisfies the differential equation $a_2W' + a_1W = 0$, with solution $W = C\exp(-\int a_1(t)/a_2(t)\,dt)$.

(c) Show that the Wronskian of the two solutions of $a_2x'' + a_1x' + a_0x = 0$ is either always zero or never zero.

(d) Show that any two solutions of $a_2x'' + a_1x' + a_0x = 0$ are linearly independent if and only if their Wronskian is nonzero.

6. Figure 8.11 shows two solutions of a second order linear differential equation with continuous coefficients. How do you know that these solutions are linearly dependent?

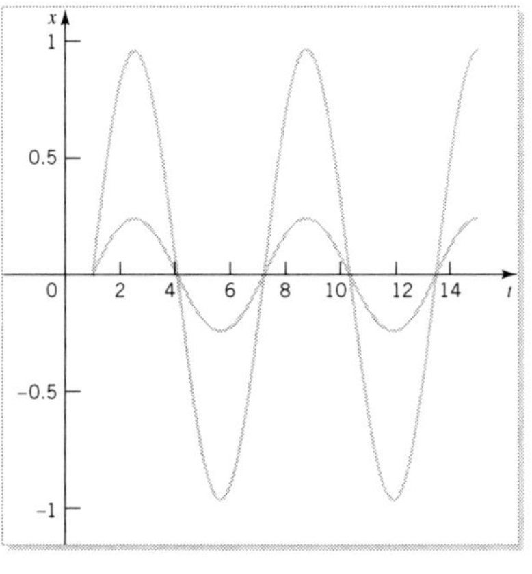

FIGURE 8.11 Two solutions

7. Figure 8.12 shows two solutions of a second order linear differential equation with continuous coefficients. How do you know that these solutions are linearly dependent?

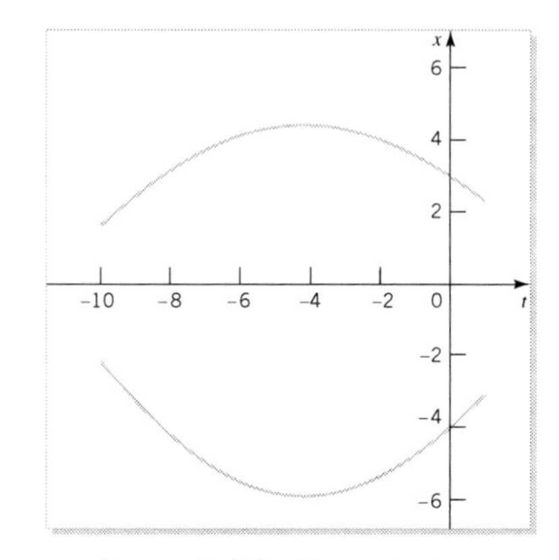

FIGURE 8.12 Two solutions

8. Show that the general solution of $a_2(t)x'' + a_1(t)x' + a_0(t)x = 0$ is $x(t) = x_1(t) + x_2(t)$, where $x_1(t)$ is the solution of the initial value problem $a_2(t)x'' + a_1(t)x' + a_0(t)x = 0$, $x(0) = 0$, $x'(0) = x_0^* \neq 0$, and $x_2(t)$ is the solution of the initial value problem $a_2(t)x'' + a_1(t)x' + a_0(t)x = 0$, $x(0) = x_0 \neq 0$, $x'(0) = 0$.

9. Show that the general solution of $a_2(t)x'' + a_1(t)x' + a_0(t)x = f(t)$ is $x(t) = x_1(t) + x_2(t) + x_p(t)$, where: $x_1(t)$ is the solution of the initial value problem $a_2(t)x'' + a_1(t)x' + a_0(t)x = 0$, $x(0) = 0$, $x'(0) = x_0^* \neq 0$; $x_2(t)$ is the solution of the initial value problem $a_2(t)x'' + a_1(t)x' + a_0(t)x = 0$, $x(0) = x_0 \neq 0$, $x'(0) = 0$; and $x_p(t)$ is the solution of the initial value problem $a_2(t)x'' + a_1(t)x' + a_0(t)x = f(t)$, $x(0) = 0$, $x'(0) = 0$. What initial conditions are satisfied by this solution?

10. For any two linearly independent functions, $x_1(t)$ and $x_2(t)$, construct the 3 by 3 determinant[6]

$$A = \begin{vmatrix} x_1 & x_2 & x \\ x_1' & x_2' & x' \\ x_1'' & x_2'' & x'' \end{vmatrix}.$$

(a) Prove that if $A = 0$, then $x(t)$ satisfies a second order linear differential equation of the type $a_2(t)x'' +$

[6]Appendix A.5 contains an introduction to determinants.

$a_1(t)x' + a_0(t)x = 0$. What are $a_2(t)$, $a_1(t)$, and $a_0(t)$ in terms of $x_1(t)$, $x_2(t)$, and their derivatives?

(b) Prove that $x_1(t)$ and $x_2(t)$ both satisfy the differential equation obtained in part (a).

(c) Explain how the results obtained in parts (a) and (b) can be used to construct a second order linear differential equation, $a_2(t)x'' + a_1(t)x' + a_0(t)x = 0$, from any two of its linearly independent solutions, $x_1(t)$ and $x_2(t)$.

(d) What happens if $x_1(t)$ and $x_2(t)$ are chosen as linearly dependent functions?

11. Use the technique outlined in Exercise 10 to construct a second order linear differential equation, $a_2(t)x'' + a_1(t)x' + a_0(t)x = 0$, that has the general solution $x(t) = C_1x_1(t) + C_2x_2(t)$, where $x_1(t)$ and $x_2(t)$ are two linearly independent functions, given as follows.

(a) t, t^2
(b) t, t^{-1}
(c) t, e^t
(d) e^{2t}, e^{-3t}
(e) e^{-t}, te^{-t}
(f) e^{rt}, te^{rt}
(g) $\sin \beta t, \cos \beta t$
(h) $e^{-t}\cos 5t, e^{-t}\sin 5t$
(i) $\sin t, \sin^2 t$
(j) $e^{1/t}, e^{-1/t}$
(k) $e^t, \sin t$

12. Use the technique outlined in Exercise 10 to show that if $x_1(t)$ and $x_2(t)$ are linearly independent functions, then $x(t) = C_1x_1(t) + C_2x_2(t)$ is the general solution of the second order linear differential equation

$$x'' - \frac{x_1x_2'' - x_1''x_2}{x_1x_2' - x_1'x_2}x' + \frac{x_1'x_2'' - x_1''x_2'}{x_1x_2' - x_1'x_2}x = 0.$$

Show that if $x_1(t)$ and $x_2(t)$ are linearly independent solutions of the second order linear differential equation $x'' + p(t)x' + q(t)x = 0$, then

$$p(t) = -\frac{x_1x_2'' - x_1''x_2}{x_1x_2' - x_1'x_2} \text{ and } q(t) = \frac{x_1'x_2'' - x_1''x_2'}{x_1x_2' - x_1'x_2}.$$

13. Use the technique outlined in Exercise 12 to show that if $w(t)$ and $1/w(t)$ are linearly independent solutions of a second order linear differential equation, then

$$x'' + \left(\frac{w'}{w} - \frac{w''}{w'}\right)x' - \left(\frac{w'}{w}\right)^2 x = 0.$$

14. Kepler's Laws.[7] If a particle moves in the xy-plane under the inverse square law, then the motion of the particle with coordinates (x, y) satisfies $x'' = -kx/r^3$ and $y'' = -ky/r^3$, where k is a positive constant and $r = \sqrt{x^2 + y^2}$. The purpose of this exercise is to show that a particle moving under the inverse square law follows a conic section. This is one of Kepler's laws.

(a) Show that the three functions, $x'(t)$, $y'(t)$, and $r'(t)$, all satisfy the second order linear differential equation $(r^3f')' = -kf$; for example, $(r^3x'')' = -kx'$.

(b) If $x'(t)$ and $y'(t)$ are linearly dependent, show that the motion is in a straight line.

(c) If $x'(t)$ and $y'(t)$ are linearly independent, explain why $r'(t) = C_1x'(t) + C_2y'(t)$, where C_1 and C_2 are constants. Integrate this to find $x^2(t) + y^2(t) = [C_1x(t) + C_2y(t) + C_3]^2$, where C_3 is a constant. Identify this family of curves.

15. Consider the differential equation

$$(\sin t \cos t)\, x' - x = -x^3.$$

(a) Show that $x_1(t) = 1$ and $x_2(t) = \sin t$ are both solutions of this differential equation.

(b) Prove that these solutions are linearly independent for $t > 0$.

(c) Is $x(t) = C_1x_1(t) + C_2x_2(t) = C_1 + C_2\sin t$ also a solution? If so, show it is; if not, explain why not.

8.5 SOLVING CAUCHY-EULER EQUATIONS

All the second order differential equations that we covered in Chapter 6 were linear with constant coefficients. These occur very often in applications, but they are not the only type of second order differential equations for which we can develop a

[7] "Inverse-square orbits: Three little-known solutions and a novel integration technique" by R. Weinstock, *The American Journal of Physics* 60, 1992, pages 615–619.

systematic method of obtaining explicit solutions. In this section we find explicit solutions of the CAUCHY-EULER DIFFERENTIAL EQUATION, namely,

$$b_2 x^2 \frac{d^2y}{dx^2} + b_1 x \frac{dy}{dx} + b_0 y = 0,$$

where b_2, b_1, and b_0 are constants.

EXAMPLE 8.13 *Electric Potential of a Charged Spherical Shell*

A differential equation that occurs in a mathematical model describing the electric potential due to a charged spherical shell is given by

$$x^2 \frac{d^2y}{dx^2} + 2x \frac{dy}{dx} - n(n+1)y = 0, \tag{8.62}$$

where x is the distance from the center of the spherical shell, y is the potential, and n is a positive constant. This differential equation does not have constant coefficients, and is one we have not yet considered. However, the Oscillation Theorem and Reduction Theorem can be used to show that nontrivial solutions of (8.62) cannot vanish more than once for $x > 0$, and cannot vanish more than once for $x < 0$ (see Exercise 6 on page 349).

In Chapter 6 we discovered a simple method for solving second order linear differential equations with constant coefficients. Thus, we look for ways to replace xy' and x^2y'' by derivatives that have constant coefficients, by using a change of variable from the independent variable, x, to a new independent variable, t, where

$$\frac{dy}{dt} = x\frac{dy}{dx}.$$

The fact that $dy/dt = (dy/dx)\,(dx/dt)$ suggests we relate x and t by the differential equation $\frac{dx}{dt} = x$,which has the solution $x = Ae^t$. Thus, we could use the change of variable

$$x = e^t, \qquad \text{if } x > 0, \qquad \text{and} \qquad x = -e^t, \qquad \text{if } x < 0.$$

Change variable In our case, $x > 0$, because distance is positive, so we use the change of variable

$$x = e^t, \tag{8.63}$$

which is equivalent to

$$t = \ln x, \tag{8.64}$$

so differentiation with respect to x gives

$$\frac{dt}{dx} = \frac{1}{x}. \tag{8.65}$$

By using the chain rule, along with (8.63), we find

$$\frac{dy}{dt} = \frac{dy}{dx}\frac{dx}{dt} = \frac{dy}{dx}e^t = x\frac{dy}{dx},$$

or

$$x\frac{dy}{dx} = \frac{dy}{dt}, \tag{8.66}$$

as desired. So, in (8.62) we can replace the $2xdy/dx$ term by $2dy/dt$. But what about the x^2d^2y/dx^2 term?

If we differentiate (8.66) with respect to x using the product rule on the left-hand side, we find

$$x\frac{d^2y}{dx^2} + \frac{dy}{dx} = \frac{d}{dx}\left(\frac{dy}{dt}\right) = \frac{d^2y}{dt^2}\frac{dt}{dx}.$$

If we multiply this equation by x and use (8.65), we find

$$x^2\frac{d^2y}{dx^2} + x\frac{dy}{dx} = \frac{d^2y}{dt^2},$$

which, from (8.66), can be written as

$$x^2\frac{d^2y}{dx^2} = \frac{d^2y}{dt^2} - \frac{dy}{dt}. \tag{8.67}$$

If we substitute (8.66) and (8.67) into (8.62), our original differential equation becomes

$$\frac{d^2y}{dt^2} - \frac{dy}{dt} + 2\frac{dy}{dt} - n(n+1)y = 0,$$

or

$$\frac{d^2y}{dt^2} + \frac{dy}{dt} - n(n+1)y = 0. \tag{8.68}$$

Because (8.68) is a linear differential equation with constant coefficients, we find the characteristic equation

$$r^2 + r - n(n+1) = (r-n)(r+n+1) = 0.$$

This has two solutions, $r = n$ and $r = -n - 1$, so our solution of (8.68) is

$$y(t) = C_1e^{nt} + C_2e^{-(n+1)t}.$$

We can use either (8.63) or (8.64) to express this solution in terms of our original variable x. In this way we get the general solution of our original differential equation as

General solution

$$y(x) = C_1x^n + C_2x^{-(n+1)}.$$

This solution is in agreement with the Oscillation Theorem and Reduction Theorem, which require that nontrivial solutions of (8.62) cannot vanish more than once for $x > 0$, and cannot vanish more than once for $x < 0$ (see Exercise 6 on page 349). ■

How to Solve a Cauchy-Euler Differential Equation

Purpose To solve differential equations of the form

$$b_2 x^2 \frac{d^2y}{dx^2} + b_1 x \frac{dy}{dx} + b_0 y = 0, \tag{8.69}$$

where b_2, b_1, and b_0 are constants.

Process

1. If $x > 0$, transform the independent variable from x to t, where

$$x = e^t, \tag{8.70}$$

so that

$$t = \ln x, \tag{8.71}$$

$$x\frac{dy}{dx} = \frac{dy}{dt}, \tag{8.72}$$

and

$$x^2\frac{d^2y}{dx^2} = \frac{d^2y}{dt^2} - \frac{dy}{dt}. \tag{8.73}$$

[If $x < 0$, use the transformation $x = -e^t$, so (8.71) becomes $t = \ln(-x)$, and (8.72) and (8.73) remain the same.]

2. Substitute (8.72) and (8.73) into (8.69) and combine like terms, obtaining

$$b_2\frac{d^2y}{dt^2} + (b_1 - b_2)\frac{dy}{dt} + b_0 y = 0, \tag{8.74}$$

which is a second order linear equation with constant coefficients.

3. Solve (8.74) for $y = y(t)$.
4. Use (8.70) or (8.71) and the solution $y = y(t)$ obtained in step 3 to find the final solution $y = y(x)$.

Comments about Solutions of the Cauchy-Euler Differential Equation

- The solution of (8.69) consists of a linear combination of two linearly independent solutions (see Exercise 4, page 349).

- The only solutions of (8.69) are linear combinations of functions of the form x^n, $x^\alpha \cos(\beta \ln x)$, $x^\alpha \sin(\beta \ln x)$, and $x^m \ln x$. (Why?)

Two Examples

EXAMPLE 8.14

Solve the initial value problem

$$x^2 \frac{d^2y}{dx^2} + 3x\frac{dy}{dx} + 5y = 0, \qquad y(1) = 0, \qquad y'(1) = 6.$$

Change variable

We change the independent variable from x to t by $x = e^t$. Substituting (8.72) and (8.73) into this differential equation yields

$$\frac{d^2y}{dt^2} + 2\frac{dy}{dt} + 5y = 0.$$

General solution

This linear differential equation with constant coefficients has the characteristic equation $r^2 + 2r + 5 = (r+1)^2 + 4 = 0$, so its general solution is $y(t) = C_1 e^{-t} \cos 2t + C_2 e^{-t} \sin 2t$. In terms of the original variables, this becomes

$$y(x) = C_1 x^{-1} \cos(2 \ln x) + C_2 x^{-1} \sin(2 \ln x).$$

Applying the initial conditions gives $C_1 = 0$, $C_2 = 3$, so our solution becomes

$$y(x) = 3x^{-1} \sin(2 \ln x).$$

■

EXAMPLE 8.15

Find a particular solution of

$$x^2 \frac{d^2y}{dx^2} + x\frac{dy}{dx} - y = \frac{2}{x+1}$$

for $x > 0$.

Our first step is to solve the associated homogeneous differential equation

$$x^2 \frac{d^2y}{dx^2} + x\frac{dy}{dx} - y = 0.$$

Change variable

This is a Cauchy-Euler equation, so we let $x = e^t$. Substituting (8.72) and (8.73) into this differential equation yields

$$\frac{d^2y}{dt^2} - y = 0,$$

with solution $y(t) = C_1 e^t + C_2 e^{-t}$. In terms of x, the solution of the associated homogeneous differential equation is thus

$$y(x) = C_1 x + C_2 x^{-1}.$$

Trial solution Because we have two linearly independent solutions (show this), we can use the method of variation of parameters, so we let

$$y_p(x) = x z_1(x) + x^{-1} z_2(x), \tag{8.75}$$

so z_1' and z_2' satisfy

$$\begin{aligned} x z_1' + x^{-1} z_2' &= 0, \\ z_1' - x^{-2} z_2' &= 2/\left[x^2(x+1)\right]. \end{aligned} \tag{8.76}$$

We solve for z_1' by multiplying the top equation in (8.76) by x^{-1} and adding the result to the bottom equation, as

$$z_1' = \frac{1}{x^2(x+1)}.$$

This gives z_2' as

$$z_2' = -x^2 z_1' = \frac{-1}{x+1},$$

which may be integrated immediately to yield

$$z_2 = -\ln(x+1).$$

To find an antiderivative for z_1, we first rewrite the form for z_1', using partial fractions, as

$$z_1' = -\frac{1}{x} + \frac{1}{x^2} + \frac{1}{x+1},$$

so

$$z_1 = -\ln x - x^{-1} + \ln(x+1).$$

Particular solution Thus, from (8.75), our particular solution is

$$y_p(x) = x \ln(1 + 1/x) - 1 - x^{-1} \ln(x+1).$$

EXERCISES

1. Find the general solution of the following differential equations, assuming $x > 0$. (Note: $y' = dy/dx$, $y'' = d^2y/dx^2$.) Wherever possible, confirm the solutions are consistent with the Oscillation Theorem and Reduction Theorem.

(a) $x^2y'' - 2y = 0$ (b) $x^2y'' + 2xy' - 6y = 0$
(c) $x^2y'' + xy' = 0$ (d) $x^2y'' + 9xy' + 2y = 0$
(e) $x^2y'' + xy' + 9y = 0$ (f) $x^2y'' - 5xy' + 25y = 0$
(g) $x^2y'' + 5xy' + 4y = 0$ (h) $2x^2y'' - xy' + y = 0$
(i) $x^2y'' - 5xy' + 5y = 0$ (j) $2x^2y'' + xy' - y = 0$
(k) $x^2y'' - xy' + y = 0$ (l) $x^2y'' - 3xy' + 4y = 0$

2. Show that $x^{-1}\cos(2\ln x)$ and $x^{-1}\sin(2\ln x)$ are linearly independent for $x > 0$.

3. Find the general solution of the following differential equations, assuming $x > 0$. (Note: $y' = dy/dx$, $y'' = d^2y/dx^2$.)

(a) $x^2y'' - 2y = \ln x$
(b) $x^2y'' - xy' + y = x\ln x$
(c) $x^2y'' + 7xy' + 5y = 10 - 4x^{-1}$
(d) $x^2y'' + xy' + y = x^3$

4. Show that if $y_1(t)$ and $y_2(t)$ are linearly independent functions of t and satisfy a second order linear differential equation with constant coefficients, then $Y_1(x) = y_1(\ln x)$ and $Y_2(x) = y_2(\ln x)$ are linearly independent functions of x for $x \neq 0$. [Hint: Obtain the Wronskian of Y_1 and Y_2 in terms of the Wronskian of y_1 and y_2.]

5. Show that, for $x > 0$, the differential equation $y'' + (a^2/x^2)y = 0$ has the general solutions that follow.

(a) $y(x) = C_1x^{r_1} + C_2x^{r_2}$, where $r_1 = \left(1 + \sqrt{1-4a^2}\right)/2$, $r_2 = \left(1 - \sqrt{1-4a^2}\right)/2$, if $a^2 < 1/4$.

(b) $y(x) = C_1\sqrt{x} + C_2\sqrt{x}\ln x$, if $a^2 = 1/4$.

(c) $y(x) = C_1\sqrt{x}\cos(\omega\ln x) + C_2\sqrt{x}\sin(\omega\ln x)$, where $\omega = \left(\sqrt{4a^2-1}\right)/2$, if $a^2 > 1/4$.

6. Show that the nontrivial solutions of

$$x^2\frac{d^2y}{dx^2} + 2x\frac{dy}{dx} - n(n+1)y = 0,$$

where n is a positive constant, cannot vanish more than once for $x > 0$, and cannot vanish more than once for $x < 0$. [Hint: See the Oscillation Theorem and Reduction Theorem.]

7. Show that trying a solution of the form x^r for

$$b_2x^2\frac{d^2y}{dx^2} + b_1x\frac{dy}{dx} + b_0y = 0, \tag{8.77}$$

where b_2, b_1, and b_0 are constants, leads to the quadratic equation $b_2r^2 + (b_1 - b_2)r + b_0 = 0$. Show that the form of the solution depends on the sign of $(b_1 - b_2)^2 - 4b_2b_0$, and find the general solution of (8.77) for the three possible cases.

8. Figure 8.13 shows various solutions of $4x^2y'' - 4xy' + 3y = 0$.

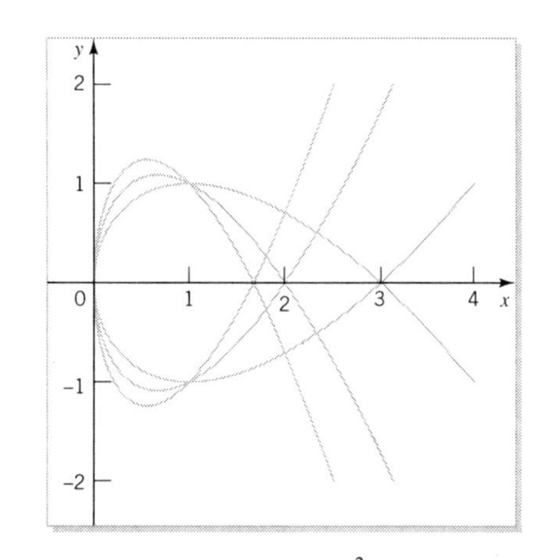

FIGURE 8.13 Solutions of $4x^2y'' - 4xy' + 3y = 0$

(a) How many solutions are shown in the figure?

(b) From the graphs you can see that the solutions have a maximum if $y > 0$ and a minimum if $y < 0$. Explain how the differential equation can be used to show that it is not possible for a solution to have a minimum if $y > 0$, nor is it possible for a solution to have a maximum if $y < 0$.

(c) Identify the solution corresponding to the initial conditions $y(1) = 1$, $y'(1) = 0$. From the graph of this solution estimate the value of $x = x_1$ $(x_1 > 0)$ for which $y(x_1) = 0$. From the graph estimate $y'(x_1) = y_1^*$.

(d) Solve the differential equation subject to $y(x_1) = 0$, $y'(x_1) = y_1^*$. Does this solution pass through the point $(1, 1)$ with horizontal slope? Should it?

(e) Solve the differential equation subject to $y(1) = 1$, $y'(1) = 0$. From this solution find the value of $x = x_1$ $(x_1 > 0)$ for which $y(x_1) = 0$. From this solution evaluate $y'(x)$. How well did it agree with your y_1^* from part (c)?

8.6 BOUNDARY VALUE PROBLEMS AND THE SHOOTING METHOD

Initial value problems have conditions on the dependent variable given at a single value of the independent variable, such as the case for $x'' + x = 0, x(0) = 0, x'(0) = 2$. With boundary value problems, conditions are given at two different values of the independent variable, usually at the end points of its domain.

Motivation

Consider the spring-mass system characterized by

$$x'' + x = 0, \tag{8.78}$$

subject to $x(0) = 0$ and $x'(0) = A$. Thus, we have a mass initially at the origin, which is given an initial velocity of A. Recall that we take x positive in the downward direction.

General solution

The general solution of (8.78) is $x(t) = C_1 \cos t + C_2 \sin t$. If we impose the initial conditions $x(0) = 0$ and $x'(0) = A$, we find

$$x(t) = A \sin t. \tag{8.79}$$

Questions

We now ask a series of questions.

- What initial velocity should we give the mass so that at time $t = \pi/2$ the mass is at 1? In other words, what value should we give A so that $x(\pi/2) = 1$? To satisfy $x(\pi/2) = 1$, we must have

$$1 = A \sin \frac{\pi}{2}; \tag{8.80}$$

  that is, $A = 1$. Thus, $x(t) = \sin t$ would be the solution we seek, and we should give the mass an initial velocity of 1 downward. In other words, the unique solution of (8.78) subject to $x(0) = 0$ and $x(\pi/2) = 1$ is $x(t) = \sin t$.

- What initial velocity should we give the mass so that at time $t = \pi/2$ the mass is at -1? In this case the counterpart of (8.80) is $-1 = A \sin \frac{\pi}{2}$, so $A = -1$ and $x(t) = -\sin t$. We should give the mass an initial velocity of 1 upward. The unique solution of (8.78) subject to $x(0) = 0$ and $x(\pi/2) = -1$ is $x(t) = -\sin t$.

- What initial velocity should we give the mass so that at time $t = \pi/2$ the mass is back at 0? In this case the counterpart of (8.80) is $0 = A \sin \frac{\pi}{2}$, so $A = 0$ and $x(t) = 0$. We should just leave the mass where it is. The unique solution of (8.78) subject to $x(0) = 0$ and $x(\pi/2) = 1$ is $x(t) = 0$.

- What initial velocity should we give the mass so that at time $t = \pi$ the mass is at 1? In this case the counterpart of (8.80) is $1 = A \sin \pi$. Here we have a problem because the left-hand side is 1 and the right-hand side is 0. Thus, there is no initial velocity we can give the mass so that at time $t = \pi$ the mass is at 1. In other words, there is no solution of (8.78) subject to $x(0) = 0$ and $x(\pi) = 1$.

- What initial velocity should we give the mass so that at time $t = \pi$ the mass is at -1? In this case the counterpart of (8.80) is $-1 = A \sin \pi$. We again have a problem because the left-hand side is -1 and the right-hand side is 0. Thus, there is no initial velocity we can give the mass so that at time $t = \pi$ the mass is at -1. In other words, there is no solution of (8.78) subject to $x(0) = 0$ and $x(\pi) = -1$.
- What initial velocity should we give the mass so that at time $t = \pi$ the mass is back at 0? In this case the counterpart of (8.80) is $0 = A \sin \pi$. Here we have a different problem because the left-hand side is 0 and the right-hand side is 0 for all values of A. Thus, any initial velocity will ensure that at time $t = \pi$ the mass is back at 0. In other words, for any A, the function $x(t) = A \sin t$ is the solution of (8.78) subject to $x(0) = 0$ and $x(\pi) = 0$. Thus, in this case there are an infinite number of solutions, one for each choice of A.

In Figure 8.14 we have plotted $x(t) = A \sin t$ for several values of A. We have also marked the points $x(\pi/2) = 1$, $x(\pi/2) = -1$, $x(\pi/2) = 0$, $x(\pi) = 1$, $x(\pi) = -1$, and $x(\pi) = 0$. This figure explains everything! If $x(0) = 0$ and the value of $x(b)$ is prescribed at $t = b$, then $x'' + x = 0$ will have a unique solution for every b if $0 < b < \pi$, no matter what value is given to $x(b)$. However, if $b = \pi$, then there will be no solution if $x(b) \neq 0$, and an infinite number of solutions if $x(b) = 0$. We see from this that $x'' + x = 0$ subject to $x(t_0) = x_0$ and $x(t_1) = x_1$ may have no solutions, one solution, or an infinite number of solutions. Lest we start thinking that we are doing something wrong, we should recall that all of our existence-uniqueness theorems to this point apply to initial value problems, not to boundary value problems.

Number of solutions

We now look at the same differential equation, $x'' + x = 0$, but change the initial condition to $x(0) = 2, x'(0) = A$. The solution of this is $x(t) = 2 \cos t + A \sin t$, which we have plotted in Figure 8.15 for several values of A. From this figure we see that if $x(0) = 2$ and the value of $x(b)$ is prescribed at $t = b$, then $x'' + x = 0$ will have a unique solution for every b if $0 < b < \pi$, no matter what value is given to $x(b)$. However, if $b = \pi$, then there will be no solution if $x(b) \neq -2$, and an infinite number of solutions if $x(b) = 2$. This is exactly the same conclusion as we found for the previous analysis.

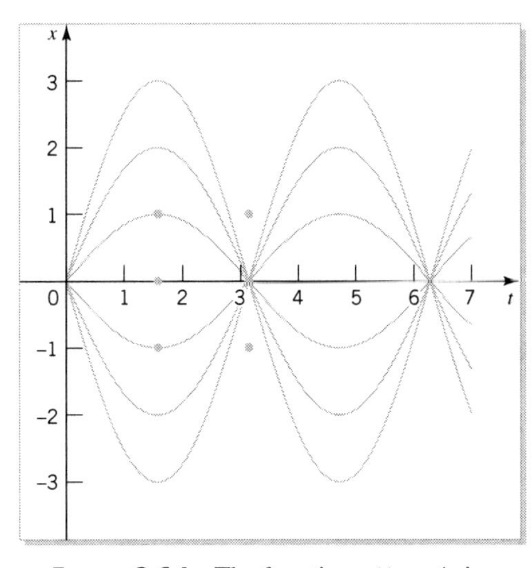

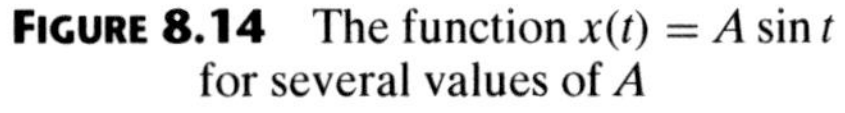

FIGURE 8.14 The function $x(t) = A \sin t$ for several values of A

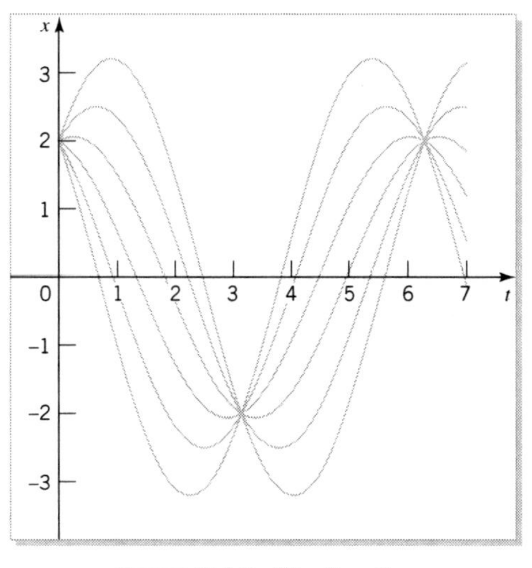

FIGURE 8.15 The function $x(t) = 2 \cos t + A \sin t$ for several values of A

The Main Result

Thus, in this example, if we do not prescribe values at both $t = 0$ and $t = \pi$, then the solution will be unique. If we do, the solution will either not exist, or will not be unique. What is so special about $t = 0$ and $t = \pi$? If we look at Figure 8.14 and compare the solution corresponding to $x(0) = 0$, $x(\pi/2) = 0$, with those corresponding to $x(0) = 0$, $x(\pi) = 0$, we see that in the first case the trivial solution $x(t) = 0$ is the only solution, whereas in the second case, not only is the trivial solution a solution, but there are also nontrivial solutions.

We have just illustrated the following result.

Theorem 8.6: The Existence Theorem for Boundary Value Problems *Consider $a_2(t)x'' + a_1(t)x' + a_0(t)x = 0$ where a_2, a_1, and a_0 are continuous in the interval $\alpha < t < \beta$ and $a_2(t) \neq 0$ in this interval. There exists a nontrivial solution $x(t)$ of the boundary value problem $a_2(t)x'' + a_1(t)x' + a_0(t)x = 0$, $x(a) = 0$, $x(b) = 0$, where $\alpha < a < b < \beta$, if and only if, for any constants x_a, x_b, the boundary value problem $a_2(t)x'' + a_1(t)x' + a_0(t)x = 0$, $x(a) = x_a$, $x(b) = x_b$ either has no solution, or, if it does have a solution, that solution is not unique.*

Proof Let $x(t) = C_1x_1(t) + C_2x_2(t)$ be the general solution of $a_2(t)x'' + a_1(t)x' + a_0(t)x = 0$, where x_1 and x_2 are linearly independent. Consider the system of algebraic equations, $x(a) = x_a$, $x(b) = x_b$; that is,

$$\begin{aligned} C_1x_1(a) + C_2x_2(a) &= x_a \\ C_1x_1(b) + C_2x_2(b) &= x_b. \end{aligned}$$

This system has a unique solution, C_1, C_2, if and only if

$$\begin{vmatrix} x_1(a) & x_2(a) \\ x_1(b) & x_2(b) \end{vmatrix} \neq 0. \tag{8.81}$$

Thus, this system does not have a unique solution (that is, either it has no solution, or, if it does have a solution, that solution is not unique) if and only if

$$\begin{vmatrix} x_1(a) & x_2(a) \\ x_1(b) & x_2(b) \end{vmatrix} = 0.$$

This condition is true if and only if

$$\begin{aligned} C_1x_1(a) + C_2x_2(a) &= 0, \\ C_1x_1(b) + C_2x_2(b) &= 0 \end{aligned}$$

has a nontrivial solution for C_1 and C_2, which is equivalent to $C_1x_1(t) + C_2x_2(t)$ being a nontrivial solution of the boundary value problem $x(a) = 0$, $x(b) = 0$. ◀

Comments about Boundary Value Problems

- The theorem says that if you want to decide whether the boundary value problem $a_2x'' + a_1x' + a_0x = 0$, $x(a) = x_a$, $x(b) = x_b$, has a unique solution, you need only consider the boundary value problem $a_2x'' + a_1x' + a_0x = 0$, $x(a) = 0, x(b) = 0$. If the second problem has only the trivial solution, then the former has a unique solution. If the second problem has a nontrivial solution, then the first either has no solution, or more than one.

- From the practical point of view of deciding whether the general solution, $x(t) = C_1x_1(t) + C_2x_2(t)$ of $a_2(t)x'' + a_1(t)x' + a_0(t)x = 0$, has a unique solution passing through $x(a) = x_a$, $x(b) = x_b$, we need only check (8.81), because this boundary value problem has a unique solution if and only if

$$\begin{vmatrix} x_1(a) & x_2(a) \\ x_1(b) & x_2(b) \end{vmatrix} \neq 0.$$

- The boundary value problem, $a_2(t)x'' + a_1(t)x' + a_0(t)x = 0$, $x(a) = x_a$, and $x(b) = x_b$, can have no solutions, one solution, or an infinite number of solutions. See Exercise 1 on page 354.
- The boundary value problem, $a_2x'' + a_1x' + a_0x = 0$, $x(a) = x_a$, $x(b) = x_b$, where a_2, a_1, and a_0 are constants, always has a unique solution if the roots of the characteristic equation $a_2r^2 + a_1r + a_0 = 0$ are real. If the roots are complex, $r = \alpha \pm i\beta$, then the boundary value problem has a unique solution if and only if $\beta = n\pi/(b-a)$, where $n = 0, \pm 1, \pm 2, \cdots$. See Exercise 2 on page 354.
- Most boundary value problems cannot be solved analytically.
- The general second order boundary value problem is to solve $x'' = R(t, x, x')$, subject to $x(a) = x_a$, $x(b) = x_b$.
- There are existence and uniqueness theorems for general second order boundary value problems $x'' = R(t, x, x')$, $x(a) = x_a$, $x(b) = x_b$, but they usually involve many hypotheses that are very specialized.

The Shooting Method

Numerical solutions

Because most boundary value problems cannot be solved analytically, we need a numerical technique. However, the techniques we have used for solving initial value problems numerically do not apply to the boundary value problem, $x'' = R(t, x, x')$, $x(a) = x_a$, $x(b) = x_b$. Nevertheless, we can adjust them by converting the boundary value problem to an initial value problem using the following idea.

Start with the initial value problem, $x'' = R(t, x, x')$, $x(a) = x_a$, $x'(a) = x_a^*$, where x_a^* is an initial guess. Use your favorite numerical technique to find a solution; let's call it $x_1(t)$, so the value of $x_1(t)$ at $t = b$ is $x_1(b)$. If $x_1(b) = x_b$, you have been very lucky. In general, $x_1(b) \neq x_b$, and based on this value of $x_1(b)$, you now change your initial guess of x_a^* and repeat the process until $x_1(b) \approx x_b$. This trial-and-error technique is called THE SHOOTING METHOD, because it is similar to the way an archer tries to hit the center of the target. If the first shot is too low, aim higher. If the second shot is too high, aim lower, but not as low as the first shot. Figure 8.16 shows the idea. However, it is wise to first make sure that a unique solution exists before using this method!

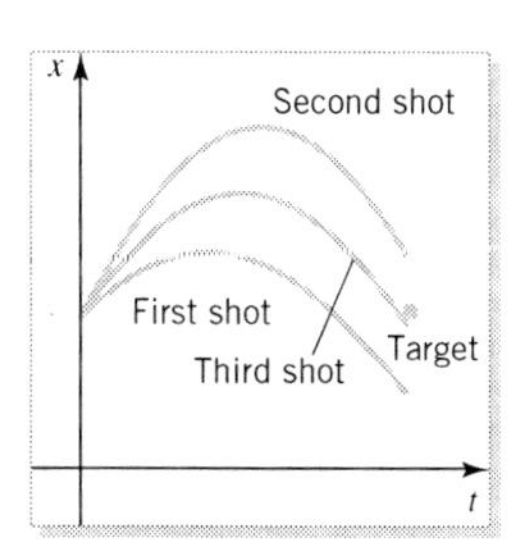

FIGURE 8.16 Pictorial representation of the shooting method

For the special boundary value problem dealt with in the theorem — namely, $a_2(t)x'' + a_1(t)x' + a_0(t)x = 0$, $x(a) = x_a$, $x(b) = x_b$ — we can use a less hit-and-miss version of the shooting method, based on the following idea. Start with the two initial value problems, $a_2x'' + a_1x' + a_0x = 0$, $x(a) = x_a$, $x'(a) = x_a^*$, and $a_2x'' + a_1x' + a_0x = 0$, $x(a) = x_a$, $x'(a) = x_a^{**}$, where x_a^* and x_a^{**} are initial guesses. Let's call the solution of the first problem $x_1(t)$, and the solution of the second $x_2(t)$. If we compute $x_1(b)$ and $x_2(b)$, unless we have been very lucky, neither of these will have the value x_b. In general we will also have $x_1(b) \neq x_2(b)$. If we form a linear combination of the two functions, $x_1(t)$ and $x_2(t)$; namely, $x(t) = C_1x_1(t) + C_2x_2(t)$, can we select C_1 and C_2 so that $x(t)$ is a solution of the boundary value problem $a_2x'' + a_1x' + a_0x = 0$, $x(a) = x_a$, $x(b) = x_b$?

For this to happen we need $x(a) = x_a$, and $x(b) = x_b$; that is,

$$\begin{cases} x_a = C_1x_1(a) + C_2x_2(a), \\ x_b = C_1x_1(b) + C_2x_2(b), \end{cases}$$

or

$$\begin{cases} x_a = C_1x_a + C_2x_a, \\ x_b = C_1x_1(b) + C_2x_2(b). \end{cases}$$

We can solve these for C_1 and C_2, and we find[8]

$$\begin{cases} C_1 = \left[x_b - x_2(b)\right] / \left[x_1(b) - x_2(b)\right], \\ C_2 = \left[x_1(b) - x_b\right] / \left[x_1(b) - x_2(b)\right]. \end{cases}$$

Thus, in this special case, we need only make two guesses, and the solution is

$$x(t) = \frac{1}{x_1(b) - x_2(b)} \left\{\left[x_b - x_2(b)\right]x_1(t) + \left[x_1(b) - x_b\right]x_2(t)\right\}.$$

In this case, the third shot in Figure 8.16 would hit the target.

EXERCISES

1. The purpose of this exercise is to show that the boundary value problem $a_2(t)x'' + a_1(t)x' + a_0(t)x = 0$, $x(a) = x_a$, and $x(b) = x_b$, can have no solutions, one solution, or an infinite number of solutions.

(a) Show that if $x_1(t)$ and $x_2(t)$ are two nontrivial solutions that satisfy the boundary value problem $a_2(t)x'' + a_1(t)x' + a_0(t)x = 0$ with boundary conditions $x(a) = x_a$, and $x(b) = x_b$, then $x(t) = \alpha x_1(t) + (1 - \alpha)x_2(t)$ satisfies the same boundary value problem for any choice of the constant α.

(b) Show that if $x_1(t)$ is the trivial solution and $x_2(t)$ is a nontrivial solution of the boundary value problem $a_2(t)x'' + a_1(t)x' + a_0(t)x = 0$ with boundary conditions $x(a) = 0$, and $x(b) = 0$, then $\alpha x_2(t)$ satisfies the same boundary value problem for any choice of the constant α.

(c) Explain why parts (a) and (b) imply that if there are two distinct solutions of the boundary value problem, then there are infinitely many.

2. Show that the boundary value problem $a_2x'' + a_1x' + a_0x = 0$, $x(a) = x_a$, $x(b) = x_b$, where a_2, a_1, and a_0 are constants, always has a unique solution if the roots of the characteristic equation $a_2r^2 + a_1r + a_0 = 0$ are real. If the roots are complex, $r = \alpha \pm i\beta$, show that the boundary value problem has a unique solution if and only if $\beta = n\pi/(b - a)$, where $n = 0, \pm1, \pm2, \cdots$. [Hint: Use (8.81).]

---

[8]If $x_a \neq 0$, then the first equation is $C_1 + C_2 = 1$. If $x_a = 0$, then we will have only one equation in two unknowns, C_1 and C_2, in which case we are free to choose an extra relation between C_1 and C_2. We choose $C_1 + C_2 = 1$.

3. Solve the following boundary value problems.

(a) $x'' + 100x = 0 \quad x(0) = 4 \quad x(\pi/4) = 500$
(b) $x'' + 10x' + 26x = 0 \quad x(0) = 0 \quad x(\pi/4) = 10$
(c) $x'' + 2x' + 5x = 0 \quad x(0) = 0 \quad x(\pi/4) = 0$

4. For what values of b will the boundary value problem, $x'' + 16x = 0, x(0) = 0, x(b) = 1$,

(a) have no solutions?
(b) have exactly one solution?
(c) have an infinite number of solutions?

5. For what values of b will the boundary value problem, $x'' + 16x = 0, x(0) = 0, x(b) = 0$,

(a) have no solutions?
(b) have exactly one solution?
(c) have an infinite number of solutions?

6. Ignoring damping, a spring-mass system is described by $x'' + \lambda^2 x = 0$, $\lambda^2 = \frac{k}{m}$. Find the values of the ratio k/m such that any motion that starts at its equilibrium position is at the same position at $t = 3$ — that is, if $x(0) = 0$, then $x(3) = 0$ also.

7. For what values of λ does $x'' + \lambda^2 x = 0$, $x(0) = 0$, $x'(\pi) = 0$ have a nontrivial solution? Give these solutions.

8. Show that if $x_1(t)$ and $x_2(t)$ satisfy boundary value problem $a_2(t)x'' + a_1(t)x' + a_0(t)x = 0$ with boundary conditions $x_1(0) = A_1$, $x_1(L) = B_1$, $x_2(0) = A_2$, $x_2(L) = B_2$, then $\alpha x_1(t) + \beta x_2(t)$ satisfies the boundary value problem $a_2(t)x'' + a_1(t)x' + a_0(t)x = 0$ with boundary conditions $x(0) = \alpha A_1 + \beta A_2$, $x(L) = \alpha B_1 + \beta B_2$.

9. Figure 8.17 shows two solutions of $a_2(t)x'' + a_1(t)x' + a_0(t)x = 0$, one satisfying the boundary conditions $x(0) = 0$, and $x(b) = 0$, and the second satisfying the boundary conditions $x(0) = -1$, and $x(b) = 1$. Is the solution of the second boundary value problem unique?

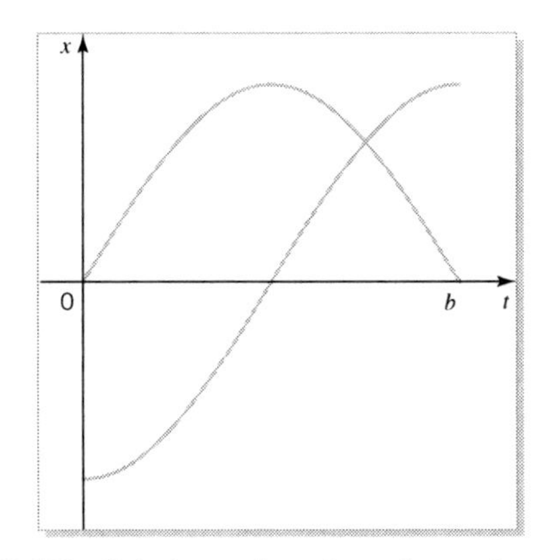

FIGURE 8.17 Solutions of two boundary value problems

10. Let $x(t) = C_1x_1(t) + C_2(t)$ be a nontrivial solution of the boundary value problem $a_2(t)x'' + a_1(t)x' + a_0(t)x = 0$, $x(a) = 0$, and $x(b) = 0$. Let this same function be the solution of the initial value problem $a_2(t)x'' + a_1(t)x' + a_0(t)x = 0$, $x(a) = 0$, $x'(a) = x_a^*$. Show that every solution of the initial value problem $a_2(t)x'' + a_1(t)x' + a_0(t)x = 0$, $x(a) = 0$, $x'(a) = mx_a^*$, where m is an arbitrary constant, has the property that $x(b) = 0$. Explain what this has to do with Figure 8.14.

11. Computer Experiment. Using software that solves initial value problems, find the value of $x'(0)$ that gives the solution of the boundary problem involving Airy's equation, $x'' + tx = 0$, $x(0) = 0$, $x(2) = 1$. In other words, experiment with the value of x_0^* until the solution of the initial value problem $x'' + tx = 0$, $x(0) = 0$, $x'(0) = x_0^*$ passes through the point $x(2) = 1$. Repeat this for $x(2) = 0$. In either case can you find a second solution?

12. Computer Experiment. Using software that solves initial value problems, find the value of $x'(\pi^2)$ that gives the solution of the boundary problem $4tx'' + 2x' + x = 0$, $x(\pi^2) = 0$, $x(4\pi^2) = 1$. Repeat this for $x(4\pi^2) = 0$. In either case can you find a second solution?

8.7 SOLVING HIGHER ORDER HOMOGENEOUS DIFFERENTIAL EQUATIONS

In this section we examine homogeneous linear differential equations in which the order is greater than two.

EXAMPLE 8.16 *Buckling of a Long Shaft*

The buckling of a long shaft, or column, of length L under an axial compressive force is modeled by the differential equation

$$\frac{d^2}{dt^2}\left(EI\frac{d^2x}{dt^2}\right) + P\frac{d^2x}{dt^2} = 0. \tag{8.82}$$

Here x is the displacement from equilibrium, t is the distance along the shaft, EI is the flexural rigidity, and P is the axial compressive force. Typical boundary conditions have the shaft supported at the ends in a manner such that there is zero deflection and zero moment. Zero deflection means that $x(0) = x(L) = 0$, and zero moment means that $x''(0) = x''(L) = 0$.

Boundary conditions

When EI and P are both constant, we may divide both sides of (8.82) by EI, and because we know that $P/(EI)$ is positive, we define λ by $\lambda^2 = P/(EI)$. This gives the resulting fourth order linear equation as

$$x'''' + \lambda^2 x'' = 0. \tag{8.83}$$

One way to solve this problem is to let $w = x''$. We then consider the resulting second order differential equation $w'' + \lambda^2 w = 0$. This differential equation has its general solution given by $w(t) = C_1 \cos \lambda t + C_2 \sin \lambda t$, and because $x'' = w$, we can integrate twice to obtain

General solution

$$x(t) = -\frac{C_1}{\lambda^2}\cos \lambda t - \frac{C_2}{\lambda^2}\sin \lambda t + C_3 t + C_4 \tag{8.84}$$

as the general solution of (8.83), where C_1, C_2, C_3, and C_4 are arbitrary constants. ■

Characteristic equation

We now look at another way to obtain the same solution. For second order linear differential equations with constant coefficients, we used the characteristic equation to find solutions of the form e^{rt}. If we try such a solution for (8.83), we obtain

$$r^4 + \lambda^2 r^2 = r^2(r^2 + \lambda^2) = 0.$$

Because this equation has a double root,[9] $r = 0$, and complex roots, $r = \pm i\lambda$, we have the solution in (8.84) — if we absorb the constant $-1/\lambda^2$ into C_1 and C_2.

Notice that we have polynomials and sinusoidal functions in our solution given by (8.84). In earlier chapters we noticed that all solutions of second order linear differential equations with constant coefficients have the form of exponential functions, sine or cosine functions, or products of these two types of functions (for nonrepeated roots of the characteristic equation). (See *How to Solve Homogeneous Second Order Linear Differential Equations with Constant Coefficients* on page 244.) Because the sine and cosine functions, using Euler's formula,[10] $e^{it} = \cos t + i \sin t$, may be written

[9] A root of multiplicity two.

[10] Appendix A.4 contains a discussion of complex numbers.

in terms of exponential functions, we want to see if this is true for solutions of higher order linear differential equations with constant coefficients. Thus, we consider the nth order linear differential equation

$$a_n \frac{d^n x}{dt^n} + a_{n-1} \frac{d^{n-1} x}{dt^{n-1}} + \cdots + a_2 \frac{d^2 x}{dt^2} + a_1 \frac{dx}{dt} + a_0 x = 0, \tag{8.85}$$

where the coefficients $a_n, a_{n-1}, \cdots, a_2, a_1$, and a_0 are constants. In summation notation this becomes

$$\sum_{m=0}^{n} a_m \frac{d^m x}{dt^m} = 0,$$

where the zeroth derivative is just the function itself. Because $de^{rt}/dt = re^{rt}$ for real or complex values of r, it follows that $d^m e^{rt}/dt^m = r^m e^{rt}$ for any positive integer m. Thus, we seek solutions of (8.85) in the form $x = e^{rt}$. Substituting this exponential function into (8.85), using the preceding result, and factoring out the common factor of e^{rt} yields

$$\left(a_n r^n + a_{n-1} r^{n-1} + \cdots + a_2 r^2 + a_1 r + a_0\right) e^{rt} = 0.$$

The exponential function is never zero, so we have

$$a_n r^n + a_{n-1} r^{n-1} + \cdots + a_2 r^2 + a_1 r + a_0 = 0.$$

Characteristic equation

This nth order algebraic equation is the characteristic equation associated with (8.85) and will have n roots. Associated with each root, designated by r_i, is the solution $e^{r_i t}$. To form our solution of the differential equation, we take linear combinations of these solutions.

Notice that if we write the differential equation in operator form,

$$\left(a_n D^n + a_{n-1} D^{n-1} + \cdots + a_2 D^2 + a_1 D + a_0\right) x = 0,$$

(D represents d/dt), then the characteristic equation may be simply obtained by replacing the derivative operator D with the parameter r.

We now illustrate this technique with an example.

EXAMPLE 8.17

Characteristic equation

Consider the fourth order linear differential equation $\left(D^4 - 8D^2 + 16\right) x = 0$, where the characteristic equation is $r^4 - 8r^2 + 16 = 0$. This equation may be factored as

$$(r^2 - 4)(r^2 + 4) = (r - 2)(r + 2)(r - 2i)(r + 2i) = 0,$$

General solution

so our solution is given by

$$x(t) = C_1 e^{2t} + C_2 e^{-2t} + C_3 \cos 2t + C_4 \sin 2t. \tag{8.86}$$

Initial conditions

If we had the initial conditions $x(0) = 4$, $x'(0) = -2$, $x''(0) = -8$, $x'''(0) = -8$, we would take derivatives of our solution in (8.86) before substituting in the value of $t = 0$. This gives the four algebraic equations

$$\begin{aligned} C_1 + C_2 + C_3 &= 4, \\ 2C_1 - 2C_2 + 2C_4 &= -2, \\ 4C_1 + 4C_2 - 4C_3 &= -8, \\ 8C_1 - 8C_2 - 8C_4 &= -8, \end{aligned}$$

in four unknowns. If we multiply the first equation by -4 and add the result to the third equation, we obtain $C_3 = 3$, whereas if we multiply the second equation by -4 and add the result to the fourth equation, we have $C_4 = 0$. Using these values in the first two equations gives $C_1 = 0$ and $C_2 = 1$, and the solution of our initial value problem is

$$x(t) = e^{-2t} + 3\cos 2t.$$

The graphs of this solution and $3\cos 2t$ are shown in Figure 8.18, where we see that the effect of the exponential term diminishes quickly with time. We would say that e^{-2t} was the transient solution and $3\cos 2t$ was the steady state solution, using the same terminology we used for first and second order linear differential equations.

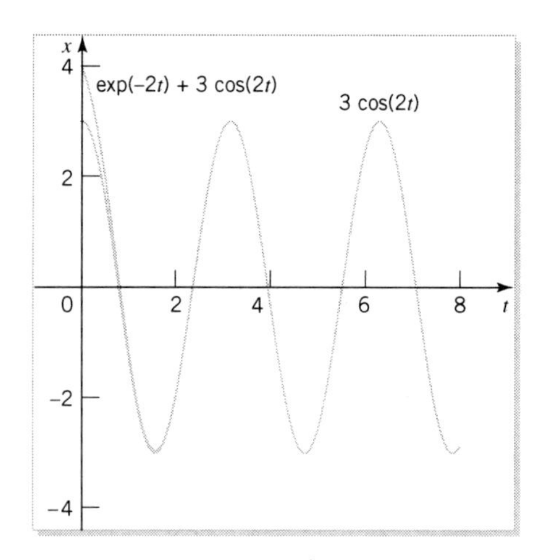

FIGURE 8.18 Graphs of $e^{-2t} + 3\cos 2t$ and $3\cos 2t$

This technique works for linear differential equations with constant coefficients of any order.

How to Solve Higher Order Linear Differential Equations with Constant Coefficients

Purpose To find the general solution of the linear differential equation

$$\sum_{m=0}^{n} a_m \frac{d^m x}{dt^m} = a_n \frac{d^n x}{dt^n} + a_{n-1} \frac{d^{n-1} x}{dt^{n-1}} + \cdots + a_2 \frac{d^2 x}{dt^2} + a_1 \frac{dx}{dt} + a_0 x = 0,$$

where the coefficients $a_n, a_{n-1}, \cdots, a_2, a_1, a_0$ are constants.

Process

1. Write the characteristic equation as
$\sum_{m=0}^{n} a_m r^m = a_n r^n + a_{n-1} r^{n-1} + \cdots + a_2 r^2 + a_1 r + a_0 = 0.$
2. Solve the characteristic equation, calling the solutions $r_1, r_2, r_3, \cdots, r_n$, and identify
   (a) all simple roots (real or complex).[11]
   (b) all repeated roots (real or complex).
3. Write the general solution. Here we have two cases:
   (a) The roots of the characteristic equation are all simple. In this case, to each root we assign the function

$$x_m(t) = e^{r_m t}, \qquad m = 1, 2, 3, \cdots, n.$$

   The general solution is given by

$$x(t) = \sum_{m=1}^{n} C_m e^{r_m t} = C_1 e^{r_1 t} + C_2 e^{r_2 t} + \cdots + C_n e^{r_n t}. \tag{8.87}$$

   (b) The characteristic equation has one or more roots that are not simple. We take the sum of arbitrary constants multiplied by the exponential functions associated with each simple root to form part of our solution, as in (8.87). To each nonsimple root r_m of multiplicity k, we assign the function

$$x_m(t) = \left(C_0^* + C_1^* t + \cdots + C_{k-1}^* t^{k-1}\right) e^{r_m t}$$

   and add this to the sum associated with the simple roots. Doing this for all the repeated roots gives us the general solution.

[11] A simple root is a root with multiplicity one.

Comments about Solving Higher Order Linear Differential Equations with Constant Coefficients

- Notice that for complex roots, it is usually advantageous to use a linear combination of $e^{\alpha t}\cos\beta t$ and $e^{\alpha t}\sin\beta t$ instead of a linear combination of $e^{(\alpha+i\beta)t}$ and $e^{(\alpha-i\beta)t}$.
- Notice that arbitrary constants are already included in the expression for $x_m(t)$ resulting from multiple roots [in step 3(b)] and that we will not need to multiply by an additional arbitrary constant when forming the solution, as in (8.87).
- This procedure produces n linearly independent functions (see Exercises 4 through 10 on page 365).
- If we choose the arbitrary constants in our general solution so it takes on specified initial values for the function and its first $n-1$ derivatives [that is, $x(0), x'(0), x''(0), \cdots, x^{(n-1)}(0)$ are specified], the resulting solution is unique.
- Higher order Cauchy-Euler differential equations have the form

$$\sum_{m=0}^{n} b_m x^m \frac{d^m y}{dx^m} = b_n x^n \frac{d^n y}{dx^n} + b_{n-1}x^{n-1}\frac{d^{n-1}y}{dx^{n-1}} + \cdots + b_2 x^2 \frac{d^2 y}{dx^2} + b_1 x \frac{dy}{dx} + b_0 y = 0,$$

where b_m, $m = 0, 1, 2, \cdots, n$ are all constants. In Section 8.5 we used the transformation $x = e^t$ to transform second order Cauchy-Euler differential equations into ones with constant coefficients. [12] In that case we replaced $x d/dx$ with d/dt and $x^2 d^2/dx^2$ with $d^2/dt^2 - d/dt$. If we let D represent the operator d/dt, we replace $x d/dx$ with D and $x^2 d^2/dx^2$ with $D^2 - D = D(D-1)$. For higher order derivatives, these replacements become

$$x^3 \frac{d^3}{dx^3} = D(D-1)(D-2),$$
$$x^4 \frac{d^4}{dx^4} = D(D-1)(D-2)(D-3), \cdots,$$
$$x^n \frac{d^n}{dx^n} = D(D-1)(D-2)(D-3)\cdots[D-(n-1)].$$

Thus, if we make these substitutions in Cauchy-Euler differential equations, we may obtain their solution by the technique previously outlined (see Example 8.20).

EXAMPLE 8.18

Solve the differential equation

$$\left(D^4 + 8D^2 + 16\right)x = 0, \tag{8.88}$$

subject to the initial conditions $x(0) = 0$, $x'(0) = 5$, $x''(0) = 0$, $x'''(0) = -28$.

---

[12] This assumes that $x > 0$. If $x < 0$, we used $x = -e^t$.

Characteristic equation

Using our previous technique, we note that the characteristic equation associated with (8.88) is $r^4 + 8r^2 + 16 = (r^2 + 4)^2 = 0$, so both $2i$ and $-2i$ are double roots.

Initial conditions

This gives our general solution of (8.88) as $x(t) = (C_1 + C_2 t)\cos 2t + (C_3 + C_4 t)\sin 2t$. Satisfying the initial conditions gives the following system of four algebraic equations:

$$\begin{aligned} x(0) &= C_1 &&= 0, \\ x'(0) &= C_2 + 2C_3 &&= 5, \\ x''(0) &= -4C_1 + 4C_4 &&= 0, \\ x'''(0) &= -12C_2 - 8C_3 &&= -28. \end{aligned}$$

Explicit solution

These equations are readily solved, yielding $C_1 = 0$, $C_2 = 1$, $C_3 = 2$, and $C_4 = 0$. Thus, the solution to this initial value problem is

$$x(t) = t\cos 2t + 2\sin 2t.$$

EXAMPLE 8.19

Find the general solution of the following differential equation, which for convenience we give in factored operator form as

$$\left[\left(D^2 - 9\right)\left(D^2 - 9\right)\left(D^2 - 4\right)\right]x = 0.$$

Characteristic equation

General solution

The characteristic equation for this sixth order equation is $(r^2 - 9)(r^2 - 9)(r^2 - 4) = 0$, which has roots at $r = 2, -2, 3$, and -3. The roots at 3 and -3 are both double roots, so our general solution is given by

$$x(t) = C_1 e^{2t} + C_2 e^{-2t} + (C_3 + C_4 t)e^{3t} + (C_5 + C_6 t)e^{-3t}.$$

EXAMPLE 8.20

Solve the differential equation

$$x^3 \frac{d^3y}{dx^3} - 6x^2 \frac{d^2y}{dx^2} + 18x\frac{dy}{dx} - 24y = 0, \tag{8.89}$$

subject to the initial conditions $y(1) = 0$, $y'(1) = 0$, $y''(1) = 3$.

Change variable

Because the initial conditions for this third order Cauchy-Euler differential equation are given at $x = 1$, we are interested in values of $x > 0$. Thus, we change our independent variable by $x = e^t$, giving $t = \ln x$. Our rules for replacing derivatives with respect to x in (8.89) to ones with respect to t, where we use D to represent d/dt, are $x^3 d^3/dx^3 = D(D-1)(D-2)$, $x^2 d^2/dx^2 = D(D-1)$, and $x\,d/dx = D$. These substitutions result in the differential equation with constant coefficients,

$$[D(D-1)(D-2) - 6D(D-1) + 18D - 24]y = 0,$$

Characteristic equation with an associated characteristic equation,

$$r(r-1)(r-2)-6r(r-1)+18r-24=0.$$

The first two terms have a common factor of $r-1$, but because the last term does not have this factor, it seems that we must expand this cubic polynomial. Doing so gives $r^3-9r^2+26r-24=0$, where a root is not obvious. We have two alternatives for finding the roots, one graphical and the other analytical.

In the graphical approach, we examine the graph of this cubic polynomial, Figure 8.19. Here it appears that this polynomial has roots at 2, 3, and 4. If so, then the product $(r-2)(r-3)(r-4)$, when expanded, will be proportional to the cubic polynomial. Expansion of these three products shows that $(r-2)(r-3)(r-4)=r^3-9r^2+26r-24$.

To follow an algebraic approach that yields rational roots, we recall that our possible integer roots of polynomials of this type (where the coefficient of the highest term is 1) must be factors of the constant term -24. This means our choices are ±1, ±2, ±3, ±4, ±6, ±8, ±12, and ±24, and it turns out that 2 is a root. If we factor out that root, we obtain

$$r^3-9r^2+26r-24=(r-2)(r^2-7r+12)=(r-2)(r-3)(r-4)=0,$$

giving 3 and 4 as the other two roots.

General solution Thus, the general solution of (8.89) is $C_1e^{2t}+C_2e^{3t}+C_3e^{4t}$, or, in terms of our original independent variable,

$$y(x)=C_1x^2+C_2x^3+C_3x^4.$$

Initial conditions To satisfy our initial conditions, we substitute $x=1$ into the above solution and its derivatives to obtain

$$\begin{aligned} y(1) &= C_1+C_2+C_3 &&= 0,\\ y'(1) &= 2C_1+3C_2+4C_3 &&= -4,\\ y''(1) &= 2C_1+6C_2+12C_3 &&= -18. \end{aligned}$$

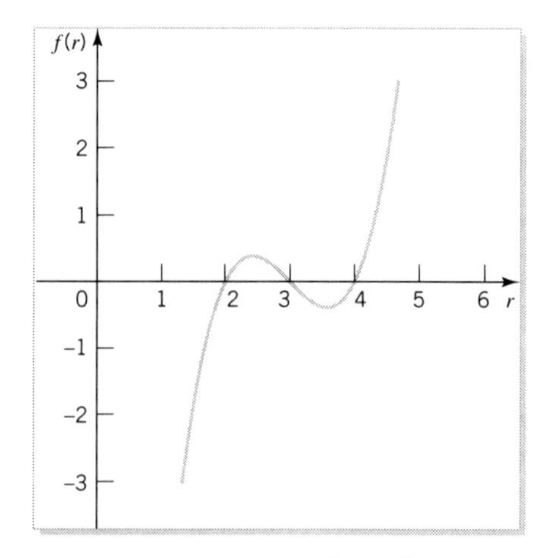

FIGURE 8.19 The cubic $r^3-9r^2+26r-24$

If we subtract the last two equations, we obtain $-3C_2 - 8C_3 = 14$, whereas if we multiply the top equation by -2 and add the result to the second equation, we obtain $C_2 + 2C_3 = -4$. Solving these two equations gives $C_2 = -2$ and $C_3 = -1$. If we use these two values in the top equation, we obtain $C_1 = 3$, giving the solution of our initial value problem as

Explicit solution

$$y(x) = 3x^2 - 2x^3 - x^4.$$

EXAMPLE 8.21 *Two Spring-Masses*

The physical example we consider now is that of two masses suspended on two springs, as shown on the left-hand side of Figure 8.20.

We treat the two masses, m_1 and m_2, as point masses and assume that the two springs obey Hooke's law with respective spring constants k_1 and k_2. If there is no motion, the system is said to be in equilibrium. We consider vertical oscillations of this system and let x and y be the displacements of the upper and lower masses, respectively, from equilibrium. Positive values of x and y are in the downward direction. If we displace each mass from its equilibrium position, as shown on the right-hand side of Figure 8.20, then the net force acting on m_1 is $-k_1x + k_2(y - x)$, and the net force acting on m_2 is $-k_2(y - x)$. Using these values in Newton's second law of motion gives

$$\begin{cases} m_1x'' = -k_1x + k_2(y - x), \\ m_2y'' = -k_2(y - x). \end{cases}$$

We now solve this system of equations subject to the initial conditions $x(0) = 0$, $x'(0) = 0$, $y(0) = y_0$, and $y'(0) = 0$. These initial conditions model holding the top mass at its equilibrium position and the bottom mass at a distance y_0 below its equilibrium position, and then releasing the two masses from rest. To simplify the

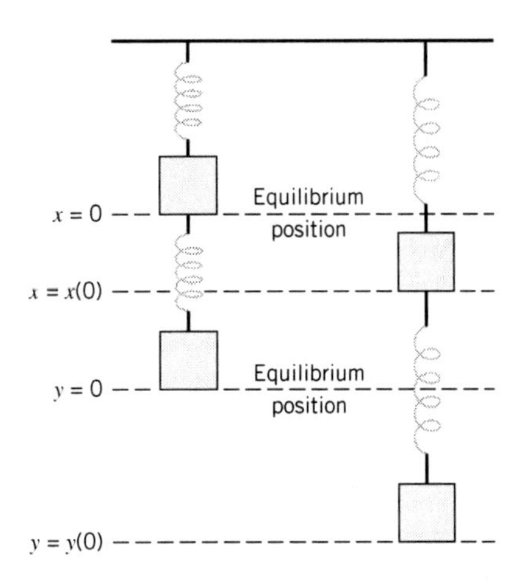

FIGURE 8.20 Two springs and two masses — in equilibrium and displaced

resulting algebra, we take the specific values of the two masses and spring constants as $m_1 = m_2 = 1$, $k_1 = 5$, and $k_2 = 6$. This gives us the system

$$\begin{cases} x'' = -5x + 6(y - x), \\ y'' = -6(y - x), \end{cases}$$

or

$$\begin{cases} x'' = -11x + 6y, \\ y'' = 6x - 6y. \end{cases} \tag{8.90}$$

Elimination method

We can solve this system by the elimination method, where we end up with an equation in only one variable. From the bottom equation of (8.90), we have

$$x = \frac{1}{6}y'' + y, \tag{8.91}$$

which when substituted in the top equation gives $[y''/6 + y]'' = -11(y'' + 6y)/6 + 6y$, or

$$y'''' + 17y'' + 30y = 0,$$

Characteristic equation

a fourth order differential equation. The characteristic equation in this case is $r^4 + 17r^2 + 30 = (r^2 + 15)(r^2 + 2) = 0$, with roots $r = \pm i\sqrt{2}$ and $\pm i\sqrt{15}$. Thus,

$$y(t) = C_1 \sin\sqrt{2}t + C_2 \cos\sqrt{2}t + C_3 \sin\sqrt{15}t + C_4 \cos\sqrt{15}t. \tag{8.92}$$

We substitute this expression for $y(t)$ into (8.91) to find $x(t)$. This means we require $y'(t)$ and $y''(t)$, which, from (8.92), are

$$y'(t) = \sqrt{2}C_1 \cos\sqrt{2}t - \sqrt{2}C_2 \sin\sqrt{2}t + \sqrt{15}C_3 \cos\sqrt{15}t - \sqrt{15}C_4 \sin\sqrt{15}t,$$
$$y''(t) = -2C_1 \sin\sqrt{2}t - 2C_2 \cos\sqrt{2}t - 15C_3 \sin\sqrt{15}t - 15C_4 \cos\sqrt{15}t.$$

Substituting these into (8.91) gives

$$x(t) = \frac{2}{3}C_1 \sin\sqrt{2}t + \frac{2}{3}C_2 \cos\sqrt{2}t - \frac{3}{2}C_3 \sin\sqrt{15}t - \frac{3}{2}C_4 \cos\sqrt{15}t.$$

Initial conditions

We now use the initial conditions $y(0) = y_0, y'(0) = 0$, on $y(t)$ and $y'(t)$ (which we have already calculated), and the initial conditions $x(0) = 0, x'(0) = 0$ on $x(t)$ and $x'(t)$ (which must be calculated) to find $y(0) = C_2 + C_4 = y_0$, $y'(0) = \sqrt{2}C_1 + \sqrt{15}C_3 = 0$, $x(0) = \frac{2}{3}C_2 + \frac{3}{2}C_4 = 0$, and $x'(0) = \frac{2}{3}\sqrt{2}C_1 - \frac{3}{2}\sqrt{15}C_3 = 0$. Solving these gives $C_1 = 0$, $C_2 = 9y_0/13$, $C_3 = 0$, and $C_4 = 4y_0/13$.

Explicit solution

Thus, the displacement of the two springs is given by

$$\begin{aligned} x(t) &= (6/13)\left(\cos\sqrt{2}t - \cos\sqrt{15}t\right) y_0, \\ y(t) &= \left[(9/13)\cos\sqrt{2}t + (4/13)\cos\sqrt{15}t\right] y_0. \end{aligned} \tag{8.93}$$

The functions $x(t)$ and $y(t)$ in (8.93) are graphed in Figure 8.21 for $y_0 = 8$.

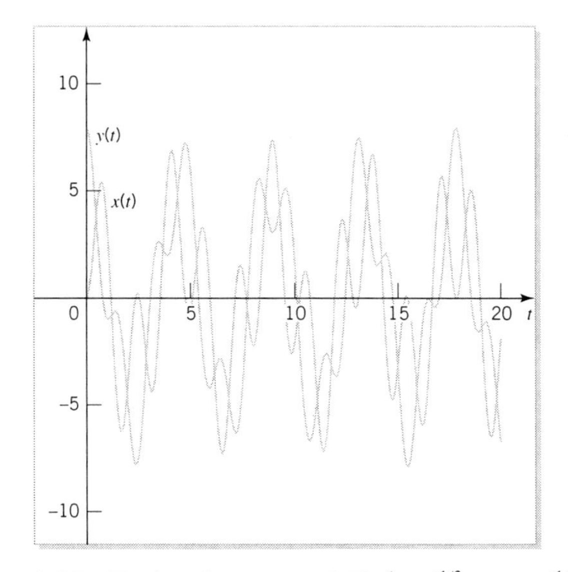

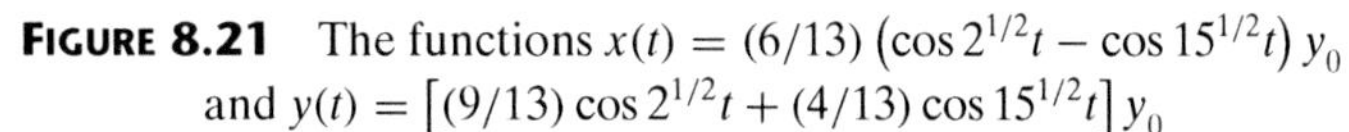
FIGURE 8.21 The functions $x(t) = (6/13)\left(\cos 2^{1/2}t - \cos 15^{1/2}t\right) y_0$ and $y(t) = \left[(9/13)\cos 2^{1/2}t + (4/13)\cos 15^{1/2}t\right] y_0$

EXERCISES

1. Find the general solution of the following differential equations.

(a) $(D^3 + D)x = 0$
(b) $(D^3 + 1)x = 0$
(c) $(D^4 + 4D^2 + 4)x = 0$
(d) $(D^4 + 2D^2 - 15)x = 0$
(e) $(D^4 + 2D^2 - 8)x = 0$
(f) $(D^3 - 7D^2 + 19D - 13)x = 0$
(g) $x''' - 2x'' - x' + 2x = 0$
(h) $x'''' + 8x'' - 9x = 0$
(i) $(D^3 + D^2 + 3D - 5)x = 0$
(j) $(D^4 - 5D^2 + 4)x = 0$

2. Find the general solution of the following differential equations assuming $x > 0$. [Note: $y' = dy/dx$.]

(a) $x^3y''' + 3xy' + y = 0$
(b) $x^3y''' + 2x^2y'' - xy' + y = 0$

3. Solve the following initial value problems.

(a) $(D^4 + 3D^3 + 2D^2)x = 0$
$x(0) = x'(0) = x''(0) = 0, x'''(0) = 8$
(b) $(D^4 + 6D^2 + 9)x = 0$
$x(0) = x'(0) = x''(0) = 0, x'''(0) = 6$
(c) $(D^3 + 6D^2 + 5D - 12)x = 0$
$x(0) = 0,\ x'(0) = 4, x''(0) = -8$
(d) $(D^4 - 16)x = 0$
$x(0) = x'(0) = x''(0) = 0, x'''(0) = 8$

[Hint for part (d): Before evaluating the arbitrary constants, write the solution of the differential equation in terms of hyperbolic and trigonometric functions instead of exponentials.]

4. Our earlier ideas of linear independence and dependence for two functions extend to a set of n functions as follows: The set of functions $\{f_1(t), f_2(t), \cdots, f_n(t)\}$ is said to be linearly independent for all t in some common domain (say, $a < t < b$) of these functions if the only way a linear combination of these n functions may equal zero for all t in this domain,

$$\sum_{m=1}^{n} b_m f_m(t) = b_1 f_1(t) + b_2 f_2(t) + \cdots + b_n f_n(t) = 0, \qquad (8.94)$$

is for all the constants in (8.94) to equal zero; that is,

$$b_1 = b_2 = b_3 = \cdots = b_n = 0.$$

If there is a way to have (8.94) satisfied with at least one of the $b_m \neq 0$, the set of functions is said to be linearly dependent for t in that domain. We define a Wronskian for a set of n functions by

$$W[f_1, f_2, \cdots, f_n] = \begin{vmatrix} f_1 & f_2 & \cdots & f_n \\ f_1' & f_2' & \cdots & f_n' \\ \vdots & \vdots & \ddots & \vdots \\ f_1^{(n-1)} & f_2^{(n-1)} & \cdots & f_n^{(n-1)} \end{vmatrix}.$$

Now use the Wronskian to show linear independence of a set of functions, $\{f_1(t), f_2(t), f_3(t)\}$, by starting with the linear combination

$$b_1 f_1(t) + b_2 f_2(t) + b_3 f_3(t) = 0, \qquad (8.95)$$

and see if you can find nonzero values of the constants that make (8.95) an identity. By differentiating this identity twice, you obtain three equations in the three unknowns b_1, b_2, and b_3, and use this to prove that if the Wronskian $W[f_1, f_2, f_3] \neq 0$ at some point in the common domain of the three functions, then the functions are linearly independent there.

5. Are the following sets of functions linearly independent or dependent? (Compute the Wronskian first.)

(a) $3, \sin t, \cos t$ (b) $3, \sin^2 t, \cos^2 t$
(c) $1, 1+t, t+3$ (d) $1-t, 1+t, t^2$
(e) e^t, te^t, t^2e^t (f) $e^{rt}, te^{rt}, t^2e^{rt}$
(g) e^t, e^{-t}, e^{2t}

6. The determinant

$$\begin{vmatrix} 1 & 1 & 1 & \cdots & 1 \\ r_1 & r_2 & r_3 & \cdots & r_n \\ r_1^2 & r_2^2 & r_3^2 & \cdots & r_n^2 \\ \vdots & \vdots & \vdots & \ddots & \vdots \\ r_1^{n-1} & r_2^{n-1} & r_3^{n-1} & \cdots & r_n^{n-1} \end{vmatrix}$$

is called VANDERMONDE'S DETERMINANT.

(a) Show that this determinant equals $r_2 - r_1$ if $n = 2$.

(b) Show that this determinant equals $(r_2 - r_1)\left[(r_3 - r_1)(r_3 - r_2)\right]$ if $n = 3$.

(c) Use induction to prove that for any positive integer n, this determinant equals

$$(r_2 - r_1)\left[(r_3 - r_1)(r_3 - r_2)\right]\left[(r_4 - r_1)(r_4 - r_2)(r_4 - r_3)\right] \times \\ \times [\cdots]\left[(r_n - r_1)(r_n - r_2)\cdots(r_n - r_{n-1})\right].$$

7. Use the results of Exercise 6 to prove that the following sets of functions are linearly independent.

(a) e^t, e^{2t}, e^{4t}.

(b) $e^t, e^{2t}, e^{3t}, e^{4t}, e^{5t}$.

(c) $e^t, e^{2t}, e^{3t}, \cdots, e^{nt}$.

8. Prove that if the Wronskian $W[f_1, f_2, f_3, f_4] \neq 0$ at some point in the common domain of the four functions $\{f_1(t), f_2(t), f_3(t), f_4(t)\}$, then the functions are linearly independent there.

9. The result we use to show that a set of solutions of a differential equation is linearly independent or dependent is as follows: Let $\{f_i(t),\ i = 1, 2, 3, \cdots n\}$ be a set of solutions of the differential equation

$$\left[\sum_{j=0}^{n} a_j(t) D^j\right] x = 0, \qquad a < t < b. \qquad (8.96)$$

Then this set of solutions is linearly independent if and only if the Wronskian

$$W[f_1, f_2, \cdots, f_n] \neq 0$$

for some t in $a < t < b$. Prove this result for $n = 3$ using the following steps.

(a) Show that

$$\frac{d}{dt} W[f_1, f_2, f_3] = \begin{vmatrix} f_1 & f_2 & f_3 \\ f_1' & f_2' & f_3' \\ f_1''' & f_2''' & f_3''' \end{vmatrix}.$$

[Hint: Recall that the derivative of a determinant is the sum of the determinants obtained by differentiating each row in turn. See Appendix A.5.]

(b) Use the fact that the f_i, $i = 1, 2, 3$, satisfy (8.96) with $n = 3$ to show that

$$\frac{d}{dt} W[f_1, f_2, f_3] = -\frac{a_2(t)}{a_3(t)} W[f_1, f_2, f_3].$$

(c) Integrate the differential equation in part (b), and derive Abel's formula,

$$W[f_1, f_2, f_3] = C \exp\left[-\int \frac{a_2(t)}{a_3(t)}\, dt\right].$$

(d) Show that if the integral in part (c) is written as a definite integral from t_0 to t, the constant C becomes the Wronskian evaluated at t_0.

(e) Use the expression for $W[f_1, f_2, f_3]$ to complete the proof.

10. Determine if the following solutions of a linear differential equation form a linearly independent set.

(a) $1, \ln t, (\ln t)^2$.

(b) $e^t, e^{2t}, e^t - e^{2t}$.

(c) $1, t, t^2, t^3, \cdots, t^n$.

(d) $\sin t, \cos t, t\sin t, t\cos t$.

(e) $e^{r_1 t}, e^{r_2 t}, e^{r_3 t}, \cdots, e^{r_n t}$, where $r_1, r_2, r_3, \cdots, r_n$ are all distinct real numbers.

11. A Two-Body Problem.[13] A linear spring with spring constant k has two masses, m and M, attached at either end. The system lies in a line on a horizontal surface in equilibrium. All motion takes place along this line and is assumed frictionless. The two masses are now moved to initial positions along the line and given initial velocities along the line. If $x(t)$ is the distance traveled by the mass m from its equilibrium position, and $y(t)$ is the corresponding position for M, then the motion of this system is governed by $mx'' = k(y - x)$, $My'' = -k(y - x)$. Solve these equations for $x(t)$ and $y(t)$.

12. Coupled Pendulums — A Simple Experiment.[14] Construct the following piece of apparatus, which is a coupled pendulum. Suspend two identical pendulums (same length wire, same mass) and join the vertical wires with a horizontal light rod. Make sure that the points of suspension of the pendulums are the same distance apart as the length of the rod so that the wires hang vertically.[15] We are first going to perform two experiments, and then we are going to explain what we have found. All motion is assumed to be in the plane of the wires.

(a) Perform the following experiment. Displace the two pendulums through the same small initial angle and simultaneously release both from rest. Describe what happens.

(b) Perform the following experiment. Displace the two pendulums through the same small initial angle but in opposite directions, and simultaneously release both from rest. Describe what happens.

(c) Perform the following experiment. Displace one of the pendulums through a small initial angle while holding the other pendulum vertical. Now simultaneously release both from rest. Describe what happens.

(d) It can be shown that if $x(t)$ and $y(t)$ are the horizontal displacements of the two pendulums from vertical, then the linearized differential equations governing the subsequent motion, neglecting air resistance, are $x'' + \lambda^2 x = \lambda^2\Delta$, $y'' + \lambda^2 y = \lambda^2\Delta$, where $\lambda^2 = g/H$ and $\Delta = \frac{1}{2}(x + y)\left(1 - \frac{H}{L}\right)$. Here L is the length of the pendulums from the pivot and H is the length of the pendulums from the rod. Solve these differential equations subject to $x(0) = x_0, x'(0) = u_0, y(0) = y_0, y'(0) = v_0$.

(e) What are the solutions of the differential equations in part (d) subject to the initial conditions $x(0) = y(0) = x_0$ and $x'(0) = y'(0) = 0$? How does this agree with part (a)?

(f) What are the solutions of the differential equations in part (d) subject to the initial conditions $x(0) = x_0$, $y(0) = -x_0$, and $x'(0) = y'(0) = 0$? How does this agree with part (b)?

(g) What are the solutions of the differential equations in part (d) subject to the initial conditions $x(0) = x_0$, $y(0) = 0$, and $x'(0) = y'(0) = 0$? How does this agree with part (c)? Show that, according to this model, the period T of the beats — that is, the time it takes for the mass at rest to return to rest — is $T = 2\pi/(\lambda - \mu)$, which can be written as $1/T = \sqrt{g}(1/\sqrt{H} - 1/\sqrt{L})/(2\pi)$. Thus, a plot of $1/T$ (the frequency) versus $1/\sqrt{H}$ should yield a straight line of slope $\sqrt{g}/(2\pi)$ and intercept $-\sqrt{g}/(\sqrt{L}2\pi)$. Conduct an experiment where the beat period, T, is measured for various values of H. Then plot $1/T$ versus $1/\sqrt{H}$. Do you get a straight line? Table 8.1 and Figure 8.22 show the results of such an experiment.[16]

Table 8.1 Frequency $1/T$ of coupled pendulums as a function of $H^{-1/2}$

$1/\sqrt{H}$ ($cm^{-1/2}$)	$1/T$ (sec^{-1})	$1/\sqrt{H}$ ($cm^{-1/2}$)	$1/T$ (sec^{-1})
0.247	0.044	0.295	0.276
0.256	0.097	0.315	0.375
0.266	0.141	0.340	0.502
0.280	0.209	0.364	0.633

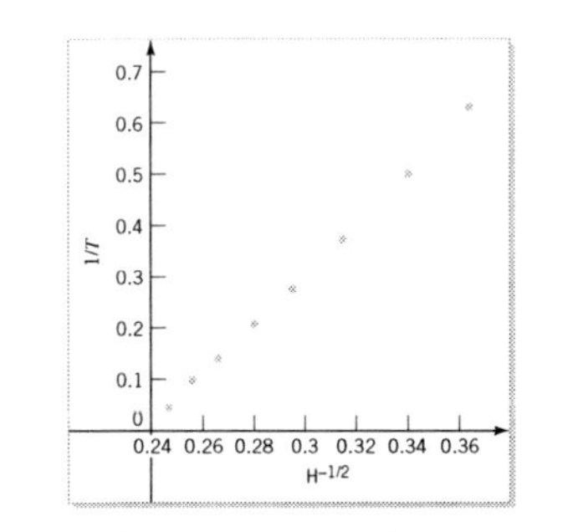

FIGURE 8.22 Frequency $1/T$ of coupled pendulums as a function of $H^{-1/2}$

[13] "A simple example for the two body problem" by E. Maor, *The Physics Teacher*, February 1973, pages 104–105.

[14] "Teaching physics with coupled pendulums" by J. Priest and J. Poth, *The Physics Teacher*, February 1982, pages 80–85.

[15] An interesting in-class version of this pendulum can be constructed from two bowling balls suspended with two steel wires about 7 feet long. For the horizontal rod, a 4-foot dowel rod attached to the wires by rubber bands can be used.

[16] "Teaching physics with coupled pendulums" by J. Priest and J. Poth, *The Physics Teacher*, February 1982, pages 80–85, Figure 6.

8.8 SOLVING HIGHER ORDER NONHOMOGENEOUS DIFFERENTIAL EQUATIONS

In the previous section we considered higher order homogeneous linear differential equations and found that our techniques from earlier chapters for solving second order linear differential equations applied there as well. Now we consider the case in which these higher order equations are nonhomogeneous, and we will discover that our previous techniques from earlier sections of this chapter need very little modification.

EXAMPLE 8.22 *Response of a Beam to a Distributed Oscillatory Force*

We now consider the situation in which a beam of length L extends horizontally from its support and is subject to an oscillatory force distributed along its length (see Figure 8.23). The vertical deflection, x, of the resulting oscillatory motion of the beam is governed by the fourth order differential equation

$$\left(EIx''\right)'' - m\omega^2 x = f(t). \tag{8.97}$$

Here t is the distance from the support (so $0 < t < L$), m is the mass per unit length of the beam, ω is the frequency of the oscillatory force, EI is the flexural rigidity, and $f(t)$ gives the vertical force as a function of time.

If we divide (8.97) by EI and define k by $k^4 = m\omega^2/(EI)$, we may rewrite the differential equation as

$$x'''' - k^4 x = \frac{1}{EI} f(t). \tag{8.98}$$

From Section 7.1 we know that general solutions of second order linear differential equations have the form

$$x(t) = x_h(t) + x_p(t),$$

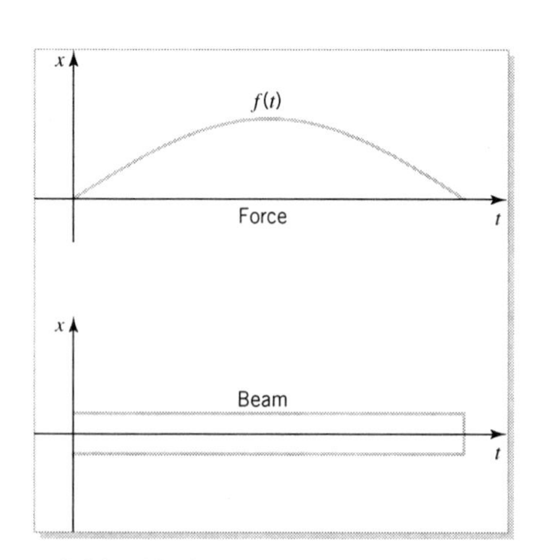

FIGURE 8.23 Horizontal beam and distributed force

where $x_h(t)$ is the general solution of the associated homogeneous differential equation and $x_p(t)$ is a particular solution. The same result is true for higher order linear differential equations. In the last section we developed ways of determining $x_h(t)$, so we now need a method of finding a particular solution. To illustrate this entire procedure, we consider the preceding situation in which the vertical distribution on the beam is such that $f(t)/(EI) = \sin(\pi t/L)$.

Characteristic equation

Our first task in solving (8.98) is to find the general solution of the associated homogeneous differential equation, which has as its characteristic equation

$$r^4 - k^4 = (r-k)(r+k)(r-ki)(r+ki) = 0.$$

Because this equation has four simple roots, we may write the solution of the associated differential equation as

$$x_h(t) = C_1 \cosh kt + C_2 \sinh kt + C_3 \cos kt + C_4 \sin kt,$$

where we have used the hyperbolic functions $\cosh kt = (e^{kt} + e^{-kt})/2$ and $\sinh kt = (e^{kt} - e^{-kt})/2$ instead of our usual exponentials.

We now need to determine a particular solution. Recall that in Section 7.2 we discussed the family of functions associated with forcing functions, consisting of exponential functions, polynomials, sine and cosine functions, and products of these functions. There we discovered that a linear combination of members of this family of functions was the proper trial solution for x_p for nonhomogeneous linear differential equations that had constant coefficients. The same reasons used for developing that family for second order equations carry over to the higher order equations as well. Because we will be using Table 7.1 on page 287 in the following development, the table is repeated here for our convenience as Table 8.2.

Trial solution

The forcing function in (8.98) is $\sin(\pi t/L)$, so Table 8.2 gives the proper form for a particular solution (for $k \neq \pi/L$) as

$$x_p(t) = A\cos(\pi t/L) + B\sin(\pi t/L)\,.$$

Table 8.2 **Trial solution for special forcing functions**

$f(t)$	Trial Solution
ae^{ct}	Ae^{ct}
$b_0 + b_1 t + b_2 t^2 + b_3 t^3 + \cdots + b_n t^n$	$A_0 + A_1 t + A_2 t^2 + A_3 t^3 + \cdots + A_n t^n$
$a \sin wt + b \cos wt$	$A \sin wt + B \cos wt$
$e^{ct}(a \sin wt + b \cos wt)$	$e^{ct}(A \sin wt + B \cos wt)$
$e^{ct} \sum_{k=0}^{k=n} a_k t^k$	$e^{ct} \sum_{k=0}^{k=n} A_k t^k$
$\sum_{k=0}^{k=n} a_k t^k (a \sin wt + b \cos wt)$	$\sum_{k=0}^{k=n} A_k t^k \sin wt + \sum_{k=0}^{k=n} B_k t^k \cos wt$
$e^{ct} \sum_{k=0}^{k=n} a_k t^k (a \sin wt + b \cos wt)$	$e^{ct}(\sum_{k=0}^{k=n} A_k t^k \sin wt + \sum_{k=0}^{k=n} B_k t^k \cos wt)$

General solution

Substituting this expression into (8.98) and equating coefficients of like terms gives $A = 0$ and $B = 1/[(\pi/L)^4 - k^4]$. This means that the general solution of our original differential equation is given by

$$x(t) = C_1 \cosh kt + C_2 \sinh kt + C_3 \cos kt + C_4 \sin kt + \frac{1}{(\pi/L)^4 - k^4} \sin(\pi t/L),$$

where C_1, C_2, C_3, and C_4 are arbitrary constants. ■

The fact that we could use Table 8.2 to solve this fourth order differential equation was no accident. The method we used to find particular solutions for forcing functions of the form appearing on the left-hand side of Table 8.2 is valid for linear differential equations with constant coefficients regardless of the order. We outline this method.

How to Find a Particular Solution for Higher Order Linear Differential Equations with Constant Coefficients

Purpose To find a particular solution of the linear differential equation

$$\sum_{m=0}^{n} a_m \frac{d^m x}{dt^m} = a_n \frac{d^n x}{dt^n} + a_{n-1} \frac{d^{n-1} x}{dt^{n-1}} + \cdots + a_2 \frac{d^2 x}{dt^2} + a_1 \frac{dx}{dt} + a_0 x = f(t),$$

where the coefficients $a_n, a_{n-1}, \cdots, a_2, a_1, a_0$ are constants and $f(t)$ is a polynomial, an exponential function, a sine or cosine function, or a product or sum of any of these three types of functions.

Process

1. Solve the associated homogeneous differential equation and obtain n linearly independent solutions, labeled $x_1(t), x_2(t), x_3(t), \ldots, x_n(t)$. See *How to Solve Higher Order Linear Differential Equations with Constant Coefficients* on page 359.
2. Consider one term in $f(t)$ and compare it with the functions listed on the left-hand side of Table 8.2. For this term, write down the proper particular solution as given on the right-hand side of the table.
3. Compare the form of this particular solution with the n linearly independent solutions listed in step 1. If this particular solution contains any of $x_1(t), x_2(t), x_3(t), \ldots, x_n(t)$, multiply this particular solution by the lowest power of t that makes every term in the result different from every one of these linearly independent solutions.
4. If there is more than one term in $f(t)$, repeat steps 2 and 3 for each term.
5. Examine all the expressions for $x_p(t)$ resulting from the preceding four steps and eliminate all duplicates. The proper form for $x_p(t)$ will be the result of this operation.

EXAMPLE 8.23

Find the general solution of the fourth order differential equation

$$x'''' - 5x'' + 4x = 80e^{3t} - 36e^{2t}. \tag{8.99}$$

If we follow the preceding procedure, we first find the general solution of the associated homogeneous differential equation

$$x'''' - 5x'' + 4x = 0. \tag{8.100}$$

Characteristic equation

Now the characteristic equation may be factored as

$$r^4 - 5r^2 + 4 = (r^2 - 4)(r^2 - 1) = (r - 2)(r + 2)(r - 1)(r + 1) = 0,$$

so the general solution of (8.100) is

$$x_h(t) = C_1e^{2t} + C_2e^{-2t} + C_3e^{t} + C_4e^{-t}, \tag{8.101}$$

where C_1, C_2, C_3, and C_4 are arbitrary constants.

To find the proper form to try for a particular solution, we first consider the forcing function $80e^{3t}$ and note from Table 8.2 that the proper trial solution is Ae^{3t}. Because this function does not appear as one of the linearly independent solutions of the homogeneous differential equation as given in (8.101), we keep this as part of our trial solution and consider the term $-36e^{2t}$. The proper trial solution for this term is Be^{2t}, but because this term occurs in (8.101), we multiply the term by t and note that Bte^{2t} does not occur in (8.101). Thus, we have our proper trial solution as

Trial solution

$$x_p(t) = Ae^{3t} + Bte^{2t}.$$

Substituting this expression into (8.99) results in

$$40Ae^{3t} + 36Be^{2t} = 80e^{3t} - 36e^{2t},$$

General solution

from which we see that we must choose $A = 2$ and $B = -1$. Thus, the general solution of (8.99) is

$$x(t) = C_1e^{2t} + C_2e^{-2t} + C_3e^{t} + C_4e^{-t} + 2e^{3t} - te^{2t}. \tag{8.102}$$

There are two terms in (8.102) that could be classified as transient terms, with the other four terms being unbounded as $t \to \infty$. (Which of these four terms dominates as $t \to \infty$?) ■

The use of Table 8.2 is obviously limited to the types of functions listed on the left-hand side of the table. For other forcing functions, or for differential equations that do not have constant coefficients, we will use a modified version of the variation of parameters technique that we discussed in Section 8.3. Because of the great similarity in the procedure for using variation of parameters for second order and higher order differential equations, we will simply write down the procedure and then give examples.

How to Find a Particular Solution for Higher Order Differential Equations Using Variation of Parameters

Purpose To find a particular solution of the linear differential equation,

$$\sum_{m=0}^{n} a_m \frac{d^m x}{dt^m} = a_n \frac{d^n x}{dt^n} + a_{n-1} \frac{d^{n-1} x}{dt^{n-1}} + \cdots + a_2 \frac{d^2 x}{dt^2} + a_1 \frac{dx}{dt} + a_0 x = f(t), \tag{8.103}$$

where the coefficients $a_n, a_{n-1}, \cdots, a_2, a_1$, and a_0 may be functions of t.

Process

1. Solve the associated homogeneous differential equation and obtain n linearly independent solutions, labeled $x_1(t), x_2(t), x_3(t), ..., x_n(t)$.
2. Assume a particular solution of the form

$$\sum_{m=1}^{n} x_m(t) z_m(t) = x_1(t) z_1(t) + x_2(t) z_2(t) + \cdots + x_n(t) z_n(t), \tag{8.104}$$

and write down the system of equations satisfied by $z'_m(t), m = 1, 2, 3, \cdots, n$,

$$\begin{aligned}
x_1(t)z'_1(t) + x_2(t)z'_2(t) + \cdots + x_n(t)z'_n(t) &= 0,\\
x'_1(t)z'_1(t) + x'_2(t)z'_2(t) + \cdots + x'_n(t)z'_n(t) &= 0,\\
x''_1(t)z'_1(t) + x''_2(t)z'_2(t) + \cdots + x''_n(t)z'_n(t) &= 0,\\
\vdots \quad \ddots \quad & \quad \vdots\\
x_1^{(n-1)}(t)z'_1(t) + x_2^{(n-1)}(t)z'_2(t) + \cdots + x_n^{(n-1)}(t)z'_n(t) &= f(t)/a_n(t).
\end{aligned}$$

3. Solve this system for $z'_m(t), m = 1, 2, 3, \cdots, n$, and integrate to obtain $z_m(t)$, $m = 1, 2, 3, \cdots, n$, where we may set the arbitrary constants equal to zero. The final form of the particular solution is obtained by substituting back into (8.104) these values of $z_m(t), m = 1, 2, 3, \cdots, n$.

Comments about Using Variation of Parameters for Higher Order Differential Equations

- If we include the arbitrary constants when integrating the expressions for the derivatives $z'_m(t), m = 1, 2, 3, \cdots, n$, the expression for the particular solution will also be the general solution of (8.103).
- If we are solving an initial value problem and cannot evaluate the integrals giving $z_m(t), m = 1, 2, 3, \cdots, n$, in terms of elementary functions, it is often convenient to write the integrals with the independent variable as the upper limit and the value of t giving the initial condition as the lower limit.
- A common mistake in writing down the system of equations in step 2 is to forget to divide by $a_n(t)$ on the right-hand side of the bottom equation.

EXAMPLE 8.24

Find the general solution of the differential equation

$$x''' + x' = \sec t. \tag{8.105}$$

We follow the preceding steps by first determining the general solution of the associated differential equation by considering the characteristic equation

Characteristic equation

$$r^3 + r = r(r^2 + 1) = r(r + i)(r - i) = 0.$$

Thus, the general solution of this homogeneous differential equation is

$$x_h(t) = C_1 + C_2 \sin t + C_3 \cos t, \tag{8.106}$$

where C_1, C_2, and C_3 are arbitrary constants.

Because the forcing function on the right-hand side of (8.105) does not appear on the left side of Table 8.2, we use the method of variation of parameters. Thus, we write our particular solution as

Trial solution

$$x_p(t) = z_1 + \sin t\, z_2 + \cos t\, z_3,$$

where z_1, z_2, and z_3 are functions of t to be determined. The differential equations satisfied by the derivatives of the three functions are

$$\begin{aligned} z_1' + \sin t\, z_2' + \cos t\, z_3' &= 0, \\ \cos t\, z_2' - \sin t\, z_3' &= 0, \\ -\sin t\, z_2' - \cos t\, z_3' &= \sec t. \end{aligned}$$

To solve this system of equations for z_1', z_2', and z_3', we multiply the last equation by $\cos t$ and add the result to what we obtain by multiplying the middle equation by $\sin t$. This gives $z_3' = -1$, so from the middle equation we discover that $z_2' = -\sin t/\cos t$, and from the top equation we have that $z_1' = \sin^2 t/\cos t + \cos t = 1/\cos t$. With help from a table of integrals, these three functions may be integrated to obtain

$$z_1(t) = \ln|\sec t + \tan t|, \qquad z_2(t) = \ln|\cos t|, \qquad z_3(t) = -t.$$

Particular solution

Thus, we have our particular solution as

$$x_p(t) = \ln|\sec t + \tan t| + \sin t \ln|\cos t| - t\cos t, \tag{8.107}$$

and the general solution of (8.105) as the sum of the functions given in (8.106) and (8.107). ■

EXAMPLE 8.25

Find the general solution of the nonhomogeneous Cauchy-Euler differential equation

$$x^3\frac{d^3y}{dx^3} - x^2\frac{d^2y}{dx^2} + 2x\frac{dy}{dx} - 2y = x^2, \qquad x > 0. \tag{8.108}$$

We start by finding the general solution of the associated homogeneous differential equation, which, with the change of independent variable $x = e^t$, becomes

$$[D(D-1)(D-2) - D(D-1) + 2D - 2]y = 0.$$

Characteristic equation

Because this is a linear differential equation with constant coefficients, we first find our characteristic equation as

$$r(r-1)(r-2) - r(r-1) + 2(r-1) = 0.$$

We notice that each term in this equation has a common factor of $r-1$, so we rewrite it as

$$(r-1)(r^2 - 2r - r + 2) = (r-1)(r^2 - 3r + 2) = (r-1)(r-1)(r-2) = 0.$$

Thus, $r = 1$ is a double root, and $r = 2$ is a single root. This means we may write our solution of the transformed differential equation as $y_h = C_1e^{2t} + (C_2 + C_3t)e^t$, which, in terms of our original independent variable, becomes

$$y_h(x) = C_1x^2 + (C_2 + C_3 \ln x)x.$$

Trial solution

To find a particular solution with the method of variation of parameters, we let

$$y_p(x) = x^2z_1(x) + xz_2(x) + x\ln x\ z_3(x).$$

Thus, our three equations for determining these three functions are

$$\begin{aligned} x^2z_1'(x) + xz_2'(x) + x\ln x\ z_3'(x) &= 0, \\ 2xz_1'(x) + z_2'(x) + (1 + \ln x)\, z_3'(x) &= 0, \\ 2z_1'(x) + \tfrac{1}{x}z_3'(x) &= \tfrac{1}{x}. \end{aligned} \tag{8.109}$$

Although there are several ways to solve this system, the way we chose takes advantage of the fact that the $z_2'(x)$ term is missing in the bottom equation of (8.109). Thus, we eliminate $z_2'(x)$ from the other two equations by multiplying the top equation by $-1/x$ and adding the result to the middle equation. This gives

$$xz_1'(x) + z_3'(x) = 0. \tag{8.110}$$

We now multiply this equation by $-1/x$ and add the result to the bottom equation in (8.109) to obtain

$$z_1'(x) = \frac{1}{x}.$$

We now use (8.110) to find that

$$z_3'(x) = -1.$$

To find $z_2'(x)$, we return to the top equation in (8.109) and obtain

$$z_2'(x) = -\ln x\, z_3'(x) - xz_1'(x) = \ln x - 1,$$

so by integration we have $z_2(x) = x\ln x - 2x$. Integrating our other two expressions gives $z_1(x) = \ln x$ and $z_3'(x) = -x$. We now can write a particular solution as

$$\begin{aligned} y_p(x) &= x^2 z_1(x) + xz_2(x) + x\ln x\, z_3(x) \\ &= x^2\ln x + x(x\ln x - 2x) - (x\ln x)x \\ &= x^2\ln x - 2x^2. \end{aligned}$$

General solution Thus, the general solution of (8.108) is

$$y(x) = C_1x^2 + (C_2 + C_3\ln x)x + x^2\ln x - 2x^2.$$

Notice that part of the particular solution $(-2x^2)$ is a constant times part of the general solution of the associated homogeneous equation. It may therefore be incorporated into the term C_1x^2, yielding

$$y(x) = C_1x^2 + (C_2 + C_3\ln x)x + x^2\ln x.$$

EXERCISES

1. Find the general solution of the following differential equations.

(a) $(D^3 + D)x = \sin 2t$
(b) $(D^3 + D)x = t$
(c) $(D^3 + D)x = \cos t$
(d) $(D^4 + 4D^2 + 4)x = 6 - t - e^t$
(e) $(D^4 + 4D^2 + 4)x = 4e^{-2t}$
(f) $(D^4 + 3D^3 + 2D^2)x = t + \sin t$
(g) $(D^4 + 3D^3 + 2D^2)x = e^{-t} + e^t$
(h) $(D^4 + 6D^2 + 9)x = \cos t$
(i) $(D^4 + 6D^2 + 9)x = \cos t + e^{-3t}$
(j) $(D^3 + 6D^2 + 5D - 12)x = e^t + e^{4t}$

2. Lift. When a projectile flies through the air, it experiences three forces — gravitation, drag, and lift — the latter due to backspin in the case of a golf ball, and the position of the skis in the case of a ski jumper.[17] If this force is linear in the velocity, then the equations of motion — including the linear drag terms — are $mx'' = -kx' - cy'$ and $my'' = -mg - ky' + cx'$, where k and c are positive constants, k is associated with the drag, and c is a measure of the lift. Solve these equations subject to the initial conditions $x(0) = y(0) = 0$ and $x'(0) = w_0\cos\alpha$, $y'(0) = w_0\sin\alpha$, to find

$$\begin{aligned} x(t) &= e^{-\lambda t}(A\sin\mu t - D\cos\mu t) + Ct + D, \\ y(t) &= e^{-\lambda t}(-D\sin\mu t - A\cos\mu t) + \\ &\quad + \tfrac{1}{\mu}\left[-C - \lambda(Ct + D) + w_0\cos\alpha\right], \end{aligned}$$

where $\lambda = k/m$, $\mu = c/g$, $A = \left(w_0\cos\alpha - D\lambda - C\right)/\mu$, $C = \mu g/(\mu^2 + \lambda^2)$, and $D = [w_0(\lambda\cos\alpha - \mu\sin\alpha)(\mu^2 + \lambda^2) - 2\lambda\mu g]/(\mu^2 + \lambda^2)^2$.

[17] "Maximum projectile range with drag and lift, with particular application to golf" by H. Erichson, *American Journal of Physics* 51, 1983, pages 357–362; and "The Flight of a Ski Jumper" by E. True, *C∘ODE∘E*, Spring 1993, pages 5–8.

(a) Does the projectile experience a terminal velocity in the x- and y-directions? Does the projectile experience a terminal velocity?

(b) **Computer Experiment.** Does the projectile experience a velocity lower than the terminal velocity?

(c) **The Flight of a Golf Ball with Drag and Lift.**[18] For a golf ball, the values of λ and μ are $\lambda = 0.25\ \text{sec}^{-1}$ and $\mu = 0.247\ \text{sec}^{-1}$. These are the same to two decimal places, so we will consider the case where $\lambda = \mu = 0.25$, $w_0 = 200$ ft/sec, and $g = 32$ ft/sec^2. In this case,

$$x(t) = e^{-t/4}\,(A \sin t/4 - D \cos t/4) + Ct + D,$$

$$y(t) = e^{-t/4}\,(-D \sin t/4 - A \cos t/4) + $$
$$- 4C - D - Ct + 800 \cos \alpha,$$

where $A = 400\,(\cos\alpha + \sin\alpha)$, $C = 64$, and $D = 400\,(\cos\alpha - \sin\alpha) - 256$. In Figure 8.24, we show the (x, y) trajectory for $\alpha = 5^\circ$ to $\alpha = 30^\circ$ in intervals of 5°. Notice that for these values of α, the greatest range occurs when $\alpha = 10^\circ$ and is a little over 600 ft.

i. **Computer Experiment.** Plot the trajectory of a golf ball for $\alpha = 5^\circ$ to $\alpha = 15^\circ$ in intervals of 1°. Which value of α gives the greatest range?

ii. Consider what happens to the flight of a golf ball with lift but no drag. Experiment with different values of μ to see whether it is possible to obtain unrealistic trajectories.

3. **A Falling Spring.**[19] A linear spring with spring constant k has two masses, m and M, attached at either end. The end with mass m is held so the spring hangs vertically. When it has reached its equilibrium position, mass M is a distance d below m. The spring is now allowed to fall under gravity. If $x(t)$ is the distance traveled by m and $y(t)$ is the distance traveled by M, then the motion of this system is governed by $mx'' = mg + \frac{k}{\ell}(y - x - \ell)$, $My'' = Mg - \frac{k}{\ell}(y - x - \ell)$, where $\ell = d/(1 + Mg/k)$. Solve these equations for $x(t)$ and $y(t)$.

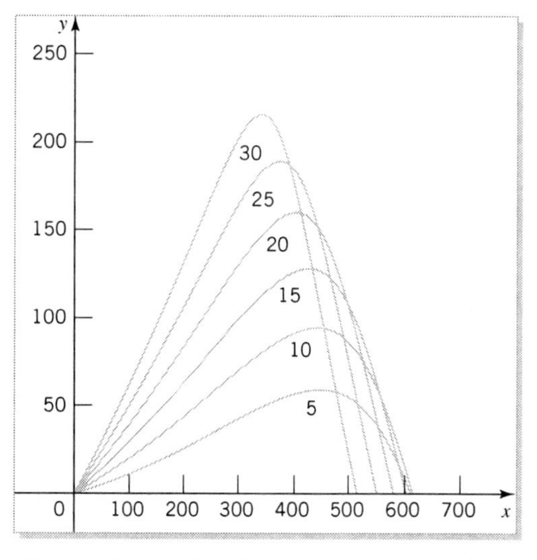

FIGURE 8.24 Golf ball trajectories with drag and lift for $\alpha = 5^\circ$ to $\alpha = 30^\circ$

[18] "Maximum projectile range with drag and lift, with particular application to golf" by H. Erichson, *American Journal of Physics* 51, 1983, pages 357–362.

[19] "Oscillations of a falling spring" by P. Glaster, *Physics Education* 20, 1993, pages 329–331.

What Have We Learned?

Main Ideas

- Two functions, $f(t)$ and $g(t)$, are linearly independent on an interval if the only way $c_1f(t) + c_2g(t) = 0$ for all values of t in this interval is for $c_1 = c_2 = 0$. See the Wronskian Theorem on page 338 for a test for independence.

The following comments refer to the linear differential equation

$$a_2(t)x'' + a_1(t)x' + a_0(t)x = 0. \tag{8.111}$$

- The general solution of (8.111) is a linear combination of two of its linearly independent solutions.
- There are ways of determining whether solutions of (8.111) oscillate without constructing the explicit solution. See the Oscillation Theorem (page 317) and Relation Theorem (page 320).
- If we know one solution of (8.111), we may always obtain a second linearly independent solution. See *How to Reduce the Order* on page 326.
- If the coefficients in (8.111) have the form of a constant times x raised to the power of the subscript [that is, $a_2(x) = b_2x^2$, $a_1(x) = b_1x$, and $a_0(x) = b_0$], we have a Cauchy-Euler equation, which may always be solved. See *How to Solve a Cauchy-Euler Differential Equation* on page 346.
- Boundary value problems consist of differential equations with conditions given at two points in the domain (usually at the boundaries of the domain), whereas initial value problems prescribe conditions at a single point in the domain.
- Boundary value problems are usually more difficult to solve than initial value problems. See Section 8.6.
- The general solution of $a_2(t)x'' + a_1(t)x' + a_0(t)x = f(t)$ is the sum of $x_h(t)$, the general solution of the associated homogeneous differential equation (8.111), and $x_p(t)$, a particular solution.

  All of the preceding comments with appropriate modifications also apply to linear differential equations of order greater than two.

How to Solve Nonhomogeneous Linear Differential Equations

Purpose To find the general solution of the nonhomogeneous linear differential equation

$$\sum_{m=0}^{n} a_m \frac{d^m x}{dt^m} = a_n \frac{d^n x}{dt^n} + a_{n-1}\frac{d^{n-1}x}{dt^{n-1}} + \cdots + a_2\frac{d^2x}{dt^2} + a_1\frac{dx}{dt} + a_0x = f(t), \tag{8.112}$$

where the coefficients $a_n, a_{n-1}, \cdots, a_2, a_1$, and a_0 may be functions of t.

Process

1. Find the general solution of the associated homogeneous differential equation. See *How to Solve Linear Differential Equations with Constant Coefficients* on page 244, or *How to Solve Higher Order Linear Differential Equations with Constant Coefficients* on page 359, which also deals with Cauchy-Euler equations in the attached comments.
2. Find a particular solution of (8.112). Here we have a choice.
   (a) See Section 7.2, *Method of Undetermined Coefficients*, or *How to Find a Particular Solution for Higher Order Linear Differential Equations with Constant Coefficients* on page 370.
   (b) See *How to Find a Particular Solution Using Reduction of Order* on page 328.
   (c) See *How to Find a Particular Solution Using Variation of Parameters* on page 334, or *How to Find a Particular Solution Using Variation of Parameters for Higher Order Differential Equations* on page 372.
3. Add the solutions found in steps 1 and 2 to obtain the general solution of (8.112).

Comments about How to Solve Nonhomogeneous Linear Differential Equations

- This three-step procedure will work for other types of linear differential equations not covered in this section. The necessary ingredient for success is to find n linearly independent solutions of the associated homogeneous differential equation and then use variation of parameters.

CHAPTER 9

LINEAR AUTONOMOUS SYSTEMS

Where We Are Going — and Why

In this chapter we return to the analysis of systems of two linear autonomous differential equations that we began in Chapter 6. For these autonomous systems, in which the coefficients in the differential equations are all constants, we obtain explicit solutions for all the possible cases and use these solutions to analyze their behavior in the phase plane. In order to characterize this behavior, we introduce the terms **node**, **center**, **focus**, and **saddle point** to indicate the behavior of solutions near equilibrium points.

We use our graphical analysis to discover how nullclines, which are isoclines for horizontal and vertical tangents in the phase plane, can aid us in determining the stability of equilibrium points. These ideas are important for Chapter 10, where we investigate nonlinear autonomous systems.

We give an alternative method for finding the explicit solutions, one that involves simple properties of 2 by 2 matrices. This alternative method introduces the notions of eigenvalues, eigenvectors, and a fundamental matrix.

Use of these differential equations is illustrated by models of solute movement between two containers, behavior of populations of two countries with mutual emigration, and compartmental models.

9.1 SOLVING LINEAR AUTONOMOUS SYSTEMS

In Chapters 4 and 5 we determined the concentration of salt in a single container in which the rates of input and output, together with the initial concentration, were known. In Chapter 6 we had several examples of systems of two first order linear autonomous differential equations for which we found explicit solutions by changing

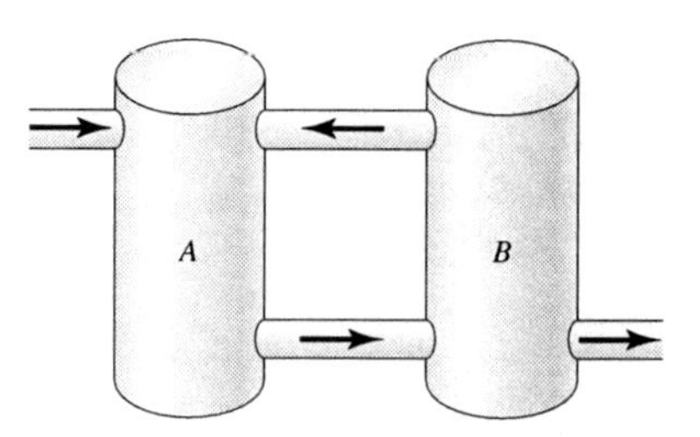

FIGURE 9.1 Two connected containers

each system to a single second order differential equation.[1] These two ideas now come together as we consider a situation with two containers, connected as shown in Figure 9.1.

Two-Container Mixture Problem

EXAMPLE 9.1 *Two-Container Mixture Problem*

Pure water is entering container A at a rate of 3 gallons per minute while the well-stirred mixture is leaving container B at the same rate. There are 100 gallons in each container, and the well-stirred mixture flows from container A to container B at a rate of 4 gallons per minute and leaks back from container B to container A at a rate of 1 gallon per minute. Predict the amount of salt in each container at any time if initially there are 16 pounds of salt in container A and 4 pounds of salt in container B. Find out how long it takes for the two containers to hold an equal amount of salt.

We let $x(t)$ be the amount of salt in container A at time t, and $y(t)$ be the amount of salt in container B at time t. Both x and y have units of pounds, and t has units of minutes. We may derive the differential equations governing this system by using a conservation equation. For container A we have

Conservation equation

$$\begin{aligned} x' &= \textit{input} - \textit{output} \\ &= (y/100)(1) - (x/100)(4), \end{aligned}$$

or

$$x' = \frac{1}{100}y - \frac{1}{25}x.$$

Balancing input and output for container B gives

$$y' = \frac{x}{100}4 - \frac{y}{100}1 - \frac{y}{100}3 = \frac{1}{25}x - \frac{1}{25}y.$$

This gives the system of equations

$$\begin{cases} x' = \frac{1}{100}y - \frac{1}{25}x, \\ y' = \frac{1}{25}x - \frac{1}{25}y. \end{cases} \tag{9.1}$$

[1] Remember that the system of differential equations, $x' = P(x, y)$, $y' = Q(x, y)$, is autonomous because both P and Q do not contain t explicitly.

To find an explicit solution of this system of equations, we differentiate the top equation, and eliminate the term involving y' by using the bottom equation, giving

Elimination method

$$x'' = \frac{1}{100}y' - \frac{1}{25}x' = \frac{1}{100}\left(\frac{1}{25}x - \frac{1}{25}y\right) - \frac{1}{25}x' = \frac{1}{100}\frac{1}{25}x - \frac{1}{100}\frac{1}{25}y - \frac{1}{25}x'.$$

From the top equation we see that $\frac{1}{100}y = x' + \frac{1}{25}x$, so the previous equation can be written as

$$x'' = \frac{1}{100}\frac{1}{25}x - \frac{1}{25}\left(x' + \frac{1}{25}x\right) - \frac{1}{25}x',$$

or

$$x'' + \frac{2}{25}x' + \frac{1}{100}\frac{3}{25}x = 0.$$

This is a second order linear differential equation with constant coefficients, and has the solution $x(t) = C_1 e^{-\frac{3}{50}t} + C_2 e^{-\frac{1}{50}t}$.

Explicit solution

From this expression for $x(t)$ we can find $y(t)$ by rearranging the top equation in (9.1) as $\frac{1}{100}y = x' + \frac{1}{25}x$, giving

$$\frac{1}{100}y = -\frac{3}{50}C_1 e^{-\frac{3}{50}t} - \frac{1}{50}C_2 e^{-\frac{1}{50}t} + \frac{1}{25}\left(C_1 e^{-\frac{3}{50}t} + C_2 e^{-\frac{1}{50}t}\right),$$

so that $y(t) = -2C_1 e^{-\frac{3}{50}t} + 2C_2 e^{-\frac{1}{50}t}$.

Initial conditions

We can evaluate C_1 and C_2 from the initial conditions, $x(0) = 16$, $y(0) = 4$, giving the two algebraic equations

$$\begin{aligned} 16 &= C_1 + C_2, \\ 4 &= -2C_1 + 2C_2. \end{aligned}$$

If we divide the bottom equation by 2 and add it to the top equation, we obtain $2C = 18$, so $C_2 = 9$. Substituting this value into either of the original equations gives $C_1 = 7$. Thus, the solution to our initial value problem is

$$\begin{cases} x(t) = 9e^{-t/50} + 7e^{-3t/50}, \\ y(t) = 18e^{-t/50} - 14e^{-3t/50}. \end{cases}$$

To determine the time, t, when we have equal amounts of salt in the two tanks, we set $x(t) = y(t)$ and solve for t. This gives

$$9e^{-t/50} + 7e^{-3t/50} = 18e^{-t/50} - 14e^{-3t/50},$$

which, when rearranged, becomes $21e^{-3t/50} = 9e^{-t/50}$. This yields $e^{2t/50} = 7/3$, so the required time is $t = 25\ln(7/3) \approx 21.18$ minutes. At that time there will be $12\sqrt{3/7} \approx 7.86$ pounds of salt in each container. (How do we know this?)

Figure 9.2 shows the graphs of $x(t) = 9e^{-t/50} + 7e^{-3t/50}$ and $y(t) = 18e^{-t/50} - 14e^{-3t/50}$. The time when these two curves cross and the value at which they cross are consistent with the preceding analysis.

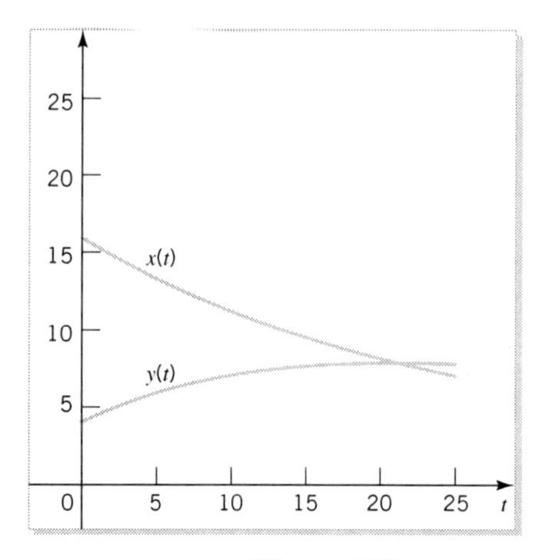

FIGURE 9.2 The functions $x(t) = 9e^{-t/50} + 7e^{-3t/50}$ and $y(t) = 18e^{-t/50} - 14e^{-3t/50}$

Explicit Solutions

We have seen several examples of systems of two first order linear autonomous differential equations for which we found explicit solutions by changing the system to a single second order differential equation. Now we consider the general situation in which the system of autonomous equations has the form

$$\begin{cases} x' = ax + by, \\ y' = cx + dy, \end{cases} \tag{9.2}$$

where $a, b, c,$ and d are constants and $x' = dx/dt, y' = dy/dt$. This system has $x(t) = 0$ and $y(t) = 0$ as an equilibrium solution. A solution where all the dependent variables are identically zero is called a TRIVIAL SOLUTION. The technique developed now is for finding nontrivial solutions of (9.2).

Trivial solution

If $b = c = 0$, the two equations in (9.2) uncouple, and each may be solved separately to obtain a solution as

$$\begin{cases} x(t) = C_1 e^{at}, \\ y(t) = C_2 e^{dt}. \end{cases} \tag{9.3}$$

If the two equations in (9.2) are to be coupled, then at least one of b or c must be nonzero. We first consider $b \neq 0$. We can differentiate the top equation in (9.2) and substitute into the result the expression for y' from the bottom equation in (9.2) to obtain

$$x'' = ax' + by' = ax' + b\,(cx + dy) = ax' + bcx + d\,(by)\,.$$

Substituting the expression for $by = x' - ax$ from the top equation in (9.2) allows the previous equation to contain only x, as $x'' = ax' + bcx + d\left(x' - ax\right)$, or

$$x'' - (a + d)x' + (ad - bc)x = 0. \tag{9.4}$$

Had we differentiated the bottom equation in (9.2) and eliminated x and x' as we did to develop (9.4), the resulting equation for y would have the same coefficients; namely,

$$y'' - (a+d)y' + (ad - bc)y = 0. \tag{9.5}$$

(See Exercise 2 on page 388.)

Characteristic equation

In Section 6.3 we discovered that to find solutions of (9.4), we should consider the characteristic equation

$$r^2 - (a+d)r + ad - bc = 0. \tag{9.6}$$

This quadratic equation has three types of solutions: two real, distinct roots; one real, repeated root; or two complex roots, one the complex conjugate of the other. Thus, the differential equation in (9.4) has three possible forms for its general solution, depending on the relative values of a, b, c, and d, and on the sign of the discriminant[2] of the quadratic equation (9.6); namely, $(a+d)^2 - 4(ad - bc) = (a-d)^2 + 4bc$. We now consider these three possibilities in turn.

Two real, distinct roots

1. **Two real, distinct roots.** If $(a-d)^2 + 4bc > 0$, and we label the roots of (9.6) r_1 and r_2, then these roots are

$$r_1 = \frac{1}{2}\left[a + d - \sqrt{(a-d)^2 + 4bc}\right], \qquad r_2 = \frac{1}{2}\left[a + d + \sqrt{(a-d)^2 + 4bc}\right],$$

and the general solution for $x(t)$ is

$$x(t) = C_1 e^{r_1 t} + C_2 e^{r_2 t}.$$

If we now substitute this expression into the top equation of (9.2), we obtain

$$\begin{aligned} by &= x' - ax \\ &= r_1 C_1 e^{r_1 t} + r_2 C_2 e^{r_2 t} - a\left(C_1 e^{r_1 t} + C_2 e^{r_2 t}\right) \\ &= C_1(r_1 - a)e^{r_1 t} + C_2(r_2 - a)e^{r_2 t}. \end{aligned}$$

This means we can write the general solution of (9.2) in this case as

$$\begin{cases} x(t) = C_1 e^{r_1 t} + C_2 e^{r_2 t}, \\ y(t) = C_1(r_1 - a)e^{r_1 t}/b + C_2(r_2 - a)e^{r_2 t}/b. \end{cases} \tag{9.7}$$

One real, repeated root

2. **One real, repeated root.** If $(a-d)^2 + 4bc = 0$, then $r = (a+d)/2$, and the general solution for $x(t)$ is

$$x(t) = C_1 e^{rt} + C_2 t e^{rt}.$$

Substituting this expression into $by = x' - ax$ gives

$$by = C_1 r e^{rt} + (1 + rt)C_2 e^{rt} - a\left(C_1 e^{rt} + C_2 t e^{rt}\right),$$

[2]The discriminant of the quadratic equation $\alpha x^2 + \beta x + \gamma = 0$ is $\beta^2 - 4\alpha\gamma$.

which means that the general solution of (9.2) in this case may be written as

$$\begin{cases} x(t) = C_1 e^{rt} + C_2 t e^{rt}, \\ y(t) = \left[C_1(r-a) + C_2\right] e^{rt}/b + C_2(r-a) t e^{rt}/b. \end{cases} \tag{9.8}$$

Two complex roots

3. Two complex roots. If $(a-d)^2 + 4bc < 0$, then we can express the roots as

$$r_1 = \frac{1}{2}\left[a + d - i\sqrt{-(a-d)^2 - 4bc}\right] = \alpha - i\beta,$$

$$r_2 = \frac{1}{2}\left[a + d + i\sqrt{-(a-d)^2 - 4bc}\right] = \alpha + i\beta.$$

Here the solution for $x(t)$ is

$$x(t) = e^{\alpha t}\left(C_1 \sin\beta t + C_2 \cos\beta t\right),$$

and by is

$$\begin{aligned} by &= x' - ax \\ &= e^{\alpha t}\left(\beta C_1 \cos\beta t - \beta C_2 \sin\beta t\right) + \alpha e^{\alpha t}\left(C_1 \sin\beta t + C_2 \cos\beta t\right) + \\ &\qquad -a e^{\alpha t}\left(C_1 \sin\beta t + C_2 \cos\beta t\right) \\ &= e^{\alpha t}\left\{\left[(\alpha - a)\, C_1 - \beta C_2\right] \sin\beta t + \left[\beta C_1 + (\alpha - a)\, C_2\right] \cos\beta t\right\}. \end{aligned}$$

This allows us to write the general solution in this case as

$$\begin{cases} x(t) = e^{\alpha t}\left(C_1 \sin\beta t + C_2 \cos\beta t\right), \\ y(t) = e^{\alpha t}\left(\left\{\left[(\alpha - a)\, C_1 - \beta C_2\right]/b\right\} \sin\beta t + \left\{\left[\beta C_1 + (\alpha - a)\, C_2\right]/b\right\} \cos\beta t\right). \end{cases} \tag{9.9}$$

The previous argument assumed that $b \neq 0$. If $c \neq 0$, then a similar analysis (see Exercise 2 on page 388) gives the solutions as follows.

1. Two real, distinct roots:

$$\begin{cases} x(t) = C_1(r_1 - d) e^{r_1 t}/c + C_2(r_2 - d) e^{r_2 t}/c, \\ y(t) = C_1 e^{r_1 t} + C_2 e^{r_2 t}. \end{cases} \tag{9.10}$$

2. One real, repeated root:

$$\begin{cases} x(t) = \left[C_1 (r-d) + C_2\right] e^{rt}/c + C_2(r-d) t e^{rt}/c, \\ y(t) = C_1 e^{rt} + C_2 t e^{rt}. \end{cases} \tag{9.11}$$

3. Two complex roots:

$$\begin{cases} x(t) = e^{\alpha t}\left(\left\{\left[(\alpha - d)\, C_1 - \beta C_2\right]/c\right\} \sin\beta t + \left\{\left[\beta C_1 + (\alpha - d)\, C_2\right]/c\right\} \cos\beta t\right), \\ y(t) = e^{\alpha t}\left(C_1 \sin\beta t + C_2 \cos\beta t\right). \end{cases} \tag{9.12}$$

Even though, for all possible situations where r satisfies (9.6), we have formulas for the general solution of (9.2) — namely, (9.3), (9.7), (9.8), (9.9), (9.10), (9.11), and

(9.12) — we do not recommend that you commit any of these to memory. Instead, we use the fact that we know the general form of the solution for the three possible cases. This allows us to write down the solution for one of the variables — say, $x(t)$ if $b \neq 0$ — and then use the top equation in (9.2) to obtain $y(t)$.

To take advantage of this knowledge, we need a simple way to obtain the characteristic equation $r^2 - (a + d)r + ad - bc = 0$ without having to commit it to memory. If we rewrite the characteristic equation in the form $(a - r)(d - r) - bc = 0$, we see that it is equivalent to the 2 by 2 determinant

Characteristic equation

$$\begin{vmatrix} a - r & b \\ c & d - r \end{vmatrix} = 0.$$

We can construct this determinant directly from (9.2) by adding the term $-r$ to the coefficients of x and y on the diagonal of the right-hand side. This gives us a simple way to obtain the characteristic equation.

How to Find the General Solution of $x' = ax + by$, $y' = cx + dy$

Purpose To solve the autonomous system

$$\begin{cases} x' = ax + by, \\ y' = cx + dy, \end{cases} \tag{9.13}$$

where a, b, c, and d are constants, for $x = x(t)$, $y = y(t)$.

Process

1. If $b = c = 0$, the equations are not coupled, and the general solution is

$$\begin{cases} x(t) = C_1 e^{at}, \\ y(t) = C_2 e^{dt}. \end{cases}$$

In this case there is nothing else to do.

2. Otherwise, solve the characteristic equation

$$\begin{vmatrix} a - r & b \\ c & d - r \end{vmatrix} = 0$$

for r, obtaining the roots r_1 and r_2.

3. If $b \neq 0$, we do the following.

   (a) Write down the solution for $x(t)$ based on r_1 and r_2 as follows.

      i. If $r_1 \neq r_2$, where r_1 and r_2 are both real, then the general solution for $x(t)$ is

$$x(t) = C_1 e^{r_1 t} + C_2 e^{r_2 t}.$$

ii. If $r_1 = r_2 = r$, then the general solution for $x(t)$ is

$$x(t) = C_1 e^{rt} + C_2 t e^{rt}.$$

iii. If r_1 and r_2 are complex, where $r_1 = \alpha - i\beta$ and $r_2 = \alpha + i\beta$, then the general solution for $x(t)$ is

$$x(t) = e^{\alpha t}\left(C_1 \sin \beta t + C_2 \cos \beta t\right).$$

(b) Calculate the general solution for $y(t)$ from the general solution for $x(t)$ and its derivative x' by rearranging the top equation in (9.13) as $y = \left(x' - ax\right)/b$.

4. If $b = 0$ and $c \neq 0$, we do the following.

(a) Write down the solution for $y(t)$ based on r_1 and r_2 as follows.

i. If $r_1 \neq r_2$, where r_1 and r_2 are both real, then the general solution for $y(t)$ is

$$y(t) = C_1 e^{r_1 t} + C_2 e^{r_2 t}.$$

ii. If $r_1 = r_2 = r$, then the general solution for $y(t)$ is

$$y(t) = C_1 e^{rt} + C_2 t e^{rt}.$$

iii. If r_1 and r_2 are complex, where $r_1 = \alpha - i\beta$ and $r_2 = \alpha + i\beta$, then the general solution for $y(t)$ is

$$y(t) = e^{\alpha t}\left(C_1 \sin \beta t + C_2 \cos \beta t\right).$$

(b) Calculate the general solution for $x(t)$ from the general solution for $y(t)$ and its derivative y' by rearranging the bottom equation in (9.13) as $x = \left(y' - dy\right)/c$.

Comments about the General Solution of $x' = ax + by$, $y' = cx + dy$

- It is possible to extend the preceding analysis to systems of equations of the form

$$\begin{cases} x' = ax + by + e, \\ y' = cx + dy + f, \end{cases}$$

where e and f are constants, if $ad - bc \neq 0$. The major difference between this case and the case when $e = f = 0$ is the fact that $x(t) = 0$ and $y(t) = 0$ is no longer a solution. Here we have the solution $x(t) = x_0$, $y(t) = y_0$, where x_0 and y_0 are constants that satisfy the two equations $ax_0 + by_0 + e = 0$ and $cx_0 + dy_0 + f = 0$. The change of variables from (x, y) to (u, v), where $u = x - x_0$ and $v = y - y_0$, will convert the preceding system in terms of x and y into an equivalent linear system in terms of u and v, where $u(t) = 0$ and $v(t) = 0$ is a solution.

EXAMPLE 9.2

Solve

$$\begin{cases} x' = -x + y, \\ y' = -kx - ky, \end{cases} \tag{9.14}$$

where (a) $k = 6$, (b) $k = 1$, (c) $k = 3 + 2\sqrt{2}$.

Characteristic equation

In all three cases the characteristic equation is

$$\begin{vmatrix} -1 - r & 1 \\ -k & -k - r \end{vmatrix} = 0,$$

or $(-1 - r)(-k - r) + k = r^2 + (1 + k)\, r + 2k = 0$, with roots

$$r = \frac{1}{2}\left[-(1 + k) \pm \sqrt{(1 + k)^2 - 8k}\,\right].$$

(a) With $k = 6$ the solutions of the characteristic equation are $r_1 = -4$ and $r_2 = -3$. Thus, $x(t) = C_1 e^{-4t} + C_2 e^{-3t}$. From this and the top equation in (9.14), we have

$$y(t) = x' + x = -4C_1 e^{-4t} - 3C_2 e^{-3t} + C_1 e^{-4t} + C_2 e^{-3t} = -3C_1 e^{-4t} - 2C_2 e^{-3t}.$$

(b) With $k = 1$ the solutions of the characteristic equation are $r_1 = -1 - i$ and $r_2 = -1 + i$. Thus, $x(t) = e^{-t}\left(C_1 \sin t + C_2 \cos t\right)$. From this and the top equation in (9.14), we have

$$\begin{aligned} y(t) &= x' + x \\ &= e^{-t}\left(-C_1 \sin t - C_2 \cos t + C_1 \cos t - C_2 \sin t\right) + e^{-t}\left(C_1 \sin t + C_2 \cos t\right), \end{aligned}$$

or $y(t) = e^{-t}\left(C_1 \cos t - C_2 \sin t\right)$.

(c) With $k = 3 + 2\sqrt{2}$ the solutions of the characteristic equation are $r = r_1 = r_2 = -(1 + k)/2 = -2 - \sqrt{2}$. Thus, $x(t) = C_1 e^{rt} + C_2 t e^{rt}$. From this and the top equation in (9.14), we have

$$y(t) = x' + x = rC_1 e^{rt} + rC_2 t e^{rt} + C_2 e^{rt} + C_1 e^{rt} + C_2 t e^{rt},$$

or $y(t) = \left[(r + 1)\, C_1 + C_2\right] e^{rt} + (r + 1)\, C_2 t e^{rt}$, where $r = -2 - \sqrt{2}$. ■

EXERCISES

1. Solve the following systems of differential equations.

(a) $x' = 2x + 5y$, $y' = x + 6y$ (b) $x' = 2x - y$, $y' = -6x + y$

(c) $x' = 5x + 6y$, $y' = x + 4y$ (d) $x' = 5x - 3y$, $y' = -x - y$

(e) $x' = 2x + 9y$, $y' = -x - 4y$ (f) $x' = 2x + y$, $y' = x - 4y$

(g) $x' = 4x - 2y$, $y' = 5x - 2y$ (h) $x' = 2x + y$, $y' = -4x + 2y$

(i) $x' = 3x + 5y$, $y' = -5x + 3y$ (j) $x' = -3x + 2y$, $y' = -4x + y$

(k) $x' = x + y$, $y' = -x + 3y$ (l) $x' = x + y$, $y' = -2x - y$

2. Differentiate the second equation of $x' = ax + by$, $y' = cx + dy$, and substitute for $x' = ax + by$ from the first and $cx = y' - dy$ from the second (assuming $c \neq 0$), to obtain (9.5); namely, $y'' - (a + d)y' + (ad - bc)y = 0$. Solve $y'' - (a + d)y' + (ad - bc)y = 0$ for $c \neq 0$, and obtain (9.10), (9.11), and (9.12).

3. Solve the following systems of differential equations.

(a) $x' = 2x + 5y - 7$, $y' = x + 6y + 14$ (b) $x' = 5x - y + 5$, $y' = 6x + y + 6$

(c) $x' = x - 2y + 1$, $y' = 2x + 5y - 7$ (d) $x' = 2x + 5y + 3$, $y' = x + 6y - 1$

(e) $x' = 5x - y + 6$, $y' = 6x + y + 5$ (f) $x' = x - 2y + 4$, $y' = 2x + 5y - 1$

4. Consider the two containers of Figure 9.1, each of which holds 100 gallons of liquid.

(a) Initially, container A has 10 pounds of salt and container B has none. How long after all the valves are opened will the number of pounds of salt in the two containers be equal?

(b) Determine the amount of salt in each container after 25 minutes.

(c) Determine the amount of salt in each container as time approaches infinity.

5. Consider the two containers of Figure 9.1, each of which holds 100 gallons of liquid. A solution with concentration of 4 pounds of salt per gallon is entering container A at a rate of 3 gallons per minute, and the well-stirred mixture is leaving container B at a rate of 3 gallons per minute. The well-stirred mixture flows from container A to container B at a rate of 4 gallons per minute and leaks back from container B to container A at a rate of 1 gallon per minute. Predict the amount of salt in each container at any time if initially

(a) there is no salt in either container.

(b) there is no salt in container A and 300 pounds of salt in container B.

9.2 CLASSIFICATION OF SOLUTIONS VIA STABILITY

In the previous section we found the general solution of the linear autonomous system of differential equations,

$$\begin{cases} x' = ax + by, \\ y' = cx + dy, \end{cases} \tag{9.15}$$

where a, b, c, and d are constants. These solutions fall into three general categories according to the roots r_1 and r_2 of the characteristic equation

$$\begin{vmatrix} a - r & b \\ c & d - r \end{vmatrix} = r^2 - (a + d)r + ad - bc = 0.$$

The relationship between a, b, c, and d, and r_1 and r_2, is

$$\begin{aligned} a+d &= r_1 + r_2, \\ ad - bc &= r_1 r_2. \end{aligned} \tag{9.16}$$

(Verify this.)

The three general categories, and typical solutions, are as follows.

1. **Two real, distinct roots r_1, r_2:**

$$\begin{cases} x(t) = C_1 e^{r_1 t} + C_2 e^{r_2 t}, \\ y(t) = K_1 e^{r_1 t} + K_2 e^{r_2 t}. \end{cases} \tag{9.17}$$

2. **One real, repeated root r:**

$$\begin{cases} x(t) = C_1 e^{rt} + C_2 t e^{rt}, \\ y(t) = K_1 e^{rt} + K_2 t e^{rt}. \end{cases} \tag{9.18}$$

3. **Two complex roots $\alpha \pm i\beta$:**

$$\begin{cases} x(t) = e^{\alpha t}(C_1 \sin \beta t + C_2 \cos \beta t), \\ y(t) = e^{\alpha t}(K_1 \sin \beta t + K_2 \cos \beta t). \end{cases} \tag{9.19}$$

Caution! **In these solutions, C_1, C_2, K_1, and K_2 are constants related to one another in various ways, so that only two of these four constants are independent.**

In this section we want to discuss the stability and long-term behavior of these solutions by looking at orbits in the phase plane as $t \to \infty$. The reason we are concerned with the limit as $t \to \infty$ (and ignore the limit as $t \to -\infty$) is that in applications, the independent variable is often time. We usually use mathematical models to predict the future, not to recover the past.

In the analysis that follows, it is important to remember that if $r < 0$, then $e^{rt} \to 0$ and $te^{rt} \to 0$ as $t \to \infty$, while if $r > 0$, then $e^{rt} \to \infty$ and $te^{rt} \to \infty$ as $t \to \infty$. Finally, if $r = 0$, then $e^{rt} = 1$.

1. In the case of **two real, distinct roots,** we may assume that $r_1 < r_2$. If we look at (9.17), we see that the long-term behavior of $x(t)$ and $y(t)$ will depend on the signs of r_1 and r_2. There are five possibilities: (a) $r_1 < r_2 < 0$, (b) $r_1 < r_2 = 0$, (c) $r_1 < 0$, $r_2 > 0$, (d) $r_1 = 0$, $r_2 > 0$, and (e) $0 < r_1 < r_2$.

2. In the case of **one real, repeated root,** we have $r = r_1 = r_2$. If we look at (9.18), we see that the long-term behavior of $x(t)$ and $y(t)$ will depend on the sign of r. There are three possibilities: (a) $r < 0$, (b) $r = 0$, and (c) $r > 0$.

3. In the case of **two complex roots,** we have $r_1 = \alpha - i\beta$, $r_2 = \alpha + i\beta$. If we look at (9.19), we see that the long-term behavior of $x(t)$ and $y(t)$ will depend on the sign of α. There are three possibilities: (a) $\alpha < 0$, (b) $\alpha = 0$, and (c) $\alpha > 0$.

Equilibrium points From our previous work in Chapter 6, we know that equilibrium points are points in the phase plane corresponding to constant solutions

$$x(t) = x_0, \qquad y(t) = y_0, \tag{9.20}$$

of (9.15). If we substitute (9.20) into (9.15), we find

$$\begin{aligned} ax_0 + by_0 &= 0, \\ cx_0 + dy_0 &= 0. \end{aligned} \tag{9.21}$$

The solution of this equation falls into two cases, depending on the value of $ad - bc$.

1. If $ad - bc \neq 0$, then (9.21) has the unique solution $x_0 = 0$, $y_0 = 0$, and so, in this case, there is exactly one equilibrium point, $(0, 0)$. From (9.16) we see that $ad - bc \neq 0$ requires that neither r_1 nor r_2 can be zero.
2. If $ad - bc = 0$, then the two equations in (9.21) are linearly dependent — that is, $cx_0 + dy_0$ is proportional to $ax_0 + by_0$ — and so the solution of (9.21) is

$$ax_0 + by_0 = 0. \tag{9.22}$$

   In (9.22) either x_0 or y_0 may be chosen arbitrarily and the other must be such that the equation is satisfied. Equation (9.22) represents a straight line, which might be vertical ($b = 0$), horizontal ($a = 0$), or slanted ($a \neq 0$ and $b \neq 0$) with slope $m = -a/b$. Thus, if $ad - bc = 0$, we have an infinite number of equilibrium points, none of which are isolated.[3] From (9.16) we see that $ad - bc = 0$ requires that at least one of r_1 and r_2 must be zero.

Representative Systems

Rather than consider (9.15) in full generality, we will look at special cases that illustrate representative long-term behaviors. We will select examples where the origin $(0, 0)$ is the only equilibrium point, so $r_1 \neq 0$ and $r_2 \neq 0$. Thus, we shall look at examples corresponding to cases 1(a), 1(c), 1(e), 2(a), 2(c), 3(a), 3(b), and 3(c). The other cases are dealt with in Exercise 2 on page 403.

EXAMPLE 9.3 *Stable Node or Sink*

First, we want an example of case 1(a), where $r_1 < r_2 < 0$, (say, $r_1 = -2$, $r_2 = -1$).[4] From (9.16) we see that to have these roots, a, b, c, and d must satisfy $a + d = -3$, with $bc = 0$. These will be satisfied if we consider $a = -2$, $d = -1$, and $b = c = 0$; that is,

$$\begin{cases} x' = -2x, \\ y' = -y. \end{cases} \tag{9.23}$$

Here the equilibrium point is $(0, 0)$.

---

[3]An equilibrium point (x_0, y_0) is ISOLATED if we can find a circle with center (x_0, y_0) inside which there is no other equilibrium point. The radius of the circle can be as small as we please.

[4]Of course there are many other possibilities.

Explicit solution

The explicit solution of the system of differential equations in (9.23) is

$$x(t) = C_1 e^{-2t}, \qquad y(t) = C_2 e^{-t}. \tag{9.24}$$

Phase plane

The orbits of (9.23) in the phase plane will satisfy

$$\frac{dy}{dx} = \frac{-y}{-2x} = \frac{y}{2x}.$$

Orbits

This is a first order separable differential equation that may be solved to yield $y^2 = Cx$, where C is an arbitrary constant.[5] [Notice that we could have also obtained this equation by eliminating t in (9.24).] If $C \neq 0$, the orbits are parabolic, opening to the right if $C > 0$ and to the left if $C < 0$. If $C = 0$, the orbit is $y = 0$. From (9.24) we see that as $t \to \infty$, both x and y approach the equilibrium point $(0, 0)$. Figure 9.3 shows the phase plane and some orbits of (9.23), including $x = 0$. Because all orbits approach the equilibrium point $(0, 0)$ as $t \to \infty$, the equilibrium point is stable. This equilibrium point is called a STABLE NODE, or a SINK.

Stable node or sink

Notice that in the vicinity of the equilibrium point all orbits appear to be asymptotic to the line $x = 0$, except for the orbit $y = 0$. We can see that $x(t)$ goes to zero faster than $y(t)$ as $t \to \infty$, by observing, from (9.24), that if $C_2 \neq 0$, then

$$\frac{x(t)}{y(t)} = \frac{C_1 e^{-2t}}{C_2 e^{-t}} = \frac{C_1}{C_2} e^{-t} \to 0, \text{ as } t \to \infty.$$

If $C_2 = 0$ then $y(t) = 0$, and this is the exceptional orbit.

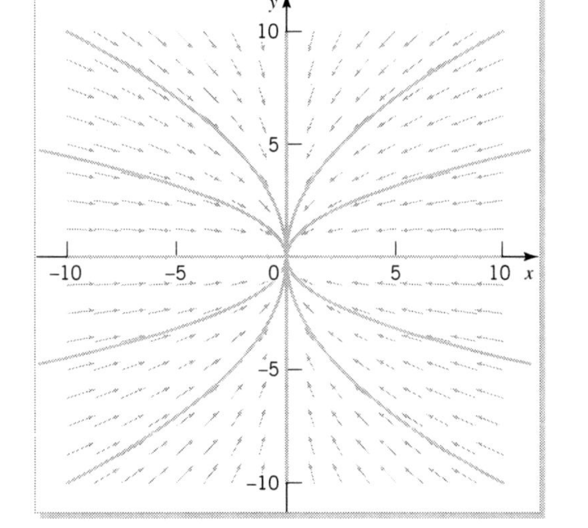

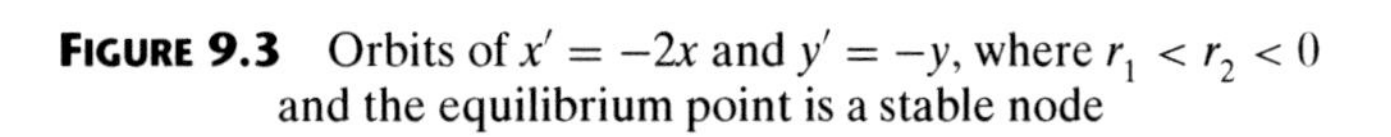

FIGURE 9.3 Orbits of $x' = -2x$ and $y' = -y$, where $r_1 < r_2 < 0$ and the equilibrium point is a stable node

[5] When you do this, don't forget the equilibrium solution $y(x) = 0$.

Figure 9.4 shows another possible example of case 1(a) — namely, $a = -1$, $d = -2$, and $b = c = 0$ — so that

$$\begin{cases} x' = -x, \\ y' = -2y. \end{cases} \tag{9.25}$$

(See Exercise 3 on page 403.)

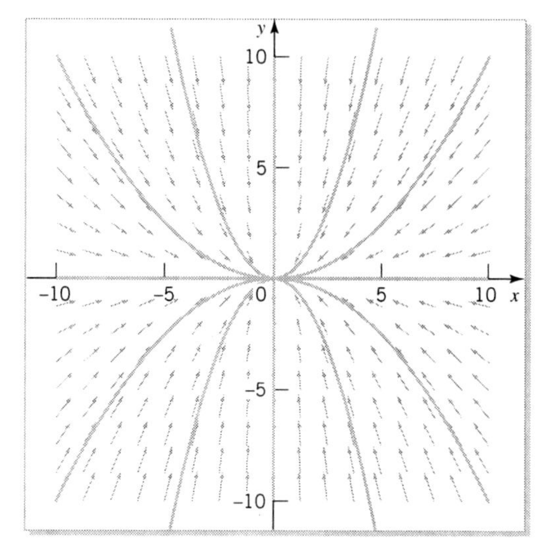

FIGURE 9.4 Orbits of $x' = -x$ and $y' = -2y$, where $r_1 < r_2 < 0$ and the equilibrium point is a stable node

EXAMPLE 9.4 *Saddle Point*

To construct an example corresponding to case 1(c), we want $r_1 < 0$ and $r_2 > 0$, so we consider $r_1 = -1$, $r_2 = 1$. From (9.16) we see that selecting $a = -1$ and $d = 1$, with $b = c = 0$, gives such an example; that is,

$$\begin{cases} x' = -x, \\ y' = y. \end{cases} \tag{9.26}$$

Here the equilibrium point is again (0, 0).

Explicit solution

The explicit solution of the system of differential equations (9.26) is

$$x(t) = C_1 e^{-t}, \qquad y(t) = C_2 e^{t}. \tag{9.27}$$

Orbits

The orbits of (9.26) will satisfy

$$\frac{dy}{dx} = \frac{y}{-x},$$

which may be solved to yield

$$y = \frac{C}{x}, \tag{9.28}$$

where C is an arbitrary constant. [Notice that we could have also obtained (9.28) by eliminating t in (9.27).] If $C \neq 0$, the orbits given by (9.28) are hyperbolas. If $C = 0$, the orbit is $y = 0$. From (9.27) we see that if $t \to \infty$, then $x \to 0$. At the same time, either $y \to \infty$ (if $C_2 \neq 0$) or $y \to 0$ (if $C_2 = 0$). Figure 9.5 shows the phase plane of (9.26) and some orbits. Because the orbit $y = 0$ approaches the equilibrium point $(0, 0)$ as $t \to \infty$, but other orbits move away from the equilibrium point, the equilibrium point is unstable. The equilibrium point is called a SADDLE POINT.

Phase plane

Saddle point

The curve $y = 0$ is special because it separates the phase plane into two regions. Orbits that start above $y = 0$ have the property that $y \to \infty$ as $t \to \infty$, whereas those that start below $y = 0$ have the property that $y \to -\infty$ as $t \to \infty$. The curve $y = 0$ is called a SEPARATRIX, and consists of the two orbits $y = 0$, $x < 0$, and $y = 0$, $x > 0$, and the equilibrium point $(0, 0)$.

Separatrix

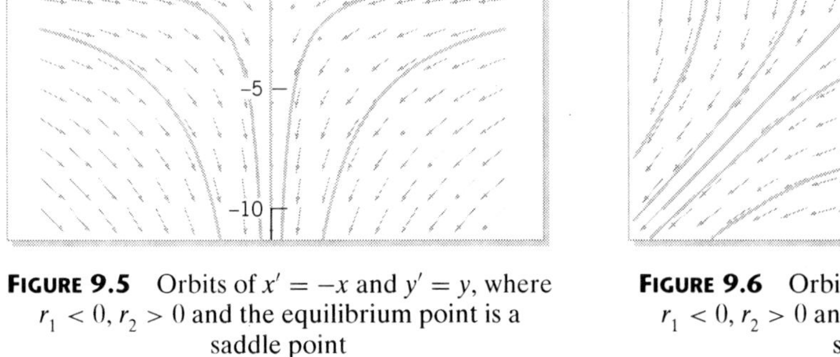

FIGURE 9.5 Orbits of $x' = -x$ and $y' = y$, where $r_1 < 0$, $r_2 > 0$ and the equilibrium point is a saddle point

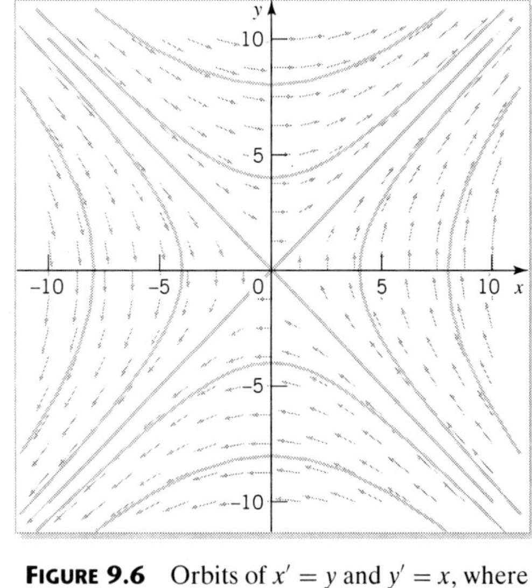

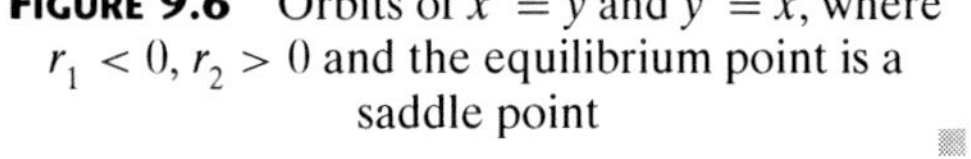

FIGURE 9.6 Orbits of $x' = y$ and $y' = x$, where $r_1 < 0$, $r_2 > 0$ and the equilibrium point is a saddle point

Figure 9.6 shows another possible example of case 1(c) with $r_1 = -1$, and $r_2 = 1$; namely, $a = d = 0$ and $b = c = 1$, so that

$$\begin{cases} x' = y, \\ y' = x. \end{cases}$$

We see that if we rotate Figure 9.6 counterclockwise through 45°, we have Figure 9.5, so the equilibrium point is again a saddle point. Here the lines $y = -x$ is a separatrix.

EXAMPLE 9.5 *Unstable Node or Source*

To construct an example corresponding to case 1(e), we want $0 < r_1 < r_2$ — say, $r_1 = 1$ and $r_2 = 2$. From (9.16) we see that selecting $a = 2$, $d = 1$, and $b = c = 0$ gives such an example, namely,

$$\begin{cases} x' = 2x, \\ y' = y. \end{cases} \tag{9.29}$$

Here the equilibrium point is again $(0, 0)$. If, in (9.29), we replace t with $-t$ (called "running the differential equation backward in time"), we see that (9.29) becomes (9.23). Thus, we can construct the phase plane for (9.29) from the phase plane for (9.23) by reversing the arrows, in which case the equilibrium point $(0, 0)$ is an UNSTABLE NODE or a SOURCE. This is shown in Figure 9.7.

Unstable node or source

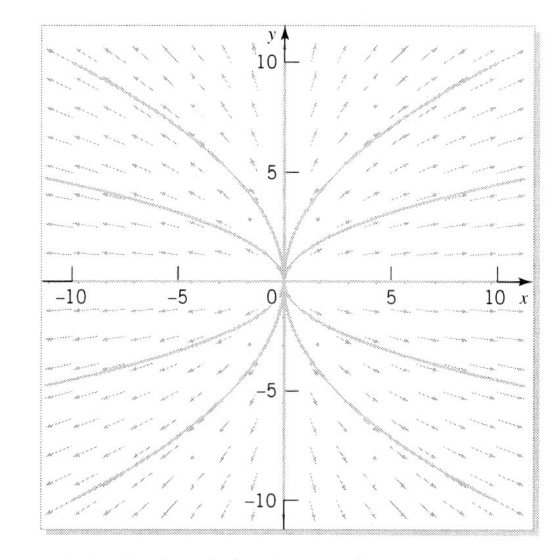

FIGURE 9.7 Orbits of $x' = 2x$ and $y' = y$, where $0 < r_1 < r_2$ and the equilibrium point is an unstable node

EXAMPLE 9.6

To construct an example corresponding to case 2(a), we want $r_1 = r_2 < 0$, so we choose $r_1 = r_2 = -1$. From (9.16) we see that $a = -1$ and $d = -1$, with $b = c = 0$, gives such an example; that is,

$$\begin{cases} x' = -x, \\ y' = -y. \end{cases} \tag{9.30}$$

Here the equilibrium point is again $(0, 0)$.

Explicit solution

The solution of (9.30) is

$$x(t) = C_1 e^{-t}, \qquad y(t) = C_2 e^{-t}. \tag{9.31}$$

Orbits The orbits of (9.30) for $x \neq 0$ will satisfy

$$\frac{dy}{dx} = \frac{y}{x},$$

which may be solved to yield

$$y = Cx, \tag{9.32}$$

where C is an arbitrary constant. The orbits, (9.32) or $x = 0$, are straight lines through the origin. From (9.31) we see that if $t \to \infty$, then $(x, y) \to (0, 0)$. Figure 9.8 shows the phase plane and some orbits of (9.30). Because all orbits approach the equilibrium point $(0, 0)$ as $t \to \infty$, the equilibrium point is a stable node.

Phase plane

Stable node

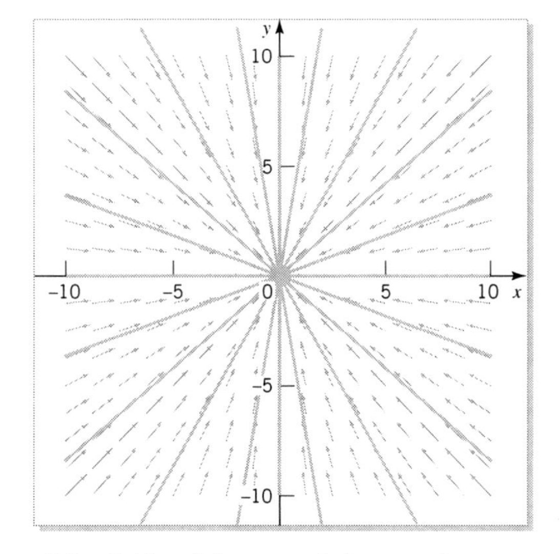

FIGURE 9.8 Orbits of $x' = -x$ and $y' = -y$, where $r_1 = r_2 < 0$ and the equilibrium point is a stable node

---

Figure 9.9 shows another possible example of case 2(a) with $r_1 = r_2 = -1$; namely, $a = d = -1$, $b = 1$, and $c = 0$, so that

$$\begin{cases} x' = -x + y, \\ y' = -y, \end{cases} \tag{9.33}$$

and the equilibrium point is another example of a stable node. The behavior of the orbits near the origin is not clear from Figure 9.9. We might consider zooming in on the origin. However, because the differential equation for the orbits in the phase plane,

$$\frac{dy}{dx} = \frac{-y}{-x + y},$$

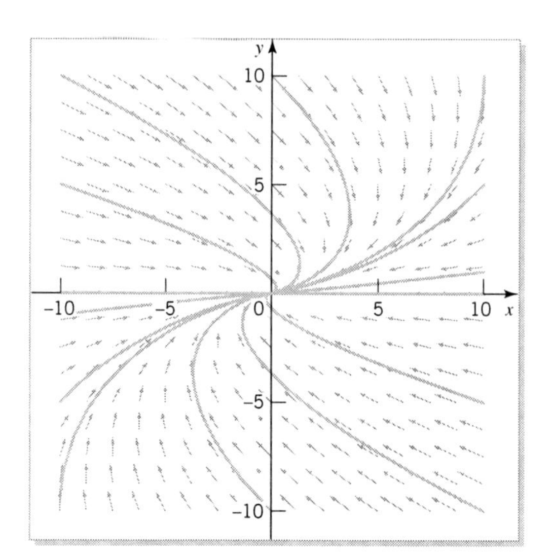

FIGURE 9.9 Orbits of $x' = -x + y$ and $y' = -y$, where $r_1 = r_2 < 0$ and the equilibrium point is a stable node

is one with homogeneous coefficients, rescaling the x- and y-axes by equal amounts in Figure 9.9 will not change the slope field. Thus, to investigate the orbits near the equilibrium point, we consider the explicit solution of (9.33), namely,

$$x(t) = C_1 e^{-t} + C_2 t e^{-t}, \qquad y(t) = C_2 e^{-t}.$$

(Show this.) From this, we see that

$$\frac{y(t)}{x(t)} = \frac{C_2 e^{-t}}{C_1 e^{-t} + C_2 t e^{-t}} = \frac{C_2}{C_1 + C_2 t} \to 0 \text{ as } t \to \infty,$$

for all choices of C_1, C_2. Thus, all orbits are asymptotic to the line $y = 0$. (Why?)

Case 2(c), with $r > 0$, can be handled in a similar way to case 2(a), by "reversing the time." The equilibrium point is an unstable node.

EXAMPLE 9.7 *Stable Focus or Spiral*

To construct an example corresponding to case 3(a), we want $a + d = 2\alpha < 0$ and $ad - bc = \alpha^2 + \beta^2$. These will be satisfied if we consider $a = \alpha < 0$, $d = \alpha$, $b = -\beta$, and $c = \beta$, namely,

$$\begin{cases} x' = \alpha x - \beta y, \\ y' = \beta x + \alpha y. \end{cases} \tag{9.34}$$

Here the equilibrium point is again $(0, 0)$.

Explicit solution

The solution of (9.34) is

$$x(t) = e^{\alpha t}\left(C_1 \cos \beta t - C_2 \sin \beta t\right), \qquad y(t) = e^{\alpha t}\left(C_2 \cos \beta t + C_1 \sin \beta t\right), \tag{9.35}$$

where $\alpha < 0$, which can be rewritten in the form (see Exercise 4 on page 403)

$$x(t) = \sqrt{C_1^2 + C_2^2}e^{\alpha t}\cos(\beta t + \phi), \qquad y(t) = \sqrt{C_1^2 + C_2^2}e^{\alpha t}\sin(\beta t + \phi). \tag{9.36}$$

As $t \to \infty$ we see that $x(t) \to 0$ and $y(t) \to 0$, because $\alpha < 0$. Thus, the equilibrium point is stable. The orbits of (9.34) in the phase plane will satisfy

$$\frac{dy}{dx} = \frac{\beta x + \alpha y}{\alpha x - \beta y}.$$

Orbits

Although this can be solved, the final solution is not very illuminating. However, (9.36) represents the parametric form of these orbits, from which we can obtain a great deal of information. From (9.36) we have

$$x^2 + y^2 = \left(C_1^2 + C_2^2\right)e^{2\alpha t}, \tag{9.37}$$

Stable focus or spiral

so as $t \to \infty$ the orbits locally follow "circles" whose radii decrease with time, because $\alpha < 0$. In fact, these orbits are spirals in the xy-plane (see Exercise 4 on page 403). Figure 9.10 shows the phase plane and an orbit of (9.34) for the case $\alpha = -0.1$ and $\beta = 1$. This equilibrium point is called a STABLE FOCUS or SPIRAL.

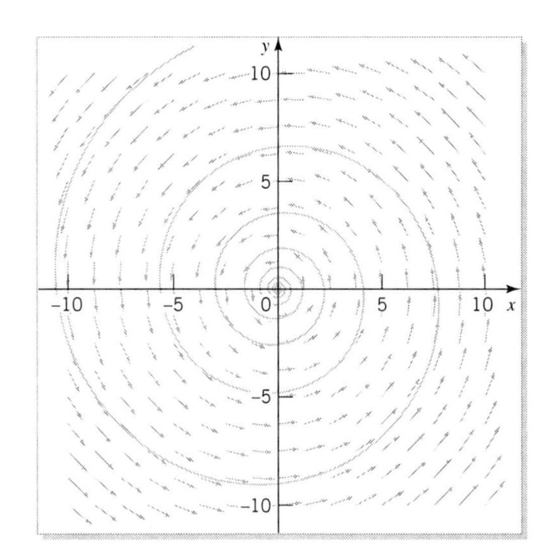

FIGURE 9.10 Orbits of $dy/dx = (\beta x + \alpha y)/(\alpha x - \beta y)$, where r_1 and r_2 are complex with negative real parts and the equilibrium point is a stable focus

Center

A similar analysis applies to case 3(b), with $\alpha = 0$, where we find the orbits are circles. Here the equilibrium point is stable (see Figure 9.11) and called a CENTER.[6]

[6] This equilibrium point is stable in the sense that orbits that start near the point stay near the point.

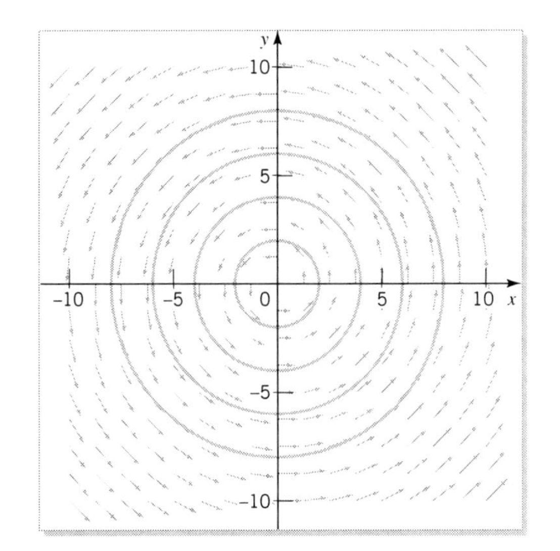

FIGURE 9.11 Orbits of $dy/dx = -x/y$, where r_1 and r_2 are complex with real parts zero and the equilibrium point is a (stable) center

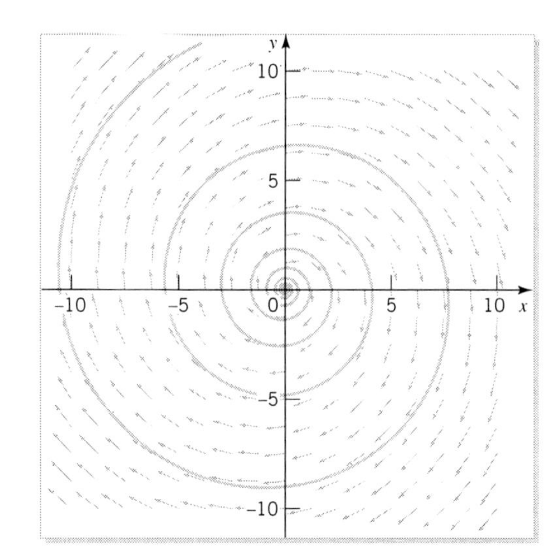

FIGURE 9.12 Orbits of $dy/dx = (\beta x + \alpha y)/(\alpha x - \beta y)$, where r_1 and r_2 are complex with positive real parts and the equilibrium point is an unstable focus

Case 3(c), with $\alpha > 0$, can be handled similarly to case 3(a) (by "reversing the time") so the orbits spiral out. This equilibrium point is unstable (see Figure 9.12 where $\alpha = 0.1$ and $\beta = -1$) and is again called a focus or spiral.

Classifying Equilibrium Solutions

We summarize these results in the following way.

▶ **Theorem 9.1: The Classification Theorem** *Consider the system of differential equations*

$$\begin{cases} x' = ax + by, \\ y' = cx + dy, \end{cases}$$

with characteristic equation $r^2 - (a + d)r + ad - bc = 0$.

1. *If the roots of the characteristic equation are real and distinct, with $r_1 < r_2$, then*

(a) *If $r_1 < r_2 < 0$, the origin is a stable node or sink.*

(b) *If $0 < r_1 < r_2$, the origin is an unstable node or source.*

(c) *If $r_1 < 0$, $r_2 > 0$, then the origin is a saddle point.*

2. *If $r_1 = r_2$, then*

(a) *If $r_1 = r_2 < 0$, the origin is a stable node or sink.*

(b) *If $0 < r_1 = r_2$, the origin is an unstable node or sink.*

3. *If $r_{1,2} = \alpha \pm i\beta$, then*

(a) *If $\alpha < 0$, the origin is a stable focus or spiral.*

(b) *If* $\alpha > 0$, *the origin is an unstable focus or spiral.*

(c) *If* $\alpha = 0$, *the origin is a center and is stable.*

EXAMPLE 9.8

Identify and classify the equilibrium point of the system

$$\begin{cases} x' = -x + y, \\ y' = -x - y. \end{cases} \tag{9.38}$$

Stable focus

The characteristic equation corresponding to (9.38) is $r^2 + 2r + 2 = 0$ with roots $r_{1,2} = -1 \pm i$. These are complex roots with a negative real part, so the equilibrium point (0, 0) is a stable focus. The phase plane and some of its orbits are shown in Figure 9.13.

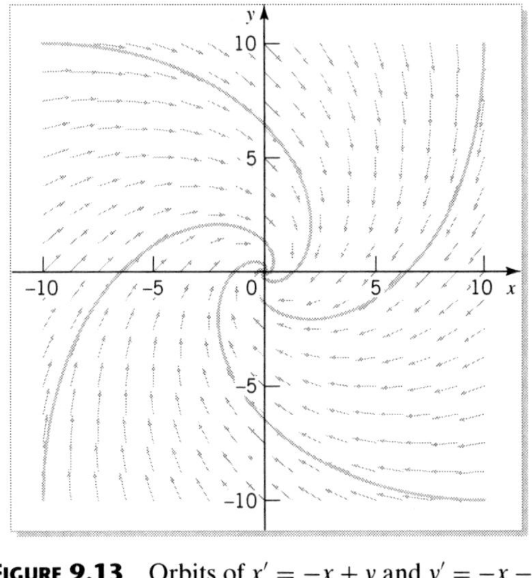

FIGURE 9.13 Orbits of $x' = -x + y$ and $y' = -x - y$

EXAMPLE 9.9

Identify and classify the equilibrium point of the system

$$\begin{cases} x' = -x + y, \\ y' = -6x - 6y. \end{cases} \tag{9.39}$$

Stable node

The characteristic equation corresponding to (9.39) is $r^2 + 7r + 12 = 0$ with roots $r_1 = -3$, $r_2 = -4$. These are negative distinct real roots, so the equilibrium point (0, 0) is a stable node. The phase plane and some of its orbits are shown in Figure 9.14.

Notice that in Figure 9.14 the orbits appear to be asymptotic to a line with negative slope. We can determine the equation of this line from the solution of (9.39) — namely, $x(t) = C_1e^{-4t} + C_2e^{-3t}$, $y(t) = -3C_1e^{-4t} - 2C_2e^{-3t}$ — by computing

$$\frac{y(t)}{x(t)} = \frac{-3C_1e^{-4t} - 2C_2e^{-3t}}{C_1e^{-4t} + C_2e^{-3t}} = \frac{-3C_1e^{-t} - 2C_2}{C_1e^{-t} + C_2} \to -2 \text{ as } t \to \infty,$$

if $C_2 \neq 0$. Thus, all orbits corresponding to $C_2 \neq 0$ are asymptotic to the line $y = -2x$. (The line $y = -2x$ is also an orbit. Why?) But what happens if $C_2 = 0$? In this case,

$$\frac{y(t)}{x(t)} = \frac{-3C_1e^{-4t}}{C_1e^{-4t}} = -3,$$

which gives the line $y = -3x$. (The line $y = -3x$ is also an orbit. Why?) Notice that it is difficult, if not impossible, to spot the orbits $y = -2x$ and $y = -3x$ in Figure 9.14. Figure 9.15 shows the lines $y = -2x$ and $y = -3x$ superimposed on Figure 9.14.

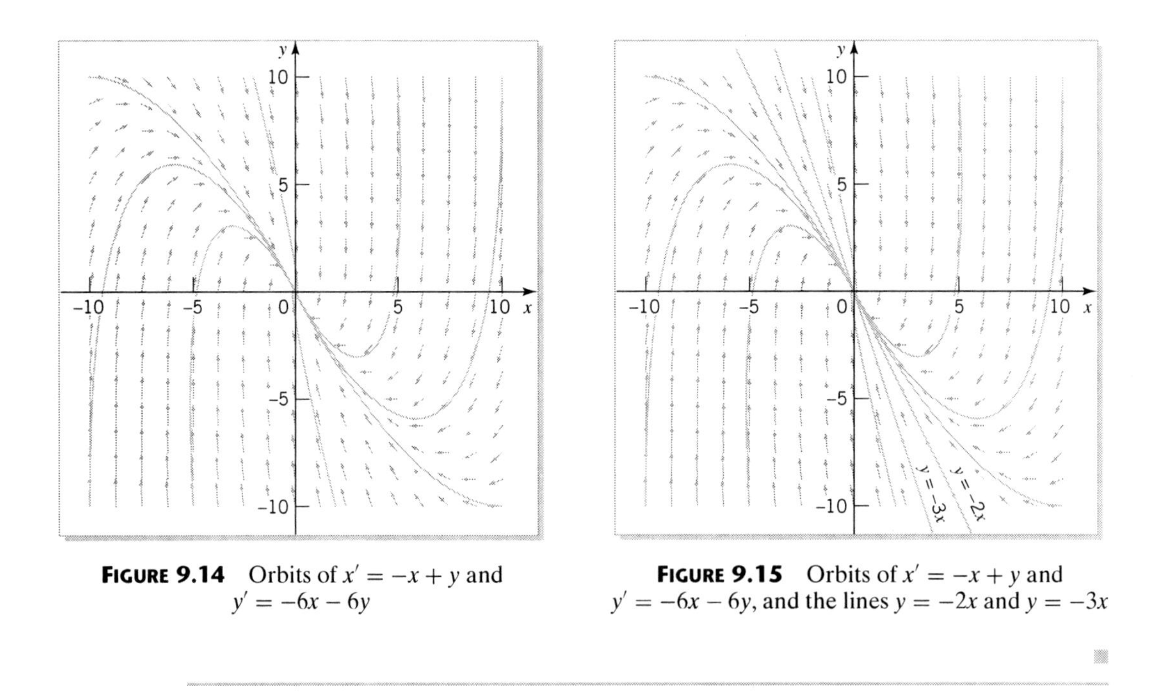

FIGURE 9.14 Orbits of $x' = -x + y$ and $y' = -6x - 6y$

FIGURE 9.15 Orbits of $x' = -x + y$ and $y' = -6x - 6y$, and the lines $y = -2x$ and $y = -3x$

Parabolic Classification Scheme

There is a way to classify the equilibrium points of

$$\begin{cases} x' = ax + by, \\ y' = cx + dy, \end{cases}$$

by introducing the two quantities $D = ad - bc = r_1 r_2$, and $T = a + d = r_1 + r_2$. We chose the letter D because it is the determinant of the 2 by 2 matrix

Determinant

$$\begin{bmatrix} a & b \\ c & d \end{bmatrix},$$

Trace

and T is the trace (the sum of the diagonal elements, a and d, of this matrix). The characteristic equation $r^2 - (a + d)r + (ad - bc) = (r - r_1)(r - r_2) = r^2 - (r_1 + r_2)r + r_1 r_2 = 0$ can be written as $r^2 - Tr + D = 0$, with roots $r = \frac{1}{2}\left(T \pm \sqrt{T^2 - 4D}\right)$.

The characteristic equation has equal roots when the discriminant $\Delta = T^2 - 4D$ is zero; that is, $T^2 - 4D = 0$. If $\Delta > 0$, then the roots will be real and distinct, and if $\Delta < 0$, the roots will be complex conjugates. Thus, if we plot D versus T, the parabola $D = \frac{1}{4}T^2$ will represent the location of equal roots — the "inside" of the parabola is where $\Delta < 0$ and thus corresponds to complex roots, while the "outside" is where $\Delta > 0$ and corresponds to real, distinct roots.

Centers, foci

If we concentrate on $\Delta < 0$, so that the roots are $\alpha \pm i\beta$, where $\alpha = \frac{1}{2}T$ and $\beta = \frac{1}{2}\sqrt{4D - T^2}$, the only cases that can occur are centers ($\alpha = 0$), stable foci ($\alpha < 0$), and unstable foci ($\alpha > 0$). Thus, when $\Delta < 0$, the region $T = 0$ corresponds to centers, the region $T < 0$ to stable foci, and the region $T > 0$ to unstable foci.

Saddle points

Nodes

If we concentrate on $\Delta > 0$, the roots are real and distinct. If they have opposite signs, then $D = r_1 r_2 < 0$. Thus, the region $\Delta > 0$, $D < 0$ represents the location of saddle points. The region $\Delta \geq 0$, $D > 0$ corresponds to a double root or roots of the same sign. If positive, then $T > 0$, so this represents unstable nodes. In the same way, $T < 0$ represents stable nodes.

The only remaining region to classify is $D = 0$; that is, $r_1 r_2 = 0$, so at least one of the roots is zero. This represents the region where equilibrium points are not isolated.

We have put these ideas together in Figure 9.16. Given a system of equations, we can immediately calculate T (the trace) and D (the determinant), locate them

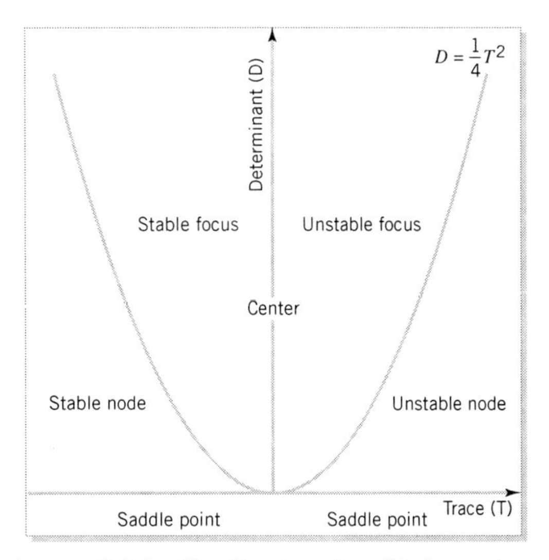

FIGURE 9.16 Classification of equilibrium points

on the D versus T plot, and classify the equilibrium point. If $T < 0$, the equilibrium point is stable, and if $T > 0$, it is unstable. If necessary, we can also calculate the roots from $r = \frac{1}{2}\left(T \pm \sqrt{T^2 - 4D}\right)$.

If we construct a 2 by 2 matrix in a random fashion, and then ask what the chances are that this matrix will generate a saddle point, a node, or focus, the answer would be very good. To see this, think about throwing darts randomly at Figure 9.16. Your chances of hitting anything except the three curves (the parabola $D = \frac{1}{4}T^2$, the positive D-axis, or the T-axis) are very good. Thus, if we randomly construct systems of linear equations, it will be **rare** to find roots of the characteristic equation that fall exactly on any of these three curves; that is, into the following categories:

Rare

- Real equal roots; that is, $r_1 = r_2$, so that $D = \frac{1}{4}T^2$.
- Real roots, one being zero, so that $D = 0$
- Complex roots with no real part; that is, $r = \pm i\beta$, so that $T = 0$ and $D > 0$.

Common

It will be **common** to find roots of the characteristic equation that fall between these three curves; that is, into the following categories:

- Real, distinct, nonzero roots.
- Complex roots with nonzero real parts.

This parabolic classification scheme will be used now and in the next chapter.

EXAMPLE 9.10

Identify and classify the equilibrium point of the system

$$\begin{cases} x' = -x + y, \\ y' = -kx - ky, \end{cases}$$

when $k = 1$ and when $k = 6$.

The matrix associated with this system is

$$\begin{bmatrix} -1 & 1 \\ -k & -k \end{bmatrix},$$

with determinant $D = 2k$ and trace $T = -1 - k$. When $k = 1$, this gives $D = 2 > 0$ and $T = -2 < 0$, so $\frac{1}{4}T^2 = 1 < D$ — a stable focus. When $k = 6$, we have $D = 12 > 0$ and $T = -7 < 0$, so $\frac{1}{4}T^2 = 12\frac{1}{4} > D$ — a stable node. ▪

EXERCISES

1. For each of the following differential equations, identify whether the equilibrium points are stable or unstable and whether each is a node, saddle point, center, or focus. Compare your answers with the ones you found in Exercise 1 on page 388.

(a) $x' = 2x + 5y$, $y' = x + 6y$ (b) $x' = 2x - y$, $y' = -6x + y$

(c) $x' = 5x + 6y$, $y' = x + 4y$ (d) $x' = 5x - 3y$, $y' = -x - y$

(e) $x' = 2x + 9y$, $y' = -x - 4y$ (f) $x' = 2x + y$, $y' = x - 4y$

(g) $x' = 4x - 2y$, $y' = 5x - 2y$ (h) $x' = 2x + y$, $y' = -4x + 2y$

(i) $x' = 3x + 5y$, $y' = -5x + 3y$ (j) $x' = -3x + 2y$, $y' = -4x + y$

(k) $x' = x + y$, $y' = -x + 3y$ (l) $x' = x + y$, $y' = -2x - y$

2. Show that the phase plane of the system of differential equations $x' = ax + by$, $y' = cx + dy$, where $ad - bc = 0$, falls into one of the following three categories.

(a) Every point is an equilibrium point.

(b) There is a line of equilibrium points, with every orbit being a straight line approaching an equilibrium point.

(c) There is a line of equilibrium points, with every orbit being a straight line parallel to the line of equilibrium points.

3. Show that the origin is a stable node of (9.25), $x' = -x$, $y' = -2y$,

(a) by finding the explicit solution.

(b) by finding the orbits in the phase plane.

4. Use the trigonometric identities for $\cos(\beta t + \phi)$ and $\sin(\beta t + \phi)$ to show that (9.35) and (9.36) are true. Then from (9.36) show that $t = \frac{1}{\beta}[\arctan(y/x) - \phi]$, so that (9.37) can be written as $x^2 + y^2 = c^2 e^{(2\alpha/\beta)\arctan(y/x)}$, where $c^2 = (C_1^2 + C_2^2)e^{-2\alpha\phi/\beta}$ is chosen appropriately. Now change to polar coordinates r and θ, where $r = \sqrt{x^2 + y^2}$ and $\theta = \arctan(y/x)$, to obtain the spiral equation $r = ce^{\alpha\theta/\beta}$.

9.3 WHEN DO STRAIGHT-LINE ORBITS EXIST?

We know that orbits in the phase plane corresponding to the system of equations

$$\begin{cases} x' = ax + by, \\ y' = cx + dy, \end{cases}$$

satisfy

$$\frac{dy}{dx} = \frac{cx + dy}{ax + by}.$$

In general, this is a first order equation with homogeneous coefficients. We know the isoclines of this equation are all straight lines through the origin, and that there is a possibility that some of these lines may be orbits. In fact, if we look carefully at Figures 9.3 through 9.14, we see that in some cases there appear to be straight-line orbits through the origin of the phase plane (Figures 9.3 through 9.9, and 9.14), whereas in others there are none (Figures 9.10 through 9.13). In this section we want to find under what circumstances straight-line orbits exist.[7]

[7]Actually, each straight line consists of two orbits that meet at the equilibrium point. If the equilibrium point is stable, then these orbits take an infinite time to reach the equilibrium point. If the equilibrium point is unstable, then those orbits going to infinity take an infinite time to reach there.

To start, we look at Figure 9.3 and notice that the vertical line $x = 0$ is an orbit. A natural question to ask is under what circumstances $x(t) = 0$ is a solution of $x' = ax + by$, $y' = cx + dy$. If we substitute $x(t) = 0$ into this, we are forced to the condition that $b = 0$. Conversely, if $b = 0$, then $x' = ax$ contains $x(t) = 0$ as one of its solutions. Thus, $b = 0$ is equivalent to the existence of the straight-line orbit $x = 0$ in the phase plane. We also notice that if $b = 0$, then the characteristic equation reduces to $r^2 - (a + d)r + ad = (r - a)(r - d) = 0$. Thus, the vertical phase-plane orbit $x = 0$ can occur only if $r_1 = a$ and $r_2 = d$. Of course, there may be other straight-line orbits for $b = 0$. Confirm these observations by looking at Figures 9.3 through 9.14.

Now we look at nonvertical straight-line orbits through the origin in the phase plane. We denote these orbits by $y = mx$ and try to determine m. If we substitute $y = mx$ into the equation for the phase plane orbits

$$\frac{dy}{dx} = \frac{cx + dy}{ax + by},$$

we find

$$m = \frac{c + dm}{a + bm},$$

or $bm^2 + (a - d)m - c = 0$.

If $b \neq 0$, then we can solve this quadratic equation to give

$$m = \frac{1}{2b}\left[-(a - d) \pm \sqrt{\Delta}\right],$$

where $\Delta = (a - d)^2 + 4bc$. Thus, there will be two values for the slope m if $\Delta > 0$, one value if $\Delta = 0$, and no values if $\Delta < 0$. But wait a minute! We have seen the quantity $\Delta = (a - d)^2 + 4bc$ before. It is exactly the discriminant of the characteristic equation, where we have real and distinct roots if $\Delta > 0$, repeated roots if $\Delta = 0$, and complex roots if $\Delta < 0$. (Confirm these observations by looking at Figures 9.3 through 9.14.) Thus, we have the following result.

- If $b \neq 0$, then
  - when we have two real distinct roots of the characteristic equation, there are two distinct straight-line orbits through the origin.
  - when we have a repeated root of the characteristic equation, there is one straight-line orbit through the origin.
  - when we have no real roots of the characteristic equation, there are no straight-line orbits through the origin.

Now let's look at the situation when $b = 0$, remembering that we already know that there is always a vertical straight-line orbit through the origin in this case. To decide if there are any other straight-line orbits through the origin, we note that because $b = 0$, the quadratic for m reduces to $(a - d)m - c = 0$. If $a \neq d$ (that is, $r_1 \neq r_2$), then there is another straight-line orbit with slope $m = c/(a - d)$. If $a = d$ (that is, $r_1 = r_2$) and $c \neq 0$, there are no more straight-line orbits. If $a = d$ (that is, $r_1 = r_2$) and $c = 0$, then there is no restriction on m, and so there are an infinite number of straight-line orbits. (Confirm these observations by looking at Figures 9.3 through 9.14.) We combine these results in the following statements.

- If $b = 0$, then
  - when we have two real distinct roots of the characteristic equation, there are two distinct straight-line orbits through the origin.
  - when we have a repeated root of the characteristic equation, there is one straight-line orbit through the origin if $c \neq 0$, and an infinite number of straight-line orbits through the origin if $c = 0$.

We thus have the following theorem.

Theorem 9.2: The Straight Line Theorem *If the characteristic equation corresponding to*

$$\begin{cases} x' = ax + by, \\ y' = cx + dy, \end{cases}$$

has

(a) two real distinct roots, then there are two distinct straight-line orbits through the origin in the phase plane.

(b) no real roots, then there are no straight-line orbits through the origin in the phase plane.

(c) a repeated root, then either there are an infinite number of straight-line orbits through the origin when $b = c = 0$, or there is one straight-line orbit through the origin.

EXAMPLE 9.11

Find the straight-line orbits through the origin of the phase plane for

$$\begin{cases} x' = -x + y, \\ y' = -kx - ky, \end{cases} \tag{9.40}$$

where (a) $k = 6$, (b) $k = 3 + 2\sqrt{2}$, (c) $k = 1$.

The characteristic equation corresponding to this system is $r^2 + (1 + k)r + 2k = 0$, with two real distinct roots if $k = 6$, a repeated real root if $k = 3 + 2\sqrt{2}$, and two complex roots if $k = 1$. Because $b \neq 0$, we expect two straight-line orbits in case (a), one in case (b), and none in case (c).

Substituting $y = mx$ into the equation for the orbits in the phase plane,

$$\frac{dy}{dx} = \frac{-kx - ky}{-x + y},$$

gives

$$m = \frac{-k - km}{-1 + m},$$

or $m^2 + (k - 1)m + k = 0$. The discriminant of this quadratic equation in m is $\Delta = (k - 1)^2 - 4k = k^2 - 6k + 1$, and the solution is $m = \frac{1}{2}\left(1 - k \pm \sqrt{\Delta}\right)$.

In case (a), where $k = 6$, we see that $\Delta = 1$, so $m = -3$ and $m = -2$, in agreement with Figure 9.15.

In case (b), where $k = 3 + 2\sqrt{2}$, we see that $\Delta = 0$, so $m = (1 - 3 - 2\sqrt{2})/2 = -1 - \sqrt{2}$. Figure 9.17 shows this.

In case (c), where $k = 1$, we see that $\Delta = -4$, so there are no real solutions for m. This agrees with Figure 9.13, which has no straight-line solutions.

In fact, if we start with $k = 6$ (Figure 9.15) and gradually decrease the value of k, the two straight-line orbits gradually converge to one line when $k = 3 + 2\sqrt{2} \approx 5.828$ (Figure 9.17). If we decrease k further, the line immediately vanishes (Figure 9.13).

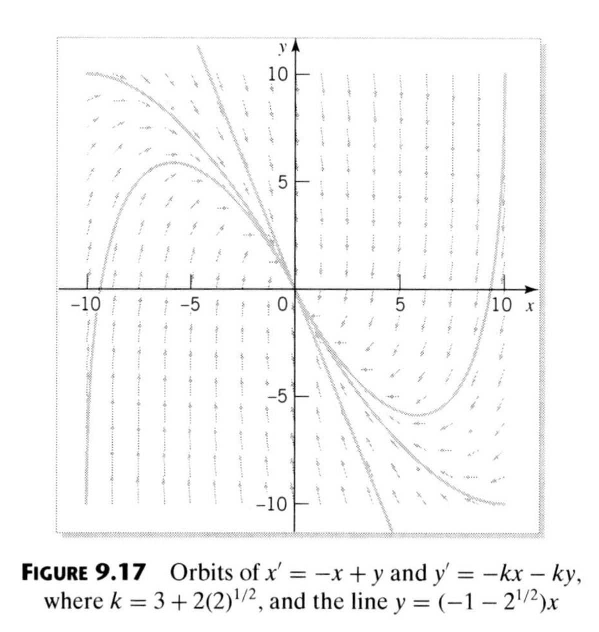

FIGURE 9.17 Orbits of $x' = -x + y$ and $y' = -kx - ky$, where $k = 3 + 2(2)^{1/2}$, and the line $y = (-1 - 2^{1/2})x$

In the previous example we found the straight-line orbits through the origin of the phase plane. However, we have already found the explicit solutions of the system of differential equations (9.40) in Example 9.2 on page 387.

In case (a), where $k = 6$, we found

$$\begin{cases} x(t) = C_1 e^{-4t} + C_2 e^{-3t}, \\ y(t) = -3C_1 e^{-4t} - 2C_2 e^{-3t}. \end{cases}$$

We can see that one of the straight-line orbits through the origin of the phase plane, $y = -3x$, corresponds to $C_2 = 0$. The other straight-line orbit, $y = -2x$, corresponds to $C_1 = 0$. If we rewrite the explicit solution in vector notation; that is,

$$\begin{bmatrix} x(t) \\ y(t) \end{bmatrix} = C_1 \begin{bmatrix} 1 \\ -3 \end{bmatrix} e^{-4t} + C_2 \begin{bmatrix} 1 \\ -2 \end{bmatrix} e^{-3t}, \tag{9.41}$$

we see that the straight-line orbits in the phase plane have directions given by the two vectors

$$\begin{bmatrix} 1 \\ -3 \end{bmatrix} \text{ and } \begin{bmatrix} 1 \\ -2 \end{bmatrix}.$$

In case (b), where $k = 3 + 2\sqrt{2}$, we found the explicit solution of the system of differential equations (9.40) as

$$\begin{cases} x(t) = C_1 e^{rt} + C_2 t e^{rt}, \\ y(t) = \left[(r+1)C_1 + C_2\right] e^{rt} + (r+1)C_2 t e^{rt}, \end{cases}$$

where $r = -2 - \sqrt{2}$. We see that the straight-line orbit through the origin of the phase plane $y = \left(-1 - \sqrt{2}\right)x$ — that is, $y = (r+1)\,x$ — corresponds to $C_2 = 0$, which is apparent if we rewrite the explicit solution in vector notation as

$$\begin{bmatrix} x(t) \\ y(t) \end{bmatrix} = \begin{bmatrix} C_1 \\ (r+1)C_1 + C_2 \end{bmatrix} e^{rt} + C_2 \begin{bmatrix} 1 \\ r+1 \end{bmatrix} t e^{rt}.$$

In case (c), where $k = 1$, we found explicit solutions of the system of differential equations (9.40) as

$$\begin{cases} x(t) = e^{-t}(C_1 \sin t + C_2 \cos t), \\ y(t) = e^{-t}(C_1 \cos t - C_2 \sin t), \end{cases}$$

which can be rewritten in the form

$$\begin{bmatrix} x(t) \\ y(t) \end{bmatrix} = C_1 \begin{bmatrix} \sin t \\ \cos t \end{bmatrix} e^{-t} + C_2 \begin{bmatrix} \cos t \\ -\sin t \end{bmatrix} e^{-t}.$$

Consequently there is an advantage in writing the explicit solutions of systems of differential equations in vector form. In that form it is possible to read off the straight-line orbits immediately.

EXERCISES

1. For each of the following differential equations, identify all straight-line orbits through the origin of the phase plane. Compare your answers with the ones you found in Exercise 1 on page 388, and Exercise 1 on page 403.

(a) $x' = 2x + 5y$, $y' = x + 6y$ (b) $x' = 2x - y$, $y' = -6x + y$

(c) $x' = 5x + 6y$, $y' = x + 4y$ (d) $x' = 5x - 3y$, $y' = -x - y$

(e) $x' = 2x + 9y$, $y' = -x - 4y$ (f) $x' = 2x + y$, $y' = x - 4y$

(g) $x' = 4x - 2y$, $y' = 5x - 2y$ (h) $x' = 2x + y$, $y' = -4x + 2y$

(i) $x' = 3x + 5y$, $y' = -5x + 3y$ (j) $x' = -3x + 2y$, $y' = -4x + y$

(k) $x' = x + y$, $y' = -x + 3y$ (l) $x' = x + y$, $y' = -2x - y$

2. For the following systems of differential equations (from Chapter 6), identify all straight-line orbits through the origin of the phase plane. (The quantities a, b, c, k, and m are all positive constants.)

(a) $x' = -ay$, $y' = -ax$ (b) $x' = y$, $y' = -(k/m)x - (b/m)y$

(c) $x' = ay$, $y' = -bx$ (d) $x' = ax - cy$, $y' = -bx$

3. Three mystery direction fields of phase planes associated with the system of differential equations $x' = ax + by$, $y' = cx + dy$, are given in Figures 9.18, 9.19, and 9.20. By looking for straight-line orbits, decide which of these direction fields has the following properties. Explain the reasons behind each choice.

   (a) The characteristic equation has real, distinct roots.

   (b) The characteristic equation has a real repeated root.

   (c) The characteristic equation has complex roots.

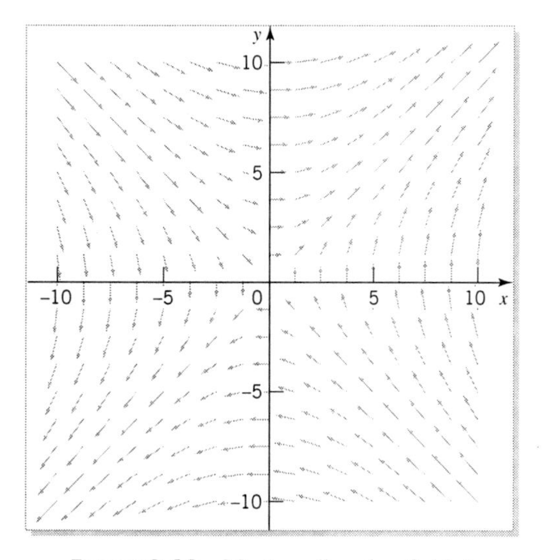

FIGURE 9.18 Mystery direction field A

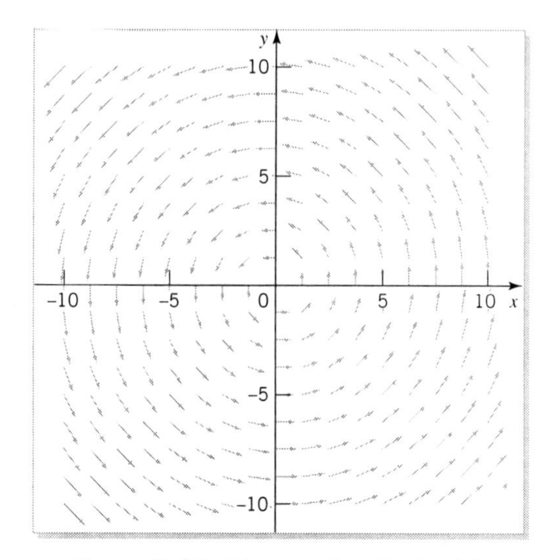

FIGURE 9.19 Mystery direction field B

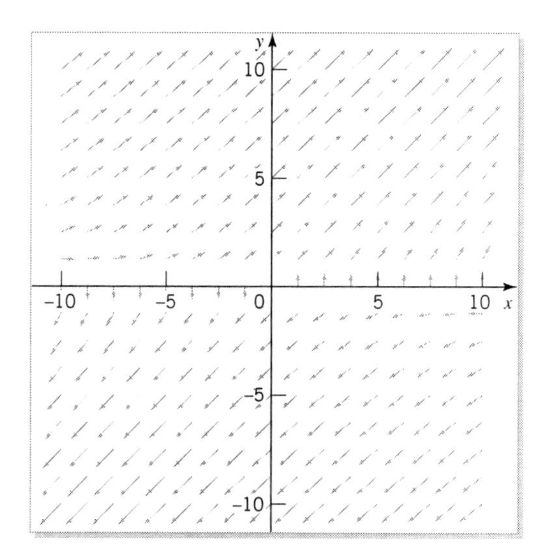

FIGURE 9.20 Mystery direction field C

9.4 QUALITATIVE BEHAVIOR USING NULLCLINES

We can frequently determine the qualitative behavior of the orbits of the system of equations

$$\begin{cases} x' = ax + by, \\ y' = cx + dy, \end{cases}$$

without actually solving them. We now show how this can be done using geometrical arguments.

EXAMPLE 9.12

We start with a system of differential equations that we discussed on page 393, namely,

$$\begin{cases} x' = y, \\ y' = x, \end{cases} \tag{9.42}$$

and its phase plane counterpart,

$$\frac{dy}{dx} = \frac{x}{y}.$$

Monotonicity To analyze (9.42) graphically, we look for regions where $x' > 0$ or $x' < 0$ (because this will tell us where x is increasing or decreasing with time), and where $y' > 0$ or $y' < 0$ (because this will tell us where y is increasing or decreasing with time). These will generally be separated by places where $x' = 0$, $y' = 0$, so we should find these regions; they will usually mark the boundaries of increasing and decreasing solutions. In the phase plane, this suggests we construct the special isoclines corresponding to horizontal ($y' = 0$) and vertical ($x' = 0$) tangent lines. These correspond to places where $x' = 0$ and $y' = 0$ and are called x-NULLCLINES and y-NULLCLINES. ***Nullclines*** The y-nullcline for (9.42) is the straight line $x = 0$ (horizontal tangents) and the x-nullcline for (9.42) is the straight line $y = 0$ (vertical tangents). We sketch these in the phase plane and add short horizontal line segments along the y-nullcline and short vertical line segments along the x-nullcline. See Figure 9.21. The point where the nullclines intersect, $(0, 0)$, is the equilibrium point.

Direction We now want to decide the direction the orbits will be traveling when they cross a nullcline, and then indicate this by adding arrowheads to the short horizontal and vertical line segments. Let's consider the y-nullcline, which is $x = 0$. We want to know how x will change with t. From $x' = y$, we see that x increases for $y > 0$ and decreases for $y < 0$. Thus, the arrows along the y-nullcline, $x = 0$, should point from left to right for $y > 0$ and from right to left for $y < 0$. In the same way the arrows along the x-nullcline, $y = 0$, will point up for $x > 0$ and down for $x < 0$. See Figure 9.22.

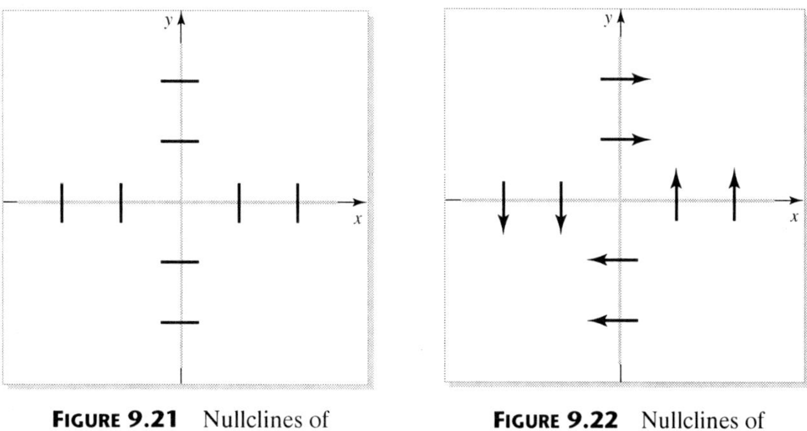

FIGURE 9.21 Nullclines of $dy/dx = x/y$ — step 1

FIGURE 9.22 Nullclines of $dy/dx = x/y$ — step 2

Thus, any orbit that enters the first quadrant cannot escape from it. Furthermore, orbits in this quadrant will move away from the equilibrium point (0, 0). Similar remarks apply to the third quadrant. In the second and fourth quadrants, orbits are being pulled toward the origin. This is indicative of a saddle point. Figure 9.6 on page 393 shows the orbits of (9.42). ∎

Saddle point

◆ *Definition 9.1:* **The x-NULLCLINES of the system of autonomous differential equations**

$$\begin{cases} x' = P(x,y), \\ y' = Q(x,y), \end{cases}$$

are the curves along which the orbits have vertical tangent lines. The y-NULLCLINES are the curves along which the orbits have horizontal tangent lines. ◆

How to Perform a Nullcline Analysis on a Linear Autonomous System

Purpose To use nullclines to obtain the qualitative behavior of the solutions of

$$\begin{cases} x' = ax + by, \\ y' = cx + dy. \end{cases}$$

Process

1. **Horizontal tangents.** Sketch the line $cx + dy = 0$ in the phase plane. This nullcline is where the orbits will have horizontal tangents — the y-nullcline. This is indicated by adding short horizontal lines along the nullcline.

2. **Vertical tangents.** Sketch the line $ax + by = 0$ in the phase plane. This nullcline is where the orbits will have vertical tangents — the x-nullcline. This is indicated by adding short vertical lines along the nullcline.

3. **Horizontal direction.** Add arrowheads to the short horizontal line segments of the y-nullcline $cx + dy = 0$ as follows. Use the equation $x' = ax + by$ to determine where $x' > 0$ along the y-nullcline. Because x is increasing with time on this part of the nullcline, the arrows should point from left to right. Now determine where $x' < 0$ along the y-nullcline. Because x is decreasing with time on this part of the nullcline, the arrows should point from right to left.

4. **Vertical direction.** Add arrowheads to the short vertical line segments of the x-nullcline $ax + by = 0$ as follows. Use the equation $y' = cx + dy$ to determine where $y' > 0$ along the x-nullcline. Because y is increasing with time on this part of the nullcline, the arrows should point up. Now determine where $y' < 0$ along the x-nullcline. Because y is decreasing with time on this part of the nullcline, the arrows should point down.

5. **Conclusion.** The phase plane is now divided into four regions. If a region has arrows only pointing into it, then orbits are trapped in that region. If a region has arrows only pointing out of it, then orbits must leave that region.

Comments about Nullcline Analysis

- The point where the x- and y-nullclines intersect, $(0, 0)$, is the equilibrium point.
- This analysis depends on the existence of distinct nullclines through the equilibrium point $(0, 0)$. The lines $ax + by = 0$ and $cx + dy = 0$ degenerate into a single line if and only if $ad - bc = 0$ — that is, if and only if at least one of the real roots, r_1 or r_2, of the characteristic equation is zero. This is the same condition that guarantees the origin is not an isolated equilibrium point.
- We must be careful when using a nullcline analysis, as can be seen by returning to

$$\begin{cases} x' = -x + y, \\ y' = -kx - ky, \end{cases}$$

where $k > 0$ is a constant. The x-nullclines are $y = x$ and the y-nullclines are $y = -x$ for all $k > 0$. Furthermore, the arrows point in a clockwise direction. It is tempting to think that this implies that the origin is either a focus or a center because of the suggested spiraling nature. **Unfortunately this is not true**; for $k \geq 3 + 2\sqrt{2}$, the origin is a node. Figures 9.23 and 9.24 show the situations for $k = 1$ and $k = 6$, respectively. We notice that although the nullclines in both figures are identical, the orbits are very different. Figure 9.24 is characteristic of a node, not a focus or center.

Important point!

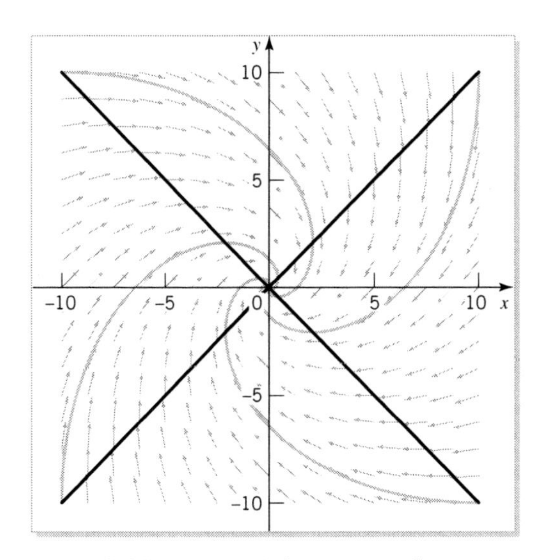

FIGURE 9.23 Orbits of $x' = -x + y$, $y' = -x - y$, and the nullclines $y = x$ and $y = -x$

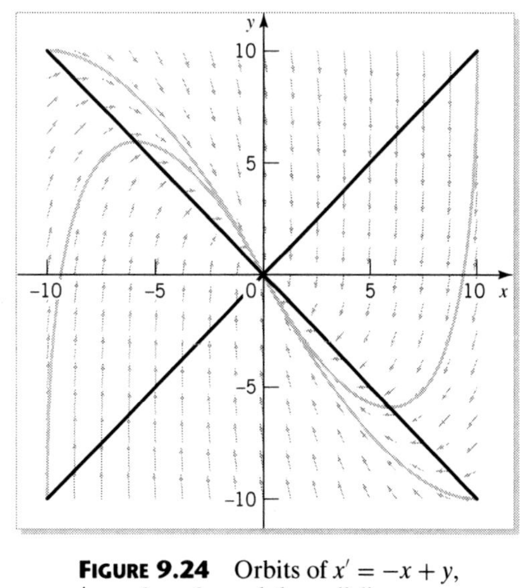

FIGURE 9.24 Orbits of $x' = -x + y$, $y' = -6x - 6y$, and the nullclines $y = x$ and $y = -x$

EXAMPLE 9.13

Use a nullcline analysis to obtain the qualitative behavior of the solutions of

$$\begin{cases} x' = -x + y, \\ y' = -2y. \end{cases} \tag{9.43}$$

Nullclines

The x-nullclines of (9.43) are the straight lines $y = x$, and the y-nullclines are $y = 0$. The equilibrium point is at $(0, 0)$, the intersection of these two lines. The arrows along the line $y = x$ point down for $y > 0$ (because $y' = -2y < 0$ for $y > 0$) and point up for $y < 0$. The arrows on the line $y = 0$ point to the left for $x > 0$ (because $x' = -x + y = -x < 0$ on the positive x-axis) and point to the right for $x < 0$. This is shown in Figure 9.25.

Stable node

Notice that all orbits that enter the region between the positive x-axis and the line $y = x$ are trapped and have to approach $(0, 0)$ without spiraling. The same is true for the region between the negative x-axis and the line $y = x$. Orbits in the other two regions are also drawn toward $(0, 0)$. The only way to reach the equilibrium point is asymptotic to the x-axis, so the equilibrium point is a stable node. This can be checked by computing the characteristic equation and finding that its roots are $r_1 = -2$ and $r_2 = -1$, which are real, distinct, and negative. Figure 9.26 shows the phase plane, some orbits of (9.43), and the lines $y = x$ and $y = 0$.

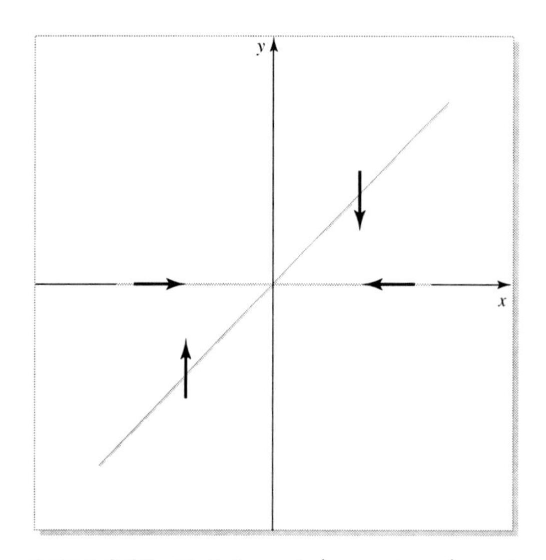

FIGURE 9.25 Nullclines of $x' = -x + y$, $y' = -2y$

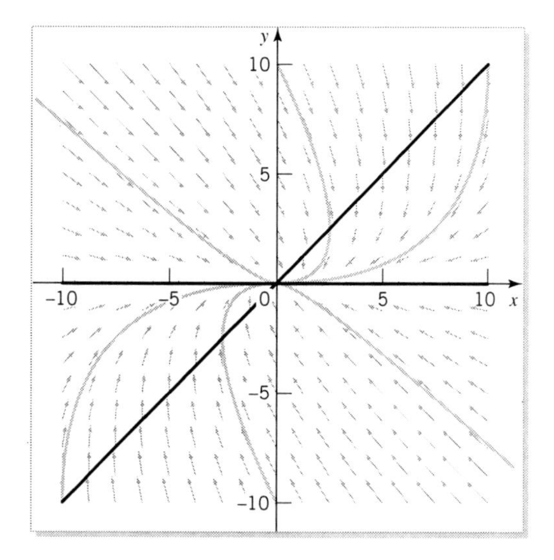

FIGURE 9.26 Orbits of $x' = -x + y$, $y' = -2y$, and the nullclines $y = x$ and $y = 0$

EXERCISES

1. For each of the following differential equations, use a nullcline analysis to identify whether the equilibrium points are stable or unstable and whether each is a node, saddle point, center, or focus. If the analysis fails, explain why. Compare your answers with the ones you found in Exercise 1 on page 388, Exercise 1 on page 403, and Exercise 1 on page 407.

   (a) $x' = 2x + 5y$, $y' = x + 6y$ (b) $x' = 2x - y$, $y' = -6x + y$
   (c) $x' = 5x + 6y$, $y' = x + 4y$ (d) $x' = 5x - 3y$, $y' = -x - y$
   (e) $x' = 2x + 9y$, $y' = -x - 4y$ (f) $x' = 2x + y$, $y' = x - 4y$
   (g) $x' = 4x - 2y$, $y' = 5x - 2y$ (h) $x' = 2x + y$, $y' = -4x + 2y$
   (i) $x' = 3x + 5y$, $y' = -5x + 3y$ (j) $x' = -3x + 2y$, $y' = -4x + y$
   (k) $x' = x + y$, $y' = -x + 3y$ (l) $x' = x + y$, $y' = -2x - y$

2. For the following systems of differential equations (from Chapter 6), use a nullcline analysis to identify whether the equilibrium points are stable or unstable and whether each is a node, saddle point, center, or focus. If the analysis fails, explain why. (The quantities a, b, c, k, and m are all positive constants.)

   (a) $x' = -ay$, $y' = -ax$ (b) $x' = y$, $y' = -(k/m)x - (b/m)y$
   (c) $x' = ay$, $y' = -bx$ (d) $x' = ax - cy$, $y' = -bx$

9.5 MATRIX FORMULATION OF SOLUTIONS

In this section we will develop an alternative method for solving

$$\begin{cases} x' = ax + by, \\ y' = cx + dy, \end{cases} \tag{9.44}$$

using matrices.[8]

On page 405 we solved (9.44) for the specific choices $a = -1$, $b = 1$, and $c = d = -6$, namely,

$$\begin{cases} x' = -x + y, \\ y' = -6x - 6y. \end{cases} \tag{9.45}$$

We found that the solution, (9.41), could be written in vector form as

$$\begin{bmatrix} x(t) \\ y(t) \end{bmatrix} = C_1 \begin{bmatrix} 1 \\ -3 \end{bmatrix} e^{-4t} + C_2 \begin{bmatrix} 1 \\ -2 \end{bmatrix} e^{-3t}. \tag{9.46}$$

The fact that we may write the solution in terms of vectors suggests that we write the system of differential equations in terms of vectors. In fact, we can rewrite (9.44) in matrix form as

$$\begin{bmatrix} x' \\ y' \end{bmatrix} = \begin{bmatrix} a & b \\ c & d \end{bmatrix} \begin{bmatrix} x \\ y \end{bmatrix} \tag{9.47}$$

[8] A summary of some elementary properties of matrices is in Appendix A.5.

or equivalently as

$$\frac{d\mathbf{X}}{dt} = \mathbf{MX},$$

Coefficient matrix

where **M** is the COEFFICIENT MATRIX given by

$$\mathbf{M} = \begin{bmatrix} a & b \\ c & d \end{bmatrix},$$

and **X** is the vector

$$\mathbf{X} = \begin{bmatrix} x \\ y \end{bmatrix}.$$

On page 389 we found that the general solution of (9.44) depended on the roots of the characteristic equation

$$\begin{vmatrix} a-r & b \\ c & d-r \end{vmatrix} = 0,$$

with typical solutions as follows.

1. Two real, distinct roots r_1 and r_2:

$$\begin{cases} x(t) = A_1 e^{r_1 t} + A_2 e^{r_2 t}, \\ y(t) = B_1 e^{r_1 t} + B_2 e^{r_2 t}. \end{cases}$$

2. One real, repeated root r:

$$\begin{cases} x(t) = A_1 e^{rt} + A_2 t e^{rt}, \\ y(t) = B_1 e^{rt} + B_2 t e^{rt}. \end{cases}$$

3. Two complex roots $\alpha \pm i\beta$:

$$\begin{cases} x(t) = e^{\alpha t}\left(A_1 \sin \beta t + A_2 \cos \beta t\right), \\ y(t) = e^{\alpha t}\left(B_1 \sin \beta t + B_2 \cos \beta t\right). \end{cases}$$

In these solutions the constants A_1, A_2, B_1, and B_2 are related to one another in various ways, so that only two of these four constants are independent. Thus, once we have determined the roots of the characteristic equation, all that is left is to determine these constants.

Each of these solutions can be written as follows.

1. Two real, distinct roots r_1 and r_2:

$$\mathbf{X}(t) = \begin{bmatrix} x(t) \\ y(t) \end{bmatrix} = \begin{bmatrix} A_1 \\ B_1 \end{bmatrix} e^{r_1 t} + \begin{bmatrix} A_2 \\ B_2 \end{bmatrix} e^{r_2 t}.$$

2. One real, repeated root r:

$$\mathbf{X}(t) = \begin{bmatrix} x(t) \\ y(t) \end{bmatrix} = \left(\begin{bmatrix} A_1 \\ B_1 \end{bmatrix} + \begin{bmatrix} A_2 \\ B_2 \end{bmatrix} t\right) e^{rt}.$$

3. Two complex roots $\alpha \pm i\beta$:

$$\mathbf{X}(t) = \begin{bmatrix} x(t) \\ y(t) \end{bmatrix} = \begin{bmatrix} A_1 \\ B_1 \end{bmatrix} e^{\alpha t} \sin \beta t + \begin{bmatrix} A_2 \\ B_2 \end{bmatrix} e^{\alpha t} \cos \beta t.$$

Thus, to determine the constants, all we need to do is substitute the appropriate solution into the differential equation $\mathbf{X}' = \mathbf{MX}$. We deal with these three cases in turn.

Two Real, Distinct Roots

We will apply this suggestion to the system $\mathbf{X}' = \mathbf{MX}$, which has the coefficient matrix

$$\mathbf{M} = \begin{bmatrix} a & b \\ c & d \end{bmatrix}.$$

This gives the characteristic equation

$$\begin{vmatrix} a - r & b \\ c & d - r \end{vmatrix} = 0,$$

which, in this case will have two real, distinct roots, r_1 and r_2, and solution

$$\mathbf{X}(t) = \begin{bmatrix} x(t) \\ y(t) \end{bmatrix} = \begin{bmatrix} A_1 \\ B_1 \end{bmatrix} e^{r_1 t} + \begin{bmatrix} A_2 \\ B_2 \end{bmatrix} e^{r_2 t}.$$

If we substitute this expression into both sides of $\mathbf{X}' = \mathbf{MX}$, realizing that

$$\frac{d}{dt} \begin{bmatrix} A \\ B \end{bmatrix} e^{rt} = \begin{bmatrix} rA \\ rB \end{bmatrix} e^{rt},$$

and that the derivative of the sum of vectors is the sum of their derivatives (see Exercise 1 on page 428), we obtain

$$\begin{bmatrix} r_1 A_1 \\ r_1 B_1 \end{bmatrix} e^{r_1 t} + \begin{bmatrix} r_2 A_2 \\ r_2 B_2 \end{bmatrix} e^{r_2 t} = \begin{bmatrix} a & b \\ c & d \end{bmatrix} \left(\begin{bmatrix} A_1 \\ B_1 \end{bmatrix} e^{r_1 t} + \begin{bmatrix} A_2 \\ B_2 \end{bmatrix} e^{r_2 t} \right).$$

Because this has to be an identity in t, and because $e^{r_1 t}$ and $e^{r_2 t}$ are linearly independent, this identity gives the two conditions

$$\begin{bmatrix} r_1 A_1 \\ r_1 B_1 \end{bmatrix} = \begin{bmatrix} a & b \\ c & d \end{bmatrix} \begin{bmatrix} A_1 \\ B_1 \end{bmatrix},$$

$$\begin{bmatrix} r_2 A_2 \\ r_2 B_2 \end{bmatrix} = \begin{bmatrix} a & b \\ c & d \end{bmatrix} \begin{bmatrix} A_2 \\ B_2 \end{bmatrix}.$$

We can rearrange the preceding equations to obtain

$$\begin{bmatrix} a & b \\ c & d \end{bmatrix}\begin{bmatrix} A_1 \\ B_1 \end{bmatrix} - \begin{bmatrix} r_1A_1 \\ r_1B_1 \end{bmatrix} = \begin{bmatrix} a-r_1 & b \\ c & d-r_1 \end{bmatrix}\begin{bmatrix} A_1 \\ B_1 \end{bmatrix} = \begin{bmatrix} 0 \\ 0 \end{bmatrix},$$

$$\begin{bmatrix} a & b \\ c & d \end{bmatrix}\begin{bmatrix} A_2 \\ B_2 \end{bmatrix} - \begin{bmatrix} r_2A_2 \\ r_2B_2 \end{bmatrix} = \begin{bmatrix} a-r_2 & b \\ c & d-r_2 \end{bmatrix}\begin{bmatrix} A_2 \\ B_2 \end{bmatrix} = \begin{bmatrix} 0 \\ 0 \end{bmatrix}.$$

Notice that both of these equations have the same structure, namely,

$$\begin{bmatrix} a-r & b \\ c & d-r \end{bmatrix}\begin{bmatrix} A \\ B \end{bmatrix} = \begin{bmatrix} 0 \\ 0 \end{bmatrix}.$$

This system of algebraic equations will have a nontrivial solution only if the determinant of the coefficients is zero, so

$$\begin{vmatrix} a-r & b \\ c & d-r \end{vmatrix} = 0.$$

This is just the characteristic equation! Thus, in the future we can bypass the previous calculation and go straight to

$$\begin{bmatrix} a-r_1 & b \\ c & d-r_1 \end{bmatrix}\begin{bmatrix} A_1 \\ B_1 \end{bmatrix} = \begin{bmatrix} 0 \\ 0 \end{bmatrix},$$

$$\begin{bmatrix} a-r_2 & b \\ c & d-r_2 \end{bmatrix}\begin{bmatrix} A_2 \\ B_2 \end{bmatrix} = \begin{bmatrix} 0 \\ 0 \end{bmatrix},$$

to find the relations among the four constants. Let's see how this works in practice.

EXAMPLE 9.14

Solve the system (9.45)

$$\begin{cases} x' = -x + y, \\ y' = -6x - 6y. \end{cases}$$

In this case,

$$\mathbf{M} = \begin{bmatrix} -1 & 1 \\ -6 & -6 \end{bmatrix},$$

and the characteristic equation is

$$\begin{vmatrix} -1-r & 1 \\ -6 & -6-r \end{vmatrix} = 0,$$

with roots $r_1 = -4$ and $r_2 = -3$. The constants A_1, A_2, B_1, and B_2 must satisfy

$$\begin{bmatrix} -1-(-4) & 1 \\ -6 & -6-(-4) \end{bmatrix}\begin{bmatrix} A_1 \\ B_1 \end{bmatrix} = \begin{bmatrix} 3 & 1 \\ -6 & -2 \end{bmatrix}\begin{bmatrix} A_1 \\ B_1 \end{bmatrix} = \begin{bmatrix} 0 \\ 0 \end{bmatrix},$$

$$\begin{bmatrix} -1-(-3) & 1 \\ -6 & -6-(-3) \end{bmatrix}\begin{bmatrix} A_2 \\ B_2 \end{bmatrix} = \begin{bmatrix} 2 & 1 \\ -6 & -3 \end{bmatrix}\begin{bmatrix} A_2 \\ B_2 \end{bmatrix} = \begin{bmatrix} 0 \\ 0 \end{bmatrix},$$

so our conditions are that $B_1 = -3A_1$ and $B_2 = -2A_2$. In this case, the explicit solution

$$\mathbf{X}(t) = \begin{bmatrix} A_1 \\ B_1 \end{bmatrix} e^{-4t} + \begin{bmatrix} A_2 \\ B_2 \end{bmatrix} e^{-3t}$$

Explicit solution

of (9.47) is

$$\mathbf{X}(t) = \begin{bmatrix} A_1 \\ -3A_1 \end{bmatrix} e^{-4t} + \begin{bmatrix} A_2 \\ -2A_2 \end{bmatrix} e^{-3t},$$

or, relabeling the constants $A_1 = C_1$ and $A_2 = C_2$,

$$\mathbf{X}(t) = C_1 \begin{bmatrix} 1 \\ -3 \end{bmatrix} e^{-4t} + C_2 \begin{bmatrix} 1 \\ -2 \end{bmatrix} e^{-3t},$$

which is (9.46).

Although this is all we wanted, we can rewrite this solution as follows. Using the fact that

$$\begin{aligned} C_1 \begin{bmatrix} 1 \\ -3 \end{bmatrix} e^{-4t} + C_2 \begin{bmatrix} 1 \\ -2 \end{bmatrix} e^{-3t} &= C_1 \begin{bmatrix} e^{-4t} \\ -3e^{-4t} \end{bmatrix} + C_2 \begin{bmatrix} e^{-3t} \\ -2e^{-3t} \end{bmatrix} \\ &= \begin{bmatrix} e^{-4t} & e^{-3t} \\ -3e^{-4t} & -2e^{-3t} \end{bmatrix}\begin{bmatrix} C_1 \\ C_2 \end{bmatrix}, \end{aligned}$$

Matrix form of solution

we may also write this solution in matrix form as

$$\mathbf{X}(t) = \begin{bmatrix} e^{-4t} & e^{-3t} \\ -3e^{-4t} & -2e^{-3t} \end{bmatrix}\begin{bmatrix} C_1 \\ C_2 \end{bmatrix},$$

or

$$\mathbf{X}(t) = \mathbf{UC},$$

where

$$\mathbf{X}(t) = \begin{bmatrix} x(t) \\ y(t) \end{bmatrix}, \mathbf{U} = \begin{bmatrix} e^{-4t} & e^{-3t} \\ -3e^{-4t} & -2e^{-3t} \end{bmatrix}, \text{ and } \mathbf{C} = \begin{bmatrix} C_1 \\ C_2 \end{bmatrix}.$$

■

Comments about $X(t) = UC$

- The matrix **U** is called a FUNDAMENTAL MATRIX of our original system of differential equations (9.47). The columns of **U** contain our two vector solutions.
- The two solutions of the characteristic equation — in this case, $r = -3$ and $r = -4$ — are called EIGENVALUES. The two vectors of constants associated with these eigenvalues, in this case

$$C_1\begin{bmatrix} 1 \\ -3 \end{bmatrix} \text{ and } C_2\begin{bmatrix} 1 \\ -2 \end{bmatrix},$$

are called EIGENVECTORS. The eigenvalues and eigenvectors characterize the behavior of the solution of this system of equations and derive their name from the German word "eigen," which means characteristic.[9]

Phase plane

Figure 9.27 gives the direction field for the phase plane associated with (9.45), along with orbits for different initial conditions. Notice that several of the orbits seem to approach the origin along a straight line. The slope of this straight line appears to be about −2. This is the same direction as the eigenvector associated with the eigenvalue of −3, namely,

$$C_2\begin{bmatrix} 1 \\ -2 \end{bmatrix}.$$

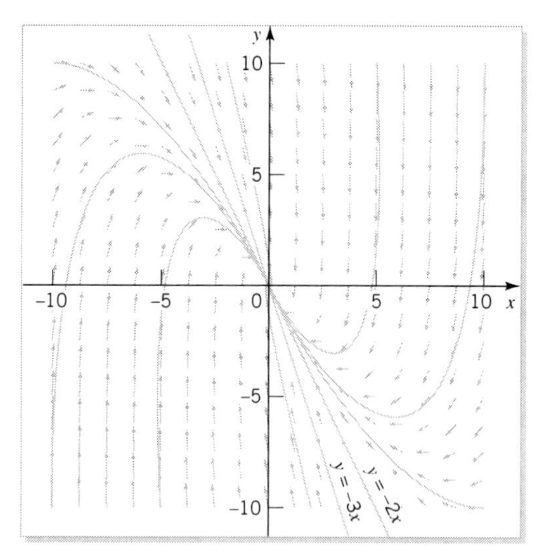

FIGURE 9.27 Direction field and several orbits for $dy/dx = (-6x - 6y)/(-x + y)$

---

[9] In fact, eigenvalues and eigenvectors are sometimes called characteristic values and characteristic vectors.

We also plot the direction of the eigenvector associated with the second eigenvalue of -4, namely,

$$C_1 \begin{bmatrix} 1 \\ -3 \end{bmatrix},$$

which gives a straight line through the origin with slope -3. We note that this line is parallel to the direction field in its vicinity. Could these straight lines also be orbits in the phase plane?

To investigate this further, we examine the differential equation for the orbits in the phase plane, given by

$$\frac{dy}{dx} = \frac{-6x - 6y}{-x + y}.$$

Straight-line orbits in the phase plane, $y = mx$, are possible if m satisfies

$$m = \frac{-6 - 6m}{-1 + m},$$

or $m^2 + 5m + 6 = 0$. This gives the directions of straight-line orbits as $m = -3$ and $m = -2$, precisely the same slopes as determined by the eigenvectors we have just found. Thus, we have discovered that the **eigenvectors give the directions of the straight-line orbits in the phase plane**. This is true for all systems of differential equations given by (9.44) that have real eigenvalues. See Exercise 8 on page 429.

Straight-line orbits

Complex Roots

We now consider a system of differential equations in which the eigenvalues are complex numbers.[10] In this case, we consider

$$\mathbf{X}' = \begin{bmatrix} a & b \\ c & d \end{bmatrix} \mathbf{X},$$

when the characteristic equation

$$\begin{vmatrix} a - r & b \\ c & d - r \end{vmatrix} = 0$$

has roots $r = \alpha \pm i\beta$, and the solution is

$$\mathbf{X}(t) = \begin{bmatrix} A_1 \\ B_1 \end{bmatrix} e^{\alpha t} \sin \beta t + \begin{bmatrix} A_2 \\ B_2 \end{bmatrix} e^{\alpha t} \cos \beta t.$$

We could substitute this solution into the differential equation $\mathbf{X}' = \mathbf{MX}$ to find conditions on the constants A_1, A_2, B_1, and B_2. However, this process is made easier

[10]Appendix A.4 contains a discussion of complex numbers.

if we first rewrite the solution entirely in terms of exponential functions using the facts that

$$\sin\beta t = \frac{1}{2i}\left(e^{i\beta t} - e^{-i\beta t}\right), \qquad \cos\beta t = \frac{1}{2}\left(e^{i\beta t} + e^{-i\beta t}\right),$$

in which case

$$\mathbf{X}(t) = \begin{bmatrix} A_1 \\ B_1 \end{bmatrix} e^{\alpha t}\frac{1}{2i}\left(e^{i\beta t} - e^{-i\beta t}\right) + \begin{bmatrix} A_2 \\ B_2 \end{bmatrix} e^{\alpha t}\frac{1}{2}\left(e^{i\beta t} + e^{-i\beta t}\right),$$

or, after some manipulation,

$$\mathbf{X}(t) = \frac{1}{2}\left(\begin{bmatrix} A_2 \\ B_2 \end{bmatrix} - i\begin{bmatrix} A_1 \\ B_1 \end{bmatrix}\right) e^{(\alpha+i\beta)t} + \frac{1}{2}\left(\begin{bmatrix} A_2 \\ B_2 \end{bmatrix} + i\begin{bmatrix} A_1 \\ B_1 \end{bmatrix}\right) e^{(\alpha-i\beta)t}.$$

If we define $A = \frac{1}{2}\left(A_2 - iA_1\right)$ and $B = \frac{1}{2}\left(B_2 - iB_1\right)$, then $\mathbf{X}(t)$ can be written in the form

$$\mathbf{X}(t) = \begin{bmatrix} A \\ B \end{bmatrix} e^{(\alpha+i\beta)t} + \begin{bmatrix} \tilde{A} \\ \tilde{B} \end{bmatrix} e^{(\alpha-i\beta)t},$$

where $\tilde{A}$ and $\tilde{B}$ are the complex conjugates of A and B, respectively. Notice that the right-hand side consists of the sum of two terms, and that the second term is the complex conjugate of the first term. Thus, once we know one of these terms, we can immediately write down the other.

If we substitute this form of $\mathbf{X}(t)$ into $\mathbf{X}' = \mathbf{MX}$, we find

$$(\alpha + i\beta)\begin{bmatrix} A \\ B \end{bmatrix} e^{(\alpha+i\beta)t} + (\alpha - i\beta)\begin{bmatrix} \tilde{A} \\ \tilde{B} \end{bmatrix} e^{(\alpha-i\beta)t} = \begin{bmatrix} a & b \\ c & d \end{bmatrix}\left(\begin{bmatrix} A \\ B \end{bmatrix} e^{(\alpha+i\beta)t} + \begin{bmatrix} \tilde{A} \\ \tilde{B} \end{bmatrix} e^{(\alpha-i\beta)t}\right).$$

Because $e^{(\alpha+i\beta)t}$ and $e^{(\alpha-i\beta)t}$ are linearly independent, we have

$$(\alpha + i\beta)\begin{bmatrix} A \\ B \end{bmatrix} = \begin{bmatrix} a & b \\ c & d \end{bmatrix}\begin{bmatrix} A \\ B \end{bmatrix},$$

and its complex conjugate,

$$(\alpha - i\beta)\begin{bmatrix} \tilde{A} \\ \tilde{B} \end{bmatrix} = \begin{bmatrix} a & b \\ c & d \end{bmatrix}\begin{bmatrix} \tilde{A} \\ \tilde{B} \end{bmatrix},$$

or, by rearranging,

$$\begin{bmatrix} a - (\alpha + i\beta) & b \\ c & d - (\alpha + i\beta) \end{bmatrix}\begin{bmatrix} A \\ B \end{bmatrix} = \begin{bmatrix} 0 \\ 0 \end{bmatrix},$$

and its complex conjugate. Notice that both of these equations have the same structure, namely,

$$\begin{bmatrix} a - r & b \\ c & d - r \end{bmatrix}\begin{bmatrix} A \\ B \end{bmatrix} = \begin{bmatrix} 0 \\ 0 \end{bmatrix},$$

where A and B are complex numbers. Let's see how this works in practice.

EXAMPLE 9.15

Solve the system

$$\frac{d}{dt}\begin{bmatrix} x \\ y \end{bmatrix} = \begin{bmatrix} 2 & 1 \\ -1 & 2 \end{bmatrix}\begin{bmatrix} x \\ y \end{bmatrix}. \tag{9.48}$$

In this case,

$$\mathbf{M} = \begin{bmatrix} 2 & 1 \\ -1 & 2 \end{bmatrix},$$

and the characteristic equation is

$$\begin{vmatrix} 2-r & 1 \\ -1 & 2-r \end{vmatrix} = 0,$$

Eigenvalues with complex eigenvalues $r = 2 \pm i$. Here the solution will be

$$\mathbf{X}(t) = \begin{bmatrix} A \\ B \end{bmatrix} e^{(2+i)t} + \begin{bmatrix} \tilde{A} \\ \tilde{B} \end{bmatrix} e^{(2-i)t},$$

where the complex numbers A and B satisfy

$$\begin{bmatrix} 2-r & 1 \\ -1 & 2-r \end{bmatrix}\begin{bmatrix} A \\ B \end{bmatrix} = \begin{bmatrix} 0 \\ 0 \end{bmatrix}. \tag{9.49}$$

If we substitute $r = 2 + i$ into (9.49), we obtain

$$\begin{bmatrix} -i & 1 \\ -1 & -i \end{bmatrix}\begin{bmatrix} A \\ B \end{bmatrix} = \begin{bmatrix} 0 \\ 0 \end{bmatrix},$$

Eigenvector so A and B must be related by $A = -iB$. This gives an eigenvector as

$$\begin{bmatrix} A \\ B \end{bmatrix} = B\begin{bmatrix} -i \\ 1 \end{bmatrix},$$

and a solution of (9.48) as

$$\mathbf{X}(t) = B\begin{bmatrix} -i \\ 1 \end{bmatrix} e^{(2+i)t} + \tilde{B}\begin{bmatrix} i \\ 1 \end{bmatrix} e^{(2-i)t},$$

or

$$\mathbf{X}(t) = B\begin{bmatrix} -i \\ 1 \end{bmatrix} e^{2t}(\cos t + i \sin t) + \tilde{B}\begin{bmatrix} i \\ 1 \end{bmatrix} e^{2t}(\cos t - i \sin t).$$

This can be rewritten in the form

$$\mathbf{X}(t) = \begin{bmatrix} -i\left(B - \tilde{B}\right) \\ B + \tilde{B} \end{bmatrix} e^{2t}\cos t + \begin{bmatrix} B + \tilde{B} \\ i\left(B - \tilde{B}\right) \end{bmatrix} e^{2t}\sin t.$$

If we define $C_1 = -i(B - \tilde{B})$ and $C_2 = B + \tilde{B}$, both $C_1$and C_2 are real (Why?), and the solution is

Explicit solution

$$\mathbf{X}(t) = \begin{bmatrix} C_1 \\ C_2 \end{bmatrix} e^{2t} \cos t + \begin{bmatrix} C_2 \\ -C_1 \end{bmatrix} e^{2t} \sin t = \begin{bmatrix} e^{2t} \cos t & e^{2t} \sin t \\ -e^{2t} \sin t & e^{2t} \cos t \end{bmatrix} \begin{bmatrix} C_1 \\ C_2 \end{bmatrix},$$

Fundamental matrix

where a fundamental matrix is

$$\mathbf{U} = \begin{bmatrix} e^{2t} \cos t & e^{2t} \sin t \\ -e^{2t} \sin t & e^{2t} \cos t \end{bmatrix}.$$

In this example we do not seek straight-line orbits in the phase plane because they do not exist for systems of differential equations with complex eigenvalues. (See Exercise 8 on page 429.) The differential equation for orbits in the phase plane associated with (9.48) is

$$\frac{dy}{dx} = \frac{-x + 2y}{2x + y}.$$

The direction field for this phase plane is shown in Figure 9.28, where we see that there are no straight-line orbits. The direction of the arrows also indicates that the origin is an unstable focus because all solutions spiral away from the origin.

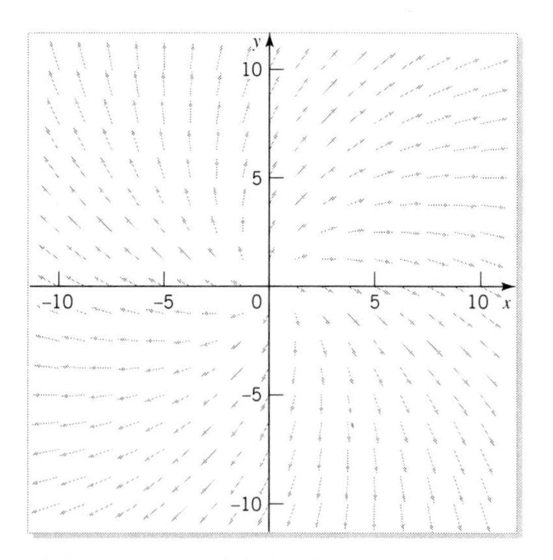

FIGURE 9.28 Direction field for $dy/dx = (-x + 2y)/(2x + y)$

Repeated Roots

Our first two examples in this section had distinct eigenvalues, the first real and the second complex. Now we examine the remaining possibility, in which the characteristic equation has a repeated root. Thus, the eigenvalue is real, but repeated. In this case, we consider

$$\mathbf{X}' = \begin{bmatrix} a & b \\ c & d \end{bmatrix}\mathbf{X},$$

when the characteristic equation

$$\begin{vmatrix} a-r & b \\ c & d-r \end{vmatrix} = 0$$

has a repeated root, r, and the solution is

$$\mathbf{X}(t) = \begin{bmatrix} x(t) \\ y(t) \end{bmatrix} = \left(\begin{bmatrix} A_1 \\ B_1 \end{bmatrix} + \begin{bmatrix} A_2 \\ B_2 \end{bmatrix} t \right) e^{rt}.$$

We demonstrate the technique in the following example.

EXAMPLE 9.16

Solve the system of differential equations

$$\frac{d}{dt}\begin{bmatrix} x \\ y \end{bmatrix} = \begin{bmatrix} 1 & 1 \\ -1 & 3 \end{bmatrix}\begin{bmatrix} x \\ y \end{bmatrix}. \tag{9.50}$$

Here the characteristic equation is $(1-r)(3-r)+1=(r-2)^2=0$, so we seek a solution of the form

$$\mathbf{X}(t) = \left(\begin{bmatrix} A_1 \\ B_1 \end{bmatrix} + \begin{bmatrix} A_2 \\ B_2 \end{bmatrix} t \right) e^{2t}.$$

Substituting this expression into (9.50) gives

$$\begin{bmatrix} 2A_1 \\ 2B_1 \end{bmatrix} e^{2t} + \begin{bmatrix} (2t+1)A_2 \\ (2t+1)B_2 \end{bmatrix} e^{2t} = \begin{bmatrix} A_1 + B_1 + t(A_2+B_2) \\ -A_1 + 3B_1 + t(-A_2 + 3B_2) \end{bmatrix} e^{2t}.$$

Because we want this equation to be valid for all values of t, we equate coefficients of e^{2t} and te^{2t} and obtain the system of algebraic equations,

$$\begin{aligned} 2A_1 + A_2 &= A_1 + B_1, \\ 2B_1 + B_2 &= -A_1 + 3B_1, \\ 2A_2 &= A_2 + B_2, \\ 2B_2 &= -A_2 + 3B_2. \end{aligned}$$

This system of equations has the solution

$$\begin{aligned} A_1 &= B_1 - B_2, \\ A_2 &= \quad B_2, \end{aligned}$$

where B_1 and B_2 may be chosen arbitrarily, so we denote them by C_1 and C_2. (See Exercise 3 on page 428.) This means that the solution of (9.50) may be written in the form

$$\mathbf{X}(t) = \begin{bmatrix} C_1 - C_2 \\ C_1 \end{bmatrix} e^{2t} + \begin{bmatrix} C_2 \\ C_2 \end{bmatrix} te^{2t},$$

Explicit solution or

$$\mathbf{X}(t) = \begin{bmatrix} e^{2t} & (t-1)e^{2t} \\ e^{2t} & te^{2t} \end{bmatrix} \begin{bmatrix} C_1 \\ C_2 \end{bmatrix}.$$

Fundamental matrix Here a fundamental matrix is

$$\mathbf{U} = \begin{bmatrix} e^{2t} & (t-1)e^{2t} \\ e^{2t} & te^{2t} \end{bmatrix}.$$

The straight line $y = x$ is an orbit in the phase plane. (Why?) This orbit is apparent in Figure 9.29, which gives the direction field for the associated phase plane. (Is there another straight-line orbit? Should there be?)

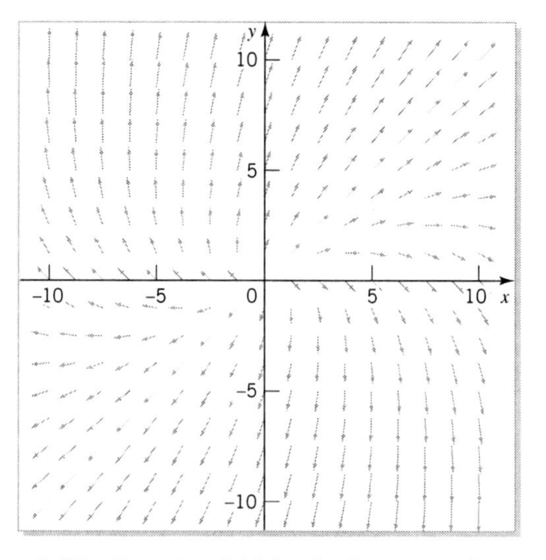

FIGURE 9.29 Direction field for $dy/dx = (-x + 3y)/(x + y)$

We now summarize the results of this section.

How to Solve $\mathbf{X}' = \mathbf{MX}$

Purpose To find the general solution of the system of differential equations $\mathbf{X}' = \mathbf{MX}$, where

$$\mathbf{M} = \begin{bmatrix} a & b \\ c & d \end{bmatrix} \text{ and } \mathbf{X} = \begin{bmatrix} x \\ y \end{bmatrix}.$$

Process

1. Write down

$$\begin{bmatrix} a-r & b \\ c & d-r \end{bmatrix} \begin{bmatrix} A \\ B \end{bmatrix} = \begin{bmatrix} 0 \\ 0 \end{bmatrix}. \tag{9.51}$$

2. Solve the characteristic equation

$$\begin{vmatrix} a-r & b \\ c & d-r \end{vmatrix} = 0$$

for r, obtaining the roots (eigenvalues) r_1 and r_2.

3. Find a fundamental matrix for this system of equations. The process in this step depends on the nature of the eigenvalues: whether they are distinct — either real or complex — or repeated.

(a) For **real, distinct eigenvalues**, r_1 and r_2, substitute one of the eigenvalues — say, r_1 — into (9.51) and solve the resulting algebraic equations for the associated eigenvector. Repeat this process for the other eigenvalue, r_2. The columns in a fundamental matrix are formed by these eigenvectors times e^{rt}, for the appropriate value of r.

(b) For **complex eigenvalues**, r_1 and r_2, substitute one of them — say, $r_1 = \alpha + i\beta$ — into (9.51) and solve the resulting algebraic equations. This gives a complex eigenvector, which when multiplied by $e^{(\alpha+i\beta)t}$ will result in a complex valued solution. The sum of this solution and its complex conjugate is the general solution.

(c) For a **repeated eigenvalue**, r, substitute

$$\mathbf{X}(t) = \begin{bmatrix} x(t) \\ y(t) \end{bmatrix} = \begin{bmatrix} A_1 \\ B_1 \end{bmatrix} e^{rt} + \begin{bmatrix} A_2 \\ B_2 \end{bmatrix} te^{rt} \tag{9.52}$$

into (9.51) and solve the resulting algebraic equations for A_1, B_1, A_2, and B_2. The result of this operation will contain two arbitrary constants. The columns of a fundamental matrix in this case are composed of the vector functions that multiply the two arbitrary constants.

4. For each of these cases the general solution of the system of differential equations is $\mathbf{X}(t) = \mathbf{UC}$, where $\mathbf{U}$ is a fundamental matrix and $\mathbf{C}$ is a vector of arbitrary constants,

$$\mathbf{C} = \begin{bmatrix} C_1 \\ C_2 \end{bmatrix}.$$

5. For initial value problems — say, $x(0) = x_0$, $y(0) = y_0$ — evaluate the arbitrary constants by solving the system

$$\mathbf{U}(0)\mathbf{C} = \begin{bmatrix} x_0 \\ y_0 \end{bmatrix}$$

for C_1 and C_2.

Comments about Solving X′ = MX

- The arbitrary constants that occur in the eigenvectors may be set equal to any convenient nonzero value. This is because in performing the matrix multiplication **UC**, the columns of **U** — containing the eigenvectors — are multiplied by the arbitrary constants contained in **C**.
- The order in which the vectors are selected to construct the matrix **U** is not important.
- For complex eigenvalues, a fundamental matrix is formed after adding a complex valued function and its complex conjugate. When done correctly, all terms involving i cancel.
- An alternative method for obtaining a fundamental matrix for complex eigenvalues is to use the real and imaginary parts of the complex valued solution as its columns.
- The eigenvectors for distinct roots give the direction of the straight-line orbits in the phase plane.
- There are no straight-line orbits in the phase plane for systems of equations with complex eigenvalues.
- The vector multiplying e^{rt} in (9.52) will be the eigenvector associated with the repeated eigenvalue r. It will give the slope of the only straight-line orbit in the phase plane determined by

$$\frac{dy}{dx} = \frac{cx + dy}{ax + by}.$$

- The reason we use the expression **a** fundamental matrix rather than **the** fundamental matrix is because fundamental matrices are not unique. They may contain arbitrary constants that can be set equal to any convenient nonzero value.
- Eigenvalues and eigenvectors arise in matrix theory in a manner that is completely independent of differential equations. If we are given a matrix **M**, the system of algebraic equations in (9.51) may be obtained without reference to differential equations. In matrix notation, it is simply finding values of r and **X** for which $\mathbf{MX} = r\mathbf{X}$ or $(\mathbf{M} - r\,\mathbf{I})\,\mathbf{X} = 0$, where **I** is the identity matrix

$$\mathbf{I} = \begin{bmatrix} 1 & 0 \\ 0 & 1 \end{bmatrix}.$$

In this setting, the characteristic equation is called the characteristic polynomial.

We conclude this section by revisiting the first example in this chapter to illustrate the use of this technique.

EXAMPLE 9.17 *Two-Container Mixture Problem Revisited*

Pure water is entering container A at a rate of 3 gallons per minute while the well-stirred mixture is leaving container B at the same rate. There are 100 gallons in each container, and the well-stirred mixture flows from container A to container B at a rate of 4 gallons per minute and leaks back from container B to container A at a rate of 1 gallon per minute. Predict the amount of salt in each container at any time.

We let $x(t)$ be the amount of salt in container A at time t, and $y(t)$ be the amount of salt in container B at time t. Both x and y have units of pounds, and t has units of minutes. As we have seen, the differential equations governing this system is (9.1), namely,

$$\begin{cases} x' = -x/25 + y/100, \\ y' = x/25 - y/25. \end{cases} \tag{9.53}$$

To find an explicit solution for this system of equations, we recast them in matrix form as $\mathbf{X}' = \mathbf{MX}$, where

$$\mathbf{X} = \begin{bmatrix} x \\ y \end{bmatrix} \text{ and } \mathbf{M} = \begin{bmatrix} -1/25 & 1/100 \\ 1/25 & -1/25 \end{bmatrix}.$$

Characteristic equation The characteristic equation is

$$\begin{vmatrix} -1/25 - r & 1/100 \\ 1/25 & -1/25 - r \end{vmatrix} = \left(-\frac{1}{25} - r\right)\left(-\frac{1}{25} - r\right) - \frac{1}{2500} = 0,$$

Eigenvalues which gives the eigenvalues as $-1/50$ and $-3/50$. These eigenvalues are real and distinct, so we seek a solution of the form

$$\mathbf{X}(t) = \begin{bmatrix} x(t) \\ y(t) \end{bmatrix} = \begin{bmatrix} A_1 \\ B_1 \end{bmatrix} e^{-t/50} + \begin{bmatrix} A_2 \\ B_2 \end{bmatrix} e^{-3t/50},$$

where

$$\begin{bmatrix} -1/25 - (-1/50) & 1/100 \\ 1/25 & -1/25 - (-1/50) \end{bmatrix}\begin{bmatrix} A_1 \\ B_1 \end{bmatrix} = \begin{bmatrix} -1/50 & 1/100 \\ 1/25 & -1/50 \end{bmatrix}\begin{bmatrix} A_1 \\ B_1 \end{bmatrix} = \begin{bmatrix} 0 \\ 0 \end{bmatrix},$$

$$\begin{bmatrix} -1/25 - (-3/50) & 1/100 \\ 1/25 & -1/25 - (-3/50) \end{bmatrix}\begin{bmatrix} A_2 \\ B_2 \end{bmatrix} = \begin{bmatrix} 1/50 & 1/100 \\ 1/25 & 1/50 \end{bmatrix}\begin{bmatrix} A_2 \\ B_2 \end{bmatrix} = \begin{bmatrix} 0 \\ 0 \end{bmatrix}.$$

Fundamental matrix Thus, $B_1 = 2A_1$ and $B_2 = -2A_2$, so a fundamental matrix is

$$\mathbf{U} = \begin{bmatrix} A_1 e^{-t/50} & A_2 e^{-3t/50} \\ 2A_1 e^{-t/50} & -2A_2 e^{-3t/50} \end{bmatrix},$$

or, by choosing $A_1 = A_2 = 1$,

$$\mathbf{U} = \begin{bmatrix} e^{-t/50} & e^{-3t/50} \\ 2e^{-t/50} & -2e^{-3t/50} \end{bmatrix}.$$

The general solution of (9.53) is then

$$\mathbf{X}(t) = \begin{bmatrix} x(t) \\ y(t) \end{bmatrix} = \mathbf{UC} = \begin{bmatrix} e^{-t/50} & e^{-3t/50} \\ 2e^{-t/50} & -2e^{-3t/50} \end{bmatrix} \begin{bmatrix} C_1 \\ C_2 \end{bmatrix}, \tag{9.54}$$

which agrees with Example 9.1 on page 380.

EXERCISES

1. The derivative of a vector function $\mathbf{V} = \begin{bmatrix} x(t) \\ y(t) \end{bmatrix}$ is defined as $\mathbf{V}' = \begin{bmatrix} x'(t) \\ y'(t) \end{bmatrix}$.

   (a) If $\mathbf{V} = \begin{bmatrix} A \\ B \end{bmatrix} f(t)$, where A and B are constants, show that $\mathbf{V}' = \begin{bmatrix} A \\ B \end{bmatrix} f'(t)$.

   (b) Show that the usual rules for differentiation apply to vector functions, namely,

$$(\mathbf{V}_1 + \mathbf{V}_2)' = \mathbf{V}_1' + \mathbf{V}_2',$$

$$(z(t)\mathbf{V})' = z'(t)\mathbf{V} + z(t)\mathbf{V}'.$$

2. Show that if $\mathbf{X} = \mathbf{V} + i\mathbf{W}$ is a solution of $\mathbf{X}' = \mathbf{MX}$, where $\mathbf{M}$ is a real 2 by 2 matrix and $\mathbf{V}$ and $\mathbf{W}$ are real vectors, then $\mathbf{V}$ and $\mathbf{W}$ are also solutions.

3. Consider the situation where the eigenvalue of

$$\mathbf{X}' = \mathbf{MX}, \qquad \mathbf{M} = \begin{bmatrix} a & b \\ c & d \end{bmatrix}, \qquad \mathbf{X} = \begin{bmatrix} x \\ y \end{bmatrix} \tag{9.55}$$

   is repeated, so $r = (a + d)/2$.

   (a) Write down the algebraic equation satisfied by the components of the associated eigenvector

$$\mathbf{Z}_2 = \begin{bmatrix} A_2 \\ B_2 \end{bmatrix}.$$

   (b) Show that trying a solution of (9.55) in the form

$$\mathbf{X}(t) = \begin{bmatrix} A_1 \\ B_1 \end{bmatrix} e^{rt} + \mathbf{Z}_2 t e^{rt}$$

   results in the same equations you found for A_2 and B_2 in part (a), and that A_1 and B_1 satisfy

$$\begin{bmatrix} a - r & b \\ c & d - r \end{bmatrix} \begin{bmatrix} A_1 \\ B_1 \end{bmatrix} = \begin{bmatrix} A_2 \\ B_2 \end{bmatrix}.$$

4. Show that the system of equations

$$\mathbf{X}' = \mathbf{MX} = \begin{bmatrix} a & 0 \\ 0 & a \end{bmatrix} \mathbf{X}$$

   has repeated eigenvalues. Show that

$$\begin{bmatrix} C_1 \\ 0 \end{bmatrix} \text{ and } \begin{bmatrix} 0 \\ C_2 \end{bmatrix}$$

   are eigenvectors for this case, and that a fundamental matrix may have the form

$$\mathbf{U} = \begin{bmatrix} e^{at} & 0 \\ 0 & e^{at} \end{bmatrix}.$$

5. Solve the following systems of differential equations by finding a fundamental matrix. For real eigenvalues find straight-line orbits in the phase plane and verify their agreement with the eigenvectors you found. For

complex eigenvalues show that there are no straight-line orbits in the phase plane. Verify your answers by considering the direction field in the phase plane.

(a) $\frac{d}{dt}\begin{bmatrix} x \\ y \end{bmatrix} = \begin{bmatrix} 2 & -1 \\ -6 & 1 \end{bmatrix}\begin{bmatrix} x \\ y \end{bmatrix}$

(b) $\frac{d}{dt}\begin{bmatrix} x \\ y \end{bmatrix} = \begin{bmatrix} 2 & 5 \\ 1 & 6 \end{bmatrix}\begin{bmatrix} x \\ y \end{bmatrix}$

(c) $\frac{d}{dt}\begin{bmatrix} x \\ y \end{bmatrix} = \begin{bmatrix} 5 & -3 \\ -1 & -1 \end{bmatrix}\begin{bmatrix} x \\ y \end{bmatrix}$

(d) $\frac{d}{dt}\begin{bmatrix} x \\ y \end{bmatrix} = \begin{bmatrix} 5 & 6 \\ 1 & 4 \end{bmatrix}\begin{bmatrix} x \\ y \end{bmatrix}$

(e) $\frac{d}{dt}\begin{bmatrix} x \\ y \end{bmatrix} = \begin{bmatrix} 2 & 9 \\ -1 & -4 \end{bmatrix}\begin{bmatrix} x \\ y \end{bmatrix}$

(f) $\frac{d}{dt}\begin{bmatrix} x \\ y \end{bmatrix} = \begin{bmatrix} 2 & 1 \\ 1 & -4 \end{bmatrix}\begin{bmatrix} x \\ y \end{bmatrix}$

(g) $\frac{d}{dt}\begin{bmatrix} x \\ y \end{bmatrix} = \begin{bmatrix} 2 & 1 \\ -4 & 2 \end{bmatrix}\begin{bmatrix} x \\ y \end{bmatrix}$

(h) $\frac{d}{dt}\begin{bmatrix} x \\ y \end{bmatrix} = \begin{bmatrix} 4 & -2 \\ 5 & -2 \end{bmatrix}\begin{bmatrix} x \\ y \end{bmatrix}$

(i) $\frac{d}{dt}\begin{bmatrix} x \\ y \end{bmatrix} = \begin{bmatrix} 3 & 5 \\ -5 & 3 \end{bmatrix}\begin{bmatrix} x \\ y \end{bmatrix}$

(j) $\frac{d}{dt}\begin{bmatrix} x \\ y \end{bmatrix} = \begin{bmatrix} 1 & 1 \\ -1 & 3 \end{bmatrix}\begin{bmatrix} x \\ y \end{bmatrix}$

6. In Example 9.15 on page 421 we found that the solution of our system of equations was given by $\mathbf{X}(t) = \mathbf{UC}$, where

$$\mathbf{U} = \begin{bmatrix} e^{2t}\cos t & e^{2t}\sin t \\ -e^{2t}\sin t & e^{2t}\cos t \end{bmatrix}. \tag{9.56}$$

(a) Evaluate the constant vector $\mathbf{C}$ so this solution satisfies the initial conditions $x(0) = x_0$, $y(0) = y_0$.

(b) Now choose a different fundamental matrix — perhaps just interchange the columns of $\mathbf{U}$ in (9.56) — and evaluate the constant vector $\mathbf{C}$ by having $\mathbf{X}(t)$ satisfy the initial conditions. Show that your resulting solution is the same as the one you obtained in part (a).

7. Consider the two containers of Figure 9.1, each of which holds 100 gallons of liquid.

(a) Initially, container A has 10 pounds of salt and container B has none. How long after all the valves are opened will the number of pounds of salt in the two containers be equal?

(b) Determine the amount of salt in each container after 25 minutes.

(c) Determine the amount of salt in each container as time approaches infinity. Could you determine this result by knowing the eigenvalues? Explain fully.

8. Consider the system of equations

$$\mathbf{X}' = \mathbf{MX}, \qquad \mathbf{M} = \begin{bmatrix} a & b \\ c & d \end{bmatrix}, \qquad \mathbf{X} = \begin{bmatrix} x \\ y \end{bmatrix}. \tag{9.57}$$

(a) Find the equation for the eigenvalues for solutions of the form

$$\mathbf{X}(t) = \begin{bmatrix} A \\ B \end{bmatrix} e^{rt}.$$

(b) For the case of a real, repeated eigenvalue, $(a + d)^2 = 4bc$, find the associated eigenvector.

(c) Find the equation for straight-line orbits in the phase plane associated with (9.57), and show that they are consistent with the direction given by the eigenvector in part (b).

(d) Repeat parts (b) and (c) for real, distinct eigenvalues.

(e) Show that there are no straight-line orbits in the phase plane associated with (9.57) when the eigenvalues are complex numbers.

9. Three mystery direction fields of phase planes associated with the system of differential equations $x' = ax + by, y' = cx + dy$, are given in Figures 9.30, 9.31, and 9.32. By looking for straight-line orbits, decide which of these direction fields has the following properties. Explain the reasons behind each choice, and, wherever possible, estimate the components of the eigenvalues.

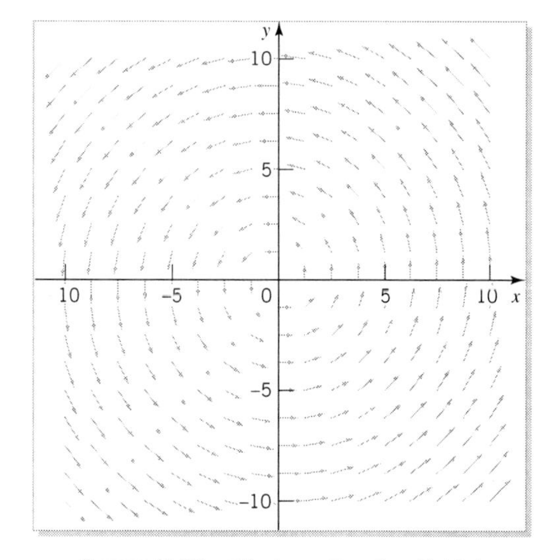

FIGURE 9.30 Mystery direction field A

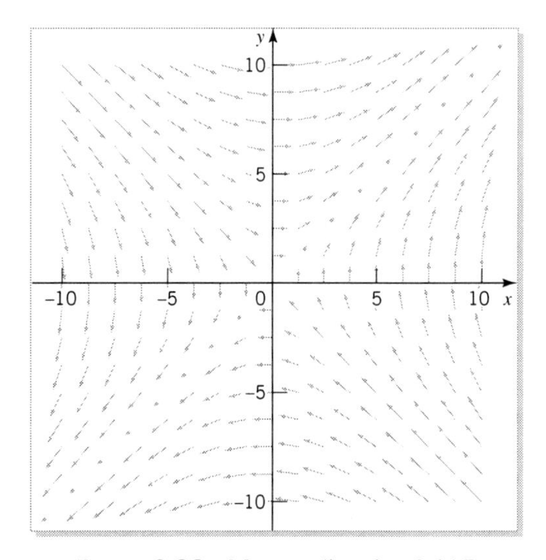

FIGURE 9.31 Mystery direction field B

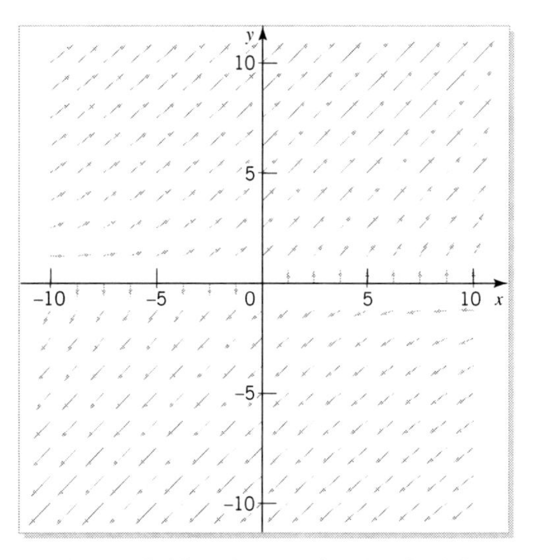

FIGURE 9.32 Mystery direction field C

(a) Real, distinct eigenvalues.

(b) A real, repeated eigenvalue.

(c) Complex eigenvalues.

10. Suppose there are two countries with populations growing exponentially with different growth rates, and a specific fraction of each population emigrates to the other country each year. Let S represent the population of country A that is growing at a per capita rate of a, and F represent the population of country B that is growing at a per capita rate of b. Suppose that the emigration from country A to country B takes place with a per capita rate of c (so the number of people leaving is cS per year), and the rate from country B to country A is k.

(a) Provide the reasoning as to why this situation could be modeled by the system of differential equations

$$\begin{cases} S' = aS + kF - cS, \\ F' = bF + cS - kF. \end{cases} \tag{9.58}$$

(b) Find all the equilibrium solutions and discuss the conditions under which they can occur. Are any of these situations realistic?

(c) Under what conditions will the populations of both countries grow for all time?

(d) For the conditions of part (c), is there a limiting ratio of the two populations as $t \to \infty$? [Hint: Try a solution of the form $F = mS$ of the differential equation in the phase plane.]

(e) For the situation where $a = 0.1$, $b = 0.2$, $k = c = 0.01$, find the equation for the orbits in the phase plane (note that you must solve a differential equation with homogeneous coefficients).

(f) Find the characteristic equation and eigenvalues associated with an exponential solution of this system of differential equations.

(g) Find the solution of the initial value problem of (9.58), with values of the constants from part (e), subject to $S(0) = 16$ and $F(0) = 8$. Will the population of country B ever surpass that of country A?

9.6 COMPARTMENTAL MODELS

Compartmental models are used to describe the dynamics of many processes in the body. The compartments in models that simulate a process are not chosen randomly, but take into account anatomical locations in the body and the function of various organs. In each compartment a differential equation is derived using conservation arguments — equating the net change of a substance to the difference between what is entering the compartment and what is leaving the compartment. A schematic drawing of a two-compartment model for such a system is shown in Figure 9.33.

One use of a compartmental model is to examine the movement of an injected substance, such as an antibiotic, or a tracer, such as inulin, within the body (often called tracer kinetics). Experiments have shown that their rate of elimination may be modeled by a system of linear differential equations. Usually the purpose of such a model is to determine body fluid sizes and rates of clearance by the kidneys.

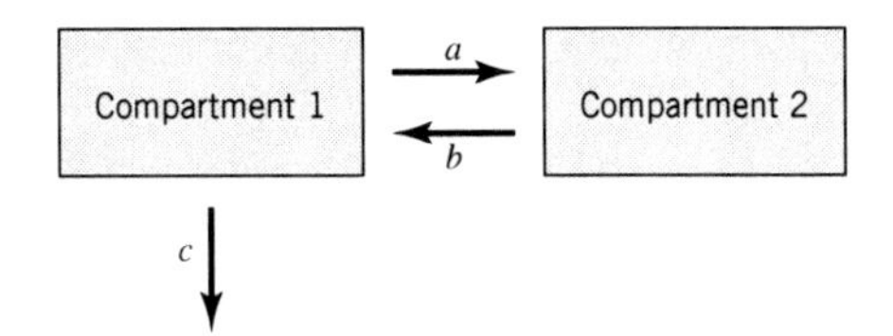

FIGURE 9.33 A two-compartment model

EXAMPLE 9.18 *Movement of Inulin*

Here we consider the movement of inulin, which is administered intravenously into the bloodstream of a rabbit, where the two compartments are the blood plasma (compartment 1) and the extracellular fluid (compartment 2). We let $x(t)$ be the concentration of inulin in the plasma and $y(t)$ be the concentration of inulin in the extracellular fluid, and use the differential equations

$$\begin{cases} x' = -ax + by - cx, \\ y' = ax - by. \end{cases}$$

In this model, the inulin is transported back and forth between the plasma and the extracellular fluid at constant rates, a and b, with the term $-cx$ representing the excretion of the inulin via the urinary tract. Initial conditions appropriate to giving an intravenous injection are $x(0) = x_0$, $y(0) = 0$.

We now find the solution of this initial value problem for the experimentally determined values of the rate constants $a = b = 11$ and $c = 12$. This gives the system of differential equations as

$$\mathbf{X}' = \mathbf{MX}, \text{ where } \mathbf{M} = \begin{bmatrix} -23 & 11 \\ 11 & -11 \end{bmatrix}. \tag{9.59}$$

To find the general solution of (9.59), we use

$$\begin{vmatrix} -23 - r & 11 \\ 11 & -11 - r \end{vmatrix} = 0$$

Eigenvalues

and find eigenvalues of -29.5 and -4.5. Thus, the solution of (9.59) will be

$$\mathbf{X}(t) = \begin{bmatrix} x(t) \\ y(t) \end{bmatrix} = \begin{bmatrix} A_1 \\ B_1 \end{bmatrix} e^{-29.5t} + \begin{bmatrix} A_2 \\ B_2 \end{bmatrix} e^{-4.5t}.$$

Using the eigenvalues in

$$\begin{bmatrix} -23-r & 11 \\ 11 & -11-r \end{bmatrix}\begin{bmatrix} A \\ B \end{bmatrix} = \begin{bmatrix} 0 \\ 0 \end{bmatrix},$$

Eigenvectors

in turn, gives eigenvectors of

$$\begin{bmatrix} -11 \\ 6.5 \end{bmatrix} \text{ and } \begin{bmatrix} 6.5 \\ 11 \end{bmatrix},$$

respectively. Thus, our general solution of (9.59) is $\mathbf{X}(t) = \mathbf{UC}$, where

$$\mathbf{U} = \begin{bmatrix} -11e^{-29.5t} & 6.5e^{-4.5t} \\ 6.5e^{-29.5t} & 11e^{-4.5t} \end{bmatrix}$$

and $\mathbf{C}$ is a vector of arbitrary constants.

If we now use our initial conditions to evaluate these arbitrary constants, we have that

$$\begin{bmatrix} x_0 \\ 0 \end{bmatrix} = \begin{bmatrix} -11 & 6.5 \\ 6.5 & 11 \end{bmatrix}\begin{bmatrix} C_1 \\ C_2 \end{bmatrix}.$$

Explicit solution

Solving these equations gives $C_1 = -0.068x_0$, $C_2 = 0.040x_0$, and the solution of our initial value problem is

$$\mathbf{X}(t) = \begin{bmatrix} x(t) \\ y(t) \end{bmatrix} = \begin{bmatrix} \left(0.74e^{-29.5t} + 0.26e^{-4.5t}\right)x_0 \\ \left(-0.44e^{-29.5t} + 0.44e^{-4.5t}\right)x_0 \end{bmatrix}.$$

Figure 9.34 shows the concentrations in the two compartments as a function of time for $x_0 = 1$. The curve that starts at the origin and increases to a maximum corresponds to compartment 2 (the y value). The fact that both functions decay to zero for any initial condition means that the equilibrium point at the origin is a stable node. (Verify this from the fundamental matrix.)

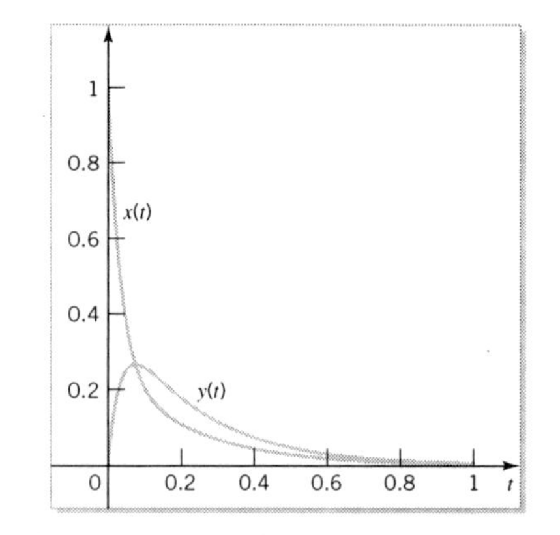

FIGURE 9.34 Concentrations of inulin for a two-compartment model

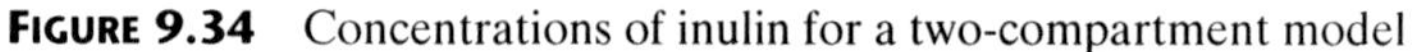

EXAMPLE 9.19 *The Compartmental Model Applied to Calves*[11]

Table 9.1 and Figure 9.35 shows the results of an experiment where the decay of the concentration of a dye in plasma in a calf (in mg/ml $\times 10^2$) was measured as a function of time (in minutes).

This can be modeled by the two-compartment model described in Example 9.18, namely,

$$\begin{cases} x' = -ax + by - cx, \\ y' = ax - by, \end{cases}$$

where a, b, and c are positive constants. Here $x(t)$ is the concentration of the dye in the plasma and $y(t)$ is the concentration of the dye in the hepatocytes. The solution for $x(t)$ will have the general form $x(t) = C_1 e^{-r_1 t} + C_2 e^{-r_2 t}$, where r_1 and r_2 are positive.

Table 9.1 The decay of the concentration of a dye in plasma in a calf as a function of time

Time (min)	Concentration (mg/ml $\times 10^2$)	Time (min)	Concentration (mg/ml $\times 10^2$)
1	10.20	10	1.26
2	7.67	13	0.77
3	5.76	16	0.52
4	4.50	19	0.39
5	3.56	22	0.28
6	2.77	25	0.22
7	2.30	28	0.17
8	1.84	31	0.14
9	1.46		

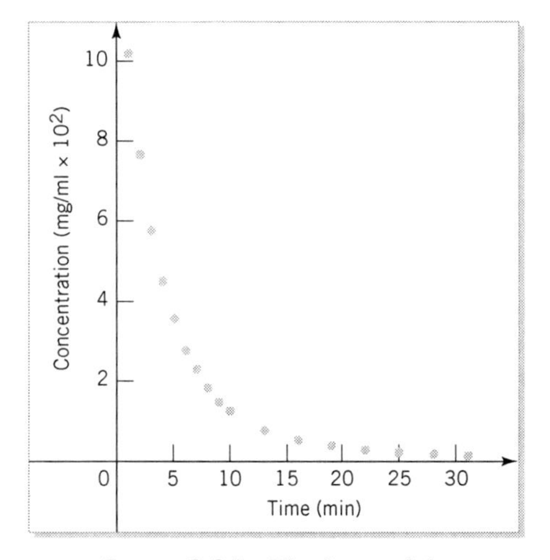

FIGURE 9.35 The decay of the concentration of a dye in plasma in a calf as a function of time

If r_1 is very much larger than r_2, then when t is small the decay will behave like $C_1 e^{-r_1 t}$, whereas when t is large it will behave like $C_2 e^{-r_2 t}$. [To convince yourself of this, consider Figure 9.36, which shows the three functions $x(t) = C_1 e^{-r_1 t} + C_2 e^{-r_2 t}$, $x_1(t) = C_1 e^{-r_1 t}$, and $x_2(t) = C_2 e^{-r_2 t}$. Notice that the shapes of $C_1 e^{-r_1 t} + C_2 e^{-r_2 t}$ and $C_1 e^{-r_1 t}$ are similar for small t, while the shapes of $C_1 e^{-r_1 t} + C_2 e^{-r_2 t}$ and $C_2 e^{-r_2 t}$, are similar for large t.] Thus, if we plot $\ln x$ versus t for the data, when t is small we expect to see a straight line with slope $-r_1$, whereas when t is large we expect to see a straight line with slope $-r_2$. Figure 9.37 shows such a plot together with the two

[11] "The Measurement of Liver Blood Flow in Conscious Calves" by F.A. Harrison, F. Hills, J.Y.F. Patterson, and R.C. Saunders, *Quarterly Journal of Experimental Physiology* 74, 1988, pages 235–247, Figure 1, as reported in *Models in Biology* by D. Brown and P. Rothery, Wiley, 1993, page 365.

straight lines, $-0.3\,(t-1)+\ln 10.2$ and $-0.07(t-1)+\ln 0.14$. Thus, we estimate $r_1 \approx 0.3$ and $r_2 \approx 0.07$. From this we can estimate that

$$x(t) = 8.8\exp\left[-0.3\,(t-1)\right] + 1.2\exp\left[-0.07\,(t-1)\right].$$

Figure 9.38 shows this function and the data set.

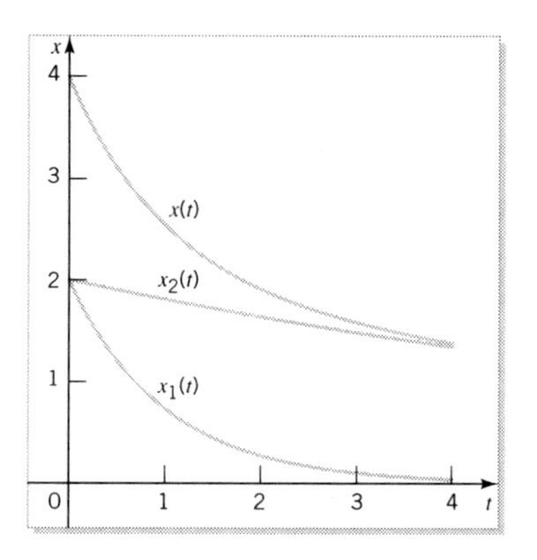

FIGURE 9.36 The functions $x(t) = 2e^{-r_1 t} + 2e^{-r_2 t}$, $x_1(t) = 2e^{-r_1 t}$, and $x_2(t) = 2e^{-r_2 t}$

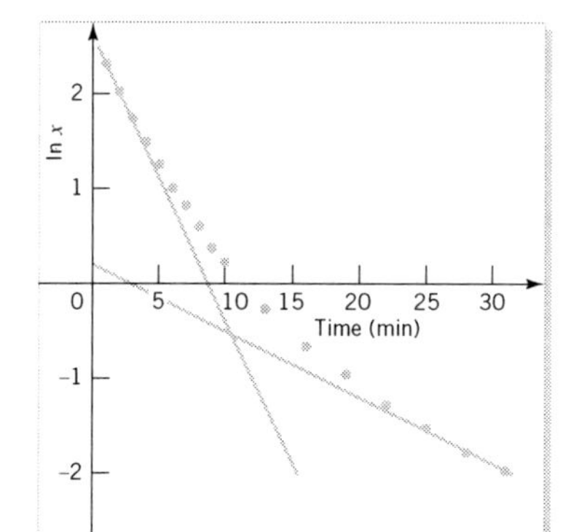

FIGURE 9.37 The ln of the concentration as a function of time and the lines $-0.3\,(t-1)+\ln 10.2$ and $-0.07(t-1)+\ln 0.14$

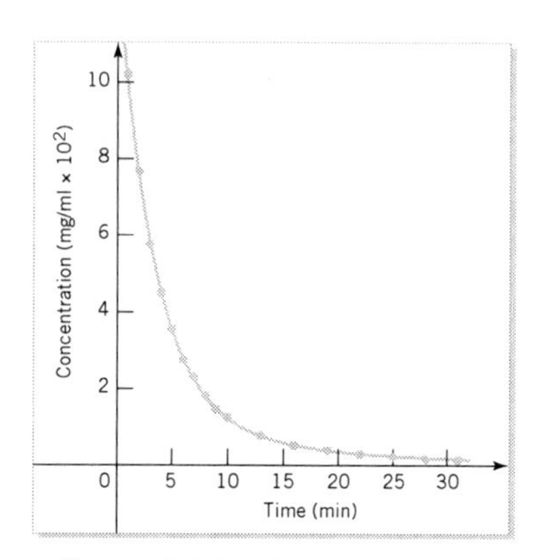

FIGURE 9.38 The decay of the concentration of a dye in plasma in a calf as a function of time and $x(t) = 8.8\exp\left[-0.3\,(t-1)\right] +$ $+1.2\exp\left[-0.07\,(t-1)\right]$

EXERCISES

1. **Kidney Machine.**[12] Patients with severe kidney disease sometimes use a dialyzer, which removes wastes from their blood. A simple model for a dialyzer has the patient's blood flowing along a membrane at a fixed rate while a purifying liquid is simultaneously flowing in the opposite direction on the other side of this membrane at a different rate. This purifying liquid has an affinity for the impurities in the blood, and the rate of change of the impurities across the membrane is proportional to the difference in concentrations. If we let x represent the concentration of impurity in the blood and y represent the concentration of impurity in the dialyzer liquid at a distance along the membrane, given by t, we have the following system of differential equations, $x' = a(y-x)/v$, $y' = a(x-y)/V$. Here a is a positive constant, and v and V are the volume flow rates of the blood and dialyzer liquid, respectively. The constant a measures the effectiveness of the dialyzer liquid: larger a values correspond to a more rapid removal of impurities.

   (a) Observe that the direction field in the phase plane is independent of a but depends strongly on the ratio v/V. Comment on the effect on the direction field of changing this ratio.

   (b) The usual initial condition is that at the beginning of the membrane, where the distance t is 0, the blood is saturated with impurities. At the other end of the membrane the dialyzer liquid is pure. This gives $x(0) = x_0$, and $y(L) = 0$, where L is the length of the membrane. Find the equation of the orbit in the phase plane for these initial conditions. Which direction along this orbit will the solution take as the distance increases?

   (c) Find the explicit solution of this system of differential equations, subject to the initial conditions given in part (b).

[12] Based on "Mathematical Model of a Kidney Machine" by D.M. Burley, *Mathematical Spectrum* 8, 1975/76, pages 69–75.

What Have We Learned?

Main Ideas

- We may always find explicit solutions for second order autonomous systems of linear differential systems of the form

$$\begin{cases} x' = ax + by, \\ y' = cx + dy, \end{cases} \tag{9.60}$$

where a, b, c, and d are constants. These solutions are given in terms of roots of the characteristic equation, obtained by expanding the determinant

$$\begin{vmatrix} a-r & b \\ c & d-r \end{vmatrix} = r^2 - (a+d)r + ad - bc = 0.$$

See *How to Find the General Solution of* $x' = ax + by, y' = cx + dy$ on page 385.

- The stability of equilibrium solutions of linear autonomous systems may be determined from the explicit solution or from the behavior of the roots of the characteristic equation (see the Classification Theorem on page 398). A graphical interpretation of stability is shown by the Parabolic Classification Scheme on page 401. Terms used to describe the various types of solution behavior include **node**, **center**, **focus**, and **saddle point**. Examples are given in Section 9.2 of stable and unstable nodes as well as stable and unstable foci. Saddle points are always unstable, and centers are always stable. Straight-line orbits in the phase plane for linear systems occur when the roots of the characteristic equation are real (see the Straight Line Theorem on page 405).

- Nullclines are isoclines in the phase plane along which orbits have either vertical or horizontal tangents. They provide an alternative approach to determine the stability of equilibrium points of a linear autonomous system of first order differential equations. See *How to Perform a Nullcline Analysis on a Linear Autonomous System* on page 410.

- Solutions of (9.60), written in matrix form as $\mathbf{X}' = \mathbf{MX}$, may also be expressed in terms a fundamental matrix, $\mathbf{U}$, as $\mathbf{X} = \mathbf{UC}$, where

$$\mathbf{C} = \begin{bmatrix} C_1 \\ C_2 \end{bmatrix}$$

is a vector of arbitrary constants. See *How to Solve* $\mathbf{X}' = \mathbf{MX}$ on page 425.

- The eigenvalues of the matrix $\mathbf{M}$ are solutions of the characteristic equation

$$\begin{vmatrix} a-r & b \\ c & d-r \end{vmatrix} = 0.$$

- The eigenvectors of the matrix $\mathbf{M}$ are vectors associated with the eigenvalues that satisfy the homogeneous algebraic equation $\mathbf{MX} = r\mathbf{X}$.

- For the case of real eigenvalues, the eigenvectors give the directions of the straight-line solutions in the phase plane.

- It is possible to extend the preceding analysis to systems of equations of the form

$$\begin{cases} x' = ax + by + e, \\ y' = cx + dy + f, \end{cases}$$

where e and f are constants and $ad - bc \neq 0$. The major difference between this case and the case when $e = f = 0$ is that the equilibrium point is no longer $x(t) = 0$, $y(t) = 0$. The equilibrium point will now be $x(t) = x_0$, $y(t) = y_0$, where x_0 and y_0 are constants that satisfy the two equations $ax_0 + by_0 + e = 0$ and $cx_0 + dy_0 + f = 0$. These are two straight lines in the phase plane whose point of intersection will be (x_0, y_0). We can translate coordinates so that this equilibrium point is at the origin if we change variables from (x, y) to (u, v), where $u = x - x_0$ and $v = y - y_0$. This will convert the preceding system in terms of x and y into an equivalent linear system in terms of u and v.

CHAPTER 10

NONLINEAR AUTONOMOUS SYSTEMS

Where We Are Going — and Why

In Chapter 9 we developed methods for discovering the behavior of linear autonomous systems of differential equations. While not all methods may have worked on a specific system, we could always determine the nature of their equilibrium points. However, models of many phenomena, such as spring-mass systems, simple pendulums, epidemics, population growth, bungee jumping, and predator-prey systems, often involve nonlinear autonomous systems. These nonlinear systems are more difficult to analyze than linear systems. In this chapter we examine several examples of what may occur and develop a number of approaches that give us qualitative information about the behavior of solutions. We collect these together in a catalog of techniques to help construct a phase portrait for nonlinear systems.

To relate the material in this chapter to that from prior chapters, we discuss some differences between linear and nonlinear systems and between autonomous and nonautonomous systems.

10.1 INTRODUCTION TO NONLINEAR AUTONOMOUS SYSTEMS

We now turn to nonlinear systems of autonomous equations of the form

$$\begin{cases} x' = P(x, y), \\ y' = Q(x, y), \end{cases} \tag{10.1}$$

where $x' = dx/dt, y' = dy/dt$, and the functions $P(x, y)$ and $Q(x, y)$ contain no explicit dependence on t. We know from the Existence-Uniqueness Theorem of Chapter 6 that if $P(x, y)$ and $Q(x, y)$ are continuously differentiable in the vicinity of the initial point $x(t_0) = x_0, y(t_0) = y_0$, then (10.1) has a unique solution with these initial conditions. This means that in the phase plane, different orbits cannot intersect. However, the same orbit might intersect itself if it returns to the same initial point (x_0, y_0) at a later time.

To gain some feeling for (10.1), in this section we will look at four examples, each of which can be solved exactly in one form or another, and whose solutions exhibit different characteristics. Our object is to gain an intuitive understanding of some of the properties of systems of differential equations with particular reference to their long-term behavior.

EXAMPLE 10.1

Discuss the behavior of the solutions of the system

$$\begin{cases} x' = -x, \\ y' = y - y^2, \end{cases} \tag{10.2}$$

subject to $x(0) = x_0$, $y(0) = y_0$.

One thing we can do is find any constant solutions. These will be equilibrium points in the phase plane. They must satisfy $-x = 0$ and $y - y^2 = 0$, so the only equilibrium points are $(0, 0)$ and $(0, 1)$. Thus, one solution of (10.2) is $x(t) = 0$, $y(t) = 0$ [which corresponds to the initial condition $x(0) = 0, y(0) = 0$], and a second is $x(t) = 0$, $y(t) = 1$ [which corresponds to the initial condition $x(0) = 0, y(0) = 1$].

Equilibrium points

We now try to find other solutions of (10.2). These equations can be treated separately, because they are not coupled to each other. In fact, we can find their explicit solutions. The top equation in (10.2) is an exponential decay equation and has solution $x(t) = C_1e^{-t}$, where $C_1 = x_0$. The bottom is a logistic equation and has solutions $y(t) = \frac{1}{1+C_2e^{-t}}$, where $C_2 = (1 - y_0)/y_0)$, and $y(t) = 0$.

Explicit solution

To investigate the long-term behavior, we consider $t \to \infty$. As $t \to \infty$ we see that $x(t) \to 0$ and that either $y(t) \to 1$, in the case $y(t) = \frac{1}{1+C_2e^{-t}}$, or $y(t) = 0$, in the case $y(t) = 0$. However, as we observed when we first discussed the logistic equation in Chapter 2, a solution of the logistic equation may have a vertical asymptote. In particular, if $y_0 < 0$ then $C_2 < 0$, so $y(t) \to -\infty$ as $t \to \ln\left(-C_2\right)$ from the left. Thus, if $y_0 < 0$, then $y(t) \to -\infty$ in finite time.

Long-term behavior

Putting these together we see that as $t \to \infty$, $x(t) \to 0$, and if $y_0 \geq 0$, either $y(t) \to 1$ or $y(t) \to 0$. Also, as t increases, $y(t) \to -\infty$ in finite time for those orbits for which $y_0 < 0$.

Orbits

We now find the orbits in the phase plane by eliminating t from the solutions $x(t)$, $y(t)$. If $C_1 \neq 0$, we can eliminate t between the equations $x(t) = C_1e^{-t}$ and $y(t) = \frac{1}{1+C_2e^{-t}}$, by writing $x = C_1e^{-t}$ in the form $e^{-t} = x/C_1$ and substituting this in $y = \frac{1}{1+C_2e^{-t}}$, to find $y = \frac{1}{1+Cx}$, where $C = C_2/C_1$. The curves $y = \frac{1}{1+Cx}$ are orbits in the phase plane. If $C_1 = 0$, the curve $x = 0$ is an orbit. Finally, the remaining solution of the logistic equation, $y = 0$, is also an orbit.

We could sketch these curves in the phase plane, but we still have to decide on the direction of travel; that is, which way the arrows point in the direction field. One way to do this is to look at the nullclines of (10.2). The x-nullcline is $x = 0$ (the y-axis) and the y-nullclines are $y = 0$ and $y = 1$. Any orbit that touches the x-nullcline must do so vertically, while orbits that touch the lines $y = 0$ or $y = 1$ must do so horizontally. From (10.2) we see that the vertical arrows along the x-nullcline point up if $0 < y < 1$ (because $y' > 0$) and down if either $y < 0$ or $y > 1$. The horizontal arrows along the y-nullclines point left if $x > 0$ (because $x' < 0$) and right if $x < 0$.

Nullclines

Phase portrait

Thus, we have a good idea what the phase plane will look like, and in Figure 10.1 we have sketched a PHASE PORTRAIT — a collection of representative orbits — of (10.2).

Look at Figure 10.1 carefully and notice the following.

- Orbits that appear to intersect take an infinite time to reach $(0, 1)$.
- If $y > 0$, all orbits are attracted to the equilibrium point $(0, 1)$. Another way of saying this is that any orbit with initial condition (x_0, y_0), where $y_0 > 0$, is attracted to the equilibrium point $(0, 1)$. The equilibrium point $(0, 1)$ is stable, and near the equilibrium point the phase portrait looks like a node.
- If $y < 1$, all orbits are repelled from the equilibrium point $(0, 0)$ except for the two orbits, $y = 0$ $(x < 0)$ and $y = 0$ $(x > 0)$, which are attracted to it. The equilibrium point $(0, 0)$ is unstable, and near the equilibrium point the phase portrait looks like a saddle point.
- Previously we observed that as $t \to \infty$, we have $x(t) \to 0$, and either $y(t) \to 1$ or $y(t) = 0$. However, according to Figure 10.1, all orbits for which $y < 0$ appear to have the property that $y(t) \to -\infty$. These are the orbits that go to $-\infty$ in finite time.
- The curve $y = 0$ — which consists of three parts, the two orbits $y = 0$ $(x < 0)$ and $y = 0$ $(x > 0)$, and the equilibrium point $(0, 0)$ — separates the phase plane into two regions. If an initial point is above the curve $y = 0$, the orbit through this point behaves in one way [it is attracted to $(0, 1)$], whereas if an initial point is below the curve $y = 0$, the orbit through this point behaves in a completely different way [it is repelled from $(0, 0)$]. Thus, the curve $y = 0$ is a separatrix.

Separatrix

Basin of attraction

- Because all orbits that are in the region $y > 0$ are attracted to the equilibrium point $(0, 1)$, this region is called the BASIN OF ATTRACTION of the equilibrium point $(0, 1)$.
- All orbits either approach $-\infty$ with increasing t, or approach an equilibrium point as $t \to \infty$.

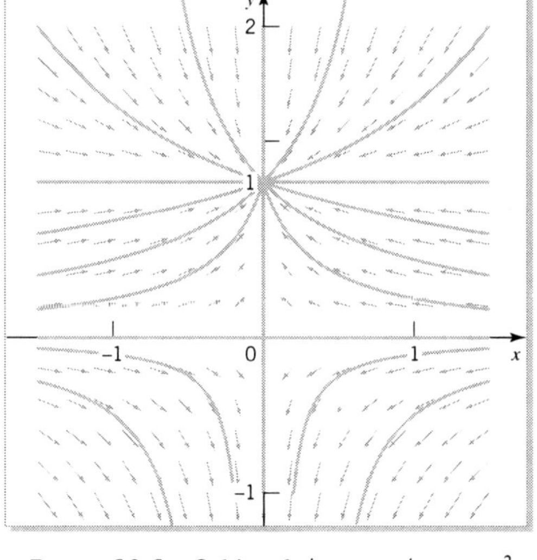

FIGURE 10.1 Orbits of $x' = -x$, $y' = y - y^2$

■

EXAMPLE 10.2 *Duffing's Equation*

Discuss the behavior of the solutions of Duffing's equation, given by the system

$$\begin{cases} x' = y, \\ y' = 2x - 2x^3. \end{cases} \tag{10.3}$$

Equilibrium points The equilibrium points must satisfy $y = 0$ and $2x - 2x^3 = 0$, so the only equilibrium points are $(-1, 0)$, $(0, 0)$, and $(1, 0)$. Thus, one solution of (10.3) is $x(t) = -1$, $y(t) = 0$ [which corresponds to the initial condition $x(0) = -1$, $y(0) = 0$], a second is $x(t) = 0$, $y(t) = 0$ [which corresponds to the initial condition $x(0) = 0$, $y(0) = 0$], and a third is $x(t) = 1$, $y(t) = 0$ [which corresponds to the initial condition $x(0) = 1$, $y(0) = 0$].

We now try the elimination method on this equation. Differentiating the top equation gives $x'' = y'$, which from the bottom equation can be written $x'' = 2x - 2x^3$. This is a second order equation for $x(t)$. In spite of its innocent appearance, ***Explicit solution*** this equation cannot be solved for $x(t)$ in terms of familiar functions. Thus, this approach fails. Nevertheless, we could solve this system using a numerical method. This requires that we supply an initial condition. Will the numerical solution we then obtain be representative of all solutions, or will different initial conditions give different behaviors?

To suggest where to look for initial conditions, we turn to a phase plane analysis. ***Orbits*** Orbits in the phase plane will satisfy

$$\frac{dy}{dx} = \frac{2x - 2x^3}{y}. \tag{10.4}$$

This is a separable differential equation with solution $y^2(x) = 2x^2 - x^4 + C$. This can be rewritten in the form $y^2 = -(x^2 - 1)^2 + C + 1$. For the right-hand side to be nonnegative, we must have $C + 1 \geq 0$. This suggests we introduce the constant $c \geq 0$ where $c^2 = C + 1$, so the orbits are $y^2 = c^2 - (x^2 - 1)^2$, or $y = \pm\sqrt{c^2 - (x^2 - 1)^2} = \pm\sqrt{(c - x^2 + 1)(c + x^2 - 1)}$. What do these orbits look like?

First, applying the techniques of calculus to (10.4), we see that if $y > 0$, then y will have local maxima at $x = \pm 1$, and local minima at $x = 0$. (Why?) Second, we note that the graph of the function $-\sqrt{(c - x^2 + 1)(c + x^2 - 1)}$ can be obtained from the graph of the function $+\sqrt{(c - x^2 + 1)(c + x^2 - 1)}$ by flipping the latter across the x-axis. (Why?) Third, for given $c > 0$, the quantity y will be defined for values of x satisfying $c^2 - (x^2 - 1)^2 \geq 0$. This means that y is defined for $-\sqrt{1 + c} \leq x \leq -\sqrt{1 - c}$ or $\sqrt{1 - c} \leq x \leq \sqrt{1 + c}$ if $0 < c < 1$, and for $-\sqrt{1 + c} \leq x \leq \sqrt{1 + c}$ if $c \geq 1$. (Why?) Thus, there are three cases to consider: $0 < c < 1$, $c = 1$, and $c > 1$.

1. We first look at the dividing case of $c = 1$, which means that $y = \pm\sqrt{x^2(2 - x^2)}$ and $-\sqrt{2} \leq x \leq \sqrt{2}$. We see that $y = 0$ in three places; namely, when $x = 0$ or $x = \pm\sqrt{2}$. The function $+\sqrt{x^2(2 - x^2)}$ is positive everywhere else with local maxima at $(\pm 1, 1)$. (Why?) Thus, the graph of this function looks like a cursive letter M. Because the function $-\sqrt{x^2(x^2 - 2)}$ can be obtained from the function $+\sqrt{x^2(2 - x^2)}$ by flipping the latter across the x-axis, the curve $y = \pm\sqrt{x^2(x^2 - 2)}$

looks like the number 8 on its side. Its maxima will occur at $(\pm 1, 1)$ and minima at $(\pm 1, -1)$.

2. If we look at a representative case for $c \geq 1$ — say, $c = 3$ — we find $y = \pm\sqrt{(4 - x^2)(2 + x^2)}$ and $-2 \leq x \leq 2$. In this case $y = 0$ at the end points $x = \pm 2$. Local maxima of $y = +\sqrt{(4 - x^2)(2 + x^2)}$ occur at $(\pm 1, 3)$, and local minima at $(0, \sqrt{8})$. Thus, the curve $\pm\sqrt{(4 - x^2)(2 + x^2)}$ looks like the number 0 with "dips" at the top and bottom, and "center" in the interval $-2 < x < 2$.

3. If we look at a representative case for $0 < c < 1$ — say, $c = \frac{3}{4}$ — we find $y = \pm\sqrt{\left(\frac{7}{4} - x^2\right)\left(x^2 - \frac{1}{4}\right)}$ and $-\sqrt{\frac{7}{4}} \leq x \leq -\frac{1}{2}$ or $\frac{1}{2} \leq x \leq \sqrt{\frac{7}{4}}$. In the first interval, $-\sqrt{\frac{7}{4}} \leq x \leq -\frac{1}{2}$, the quantity y is zero at the end points $x = -\sqrt{\frac{7}{4}}$ and $x = -\frac{1}{2}$, so the curve $y = \pm\sqrt{\left(\frac{7}{4} - x^2\right)\left(x^2 - \frac{1}{4}\right)}$ looks like the number 0 with "center" in the interval $-\sqrt{\frac{7}{4}} < x < -\frac{1}{2}$. It has a local maximum at $\left(-1, \frac{3}{4}\right)$ and local minimum at $\left(-1, -\frac{3}{4}\right)$. In the second case, $\frac{1}{2} \leq x \leq \sqrt{\frac{7}{4}}$, the quantity y is zero at the end points $x = \frac{1}{2}$ and $x = \sqrt{\frac{7}{4}}$, so the curve $y = \pm\sqrt{\left(\frac{7}{4} - x^2\right)\left(x^2 - \frac{1}{4}\right)}$ looks like the number 0 with "center" in the interval $\frac{1}{2} < x < \sqrt{\frac{7}{4}}$. It has a local maximum at $\left(1, \frac{3}{4}\right)$ and local minimum at $\left(1, -\frac{3}{4}\right)$. Thus, for $0 < c < 1$ we find two disconnected "circles."

We still have to decide on the direction of travel; that is, which way the arrows point in the direction field. This can be done by returning to (10.3) and looking at the nullclines. The x-nullcline of (10.3) is $y = 0$ (the x-axis) and the y-nullclines are $x = 0$ and $x = \pm 1$. Any orbit that touches $y = 0$ must do so vertically, whereas orbits that touch the lines $x = 0$ and $x = \pm 1$ must do so horizontally. From (10.3) we see that the vertical arrows along $y = 0$ point up if either $x < -1$ or $0 < x < 1$ (because $y' > 0$) and down if either $-1 < x < 0$ or $1 < x$. The horizontal arrows along the lines $x = 0$ and $x = \pm 1$ point right if $y > 0$ (because $x' > 0$) and left if $y < 0$. Thus, we have a good idea what the phase plane will look like, and also we know that to see a representative phase portrait by numerical means, we should start with initial conditions in each of the eight regions bounded by $x = 0, x = \pm 1$, and $y = 0$.

Nullclines

Figure 10.2 shows the direction field and some orbits of (10.4) for various values of c. Look at Figure 10.2 carefully and notice the following.

- Orbits that start near the equilibrium point $(-1, 0)$ stay near $(-1, 0)$, and so $(-1, 0)$ is a stable equilibrium point that behaves like a center. By symmetry, the equilibrium point $(1, 0)$ is stable and behaves like a center.
- The equilibrium point $(0, 0)$ is unstable and behaves like a saddle point.
- Because there are no attracting stable equilibrium points, there is no basin of attraction.
- The only orbits that intersect are closed curves; that is, curves that are self-intersecting. For example, the innermost curve centered at $(1, 0)$ is a closed curve, as is the outermost one centered at $(0, 0)$. We have already seen that if a solution is periodic, then the orbit is closed. The converse is also true — **a closed orbit represents periodic motion**. Orbits that are closed curves are called CYCLES. Notice that every cycle contains at least one equilibrium point.

Closed curves or cycles

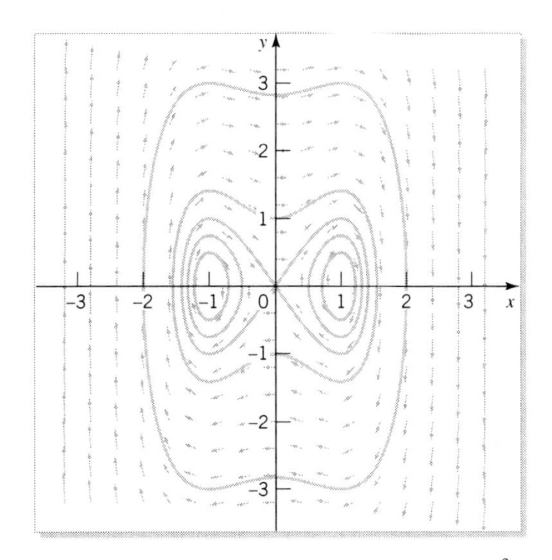

FIGURE 10.2 Orbits of $x' = y$, $y' = 2x - 2x^3$

- The curve that looks like the number 8 on its side consists of three orbits — the orbit $y = \pm\sqrt{x^2(2 - x^2)}$ with $x < 0$, the orbit $y = \pm\sqrt{x^2(2 - x^2)}$ with $x > 0$, and the equilibrium point $(0, 0)$. This curve is a separatrix. It is not a closed curve. (Why?)

Separatrix

- The right-hand orbit that forms part of this separatrix approaches the equilibrium point $(0, 0)$ after initially being repelled by it. A similar comment is true for the left-hand orbit.

Interpretation

It is possible to interpret this phase portrait in terms of the horizontal location, $x(t)$, of a ball that rolls on a curved surface with two local minima of equal depth at $x = \pm 1$, and a local maximum at $x = 0$, as its cross-sectional profile. In this interpretation, the vertical axis in the phase plane will represent the velocity x', so $y = x'$, and we assume that the ball rolls without friction or slipping. We will now look at some typical motions of the ball using Figure 10.3, and interpret them in terms of the orbits in Figure 10.2.

A ball at rest for all time at the point marked A will have velocity $x' = 0$ and horizontal location $x = -1$. In the phase plane this stationary motion is represented by the equilibrium point $(-1, 0)$.

A ball released from rest at point B will fall, moving to the right until it reaches point C, the same height as point B. Then the ball will roll back to B, then back to C, oscillating forever. In the phase plane this oscillating motion could be represented by the inner closed curve with "center" at $(-1, 0)$.

Now imagine we again start with the ball at B but now give it a small positive initial velocity. It will pass through A and rise higher than C to D, and return through B to a point the same height as D and continue oscillating like the previous case, but now with a larger amplitude. In the phase plane this oscillating motion could be represented by the second innermost closed curve with "center" at $(-1, 0)$.

Now start the ball at B with a slightly larger velocity, just enough to reach the point E and come to rest there in infinite time; that is, ending at the point $x = 0$,

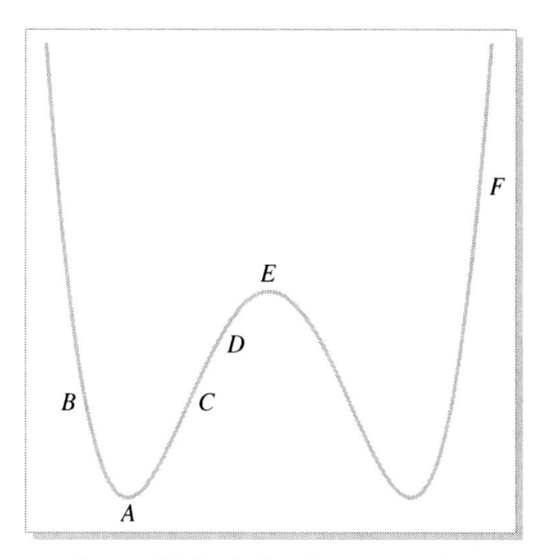

FIGURE 10.3 Ball rolling on a surface

$y = 0$ in the phase plane. In the phase plane this motion would be represented by the left-hand portion of the curve shaped like the number 8 on its side.

Now we restart the ball from B with a larger velocity than before. The ball will roll past E and come to rest at F, higher than E. (Why?) It then returns, passing E and B and coming to rest at a point the same height as F. It then oscillates back and forth between this point and F. In the phase plane this oscillating motion could be represented by one of the two outermost closed curves with "center" at $(0, 0)$.

To see whether you have understood this interpretation, do Exercise 2 on page 446.

EXAMPLE 10.3

Discuss the behavior of the solutions of the system

$$\begin{cases} x' = -y + ax\left(x^2 + y^2\right), \\ y' = x + ay\left(x^2 + y^2\right), \end{cases} \tag{10.5}$$

for $a = 0$, $a > 0$, and $a < 0$.

Equilibrium points

The equilibrium points must satisfy the two equations $-y + ax\left(x^2 + y^2\right) = 0$ and $x + ay\left(x^2 + y^2\right) = 0$. To solve these two equations, we multiply the first by $-y$ and the second by x and add the resulting equations, to find $x^2 + y^2 = 0$, which requires that $x = y = 0$. (Why?) Therefore, the only equilibrium point is $(0, 0)$ for any choice of a. Thus, one solution of (10.5) is $x(t) = 0$, $y(t) = 0$, which corresponds to the initial condition $x(0) = 0$, $y(0) = 0$.

Explicit solution

Trying to solve this system by the elimination method does not seem to lead anywhere, except the algebra makes us wish that the term $a\left(x^2 + y^2\right)$ was either absent from an equation, or appeared throughout it. If we multiply the top equation of (10.5) by y, the bottom by x, and subtract the resulting equations, we can eliminate the term $a\left(x^2 + y^2\right)$ in (10.5), and find $yx' - xy' = -x^2 - y^2$. If we multiply the top equation by x, the bottom by y, and add the resulting equations, we retain the term $a\left(x^2+y^2\right)$, and find $xx'+yy'=a\left(x^2+y^2\right)^2$. We are thus led to the equations

$$\begin{cases} xy'-yx' = x^2+y^2, \\ xx'+yy' = a\left(x^2+y^2\right)^2. \end{cases} \tag{10.6}$$

This set of equations suggests that we introduce polar coordinates $x=r\cos\theta$, $y=r\sin\theta$ — from which $r^2=x^2+y^2$ and $\tan\theta=y/x$ follow. Because $\left(r^2\right)'=2xx'+2yy'$, the bottom equation in (10.6) becomes $r'=ar^3$, and the first ultimately gives $\theta'=1$. (Do this.) This last equation tells us that $\theta=t+C$; that is, all orbits proceed counterclockwise, so a nullcline analysis is not necessary. Although we could solve $r'=ar^3$ for $r(t)$, we need only perform a phase line analysis for $r\geq 0$ to understand the solution. We know that $r=0$ is an unstable equilibrium solution if $a>0$ and a stable equilibrium solution if $a<0$. Thus, for $a>0$ all solutions increase monotonically in r to ∞ as t increases, whereas if $a<0$ they decrease to $r=0$. If $a=0$ the solution is $r=c$. In terms of the phase plane, $r=0$ is the origin and $r=c$ is the circle $x^2+y^2=c^2$. So, if $a>0$ all orbits spiral outward to infinity; if $a=0$ all orbits are circles with the equilibrium point as the center; and if $a<0$ all orbits spiral inward to the origin. These motions are counterclockwise. Phase portraits for this example are shown in Figures 10.4 (with $a=0.3$) and 10.5 ($a=-0.3$). Notice that a small change in the value of a from $a=0$ makes an enormous difference in the ultimate behavior of a solution.

Orbits

Long-term behavior

Phase portrait

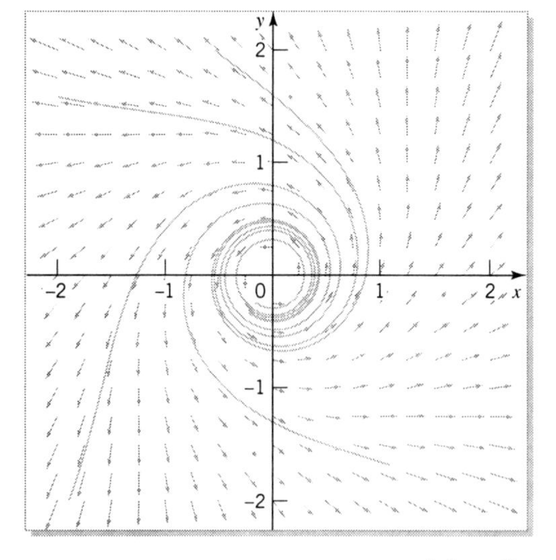

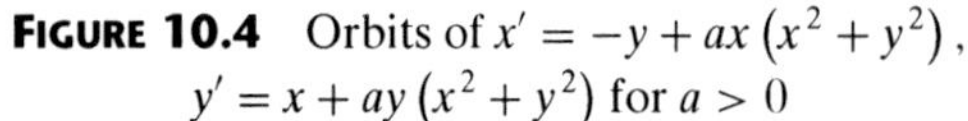

FIGURE 10.4 Orbits of $x'=-y+ax\left(x^2+y^2\right)$, $y'=x+ay\left(x^2+y^2\right)$ for $a>0$

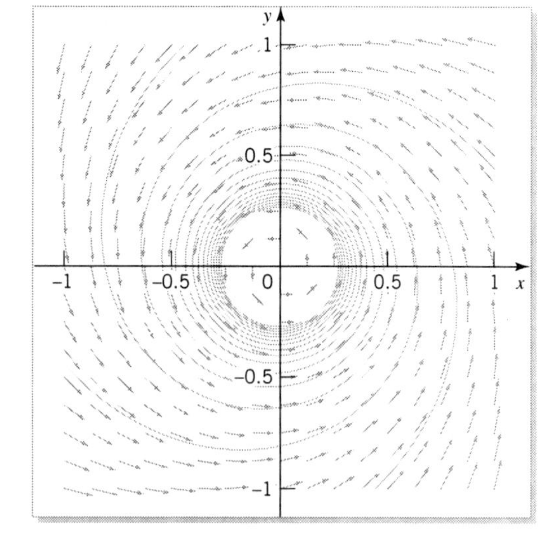

FIGURE 10.5 Orbits of $x'=-y+ax\left(x^2+y^2\right)$, $y'=x+ay\left(x^2+y^2\right)$ for $a<0$

EXAMPLE 10.4

Discuss the behavior of the solutions of the system

$$\begin{cases} x' = -4y + x\left(1 - x^2 - y^2\right), \\ y' = 4x + y\left(1 - x^2 - y^2\right). \end{cases} \tag{10.7}$$

Equilibrium points The equilibrium points must satisfy $-4y + x\left(1 - x^2 - y^2\right) = 0$ and $4x + y\left(1 - x^2 - y^2\right) = 0$. To solve these two equations, we multiply the first by $-y$ and the second by x and add the resulting equations, to find $4\left(x^2 + y^2\right) = 0$, which requires that $x = y = 0$. Thus, the only equilibrium point is $(0, 0)$, so one solution of (10.7) is $x(t) = 0$, $y(t) = 0$, which corresponds to the initial condition $x(0) = 0$, $y(0) = 0$.

Explicit solution Trying to solve this system by the elimination method does not seem to lead anywhere, but is reminiscent of the previous example, so we follow that pattern. If we multiply the top equation of (10.7) by y, the bottom by x, and subtract the resulting equations, we can eliminate the term $\left(1 - x^2 - y^2\right)$, and find $yx' - xy' = -4\left(x^2 + y^2\right)$. If we multiply the top equation by x, the bottom by y, and add the resulting equations, we retain the term $\left(1 - x^2 - y^2\right)$, and find $xx' + yy' = \left(x^2 + y^2\right)\left(1 - x^2 - y^2\right)$. We are thus led to the equations

$$\begin{cases} xy' - yx' = 4\left(x^2 + y^2\right), \\ xx' + yy' = \left(x^2 + y^2\right)\left(1 - x^2 - y^2\right). \end{cases} \tag{10.8}$$

Again we introduce polar coordinates $x = r\cos\theta$, $y = r\sin\theta$. Because $\left(r^2\right)' = 2xx' + 2yy'$, the bottom equation in (10.8) becomes $\left(r^2\right)' = 2r^2\left(1 - r^2\right)$, while the first ultimately gives $\theta' = 4$. (Do this.) This last equation tells us that $\theta = 4t + C$; that is, all

Orbits orbits proceed counterclockwise, so a nullcline analysis is not necessary. If we define

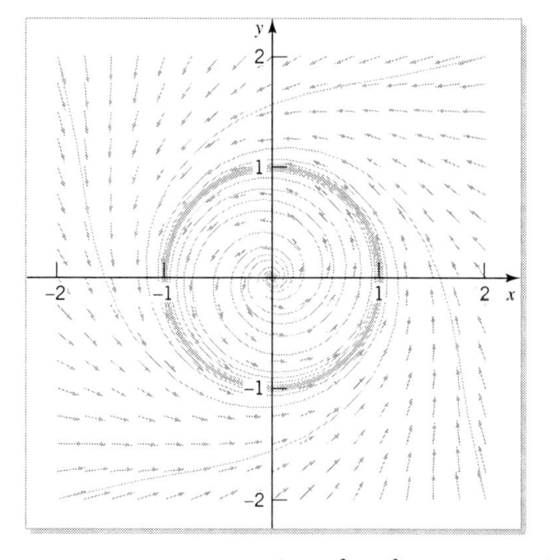

FIGURE 10.6 Orbits of $x' = -4y + x\left(1 - x^2 - y^2\right)$, $y' = 4x + y\left(1 - x^2 - y^2\right)$

$R = r^2$, we see that $(r^2)' = 2r^2(1 - r^2)$ is the logistic equation $R' = 2R(1 - R)$. Although we could solve this for $R(t)$, we need only perform a phase line analysis for $R \geq 0$ to understand the solution. We know that $R = 0$ is an unstable equilibrium and $R = 1$ is a stable equilibrium. Thus, for $R > 0$ all solutions tend monotonically to $R = 1$ as $t \to \infty$ — if $0 < R < 1$ they increase to $R = 1$, and if $R > 1$ they decrease to $R = 1$. In terms of the phase plane, the solution $R = 0$ is the origin and $R = 1$ is the circle $x^2 + y^2 = 1$. So all orbits, except the equilibrium point $(0, 0)$ and the circle $x^2 + y^2 = 1$, spiral toward the unit circle, from within and without, in a counterclockwise fashion. This is shown in Figure 10.6. The circle $x^2 + y^2 = 1$ is called a LIMIT CYCLE. It is a stable limit cycle.

Long-term behavior

Limit cycle

Summary

Thus, in this section we have seen examples of orbits of nonlinear systems with the following properties:

- Orbits can self-intersect, generating closed curves, but cannot intersect other orbits.
- Orbits can approach an equilibrium point, possibly by spiraling.
- Orbits can approach infinity as t increases, possibly by spiraling.
- Orbits can spiral out or in and approach a closed curve.
- Orbits that are closed curves contain an equilibrium point.

EXERCISES

1. From (10.2) write down the differential equation that the orbits in the phase plane must satisfy. Solve this differential equation. Confirm that the orbits you obtained agree with those found in Example 10.1.

2. This question refers to Figures 10.2 and 10.3.

(a) What does the dip in the outermost closed curve in Figure 10.2 correspond to in Figure 10.3?

(b) A ball is placed at rest at E and stays there for all time. What is its orbit in the phase plane?

(c) A ball is released from point B with an initial velocity to the left. What are its possible orbits?

(d) What do the stable and unstable equilibrium points correspond to in Figure 10.3? Why are stable and unstable good descriptions?

3. Consider the system of equations

$$\begin{cases} x' = -x, \\ y' = y - x^2. \end{cases} \tag{10.9}$$

(a) Find all the constant solutions of (10.9).

(b) Find all the solutions of (10.9) in the following two distinct ways, and then make sure your answers agree.

i. Integrate the top equation in (10.9) to find $x(t)$. Substitute this solution into the bottom equation in (10.9) and integrate the result to find $y(t)$.

ii. Differentiate the bottom equation of (10.9) and use the top equation to write it as $y'' = y' + 2x^2$. Now use the bottom equation again — in the form $x^2 = -y' + y$ — to eliminate the x^2 term from $y'' = y' + 2x^2$, obtaining $y'' + y' - 2y = 0$. Solve this second order linear equation for $y(t)$ and then find $x(t)$ from $x^2 = -y' + y$.

(c) Use the results from part (b) to decide what happens to $x(t)$ and $y(t)$ as $t \to \infty$.

(d) Find the orbits in the phase plane in the following two distinct ways and then make sure your answers agree.

i. Use the results from part (b) to eliminate t between the equations for $x(t)$ and $y(t)$.
ii. From (10.9) write down the differential equation for the orbits in the phase plane, and then solve it.

(e) Sketch the nullclines of (10.9), making sure you include the direction of travel. Add several orbits to create a phase portrait for (10.9).

(f) According to your phase portrait, describe what happens to x and y as t increases.

(g) Does this system have a separatrix? If so, identify it. Does this system have a basin of attraction? If so, identify it.

4. Repeat Exercise 3 for the system of equations

$$\begin{cases} x' = 1 - x^2, \\ y' = xy, \end{cases} \tag{10.10}$$

replacing part (b) ii with the following. Differentiate the bottom equation of (10.10) and use the top equation to write it as $y'' = xy' + (1 - x^2)\,y$. Now use the bottom equation again — in the form $x = y'/y$ — to eliminate the x terms from $y'' = xy' + (1 - x^2)\,y$, obtaining $y'' + y = 0$. Solve this second order linear equation for $y(t)$ and then find $x(t)$ from $x = y'/y$.

5. Consider the system of equations

$$\begin{cases} x' = y, \\ y' = 2x - x^2. \end{cases} \tag{10.11}$$

(a) Find all the constant solutions of (10.11).

(b) From (10.11) write down the differential equation for the orbits in the phase plane, and then solve it.

(c) Sketch the nullclines of (10.11), making sure you include the direction of travel. Add several of the orbits you found in part (b) to create a phase portrait for (10.11). [Hint: By considering the graph of the function $2x^2 - \frac{2}{3}x^3$, show that the function $2x^2 - \frac{2}{3}x^3 + C$ has three distinct real roots if $C < 0$, two distinct real roots (one repeated) if $C = 0$, and exactly one real root if $C > 0$.]

(d) According to your phase portrait, describe what happens to x and y as t increases.

(e) Does this system have a separatrix? If so, identify it. Does this system have a basin of attraction? If so, identify it.

6. Consider the system of equations

$$\begin{cases} x' = 2xy, \\ y' = y^2 - x^2. \end{cases} \tag{10.12}$$

(a) Find all the constant solutions of (10.12).

(b) From (10.12) write down the differential equation for the orbits in the phase plane, and then solve it. Are these orbits circles?

(c) Sketch the nullclines of (10.12), making sure you include the direction of travel. Add several of the orbits you found in part (b) to create a phase portrait for (10.12).

(d) According to your phase portrait, describe what happens to x and y as t increases.

(e) Does this system have any separatrices? If so, identify them. Does this system have a basin of attraction? If so, identify it.

(f) Solve the equation you found for the orbits in part (b) for $y = y(x)$. Substitute this result in the top equation of (10.12), obtaining the first order differential equation $x' = \pm 2x\sqrt{cx - x^2}$. Solve this equation for $x = x(t)$ and substitute the result in $y = y(x)$ to find $y = y(t)$. What happens to x and y as t increases? Does this agree with the result you found in part (d)?

7. Consider the system of equations

$$\begin{cases} x' = x(x - 2y), \\ y' = y(y - 2x). \end{cases} \tag{10.13}$$

(a) Find all the constant solutions of (10.13).

(b) From (10.13) write down the differential equation for the orbits in the phase plane, and then solve it.

(c) Sketch the nullclines of (10.13), making sure you include the direction of travel. Add several of the orbits you found in part (b) to create a phase portrait for (10.13).

(d) According to your phase portrait, describe what happens to x and y as t increases.

(e) Does this system have any separatrices? If so, identify them. Does this system have a basin of attraction? If so, identify it.

8. Sketch the phase portrait of an unstable limit cycle. How many possibilities are there? Locate the equilibrium point in your sketch.

9. A wire is bent into the shape $y = f(x)$ and is placed so that the positive y-axis points upward. A bead is at the point $(x(t), y(t))$ at time t and slides without friction on this wire under the downward force of gravity. Use your intuition as to how such a bead might move to answer the following questions.

(a) Sketch a phase portrait of x' versus x for each of the following wire shapes $y = f(x)$. In each case, identify the equilibrium points, separatrices, and basins of attraction.

(i) x^2 (ii) $-x^2$ (iii) $-x^2(x-1)$ (iv) $x^2(x-1)$

(b) It is claimed that Figure 10.7 is a phase portrait for a particular shaped wire. Describe the wire's shape.

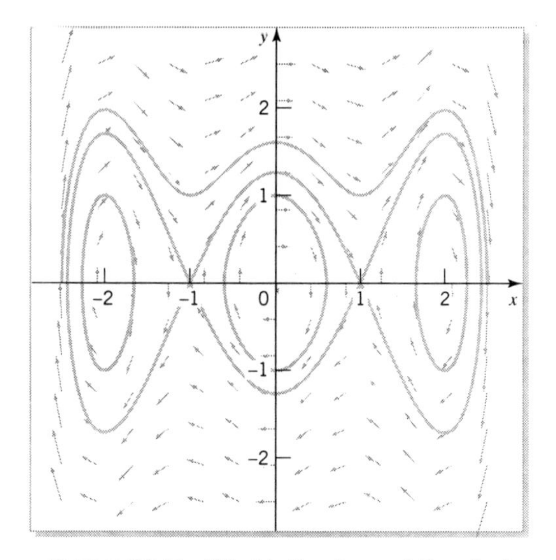

FIGURE 10.7 What is the shape of the wire?

10.2 QUALITATIVE BEHAVIOR USING NULLCLINE ANALYSIS

Unlike the linear autonomous systems discussed in the previous chapter, and the carefully selected examples in the previous section, it is usually impossible to find explicit solutions of nonlinear systems even in the phase plane. We could try to solve a nonlinear system using a numerical method and obtain quantitative information of a solution. This requires that we supply an initial condition. Will the numerical solution be accurate? Will the numerical solution be characteristic of all solutions, or will different initial conditions give different behaviors? To answer these questions we require qualitative information on the long-term behavior of the solutions. In the next two sections we discuss how to obtain such information.

Numerical solution

In this section we discuss the use of nullclines. Until now, their primary use has been to establish the direction of the arrows in the phase plane. However, frequently they can tell us more than that. To demonstrate this, think about the following problem. You are told that a particular system of differential equations has the following properties.

- It has exactly one x-nullcline at $x = 0$.
- It has exactly two y-nullclines at $y = 0$ and $y = 1$.
- The vertical arrows on the x-nullcline point down if $y > 1$ or $y < 0$ and up if $0 < y < 1$.
- The horizontal arrows on both y-nullclines point left if $x > 0$ and right if $x < 0$.

From this information alone, use common sense to identify the equilibrium points and sketch a reasonable phase portrait for the system of equations. It is important that you try this now. (What do you find?[1])

Law of Competitive Exclusion

EXAMPLE 10.5 *A Simple Two-Population Model*

In earlier chapters we discussed the situation in which a population, $x(t)$, grows with time according to the differential equation $x'/x = k$, where k is a positive constant. This is exponential growth. We now look at the situation where we have two populations, $x(t)$ and $y(t)$, both growing in this way with the same constant, k, but competing for the same finite resources. We use a very simple model where we assume that for the first population, x, the time rate of change of x per unit population, k, is reduced by an amount proportional to the second population, y, so that

$$\frac{1}{x}x' = k - ay,$$

where a is a positive constant. A similar assumption is made for the rate of change of y. Thus, we have

$$\begin{cases} x' = x(k - ay), \\ y' = y(k - bx), \end{cases} \tag{10.14}$$

where a and b are positive constants. Notice that the terms multiplying a and b are both $-xy$, and so the system of equations (10.14) is nonlinear. In order to avoid unnecessary algebra, we will consider the case where $k = 5$ and $a = b = 1$, so that (10.14) becomes

$$\begin{cases} x' = x(5 - y), \\ y' = y(5 - x). \end{cases} \tag{10.15}$$

We are concerned with $x \geq 0$ and $y \geq 0$. (Why?)

Explicit solution

To find an explicit solution, the first thing we might try is to find a differential equation involving only one of the variables x or y. We can do this by the same technique that we used in earlier sections. We differentiate the top equation in (10.15), and into the result substitute the expression for y' from the bottom equation in (10.15) — and y from the top equation in (10.15) — to obtain

$$\begin{aligned} x'' &= x'(5 - y) - xy' \\ &= x'\left(x'/x\right) - xy(5 - x) \\ &= \left(x'\right)^2/x - x\left(5 - x'/x\right)(5 - x) \\ &= \left(x'\right)^2/x + (5 - x)x' - 5x(5 - x), \end{aligned}$$

[1] You should find something similar to Figure 10.1 on page 439. The equilibrium points are at (0, 0) and (0, 1), where the x- and y-nullclines cross.

or $x'' - \frac{1}{x}(x')^2 - (5 - x)x' + 5x(5 - x) = 0$, which is a second order nonlinear differential equation. We have no techniques for solving this equation, so the elimination method does not help. As mentioned earlier, this is usually what happens if we try this method on the general equation (10.1).

Orbits The next thing we might try is to find the orbits by solving

$$\frac{dy}{dx} = \frac{y(5 - x)}{x(5 - y)}. \tag{10.16}$$

This is a separable differential equation with solutions

$$5 \ln y - y = 5 \ln x - x + C. \tag{10.17}$$

(Show this.) This cannot be solved for $y = y(x)$ or $x = x(y)$ explicitly. We could use the technique developed in Section 4.1 for drawing solutions of separable differential equations; in this case, for various representative values of C. However, this is very lengthy because we have to repeat the process for each value of C that we select (see Exercise 4 on page 455). Nevertheless, from (10.17) we see that if $C = 0$, then $y = x$ is a solution of (10.17).

Equilibrium points The third method is to find the equilibrium points and then use a nullcline analysis. In the case of (10.15), the equilibrium points will occur when

$$\begin{aligned} x(5 - y) &= 0, \\ y(5 - x) &= 0; \end{aligned}$$

that is, when $x = 0$ or $y = 5$, and $y = 0$ or $x = 5$. This gives two equilibrium points; namely, $(0, 0)$ and $(5, 5)$.

Nullclines If we use a nullcline analysis on (10.15), we see that there are four lines to consider:

- the x-nullclines $x = 0$ and $y = 5$, which have vertical arrows in the phase plane, and
- the y-nullclines $y = 0$ and $x = 5$, which have horizontal arrows.

The directions of the arrows are determined from (10.15) as follows.

To decide on the direction of the vertical arrows we use $y' = y(5 - x)$. When $x = 0$ we have $y' = 5y$, which is positive if $y > 0$ and negative if $y < 0$. Thus, along $x = 0$, the vertical arrows point up if $y > 0$ and down if $y < 0$. When $y = 5$ we have $y' = 5(5 - x)$, which is positive if $x < 5$ and negative if $x > 5$. Thus, along $y = 5$, the vertical arrows point up if $x < 5$ and down if $x > 5$.

To decide on the direction of the horizontal arrows we use $x' = x(5 - y)$. When $y = 0$ we have $x' = 5x$, which is positive if $x > 0$ and negative if $x < 0$. Thus, along $y = 0$, the horizontal arrows point right if $x > 0$ and left if $x < 0$. When $x = 5$ we have $x' = 5(5 - y)$, which is positive if $y < 5$ and negative if $y > 5$. Thus, along $x = 5$, the horizontal arrows point right if $y < 5$ and left if $y > 5$.

These nullclines are shown in Figure 10.8 for $x \geq 0$, $y \geq 0$.

Notice that the equilibrium points occur only where an x-nullcline (a nullcline with vertical arrows) intersects a y-nullcline (a nullcline with horizontal arrows); namely, at $(0, 0)$ and $(5, 5)$. The points $(5, 0)$ and $(0, 5)$ are not equilibrium points. (Why?)

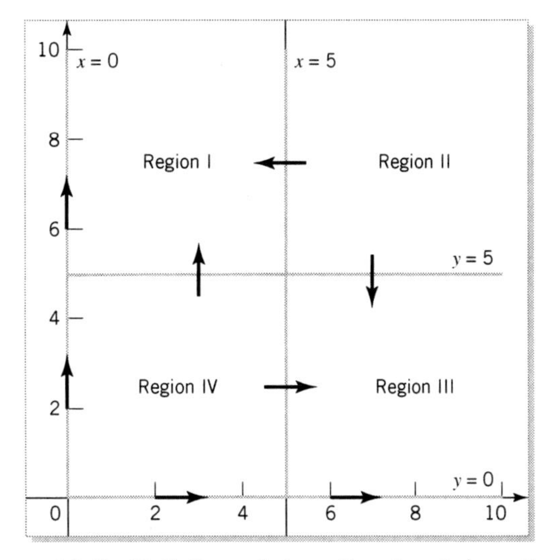

FIGURE 10.8 Nullclines of $x' = x(5 - y)$ and $y' = y(5 - x)$

In order to facilitate our discussion, we have labeled the regions I, II, III, and IV. Notice that if an orbit enters region I, it cannot escape and it appears to satisfy $x \to 0$ and $y \to \infty$ as t increases. Similarly, orbits entering region III cannot escape and appear to satisfy $x \to \infty$ and $y \to 0$ as t increases. Orbits in region II are either drawn to the equilibrium point $(5, 5)$ or cross a nullcline into regions I or III. Once they enter those regions, we already know what happens to them. Orbits in region IV move away from the equilibrium point $(0, 0)$ and are either drawn to the equilibrium point $(5, 5)$ or cross a nullcline into regions I or III. Thus, the equilibrium point $(0, 0)$ behaves like an unstable node, and the equilibrium point $(5, 5)$ behaves like a saddle point. Figure 10.9 shows a computer-drawn phase portrait of (10.15), which is consistent with this information.

Numerical solution
Phase portrait

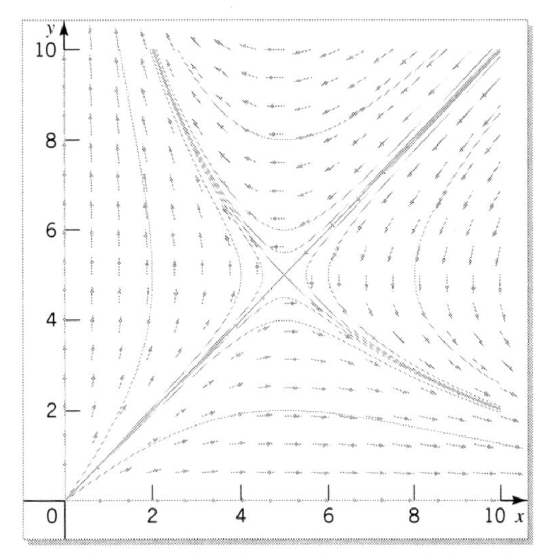

FIGURE 10.9 Orbits of $x' = x(5 - y)$ and $y' = y(5 - x)$

Straight-line orbits

Notice that there appear to be straight-line orbits approaching (5, 5) from both the left and the right. This is the line $y = x$, the solution of (10.17) with $C = 0$, and consists of three orbits: $y = x$ with $x < 5$; $y = x$ with $x > 5$; and the equilibrium point (5, 5). This particular line divides the xy plane into two regions. An initial condition that starts an orbit above this line has one type of behavior ($x \to 0, y \to \infty$), whereas orbits that start below this line have another type ($x \to \infty$, $y \to 0$). This particular curve, which separates the behavior of other orbits, is another example of a separatrix.

Separatrix

Long-term behavior

So, what is the ultimate fate of these populations? If initially the x population exceeds the y population, then the y population becomes extinct, and vice versa. This type of behavior is sometimes called the LAW OF COMPETITIVE EXCLUSION. If it happens that initially $x = y$, then they will tend to coexist and tend to the equilibrium point (5, 5) as t increases. However, notice that if by some random event $x \neq y$, then either $x \to \infty$ or $y \to \infty$. (Is this a realistic model?) ■

So far, all our nullclines have been lines. This need not be the case, as the next example shows.

EXAMPLE 10.6

Discuss the behavior of the solutions of

$$\begin{cases} x' = x^3 - y, \\ y' = -x^2 y. \end{cases}$$

Equilibrium points

The only equilibrium point is (0, 0). (Why?) Neither the second order equation arising from the elimination method, nor the differential equation for orbits in the phase plane have explicit solutions in this case.

Nullclines

We try a nullcline analysis. The x-nullcline is the curve $y = x^3$, and the y-nullcline consists of two lines, $x = 0$ and $y = 0$. Sketch these now.

To decide on the direction of the vertical arrows (those along $y = x^3$), we look at $y' = -x^2 y$, which is negative if $y > 0$ and positive if $y < 0$. Thus, when $y > 0$, solutions are decreasing in the y direction, so the arrows point down for $y > 0$. In the same way, the arrows point up for $y < 0$.

To decide on the direction of the horizontal arrows (those along $x = 0$ and $y = 0$), we look at $x' = x^3 - y$. Along $x = 0$ we have $x' = -y$, which is negative if $y > 0$ and positive if $y < 0$. Thus, when $y > 0$, solutions are decreasing in the x direction, so the arrows on $x = 0$ point left for $y > 0$. They point right for $y < 0$. Along the second y-nullcline, $y = 0$, the equation $x' = x^3 - y$ becomes $x' = x^3$, which is negative if $x < 0$ and positive if $x > 0$. Thus, when $x < 0$, solutions are decreasing in the x direction, so the arrows on $y = 0$ point left for $x < 0$. They point right for $x > 0$.

Phase portrait

Numerical solution

Figure 10.10 summarizes this information. Sketch by hand some typical orbits consistent with this figure. You should find that the equilibrium point (0, 0) behaves like a saddle point. Figure 10.11 shows a phase portrait generated from numerical solutions. (Can you see a separatrix beginning to appear?)

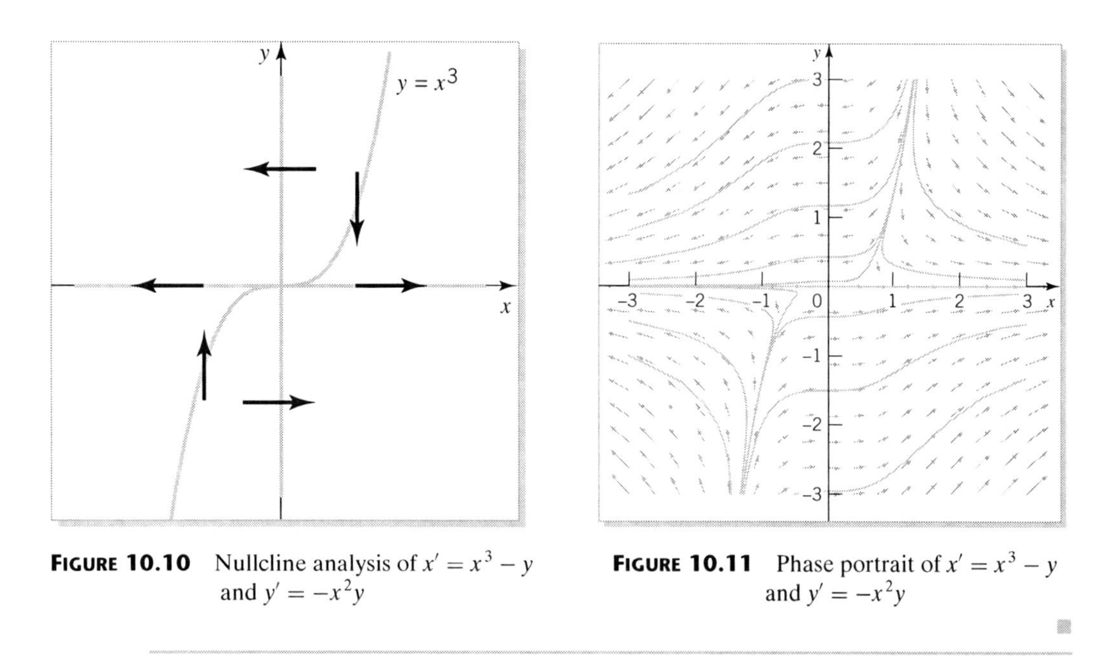

FIGURE 10.10 Nullcline analysis of $x' = x^3 - y$ and $y' = -x^2y$

FIGURE 10.11 Phase portrait of $x' = x^3 - y$ and $y' = -x^2y$

Important point As we pointed out in Section 9.4 on page 411, **we must be careful when using a nullcline analysis if all the arrows rotate in the same direction around an equilibrium point.** In such cases the equilibrium point could behave like a node, a center, or a focus, and the nullcline analysis is inconclusive. We demonstrate this in the next example.

EXAMPLE 10.7

Discuss the behavior of the solutions of

$$\begin{cases} x' = y^2 - (x+1)^2, \\ y' = -x. \end{cases}$$

Neither the second order equation arising from the elimination method, nor the differential equation for orbits in the phase plane have explicit solutions in this case.

Nullclines We try a nullcline analysis. The x-nullcline is characterized by $y^2 - (x+1)^2 = 0$, which consists of the two straight lines $y = x + 1$ and $y = -x - 1$. The y-nullcline consists of the line $x = 0$. Sketch these now and identify the equilibrium points. You should have found two.

Equilibrium points

To decide on the direction of the vertical arrows (those along $y = x + 1$ and $y = -x - 1$), we look at $y' = -x$, which is negative if $x > 0$ and positive if $x < 0$. Thus, when $x > 0$, solutions are decreasing in the y direction, so the arrows point down for $x > 0$. In the same way, the arrows point up for $x < 0$.

To decide on the direction of the horizontal arrows (those along $x = 0$), we look at $x' = y^2 - (x+1)^2$. Along $x = 0$ we have $x' = y^2 - 1$, which is negative if $-1 < y < 1$ and positive if $y < -1$ or $y > 1$. Thus, when $-1 < y < 1$, solutions are decreasing in the x direction, so the arrows on $x = 0$ point left for $-1 < y < 1$. They point right if $y < -1$ or $y > 1$.

Figure 10.12 summarizes this information. Sketch by hand some typical orbits consistent with this figure. You should find that one of the equilibrium points behaves like a saddle point, but the behavior near the other could be one of many possibilities. Here the nullcline analysis is inconclusive. We will return to this example in the next section.

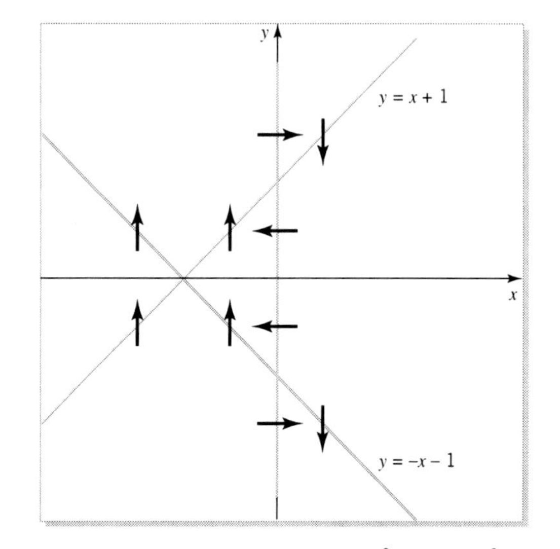

FIGURE 10.12 Nullcline analysis of $x' = y^2 - (x+1)^2$ and $y' = -x$

How to Perform a Nullcline Analysis on a Nonlinear Autonomous System

Purpose To use nullclines to obtain the qualitative behavior of the solutions of

$$\begin{cases} x' = P(x, y), \\ y' = Q(x, y). \end{cases}$$

Process

1. Determine the equilibrium points by simultaneously solving $P(x, y) = 0$ and $Q(x, y) = 0$ for x and y.
2. Determine the curves along which the orbits in the phase plane have horizontal or vertical tangents. These are the y-nullclines and x-nullclines, respectively.
3. Graph these curves in the phase plane, and note that they divide the plane into several regions. Add arrows to indicate the direction of the orbits as they cross these nullclines.
4. Confirm that the equilibrium points occur where the nullclines with horizontal arrows intersect those with vertical arrows.

5. Analyze the stability by considering the direction of orbits crossing the nullclines. If two nullclines have arrows pointing into a region, then orbits are trapped in that region.
6. Keep alert for the presence of a separatrix.

Comments about Nullcline Analysis

- We must be careful when using a nullcline analysis. If all the arrows appear to rotate in the same direction around an equilibrium point, then the equilibrium point could behave like a node, a center, or a focus. In this case, the nullcline analysis is inconclusive.

EXERCISES

1. Use a nullcline analysis to sketch the phase plane orbits of the following systems of nonlinear equations. Use a computer/calculator program to obtain numerical solutions for the orbits in the phase plane and then check these against your expectations.

   (a) $x' = x + y$, $y' = x^2 + y$
   (b) $x' = x(y + 1)$, $y' = y(x + 1)$
   (c) $x' = -y + x^2 - 1$, $y' = -y - x^2 + 1$
   (d) $x' = 1 - y^2$, $y' = y - x$

2. Each of the following systems occurred as examples in the previous section. Using only nullclines, try to determine the qualitative behavior of the corresponding orbits in the phase plane. In the case when the nullcline analysis is inconclusive, give two different possibilities. Use a computer/calculator program to obtain numerical solutions for the orbits in the phase plane and then check these against your expectations. In each case compare your answers with those you previously obtained.

   (a) $x' = -x$, $y' = y - y^2$ [Eq. (10.2) on p. 438.]
   (b) $x' = y$, $y' = 2x - 2x^3$ [Eq. (10.3) on p. 440.]
   (c) $x' = -y + ax(x^2 + y^2)$, $y' = x + ay(x^2 + y^2)$ [Eq. (10.5) on p. 443.] for $a = 0$, $a > 0$, and $a < 0$.
   (d) $x' = -4y + x(1 - x^2 - y^2)$, $y' = 4x + y(1 - x^2 - y^2)$ [Eq. (10.7) on p. 445.]

3. Each of the following systems occurred as exercises in the previous section. Using only nullclines, try to determine the qualitative behavior of the corresponding orbits in the phase plane. In the case when the nullcline analysis is inconclusive, give two different possibilities. Use a computer/calculator program to obtain numerical solutions for the orbits in the phase plane and then check these against your expectations. In each case compare your answers with those you previously obtained.

   (a) $x' = -x$, $y' = y - x^2$ [Eq. (10.9) on p. 446.]
   (b) $x' = 1 - x^2$, $y' = xy$ [Eq. (10.10) on p. 447.]
   (c) $x' = y$, $y' = 2x - x^2$ [Eq. (10.11) on p. 447.]
   (d) $x' = 2xy$, $y' = y^2 - x^2$ [Eq. (10.12) on p. 447.]
   (e) $x' = x(x - 2y)$, $y' = y(y - 2x)$ [Eq. (10.13) on p. 447.]

4. Graph the implicit solution (10.17) of Example 10.5 on page 449 — namely, $5 \ln y - y = 5 \ln x - x + C$ — using the technique from Section 4.1. Figure 10.13 shows the function $5 \ln x - x$.

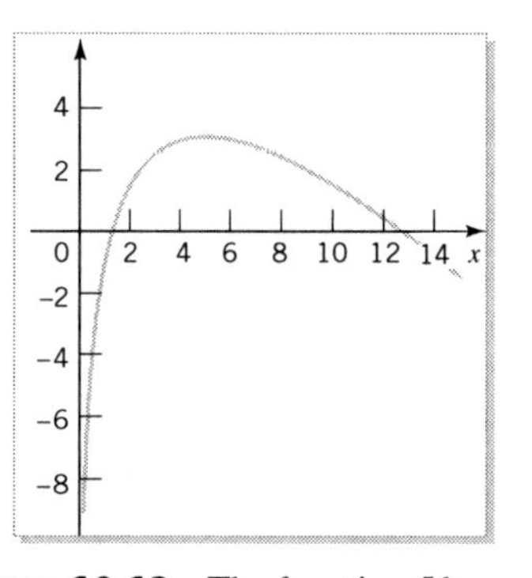

FIGURE 10.13 The function $5 \ln x - x$

10.3 QUALITATIVE BEHAVIOR USING LINEARIZATION

In addition to the nullcline analysis discussed in the previous section, we can sometimes obtain qualitative information about the long-term behavior of the solutions by using a technique that relates the solutions in the nonlinear case to the corresponding solutions in the linear case. This technique is called LINEARIZATION.

Linearization

In order to motivate the technique introduced in this section, first think about the following problem. You are told that a particular system of differential equations has the following properties.

- It has exactly two equilibrium points located at $(0, 0)$ and $(0, 1)$ in the phase plane.
- Near the equilibrium point $(0, 0)$ the phase portrait looks like a saddle point.
- Near the equilibrium point $(0, 1)$ the phase portrait looks like a stable node.

From this information alone, use common sense to sketch a reasonable phase portrait for the system of equations. It is important that you try this now. (What do you find?[2])

This is reminiscent of the corresponding situation for first order autonomous differential equations, where knowledge of the type of equilibrium solutions allowed us to sketch typical solutions. We briefly look at that situation again to gain insight into what to do in the current situation.

Linear

The general linear first order autonomous differential equation is $x' = cx$, where c is a constant. This linear equation has the equilibrium solution $x = 0$. If $c \neq 0$, this is the only equilibrium solution, stable if $c < 0$ and unstable if $c > 0$. If $c = 0$, there are an infinite number of equilibrium solutions, $x = C$.

Nonlinear

The prototype of a first order nonlinear autonomous differential equation is the logistic equation $x' = ax(b - x)$, where a and b are positive constants. A phase line analysis of the logistic equation determines the qualitative behavior of all solutions by first locating all equilibrium solutions ($x = 0$ and $x = b$), then determining their stability (unstable and stable respectively), and finally using common sense to obtain the global behavior from the local behavior. Locating the equilibrium solutions for nonlinear equations is usually easy — determining their stability is more complicated. There is a way of determining the stability of nonlinear equations in terms of linear equations, by linearizing the nonlinear equation about each of its equilibrium solutions in turn. We now demonstrate this idea.

First, consider the equilibrium solution $x = 0$, and concentrate on values of x near $x = 0$. In the expression $ax(b - x) = abx - ax^2$, if x is near zero (that is, x is small), then we can neglect terms in x^2 compared to terms in x, because the magnitude of x^2 will be much smaller than the magnitude of x. Thus, near $x = 0$, $ax(b - x) \approx abx$. In this case, near $x = 0$, we would expect the solutions of the logistic equation $x' = ax(b - x)$ to behave like the solutions of the linear equation $x' = abx$. Because $ab > 0$, the equilibrium solution of this equation, $x = 0$, is unstable. Thus, we would expect the equilibrium solution $x = 0$ of the logistic equation to be unstable — that is, we would expect the equilibrium solution of the nonlinear case to inherit the instability of the linear case.

Linearization

[2]You should find something similar to Figure 10.1 on page 439.

Second, consider the remaining equilibrium solution, $x = b$, of the logistic equation. We would like to deal with this solution in the same way as we dealt with $x = 0$. One way to do this is to use a translation to make $x = b$ the origin. This can be done by introducing the variable $u = x - b$, so that $x = u + b$, and the equilibrium solution $x = b$ is $u = 0$. In this case the equation $x' = ax(b - x)$ becomes $u' = a(u + b)(-u)$, or $u' = -au(u + b)$. In the expression $-au(u + b) = -au^2 - abu$, if u is near zero (that is, u is small so x is near b), then we can neglect terms in u^2 compared to terms in u, so that, near $u = 0$, $-au(u + b) \approx -abu$. Thus, near $u = 0$, we would expect the solutions of the equation $u' = -au(u + b)$ to behave like the solutions of the linear equation $u' = -abu$. Because $-ab < 0$, the equilibrium solution, $u = 0$, of this equation is stable. Thus, we would expect the equilibrium solution $u = 0$ of the equation $u' = -au(u + b)$ to be stable, and in turn the equilibrium solution $x = b$ of the equation $x' = ax(b - x)$ to be stable.

Linearization

With an unstable equilibrium at $x = 0$ and a stable one at $x = b$, the only possible phase line is shown in Figure 10.14, and typical solutions are shown in Figure 10.15 for $a = 1, b = 1.5$. Notice that the region $x > 0$ is the basin of attraction of the equilibrium point $x = b$.

FIGURE 10.14 Phase line analysis for $x' = ax(b - x)$

This, then, is the crux of the idea; namely, find the equilibrium solutions and then analyze each of them in turn, in every case linearizing the nonlinear equation about each of its equilibrium solutions. Finally, obtain the global behavior from this local behavior. Let's consider what we could do in the general case of $x' = g(x)$. To fix ideas, let's assume that $g(x_0) = 0$, so that $x = x_0$ is an equilibrium solution. How can we determine the nature of its stability? The answer lies in the proof of

Linearization

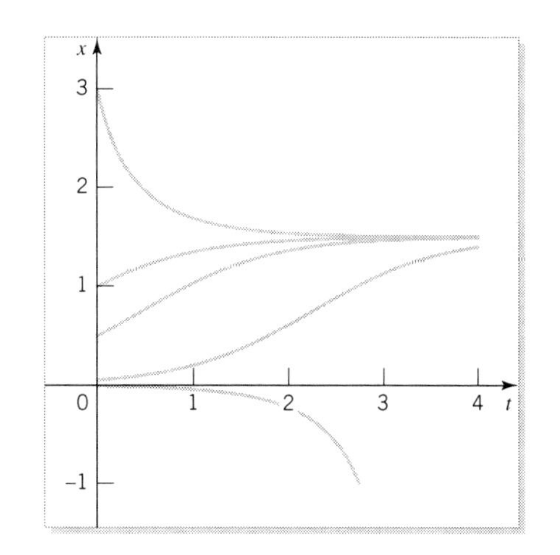

FIGURE 10.15 Typical solution curves of $x' = ax(b - x)$

Theorem 2.2 — the Derivative Test for Stable or Unstable Equilibrium Solutions — in Section 2.5. If $g(x)$ has a continuous derivative, then

$$g(x) \approx g(x_0) + c(x - x_0),$$

where $c = \frac{dg}{dx}(x_0)$, and, because $g(x_0) = 0$,

$$g(x) \approx c(x - x_0).$$

Thus, near the equilibrium solution $x = x_0$, we would expect the solutions of $x' = g(x)$ to behave like the solutions of $x' = c(x - x_0)$. If we introduce the variable $u = x - x_0$, then $x' = c(x - x_0)$ becomes $u' = cu$. Thus, we expect the solutions of $x' = g(x)$ to behave like the solutions of $u' = cu$. Thus, c determines the stability information of x_0 as long as $c \neq 0$. If $c = 0$, the technique is inconclusive. (For example, the two equations, $x' = ax^2$ and $x' = ax^3$, where a is a constant, both have $x' = 0$ as their linearized counterpart, yet they exhibit every type of stability — stable, unstable, and semistable.)

We now generalize this idea to (10.1). First, we assume that $P(x_0, y_0) = 0$ and $Q(x_0, y_0) = 0$ so that (x_0, y_0) is an equilibrium solution of the system. If $P(x, y)$ and $Q(x, y)$ have continuous derivatives in the vicinity of (x_0, y_0), then[3]

Local linearity

$$\begin{aligned} P(x, y) &\approx P(x_0, y_0) + P_x(x_0, y_0)(x - x_0) + P_y(x_0, y_0)(y - y_0), \\ Q(x, y) &\approx Q(x_0, y_0) + Q_x(x_0, y_0)(x - x_0) + Q_y(x_0, y_0)(y - y_0), \end{aligned}$$

where $P_x = \frac{\partial P}{\partial x}$, $P_y = \frac{\partial P}{\partial y}$, $Q_x = \frac{\partial Q}{\partial x}$, and $Q_y = \frac{\partial Q}{\partial y}$. However, $P(x_0, y_0) = 0$ and $Q(x_0, y_0) = 0$, so that

$$\begin{aligned} P(x, y) &\approx a(x - x_0) + b(y - y_0), \\ Q(x, y) &\approx c(x - x_0) + d(y - y_0), \end{aligned}$$

where $a = P_x(x_0, y_0)$, $b = P_y(x_0, y_0)$, $c = Q_x(x_0, y_0)$, and $d = Q_y(x_0, y_0)$. Thus, near the equilibrium point (x_0, y_0), we can associate the linear system

Linearization

$$\begin{cases} x' = a(x - x_0) + b(y - y_0), \\ y' = c(x - x_0) + d(y - y_0), \end{cases}$$

with the nonlinear one,

$$\begin{cases} x' = P(x, y), \\ y' = Q(x, y). \end{cases}$$

The translation, $u = x - x_0$, $v = y - y_0$, converts the equilibrium point from $x = x_0$, $y = y_0$, to the origin of the uv system, and converts the linearized system into

$$\begin{cases} u' = au + bv, \\ v' = cu + dv. \end{cases}$$

[3] These equations could be obtained by using local linearity twice, first on x and second on y.

The classification of the equilibrium solution (x_0, y_0) of this system is determined by the matrix

$$\begin{bmatrix} a & b \\ c & d \end{bmatrix} = \begin{bmatrix} P_x(x_0, y_0) & P_y(x_0, y_0) \\ Q_x(x_0, y_0) & Q_y(x_0, y_0) \end{bmatrix},$$

where the matrix

$$J(x, y) = \begin{bmatrix} P_x(x, y) & P_y(x, y) \\ Q_x(x, y) & Q_y(x, y) \end{bmatrix}$$

Jacobian matrix

is called the JACOBIAN MATRIX. Thus, to classify the equilibrium point (x_0, y_0) of the linearized system, we need only construct the Jacobian matrix at (x_0, y_0) and analyze it using the parabolic classification scheme on page 401, by computing its trace and determinant. We then hope that the nature of the equilibrium point of the linear system is bequeathed to its nonlinear counterpart.

To explore this idea, we consider the following example.

EXAMPLE 10.8

Discuss the behavior of the solutions of

$$\begin{cases} x' = x - y, \\ y' = x - x^3. \end{cases}$$

Equilibrium points

Here, the equilibrium points are $(0, 0)$, $(1, 1)$, and $(-1, -1)$.

Jacobian matrix

Because $P(x, y) = x - y$ and $Q(x, y) = x - x^3$, we have $P_x = 1$, $P_y = -1$, $Q_x = 1 - 3x^2$, and $Q_y = 0$, so the Jacobian matrix is

$$J(x, y) = \begin{bmatrix} 1 & -1 \\ 1 - 3x^2 & 0 \end{bmatrix}.$$

At $(0, 0)$ we find

$$J(0, 0) = \begin{bmatrix} 1 & -1 \\ 1 & 0 \end{bmatrix},$$

with determinant $D = 1$ and trace $T = 1$, so $D - \frac{1}{4}T^2 > 0$ which is characteristic of an unstable focus. (Why?) Thus, we hope that in the nonlinear case that $(0, 0)$ behaves like an unstable focus.

At $(1, 1)$ and $(-1, -1)$ we find

$$J(1, 1) = J(-1, -1) = \begin{bmatrix} 1 & -1 \\ -2 & 0 \end{bmatrix},$$

with determinant -2 and trace 1. This characteristic of a saddle point because the determinant is negative. Thus, we hope that in the nonlinear case that $(1, 1)$ and $(-1, -1)$ behave like saddle points.

Phase portrait

Figure 10.16 shows a phase portrait generated numerically, which is consistent with this linearization technique.

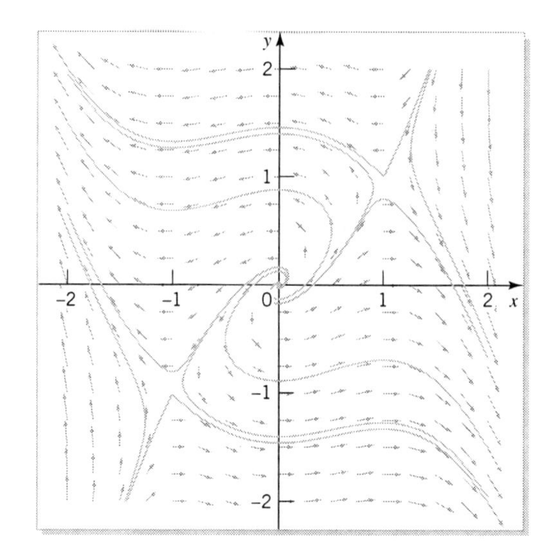

FIGURE 10.16 Phase portrait of $x' = x - y$ and $y' = x - x^3$

If we apply this technique to (10.5), namely,

$$\begin{cases} x' = -y + ax(x^2 + y^2), \\ y' = x + ay(x^2 + y^2), \end{cases}$$

Jacobian matrix which we know has $(0, 0)$ as the only equilibrium solution, we find the Jacobian matrix

$$J(x, y) = \begin{bmatrix} a(3x^2 + y^2) & -1 + 2axy \\ 1 + 2axy & a(x^2 + 3y^2) \end{bmatrix},$$

which gives

$$J(0, 0) = \begin{bmatrix} 0 & -1 \\ 1 & 0 \end{bmatrix}.$$

This matrix has determinant 1 and trace 0, which is characteristic of a center. However, as we have already seen, the behavior of this nonlinear system depends critically on a. The equilibrium point is a stable focus if $a < 0$, an unstable focus if $a > 0$, and a center if $a = 0$.

Important point Thus, **the linearization technique is inconclusive if the linearized system gives a center.**

So when does the linearization technique work? This can be answered by reference to the parabolic classification scheme for equilibrium points of linear systems, Figure 10.17.

▶ **Theorem 10.1: The Linearization Theorem** *Consider the nonlinear system*

$$\begin{cases} x' = P(x, y), \\ y' = Q(x, y). \end{cases}$$

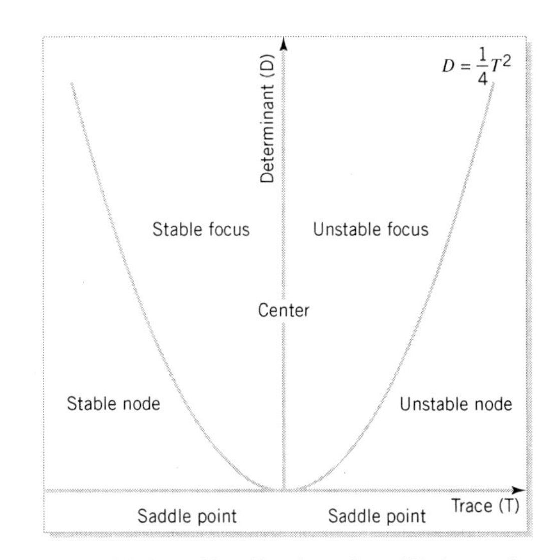

FIGURE 10.17 Classification of equilibrium points

Let (x_0, y_0) be an equilibrium point of this system, and let $P(x, y)$ and $Q(x, y)$ be continuously differentiable in the vicinity of (x_0, y_0). The linear system associated with the nonlinear one is

$$\begin{cases} u' = au + bv, \\ v' = cu + dv, \end{cases}$$

where $u = x - x_0$, $v = y - y_0$, $a = P_x(x_0, y_0)$, $b = P_y(x_0, y_0)$, $c = Q_x(x_0, y_0)$, *and* $d = Q_y(x_0, y_0)$.

1. *If the equilibrium point (x_0, y_0) of the linear system falls into the common category according to the parabolic classification scheme — that is, $r_1 \neq r_2$ or $r = \alpha \pm i\beta$ with $\alpha \neq 0$ — then the equilibrium point (x_0, y_0) of the nonlinear system inherits the properties of the equilibrium point of the linear system.*

2. *If the equilibrium point (x_0, y_0) of the linear system is a center, then the nonlinear equilibrium point may be a center, a stable focus, or an unstable focus.*

3. *If the equilibrium point (x_0, y_0) of the linear system is a stable (unstable) node with $r_1 = r_2 \neq 0$, then it is either a stable (unstable) node or a stable (unstable) focus.* ◀

This theorem verifies the behavior of solutions near the equilibrium points that we predicted in Example 10.8.

We now apply this theorem to an example from the previous section where the nullcline analysis was inconclusive.

EXAMPLE 10.9

Discuss the behavior of the solutions of

$$\begin{cases} x' = y^2 - (x+1)^2, \\ y' = -x. \end{cases}$$

Jacobian matrix The equilibrium points are (0, 1) and (0, −1), and the Jacobian matrix is

$$J(x, y) = \begin{bmatrix} -2(x+1) & 2y \\ -1 & 0 \end{bmatrix}.$$

First, we concentrate on (0, 1), where

$$J(0, 1) = \begin{bmatrix} -2 & 2 \\ -1 & 0 \end{bmatrix}.$$

Nullclines The nullcline analysis was inconclusive in this case. This matrix has determinant 2 and trace −2, so according to the parabolic classification scheme, the equilibrium point of the linearized system is a stable focus. (Why?) Thus, the equilibrium point (0, 1) of the nonlinear system behaves like a stable focus.

Second, we concentrate on (0, −1), where

$$J(0, -1) = \begin{bmatrix} -2 & -2 \\ -1 & 0 \end{bmatrix}.$$

This matrix has determinant −2 and trace −2 so, according to the parabolic classification scheme, the equilibrium point of the linearized system is a saddle point. (Why?) Thus, the equilibrium point (0, −1) of the nonlinear system behaves like a saddle point. This agrees with the previous nullcline analysis.

Phase portrait Figure 10.18 shows a phase portrait generated numerically, which is consistent with this linearization analysis.

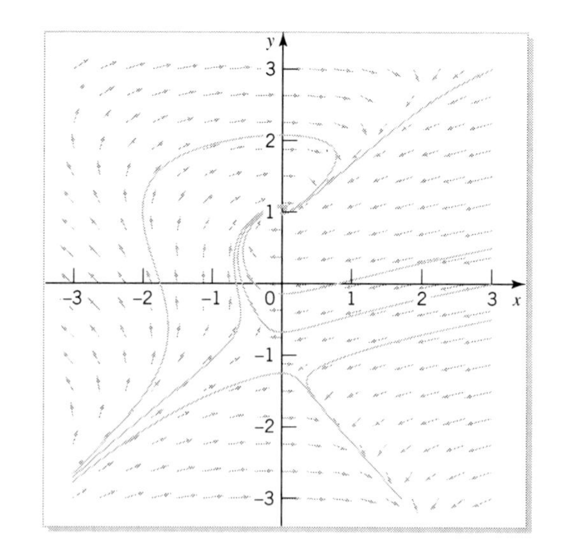

FIGURE 10.18 Phase portrait of $x' = y^2 - (x+1)^2$ and $y' = -x$

■

The previous example is one where the linearization gives information that the nullclines did not. However, the system

$$\begin{cases} x' = x^3 - y, \\ y' = -x^2 y, \end{cases}$$

Jacobian matrix

which we analyzed successfully using nullclines, has Jacobian matrix

$$J(x, y) = \begin{bmatrix} 3x^2 & -1 \\ -2xy & -2x^2 \end{bmatrix},$$

which at the equilibrium point $(0, 0)$ gives

$$J(0, 0) = \begin{bmatrix} 0 & -1 \\ 0 & 0 \end{bmatrix}$$

with determinant and trace zero, which, according to the parabolic classification, is a rare case, so the linearization technique is inconclusive.

How to Perform a Linearization Analysis on a Nonlinear Autonomous System

Purpose To use a linearization analysis to obtain the qualitative behavior of the solutions of

$$\begin{cases} x' = P(x, y), \\ y' = Q(x, y). \end{cases}$$

Process

1. Find all the equilibrium points (x_0, y_0) by simultaneously solving $P(x_0, y_0) = 0$ and $Q(x_0, y_0) = 0$.
2. Confirm that $P(x, y)$ and $Q(x, y)$ are continuously differentiable in the vicinity of each (x_0, y_0).
3. Compute the Jacobian matrix

$$J(x, y) = \begin{bmatrix} P_x(x, y) & P_y(x, y) \\ Q_x(x, y) & Q_y(x, y) \end{bmatrix}.$$

4. Evaluate $J(x, y)$ at each of the equilibrium points (x_0, y_0). Determine the nature of the equilibrium point of the linearized system by analyzing this matrix in terms of the parabolic classification scheme.
5. The equilibrium point in the nonlinear system inherits the stability properties of the linearized one if the linearized one falls either into the common category or into the rare category with real repeated nonzero roots.

We use these ideas on an example we looked at in Chapter 6.

EXAMPLE 10.10 *The Simple Pendulum with Damping*

The differential equation that governs the motion of a simple pendulum experiencing damping is $x'' + kx' + \lambda^2 \sin x = 0$, where $k > 0$. If we define $y = x'$, then the corresponding system is

$$\begin{cases} x' = y, \\ y' = -ky - \lambda^2 \sin x. \end{cases}$$

Equilibrium points The equilibrium points occur where $y = 0$ and $\sin x = 0$ — that is, at $(n\pi, 0)$, where $n = 0, \pm 1, \pm 2, \cdots$ — so there are an infinite number of equilibrium points.

Jacobian matrix The Jacobian matrix is

$$J(x, y) = \begin{bmatrix} 0 & 1 \\ -\lambda^2 \cos x & -k \end{bmatrix},$$

with determinant $\lambda^2 \cos x$ and trace $-k$. The trace is always negative, whereas the value of the determinant will be λ^2 when $x = 2n\pi$ and $-\lambda^2$ when $x = (2n+1)\pi$. This tells us that the equilibrium points $(2n\pi, 0)$ are stable foci if $\lambda^2 > k^2/4$, and stable nodes if $\lambda^2 < k^2/4$. The equilibrium points $((2n+1)\pi, 0)$ are saddle points. Figures 10.19 and 10.20 show numerically generated phase portraits for the case $\lambda^2 > k^2/4$ $(\lambda = 3, k = \frac{1}{4})$ and $\lambda^2 < k^2/4$ $(\lambda = 3, k = 7)$, respectively, which are consistent with this linearization analysis.

Phase portrait

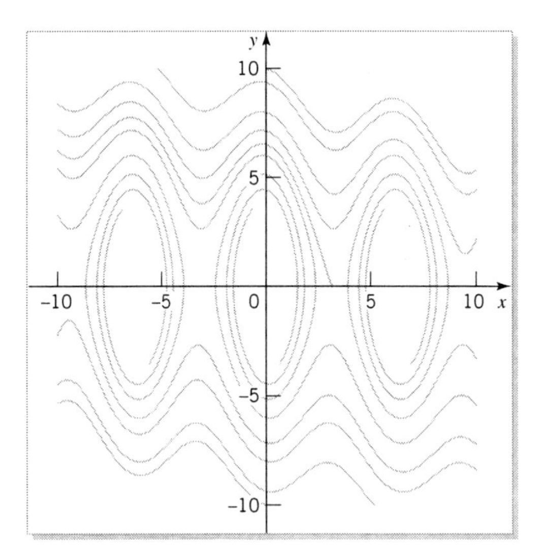

FIGURE 10.19 Phase portrait of pendulum with friction for the case $\lambda^2 > k^2/4$

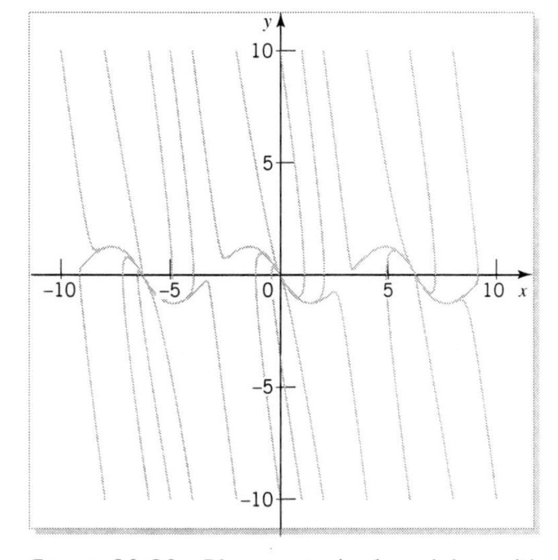

FIGURE 10.20 Phase portrait of pendulum with friction for the case $\lambda^2 < k^2/4$

We now construct a checklist of techniques that we can use when trying to construct a phase portrait of a nonlinear system of equations.

How to Construct a Phase Portrait of a Nonlinear Autonomous System

Purpose To construct a phase portrait of

$$\begin{cases} x' = P(x, y), \\ y' = Q(x, y). \end{cases}$$

Process

1. **Explicit Solution.** Try to solve the system for $x(t)$ and $y(t)$.
2. **Orbits.** Try to solve the system for orbits in the phase plane.
3. **Equilibrium Points.** Find the equilibrium points; that is, find all solutions (x_0, y_0), where x_0 and y_0 are constants, of $P(x_0, y_0) = 0$, $Q(x_0, y_0) = 0$.
4. **Nullcline Analysis.** Perform a nullcline analysis. See *How to Perform a Nullcline Analysis on a Nonlinear Autonomous System* on page 454.
5. **Direction Field.** Draw the slope field and then use the nullcline analysis to determine the direction of the slopes.
6. **Linearization Analysis.** Perform a linearization analysis. See *How to Perform a Linearization Analysis on a Nonlinear Autonomous System* on page 463.
7. **Numerical Solutions.** Use a computer program to numerically solve the system of equations for $x(t)$ and $y(t)$, and for orbits in the phase plane.
8. **Common Sense.** Use common sense and the information you have obtained from parts (1) through (7) to try to construct the global behavior of the phase portrait from the local behavior.

Comments about Constructing a Phase Portrait

- The order in which we have presented these processes is not important.
- The information from these processes should reinforce each other.
- It is usually not possible to solve either for $x(t)$, $y(t)$, or for the orbits in the phase plane. Even if you can, the results are seldom very useful.

Important point

It is important to realize that **if the equilibrium point behaves like a center — that is, the orbits are closed curves, so the motion is periodic — then neither the nullcline analysis nor the linearization analysis is conclusive**. In this case, other techniques have to be used. There is no simple way to do this.

EXERCISES

1. A particular system has exactly two equilibrium solutions, which on linearization appear to behave as indicated by each of the cases (a) through (f). In each case sketch all phase portraits (up to the location of the equilibrium points and distortion of the orbits) that are consistent with this information.

(a) A saddle point and an unstable node.

(b) A stable focus and an unstable focus.

(c) Two saddle points.

(d) A stable focus and a saddle point.

(e) Two stable nodes.

(f) Two centers.

2. Can a system have a phase plane with a closed curve orbit and exactly one equilibrium point that behaves like a saddle point on linearization?

3. Use a linearization analysis to sketch the phase plane orbits of the following systems of nonlinear equations. Use a computer/calculator program to obtain numerical solutions for the orbits in the phase plane and then check these against your expectations. In each case compare your answers with those you obtained in Example 1 on page 455.

(a) $x' = x + y$, $y' = x^2 + y$

(b) $x' = x(y+1)$, $y' = y(x+1)$

(c) $x' = -y + x^2 - 1$, $y' = -y - x^2 + 1$

(d) $x' = 1 - y^2$, $y' = y - x$

4. Each of the following systems occurred as examples in the previous sections. By linearizing these equations about each of the equilibrium points, try to determine the qualitative behavior of the corresponding orbits in the phase plane. In the case where the linearization analysis is inconclusive, give two different possibilities. Use a computer/calculator program to obtain numerical solutions for the orbits in the phase plane and then check these against your expectations. In each case compare your answers with those you previously obtained.

(a) $x' = -x$, $y' = y - y^2$ [Eq. (10.2) on p. 438.]

(b) $x' = y$, $y' = 2x - 2x^3$ [Eq. (10.3) on p. 440.]

(c) $x' = -y + ax(x^2 + y^2)$, $y' = x + ay(x^2 + y^2)$ for $a = 0$, $a > 0$, and $a < 0$. [Eq. (10.5) on p. 443.]

(d) $x' = -4y + x(1 - x^2 - y^2)$, $y' = 4x + y(1 - x^2 - y^2)$ [Eq. (10.7) on p. 445.]

(e) $x' = x(5 - y)$, $y' = y(5 - x)$ [Eq. (10.15) on p. 449.]

(f) $x' = x^3 - y$, $y' = -x^2 y$ [Ex. (10.6) on p. 452.]

(g) $x' = y^2 - (x+1)^2$, $y' = -x$ [Ex. (10.7) on p. 453.]

5. Each of the following systems occurred as exercises in the previous sections. By linearizing these equations about each of the equilibrium points, try to determine the qualitative behavior of the corresponding orbits in the phase plane. In the case where the linearization analysis is inconclusive, give two different possibilities. Use a computer/calculator program to obtain numerical solutions for the orbits in the phase plane and then check these against your expectations. In each case compare your answers with those you previously obtained.

(a) $x' = -x$, $y' = y - x^2$ [Eq. (10.9) on p. 446.]

(b) $x' = 1 - x^2$, $y' = xy$ [Eq. (10.10) on p. 447.]

(c) $x' = y$, $y' = 2x - x^2$ [Eq. (10.11) on p. 447.]

(d) $x' = 2xy$, $y' = y^2 - x^2$ [Eq. (10.12) on p. 447.]

(e) $x' = x(x - 2y)$, $y' = y(y - 2x)$ [Eq. (10.13) on p. 447.]

6. Computer Experiment — The Simple Pendulum with Damping. Use a computer program to plot a phase portrait of

$$\begin{cases} x' = y, \\ y' = -\frac{1}{4}y - 9\sin x, \end{cases}$$

in the window $-10 \le x \le 10$, $-10 \le y \le 10$. Sketch a number of orbits satisfying the initial conditions $x = 10$, $y = y_0$, where $4.4 \le y_0 \le 6.2$, and $x = -10$, $y = y_0$, where $-6.2 \le y_0 \le -4.4$. What happens to all these orbits? What is this region called?

10.4 MODELS INVOLVING NONLINEAR AUTONOMOUS EQUATIONS

In this section we look at nonlinear systems of equations that occur in population models, predator-prey models, and epidemic models.

Population Models

In Example 10.5 on page 449 we assumed that, in the absence of competition for the same finite resources, the populations would grow according to an exponential growth law. In the next two examples, we investigate what happens when the populations x and y still have to compete for the same finite resources, but each grows according to a logistic growth law so that

$$\begin{cases} x' = x\,(k - cx - ay)\,, \\ y' = y\,(k - dy - bx)\,, \end{cases} \tag{10.18}$$

where a, b, c, d, and k are positive constants. We are concerned only with $x \geq 0$ and $y \geq 0$. This system of equations (10.18) is nonlinear. In order to avoid unnecessary algebra, we will consider the case where $k = 5$, $a = b = 1$, and $d = c$, so that (10.18) becomes

$$\begin{cases} x' = x\,(5 - cx - y)\,, \\ y' = y\,(5 - cy - x)\,, \end{cases} \tag{10.19}$$

where $c > 0$. This means that in the absence of competition, the carrying capacity for both x and y is $5/c$. (Why?)

Equilibrium points Our first step is to find the equilibrium points, which in the case of (10.19) occur when

$$x\,(5 - cx - y) = 0$$

and

$$y\,(5 - cy - x) = 0;$$

that is, when $x = 0$ or $y = 5 - cx$, and $y = 0$ or $y = (5 - x)/c$. If $c \neq 1$, we have four equilibrium points; namely, $(0, 0)$, $(0, 5/c)$, $(5/c, 0)$, and $(5/(c + 1), 5/(c + 1))$. If $c = 1$, we have an infinite number of equilibrium points; namely, $(0, 0)$ and all points on the line $y = 5 - x$. In the two following examples we will consider $c < 1$ and $c > 1$.

EXAMPLE 10.11 *Another Two-Population Model (Part 1)*

We illustrate the case $c < 1$ by considering $c = 5/8$, so that (10.19) becomes

$$\begin{cases} x' = x\,(5 - 5x/8 - y)\,, \\ y' = y\,(5 - 5y/8 - x)\,, \end{cases} \tag{10.20}$$

Equilibrium points with equilibrium points at $(0, 0)$, $(0, 8)$, $(8, 0)$, and $(40/13, 40/13)$.

If we use a nullcline analysis on (10.20), we find that there are four lines to

Nullclines consider: $x = 0$ and $y = 5 - 5x/8$ (x-nullclines, which have vertical arrows in the phase plane), and $y = 0$ and $y = 8(5 - x)/5$ (y-nullclines, which have horizontal arrows). The directions of the arrows are determined from (10.20) and are shown in Figure 10.21. (Do this.)

Again we notice that the equilibrium points occur only where a nullcline with vertical arrows intersects one with horizontal arrows; namely, at $(0, 0)$, $(0, 8)$, $(8, 0)$, and $(40/13, 40/13)$. The points $(5, 0)$ and $(0, 5)$ are not equilibrium points. (Why?) In order to facilitate the discussion, we have labeled the regions I, II, III, and IV. We notice that if an orbit enters region I, it cannot escape and appears to be attracted to the equilibrium point $(0, 8)$ as t increases. Similarly, orbits entering region III cannot escape and appear to be attracted to the equilibrium point $(8, 0)$ as t increases. Orbits in region II are either drawn to the equilibrium point $(40/13, 40/13)$ or cross a nullcline into regions I or III. Once they enter those regions, we know what happens to them. Orbits in region IV move away from the equilibrium point $(0, 0)$ and are either drawn to the equilibrium point $(40/13, 40/13)$ or cross a nullcline into regions I or III. Thus, the equilibrium point $(0, 0)$ behaves like an unstable node, the equilibrium points $(0, 8)$ and $(8, 0)$ behave like stable nodes, and the equilibrium point $(40/13, 40/13)$ behaves like a saddle point.

Linearization
Jacobian matrix

We can try to confirm this behavior by linearizing the nonlinear system about each of its equilibrium points. The Jacobian matrix is

$$J(x, y) = \begin{bmatrix} 5 - 5x/4 - y & -x \\ -y & 5 - 5y/4 - x \end{bmatrix}.$$

At the equilibrium point $(0, 0)$ this gives

$$J(0, 0) = \begin{bmatrix} 5 & 0 \\ 0 & 5 \end{bmatrix},$$

which has determinant $D = 25$ and trace $T = 10$ so $D = \frac{1}{4}T^2$. This is a rare case, so the Linearization Theorem guarantees only that $(0, 0)$ is unstable because $T > 0$.

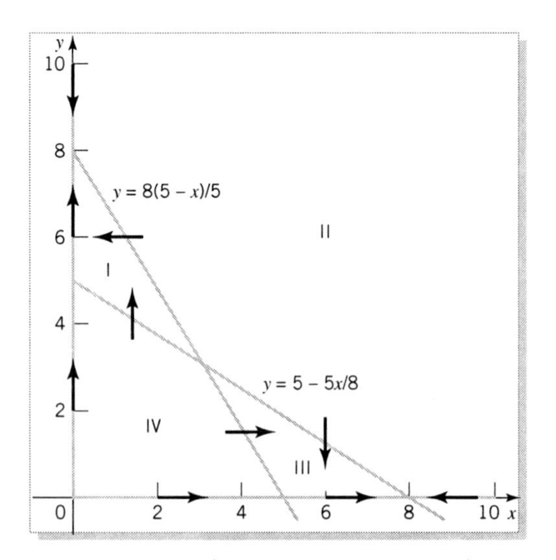

FIGURE 10.21 Nullclines of $x' = x(5 - 5x/8 - y)$ and $y' = y(5 - 5y/8 - x)$

We now turn to the equilibrium point $(40/13, 40/13)$, where

$$J(40/13, 40/13) = \begin{bmatrix} -25/13 & -40/13 \\ -40/13 & -25/13 \end{bmatrix}.$$

Here the determinant is $-65 \cdot 16/13^2$, and the trace is $-50/13$, so $(40/13, 40/13)$ behaves like a saddle point. (Why?)

We now turn to the equilibrium point $(0, 8)$, where

$$J(0, 8) = \begin{bmatrix} -3 & 0 \\ -8 & -5 \end{bmatrix}.$$

Here the determinant is 15 and the trace is -8, so the equilibrium point $(0, 8)$ behaves like a stable node. (Why?) By symmetry, a similar analysis applies to the equilibrium point $(8, 0)$.

Phase portrait

We can also evaluate these orbits numerically, and these are shown in Figure 10.22 along with the direction field. We notice that there appear to be straight-line orbits approaching $(40/13, 40/13)$ from both the left and the right. If we substitute $y = mx$ into (10.20), we find $m = 1$, so $y = x$ is a straight-line solution. (Show this.)

Separatrix

The curve $y = x$ is a separatrix.

Long-term behavior

So, what is the ultimate fate of these populations? If initially the x population exceeds the y population, then the y population becomes extinct, and $x \to 8$ as t increases. If initially the y population exceeds the x population, then the x population becomes extinct, and $y \to 8$ as t increases.

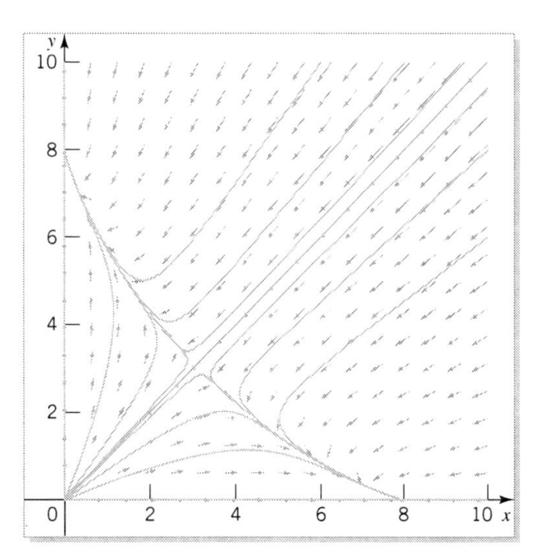

FIGURE 10.22 Numerically computed orbits of $x' = x(5 - 5x/8 - y)$ and $y' = y(5 - 5y/8 - x)$

EXAMPLE 10.12 *Another Two-Population Model (Part 2)*

We now illustrate the case $c > 1$ by considering $c = 5/4$, so that (10.19) becomes

$$\begin{cases} x' = x(5 - 5x/4 - y), \\ y' = y(5 - 5y/4 - x), \end{cases} \tag{10.21}$$

Equilibrium points with equilibrium points at (0, 0), (0, 4), (4, 0), and (20/9, 20/9).

Nullclines If we use a nullcline analysis on (10.21), we find that there are four lines to consider: $x = 0$ and $y = 5 - 5x/4$ (x-nullclines, which have vertical arrows in the phase plane), and $y = 0$ and $y = 4(5 - x)/5$ (y-nullclines, which have horizontal arrows). The directions of the arrows are determined from (10.21) and are shown in Figure 10.23. (Confirm this.)

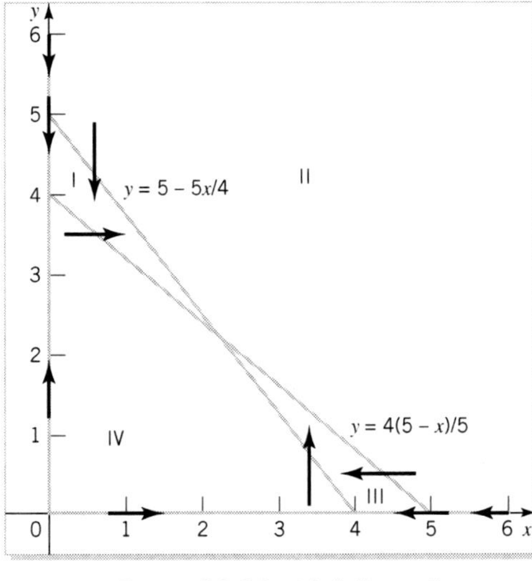

FIGURE 10.23 Nullclines of $x' = x(5 - 5x/4 - y)$ and $y' = y(5 - 5y/4 - x)$

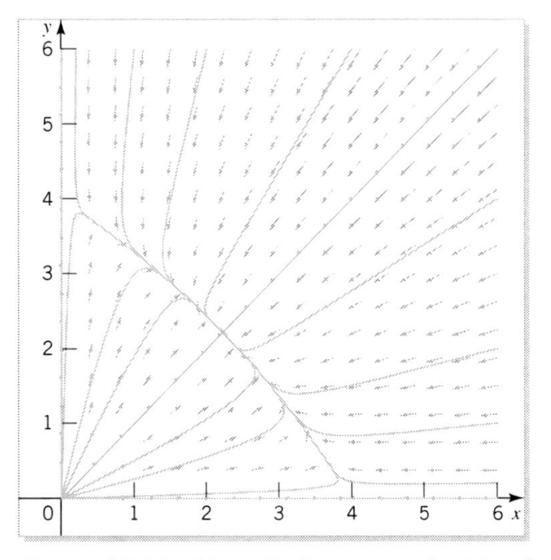

FIGURE 10.24 Numerically computed orbits of $x' = x(5 - 5x/4 - y)$ and $y' = y(5 - 5y/4 - x)$

Again we notice that the equilibrium points occur only where a nullcline with vertical arrows intersects one with horizontal arrows; namely, at (0, 0), (0, 4), (4, 0), and (20/9, 20/9). The points (5, 0) and (0, 5) are not equilibrium points. In order to facilitate the discussion, we have labeled the regions I, II, III, and IV. Notice that if an orbit enters region I, it cannot escape and appears to be attracted to the equilibrium point (20/9, 20/9) as t increases. Similarly, orbits entering region III cannot escape and appear to be attracted to the equilibrium point (20/9, 20/9) as t increases. Orbits in region II are either drawn directly to the equilibrium point (20/9, 20/9) or cross a nullcline into regions I or III and are again drawn to the equilibrium point (20/9, 20/9). Orbits in region IV move away from the equilibrium points (0, 0), (0, 4), and (4, 0) and either are drawn to the equilibrium point (20/9, 20/9) directly or cross a nullcline into regions I or III and are again drawn to the equilibrium point (20/9, 20/9). Thus, the equilibrium points (0, 4) and (4, 0) behave like saddle points, the equilibrium point (0, 0) behaves like an unstable node, and the equilibrium point (20/9, 20/9) behaves like a stable node.

Linearization
Jacobian matrix

As in the last example, we can try to convince ourselves of this by linearizing the nonlinear system about each of its equilibrium points. The Jacobian matrix for this system is

$$J(x, y) = \begin{bmatrix} 5 - 5x/2 - y & -x \\ -y & 5 - 5y/2 - x \end{bmatrix}.$$

At the equilibrium point $(0, 0)$ we find

$$J(0, 0) = \begin{bmatrix} 5 & 0 \\ 0 & 5 \end{bmatrix}.$$

We found this matrix in the previous example, so we know that $(0, 0)$ is unstable.

We now turn to the equilibrium point $(20/9, 20/9)$, where the Jacobian matrix is

$$J(20/9, 20/9) = \begin{bmatrix} -25/9 & -20/9 \\ -20/9 & -25/9 \end{bmatrix}.$$

This has determinant $25/9$ and trace $-50/9$, thus $(20/9, 20/9)$ behaves like a stable node. (Why?)

At the equilibrium point $(0, 4)$, we find

$$J(0, 4) = \begin{bmatrix} 1 & 0 \\ -4 & -5 \end{bmatrix},$$

which has determinant -5 and trace -4. Thus, $(0, 4)$ behaves like a saddle point. (Why?) A similar analysis shows that the equilibrium point $(4, 0)$ also behaves like a saddle point.

Phase portrait

We can also evaluate these orbits numerically, and these are shown in Figure 10.24. In spite of appearances, the slightly curved arc with radius 4 consists of parts of 14 distinct orbits, all of which approach $(20/9, 20/9)$ as t increases.

Long-term behavior

So, what is the ultimate fate of these populations? All populations will coexist and tend to the equilibrium point $(20/9, 20/9)$ as t increases. ■

The Lotka-Volterra Predator-Prey Model

EXAMPLE 10.13 *The Lotka-Volterra Predator-Prey Model*

Several different biologists and mathematicians contributed to the development of the following mathematical model, often called the Lotka-Volterra model. The Lotka-Volterra model for predator-prey interactions is

$$S' = -aS + bSF = S(-a + bF), \tag{10.22}$$

$$F' = kF - cSF = F(k - cS), \tag{10.23}$$

where S represents the predator population (sharks) and F the prey population (fish). The coefficients are all positive, with b and c being the interaction coefficients.

Predators gain from interaction with prey (they eat them), so the term bSF would give a positive rate of change of the predators. Another way of looking at this equation is in the factored form of (10.22), where the coefficient of S could be considered as the net growth rate: having the normal rate of demise in the absence of prey $(-a)$ augmented by a term that accounts for the positive effect of the presence of prey. A way of looking at the terms in (10.23) is that kF represents the growth rate of the prey if there were no predators, while the interaction term decreases this growth rate because of the predator-prey interaction (the prey get eaten).

Explicit solution This system of differential equations is nonlinear, and none of the methods developed thus far for obtaining an explicit time-dependent solution applies. Before looking at some numerical solutions, we seek to obtain as much information as possible about the solutions by simply analyzing this system of differential equations. For this mathematical model, both S and F will be nonnegative, so we concentrate on the first quadrant.

Equilibrium points Notice that $S = 0$ and $F = 0$ satisfy this system of equations identically, corresponding to an equilibrium point in the phase plane. This is a fairly uninteresting equilibrium point because it is not too surprising that one way for predators and prey to be in equilibrium is for the numbers of each to be zero. However, there is another equilibrium point given by $S = k/c$ and $F = a/b$.

Nullclines The nullclines are given by $S = 0$ and $F = a/b$ (x-nullclines) and by $F = 0$ and $S = k/c$ (y-nullclines). These four lines divide the first quadrant of the phase plane into four regions, as shown in Figure 10.25 for the case $a = b = c = k = 1$. The direction of the arrows is determined from (10.22) and (10.23). They imply that the orbits will be traversed in a clockwise direction. (Why?) However, this is the case where the nullcline analysis is inconclusive, because we cannot tell from this whether the equilibrium point $(k/c, a/b)$ is a center, a focus, or a node.

Jacobian matrix The Jacobian matrix of this system is given by

$$J(S, F) = \begin{bmatrix} P_S(S, F) & P_F(S, F) \\ Q_S(S, F) & Q_F(S, F) \end{bmatrix} = \begin{bmatrix} -a + bF & bS \\ -cF & k - cS \end{bmatrix},$$

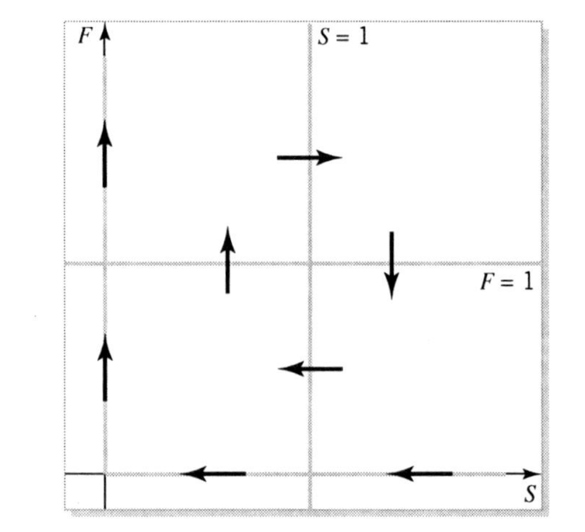

FIGURE 10.25 The nullclines for the predator-prey problem

which, when evaluated at the equilibrium point $S = k/c$, $F = a/b$, becomes

$$J(k/c, a/b) = \begin{bmatrix} 0 & bk/c \\ -ca/b & 0 \end{bmatrix}.$$

Linearization This has determinant ka and trace 0. However, this is the case where the linearization analysis

is inconclusive, so we cannot tell from this whether the equilibrium point $(k/c, a/b)$ is a center, a stable focus, or an unstable focus.

We are thus forced to look at this problem more closely.

Monotonicity From (10.22) we see that S will be an increasing function of t if F is greater than a/b (that is, the prey population is greater than this equilibrium value), and a decreasing function if F is less than a/b. In other words, if the prey supply is greater than needed to be in equilibrium with the predator, the predator will gain in number. We also observe from (10.23) that F will be an increasing function of t if S is less than k/c, and a decreasing function if S is greater than k/c.

Concavity To examine concavity, we obtain the differential equation of the phase plane as

$$\frac{dF}{dS} = \frac{F(k - cS)}{S(-a + bF)}, \tag{10.24}$$

and then compute its derivative as

$$\frac{d^2F}{dS^2} = \frac{-F\left[a(k - cS)^2 + k(-a + bF)^2\right]}{S^2(-a + bF)^3}. \tag{10.25}$$

Thus, we see that the orbits in the phase plane will be concave up if $-a + bF < 0$, or $F < a/b$ — that is, if the number of prey is less than their nonzero equilibrium value — and concave down otherwise. (Note that this is consistent with the fact we discovered earlier about the movement in the phase plane being clockwise.)

Orbits Thus, we have all the information about the phase plane for this model regarding where the orbits are increasing, decreasing, concave up, and concave down, and direction of motion along any orbit. The direction field corresponding to (10.24) and the nullclines $S = k/c$ and $F = a/b$ (for the case $a = b = c = k = 1$) are shown in Figure 10.26. Everything we have discovered so far is consistent with this.

Because (10.24) is a separable equation, we can integrate to obtain the implicit solution

$$F^a e^{-bF} = CS^{-k} e^{cS}. \tag{10.26}$$

For the initial conditions $S(0) = S_0$, $F(0) = F_0$, we find $C = F_0^a e^{-bF_0} S_0^k e^{-cS_0}$, so (10.26) becomes

$$\left(\frac{F}{F_0}\right)^a e^{-b(F-F_0)} = \left(\frac{S}{S_0}\right)^{-k} e^{c(S-S_0)}. \tag{10.27}$$

Because the differential equation is separable, we can use our technique from Section 4.2 to plot (10.27). We do this for the case $a = b = c = k = 1$.

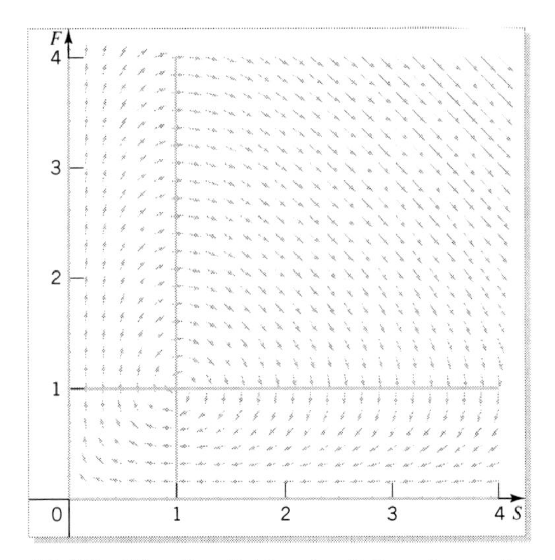

FIGURE 10.26 Direction field and nullclines for predator-prey

With the initial conditions $S_0 = 1$, $F_0 = 3.5$, we plot the function of S on the right-hand side of (10.27) — namely, $(S/S_0)^{-k}e^{c(S-S_0)}$ — in the lower left-hand box of Figure 10.27, and the function of F on the left-hand side of (10.27) — namely, $(F/F_0)^a e^{-b(F-F_0)}$ — in the lower right-hand box. Because the graph of the function of F is bounded between 0 and some maximum value, and the graph of the function of S is bounded away from 0, the orbits in the phase plane will be bounded. Notice also that the equilibrium point in the phase plane is the place where both the S curves and the F curves in the lower boxes of Figure 10.27 have their minimum and maximum values respectively. You should complete Figure 10.27. What shape do you get? What does that tell you about the long-term behavior of the shark and fish population starting from $S_0 = 1$, $F_0 = 3.5$?

Long-term behavior

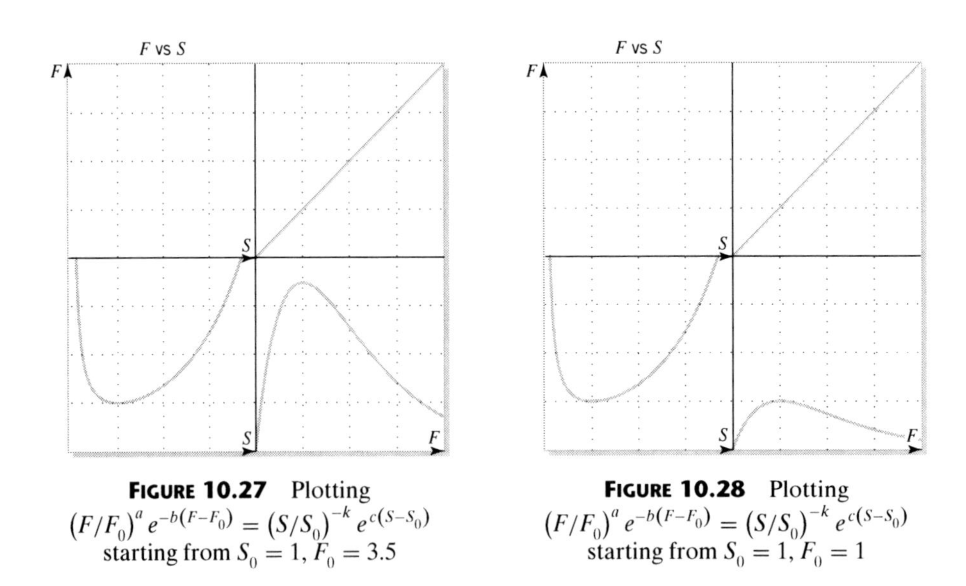

FIGURE 10.27 Plotting $(F/F_0)^a e^{-b(F-F_0)} = (S/S_0)^{-k} e^{c(S-S_0)}$ starting from $S_0 = 1$, $F_0 = 3.5$

FIGURE 10.28 Plotting $(F/F_0)^a e^{-b(F-F_0)} = (S/S_0)^{-k} e^{c(S-S_0)}$ starting from $S_0 = 1$, $F_0 = 1$

Figure 10.28 is the counterpart of Figure 10.27 with the initial conditions $S_0 = 1$, $F_0 = 1$. Complete Figure 10.28. What shape do you get? What does that tell you about the long-term behavior of the shark and fish population starting from $S_0 = 1$, $F_0 = 1$?

Figure 10.29 shows the function $(F/F_0)^a e^{-b(F-F_0)}$ (with $a = b = 1$) plotted for $F_0 = 3.5, 3.0, 2.5, 2.0$, and 1.0. What do these different initial conditions imply about the orbits in the phase plane? Figure 10.30 shows the function $(F/F_0)^a e^{-b(F-F_0)}$ (with $a = b = 1$) plotted for $F_0 = 1, 0.4, 0.25, 0.15$, and 0.11. What do these different initial conditions imply about the orbits in the phase plane? Based on these observations, what do you expect the orbits in the phase plane to look like?

Phase portrait Some numerical solutions of (10.22) and (10.23) for the case $a = b = c = k = 1$ are shown in Figure 10.31 using the initial conditions $S_0 = 1$ and $F_0 = 3.5, 3.0, 2.5$, and 2.0. These show closed orbits, which is what you should have found when analyzing Figures 10.27 through 10.30. Closed orbits in the phase plane means that F and S keep returning to their initial values. Thus, if we start at a point (S_0, F_0) and then, at some later time, the solution of (10.22) and (10.23) returns to that same point, the motion for later times will simply repeat. In Exercise 2 on page 483 we ask you to show that all orbits are closed.

Another way to state our previous result would be that if there exists a value of time (say T) such that both $S(t + T) = S(t)$ and $F(t + T) = F(t)$, then orbits in the phase plane will be closed curves, and S and F will oscillate with period T. Because this is the case for the system in (10.22) and (10.23), the equilibrium point $(k/c, a/b)$ is stable.

Let us look at a single orbit in Figure 10.31 — say, the outermost one that passes through the point $S = 1$, $F = 3.5$. At this point we have a maximum number of fish. As time progresses we follow the orbit clockwise and see that the number of fish decreases while the number of sharks increases, due to the sharks preying on the fish. This continues until the orbit reaches the point $S = 3.5$ and $F = 1$, which corresponds to the maximum number of sharks and a small number of fish. Now the sharks start decreasing in number, because there are too few fish to sustain them. The fish and shark populations continue to decrease until they reach the point $S = 1$, $F \approx 0.11$.

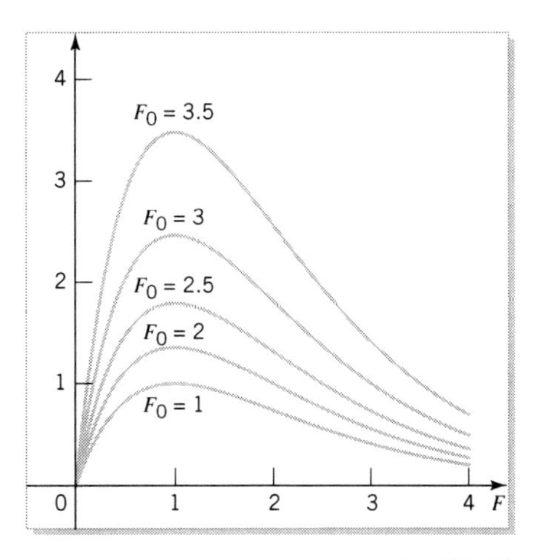

FIGURE 10.29 Plotting $(F/F_0)^a e^{-b(F-F_0)}$ for $F_0 = 3.5, 3.0, 2.5, 2.0$, and 1.0

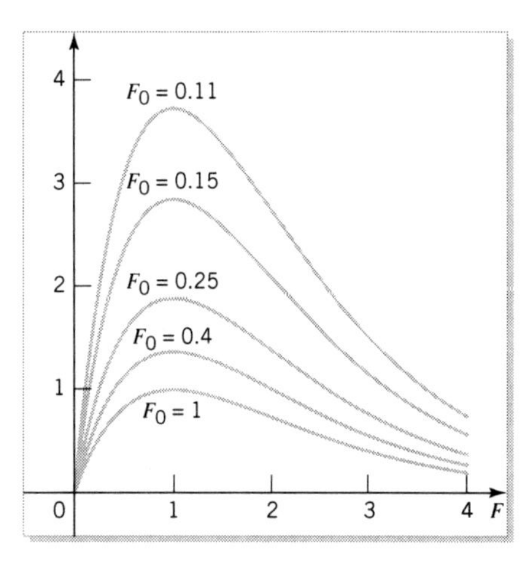

FIGURE 10.30 Plotting $(F/F_0)^a e^{-b(F-F_0)}$ for $F_0 = 1, 0.4, 0.25, 0.15$, and 0.11

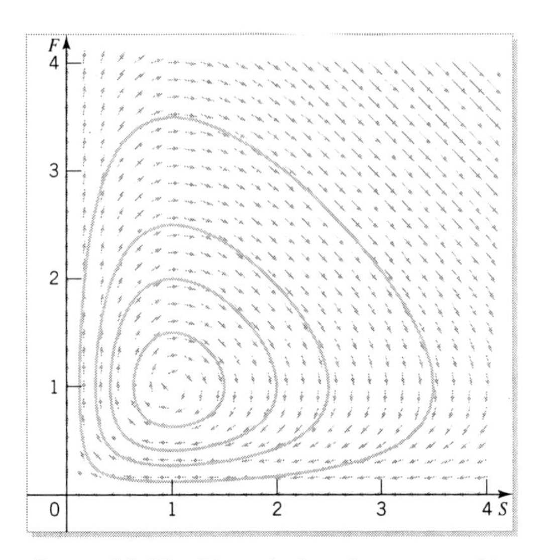

FIGURE 10.31 Numerical predator-prey orbits

At this stage the fish population is at its minimum. Because the shark population is so low, the fish have an opportunity to recover and gradually increase in numbers. The shark population continues to decline until the point $S \approx 0.11, F = 1$ is reached. Now there are sufficient fish for the sharks to increase, but not enough to cause a decline in the fish population. They eventually reach the point $S = 1, F = 3.5$ again, and the cycle repeats itself. This seems a reasonable model.

Notice that the fish population reaches a maximum before the shark population, and it is for this reason that we can think of the sharks as the predator and the fish as the prey. If the arrows in the direction field shown in Figure 10.31 were reversed, then this would mean that the shark population reaches a maximum before the fish population, and so that would mean that the fish are the predator and the sharks are the prey. The Lotka-Volterra system of equations has been used to model the Canadian lynx and snowshoe hare data from the Hudson Bay Company's records from 1875 to 1904. However, for this to be accurate, the hare would have been the predator and the lynx the prey![4]

A physically meaningful quantity associated with a periodic solution is its average value over one period.[5] We denote the average value of $S(t)$ over one period T by $\overline{S}$, and the average value of $F(t)$ by $\overline{F}$, so that

$$\overline{S} = \frac{1}{T}\int_0^T S(t)\,dt \tag{10.28}$$

[4]For a detailed analysis and additional references, see *Mathematical Biology* by J.D. Murray, Springer-Verlag, 1989, pages 66–68.

[5]The average value of a continuous function $f(t)$ over the interval $a \le t \le b$ is $\overline{f} = \frac{1}{b-a}\int_a^b f(t)\,dt$.

and

$$\overline{F} = \frac{1}{T}\int_0^T F(t)\,dt. \tag{10.29}$$

We can evaluate $\overline{S}$ if we consider the integral

$$\int_0^T \frac{F'(t)}{F(t)}\,dt \tag{10.30}$$

in two different ways.

On the one hand, we can evaluate (10.30) directly and obtain

$$\int_0^T \frac{F'(t)}{F(t)}\,dt = \ln F(T) - \ln F(0) = 0, \tag{10.31}$$

where we have used the fact that $F(t)$ is periodic of period T, so $F(T) = F(0)$. On the other hand, we can use (10.23) to evaluate (10.30) indirectly as

$$\int_0^T \frac{F'(t)}{F(t)}\,dt = \int_0^T \frac{kF - cSF}{F}\,dt = \int_0^T (k - cS)\,dt = kT - c\overline{S}T. \tag{10.32}$$

From (10.31) and (10.32), we thus have

$$\overline{S} = \frac{k}{c}. \tag{10.33}$$

In the same way, we can show that

$$\overline{F} = \frac{a}{b}. \tag{10.34}$$

Notice that the average values, (10.33) and (10.34), are just the coordinates of the equilibrium point.

These results have an interesting consequence. Imagine the prey is actually a pest being controlled by the predator. We decide to reduce the pest population further by using a pesticide, which is not discriminatory, so it also kills predators. If we look at the prey equation (10.23) — namely, $F' = kF - cSF$ — the effect of a pesticide would be to introduce a term, $-h_1F$ (h_1 a positive constant), so the modified prey equation would be

$$F' = kF - cSF - h_1F = (k - h_1)F - cSF. \tag{10.35}$$

In the same way, the predator equation (10.22) must be augmented by the term $-h_2S$ (h_2 a positive constant), so we have

$$S' = -aS + bSF - h_2S = -(a + h_2)S + bSF. \tag{10.36}$$

The average values, $\overline{S}_m$ and $\overline{F}_m$, for these modified equations are

$$\overline{S}_m = \frac{k - h_1}{c},$$

and

$$\overline{F}_m = \frac{a + h_2}{b},$$

from which it is clear that $\overline{S}_m < \overline{S}$ and $\overline{F}_m > \overline{F}$. In other words, the average pest population increases, while the average prey population decreases, which is exactly the opposite impact we wanted! This phenomenon is observed in practice. See Exercise 4 on page 484.

The S-I-R Model

EXAMPLE 10.14 *The S-I-R Model*

The following is a simple model for the evolution of an epidemic, such as the plague or measles.[6] Suppose there is a homogeneous group of N individuals, which at some time t consists of $S(t)$ individuals who are susceptible to infection (say by some virus), $I(t)$ individuals who are already infected, and $R(t)$ individuals who have been removed (by isolation, death, or recovery), and are thus immune. This model is often called the *S-I-R* model for epidemics. Note that we really only have two unknown quantities, because $S(t) + I(t) + R(t) = N$.

If we assume that the rate at which susceptible individuals become infected is proportional to both the number of infected individuals and the number of susceptible individuals, we have the differential equation

$$S' = -aSI. \tag{10.37}$$

Here we take a to be positive constant, so we need a minus sign in this equation, because the number of individuals susceptible to infection must be diminishing.

Conservation equation

The rate of change of the number of individuals who are infected is governed by a conservation equation,

$$\frac{dI}{dt} = \textit{rate of susceptibles becoming infected} - \textit{rate of infected being removed}. \tag{10.38}$$

If we assume that the rate of infected being removed is proportional to the number of individuals infected (the proportionality constant being the positive constant b), then we have

$$I' = aSI - bI. \tag{10.39}$$

Note that because $S(t) + I(t) + R(t) = N$, we have $0 \leq S(t) + I(t) = N - R(t) < N$ for all time, and that the rate of change of the removed category is given by

$$R' = -S' - I' = bI. \tag{10.40}$$

[6] "A contribution to the mathematical theory of epidemics" by W. O. Kermack and A. G. McKendrick, *Proc. Roy. Soc.* 115A, 1927, pages 700–721.

If we assume that at time $t = 0$, n people become ill and the rest are not infected, our initial conditions are

$$I(0) = n, \qquad S(0) = N - n, \tag{10.41}$$

which implies that $R(0) = 0$.

Phase plane

The differential equation governing the orbits in the SI phase plane is

$$\frac{dI}{dS} = \frac{aSI - bI}{-aSI} = -1 + \frac{c}{S},$$

Orbits

where $c = b/a$. Integration gives the orbits,

$$I(t) = -S(t) + c \ln S(t) + C. \tag{10.42}$$

Using the initial conditions, we find $C = N - c \ln S(0)$, so

$$I(t) = N - S(t) + c \ln \left[\frac{S(t)}{S(0)} \right]. \tag{10.43}$$

These are the orbits in the phase plane. In view of the fact that at time $t = 0$, $R(0) = 0$, we must have $I(0) = N - S(0)$, so that initial conditions for I and S must lie on the line $I = N - S$ in the phase plane. Some orbits and this line are plotted in Figure 10.32 for $N = 261$, $a = 2.78/159$, and $b = 2.78$, so $c = 159$.

From Figure 10.32 we see that $I(t)$ has a maximum (Why?), which, from (10.43), occurs at $S = c$. Thus, if the initial value of S is less than c, then the number of infectives decreases to zero, and the disease dies out. However, if the initial value of S is greater than c, the number of infectives increases to a maximum, and then dies to zero. This is sometimes called an epidemic.

Epidemic

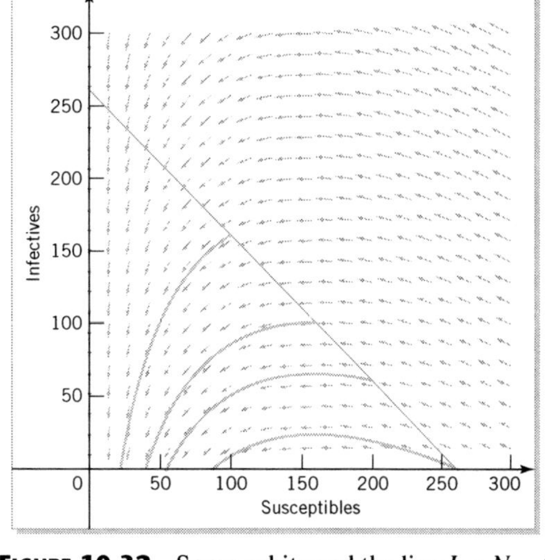

FIGURE 10.32 Some orbits and the line $I = N - S$

EXAMPLE 10.15 *The Eyam Plague*

Table 10.1 and Figure 10.33 show a data set for the number of deaths due to the plague in the small village of Eyam, England — with an initial population of 261 — from the start of the epidemic (June 18, 1666) to its end (October 20, 1666) in regular intervals.[7] The values for $S(t)$ in Table 10.1 were computed from $S(t) = N - R(t) - I(t)$, where $N = 261$. We want to see how well (10.43) fits this data set. To do that we need to estimate c, knowing that $N = 261$, and $S(0) = 254$. From Figure 10.32 we see that all orbits end on the horizontal axis, so that as the epidemic ends, $I \to 0$ as $S \to 83$, the number of susceptibles at the end of the epidemic. If we substitute this into (10.43), we find

$$0 = 261 - 83 + c \ln \left[\frac{83}{254} \right],$$

which allows us to estimate that $c \approx 159$. Figure 10.34 shows the data set and the graph of (10.43) with $c = 159$, $N = 261$, and $S(0) = 254$.

Table 10.1 The Eyam plague

Time	$R(t)$	$I(t)$	$S(t)$
0.0	0.0	7.0	254.0
0.5	11.5	14.5	235.0
1.0	38.0	22.0	201.0
1.5	78.5	29.0	153.5
2.0	120.0	20.0	121.0
2.5	145.0	8.0	108.0
3.0	156.0	8.0	97.0
3.5	167.5	4.0	89.5
4.0	178.0	0.0	83.0

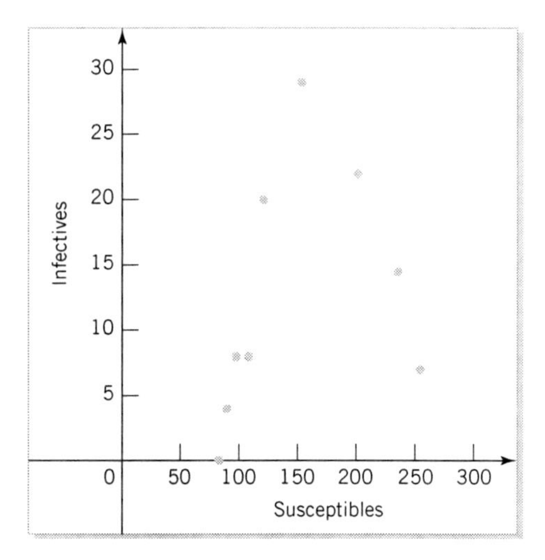

FIGURE 10.33 The infectives as a function of the susceptibles for the Eyam plague

Phase plane

If we are concerned about the number of removed, $R(t)$, then from (10.37) and (10.40) we have

$$\frac{dS}{dR} = \frac{dS}{dt} \bigg/ \frac{dR}{dt} = \frac{-aSI}{bI} = -\frac{a}{b}S,$$

[7] "Modelling the Eyam Plague" by G.F. Raggett, *Institute of Mathematics and Its Applications* 18, 1982, pages 221–226. The intervals were $15\frac{1}{2}$ days long, half of a 31-day month. Because each time interval contains a half-day, the actual data were averaged. This accounts for the fact that the table appears to contain half-people.

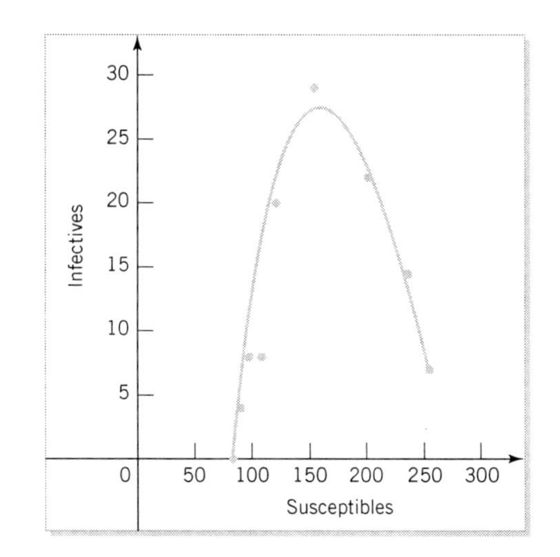

FIGURE 10.34 The infectives as a function of the susceptibles for the Eyam plague, and the theoretical curve

with solution

$$S(t) = A \exp\left[-\frac{a}{b}R(t)\right],$$

where A is a constant. Using the initial condition $R(0) = 0$, we find

$$S(t) = S(0) \exp\left[-\frac{a}{b}R(t)\right]. \tag{10.44}$$

From (10.40), (10.43), and (10.44), we thus have

$$\frac{dR}{dt} = bI = b\left\{N - S(t) + c\ln\left[\frac{S(t)}{S(0)}\right]\right\} = b\left\{N - S(0)\exp\left[-\frac{a}{b}R(t)\right] - \frac{ac}{b}R(t)\right\},$$

which, because $c = b/a$, reduces to

$$\frac{dR}{dt} = b\left\{N - S(0)\exp\left[-\frac{a}{b}R(t)\right] - R(t)\right\}. \tag{10.45}$$

We cannot solve this first order differential equation in terms of familiar functions. However, it is an autonomous differential equation, so we can analyze it using the techniques developed in Chapter 2. In fact, we have already examined (10.45) in that chapter for specific values of the parameters. We can repeat that technique, using the current parameters, and then compare the theoretical values of $R(t)$ with the experimental values. (Do this.) Figure 10.35 shows the result of this calculation together with the experimental values.

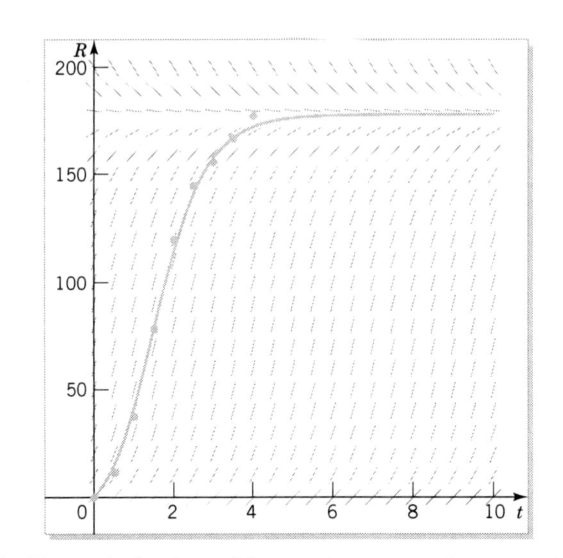

FIGURE 10.35 Theoretical values of the number removed compared to the real data

EXERCISES

1. **Nonlinear Springs.** The differential equation that governs the vertical motion of a mass m at the end of a linear spring with spring constant k is $mx'' = -F(x)$, where $F(x) = kx$ and x is the distance of the mass from its equilibrium position — the natural length of the spring. This is Hooke's law. Various generalizations of Hooke's law have been proposed — one of these is

$$F(x) = kx + ax^3, \tag{10.46}$$

where a is a constant. In this case the differential equation is $mx'' = -kx - ax^3$, where the constants m and k are positive, but a may be either positive or negative, depending on the spring. If $a > 0$ the spring is called **hard**, and if $a < 0$ it is called **soft**. Figure 10.36 shows the three different forces. By introducing the variable $y = x'$, this equation can be written as the system of equations $x' = y$, $y' = -\lambda^2 \left(x + 2bx^3\right)$, where $\lambda^2 = k/m$ and $b = a/(2k)$.

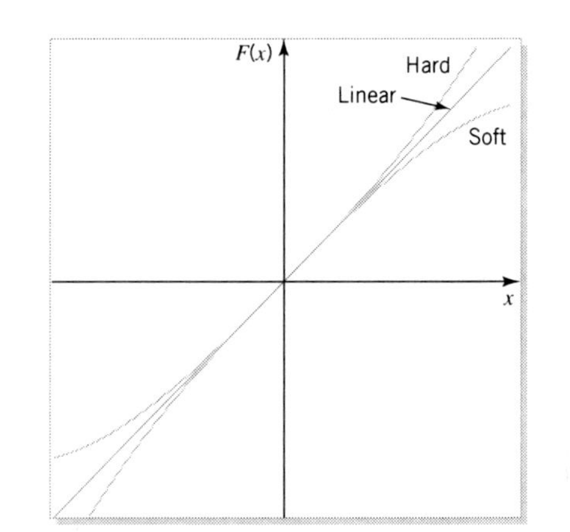

FIGURE 10.36 The forces due to a linear, hard, and soft spring

(a) Find all equilibrium points. Do they depend on whether the spring is hard or soft?

(b) By using a nullcline analysis, try to characterize the equilibrium point(s) and sketch some orbits in the phase plane.

(c) By linearizing the nonlinear system, try to characterize the equilibrium point(s) and sketch some orbits in the phase plane.

(d) Show that orbits in the phase plane are characterized by $y^2 = -\lambda^2 \left(x^2 + bx^4\right) + C$.

(e) Figures 10.37 and 10.38 show the direction fields and some orbits corresponding to a hard and a soft spring (with $\lambda = 1$ and $b = \pm 1/2$). Which is which? Explain fully.

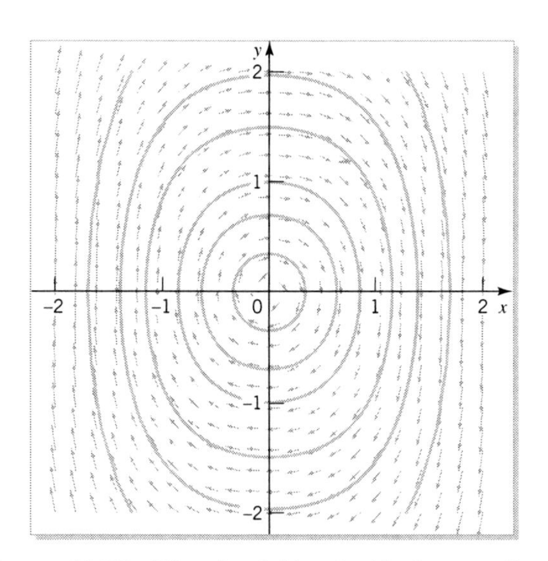

FIGURE 10.37 Direction field and orbits for a nonlinear spring. Is the spring hard or soft?

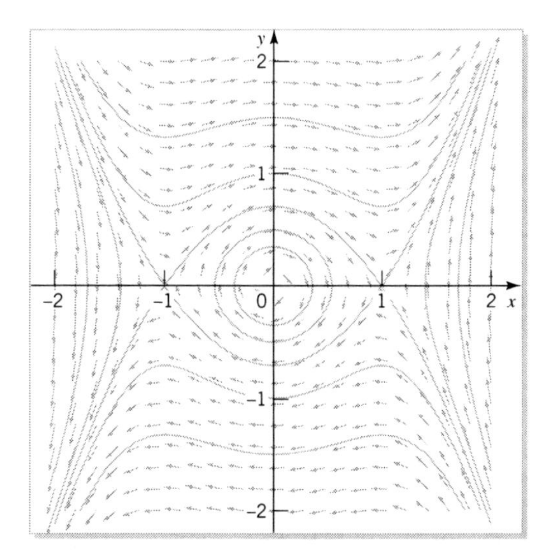

FIGURE 10.38 Direction field and orbits for a nonlinear spring. Is the spring hard or soft?

(f) The orbits in Figure 10.37 are all ovals. What does this tell you about the motion of the mass?

(g) The orbits in Figure 10.38 fall into different categories. Describe what each of these tells you about the motion of the mass. Is this a realistic model?

(h) We want to find the period, T, of periodic orbits. Use part (d) to show that if $x = x_0$ and $y = 0$ at $t = 0$, then $y^2 = \lambda^2 \left[x_0^2 - x^2 + b\left(x_0^4 - x^4\right)\right]$. Show that for periodic orbits, the period T is given by the improper integral

$$T = \frac{4}{\lambda} \int_0^{x_0} \frac{1}{\sqrt{x_0^2 - x^2 + b\left(x_0^4 - x^4\right)}}\, dx,$$

which, by the substitution $x = x_0 \sin u$, can be expressed as the proper integral

$$T = \frac{4}{\lambda} \int_0^{\pi/2} \frac{1}{\sqrt{1 + bx_0^2 + bx_0^2 \sin^2 u}}\, du.$$

Explain why this integral will exist always if the spring is hard, but will exist only when $x_0^2 \leq -1/(2b)$ if the spring is soft. Confirm that this is consistent with Figures 10.37 and 10.38. Table 10.2 shows the values of T for various x_0 for a hard and a soft spring ($b = \pm 1/2$). What do you notice as x_0 increases? Explain what this means in terms of Figures 10.37 and 10.38.

Table 10.2 Periods of hard and soft springs

x_0	Hard	Soft
0.1	$6.2598\sqrt{m/k}$	$6.3069\sqrt{m/k}$
0.2	$6.1911\sqrt{m/k}$	$6.3797\sqrt{m/k}$
0.3	$6.0818\sqrt{m/k}$	$6.5073\sqrt{m/k}$
0.4	$5.9385\sqrt{m/k}$	$6.7005\sqrt{m/k}$
0.5	$5.7688\sqrt{m/k}$	$6.9783\sqrt{m/k}$
0.6	$5.5807\sqrt{m/k}$	$7.3750\sqrt{m/k}$
0.7	$5.3811\sqrt{m/k}$	$7.9571\sqrt{m/k}$
0.8	$5.1761\sqrt{m/k}$	$8.8765\sqrt{m/k}$
0.9	$4.9701\sqrt{m/k}$	$10.6192\sqrt{m/k}$

2. In this exercise you are going to show that all orbits of the type (10.27), namely,

$$\left(\frac{F}{F_0}\right)^a e^{-b(F-F_0)} = \left(\frac{S}{S_0}\right)^{-k} e^{c(S-S_0)},$$

other than the equilibrium point $(k/c, a/b)$, are closed and contain the equilibrium point. Furthermore, for every orbit, the extreme values of F occur along the line $S = k/c$, and the extreme values of S occur along the line $F = a/b$.

(a) Consider the function $f(S) = (S/S_0)^{-k} e^{c(S-S_0)}$. Show that: $f(S) > 0$, $\lim_{S \to 0} f(S) = \infty$, $\lim_{S \to \infty} f(S) = \infty$, and $f'(S) = 0$ if $S = k/c$. Thus, the graph of $f(S)$ has a minimum $m = f(k/c)$ at $S = k/c$, and generally looks like the curve in the lower left-hand box of Figure 10.27.

(b) Consider the function $g(F) = (F/F_0)^a e^{-b(F-F_0)}$. Show that: $g(F) > 0$, $\lim_{F\to 0} g(F) = 0$, $\lim_{F\to\infty} g(F) = 0$, and $g'(F) = 0$ if $F = a/b$. Thus, the graph of $g(F)$ has a maximum $M = g(a/b)$ at $F = a/b$, and generally looks like the curve in the lower right-hand box of Figure 10.27.

(c) There are three possibilities concerning the values of m and M; namely, $m < M$, $m = M$, and $m > M$. By looking at Figure 10.27 show the following.

i. If $m < M$, then the orbits are closed and contain the point $(k/c, a/b)$. Also, the extreme values of F occur along the line $S = k/c$, and the extreme values of S occur along the line $F = a/b$.

ii. If $m = M$, then the point $(k/c, a/b)$ is the orbit.

iii. If $m > M$, then there are no orbits.

3. Computer Experiment — van der Pol's Equation. A special case of van der Pol's equation is $x'' + (x^2 - 1)x' + x = 0$. Show that, if $y = x'$, its orbits in the phase plane satisfy

$$\frac{dy}{dx} = -\frac{x + (x^2 - 1)y}{y}.$$

(a) What are the equilibrium point(s) of this system?

(b) Use a computer to plot various orbits. What long-term behavior do you notice?

(c) Draw the nullclines for this system. Do these, by themselves, explain the behavior you observed in part (b)?

(d) Linearize the nonlinear equation. Does this explain the behavior you observed in part (b)?

(e) Rewrite the phase plane differential equation in the form

$$\frac{dy}{dx} = -\frac{x}{y} - (x^2 - 1).$$

Explain why this can be interpreted as "the slope of an orbit for the van der Pol equation is $x^2 - 1$ less than the slope of an orbit for a circle." Why does this imply that if $|x| > 1$, then the orbits for the van der Pol equation move toward the origin, whereas if $|x| < 1$, then the orbits for the van der Pol equation move away from the origin?

(f) Use parts (c), (d), and (e) to explain the behavior you noticed in part (b).

(g) The full van der Pol equation is $x'' + a(x^2 - 1)x' + x = 0$, where a is a positive constant. Use the computer to plot the behavior of orbits for different values of a. Use an analysis similar to that used in parts (c), (d), and (e) to explain this behavior.

4. Data for fish predators (sharks, skates, etc.) caught as a percentage of the total fish catch for the port of Fiume, Italy, for the period 1914–1923 are given in Table 10.3. This spans World War I (1914–1918), when the amount of fishing was reduced, and shows that the percentage of predators was higher than under nonwar conditions. Assuming that this situation can be modeled by a Lotka-Volterra system of differential equations, explain why a reduced level of fishing is more beneficial to the predators than their prey.

5. The growth of two particular populations, $x(t)$ and $y(t)$, is modeled by the nonlinear system of equations

$$\begin{cases} x' = x(9 - x - y), \\ y' = y(6 - x - y). \end{cases}$$

What is the long-term behavior of these populations? Is there a straight-line orbit in the phase plane?

6. Another predator-prey model takes into consideration that the supply of food for the prey is limited. Recall that the logistic equation contained a term involving the square of the dependent variable to provide for a limiting value of population growth. Such a model would add a term such as $-hF^2$ to the right-hand side of (10.23), giving a system

$$\begin{cases} S' = S(-a + bF), \\ F' = F(k - cS - hF), \end{cases}$$

where h is a nonnegative constant.

(a) Find the carrying capacity (that is, the equilibrium solution) for the case where there are no predators, $(S = 0)$.

(b) Find all the equilibrium points, and determine their stability. Describe the effect of increasing h on the location of these points.

(c) Interpret each of these equilibrium points in terms of the numbers of predators and prey.

Table 10.3 **Percentage of sharks caught**

Year	1914	1915	1916	1917	1918	1919	1920	1921	1922	1923
%	11.9	21.4	22.1	21.2	36.4	27.3	16.0	15.9	14.8	10.7

10.5 BUNGEE JUMPING

A bungee jumper attaches one end of a piece of elastic to himself and attaches the other end to a platform above the ground. He then steps off the platform. Ideally he hurtles down until the elastic is fully stretched, and then rebounds, and continues to oscillate until he comes to rest. The elastic is then about twice its original length.

Consider an elastic bungee that is initially 10 m long, which, with a particular person attached to the end, stretches to a length of 20 m. If the platform is 35 m above the ground, is the bungee jumper going to hit the ground? We will answer this question by looking at two different models.

A Linear Model

Let x be the distance from the anchor point to the center of mass of the body of mass m at time t. If the natural length of the elastic is L, and the elastic obeys a version of Hooke's Law, namely,

$$F(x) = \begin{cases} 0 & \text{if } x \leq L, \\ k(x - L) & \text{if } x > L, \end{cases}$$

then the differential equation will be

$$mx'' = \begin{cases} mg & \text{if } x \leq L, \\ mg - k(x - L) & \text{if } x > L, \end{cases}$$

or

$$x'' = \begin{cases} g & \text{if } x \leq L, \\ g - \lambda^2(x - L) & \text{if } x > L, \end{cases}$$

where $\lambda = \sqrt{k/m}$. This is a linear differential equation. Because the body starts from rest, we have $x(0) = x'(0) = 0$. Here we have assumed that the air resistance and the mass of the elastic are negligible. [8]

The equilibrium solution will occur when $g - \lambda^2(x - L) = 0$; that is, when

$$x = L + L_E,$$

where

$$L_E = \frac{g}{\lambda^2} = \frac{gm}{k}.$$

The length of the elastic from the top to its equilibrium position is $L_E + L$, so L_E is the extension due to the mass, m. Thus, L_E is the amount by which the elastic stretches if we place the body in just the right position so it does not move. This

[8] These assumptions are unrealistic. Without any resistance, the bungee jumper will oscillate forever. Also, the assumption that the mass of the elastic is negligible will be disputed by any climber carrying a long coiled rope. Thus, this model has severe limitations.

can be used to estimate λ^2. For example, if $L = L_E = 10$, then $g/\lambda^2 = 10$, and so $\lambda^2 = g/10 \approx 0.98$.

If we introduce the change of variable $X = x - L$, this system becomes

$$X'' = \begin{cases} g & \text{if } X \leq 0, \\ g - \lambda^2 X & \text{if } X > 0, \end{cases}$$

subject to $X(0) = -L$, $X'(0) = 0$. The variable, X, is the distance of the body from the end of the natural length of the elastic.

We start by solving the top equation, $X'' = g$, in order to describe the initial fall and to find the position, time, and velocity of the body when the elastic starts to take effect at $X = 0$. These values for X, t, and X' will be the initial values for the bottom equation, $X'' = g - \lambda^2 X$. We then solve the bottom equation, which will be valid until $X = 0$ again.

The top equation can be integrated immediately, and using the initial conditions we find $X'(t) = gt$, $X(t) = gt^2/2 - L$. If T_1 is the first time that $X(T_1) = 0$, then $L = gT_1^2/2$, and so $T_1 = \sqrt{2L/g}$. Thus, $X'(T_1) = gT_1 = \sqrt{2gL}$.

The general solution of the bottom equation can be written in the form

$$X(t) = -A\cos\left[\lambda(t - T_1) + \phi\right] + L_E,$$

where A and ϕ are constants. Imposing the initial conditions $X(T_1) = 0$, $X'(T_1) = \sqrt{2gL}$, we find

$$A = \sqrt{L_E^2 + 2L_E L}$$

and

$$\phi = \arctan\left[\lambda\sqrt{\frac{2L}{g}}\right].$$

This solution is valid until the bungee jumper returns to $X = 0$ at time $T_2 = T_1 + (2\pi - 2\phi)/\lambda$. In terms of the original variables, we have

$$x(t) = \begin{cases} gt^2/2 & \text{if } t \leq T_1, \\ -A\cos\left[\lambda\left(t - T_1\right) + \phi\right] + L_E + L & \text{if } T_1 < t \leq T_1 + (2\pi - 2\phi)/\lambda. \end{cases}$$

At time $T_2 = T_1 + (2\pi - 2\phi)/\lambda$, the "free fall" part of the differential equation takes over; that is,

$$x'' = g,$$

subject to the initial conditions $x(T_2) = L$, $x'(T_2) = -gT_1$, which has solution

$$x(t) = \frac{1}{2}g\left(t - T_2\right)^2 - gT_1\left(t - T_2\right) + L.$$

This is valid until time T_3, when $X(T_3) = 0$, and so on. Thus, this model predicts that the bungee jumper oscillates indefinitely between a parabolic time dependence ("free fall") and a periodic time dependence ("simple harmonic motion"). This is shown in Figure 10.39 for the case $L = L_E = 10$ and represents the first time interval when the jumper returns to the height of the top of the tower. This same motion is then repeated indefinitely.

The maximum distance that the body will fall, x_{max}, occurs when $\cos[\lambda(t - T) + \phi] = -1$, and so

$$x_{max} = A + \frac{g}{\lambda^2} + L,$$

or

$$x_{max} = \sqrt{L_E^2 + 2L_E L} + L_E + L.$$

Notice that if we know L, the natural length of the elastic, and L_E, the amount by which it stretches to achieve equilibrium, then we can predict the maximum height of the tower top to avoid collision with the earth. This result can also be obtained by energy considerations.[9]

It is possible to analyze this motion in the phase plane by introducing $y = x'$ and considering

$$\frac{dy}{dx} = \begin{cases} g/y & \text{if } x \le L, \\ [g - \lambda^2(x - L)]/y & \text{if } x > L. \end{cases}$$

For $x \le L$ the orbits are parabolas,

$$y^2 = 2gx + C,$$

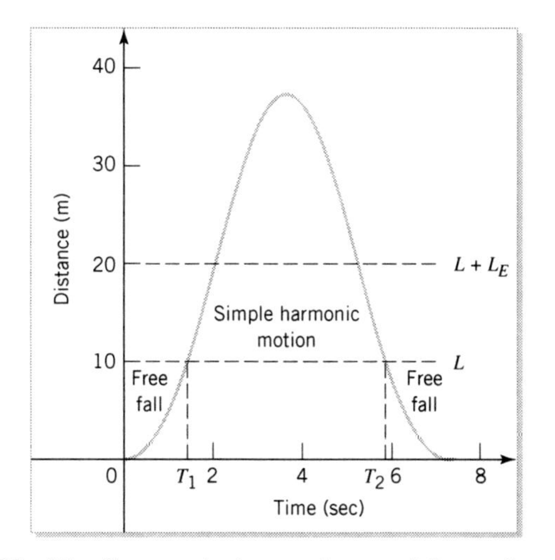

FIGURE 10.39 The distance the bungee jumper falls as a function of time

[9] "The Physics of Bungee Jumping" by P. G. Menz, *The Physics Teacher* 31, 1993, pages 483–487; and "The physics of bungee jumping" by T. Martin and J. Martin, *Physics Education* 29, 1994, pages 247–248.

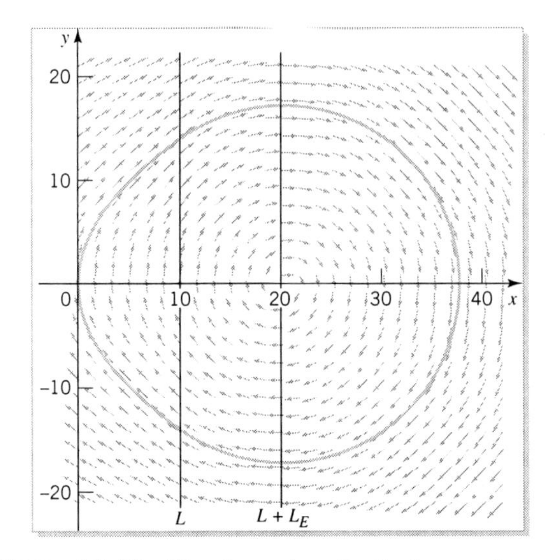

FIGURE 10.40 The phase plane for the bungee jumper

and for $x > L$ they are ellipses,

$$y^2 + \lambda^2 (x - L - L_E)^2 = C.$$

The direction field and the orbit corresponding to $x(0) = y(0) = 0$, namely,

$$y^2 = \begin{cases} 2gx & \text{if } x \leq L, \\ \lambda^2 (2L_E L + L_E^2) - \lambda^2 (x - L - L_E)^2 & \text{if } x > L, \end{cases}$$

are shown in Figure 10.40 for the case $L = L_E = 10$. We see that x_{max} occurs when $y = 0$. If we set $y = 0$ in the previous equation and solve for x_{max}, we again find

$$x_{max} = \sqrt{L_E^2 + 2L_E L} + L_E + L.$$

Because $L = L_E = 10$ m, we have $x_{max} \approx 37.32$ m, so the model predicts that the bungee jumper will hit the ground if he steps off a platform that is 35 m high.

A Nonlinear Model

The previous model does not take into account that bungee cords are made from rubber, and rubber has some complicated properties.[10] For example, take a thick flat rubber band about 3 inches long and 0.25 inches wide. Hold it without stretching between your two index fingers, and touch your lips with the flat side. It should feel slightly cool because the rubber is at room temperature, which is lower than your lip

[10] We are indebted to a colleague, John Kessler of the Department of Physics at the University of Arizona, for this information and demonstration.

temperature. Now quickly move your index fingers apart, causing the rubber band to stretch, and immediately touch your lips again. What do you notice? Keep the rubber band stretched until it attains room temperature, and now let it contract to its unstretched length, and touch it to your lips again. What do you notice?

Taking these properties into account leads to a different model than Hooke's Law for elastic; namely, the ideal rubber law[11]

$$F(x) = \begin{cases} 0 & \text{if } x \le L, \\ k\left(x - L^3/x^2\right) & \text{if } x > L. \end{cases}$$

In this case, the differential equation will be

$$x'' = \begin{cases} g & \text{if } x \le L, \\ g - \lambda^2\left(x - L^3/x^2\right) & \text{if } x > L, \end{cases}$$

where $\lambda = \sqrt{k/m}$. This is a nonlinear differential equation.

The equilibrium solution will occur when $g - \lambda^2\left(x - L^3/x^2\right) = 0$; that is, when

$$x = L + L_E,$$

where L_E satisfies the cubic equation

$$\left(L + L_E\right)^3 - \frac{g}{\lambda^2}\left(L + L_E\right)^2 - L^3 = 0.$$

The length of the elastic from the top to its equilibrium position is $L_E + L$, so L_E is the extension due to the mass, m. Thus, L_E is the amount by which the elastic stretches if we place the body in just the right position so it does not move. This can be used to estimate λ^2. For example, if $L = L_E = 10$, then $g/\lambda^2 = 17.5$, and so $\lambda^2 = g/17.5 \approx 0.56$.

Although it is not clear whether we can obtain an exact solution of the second order differential equation, we can still perform a phase plane analysis, obtaining

$$y^2 = \begin{cases} 2gx & \text{if } x \le L, \\ 2gx - \lambda^2(x^2 + 2L^3/x - 3L^2) & \text{if } x > L, \end{cases}$$

as the orbit through $x(0) = 0, y(0) = 0$. (Do this.) This orbit for the case $L = L_E = 10$ is shown in Figure 10.41. This orbit is oval — what does this tell you about the motion of the bungee jumper? Returning to our previous rubber band demonstration, place the rubber band against your lips, then stretch and contract the rubber band, and once more place it against your lips. What do you notice? How is this consistent with the oval orbit?[12]

---

[11] *Chemical Thermodynamics* by F. T. Wall, Freeman, 1974. See page 348.

[12] We are indebted to a colleague, John Robson of the Department of Physics at the University of Arizona, for this observation.

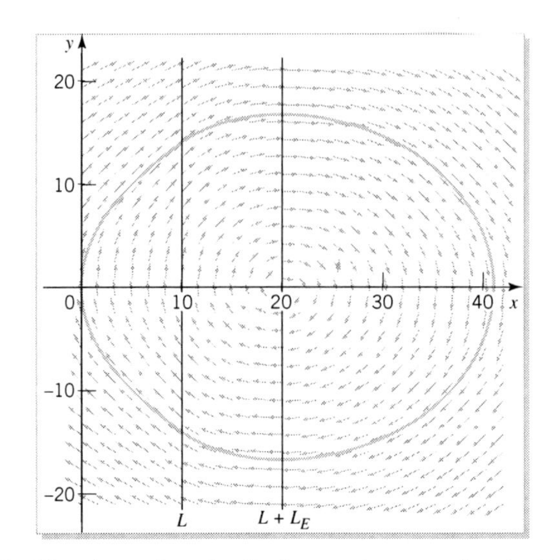

FIGURE 10.41 The phase plane for the bungee jumper using the ideal rubber law

We see that x_{max} occurs when $y = 0$. If we set $y = 0$ and $x = x_{max}$ in the previous equation, we see that x_{max} will satisfy the cubic equation

$$2gx_{max} - \lambda^2(x_{max}^2 + 2L^3/x_{max} - 3L^2) = 0.$$

In the case of $L = L_E = 10$, this equation becomes

$$x_{max}^3 - 35x_{max}^2 - 300x_{max} + 2000 = 0,$$

which we can solve numerically to find $x_{max} \approx 41.11$. That this distance is greater than we found previously, when $x_{max} \approx 37.32$, is consistent with the fact that this nonlinear model exerts less restoring force than the linear model.

EXERCISES

1. Show that if $f(x)$ is continuous in $a \leq x \leq b$; $f(x)$ has exactly two roots in $a < x < b$, namely, x_1 and x_2, where $a < x_1 < x_2 < b$; $f(x) > 0$ when $x_1 < x < x_2$; and $f(x) < 0$ when $x < x_1$ or when $x_2 < x$; then $y^2 = f(x)$ represents a closed curve in the xy-plane. [Hint: Sketch the graph of $y = +\sqrt{f(x)}$ and from it draw the graph of $y = -\sqrt{f(x)}$].[13] Explain how this shows that curves satisfying $x^{2m} + y^{2n} = 1$, where m and n are positive integers, are closed curves.

2. The purpose of this exercise is to show that the orbit shown in Figure 10.41 is in fact closed.

[13] This is based on the suggestion of our colleague, Shandelle Henson.

(a) Show that for the case $L = L_E = 10$, the orbit satisfies

$$y^2 = f(x),$$

where

$$f(x) = \begin{cases} 2gx & \text{if } x \le 10, \\ g\left(-x^3 + 35x^2 + 300x - 2000\right) / (17.5x) & \text{if } x > 10. \end{cases}$$

(b) Show that $f(x)$ is continuous for $x \ge 0$.

(c) Show that $f(x)$ has exactly two roots — namely, 0 and a where $a > 10$. [Hint: Observe that the cubic, $-x^3 + 35x^2 + 300x - 2000$, has three real roots: the first is negative, the second is positive but less that 10, and the third is a, which is greater than 10.]

(d) Show that $f(x) > 0$ if $0 < x < a$, and $f(x) < 0$ if $x < 0$ or $x > a$.

(e) Use the previous exercise to show that the orbits are closed.

(f) Repeat this for the general case

$$y^2 = f(x),$$

where

$$f(x) = \begin{cases} 2gx & \text{if } x \le L, \\ g\left(-x^3 + 2\alpha x^2 + 3L^2x - 2L^3\right) / (\alpha x) & \text{if } x > L, \end{cases}$$

where $\alpha = \left[\left(L + L_E\right)^3 - L^3\right] / \left(L + L_E\right)$.

3. Table 10.4 and Figure 10.42 show the results of an experiment where various known forces are applied to a rubber band of unstretched length 5.56 cm and its stretched length is measured.[14] How well does this fit the ideal rubber law, $F(x) = k(x - L^3/x^2)$?

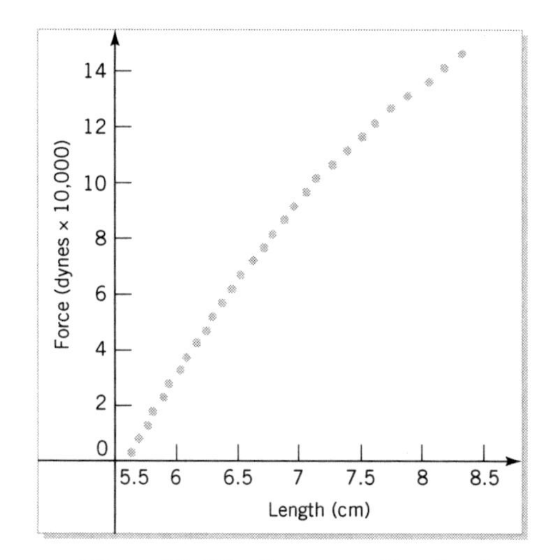

FIGURE 10.42 Force on rubber band as a function of the length

Table 10.4 Force on rubber band as a function of the length

Length (cm)	Force (dynes ×10,000)	Length (cm)	Force (dynes ×10,000)	Length (cm)	Force (dynes ×10,000)
5.63	0.3191	6.29	5.2128	7.14	10.1702
5.69	0.8085	6.37	5.7021	7.27	10.6809
5.76	1.2766	6.45	6.1915	7.39	11.1915
5.80	1.7872	6.52	6.7021	7.51	11.6809
5.89	2.2979	6.62	7.2128	7.61	12.1702
5.93	2.7660	6.71	7.6809	7.74	12.7021
6.03	3.2766	6.78	8.1702	7.88	13.1489
6.08	3.7234	6.88	8.7021	8.05	13.6383
6.16	4.2553	6.96	9.1702	8.18	14.1489
6.24	4.6809	7.06	9.6809	8.32	14.6596

[14] "A general physics laboratory investigation of the thermodynamics of a rubber band" by G. Savarino and M. R. Fisch, *The American Journal of Physics* 59, 1991, pages 141–145, figure 2.

10.6 LINEAR VERSUS NONLINEAR DIFFERENTIAL EQUATIONS

There are large differences between linear and nonlinear differential equations, whether autonomous or nonautonomous. We will demonstrate some of these differences by comparing the solutions of a linear differential equation with the solutions of a nonlinear differential equation.

We comment that under very special circumstances it is possible to convert a nonlinear differential equation into one that can be solved. See Exercises 1, 2, and 4.

Linear Versus Nonlinear Homogeneous

The linear differential equation $x'' - x = 0$ has the linearly independent functions e^t and e^{-t} as particular solutions, and the general solution is a linear combination of these functions, namely, $C_1e^t + C_2e^{-t}$.

The nonlinear differential equation $xx'' - 2(x')^2 + x^2 = 0$ has the linearly independent functions e^t and e^{-t} as particular solutions. However, a linear combination of these functions — namely, $C_1e^t + C_2e^{-t}$ — is not a solution. The general solution is $1/(C_1e^t + C_2e^{-t})$. (See Exercise 7 on page 497.)

Important point

In general, **a linear combination of two linearly independent solutions of a nonlinear differential equation is not the general solution of that differential equation. The Principle of Linear Superposition does NOT hold for nonlinear differential equations.**

The linear differential equation $a_2(t)x'' + a_1(t)x' + a_0(t)x = 0$ has the general solution $x(t) = C_1x_1(t) + C_2x_2(t)$. If $x(t)$ has a vertical asymptote at $t = t_1$, then either $x_1(t)$ or $x_2(t)$ has a vertical asymptote at $t = t_1$. The location $t = t_1$ of the vertical asymptote does not depend on C_1 and C_2 — that is, it is independent of the initial conditions — apart, possibly, from the exceptional case where either C_1 or C_2 is zero.

The nonlinear differential equation $xx'' - 2(x')^2 + x^2 = 0$ has the general solution $x(t) = 1/(C_1e^t + C_2e^{-t})$. The function $x(t)$ has a vertical asymptote at $t = \frac{1}{2}\ln(-C_2/C_1)$, which depends on C_1 and C_2 — that is, the location of the vertical asymptote is dependent on the initial conditions. For example, the initial conditions $x(0) = -1/3$, $x'(0) = 5/9$, give $C_1 = 1$ and $C_2 = -4$, and vertical asymptote at $t = \ln 2 \approx 0.6931$, whereas $x(0) = -1/3$, $x'(0) = 5/12$, give $C_1 = 3/8$ and $C_2 = -27/8$, and vertical asymptote at $t = \ln 3 \approx 1.0986$.

Important point

The location of a vertical asymptote is not dependent on the initial conditions if the differential equation is linear, but, in general, is dependent if it is nonlinear.

Linear Versus Nonlinear Nonhomogeneous

The linear differential equation $x'' - x = -2\sin t$ has the general solution $C_1e^t + C_2e^{-t} + \sin t$, which is of the form $x_h(t) + x_p(t)$, where $x_h(t) = C_1e^t + C_2e^{-t}$ — the complementary function — satisfies the associated homogeneous equation — namely, $x'' - x = 0$ — and $x_p(t) = \sin t$ is a particular solution of the original equation. The particular solution can be obtained by the method of undetermined coefficients, reduction of order, or variation of parameters.

The nonlinear differential equation $[xx'' - 2(x')^2 + x^2]/x^3 = -2\sin t$ has $x_h(t) = 1/(C_1e^t + C_2e^{-t})$ as the solution of $xx'' - 2(x')^2 + x^2 = 0$, and $x_p(t) = -\csc t$

as a particular solution of the original equation. However, the general solution is $1/(C_1 e^t + C_2 e^{-t} - \sin t)$, which is not of the form $x_h(t) + x_p(t)$. (See Exercise 8 on page 497.)

Important point

The solution of a nonhomogeneous linear differential equation is the sum of the complementary function and a particular solution. In general, this is not the case for nonhomogeneous nonlinear differential equations.

Consider the driven damped linearized simple pendulum with equation

$$x'' + \frac{1}{2}x' + x = a\cos\left(\frac{2}{3}t\right),$$

where a is a positive constant. Physically this corresponds to a damped linearized pendulum whose pivot experiences a torque of $a\cos\left(\frac{2}{3}t\right)$. The solution of this equation is

$$x(t) = e^{-t/4}\left[C_1 \sin\left(\frac{\sqrt{15}}{4}t\right) + C_2\cos\left(\frac{\sqrt{15}}{4}t\right)\right] + \frac{45}{34}a\cos\left(\frac{2}{3}t\right) + \frac{27}{34}a\sin\left(\frac{2}{3}t\right).$$

No matter what the initial conditions, the long-term behavior (steady state) is $\frac{45}{34}a\cos\left(\frac{2}{3}t\right) + \frac{27}{34}a\sin\left(\frac{2}{3}t\right)$, which is periodic with period 3π and amplitude $9a/\sqrt{34}$. Figure 10.43 shows the steady state solution for $a = 1$. Notice than any change in the value of a affects the amplitude but not the period of the long-term behavior. We contrast this with the nonlinear case.

We now consider the driven damped simple pendulum with nonlinear equation

$$x'' + \frac{1}{2}x' + \sin x = a\cos\left(\frac{2}{3}t\right), \tag{10.47}$$

where a is a positive constant. Physically this corresponds to a damped pendulum whose pivot experiences a torque of $a\cos\left(\frac{2}{3}t\right)$.

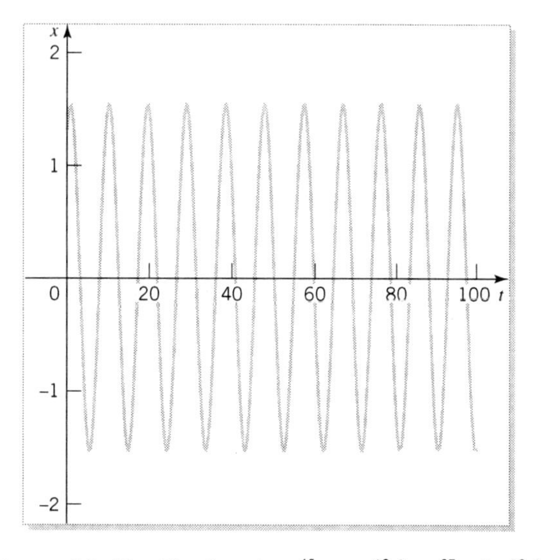

FIGURE 10.43 The function $\frac{45}{34}a\cos\left(\frac{2}{3}t\right) + \frac{27}{34}a\sin\left(\frac{2}{3}t\right)$

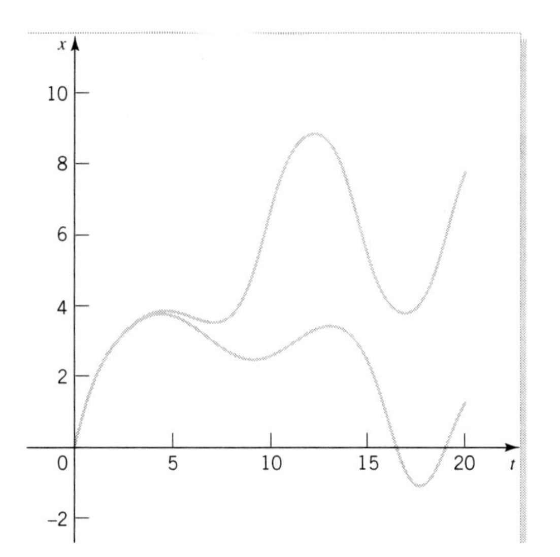

FIGURE 10.44 Two solutions of $x'' + \frac{1}{2}x' + \sin x = a\cos\left(\frac{2}{3}t\right)$

The following figures show different solutions of (10.47) where a is gradually increased through $a = 0.92$, 1.07, 1.08, and 1.15. All were produced using Runge-Kutta 4 with a step-size of 0.05.

In Figures 10.44 and 10.45, $a = 0.92, x(0) = 0$, and $x'(0)$ takes on the two values 2.20 and 2.21. The first figure shows the two solutions as functions of t, and the second shows the phase plane counterpart. This is an example of sensitivity to initial conditions.

We are now going to look at the long-term behavior of the solutions of (10.47). We do this by computing x from $t = 0$ to 200, and then plotting $x(t)$ from $t = 200$ to 300.

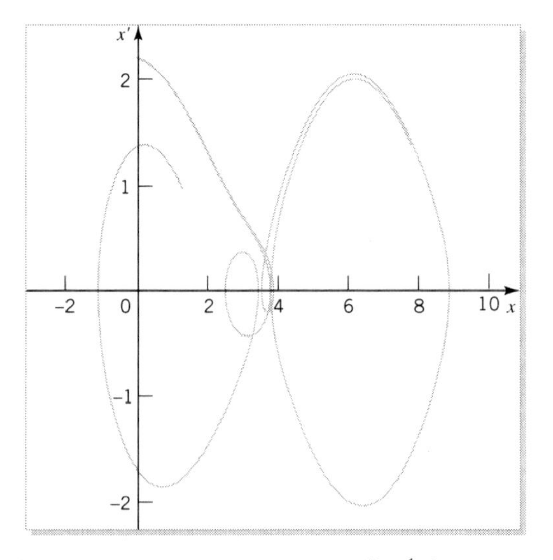

FIGURE 10.45 Orbits of two solutions of $x'' + \frac{1}{2}x' + \sin x = a\cos\left(\frac{2}{3}t\right)$

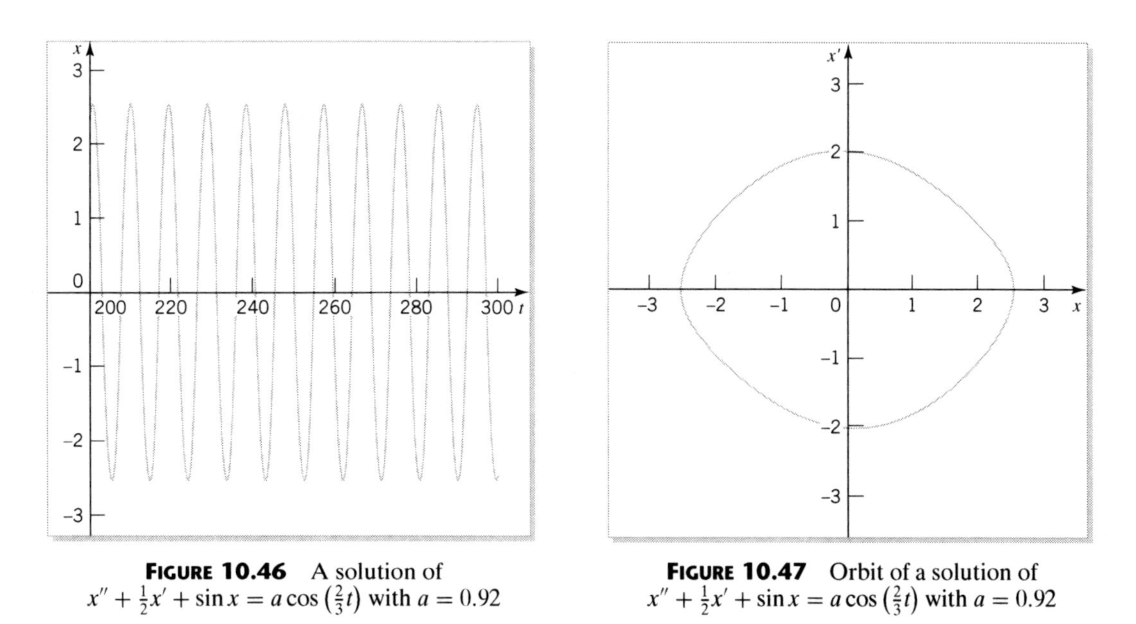

FIGURE 10.46 A solution of $x'' + \frac{1}{2}x' + \sin x = a\cos\left(\frac{2}{3}t\right)$ with $a = 0.92$

FIGURE 10.47 Orbit of a solution of $x'' + \frac{1}{2}x' + \sin x = a\cos\left(\frac{2}{3}t\right)$ with $a = 0.92$

In Figures 10.46 and 10.47, $a = 0.92$, $x(0) = 0$, and $x'(0) = 2$. The first figure shows the solution x as a function of t from $t = 200$ to $t = 300$, and the second shows the phase plane counterpart. This is periodic behavior.

In Figures 10.48 and 10.49, $a = 1.07$, $x(0) = 0$, and $x'(0) = 2$. The first figure shows the solution x as a function of t from $t = 200$ to $t = 300$, and the second shows the phase plane counterpart. This is periodic behavior, where the period is about double that in Figure 10.46.

In Figures 10.50 and 10.51, $a = 1.08$, $x(0) = 0$, and $x'(0) = 2$. The first figure shows the solution x as a function of t from $t = 200$ to $t = 300$, and the second shows

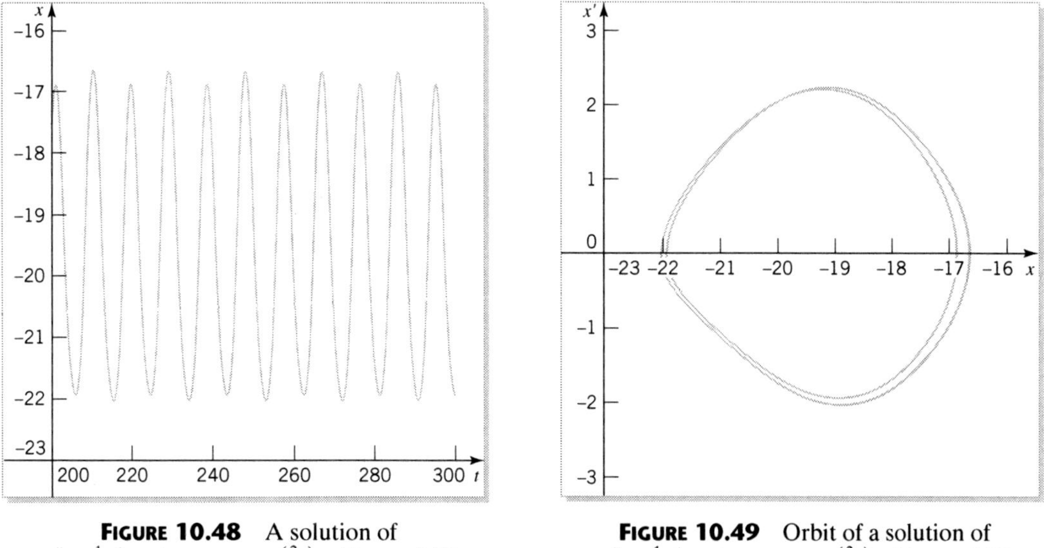

FIGURE 10.48 A solution of $x'' + \frac{1}{2}x' + \sin x = a\cos\left(\frac{2}{3}t\right)$ with $a = 1.07$

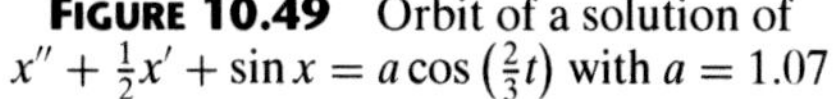

FIGURE 10.49 Orbit of a solution of $x'' + \frac{1}{2}x' + \sin x = a\cos\left(\frac{2}{3}t\right)$ with $a = 1.07$

the phase plane counterpart. This is periodic behavior, where the period is about double that in Figure 10.48 and quadruple that in Figure 10.46.

In Figures 10.52 and 10.53, $a = 1.15$, $x(0) = 0$, and $x'(0) = 2$. The first figure shows the solution x as a function of t from $t = 0$ to $t = 250$, and the second shows the phase plane continuation from $t = 250$ to $t = 500$. Chaos!

Thus, in contrast to the linear case, in the nonlinear case the period depends on the choice of the parameter a.

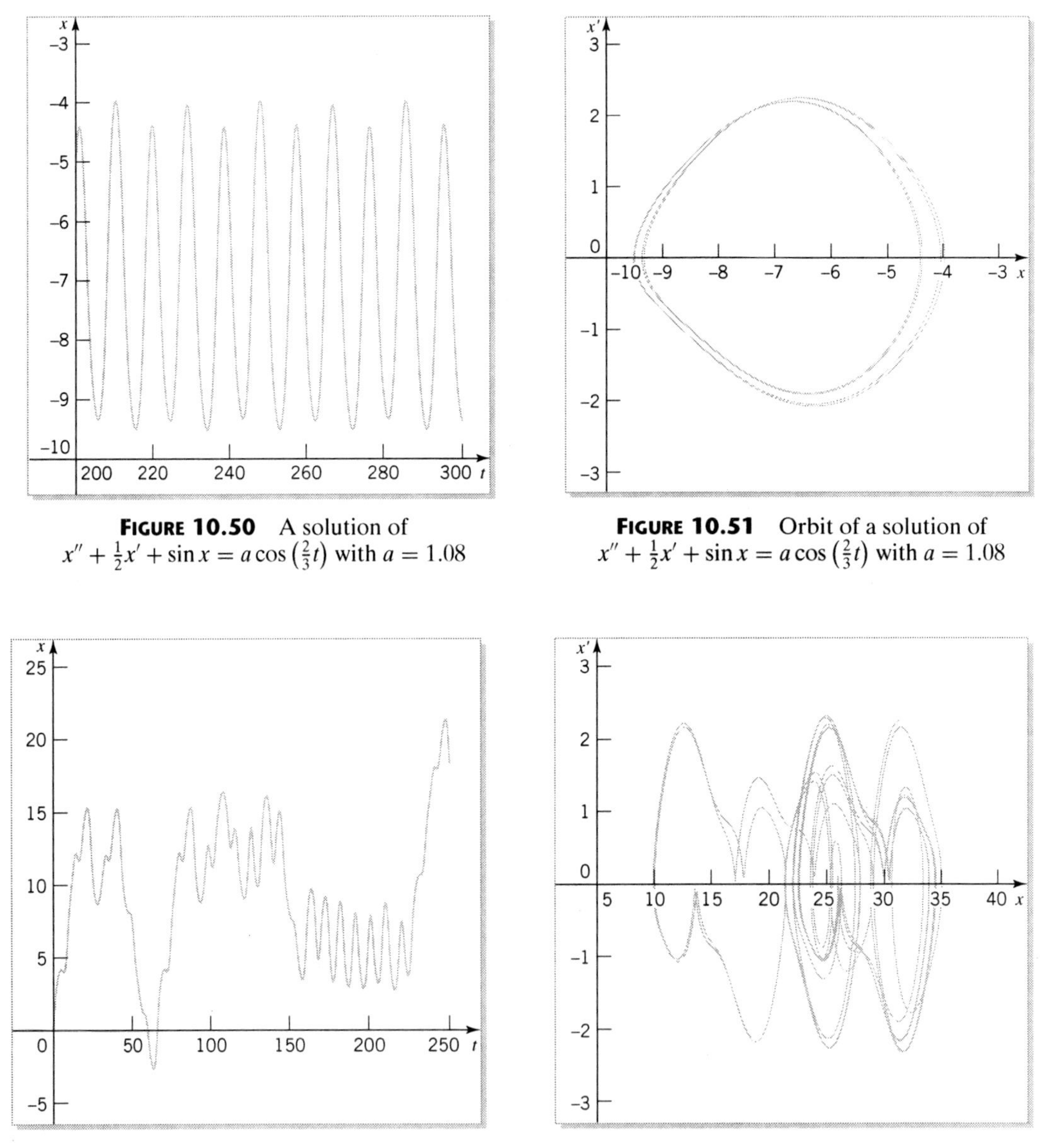

FIGURE 10.50 A solution of $x'' + \frac{1}{2}x' + \sin x = a\cos\left(\frac{2}{3}t\right)$ with $a = 1.08$

FIGURE 10.51 Orbit of a solution of $x'' + \frac{1}{2}x' + \sin x = a\cos\left(\frac{2}{3}t\right)$ with $a = 1.08$

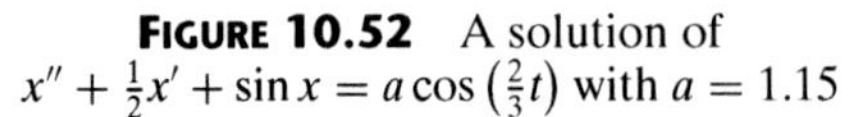

FIGURE 10.52 A solution of $x'' + \frac{1}{2}x' + \sin x = a\cos\left(\frac{2}{3}t\right)$ with $a = 1.15$

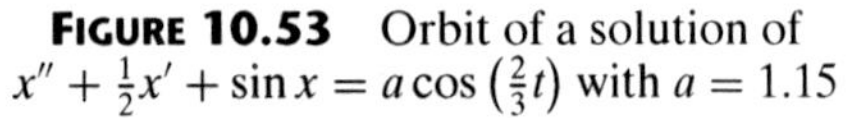

FIGURE 10.53 Orbit of a solution of $x'' + \frac{1}{2}x' + \sin x = a\cos\left(\frac{2}{3}t\right)$ with $a = 1.15$

EXERCISES

1. How to solve

$$\frac{d^2x}{dt^2} = F\left(t, \frac{dx}{dt}\right); \qquad (10.48)$$

that is, where the dependent variable x is missing, for $x = x(t)$. Make the substitution $y = \frac{dx}{dt}$, so $\frac{dy}{dt} = \frac{d^2x}{dt^2}$. The differential equation (10.48) becomes $\frac{dy}{dt} = F(t, y)$. This is a first order differential equation for $y = y(t)$. If possible, solve it for $y = y(t)$. The solution may be explicit or implicit. If it is explicit, solve $\frac{dx}{dt} = y(t)$ for $x = x(t)$. Use these ideas to solve $tx'' + x' - 4t = 0$.

2. Autonomous Equations. How to solve the autonomous equation

$$\frac{d^2x}{dt^2} = F\left(x, \frac{dx}{dt}\right); \qquad (10.49)$$

that is, where the independent variable t is missing, for $x = x(t)$. [Note that if $\frac{d^2x}{dt^2} = F\left(\frac{dx}{dt}\right)$ — that is, the dependent variable x is also missing — then the technique described in Exercise 1 also applies.] Make the substitution $y = \frac{dx}{dt}$. In this case, we compute x'' to eliminate t as the variable. Here we use the chain rule to write $\frac{d^2x}{dt^2} = \frac{d}{dt}\left(\frac{dx}{dt}\right) = \frac{d}{dt}(y) = \frac{dy}{dx}\frac{dx}{dt} = \frac{dy}{dx}y$. The differential equation (10.49) now becomes $y\frac{dy}{dx} = F(x, y)$. This is a first order differential equation for $y = y(x)$. It gives the orbits in the phase plane. If possible, solve it for $y = y(x)$. The solution may be explicit or implicit. If it is explicit, solve $\frac{dx}{dt} = y(x)$ for $x = x(t)$. Use these ideas to solve $2xx'' - (x')^2 - 1 = 0$.

3. Use exercises 1 and 2 to solve the following.

(a) $2tx'' + (x')^2 - 1 = 0$ (b) $x'' - (x')^2 = 0$
(c) $xx'' + (x')^2 = 0$ (d) $x'' + (x')^2 - 1 = 0$
(e) $(x^2 + 1)x'' - 2x(x')^2$

4. The Generalized Cauchy-Euler Equation. How to solve

$$t^2\frac{d^2x}{dt^2} = G\left(x, t\frac{dx}{dt}\right) \qquad (10.50)$$

for $x = x(t)$. Make the substitution $t = e^s$, so that $t\frac{dx}{dt} = \frac{dx}{ds}$ and $t^2\frac{d^2x}{dt^2} = \frac{d^2x}{ds^2} - \frac{dx}{ds}$. The differential equation (10.50) now becomes $\frac{d^2x}{ds^2} = \frac{dx}{ds} + G\left(x, \frac{dx}{ds}\right) = F\left(x, \frac{dx}{ds}\right)$, which is dealt with in Exercise 2. Use these ideas to solve $txx'' - t(x')^2 + xx' = 0$.

5. Curves of Constant Curvature. The curvature, $\kappa(x)$, of a curve $y = f(x)$ at the point (x, y) is

$$\kappa(x) = \frac{y''}{\left[1 + (y')^2\right]^{3/2}}.$$

For any given curve, $\kappa(x)$ measures how much the curve arches at the point (x, y). Find all curves $y = y(x)$ that have constant curvature — that is, solve $y'' = \kappa\left[1 + (y')^2\right]^{3/2}$ for $y = y(x)$ where κ is a nonzero constant.

6. If a projectile of mass m is fired vertically from the earth, the projectile's distance from the center of the earth, x, is governed by the differential equation $m\frac{d^2x}{dt^2} = -\frac{mgR^2}{x^2}$, where g is the gravitational constant and R is the radius of the earth.

(a) Letting $\frac{dx}{dt} = y$ and $\frac{d^2x}{dt^2} = y\frac{dy}{dx}$, solve the foregoing differential equation for y, subject to $y(R) = y_0$, obtaining $y = \sqrt{y_0^2 + 2gR^2\left(\frac{1}{x} - \frac{1}{R}\right)}$.

(b) Show that if $y_0^2 - 2gR > 0$, the velocity y will never be zero. (This critical value of y_0, $y_0 = \sqrt{2gR}$, is called the **escape velocity** of the earth.)

7. Show that e^t and e^{-t} are particular solutions of $xx'' - 2(x')^2 + x^2 = 0$. Show that the function $x(t) = C_1e^t + C_2e^{-t}$ is not a solution of $xx'' - 2(x')^2 + x^2 = 0$. Solve $xx'' - 2(x')^2 + x^2 = 0$ by using the substitution $x = 1/y$.

8. Solve $[xx'' - 2(x')^2 + x^2]/x^3 = -2\sin t$ by using the substitution $x = 1/y$.

9. Show that $1/t$ and $t/(1 + t^2)$ are particular solutions of $x'' + 6xx' + 4x^3 = 0$. Show that the function $x(t) = C_1/t + C_2t/(1 + t^2)$ is not a solution of $x'' + 6xx' + 4x^3 = 0$. Solve $x'' + 6xx' + 4x^3 = 0$ by using the substitution $x = y'/(2y)$.

10. Let a cable of uniformly distributed weight be suspended between two supports (see Figure 10.54). If we put the origin of our coordinate system a distance a below the lowest point of the cable, and the coordinates of any point on the cable are (x, y), the differential equation relating x and y is

$$a\frac{d^2y}{dx^2} = \left[1 + \left(\frac{dy}{dx}\right)^2\right]^{1/2}.$$

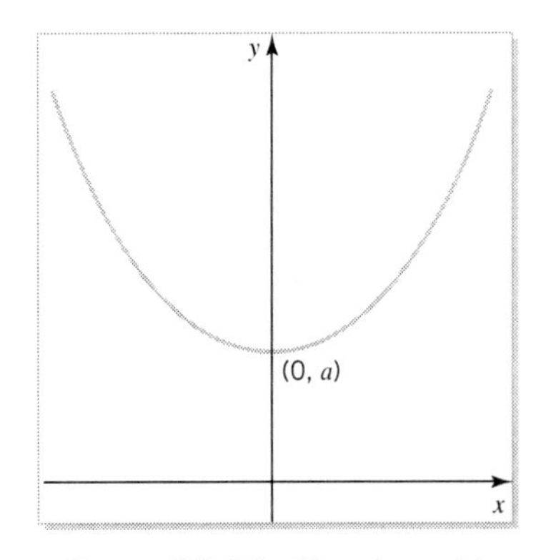

FIGURE 10.54 Hanging cable

The constant a in this equation is the ratio of the tension in the cable at the lowest point to the density of the cable. Show that the solution of this differential equation, subject to the condition $y(0) = a$, $y'(0) = 0$, may be expressed as $y = a \cosh \frac{x}{a}$. (The graph of this equation is called a **catenary**, which is derived from the Latin word for chain.)

11. **Computer Experiment — The Forced Ueda Equation.** The differential equation

$$x'' + 0.08x' + x^3 = 0.2 \cos t,$$

is a special case of the FORCED UEDA EQUATION. Use a computer program to solve this differential equation numerically subject to the initial condition $x(0) = -1$, $x'(0) = 0$, for $0 \leq t \leq 200$. Plot the continuation of your results in the window $-0.8 \leq x \leq 0.8$, $-0.8 \leq x' \leq 0.8$, of the phase plane for $200 \leq t \leq 300$. What is the period of your solution? Now change the initial condition to $x(0) = x_0$, for $x_0 = -0.9, -0.8$, and -0.7. What do you notice? Does this problem exhibit "sensitivity to initial conditions"?

12. **Computer Experiment — The Forced Duffing Equation.** The differential equation

$$x'' + x' - x + x^3 = a \cos t,$$

where a is a positive parameter, is a special case of the FORCED DUFFING EQUATION.

(a) Use a computer program to solve this differential equation numerically subject to the initial condition $x(0) = 0$, $x'(0) = 1$, for the case $a = 0.6$ for $0 \leq t \leq 100$. Plot the continuation of your results in the phase plane for $100 \leq t \leq 200$. What is the period of your solution? Now change the initial condition to $x'(0) = x_0^*$, for several different values of x_0^*. What do you notice? Does this problem exhibit "sensitivity to initial conditions"?

(b) Repeat part (a) for different values of a between 0.6 and 0.8. Can you find values of a that give limit cycles of period 2? Period 4? Chaos?

10.7 AUTONOMOUS VERSUS NONAUTONOMOUS DIFFERENTIAL EQUATIONS

There are large differences between autonomous and nonautonomous differential equations, whether linear or nonlinear. We will demonstrate some of these differences by comparing the solutions of a linear autonomous differential equation to the solutions of a linear nonautonomous differential equation.

Autonomous

Consider the linear autonomous differential equation $x'' + 3x' + 2x = 0$, which has the general solution

$$x(t) = C_1 e^{-t} + C_2 e^{-2t}.$$

If at $t = 0$ we have the initial conditions $x(0) = 0$, $x'(0) = 10$, then we find $C_1 = 10$, $C_2 = -10$. If we call this particular solution $x_1(t)$, then

$$x_1(t) = 10\left(e^{-t} - e^{-2t}\right).$$

If we want to find the time T at which $x_1' = 0$, then we would solve $x_1'(T) = 0$ for T; that is ,

$$x_1'(T) = 10\left(-e^{-T} + 2e^{-2T}\right) = 0,$$

in which case $T = \ln 2$. At this time we find $x_1(T) = 10\left(e^{-T} - e^{-2T}\right) = 10\left(\frac{1}{2} - \frac{1}{4}\right) = 2.5$.

If at $t = 0$ we have the initial conditions $x(0) = 2.5$, $x'(0) = 0$, then we find $C_1 = 5$, $C_2 = -2.5$. If we call this particular solution $x_2(t)$, then

$$x_2(t) = 5e^{-t} - 2.5e^{-2t}.$$

So if we have two identical systems, each of which obeys $x'' + 3x' + 2x = 0$, and at $t = 0$ they start with the initial conditions $x(0) = 0$, $x'(0) = 10$, and $x(0) = 2.5$, $x'(0) = 0$ respectively, then after $T = \ln 2$ the first system will be at the same place with the same velocity as the second one was when it started. Apart from the T delay, are the motions identical? A simple way to get some insight into this is to plot both orbits in the phase plane; that is, evaluate $x_1(t)$ and $x_1'(t)$ for various values of t and then plot $\left(x_1(t), x_1'(t)\right)$, and repeat this for $x_2(t)$. This is done in Figure 10.55, where the orbits are indistinguishable.

It appears that $x_1(t)$ for $t > T$ and $x_2(t)$ for $t > 0$ are the same; in other words, $x_1(T + t) = x_2(t)$. In fact,

$$x_1(T + t) = 10\left(e^{-(T+t)} - e^{-2(T+t)}\right) = 10\left(\frac{1}{2}e^{-t} - \frac{1}{4}e^{-2t}\right) = 5e^{-t} - 2.5e^{-2t} = x_2(t).$$

If we plot $x_1(t)$ and $x_2(t)$ on the same graph, we find Figure 10.56, where it appears that one graph is a translation of the other.

Thus, we are led to the following results.

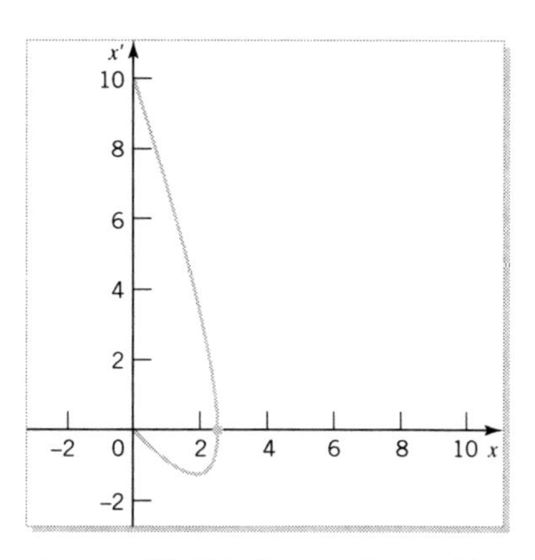

FIGURE 10.55 Phase plane orbits through (0, 10) and (2.5, 0) of $x'' + 3x' + 2x = 0$

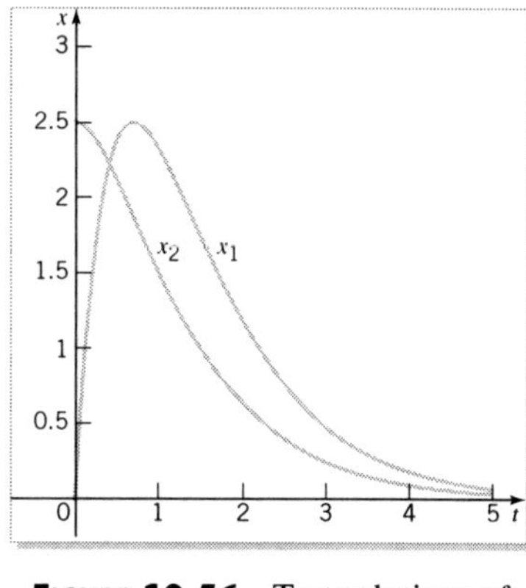

FIGURE 10.56 Two solutions of $x'' + 3x' + 2x = 0$

▶ *Theorem 10.2: The Translation Theorem*

(a) *If $x(t)$ is a solution of the autonomous differential equation $x'' = R(x, x')$, then so is $x(t + c)$ for any constant c.*

(b) *If $x(t)$, $y(t)$, are solutions of the autonomous system of differential equations*

$$\begin{aligned} x' &= P(x, y), \\ y' &= Q(x, y), \end{aligned}$$

then so are $x(t + c)$, $y(t + c)$, for any constant c. ◀

Comments about the Translation Theorem

- These solutions are obtained from each other by translating along the t-axis. There is a very simple interpretation of this in terms of how we related the solutions of a system of equations to its phase plane orbit described on page 267. We have reproduced Figure 6.32 here as Figure 10.57. A second solution, $x(t)$, $y(t)$, could be obtained from the first by sliding $x(t)$ and $y(t)$ in the right-hand boxes a distance c to the right. If we then construct the orbit corresponding to this new solution, we find it is exactly the same orbit as the one already drawn.
- One orbit in the phase plane represents an infinite number of solutions, $x(t)$, $y(t)$.
- A knowledge of all solutions subject to $x(0)$, $x'(0)$, determines all solutions subject to $x(t_0)$, $x'(t_0)$.

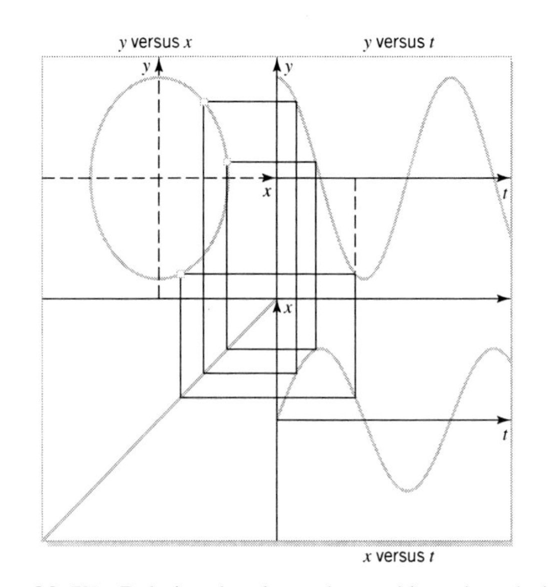

FIGURE 10.57 Relating the phase plane orbit to the solution

Nonautonomous

We now apply the same calculations and ideas to the linear nonautonomous differential equation

$$x'' + 2tx' + \left(1 + t^2\right)x = 0,$$

which has the general solution

$$x(t) = \left(C_1 + C_2 t\right)e^{-\frac{1}{2}t^2}.$$

(Check this.)

If at $t = 0$ we have the initial conditions $x(0) = 0$, $x'(0) = 10$, then we find $C_1 = 0$, $C_2 = 10$. If we call this particular solution $x_1(t)$, then

$$x_1(t) = 10te^{-\frac{1}{2}t^2}.$$

If we want to find the time T at which $x_1' = 0$, then we would solve $x_1'(T) = 0$ for T; that is,

$$x_1'(T) = 10\left(1 - t\right)e^{-\frac{1}{2}t^2} = 0,$$

in which case $T = 1$. At this time $x_1(T) = 10Te^{-\frac{1}{2}T^2} = 10e^{-\frac{1}{2}}$.

If at $t = 0$ we have the initial conditions $x(0) = 10e^{-\frac{1}{2}}$, $x'(0) = 0$, then we find $C_1 = 10e^{-\frac{1}{2}}$, $C_2 = 0$. If we call this particular solution $x_2(t)$, then

$$x_2(t) = 10e^{-\frac{1}{2}}e^{-\frac{1}{2}t^2}.$$

So if we have two identical systems, each of which obeys $x'' + 2tx' + (1 + t^2)x = 0$, and at $t = 0$ they start with the initial conditions $x(0) = 0$, $x'(0) = 10$, and $x(0) = 10e^{-1/2}$, $x'(0) = 0$ respectively, then after $T = 1$ the first system will be at the same place with the same velocity as the second one was when it started. Apart from the T delay, are the motions identical? A simple way to get some insight into this is to plot both orbits in the phase plane; that is, evaluate $x_1(t)$ and $x_1'(t)$ for various values of t and plot $(x_1(t), x_1'(t))$, and then repeat this for $x_2(t)$. See Figure 10.58. It appears that $x_1(t)$ from $t = T$ and $x_2(t)$ from $t = 0$ are different; in other words, $x_1(T + t) \neq x_2(t)$. In fact, there is no time when $x_2(t) = 0$. The curve that starts at $(0, 10)$ is the outermost curve. The one that starts at $10e^{-1/2} \approx 6$ is the innermost curve. If we were to trace this orbit backward in time from $x(0) = 10e^{-1/2}$, $x'(0) = 0$, what would you expect it to look like? See Figure 10.59.

If we plot $x_1(t)$ and $x_2(t)$ on the same graph, we find Figure 10.60.

Comments about Nonautonomous Differential Equations

- The orbits in the phase portrait can be very tangled because they depend on time.
- The phase portrait may still be useful but it does not give the global picture. Closed orbits need not correspond to periodic solutions. (See Exercise 2 on page 502).

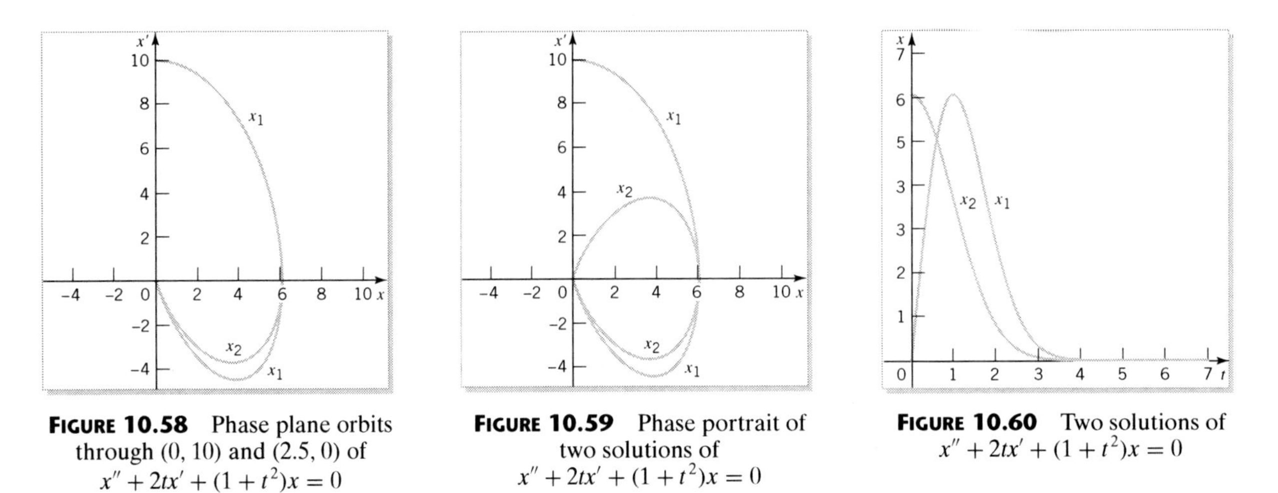

FIGURE 10.58 Phase plane orbits through (0, 10) and (2.5, 0) of $x'' + 2tx' + (1 + t^2)x = 0$

FIGURE 10.59 Phase portrait of two solutions of $x'' + 2tx' + (1 + t^2)x = 0$

FIGURE 10.60 Two solutions of $x'' + 2tx' + (1 + t^2)x = 0$

- The phase portrait may change dramatically depending on t. See Figure 10.61, which shows two solutions of $x'' + 2tx' + (1 + t^2)x = 0$, subject to the initial conditions $x(0) = 0, x'(0) = 10$, and $x(1) = 3.6, x'(1) = 10$, respectively.

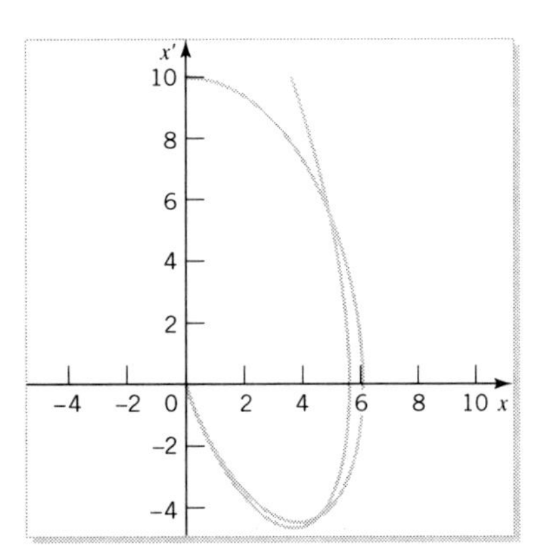

FIGURE 10.61 Phase portrait of two solutions of $x'' + 2tx' + (1 + t^2)x = 0$

EXERCISES

1. Figure 10.62 shows the graph of a solution of the autonomous differential equation $x'' = R(x, x')$. Draw the graph of another solution of the same equation.

2. Consider the system of equations

$$\begin{cases} x' = -y/t, \\ y' = x/t, \end{cases} \tag{10.51}$$

for $t > 0$.

(a) Use the elimination method on (10.51) to show that x satisfies the Cauchy-Euler equation $t^2x'' + tx' + x = 0$. Solve this for $x(t)$ and then use the top equation in (10.51) — namely, $y = -tx'$ — to find $y(t)$. Do $x(t)$ and $y(t)$ oscillate? Are $x(t)$ and $y(t)$ periodic?

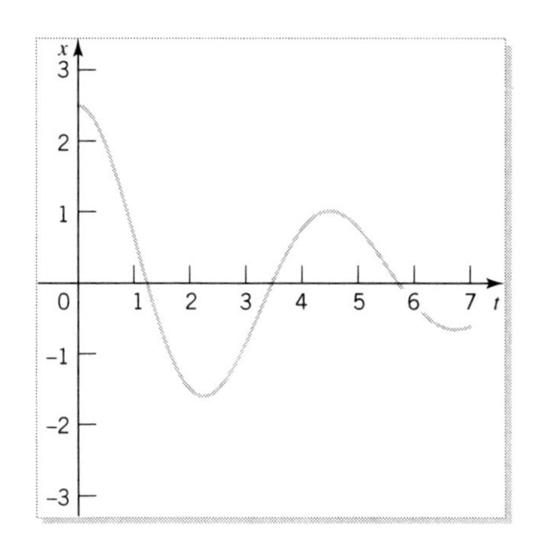

FIGURE 10.62 A solution of $x'' = R(x, x')$

(b) Show that the orbits of (10.51) in the phase plane are given by circles $x^2 + y^2 = C$. If (10.51) was autonomous, what type of motion would be characterized by a circle in the phase plane? Reconcile this result with your conclusions to part (a).

What Have We Learned?

Main Ideas

- We examined four examples where we could use some type of explicit solution, either in the time domain or in the phase plane, to see what types of long-term behavior are possible for nonlinear autonomous systems of differential equations. Nonlinear systems may have more than one isolated equilibrium point, allowing more variety in the behavior of solutions of nonlinear systems.
- We discovered that the method of nullclines from Chapter 9 applied directly to nonlinear systems — see *How to Perform a Nullcline Analysis on a Nonlinear Autonomous System* on page 454.
- We developed a linearization method to find out when the behavior of an associated linear system dictates the behavior of the nonlinear system — see the Linearization Theorem on page 460 and *How to Perform a Linearization Analysis on a Nonlinear Autonomous System* on page 463.
- To present a systematic approach to constructing a phase portrait for nonlinear autonomous systems, we developed a checklist of applicable techniques — see *How to Construct a Phase Portrait of a Nonlinear Autonomous System* on page 465.
- There is no single approach that will allow us to discover the behavior of solutions of a general autonomous system of differential equations.
- We discussed some differences between properties of linear and nonlinear systems and between autonomous and nonautonomous systems.
- Models considered in this chapter that involved nonlinear systems of autonomous differential equations pertained to populations, predator-prey situations, nonlinear springs and simple pendulums, epidemics, and bungee jumping.

CHAPTER 11

USING LAPLACE TRANSFORMS

Where We Are Going — and Why

While we already have techniques for solving all orders of nonhomogeneous linear differential equations with constant coefficients, those techniques are cumbersome for discontinuous or periodic forcing functions. A method that easily handles these situations will be introduced in this chapter — Laplace transforms. It provides an alternative method for solving the differential equations in previous chapters by reducing the problem to a purely algebraic one. It is particularly suited to initial value problems. This transform method also applies to systems of linear differential equations with constant coefficients.

Because the existence of a table of transforms and inverse transforms of many functions greatly eases the use of Laplace transforms, we develop results that help establish such tables.

Laplace transforms are the method of choice for solving differential equations in some disciplines.

11.1 MOTIVATION

We introduce Laplace transforms as another way of solving initial value problems for linear differential equations with constant coefficients. This technique allows us to convert the task of solving a differential equation to that of solving an algebraic equation. We begin this section by demonstrating the use of this technique with a basic example.

Basic Example

EXAMPLE 11.1

Consider the initial value problem

$$y' + y = e^{at}, \tag{11.1}$$

where a is a constant, subject to

$$y(0) = 0. \tag{11.2}$$

Explicit solution Using the techniques from Chapter 5, we already know that our solution is

$$y(t) = \frac{1}{a+1}\left(e^{at} - e^{-t}\right) \tag{11.3}$$

if $a \neq -1$, and

$$y(t) = te^{-t} \tag{11.4}$$

if $a = -1$. (Verify that these are the solutions.) We will now solve this same initial value problem in a completely different way, which at first will appear very arbitrary and unmotivated.

Many of the initial value problems that arise in applications have time as the independent variable. Usually the initial condition is given at $t = 0$ and the region of interest is $t \geq 0$. Thus, as we solve this problem, we focus attention on the interval $0 \leq t < \infty$.

First we multiply (11.1) by e^{-st}, where s is a positive constant, and integrate the result from 0 to ∞, obtaining

$$\int_0^\infty e^{-st}\left(y' + y\right)\,dt = \int_0^\infty e^{-st}e^{at}\,dt,$$

or

$$\int_0^\infty e^{-st}y'\,dt + \int_0^\infty e^{-st}y\,dt = \int_0^\infty e^{(a-s)t}\,dt. \tag{11.5}$$

We next evaluate the integral on the right-hand side as

$$\int_0^\infty e^{(a-s)t}\,dt = \lim_{b\to\infty}\int_0^b e^{(a-s)t}\,dt = \lim_{b\to\infty}\left[\frac{1}{a-s}e^{(a-s)t}\Bigg|_0^b\right] = \lim_{b\to\infty}\frac{1}{a-s}e^{(a-s)b} - \frac{1}{a-s},$$

or, if $s > a$,

$$\int_0^\infty e^{(a-s)t}\,dt = \frac{1}{s-a}. \tag{11.6}$$

Substituting (11.6) into (11.5) gives

$$\int_0^\infty e^{-st}y'\,dt + \int_0^\infty e^{-st}y\,dt = \frac{1}{s-a}. \tag{11.7}$$

We now look at the left-hand side of (11.7). If we use integration by parts — that is,

$$\int uv'\,dt = uv - \int vu'\,dt$$

on the first integral, by letting $u = e^{-st}$ and $v' = y'$, so $u' = -se^{-st}$ and $v = y$ — we find

$$\int e^{-st}y'\,dt = e^{-st}y + s\int e^{-st}y\,dt.$$

Inserting the limits of integration gives

$$\int_0^\infty e^{-st}y'\,dt = \lim_{b\to\infty}\left(e^{-st}y\Big|_0^b\right) + s\int_0^\infty e^{-st}y\,dt,$$

or

$$\int_0^\infty e^{-st}y'\,dt = \lim_{b\to\infty} e^{-sb}y(b) - y(0) + s\int_0^\infty e^{-st}y\,dt. \tag{11.8}$$

Using the initial condition $y(0) = 0$, we find

$$\int_0^\infty e^{-st}y'\,dt = \lim_{b\to\infty} e^{-sb}y(b) + s\int_0^\infty e^{-st}y\,dt. \tag{11.9}$$

If we look at the term $\lim_{b\to\infty} e^{-sb}y(b)$ and realize that $\lim_{b\to\infty} e^{-sb} = 0$, we would expect that $\lim_{b\to\infty} e^{-sb}y(b) = 0$ for many functions. For the time being, let's assume that this is true for our solution y, in which case (11.9) becomes

$$\int_0^\infty e^{-st}y'\,dt = s\int_0^\infty e^{-st}y\,dt. \tag{11.10}$$

We now substitute this equation into (11.7) to find

$$s\int_0^\infty e^{-st}y\,dt + \int_0^\infty e^{-st}y\,dt = \frac{1}{s-a},$$

or

$$(s+1)\int_0^\infty e^{-st}y\,dt = \frac{1}{s-a}.$$

This means that the solution $y = y(t)$ that we seek must satisfy

$$\int_0^\infty e^{-st}y\,dt = \frac{1}{(s+1)(s-a)}. \tag{11.11}$$

One way to find a solution would be to compute $\int_0^\infty e^{-st}y\,dt$ for many different functions $y(t)$, such as 1, t, $\sin t$, $\cos t$, e^t, and so on, or more generally for a, t^n, $\sin at$, $\cos at$, and e^{at}. Then, if we are lucky, we might find that for one of these functions we had (11.11). For example, in (11.6) we found that

$$\int_0^\infty e^{-st}e^{at}\,dt = \frac{1}{s-a}, \tag{11.12}$$

so that if the right-hand side of (11.11) were $1/(s-a)$, our solution would be $y(t) = e^{at}$. In the present case, we can take advantage of this knowledge as follows. For the

Partial fractions case $a \neq -1$, we use partial fractions to change the form of the right-hand side of (11.11) to[1]

$$\frac{1}{(s+1)(s-a)} = \frac{1}{a+1}\left(\frac{1}{s-a} - \frac{1}{s+1}\right).$$

Substituting this expression into (11.11) yields

$$\int_0^\infty e^{-st} y \, dt = \frac{1}{a+1}\left(\frac{1}{s-a} - \frac{1}{s+1}\right),$$

or, by using (11.12) twice — once with arbitrary a and once with $a = -1$ — we find

$$\int_0^\infty e^{-st} y \, dt = \frac{1}{a+1}\left(\int_0^\infty e^{-st} e^{at} \, dt - \int_0^\infty e^{-st} e^{-t} \, dt\right),$$

or

$$\int_0^\infty e^{-st} y \, dt = \int_0^\infty e^{-st}\left[\frac{1}{a+1}\left(e^{at} - e^{-t}\right)\right] dt.$$

Explicit solution Thus, the solution of our initial value problem is

$$y(t) = \frac{1}{a+1}\left(e^{at} - e^{-t}\right),$$

which agrees with (11.3).

Working Assumptions

Now let's look at what we have just done, first in broad outline, and then with a more critical eye. We started with (11.1) and, after multiplying by e^{-st} and integrating, obtained (11.5). Notice that there are two terms on the left-hand side of (11.5), one involving y' and the other y. We used integration by parts to obtain (11.8), which relates the term $\int_0^\infty e^{-st} y' \, dt$ to the term $\int_0^\infty e^{-st} y \, dt$. This led to (11.11), which we "solved" for y by luck. That is the broad outline.

However, there are three questions that we have avoided discussing.

- Under what circumstances do integrals such as $\int_0^\infty e^{-st} y \, dt$ exist?
- Under what circumstances can we guarantee that $\lim_{b\to\infty} e^{-sb} y(b) = 0$?
- Is $\left(e^{at} - e^{-t}\right)/(a+1)$ the only function that satisfies (11.11)? The associated general question is, if

$$\int_0^\infty e^{-st} f(t) \, dt = \int_0^\infty e^{-st} g(t) \, dt,$$

is $f(t) = g(t)$?

[1] There is a discussion of partial fractions in Appendix A.2.

We will return to these points in Section 11.8, but for the time being, in order to concentrate on the broad outline, we are going to make the following **working assumptions** for the functions that we will be using.[2]

1. $\int_0^\infty e^{-st} y \, dt$ exists.

2. $\lim_{b\to\infty} e^{-sb} y(b) = 0$.

3.
$$\text{If } \int_0^\infty e^{-st} f(t)\, dt = \int_0^\infty e^{-st} g(t)\, dt, \text{ then } f(t) = g(t). \tag{11.13}$$

The quantity $\int_0^\infty e^{-st} y(t)\, dt$ appears repeatedly and is given a special name and notation.

◆ *Definition 11.1:* **The integral**

$$Y(s) = \mathcal{L}\{y(t)\} = \int_0^\infty e^{-st} y(t)\, dt \tag{11.14}$$

is called the LAPLACE TRANSFORM of $y(t)$. ◆

Comments about Laplace Transforms

- Unstated in this definition is a qualification that the integral must exist.
- Because of the ∞ in the upper limit, the integral is an improper integral and is defined — and evaluated — by

$$\int_0^\infty e^{-st} y(t)\, dt = \lim_{b\to\infty} \int_0^b e^{-st} y(t)\, dt. \tag{11.15}$$

- The two different notations, $Y(s)$ and $\mathcal{L}\{y(t)\}$, may appear confusing at first. The $Y(s)$ notation is the standard function notation, where the only variable that remains after the right-hand side of (11.14) is integrated will be s. The $\mathcal{L}\{y(t)\}$ notation is used when we want to draw attention to the function $y(t)$ in the integrand. If we change $y(t)$, then the integral will change, which is what the $\mathcal{L}\{y(t)\}$ notation is pointing out. Sometimes $\mathcal{L}\{y(t)\}$ is written $\mathcal{L}\{y\}$.
- The pattern of $Y(s) = \mathcal{L}\{y(t)\}$ is followed for other functions. For example, $F(s) = \mathcal{L}\{f(t)\}$, $G(s) = \mathcal{L}\{g(t)\}$, and so on.
- If we use the assumption that $\lim_{b\to\infty} e^{-sb} y(b) = 0$, then (11.8) can be written as

$$\mathcal{L}\{y'\} = -y(0) + s\mathcal{L}\{y\}. \tag{11.16}$$

This allows us to relate the Laplace transform of y' to the Laplace transform of y.

[2] We stress that the full mathematical conditions under which these assumptions are justified will be established in Section 11.8.

- We can rewrite the assumption given in (11.13) as

$$\text{if } \mathcal{L}\{f(t)\} = \mathcal{L}\{g(t)\} \text{ then } f(t) = g(t). \tag{11.17}$$

- We can rewrite the equation

$$\int_0^\infty e^{-st}\left[af(t) + bg(t)\right] dt = a\int_0^\infty e^{-st} f(t)\, dt + b\int_0^\infty e^{-st} g(t)\, dt,$$

which we used to find (11.5) from the equation preceding it, as

$$\mathcal{L}\{af(t) + bg(t)\} = a\mathcal{L}\{f(t)\} + b\mathcal{L}\{g(t)\}. \tag{11.18}$$

A useful special case of this is

$$\mathcal{L}\{af(t)\} = a\mathcal{L}\{f(t)\}.$$

- We can rewrite the result (11.12) in the form

$$\mathcal{L}\left\{e^{at}\right\} = \frac{1}{s-a} \tag{11.19}$$

if $s > a$, while

$$\int_0^\infty e^{-st} 0\, dt = 0$$

is

$$\mathcal{L}\{0\} = 0. \tag{11.20}$$

EXAMPLE 11.2

Take the Laplace transform

With these changes of notation, we will now repeat the technique used to solve (11.1). We first take the Laplace transform of each side of

$$y' + y = e^{at},$$

obtaining

$$\mathcal{L}\{y' + y\} = \mathcal{L}\{e^{at}\}.$$

Using (11.18) allows us to rewrite this as

$$\mathcal{L}\{y'\} + \mathcal{L}\{y\} = \mathcal{L}\{e^{at}\}.$$

We substitute for $\mathcal{L}\{y'\}$ from (11.16) and $\mathcal{L}\{e^{at}\}$ from (11.19) to find

$$-y(0) + s\mathcal{L}\{y\} + \mathcal{L}\{y\} = \frac{1}{s-a}.$$

Solve for $\mathcal{L}\{y\}$ Using the initial condition $y(0) = 0$ and solving for $\mathcal{L}\{y\}$ leads to

$$\mathcal{L}\{y\} = \frac{1}{(s+1)(s-a)},$$

which, for the case $a \neq -1$, can be rewritten as

$$\mathcal{L}\{y\} = \frac{1}{a+1}\left(\frac{1}{s-a} - \frac{1}{s+1}\right). \tag{11.21}$$

From (11.18) and (11.19) we have

$$\mathcal{L}\{y\} = \mathcal{L}\left\{\frac{1}{a+1}\left(e^{at} - e^{-t}\right)\right\}, \tag{11.22}$$

Explicit solution which finally gives $y(t) = \left(e^{at} - e^{-t}\right)/(a+1)$ from (11.17). ■

If we look at the preceding example for the case when $a = -1$, then the counterpart of (11.21) is

$$\mathcal{L}\{y\} = \frac{1}{(s+1)^2}. \tag{11.23}$$

Here we have a problem: We don't have an equation similar to (11.19) for the function $1/(s+1)^2$.

The crucial step for the success of this analysis in the case $a \neq -1$ is going from (11.21) to (11.22). We were lucky to have the result in (11.19) at hand. The reason we are unsuccessful in the case $a = -1$ is the fact that we were not lucky enough to have at hand a result similar to (11.19) for the function $1/(s+1)^2$.

However, because we know that the solution of this initial value problem for $a = -1$ is te^{-t}, this suggests we compute the integral

$$\int_0^\infty e^{-st}te^{-t}\,dt = \int_0^\infty te^{-(s+1)t}\,dt = \lim_{b\to\infty}\int_0^b te^{-(s+1)t}\,dt,$$

which, using integration by parts, gives

$$\int_0^\infty e^{-st}te^{-t}\,dt = \lim_{b\to\infty}\left[-\frac{1}{s+1}te^{-(s+1)t}\Big|_0^b\right] + \int_0^\infty \frac{1}{s+1}e^{-(s+1)t}\,dt.$$

The first term on the right-hand of this equation is 0 for $s+1 > 0$, and, using (11.6), the second term evaluates as $1/(s+1)^2$, so this gives us

$$\mathcal{L}\left\{te^{-t}\right\} = \frac{1}{(s+1)^2},$$

as we expected.

Constructing a Table of Laplace Transforms

In order to be so lucky in the future, we need to construct a table of Laplace transforms $\mathcal{L}\{y\}$ for many different functions $y(t)$. In the next example we compute the Laplace transform of $\sin t$.

EXAMPLE 11.3 *The Laplace Transform of* sin t

To find the Laplace transform of $\sin t$, we need to calculate

$$\mathcal{L}\{\sin t\} = \int_0^\infty e^{-st} \sin t \, dt.$$

We could integrate this by parts twice to evaluate this integral (see Exercise 1 on page 515). Instead, we will use a different approach that foreshadows the techniques used in the next section — namely, use an existing result to generate new Laplace transforms. In this case, we take advantage of (11.16) — that is,

$$\mathcal{L}\{y'\} = -y(0) + s\mathcal{L}\{y\}. \tag{11.24}$$

To find $\mathcal{L}\{\sin t\}$ from this equation, we could try $y'(t) = \sin t$, in which case we should choose $y(t) = -\cos t$. In this way we find

$$\mathcal{L}\{\sin t\} = -(-1) + s\mathcal{L}\{-\cos t\} = 1 - s\mathcal{L}\{\cos t\}, \tag{11.25}$$

which relates $\mathcal{L}\{\sin t\}$ to $\mathcal{L}\{\cos t\}$ but does not find $\mathcal{L}\{\sin t\}$ explicitly. However, perhaps we can use (11.24) again to find $\mathcal{L}\{\cos t\}$. If we try $y'(t) = \cos t$, so that $y(t) = \sin t$, in (11.24), we now find

$$\mathcal{L}\{\cos t\} = s\mathcal{L}\{\sin t\}. \tag{11.26}$$

Substituting (11.26) into (11.25), we see that $\mathcal{L}\{\sin t\} = 1 - s^2\mathcal{L}\{\sin t\}$. We solve this equation for $\mathcal{L}\{\sin t\}$, finally finding the Laplace transform of $\sin t$, namely,

$$\mathcal{L}\{\sin t\} = \frac{1}{s^2+1}.$$

We notice that from (11.26), we can also write down the Laplace transform of $\cos t$ as

$$\mathcal{L}\{\cos t\} = \frac{s}{s^2+1}.$$

■

The Laplace transforms of a few other familiar functions are developed in the exercises, and the results are listed in Table 11.1. In Table 11.2 we summarize the elementary properties of Laplace transforms that we have found thus far.

Now that we have a table of Laplace transforms of several functions, let us solve another initial value problem.

Table 11.1 Laplace transforms of several functions

$f(t)$	$\mathcal{L}\{f(t)\} = F(s)$
1	$1/s$
t	$1/s^2$
e^{at}	$1/(s-a)$
te^{at}	$1/(s-a)^2$
$\sin t$	$1/(s^2+1)$
$\cos t$	$s/(s^2+1)$
$\sinh t$	$1/(s^2-1)$
$\cosh t$	$s/(s^2-1)$

Table 11.2 Simple properties of Laplace transforms

If $\mathcal{L}\{f(t)\} = F(s)$, then
$\mathcal{L}\{af(t) + bg(t)\} = a\mathcal{L}\{f(t)\} + b\mathcal{L}\{g(t)\}$
$\mathcal{L}\{af(t)\} = a\mathcal{L}\{f(t)\}$
$\mathcal{L}\{f'(t)\} = s\mathcal{L}\{f(t)\} - f(0)$

EXAMPLE 11.4

Solve the initial value problem $y' + 2y = 10 \sin t$, $y(0) = 1$.

Take the Laplace transform

Taking the Laplace transform of both sides of this differential equation and using the property given in (11.18), we have $\mathcal{L}\left\{y' + 2y\right\} = \mathcal{L}\left\{10 \sin t\right\}$, or

$$\mathcal{L}\left\{y'\right\} + 2\mathcal{L}\left\{y\right\} = 10\mathcal{L}\left\{\sin t\right\}. \tag{11.27}$$

We now use results from Table 11.1, the fact that $\mathcal{L}\left\{y'\right\} = s\mathcal{L}\{y\} - y(0)$, and the initial condition $y(0) = 1$ to write (11.27) as

$$s\mathcal{L}\{y\} - 1 + 2\mathcal{L}\{y\} = \frac{10}{s^2+1}.$$

Solve for $\mathcal{L}\{y\}$

Solving for $\mathcal{L}\{y\}$ gives

$$\mathcal{L}\{y\} = \frac{1}{s+2} + \frac{10}{(s+2)\left(s^2+1\right)}. \tag{11.28}$$

Partial fractions

In order to rearrange (11.28) to exploit Tables 11.1 and 11.2, we use partial fractions as follows:

$$\frac{10}{(s+2)\left(s^2+1\right)} = \frac{A}{s+2} + \frac{Bs+C}{s^2+1}$$

or

$$\frac{10}{(s+2)\left(s^2+1\right)} = \frac{A\left(s^2+1\right) + (Bs+C)(s+2)}{(s+2)\left(s^2+1\right)},$$

where A, B, and C are constants to be determined. Because the denominators on the two sides of this equation are the same, the numerators must also be equal, giving

$$10 = A\left(s^2+1\right) + (Bs+C)(s+2) = (A+B)s^2 + (2B+C)s + (A+2C).$$

This equation is an identity in s, so we may equate the coefficients of like powers of s to obtain $10 = A + 2C$, $0 = 2B + C$, and $0 = A + B$. Solving these equations gives $A = 2$, $B = -2$, and $C = 4$, so we have the equivalent expression

$$\frac{10}{(s+2)\left(s^2+1\right)} = \frac{2}{s+2} + \frac{-2s+4}{s^2+1}.$$

Substituting this expression into (11.28), we find

$$\mathcal{L}\{y\} = \frac{3}{s+2} - \frac{2s}{s^2+1} + \frac{4}{s^2+1}.$$

Explicit solution Using Tables 11.1 and 11.2 allows us to give the solution of our initial value problem as

$$y(t) = 3e^{-2t} - 2\cos t + 4\sin t.$$

We now formalize the procedure we have just used.

How to Solve First Order Linear Differential Equations Using Laplace Transforms

Purpose To find $y(t)$ that satisfies the initial value problem

$$a_1 y' + a_0 y = f(t), \qquad y(0) = y_0,$$

by using Laplace transforms. Here $a_1 \neq 0$, a_0 and y_0 are all constants.

Process

1. Put the linear differential equation in standard form by dividing by a_1, then multiply both sides of the result by e^{-st} and integrate with respect to t from 0 to ∞.

2. Use results from Tables 11.1 and 11.2 to change all of the integrated expressions to those involving the transformed dependent variable

$$Y(s) = \mathcal{L}\{y(t)\} = \int_0^\infty e^{-st} y(t)\, dt$$

   and known functions of s.

3. Solve for the transformed dependent variable, $Y(s)$, and rearrange the result so all the terms have the form of an expression on the right-hand side of Table 11.1.

4. Use Tables 11.1 and 11.2 to write down the solution for $y(t)$.

Comments about Using Laplace Transforms to Solve Initial Value Problems

- The process of multiplying both sides of a differential equation by e^{-st} and integrating on t from 0 to ∞ is called taking the Laplace transform of the equation.
- The broad outline for this process is as follows.
  - Transform the original differential equation to its Laplace transform counterpart.
  - Use algebraic manipulations to solve for the unknown Laplace transform.
  - Change from the Laplace transform variables back to the original differential equation variables.
- If no initial condition is given, we may leave our answer in terms of the arbitrary constant $y(0)$.
- Partial fractions are sometimes useful in step 3 of the process.
- Extensive lists of Laplace transforms are found in many books.[3]
- If the initial condition is not at $t = 0$ but at $t = t_0$, we first make the change of variable $\tau = t - t_0$, and then we use Laplace transforms on the differential equation in τ.

EXERCISES

1. Use integration by parts to evaluate $\mathcal{L}\{\sin t\}$ and $\mathcal{L}\{\cos t\}$.

2. Complete Table 11.1 by evaluating the Laplace transforms of the following functions.

(a) 1 (b) t (c) te^{at} (d) $\sinh t = (e^t - e^{-t})/2$
(e) $\cosh t = (e^t + e^{-t})/2$

3. Use (11.24) to evaluate the following.

(a) $\mathcal{L}\{\sin at\}$ (b) $\mathcal{L}\{\cos at\}$
(c) $\mathcal{L}\{\sinh at\}$ (d) $\mathcal{L}\{\cosh at\}$

4. Use Tables 11.1 and 11.2 to evaluate the Laplace transforms of the following functions.

(a) $\sin 2t + \cos 2t$ (b) $2e^t + 3e^{-t}$
(c) $4te^{at} - t$ (d) $e^{-t} + 2\sinh t$

Are you surprised at the result for part (d)?

5. Use Table 11.1, Table 11.2, and (11.17) to find functions whose Laplace transforms are as follows.

(a) $1/(s-2) - 1/(s+3)$ (b) $6/[s(s+3)]$
(c) $1/[(s+2)(s+3)]$ (d) $5/[(s+2)(s^2+1)]$
(e) $6/[(s-2)(s^2-1)]$ (f) $4/(s^4-1)$

6. In deriving $\mathcal{L}\{e^{at}\} = 1/(s-a)$, we used $\lim_{b\to\infty}[e^{(a-s)b}/(a-s)] = 0$ for $s > a$. If a and s are complex numbers, it is also true that $\lim_{b\to\infty}[e^{(a-s)b}/(a-s)] = 0$ provided that the real part of $s - a$ — namely, $Re[s-a]$ — is positive; that is, $Re[s-a] > 0$. Thus, we have that $\mathcal{L}\{e^{it}\} = 1/(s-i)$ for $Re[s-i] = Re[s] > 0$, and $\mathcal{L}\{e^{-it}\} = 1/(s+i)$ for $Re[s+i] = Re[s] > 0$. Use these results to derive the formulas for $\mathcal{L}\{\sin t\}$ and $\mathcal{L}\{\cos t\}$.

7. Solve the following initial value problems.

(a) $y' + 3y = 5e^{2t}$ $\quad y(0) = 0$
(b) $y' + 3y = 6 - e^{-2t}$ $\quad y(0) = 1$
(c) $y' + 2y = 5\sin t$ $\quad y(0) = 0$
(d) $y' - 2y = 6\sinh t$ $\quad y(0) = 2$
(e) $y' + 3y = e^{2t}$ $\quad y(2) = 0$
(f) $y' + 3y = 10(t - e^{-2t})$ $\quad y(-2) = 1$

[3] For example, *Tables of Integral Transforms: Volumes I and II* by A.C. Erdelyi et al., McGraw-Hill, 1954.

11.2 CONSTRUCTING NEW LAPLACE TRANSFORMS FROM OLD

In the previous section we found Laplace transforms of several functions. We also solved two first order linear differential equations with constant coefficients using these transforms, and our success depended in part on our ability to recast the original differential equation to an algebraic equation that contained Laplace transforms of known functions. However, at the moment we have a very modest list of Laplace transforms — Table 11.1. Thus, in this section, we will create a larger list of functions than is given in that table.

We could always use the definition of the Laplace transform of $f(t)$, namely,

$$\mathcal{L}\{f(t)\} = \int_0^\infty e^{-st} f(t)\, dt,$$

to calculate a Laplace transform using integration techniques. However, we will try to avoid this lengthy process by using Tables 11.1 and 11.2 as our starting points to generate Laplace transforms of more functions. This means that the methods we discover will be largely disjointed but will stem from relevant observations and questions. We must remember — as we pointed out in the previous section — that in all cases we will assume that we are dealing only with functions whose Laplace transforms exist.

Extending the Table of Laplace Transforms

If we look at the entries under $f(t)$ in Table 11.1, perhaps the most glaring omission is that we have no Laplace transform for t^2, or any higher power of t. We could evaluate this Laplace transform from its definition, namely,

$$\mathcal{L}\left\{t^2\right\} = \int_0^\infty e^{-st} t^2\, dt,$$

but this will involve integrating by parts twice — and n times in the case of t^n. Instead we follow our previous suggestion and try to use old results to generate new results.

EXAMPLE 11.5 *The Laplace Transform of* t^n

We know that

$$\mathcal{L}\left\{y'(t)\right\} = s\mathcal{L}\{y(t)\} - y(0),$$

so this could lead to new results if we choose $y(t)$ in different ways.

If we try $y(t) = t^2$ and note that $y(0) = 0$ and $y'(t) = 2t$, we see that

$$\mathcal{L}\{2t\} = s\mathcal{L}\left\{t^2\right\}.$$

Thus, solving for $\mathcal{L}\left\{t^2\right\}$ and using Tables 11.1 and 11.2 gives

$$\mathcal{L}\left\{t^2\right\} = \frac{1}{s}\mathcal{L}\{2t\} = \frac{2}{s}\mathcal{L}\{t\} = \frac{2}{s^3}.$$

Notice that we have found a new Laplace transform from an old one. We will add this to our extended list of Laplace transforms.

If we try the same technique on $y(t) = t^n$ for a positive integer n, we find $\mathcal{L}\left\{nt^{n-1}\right\} = s\mathcal{L}\left\{t^n\right\}$, or

$$\mathcal{L}\left\{t^n\right\} = \frac{n}{s}\mathcal{L}\left\{t^{n-1}\right\}. \tag{11.29}$$

This gives $\mathcal{L}\left\{t^n\right\}$ in terms of $\mathcal{L}\left\{t^{n-1}\right\}$. Thus, for example, if we put $n = 3$, we have

$$\mathcal{L}\left\{t^3\right\} = \frac{3}{s}\mathcal{L}\left\{t^2\right\} = \frac{3}{s}\frac{2}{s^3} = \frac{3 \cdot 2}{s^4}.$$

Continuing in this way, we find

$$\mathcal{L}\left\{t^n\right\} = \frac{n!}{s^{n+1}}. \tag{11.30}$$

We can also use the property $\mathcal{L}\left\{y'(t)\right\} = s\mathcal{L}\left\{y(t)\right\} - y(0)$ on the function $y(t) = f'(t)$ to find

$$\mathcal{L}\left\{f''(t)\right\} = s\mathcal{L}\left\{f'(t)\right\} - f'(0).$$

If we substitute $\mathcal{L}\left\{f'(t)\right\} = s\mathcal{L}\left\{f(t)\right\} - f(0)$ into this expression for $\mathcal{L}\left\{f''(t)\right\}$, we find $\mathcal{L}\left\{f''(t)\right\} = s\left[s\mathcal{L}\left\{f(t)\right\} - f(0)\right] - f'(0)$, or

$$\mathcal{L}\left\{f''(t)\right\} = s^2\mathcal{L}\left\{f(t)\right\} - \left[sf(0) + f'(0)\right].$$

We can continue this process as often as we like to find the Laplace transform of the nth derivative of $f(t)$; namely, $f^{(n)}(t)$.

Theorem 11.1: The Derivative Theorem *If the Laplace transform of $f(t)$ and its first $n-1$ derivatives exist, then*

$$\mathcal{L}\left\{f^{(n)}(t)\right\} = s^n\mathcal{L}\left\{f(t)\right\} - \left[s^{n-1}f(0) + s^{n-2}f'(0) + \cdots + sf^{(n-2)}(0) + f^{(n-1)}(0)\right]. \tag{11.31}$$

Comments about the Laplace Transform of $f^{(n)}(t)$

- We essentially used this theorem, with $n = 2$, in Example 11.3 on page 512 when we calculated the Laplace transform of $\sin t$.
- This property is crucial for solving higher order linear differential equations using Laplace transforms. It will be used repeatedly in Section 11.6.

EXAMPLE 11.6 *The Laplace Transform of* $t \sin t$

We apply (11.31), with $n = 2$, to the function $f(t) = t \sin t$. We will need $f'(t)$ and $f''(t)$, which are $f'(t) = \sin t + t \cos t$ and $f''(t) = 2 \cos t - t \sin t$, respectively. Thus, from (11.31), and because in this case $f(0) = f'(0) = 0$, we have $\mathcal{L}\{2 \cos t - t \sin t\} = s^2 \mathcal{L}\{t \sin t\}$. Solving for $\mathcal{L}\{t \sin t\}$ gives

$$\mathcal{L}\{t \sin t\} = \frac{1}{s^2+1} \mathcal{L}\{2 \cos t\}.$$

From our previous results we know that

$$\mathcal{L}\{2 \cos t\} = 2\mathcal{L}\{\cos t\} = \frac{2s}{s^2+1}.$$

This leads to

$$\mathcal{L}\{t \sin t\} = \frac{2s}{\left(s^2+1\right)^2}.$$

To appreciate the power of this technique, you should try to evaluate $\mathcal{L}\{t \sin t\}$ from its definition; namely,

$$\mathcal{L}\{t \sin t\} = \int_0^\infty e^{-st} t \sin t \, dt.$$

Good luck! ■

Changing Scale

We now turn to a different question. Our list of Laplace transforms in Table 11.1 includes that of $\sin t$, but does not include that of $\sin at$. Is there an easy way for us to use the Laplace transform of $\sin t$ to find the Laplace transform of $\sin at$?

EXAMPLE 11.7 *The Laplace Transform of* $\sin at$

To explore this idea, we write the expression for $\mathcal{L}\{\sin at\}$, namely

$$\mathcal{L}\{\sin at\} = \int_0^\infty e^{-st} \sin at \, dt. \tag{11.32}$$

Again, we could evaluate this integral directly, but that involves integrating by parts twice. Instead we will try to use our old results. Because we know that the Laplace transform of $\sin t$ is

$$F(s) = \mathcal{L}\{\sin t\} = \int_0^\infty e^{-st} \sin t \, dt,$$

Change variable we make the change of variable $at = z$ in (11.32) to find

$$\mathcal{L}\{\sin at\} = \int_0^\infty e^{-sz/a} \sin z \, dz/a = \frac{1}{a} \int_0^\infty e^{-(s/a)z} \sin z \, dz. \tag{11.33}$$

If we compare the integral in (11.33) with the integral giving the Laplace transform of $\sin t$, we see that

$$\mathcal{L}\{\sin at\} = \frac{1}{a}F\left(\frac{s}{a}\right).$$

From Table 11.1 we have

$$F(s) = \mathcal{L}\{\sin t\} = \frac{1}{s^2+1}.$$

This means that

$$\mathcal{L}\{\sin at\} = \frac{1}{a}F\left(\frac{s}{a}\right) = \frac{1}{a}\frac{1}{(s/a)^2+1} = \frac{a}{s^2+a^2}. \tag{11.34}$$

This shows that we are able to find the Laplace transform of $\sin at$ from the Laplace transform of $\sin t$. ■

This same procedure works for any two functions, $f(t)$ and $f(at)$, that have Laplace transforms as stated in the following result.

▶ **Theorem 11.2: The Scale Change Theorem** *If the Laplace transform of $f(t)$ exists — that is, $\mathcal{L}\{f(t)\} = F(s)$ — then*

$$\mathcal{L}\{f(at)\} = \frac{1}{a}F\left(\frac{s}{a}\right)$$

if $a > 0$. ◀

Comments about the Scale Change Theorem

- The primary use of this theorem is to supplement our list of Laplace transforms. For example, because we know that $\mathcal{L}\{\cos t\} = s/(s^2+1)$, we can use this theorem to conclude that

$$\mathcal{L}\{\cos at\} = \frac{1}{a}\frac{(s/a)}{(s/a)^2+1} = \frac{s}{s^2+a^2}.$$

- We restricted the value of a to be positive to guarantee the existence of integrals such as $\int_0^\infty e^{-sz/a}\sin z\,dz$. See Exercise 2 on page 522.
- Because the details in the proof of this theorem closely follow the procedure used in Example 11.7 on page 518, they are left to Exercise 2 on page 522.

Shifting

We now turn to an unrelated observation. Notice in Table 11.1 that we have $\mathcal{L}\{t\} = 1/s^2$, while $\mathcal{L}\{te^t\} = 1/(s-1)^2$. Because the forms of these two transforms are similar, we might be tempted to ask if there is some general principle lurking here.

EXAMPLE 11.8 *The Laplace Transform of* te^t

To check this out, we write down the Laplace transform of te^t,

$$\mathcal{L}\left\{te^t\right\} = \int_0^\infty e^{-st}te^t\,dt = \int_0^\infty e^{-(s-1)t}t\,dt, \tag{11.35}$$

and compare it with the Laplace transform of t,

$$\int_0^\infty e^{-st}t\,dt = \frac{1}{s^2}. \tag{11.36}$$

We see that the integrand in (11.35) differs from the one in (11.36) only in that there is a factor of $s-1$ in the exponent of e rather than just s. This means that we can evaluate (11.35) by replacing s by $s-1$ throughout (11.36), giving

$$\mathcal{L}\left\{te^t\right\} = \int_0^\infty e^{-(s-1)t}t\,dt = \frac{1}{(s-1)^2}.$$

This result may be generalized to the following theorem, with the proof left to Exercise 4 on page 522.

▶ **Theorem 11.3: The Shifting Theorem** *If the Laplace transform of $f(t)$ exists for $s > \alpha$, so that $\mathcal{L}\{f(t)\} = F(s)$ for $s > \alpha$, then*

$$\mathcal{L}\left\{e^{at}f(t)\right\} = F(s-a)$$

for $s > \alpha + a$. ◀

Comments about the Shifting Theorem

- The primary use of this theorem is to supplement our list of Laplace transforms. For example, because we know from (11.34) that $\mathcal{L}\{\sin bt\} = b/(s^2+b^2)$, we can use this theorem to conclude that

$$\mathcal{L}\left\{e^{at}\sin bt\right\} = \frac{b}{(s-a)^2+b^2}.$$

  In a similar way,

$$\mathcal{L}\left\{e^{at}\cos bt\right\} = \frac{s-a}{(s-a)^2+b^2}.$$

- The property of Laplace transforms given in this theorem is often called the SHIFTING PROPERTY. (Explain why this terminology is appropriate.)

A Special Product

We turn to a new question. A previous property dealt with what happens when we differentiate $y(t)$ with respect to t. We could ask whether there are any useful properties obtained by differentiating the Laplace transform of $f(t)$, namely,

$$F(s) = \mathcal{L}\{f(t)\} = \int_0^\infty e^{-st} f(t)\, dt,$$

with respect to s. If we differentiate both sides with respect to s (assuming it is legal to interchange the order of differentiation and integration), we obtain

$$\frac{dF(s)}{ds} = \frac{d}{ds}\left[\int_0^\infty e^{-st} f(t)\, dt\right] = \int_0^\infty \frac{d}{ds}\left[e^{-st} f(t)\right] dt = \int_0^\infty -te^{-st} f(t)\, dt,$$

or

$$\frac{dF(s)}{ds} = -\mathcal{L}\{tf(t)\}.$$

Differentiating once more yields

$$\frac{d^2F(s)}{ds^2} = \int_0^\infty (-t)^2 e^{-st} f(t)\, dt = (-1)^2 \mathcal{L}\left\{t^2 f(t)\right\}.$$

This process may be repeated as many times as needed until we find the following result.

▶ **Theorem 11.4: The Special Product Theorem** *If $\mathcal{L}\{f(t)\} = F(s)$, then for any integer $n \geq 0$, we have*

$$\mathcal{L}\{t^n f(t)\} = (-1)^n \frac{d^n F(s)}{ds^n}.$$

◀

Comments about the Special Product Theorem

- The primary use of this theorem is to supplement our list of Laplace transforms. For example, because we know from (11.34) that $\mathcal{L}\{\sin at\} = a/(s^2 + a^2)$, we can use this theorem to conclude that

$$\mathcal{L}\{t \sin at\} = -\frac{d}{ds}\left(\frac{a}{s^2 + a^2}\right) = \frac{2as}{\left(s^2 + a^2\right)^2}$$

and

$$\mathcal{L}\left\{t^2 \sin at\right\} = \frac{d^2}{ds^2}\left(\frac{a}{s^2 + a^2}\right) = \frac{2a\left(a^2 - 3s^2\right)}{\left(s^2 + a^2\right)^3}.$$

In the same way we can find

$$\mathcal{L}\{t \cos at\} = -\frac{d}{ds}\left(\frac{s}{s^2 + a^2}\right) = \frac{s^2 - a^2}{\left(s^2 + a^2\right)^2}.$$

Using this theorem to find these Laplace transforms is considerably less daunting than the prospect of evaluating the integrals $\int_0^\infty e^{-st} t \sin at\, dt$, $\int_0^\infty e^{-st} t^2 \sin at\, dt$, and $\int_0^\infty e^{-st} t \cos at\, dt$, by standard integration techniques.

- We can use this theorem as yet another way of calculating $\mathcal{L}\{te^{at}\}$. [Use it to show that $\mathcal{L}\{te^{at}\} = 1/(s-a)^2$.]

We collect some of these Laplace transforms and properties together in Tables 11.3 and 11.4.

Table 11.3 **Laplace transforms of several functions**

$f(t)$	$\mathcal{L}\{f(t)\} = F(s)$
a	a/s
t	$1/s^2$
t^n	$n!/s^{n+1}$
e^{at}	$1/(s-a)$
te^{at}	$1/(s-a)^2$
$\sin at$	$a/(s^2+a^2)$
$\cos at$	$s/(s^2+a^2)$
$\sinh at$	$a/(s^2-a^2)$
$\cosh at$	$s/(s^2-a^2)$
$t \sin at$	$2as/(s^2+a^2)^2$
$t \cos at$	$(s^2-a^2)/(s^2+a^2)^2$

Table 11.4 **Properties of Laplace transforms**

If $\mathcal{L}\{f(t)\} = F(s)$, then
$\mathcal{L}\{af(t)+bg(t)\} = a\mathcal{L}\{f(t)\} + b\mathcal{L}\{g(t)\}$
$\mathcal{L}\{af(t)\} = a\mathcal{L}\{f(t)\}$
$\mathcal{L}\{f'(t)\} = s\mathcal{L}\{f(t)\} - f(0)$
$\mathcal{L}\{f''(t)\} = s^2\mathcal{L}\{f(t)\} - [sf(0) + f'(0)]$
$\mathcal{L}\{f(at)\} = (1/a)F(s/a)$
$\mathcal{L}\{e^{at}f(t)\} = F(s-a)$
$\mathcal{L}\{t^n f(t)\} = (-1)^n F^{(n)}(s)$

EXERCISES

1. Evaluate $\mathcal{L}\{t \cos at\}$ by

(a) the method of Example 11.6 on page 518.

(b) integration by parts.

2. Prove the Scale Change Theorem by following the procedure used in Example 11.7 on page 518. If $a < 0$, what complications are introduced in trying to prove this theorem?

3. Use the Derivative Theorem to evaluate

(a) $\mathcal{L}\{t \sinh t\}$ (b) $\mathcal{L}\{t \cosh t\}$

4. Prove the Shifting Theorem by using the definition of the Laplace transform.

5. Use the Shifting Theorem to evaluate

(a) $\mathcal{L}\{e^{at} \sinh at\}$ (b) $\mathcal{L}\{e^{at} \cosh at\}$ (c) $\mathcal{L}\{t^2 e^{at}\}$

6. Another property of Laplace transforms is as follows: If $\mathcal{L}\{f(t)\} = F(s)$ for $s > \alpha$, and $f(t)/t$ is bounded for $t > 0$, then

$$\mathcal{L}\left\{\frac{1}{t} f(t)\right\} = \int_s^\infty F(z)\, dz.$$

(a) Verify this result by writing $F(z) = \int_0^\infty e^{-zt} f(t)\, dt$ and integrating both sides from $z = s$ to $z = \infty$. Interchange the order of the resulting integrations.

(b) This theorem is used to calculate Laplace transforms for which straightforward integration techniques may fail. Use this theorem to evaluate $\mathcal{L}\{(\sin t)/t\}$ and $\mathcal{L}\{(\sinh t)/t\}$.

(c) Use the definition of the Laplace transform to observe the difficulty of evaluating $\mathcal{L}\{(\sin t)/t\}$ by integration techniques.

7. Use the Special Product Theorem to evaluate

(a) $\mathcal{L}\left\{t^2 \cos at\right\}$ (b) $\mathcal{L}\{t \sinh at\}$
(c) $\mathcal{L}\{t \cosh at\}$ (d) $\mathcal{L}\left\{t^2 e^{at}\right\}$

8. In Chapter 1 we suggested that $\lim_{x\to\infty} erf(x) = 1$, so that

$$\int_0^\infty e^{-x^2}\, dx = \frac{\sqrt{\pi}}{2}.$$

Accepting this as true, use the definition of the Laplace transform to show that $\mathcal{L}\left\{t^{-1/2}\right\} = \sqrt{\pi/s}$.

11.3 THE INVERSE LAPLACE TRANSFORM AND THE CONVOLUTION THEOREM

In Section 11.1 we outlined a procedure for solving first order linear differential equations with constant coefficients using Laplace transforms. Our success at finding explicit solutions of such initial value problems depends on two things: being able to transform the original differential equation into an algebraic equation involving the Laplace transform of our explicit solution, and then being able to return to familiar functions using our tables. In the previous section we developed techniques that allow us to go from the differential equation to the transformed equation. In this section we concentrate on the inverse problem: Given the Laplace transform $F(s) = \mathcal{L}\{f(t)\}$, what is $f(t)$?

The key to making this process work is Table 11.3 on page 522. We should notice the parallel between using this table and using a typical table of integrals. With a table of integrals, the integral we want to evaluate may not appear explicitly. When that happens we try to use the properties of integrals to convert the integral we want into integrals that are already in the table. Exactly the same reasoning applies here, so the correct way to view Table 11.3 in order to answer our question – given a Laplace transform $F(s) = \mathcal{L}\{f(t)\}$, what is $f(t)$? – is from right to left. From this viewpoint, there are many functions missing from the right-hand side of Table 11.3. Do you notice anything unusual about the functions that are there? All of them are rational functions – that is, one polynomial in s divided by another. In fact, they are proper rational functions, where the degree of the numerator is always less than that of the denominator.

Finding Inverse Laplace Transforms

We now look at some examples in which we are given the Laplace transform of $f(t)$ – namely, $F(s) = \mathcal{L}\{f(t)\}$ – and want to find $f(t)$. This process is used so often that we give it a special name.

◆ *Definition 11.2:* **If $\mathcal{L}\left\{f(t)\right\} = F(s)$, then the INVERSE LAPLACE TRANSFORM of $F(s)$ is $f(t)$ and is denoted by**

$$\mathcal{L}^{-1}\{F(s)\} = f(t).$$ ◆

Comments about Inverse Laplace Transforms

- In Section 11.8 we will discuss the existence and uniqueness of inverse Laplace transforms. Until then, we will assume that any particular inverse Laplace transform exists and is unique.

- The process of solving a differential equation by Laplace transforms is to take the Laplace transform of the differential equation, manipulate the Laplace transform of the unknown solution, and then use the inverse Laplace transform to return to the original variables.

- From $\mathcal{L}\{af(t)+bg(t)\} = a\mathcal{L}\{f(t)\}+b\mathcal{L}\{g(t)\} = aF(s)+bG(s)$, we immediately have

$$\mathcal{L}^{-1}\{aF(s)+bG(s)\} = a\mathcal{L}^{-1}\{F(s)\}+b\mathcal{L}^{-1}\{G(s)\}. \tag{11.37}$$

- Table 11.3 can be rewritten in terms of inverse Laplace transforms as Table 11.5.

- The Scale Change Theorem from the previous section can be useful when we are looking for inverse Laplace transforms. We restate it using the current terminology:

$$\text{If } \mathcal{L}^{-1}\{F(s)\} = f(t), \text{ then } \mathcal{L}^{-1}\{F(s/a)\} = af(at). \tag{11.38}$$

  This will allow us to compute $\mathcal{L}^{-1}\{F(s/a)\}$ if we know $\mathcal{L}^{-1}\{F(s)\}$.

- The Shifting Theorem from the previous section can also be useful when looking for inverse Laplace transforms. We restate it using the current terminology:

$$\text{If } \mathcal{L}^{-1}\{F(s)\} = f(t), \text{ then } \mathcal{L}^{-1}\{F(s-a)\} = e^{at}f(t). \tag{11.39}$$

  This will allow us to compute $\mathcal{L}^{-1}\{F(s-a)\}$ if we know $\mathcal{L}^{-1}\{F(s)\}$.

- In many cases the quantity $F(s)$ will be a proper rational function. Consequently, in this case we could always rewrite $F(s)$ using partial fractions. The typical components of a partial fraction decomposition are $A/(s-a)^n$, $B/\left(s^2+a^2\right)^n$, and $Cs/\left(s^2+a^2\right)^n$. Some of these are already in our list of inverse Laplace transforms — Table 11.5. However, we should be cognizant of the fact that blindly applying partial fractions may make the problem more difficult than it needs to be. For

Table 11.5 The inverse Laplace transforms of several functions

$F(s)$	$\mathcal{L}^{-1}\{F(s)\} = f(t)$
$1/s$	1
$1/s^2$	t
$n!/s^{n+1}$	t^n
$1/(s-a)$	e^{at}
$1/(s-a)^2$	te^{at}
$a/(s^2+a^2)$	$\sin at$
$s/(s^2+a^2)$	$\cos at$
$a/(s^2-a^2)$	$\sinh at$
$s/(s^2-a^2)$	$\cosh at$
$2as/(s^2+a^2)^2$	$t\sin at$
$(s^2-a^2)/(s^2+a^2)^2$	$t\cos at$

example, it would be silly to decompose $1/(s^2 - a^2)$ using partial fractions, because Table 11.5 already has its inverse Laplace transform, $\mathcal{L}^{-1}\left\{1/(s^2 - a^2)\right\} = \sinh at$.

We will look at a few examples in which we compute inverse Laplace transforms.

EXAMPLE 11.9 *The inverse Laplace transform of* $1/[(s-a)(s-b)]$

If

$$F(s) = \mathcal{L}\{f(t)\} = \frac{1}{(s-a)(s-b)},$$

Partial fractions

where $a \neq b$, what is $f(t)$? That is, what is $\mathcal{L}^{-1}\{F(s)\}$? We use partial fractions and write

$$\frac{1}{(s-a)(s-b)} = \frac{1}{a-b}\left(\frac{1}{s-a} - \frac{1}{s-b}\right),$$

which from (11.37) yields

$$\mathcal{L}^{-1}\left\{\frac{1}{(s-a)(s-b)}\right\} = \frac{1}{a-b}\left(\mathcal{L}^{-1}\left\{\frac{1}{s-a}\right\} - \mathcal{L}^{-1}\left\{\frac{1}{s-b}\right\}\right).$$

The inverse Laplace transform of $1/(s-a)$ occurs in Table 11.5, so we have

$$f(t) = \mathcal{L}^{-1}\left\{\frac{1}{(s-a)(s-b)}\right\} = \frac{e^{at} - e^{bt}}{a-b}. \tag{11.40}$$

■

EXAMPLE 11.10 *The inverse Laplace transform of* $s/[(s-a)(s-b)]$

If

$$F(s) = \mathcal{L}\{f(t)\} = \frac{s}{(s-a)(s-b)},$$

where $a \neq b$, what is $f(t) = \mathcal{L}^{-1}\{F(s)\}$?

We could decompose $s/\left[(s-a)(s-b)\right]$ using partial fractions, and there is nothing wrong with that. However, we can frequently use known inverse Laplace transforms to calculate as yet unknown ones. In this example we can add and subtract a from the numerator to find

$$\frac{s}{(s-a)(s-b)} = \frac{(s-a)+a}{(s-a)(s-b)} = \frac{1}{s-b} + \frac{a}{(s-a)(s-b)}.$$

Thus, taking the inverse Laplace transform yields

$$\mathcal{L}^{-1}\left\{\frac{s}{(s-a)(s-b)}\right\} = \mathcal{L}^{-1}\left\{\frac{1}{s-b}\right\} + a\mathcal{L}^{-1}\left\{\frac{1}{(s-a)(s-b)}\right\},$$

or

$$\mathcal{L}^{-1}\left\{\frac{s}{(s-a)(s-b)}\right\} = e^{bt} + a\frac{e^{at}-e^{bt}}{a-b} = \frac{ae^{at}-be^{bt}}{a-b}. \tag{11.41}$$

EXAMPLE 11.11 *The inverse Laplace transform of* $1/(s^2+1)^2$

If

$$F(s) = \mathcal{L}\{f(t)\} = \frac{1}{(s^2+1)^2},$$

what is $f(t) = \mathcal{L}^{-1}\{F(s)\}$?

There is no point in trying to decompose $1/(s^2+1)^2$ by partial fractions — it is already decomposed. (Why?) Thus, the only method we have is to try to rewrite $1/(s^2+1)^2$ in terms of expressions that appear on the left-hand side of Table 11.5. After much thought and experimenting, we find

$$\frac{1}{(s^2+1)^2} = \frac{(s^2+1)-s^2}{(s^2+1)^2} = \frac{1}{s^2+1} - \frac{s^2}{(s^2+1)^2}.$$

The first term on the right-hand side of this equation is listed in Table 11.5, so we concentrate on the second. We notice that it is part of $(s^2-1)/(s^2+1)^2$, which is in the table, so we try to use this information by writing

$$\frac{1}{(s^2+1)^2} = \frac{1}{s^2+1} - \frac{(s^2-1)+1}{(s^2+1)^2} = \frac{1}{s^2+1} - \frac{s^2-1}{(s^2+1)^2} - \frac{1}{(s^2+1)^2}.$$

Ah! If we move the final term on the right-hand side to the left-hand side and divide by 2, we find

$$\frac{1}{(s^2+1)^2} = \frac{1}{2}\left[\frac{1}{s^2+1} - \frac{s^2-1}{(s^2+1)^2}\right].$$

Thus, we finally have

$$\mathcal{L}^{-1}\left\{\frac{1}{(s^2+1)^2}\right\} = \frac{1}{2}\left[\mathcal{L}^{-1}\left\{\frac{1}{s^2+1}\right\} - \mathcal{L}^{-1}\left\{\frac{s^2-1}{(s^2+1)^2}\right\}\right],$$

or

$$\mathcal{L}^{-1}\left\{\frac{1}{(s^2+1)^2}\right\} = \frac{1}{2}(\sin t - t\cos t). \tag{11.42}$$

It would be an understatement to say we were lucky to find this last inverse Laplace transform. In fact, while looking down Table 11.5, you will have noticed that $\mathcal{L}^{-1}\left\{1/(s^2+1)\right\} = \sin t$, and you may have been tempted to think that $\mathcal{L}^{-1}\left\{1/(s^2+1)^2\right\} = \mathcal{L}^{-1}\left\{1/(s^2+1)\right\} \times \mathcal{L}^{-1}\left\{1/(s^2+1)\right\}$. If this were true then we would have found $\mathcal{L}^{-1}\left\{1/(s^2+1)^2\right\} = \sin^2 t$, which does not agree with the correct value of $(\sin t - t\cos t)/2$, which we just obtained. Although this technique is incorrect, it does raise the following question: Is there a formula for $\mathcal{L}^{-1}\{F(s)G(s)\}$? If there is a formula, then we may be able to use it on this example, without relying on luck.

The Convolution Theorem

There is a formula, but it is very unlikely you would guess what it is. So, without motivation, we state the result.

Theorem 11.5: The Convolution Theorem *If $F(s) = \mathcal{L}\{f(t)\}$ and $G(s) = \mathcal{L}\{g(t)\}$, then*

$$\mathcal{L}^{-1}\{F(s)G(s)\} = \int_0^t f(z)g(t-z)\,dz.$$

Comments about the Convolution Theorem

- This theorem is sometimes called the Faltung Theorem. An outline of its proof is given in Exercise 6 on page 530.
- The integral $\int_0^t f(z)g(t-z)\,dz$ is called the CONVOLUTION of the two functions $f(t)$ and $g(t)$ and is denoted by $f * g$. Thus,

$$f * g = \int_0^t f(z)g(t-z)\,dz,$$

  and the theorem states that $\mathcal{L}^{-1}\{F(s)G(s)\} = f * g$.
- The convolution $f * g$ is symmetric in the sense that $f * g = g * f$. See Exercise 8 on page 530.
- This theorem may also be stated in the form $\mathcal{L}\{\int_0^t f(z)g(t-z)\,dz\} = \mathcal{L}\{f\}\mathcal{L}\{g\}$, or $\mathcal{L}\{f * g\} = \mathcal{L}\{f\}\mathcal{L}\{g\}$.
- A useful special case of this theorem is obtained by choosing $g(t) = 1$ so that $G(s) = 1/s$. We then have the following statements:

$$\text{If } F(s) = \mathcal{L}\{f(t)\}\text{, then } \mathcal{L}^{-1}\left\{\frac{1}{s}F(s)\right\} = \int_0^t f(z)\,dz, \tag{11.43}$$

  or, equivalently,

$$\text{If } \mathcal{L}\{f(t)\} = F(s), \text{ then } \mathcal{L}\left\{\int_0^t f(z)\,dz\right\} = \frac{1}{s}F(s).$$

We will now use this theorem to recalculate $\mathcal{L}^{-1}\left\{1/(s^2+1)^2\right\}$.

Example 11.12 *The inverse Laplace transform of* $1/(s^2+1)^2$ *revisited*

We apply the theorem to $f(t) = g(t) = \sin t$ and use the fact that $\mathcal{L}^{-1}\left\{1/(s^2+1)\right\} = \sin t$ to obtain

$$\mathcal{L}^{-1}\left\{\frac{1}{\left(s^2+1\right)^2}\right\} = \int_0^t \sin z \sin(t-z)\,dz.$$

Using the identity $\sin A \sin B = \frac{1}{2}\left[\cos(A-B) - \cos(A+B)\right]$ gives

$$\mathcal{L}^{-1}\left\{\frac{1}{\left(s^2+1\right)^2}\right\} = \frac{1}{2}\int_0^t \left[\cos(2z-t) - \cos t\right]\,dz$$
$$= \frac{1}{2}\int_0^t \cos(2z-t)\,dz - \frac{1}{2}\cos t \int_0^t dz.$$

(Remember we are integrating with respect to z, so t plays the role of a constant in this process.) The first integral on the right-hand side can be evaluated with the change of variable $u = 2z - t$, so that

$$\int_0^t \cos(2z-t)\,dz = \frac{1}{2}\int_{-t}^t \cos u\,du = \sin t.$$

The second integral can be evaluated immediately, so we have

$$\mathcal{L}^{-1}\left\{\frac{1}{\left(s^2+1\right)^2}\right\} = \frac{1}{2}\sin t - \frac{1}{2}t\cos t, \tag{11.44}$$

as anticipated. Notice that we can use (11.38) and (11.44) to show that

$$\mathcal{L}^{-1}\left\{\frac{1}{\left(s^2+a^2\right)^2}\right\} = \frac{1}{2a^3}\left(\sin at - at\cos at\right).$$

(See Exercise 2 on page 530.)

Whenever we have the term $1/s$ in a Laplace transform, we should consider using the special case of the Convolution Theorem; namely, (11.43). We now give an example of its use.

Example 11.13 *The inverse Laplace transform of* $1/[s(s-1)^2]$

We seek $f(t)$ for which

$$\mathcal{L}\{f(t)\} = \frac{1}{s(s-1)^2}.$$

If we write this as

$$f(t) = \mathcal{L}^{-1}\left\{\frac{1}{s(s-1)^2}\right\} = \mathcal{L}^{-1}\left\{\frac{1}{s}\frac{1}{(s-1)^2}\right\},$$

we recognize that this is in the form of (11.43). Because we know that $\mathcal{L}\{te^t\} = 1/(s-1)^2$, we have

$$f(t) = \mathcal{L}^{-1}\left\{\frac{1}{s}\frac{1}{(s-1)^2}\right\} = \int_0^t ze^z\,dz.$$

Integration by parts gives $\int_0^t ze^z\,dz = te^t - e^t + 1$, so $f(t) = te^t - e^t + 1$, and

$$\mathcal{L}^{-1}\left\{\frac{1}{s(s-1)^2}\right\} = te^t - e^t + 1. \tag{11.45}$$

We can use (11.38) and (11.45) to show that

$$\mathcal{L}^{-1}\left\{\frac{1}{s(s-a)^2}\right\} = \frac{1}{a^2}\left(ate^{at} - e^{at} + 1\right).$$

(See Exercise 3 on page 530.)

Table 11.6 contains all these results.

Table 11.6 **The inverse Laplace transform list**

$F(s)$	$\mathcal{L}^{-1}\{F(s)\} = f(t)$	
a/s	a	
$1/s^2$	t	
$n!/s^{n+1}$	t^n	
$1/(s-a)$	e^{at}	
$1/(s-a)^2$	te^{at}	
$1/[(s-a)(s-b)]$	$(e^{at} - e^{bt})/(a-b)$	$a \neq b$
$s/[(s-a)(s-b)]$	$(ae^{at} - be^{bt})/(a-b)$	$a \neq b$
$a/(s^2+a^2)$	$\sin at$	
$s/(s^2+a^2)$	$\cos at$	
$a/(s^2-a^2)$	$\sinh at$	
$s/(s^2-a^2)$	$\cosh at$	
$s/[(s^2-a^2)(s-b)]$	$[e^{at}/(a-b) - e^{-at}/(a+b) - 2be^{bt}/(a^2-b^2)]/2$	$a \neq b$
$b/[(s-a)(s^2+b^2)]$	$(be^{at} - a\sin bt - b\cos bt)/(a^2+b^2)$	
$1/(s^2+a^2)^2$	$(\sin at - at\cos at)/(2a^3)$	
$1/(s^2-a^2)^2$	$(-\sinh at + at\cosh at)/(2a^3)$	
$s^2/(s^2+a^2)^2$	$(\sin at + at\cos at)/(2a)$	
$2as/(s^2+a^2)^2$	$t\sin at$	
$(s^2-a^2)/(s^2+a^2)^2$	$t\cos at$	
$1/[s(s-a)^2]$	$(ate^{at} - e^{at} + 1)/a^2$	

EXERCISES

1. Find the inverse Laplace transform of the following functions. Here a and b are constants, and $a \neq b$.

(a) $1/(s^2 - 3s + 2)$ (b) $1/(s^2 + 2s + 1)$
(c) $1/(s^3 + 2s^2 - s - 2)$ (d) $1/\left[s^2(s+1)\right]$
(e) $s/\left[(s-1)(s^2+a^2)\right]$ (f) $1/\left[(s+1)(s^2+1)\right]$
(g) $1/\left[s(s^2+4s+13)\right]$ (h) $1/\left[s(s^2+16)\right]$
(i) $a/\left[(s-1)(s^2+a^2)\right]$ (j) $1/\left[(s-b)(s^2-a^2)\right]$
(k) $s/\left[(s-b)(s^2-a^2)\right]$ (l) $s^2/(s^2+a^2)^2$
(m) $1/(s^2-a^2)^2$

2. From (11.38) and (11.44) show that

$$\mathcal{L}^{-1}\left\{\frac{1}{\left(s^2+a^2\right)^2}\right\} = \frac{1}{2a^3}(\sin at - at\cos at).$$

3. From (11.38) and (11.45) show that

$$\mathcal{L}^{-1}\left\{\frac{1}{s(s-a)^2}\right\} = \frac{1}{a^2}\left(ate^{at} - e^{at} + 1\right).$$

4. The property $F'(s) = -\mathcal{L}\{tf(t)\}$ may be used to invert Laplace transforms if we write it as $tf(t) = -\mathcal{L}^{-1}\{F'(s)\}$. Thus, if we know the inverse Laplace transform of $F'(s)$, we also know the inverse Laplace transform of $F(s)$. Use this property to find

(a) $$\mathcal{L}^{-1}\left\{\ln\left|\frac{s-a}{s-b}\right|\right\}.$$

(b) $$\mathcal{L}^{-1}\left\{\ln\left|1+\frac{1}{s^2}\right|\right\}.$$

(c) $$\mathcal{L}^{-1}\left\{\frac{\pi}{2} - \arctan\frac{s}{2}\right\}.$$

5. The Laplace transform may also be used to solve certain types of integral equations. One type, called a VOLTERRA INTEGRAL EQUATION, has the form

$$y(t) = f(t) + \int_0^t y(z)g(t-z)\,dz,$$

where f and g are known functions, and $y(t)$ is to be determined. Volterra integral equations may be solved with the aid of the Convolution Theorem. Solve the following equations for $y(t)$.

(a) $y(t) = 4t - 3 - \int_0^t \sin(t-z)\,y(z)\,dz$
(b) $y(t) = t - \int_0^t (t-z)\,y(z)\,dz$
(c) $y(t) = 3\sin t - 2\int_0^t \cos(t-z)\,y(z)\,dz$
(d) $y(t) = \sin t + 4e^{-t} - 2\int_0^t \cos(t-z)\,y(z)\,dz$
(e) $y(t) = t - \int_0^t e^{t-z}y(z)\,dz$
(f) $y(t) = \cos t + \int_0^t e^{t-z}y(z)\,dz$

6. An outline of the proof of the Convolution Theorem uses the following steps.

(a) Show that

$$\mathcal{L}\{f(t) * g(t)\} = \int_R e^{-st}f(z)g(t-z)\,dz\,dt,$$

where R is the region shown in Figure 11.1.

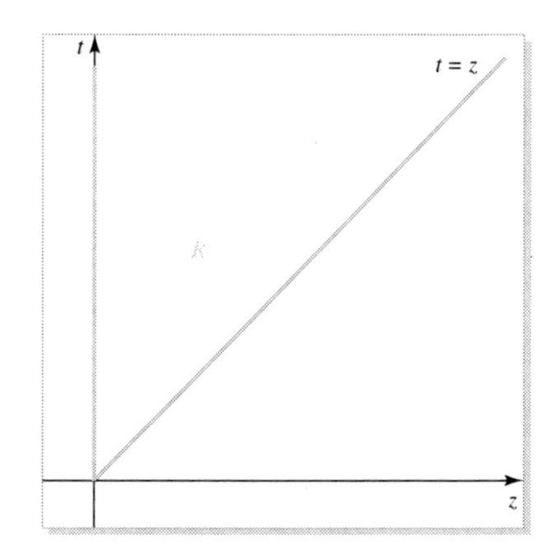

FIGURE 11.1 The region R

(b) Make the change of variables $u = t - z$, $v = z$, to show that

$$\mathcal{L}\{f(t) * g(t)\} = \int_0^\infty \int_0^\infty e^{-s(u+v)}f(v)g(u)\,du\,dv.$$

(c) Write the integral in part (b) as the product of two integrals, and note that this product contains the Laplace transform of f and g.

7. Use the Convolution Theorem to prove

$$\int_0^1 (1-t)^m t^n\,dt = \frac{n!m!}{(n+m+1)!},$$

where m and n are positive integers.

8. By making the substitution $u = t - z$, prove that $f * g = g * f$.

11.4 FUNCTIONS THAT JUMP

In the first section of this chapter we showed how the Laplace transform is used to solve first order linear differential equations with constant coefficients. Because we already have methods for solving such equations in earlier chapters, you may wonder why we introduce another. There are two reasons. The first is that using the Laplace transform process reduces the problem of solving a differential equation to a purely algebraic one. The second is the ease with which the Laplace transform handles certain types of forcing functions, especially ones that suddenly "jump" from one value to another value at a particular time.

An example of this is the situation in which a forcing function in an electrical circuit is 0 from $t = 0$ to $t = 2$ and then jumps to 12 at $t = 2$ and remains at 12 thereafter. (This would correspond to a switch being turned on at $t = 2$ when a constant voltage of 12 is applied.) If we denote such a function by $E(t)$, then

$$E(t) = \begin{cases} 0 & \text{if } 0 \le t < 2, \\ 12 & \text{if } 2 \le t. \end{cases}$$

The graph of $E(t)$ is shown in Figure 11.2. Sometimes when graphing functions with jumps, it is difficult to see the major features because of the gaps. For this reason we will always fill a gap with a vertical line, as we have done in Figure 11.3. However, these vertical lines are merely aids to the eye. They are not part of the function.

Functions with Jumps

Because functions with jumps, such as shown by $E(t)$, are common in applications, we define a special function with this property.

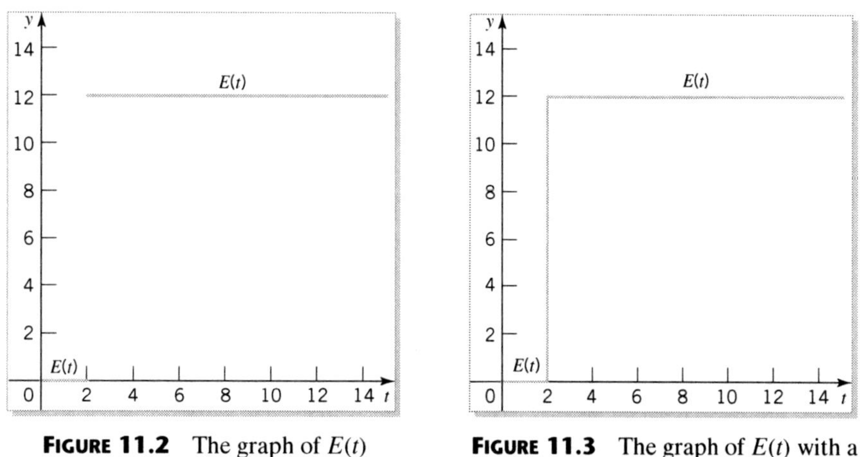

FIGURE 11.2 The graph of $E(t)$

FIGURE 11.3 The graph of $E(t)$ with a vertical line added

◆ *Definition 11.3:* **The UNIT STEP FUNCTION $u(t)$ is given by**[4]

$$u(t) = \begin{cases} 0 & \text{if } t < 0, \\ 1 & \text{if } 0 \le t. \end{cases}$$

◆

Comments about the Unit Step Function

- The graph of the unit step function is shown in Figure 11.4.
- If we are given a function, $f(t)$, then the function $f(t)u(t)$ is 0 for $t < 0$ and the original function $f(t)$ for $t \geq 0$. The graph of $tu(t)$ is shown in Figure 11.5.
- Because $u(t)$ has its jump at $t = 0$, the function $u(t - a)$ has its jump at $t = a$. (Why?) The graph of $u(t - 1)$ is shown in Figure 11.6.
- If we are given a function $f(t)$, then the function $f(t)u(t - a)$ is 0 for $t < a$ and the original function $f(t)$ for $t \geq a$. Thus, the function $E(t)$ on page 531 can be written in terms of the unit step function by $E(t) = 12u(t - 2)$.
- If we are given a function $f(t)$ that is 0 for $t < 0$, then the function $f(t - a)u(t - a)$ is the original function $f(t)$ shifted to the right by a units, if $a > 0$. If $a < 0$, the function is shifted to the left. The graph of $(t - 1)u(t - 1)$ is shown in Figure 11.7.

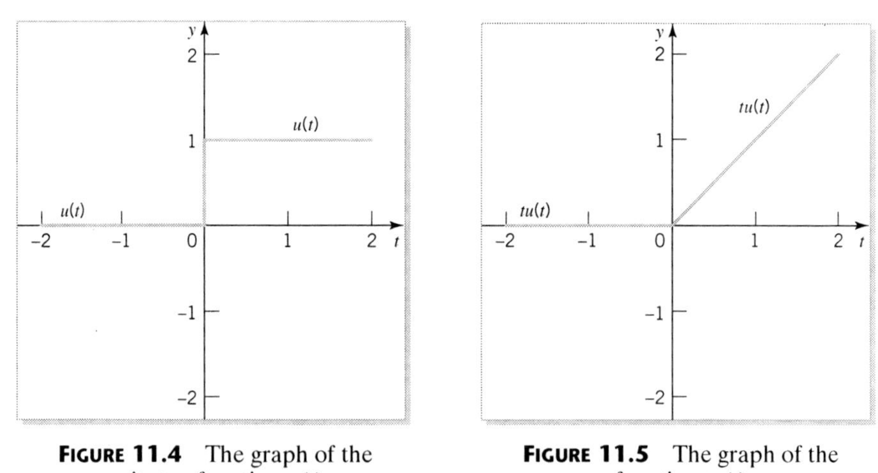

FIGURE 11.4 The graph of the unit step function $u(t)$

FIGURE 11.5 The graph of the function $tu(t)$

[4] This function is sometimes called the Heaviside function.

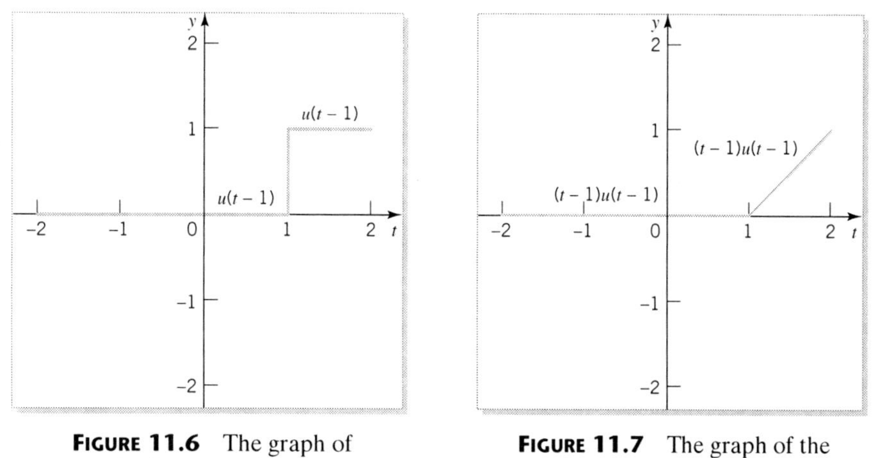

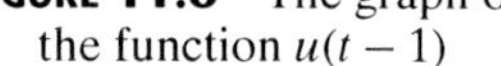

FIGURE 11.6 The graph of the function $u(t-1)$

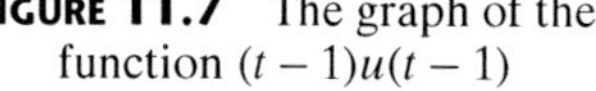

FIGURE 11.7 The graph of the function $(t-1)u(t-1)$

EXAMPLE 11.14 *The Function* $u(t-1) - u(t-3)$

To sketch the graph of $u(t-1) - u(t-3)$, we first recognize that there are two possible jumps, at $t = 1$ and $t = 3$. Thus, we should analyze $u(t-1) - u(t-3)$ in the three regions $t < 1$, $1 < t < 3$, and $3 < t$.

If $t < 1$, then both $u(t-1)$ and $u(t-3)$ contribute 0, so the function is 0 here. At $t = 1$ the contribution from $u(t-1)$ is 1. If $1 < t < 3$, then $u(t-1) = 1$, but $u(t-3) = 0$, so their difference is 1. If $3 \leq t$, both $u(t-1)$ and $u(t-3)$ contribute 1, so their difference is 0. Combining this information, we find

$$u(t-1) - u(t-3) = \begin{cases} 0 & \text{if } t < 1, \\ 1 & \text{if } 1 \leq t < 3, \\ 0 & \text{if } 3 \leq t. \end{cases}$$

The graph of the $u(t-1) - u(t-3)$ is shown in Figure 11.8. Physically this might correspond to a pulse of magnitude 1 lasting for 2 time units, starting at $t = 1$ and ending at $t = 3$.

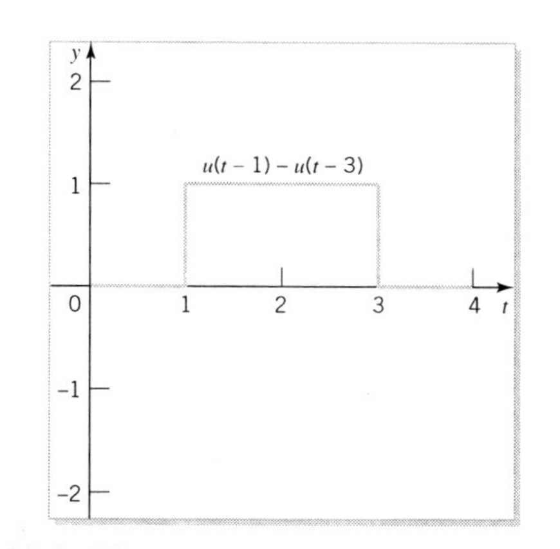

FIGURE 11.8 The graph of the function $u(t-1) - u(t-3)$

This example can be generalized to give

$$f(t)\,[u(t-a)-u(t-b)] = \begin{cases} 0 & \text{if } t < a, \\ f(t) & \text{if } a \le t < b, \\ 0 & \text{if } b \le t. \end{cases} \tag{11.46}$$

EXAMPLE 11.15

Write down a formula for the square wave function $f(t)$ whose graph is shown in Figure 11.9. The graph continues in the same fashion forever.

We see that $f(t) = 1$ if $0 < t < 1$, and that $f(t) = 0$ if $1 < t < 2$. Then the function repeats itself indefinitely — usually called PERIODIC of period 2.[5] From (11.46) we see that $u(t) - u(t-1)$ will be 1 if $0 \le t < 1$ and 0 if $1 \le t$. In the same way, the second bump can be obtained from $u(t-2) - u(t-3)$, so $u(t) - u(t-1) + u(t-2) - u(t-3)$ gives the first two bumps. From this we see that

$$f(t) = u(t) - u(t-1) + u(t-2) - u(t-3) + \cdots + u(t-2n) - u(t-2n-1) + \cdots$$

or

$$f(t) = \sum_{k=0}^{\infty} [u(t-2k) - u(t-2k-1)]\,.$$

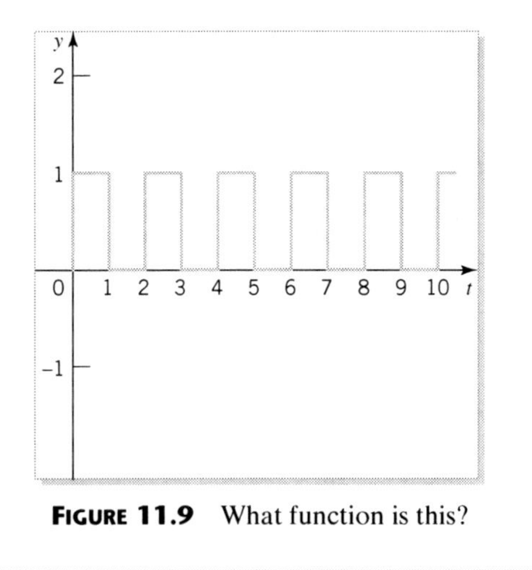

FIGURE 11.9 What function is this?

The Laplace Transform of the Unit Step Function

Because of the appearance of the unit step function in various applications, we need to compute its Laplace transform.

[5] Remember, a function $f(t)$ is periodic of period P if $f(t+P) = f(t)$ for every t for which $f(t)$ is defined, and where P is the smallest positive number with this property.

EXAMPLE 11.16 *The Laplace Transform of* $u(t)$

The Laplace transform of $u(t)$ is

$$\mathcal{L}\{u(t)\} = \int_0^\infty e^{-st}\,dt = \lim_{b\to\infty} -\frac{1}{s}e^{-st}\Big|_0^b,$$

or, for $s > 0$,

$$\mathcal{L}\{u(t)\} = \frac{1}{s}.$$

Thus, we also have

$$\mathcal{L}^{-1}\left\{\frac{1}{s}\right\} = u(t).$$

EXAMPLE 11.17 *The Laplace Transform of* $u(t-a)$

The Laplace transform of $u(t-a)$ is

$$\mathcal{L}\{u(t-a)\} = \int_0^a e^{-st}0\,dt + \int_a^\infty e^{-st}\,dt = \lim_{b\to\infty} -\frac{1}{s}e^{-st}\Big|_a^b,$$

or, for $s > 0$,

$$\mathcal{L}\{u(t-a)\} = \frac{1}{s}e^{-sa}.$$

Thus, we also have

$$\mathcal{L}^{-1}\left\{\frac{1}{s}e^{-sa}\right\} = u(t-a).$$

EXAMPLE 11.18 *The Laplace Transform of* $f(t-a)u(t-a)$

The Laplace transform of $f(t-a)u(t-a)$ is

$$\mathcal{L}\{f(t-a)u(t-a)\} = \int_0^a e^{-st}0\,dt + \int_a^\infty e^{-st}f(t-a)\,dt.$$

We now make the change of variable $t - a = z$ in the second integral to find that

$$\mathcal{L}\{f(t-a)u(t-a)\} = \int_0^\infty e^{-s(z+a)}f(z)\,dz = e^{-sa}\int_0^\infty e^{-sz}f(z)\,dz = e^{-sa}\mathcal{L}\{f(t)\}.$$

Thus, we have that if $F(s) = \mathcal{L}\{f(t)\}$ and $a > 0$, then

$$\mathcal{L}\{f(t-a)u(t-a)\} = e^{-sa}\mathcal{L}\{f(t)\}.$$

We also have

$$\mathcal{L}^{-1}\left\{e^{-sa}F(s)\right\} = f(t-a)u(t-a). \tag{11.47}$$

Although the derivation of (11.47) used $a > 0$, a different change of variables will show that (11.47) is also true for $a < 0$.

An Application

EXAMPLE 11.19

In order to illustrate the use of these results, we consider an electrical circuit that has a 12-volt battery connected in series with a resistor and a capacitor. At $t = 0$, the battery is bypassed, allowing the capacitor to begin discharging. Two seconds later the battery is inserted back into the circuit. (See Figure 11.10.) We want to discover how the charge, q, on the capacitor behaves under this situation.

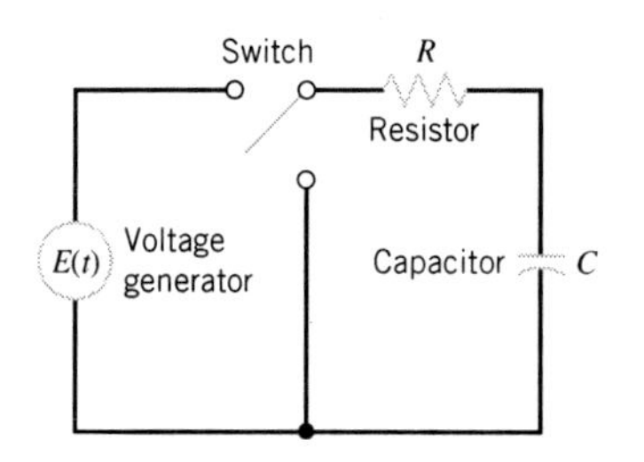

FIGURE 11.10 Electrical circuit

Our pertinent initial value problem is

$$Rq' + \frac{1}{C}q = \begin{cases} 0 & \text{if } 0 \le t < 2, \\ 12 & \text{if } 2 \le t, \end{cases}$$

In this example we will take $R = 10$, $C = 1/10$, and $q(0) = 12C$, so we want to solve

$$q' + q = \begin{cases} 0 & \text{if } 0 \le t < 2, \\ 1.2 & \text{if } 2 \le t, \end{cases}$$

with $q(0) = 1.2$.

We could solve this using the techniques of Chapter 5. We would then proceed in two steps. First solve $q' + q = 0$ subject to $q(0) = 1.2$. This solution will be valid for $0 \le t < 2$. From this we would evaluate $q(2) = \lim_{t\to 2^-} q(t)$. Then we would solve $q' + q = 1.2$, subject to the initial value of $q(2)$. This solution will be valid for $2 \le t$. Although we could solve the problem in this way, it is very clumsy.

Instead we use Laplace transforms. We first express the forcing function in terms of the unit step function. This gives our initial value problem as

$$q' + q = 1.2u(t-2), \tag{11.48}$$

with $q(0) = 1.2$.

Take the Laplace transform We take the Laplace transform of both sides of (11.48) and use our initial condition and properties of Laplace transforms to obtain

$$\mathcal{L}\{q'\} + \mathcal{L}\{q\} = 1.2\, e^{-2s}/s,$$

or

$$s\mathcal{L}\{q\} - 1.2 + \mathcal{L}\{q\} = 1.2\, e^{-2s}/s.$$

Solve for $\mathcal{L}\{q\}$ Thus, we may solve for $\mathcal{L}\{q\}$ as

$$\mathcal{L}\{q\} = \frac{1.2}{s+1} + \frac{1.2e^{-2s}}{s(s+1)}. \tag{11.49}$$

We know the function that has a Laplace transform of $1.2/(s+1)$ is $1.2e^{-t}$. But what about the second term in (11.49)? What function has a Laplace transform of $1.2e^{-2s}/[s(s+1)]$? In order to use our result in (11.47), we need to know what function has a Laplace transform of $1/[s(s+1)]$. From Table 11.6 we have that

$$\mathcal{L}\left\{\frac{1}{a}\left(e^{at} - 1\right)\right\} = \frac{1}{s(s-a)},$$

so

$$\mathcal{L}\left\{-\left(e^{-t} - 1\right)\right\} = \frac{1}{s(s+1)}.$$

Using (11.47), with $f(t) = 1 - e^{-t}$, gives our solution to (11.48) as

$$q(t) = 1.2\left\{e^{-t} + \left[1 - e^{-(t-2)}\right]u(t-2)\right\}. \tag{11.50}$$

Explicit solution This could be expressed as

$$q(t) = \begin{cases} 1.2e^{-t} & \text{if } 0 \le t < 2, \\ 1.2\left\{e^{-t} + \left[1 - e^{-(t-2)}\right]\right\} & \text{if } 2 \le t. \end{cases}$$

The graph of (11.50) is shown in Figure 11.11. (Does its behavior agree with what you anticipated?)

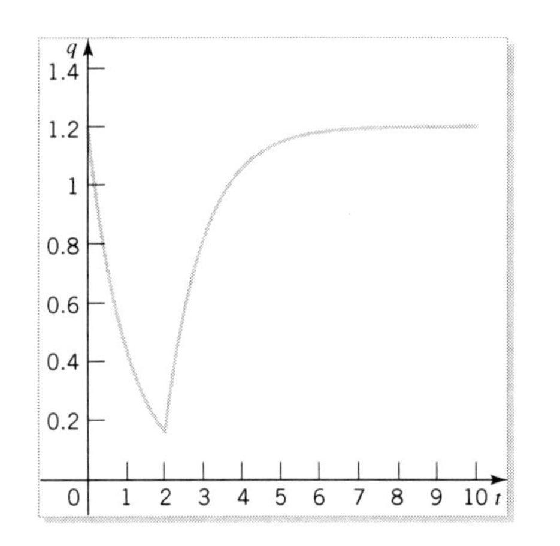

FIGURE 11.11 The graph of $q(t) = 1.2\{e^{-t} + [1 - e^{-(t-2)}]u(t-2)\}$

EXAMPLE 11.20 *The Laplace Transform of $\sum_{k=0}^{\infty}[u(t-2k) - u(t-2k-1)]$*

If a forcing function such as the one we used in our previous example turns on and off indefinitely, we will need the Laplace transform of the square wave function

$$f(t) = \sum_{k=0}^{\infty} [u(t-2k) - u(t-2k-1)];$$

namely, $\mathcal{L}\{f(t)\} = \mathcal{L}\left\{\sum_{k=0}^{\infty} [u(t-2k) - u(t-2k-1)]\right\}$. Assuming it is legal to extend (11.18) to an infinite number of terms, we find

$$\mathcal{L}\left\{\sum_{k=0}^{\infty} [u(t-2k) - u(t-2k-1)]\right\} = \sum_{k=0}^{\infty} \mathcal{L}\{u(t-2k)\} - \sum_{k=0}^{\infty} \mathcal{L}\{u(t-2k-1)\}.$$

From

$$\mathcal{L}\{u(t-a)\} = \frac{1}{s}e^{-sa}$$

we have

$$\mathcal{L}\left\{\sum_{k=0}^{\infty} [u(t-2k) - u(t-2k-1)]\right\} = \sum_{k=0}^{\infty} \frac{1}{s}e^{-s2k} - \sum_{k=0}^{\infty} \frac{1}{s}e^{-s(2k+1)}.$$

Using $e^{-s(2k+1)} = e^{-s2k}e^{-s}$ in the last term on the right-hand side allows us to combine both terms in the form

$$\mathcal{L}\left\{\sum_{k=0}^{\infty} [u(t-2k) - u(t-2k-1)]\right\} = \frac{1-e^{-s}}{s} \sum_{k=0}^{\infty} e^{-s2k}. \tag{11.51}$$

The infinite series $\sum_{k=0}^{\infty} e^{-s2k}$ is a geometric series $\sum_{k=0}^{\infty} r^k$ (where $r = e^{-s2}$ or e^{-2s}) with sum $1/(1-r)$; that is, $1/\left(1-e^{-2s}\right)$. Finally, we have the Laplace transform of the square wave as

$$\mathcal{L}\left\{\sum_{k=0}^{\infty}[u(t-2k)-u(t-2k-1)]\right\} = \frac{1-e^{-s}}{s}\frac{1}{\left(1-e^{-2s}\right)}.$$

Alternatively, using $1-e^{-2s} = \left(1-e^{-s}\right)\left(1+e^{-s}\right)$, we have

$$\mathcal{L}\left\{\sum_{k=0}^{\infty}[u(t-2k)-u(t-2k-1)]\right\} = \frac{1}{s}\frac{1}{\left(1+e^{-s}\right)}.$$

Sometimes it is useful to write this result in the form

$$\mathcal{L}\left\{\sum_{k=0}^{\infty}[u(t-2k)-u(t-2k-1)]\right\} = \frac{1}{s}\sum_{k=0}^{\infty}(-1)^k e^{-sk}. \tag{11.52}$$

(How did we do this?) ■

Notice that we were able to simplify the form of the Laplace transform of this periodic function considerably. Was this simplification due to the periodicity or to the fact that it consists of parts of horizontal lines? To investigate further we look at the transform of a general function with period P. Thus, we consider $\mathcal{L}\{f(t)\}$ where $f(t+P) = f(t)$ and are able to use the standard results from calculus (see Exercise 6 on page 544) to prove the following result.

► **Theorem 11.6: The Periodic Theorem** *If $f(t)$ is periodic with period P, then*

$$\mathcal{L}\{f(t)\} = \frac{1}{1-e^{-Ps}}\int_0^P e^{-st}f(t)\,dt.$$ ◄

Another Application

EXAMPLE 11.21

Let us now consider the RC circuit in Example 11.19 on page 536 in which the battery is connected and disconnected every 4 seconds following the pattern started in the first 4 seconds. (See Figure 11.10 on page 536.) Thus, our initial value problem for the charge $q(t)$ is now

$$q' + q = E(t), \qquad q(0) = 1.2, \tag{11.53}$$

where

$$E(t) = \begin{cases} 0 & \text{if } 0 \le t < 2, \\ 1.2 & \text{if } 2 \le t < 4, \end{cases} \quad \text{and } E(t+4) = E(t) \text{ for } t \ge 0.$$

In terms of unit step functions we have $E(t) = 1.2\sum_{k=0}^{\infty}[u(4k+2)-u(4k+4)]$.

Take the Laplace transform

We take the Laplace transform of (11.53) to obtain

$$\mathcal{L}\{q'\} + \mathcal{L}\{q\} = \mathcal{L}\{E(t)\}. \tag{11.54}$$

Rather than compute $\mathcal{L}\{E(t)\}$ as we did in Example 11.20 on page 538, we use the Periodic Theorem to find

$$\mathcal{L}\{E(t)\} = \frac{1}{1-e^{-4s}}\left(\int_0^2 e^{-st}0\,dt + \int_2^4 e^{-st}1.2\,dt\right),$$

or

$$\mathcal{L}\{E(t)\} = 1.2\frac{1}{s}\left(\frac{e^{-2s}-e^{-4s}}{1-e^{-4s}}\right).$$

We can simplify this expression once we realize that $e^{-2s}-e^{-4s} = e^{-2s}\left(1-e^{-2s}\right)$, and $1-e^{-4s} = \left(1-e^{-2s}\right)\left(1+e^{-2s}\right)$, obtaining

$$\mathcal{L}\{E(t)\} = 1.2\frac{1}{s}\frac{e^{-2s}}{\left(1+e^{-2s}\right)}.$$

Using the properties of Laplace transforms in Table 11.4, we can rewrite (11.54) as

$$s\mathcal{L}\{q(t)\} - 1.2 + \mathcal{L}\{q(t)\} = 1.2\frac{1}{s}\frac{e^{-2s}}{\left(1+e^{-2s}\right)},$$

Solve for $\mathcal{L}\{q\}$

or, solving for $\mathcal{L}\{q(t)\}$,

$$\mathcal{L}\{q(t)\} = 1.2\frac{1}{s+1} + 1.2\frac{1}{s(s+1)}\frac{e^{-2s}}{\left(1+e^{-2s}\right)}. \tag{11.55}$$

Looking at the right-hand side of (11.55), we see that we know the inverse Laplace transform of $1/(s+1)$ and $1/[s(s+1)]$. But what do we do with the term involving $e^{-2s}/\left(1+e^{-2s}\right)$? Our clue comes from Example 11.20 on page 538, in which we transformed the infinite series $\sum_{k=0}^{\infty} e^{-s2k}$ to $1/(1-e^{-2s})$. Because we have formulas for the inverse Laplace transform of products of $e^{as}F(s)$, using $1/(1-r) = \sum_{k=0}^{\infty} r^k$ allows us to write $e^{-2s}/\left(1+e^{-2s}\right)$ as a series of exponential functions. Thus, we use $1/\left(1+e^{-2s}\right) = \sum_{k=0}^{\infty}(-1)^k e^{-2sk}$ to find that

$$\frac{e^{-2s}}{1+e^{-2s}} = e^{-2s}\sum_{k=0}^{\infty}(-1)^k e^{-2sk} = \sum_{k=1}^{\infty}(-1)^{k+1} e^{-2sk}.$$

Using this result in (11.55) gives

$$\mathcal{L}\{q(t)\} = \frac{1.2}{s+1} + 1.2\frac{1}{s(s+1)}\sum_{k=1}^{\infty}(-1)^{k+1} e^{-2sk}.$$

From Table 11.6 we have $\mathcal{L}^{-1}\{1/(s+1)\} = e^{-t}$ and $\mathcal{L}^{-1}\left\{1/[s(s+1)]\right\} = 1 - e^{-t}$.

Explicit solution These results and (11.47) give us the solution of our initial value problem:

$$q(t) = 1.2e^{-t} + 1.2\sum_{k=1}^{\infty}(-1)^{k+1}\left[1 - e^{-(t-2k)}\right]u(t-2k). \tag{11.56}$$

If we expand the right-hand side of our solution for the first four time periods, we find

$$q(t) = \begin{cases} 1.2e^{-t} & \text{if } 0 \le t < 2, \\ 1.2\left\{e^{-t} + \left[1 - e^{-(t-2)}\right]\right\} & \text{if } 2 \le t < 4, \\ 1.2\left\{e^{-t} + \left[1 - e^{-(t-2)}\right] - \left[1 - e^{-(t-4)}\right]\right\} & \text{if } 4 \le t < 6, \\ 1.2\left\{e^{-t} + \left[1 - e^{-(t-2)}\right] - \left[1 - e^{-(t-4)}\right] + \left[1 - e^{-(t-6)}\right]\right\} & \text{if } 6 \le t < 8, \end{cases} \tag{11.57}$$

which agrees with our earlier results in Example 11.19 on page 536 for the common domain of $0 \le t < 4$. The graph of the charge given by (11.56) is shown in Figure 11.12 for $0 \le t < 8$.

The figure suggests that we have periodic behavior with constant amplitude. Do we? To answer this question we need to look at the solution $q(t)$ more carefully. If we simplify (11.57), we find

$$q(t) = \begin{cases} 1.2e^{-t} & \text{if } 0 \le t < 2, \\ 1.2\left[1 + e^{-t}\left(1 - e^{2}\right)\right] & \text{if } 2 \le t < 4, \\ 1.2e^{-t}\left(1 - e^{2} + e^{4}\right) & \text{if } 4 \le t < 6, \\ 1.2\left[1 + e^{-t}\left(1 - e^{2} + e^{4} - e^{6}\right)\right] & \text{if } 6 \le t < 8. \end{cases}$$

From this we can see a pattern emerge depending on whether t satisfies $4n \le t < 4n + 2$ or $4n + 2 \le t < 4n + 4$ for $n = 0, 1, 2, \cdots$. That pattern is

$$q(t) = \begin{cases} 1.2e^{-t}\left(1 - e^{2} + e^{4} - \cdots + e^{4n}\right) & \text{if } 4n \le t < 4n + 2, \\ 1.2\left[1 + e^{-t}\left(1 - e^{2} + e^{4} - \cdots - e^{4n+2}\right)\right] & \text{if } 4n + 2 \le t < 4n + 4. \end{cases} \tag{11.58}$$

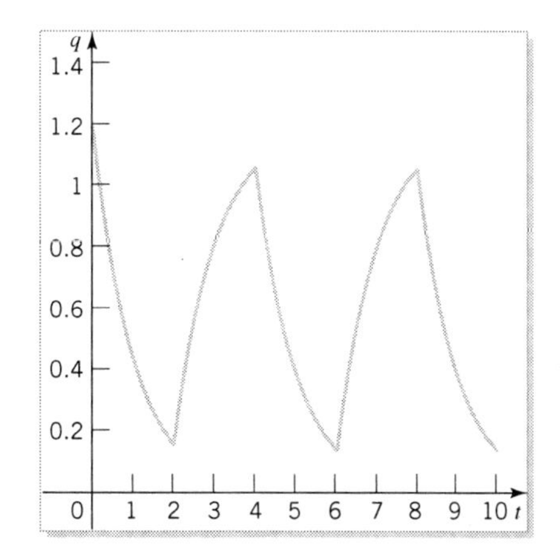

FIGURE 11.12 The charge from a square wave forcing function

We recognize that $1 - e^2 + e^4 - \cdots + e^{4n}$ can be simplified by means of the geometric series property

$$1 + z + z^2 + \cdots + z^m = \frac{1 - z^{m+1}}{1 - z}$$

as

$$1 - e^2 + e^4 - \cdots + e^{4n} = \frac{1 - (-e^2)^{2n+1}}{1 - (-e^2)} = \frac{1 + e^{4n+2}}{1 + e^2}.$$

In a similar way we find

$$1 - e^2 + e^4 - \cdots - e^{4n+2} = \frac{1 + e^{4n+4}}{1 + e^2},$$

so we can write (11.58) as

$$q(t) = \begin{cases} 1.2e^{-t}\left(1 + e^{4n+2}\right)/\left(1 + e^2\right) & \text{if } 4n \le t < 4n+2, \\ 1.2\left[1 + e^{-t}\left(1 + e^{4n+4}\right)/\left(1 + e^2\right)\right] & \text{if } 4n+2 \le t < 4n+4. \end{cases} \tag{11.59}$$

The successive maxima of $q(t)$ will occur at $t = 4n$, $n = 0, 1, 2, \cdots$. That is, they will occur at

$$q(4n) = 1.2e^{-4n}\frac{1 + e^{4n+2}}{1 + e^2} = 1.2\frac{e^{-4n} + e^2}{1 + e^2},$$

which is a decreasing function that tends to $1.2e^2/\left(1 + e^2\right)$ as $t \to \infty$. In the same way the minima will occur at $t = 4n + 2$, $n = 0, 1, 2, \cdots$, which is at

$$q(4n+2) = 1.2\left[1 + e^{-4n-2}\frac{1 - e^{4n+4}}{1 + e^2}\right] = 1.2\left[1 + \frac{e^{-4n-2} - e^2}{1 + e^2}\right] = 1.2\left[\frac{e^{-4n-2} + 1}{1 + e^2}\right].$$

This expression is a decreasing function that tends to $1.2/\left(1 + e^2\right)$ as $t \to \infty$. Thus, the maxima will lie on the curve $1.2\left(e^{-t} + e^2\right)/\left(1 + e^2\right)$ and the minima on the curve $1.2\left(e^{-t} + 1\right)/\left(1 + e^2\right)$. Figure 11.13 shows these two curves superimposed on Figure 11.12. Thus, in spite of its appearance, the amplitude of $q(t)$ is not constant.

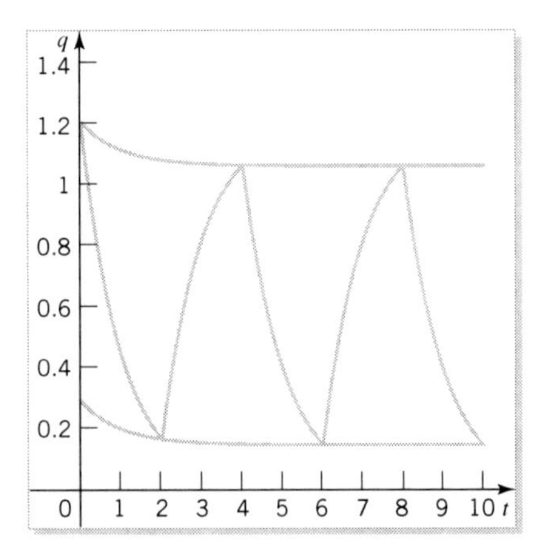

FIGURE 11.13 The functions $1.2(e^{-t} + e^2)/(1 + e^2)$ and $1.2(e^{-t} + 1)/(1 + e^2)$ and the charge from a square wave forcing function

Notice that in Examples 11.19 and 11.21 on pages 536 and 539 we had discontinuous forcing functions but continuous solutions, as shown in Figure 11.11 on page 538, and in Figure 11.12 on page 541. (How can you tell from the figures that the solutions are continuous?) We will return to this comment in Section 11.8, where we will discuss the circumstances under which this situation occurs.

EXERCISES

1. Graph the following functions for $t \geq 0$, and then write them in terms of the unit step function.

(a) $f(t) = \begin{cases} 1 & \text{if } 0 \leq t < 4, \\ 0 & \text{if } 4 \leq t. \end{cases}$

(b) $f(t) = \begin{cases} 1 & \text{if } 0 \leq t < 2, \\ 0 & \text{if } 2 \leq t < 3, \\ 3 & \text{if } 3 \leq t. \end{cases}$

(c) $f(t) = \begin{cases} t & \text{if } 0 \leq t < 2, \\ 2 & \text{if } 2 \leq t < 4, \\ 6-t & \text{if } 4 \leq t < 6, \\ 0 & \text{if } 6 \leq t. \end{cases}$

(d) $f(t) = \begin{cases} 1 & \text{if } 0 \leq t < 2, \\ -1 & \text{if } 2 \leq t < 4, \end{cases}$ where $f(t+4) = f(t)$.

(e) $f(t) = \begin{cases} 1 & \text{if } 0 \leq t < a, \\ 0 & \text{if } a \leq t < 2a, \\ -1 & \text{if } 2a \leq t < 3a, \end{cases}$ where $f(t+3a) = f(t)$.

2. Evaluate the Laplace transforms of the functions in Exercise 1.

3. The function $g(t) = \sin t$ if $0 \leq t < \pi$, where $g(t+\pi) = g(t)$ for $t \geq 0$, is shown in Figure 11.14. This is called the FULL-WAVE RECTIFICATION of $\sin t$.[6] Express this function in terms of the unit step function, and then evaluate its Laplace transform.

4. Find the equations, in terms of the unit step function, for the graphs in the following figures.

(a) Figure 11.15 (b) Figure 11.16

(c) Figure 11.17 (d) Figure 11.18

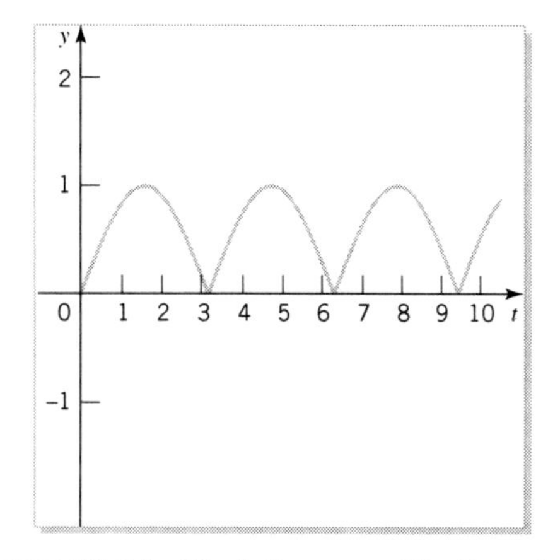

FIGURE 11.14 The full-wave rectification of $\sin t$

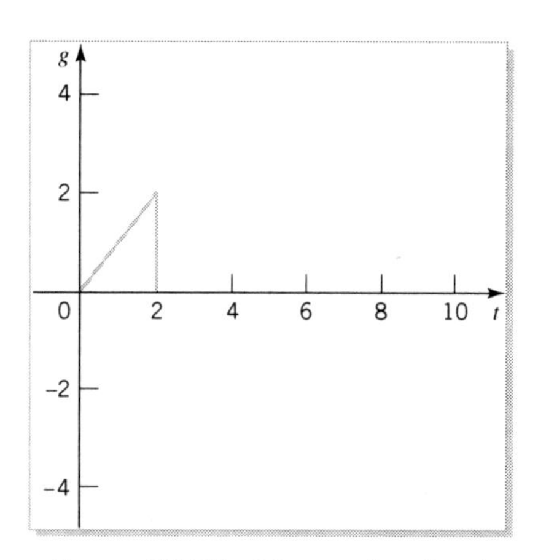

FIGURE 11.15 Mystery function A

[6] The full-wave rectification of $f(t)$ is the function $|f(t)|$. The half-wave rectification of $f(t)$ is the function $f(t)$ when $f(t) \geq 0$, and 0 when $f(t) < 0$.

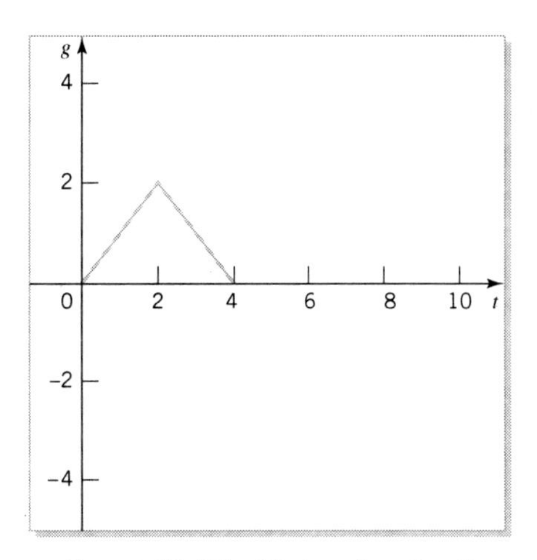

FIGURE 11.16 Mystery function B

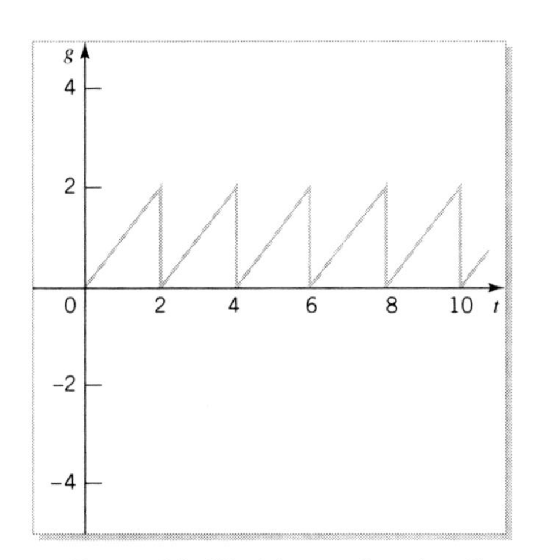

FIGURE 11.17 Mystery function C

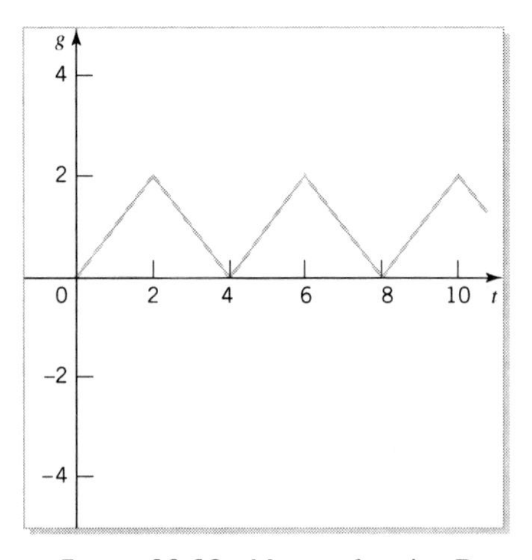

FIGURE 11.18 Mystery function D

5. Evaluate the Laplace transforms of the functions in Exercise 4.

6. Prove the Periodic Theorem using the following outline.

(a) Write the Laplace transform of the function $f(t)$ with period P in the form

$$\mathcal{L}\{f(t)\} = \int_0^\infty e^{-st} f(t)\,dt = \int_0^P e^{-st} f(t)\,dt + \int_P^\infty e^{-st} f(t)\,dt.$$

(b) Make the change of variable $z = t - P$ in the second integral to find

$$\mathcal{L}\{f(t)\} = \int_0^P e^{-st} f(t)\,dt + \int_0^\infty e^{-s(z+P)} f(z+P)\,dz.$$

(c) Use the periodicity of $f(t)$ in the form $f(z+P) = f(z)$ to factor out an e^{-sP} to remove the dependence on P in the integrand.

(d) Finish the argument.

7. Evaluate the inverse Laplace transforms of the following functions.

(a) $e^{-s}/(s+1)^2$ (b) $e^{-2s}/(s+1)$
(c) $e^{-3s}/[s(s+1)]$ (d) $e^{-4s}/(s^2-1)$
(e) $\left(e^{-s} - e^{-2s}\right)/(s-4)^2$ (f) $\left(e^{-2s} + e^{-3s}\right)/(s^2 - 3s + 2)$
(g) $e^{-3s}/(s^2 + 6s + 25)$

8. Dirac Delta Function. The Dirac Delta "function," $\delta(t)$, is sometimes used to model an impulsive force. Although it is not a function in the traditional mathematical sense, we can motivate its properties as follows.

(a) Let $d_\varepsilon(t)$ be defined by

$$d_\varepsilon(t) = \begin{cases} 0 & \text{if } t < -\varepsilon, \\ 1/(2\varepsilon) & \text{if } -\varepsilon \le t \le \varepsilon, \\ 0 & \text{if } \varepsilon < t, \end{cases}$$

where ε is a positive constant. Show that, if a is a given constant,

i. $\int_{-\infty}^{\infty} d_\varepsilon(t)\,dt = 1$, and $\int_{-\infty}^{\infty} d_\varepsilon(t-a)\,dt = 1$.

ii. $d_\varepsilon(t) = \frac{1}{2\varepsilon}\left[u(t+\varepsilon) - u(t-\varepsilon)\right]$, and $d_\varepsilon(t-a) = \frac{1}{2\varepsilon}\left[u(t-a+\varepsilon) - u(t-a-\varepsilon)\right]$.

iii. $\mathcal{L}\left\{d_\varepsilon(t)\right\} = \left(e^{\varepsilon s} - e^{-\varepsilon s}\right)/(2\varepsilon s)$, and $\mathcal{L}\left\{d_\varepsilon(t-a)\right\} = \left(e^{\varepsilon s} - e^{-\varepsilon s}\right)e^{-sa}/(2\varepsilon s)$.

iv. If $f(t)$ is a continuous function on $-\infty < t < \infty$, then there is number, c, for which $\int_{-\infty}^{\infty} d_\varepsilon(t-a) f(t)\,dt = f(c)$, where $a - \varepsilon \le c \le a + \varepsilon$. [Hint: Apply the mean value theorem for integrals, $\int_\alpha^\beta f(t)\,dt = f(c)$, where $\alpha \le c \le \beta$.]

(b) By formally letting $\delta(t-a) = \lim_{\varepsilon\to 0} d_\varepsilon(t-a)$, and assuming that the usual properties of limits are valid, show that the Dirac Delta function $\delta(t-a)$ has the following properties.

i. $\delta(t-a) = 0$ if $t \neq a$ and $\delta(t-a) = \infty$ if $t = a$.
ii. $\int_{-\infty}^{\infty} \delta(t-a)\,dt = 1$.
iii. $\delta(t-a) = u'(t-a)$.
iv. $\mathcal{L}\{\delta(t-a)\} = e^{-sa}$.
v. If $f(t)$ is a continuous function on $-\infty < t < \infty$, then $\int_{-\infty}^{\infty} \delta(t-a) f(t)\,dt = f(a)$.

11.5 MODELS INVOLVING FIRST ORDER LINEAR DIFFERENTIAL EQUATIONS

Now that we have developed several properties of Laplace transforms, and tables containing transforms of familiar functions, we move on to solve linear first order differential equations with constant coefficients. We start by revisiting two applications we covered earlier to show how using Laplace transforms simplifies some of our calculations. We end this section with two applications from forensic pathology.

EXAMPLE 11.22 *Population of Botswana, with and without Emigration*

In two different parts of this book, we considered the population growth of Botswana, once to obtain a model assuming no emigration, and later where we included hypothetical emigration. Here we combine these two into a single mathematical model by starting with a population in 1975, $t = 0$, of 0.755 million people and assuming that there is no significant emigration or immigration until 1990, $t = 15$. We suppose that at that time, emigration commences in a linear manner with 1.285 million people leaving over a 20-year period (see Example 5.1). With these assumptions, our mathematical model requires us to find the population, $P(t)$, such that

$$P' - kP = \begin{cases} 0 & \text{if } 0 < t < 15, \\ -a(t-15) & \text{if } 15 \leq t, \end{cases}$$

with $P(0) = 0.755$, $k = 0.0355$, and $a = 1.60625 \times 10^{-3}$. In order to use prior results from Laplace transforms, we first recast this equation in terms of unit step functions:

$$P' - kP = -a(t-15)u(t-15). \tag{11.60}$$

Take the Laplace transform

If we take the Laplace transform of (11.60), we obtain

$$\mathcal{L}\{P' - kP\} = \mathcal{L}\{-a(t-15)u(t-15)\},$$

or, by using the properties listed in (11.47), Table 11.3, and Table 11.4,

$$s\mathcal{L}\{P(t)\} - 0.755 - k\mathcal{L}\{P(t)\} = -\frac{a}{s^2}e^{-15s}.$$

Solve for $\mathcal{L}\{P(t)\}$

Solving for $\mathcal{L}\{P(t)\}$ gives

$$\mathcal{L}\{P(t)\} = \frac{0.755}{s-k} - \frac{a}{(s-k)s^2}e^{-15s}. \tag{11.61}$$

The inverse Laplace transform of the first term on the right-hand side of (11.61) is $0.755\,e^{kt}$ but with the term e^{-15s} present in the second term, we need to use (11.47). However, to do that we need the inverse Laplace transform of $1/[(s-k)s^2]$. Although this function does not appear on our largest list of inverse Laplace transforms, Table 11.6, both $1/(s-k)$ and $1/s^2$ do appear. Thus, we can use the Convolution Theorem with $F(s) = 1/(s-k)$, so $f(t) = e^{kt}$, and $G(s) = 1/s^2$, so $g(t) = t$. This gives us

$$\mathcal{L}^{-1}\left\{\frac{1}{(s-k)\,s^2}\right\} = \int_0^t e^{kz}(t-z)\,dz.$$

We use integration by parts to find

$$\mathcal{L}^{-1}\left\{\frac{1}{(s-k)\,s^2}\right\} = \frac{1}{k}e^{kz}(t-z)\bigg|_0^t - \int_0^t -\frac{1}{k}e^{kz}\,dz,$$

or

$$\mathcal{L}^{-1}\left\{\frac{1}{(s-k)\,s^2}\right\} = -\frac{t}{k} + \frac{e^{kt}-1}{k^2}.$$

Explicit solution

Now we use (11.47) to write our solution for the population as

$$P(t) = 0.755e^{kt} - a\left\{-\frac{t-15}{k} + \frac{1}{k^2}\left[e^{k(t-15)} - 1\right]\right\}u(t-15). \tag{11.62}$$

The graph of (11.62) is shown in Figure 11.19 for a 45-year period until 2020. (Does it look the way you expected?)

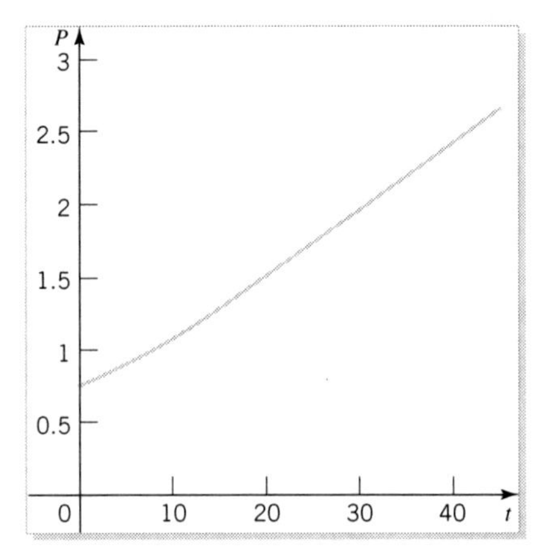

FIGURE 11.19 The population of Botswana for the period 1975 through 2020

EXAMPLE 11.23 *The Yam in the Oven Revisited*

In Example 5.7, we had another situation in which the forcing function in a differential equation had different forms for different time intervals. In that example, the oven temperature was described by

$$T_a(t) = \begin{cases} 70 + 66t & \text{if } 0 \le t < 5, \\ 400 & \text{if } 5 \le t, \end{cases}$$

or in terms of our unit step function,

$$T_a(t) = (70 + 66t)\,[u(t) - u(t-5)] + 400u(t-5) = (70 + 66t) + (330 - 66t)u(t-5).$$

We will now solve that initial value problem, namely,

$$\frac{dT}{dt} = k\left[T - T_a(t)\right] = k\left[T - (70 + 66t)u(t) - (330 - 66t)u(t-5)\right], \tag{11.63}$$

with $k = -0.04$ and $T(0) = 70$, using Laplace transforms.

Take the Laplace transform

We start by taking the Laplace transform of (11.63) as

$$\mathcal{L}\left\{T'(t)\right\} = \mathcal{L}\left\{k[T - (70 + 66t)u(t) - (330 - 66t)u(t-5)]\right\}. \tag{11.64}$$

We have results to evaluate the Laplace transform of all of the terms in (11.64) except the last one. There, in order to use (11.47), we need the term involving $u(t-5)$ to have the form $f(t-a)u(t-a)$. Thus, we write $t = (t-5) + 5$ and rewrite $(330 - 66t)u(t-5)$ as $-66(t-5)u(t-5)$. Using this result allows us to write (11.64) as

$$\mathcal{L}\left\{T'(t)\right\} = \mathcal{L}\left\{k[T - (70 + 66t)u(t) + 66(t-5)u(t-5)]\right\}. \tag{11.65}$$

Now we may use the properties in (11.47), Table 11.3, and Table 11.4 to find

$$s\mathcal{L}\{T(t)\} - 70 = k\left(\mathcal{L}\{T(t)\} - \frac{70}{s} - \frac{66}{s^2} + \frac{66}{s^2}e^{-5s}\right).$$

Solve for $\mathcal{L}\{T(t)\}$

Solving for $\mathcal{L}\{T(t)\}$ gives

$$\mathcal{L}\{T(t)\} = \frac{70}{s-k} - \frac{70k}{s(s-k)} - \frac{66k}{s^2(s-k)} + \frac{66k}{s^2(s-k)}e^{-5s}. \tag{11.66}$$

The first two terms on the right-hand side of (11.66) are included in Table 11.6, and we have just evaluated the inverse Laplace transform of the third and fourth terms in our previous example. Thus, from (11.66), we have

Explicit solution

$$T(t) = 70 - 66[f(t) - f(t-5)u(t-5)], \tag{11.67}$$

where

$$f(t) = -t + \frac{1}{k}(e^{kt} - 1).$$

The graph of $T(t)$ from (11.67) is shown in Figure 11.20.

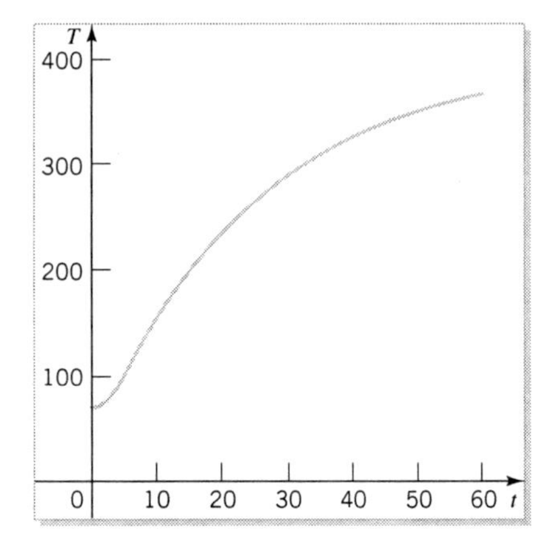

FIGURE 11.20 The temperature of the yam

EXAMPLE 11.24 *Estimating the Time of Death of a Murder Victim*

Consider a crime scene in Alaska in which the owner of a jewelry store was killed. The body was discovered in the jewelry store early in the morning, and at 7:00 A.M. the coroner measured its temperature as 72.5°F. One hour later the body temperature was 72°F. During the night the temperature of the store was a constant 70° F, and we want to find the time the person died.

We assume that the temperature of the body, $T(t)$, is governed by Newton's law of cooling,

$$T' = k\left[T(t) - T_a(t)\right], \tag{11.68}$$

where k is a negative constant, $T_a(t)$ is the ambient temperature, and the time, t, is measured in hours from the time of death. We also assume that the temperature of the person at the time of death was 98.6°F so that $T(0) = 98.6$.

The temperatures taken by the coroner give us $T(t_c) = 72.5$ and $T(t_c + 1) = 72$, where t_c is the number of hours after death. We will use these two pieces of information to evaluate k and t_c.

Take the Laplace transform

We start by taking the Laplace transform of (11.68), which gives us

$$\mathcal{L}\left\{T'(t)\right\} = \mathcal{L}\left\{k\left[T(t) - T_a(t)\right]\right\}.$$

If we use the properties of Laplace transforms in Table 11.4, we find that

$$s\mathcal{L}\{T(t)\} - T(0) = k\mathcal{L}\{T(t)\} - k\mathcal{L}\left\{T_a(t)\right\},$$

Solve for $\mathcal{L}\{T(t)\}$

and we may solve for $\mathcal{L}\{T(t)\}$ as

$$\mathcal{L}\{T(t)\} = \frac{T(0)}{s-k} - \frac{k}{s-k}\mathcal{L}\left\{T_a(t)\right\}. \tag{11.69}$$

We now use the Convolution Theorem to find the inverse Laplace transform of (11.69) as

$$T(t) = T(0)e^{kt} - \int_0^t kT_a(z)e^{k(t-z)}\,dz,$$

Explicit solution

or

$$T(t) = T(0)e^{kt} - ke^{kt}\int_0^t T_a(z)e^{-kz}\,dz. \tag{11.70}$$

The ambient temperature was a constant 70°F after the person's death. Thus, we take $T_a(t) = 70$ and integrate (11.70) to obtain $T(t) = 98.6e^{kt} + 70\left(1 - e^{-kt}\right)$, or

$$T(t) = 28.6e^{kt} + 70. \tag{11.71}$$

From the coroner's measurements we have

$$T(t_c) = 72.5 = 28.6e^{kt_c} + 70, \tag{11.72}$$

and

$$T(t_c + 1) = 72 = 28.6e^{k(t_c+1)} + 70.$$

We want to solve these two equations for t_c and k. If we subtract 70 from both equations and then divide the resulting equations, we obtain $2.5/2 = e^{-k}$, so $k = \ln 0.8 \approx -0.223$. If we substitute this value of k into (11.72), we obtain $e^{-0.223t_c} \approx 2.5/28.6$, or $t_c \approx 10.92$. Thus, the coroner thinks the person died about 10.92 hours before 7 A.M. — that is, about 8 P.M. the previous evening.[7] Figure 11.21 shows the graph of the temperature from (11.71).

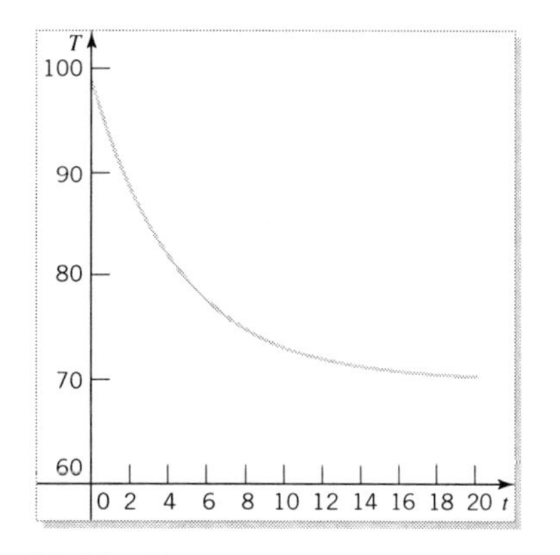

FIGURE 11.21 The temperature of the victim's body

[7] Notice that contrary to the situation frequently depicted on television crime shows, we need to know the temperature of the room and must take the temperature of the body at **two** different times to estimate the time of death.

EXAMPLE 11.25 *Reexamining the Crime Scenario*

Consider the previous crime scene through the eyes of the perpetrator, who in fact committed the crime at midnight when the owner of the jewelry store was locking the front door of the store. However, the killer left the victim's body outside the store for some time, where the temperature was a constant 30°F. Later he moved the body inside the store, where it was discovered under the circumstances described in the previous example. How long was the body outside?

The main difference between this example and the previous one is that the ambient temperature, $T_a(t)$, changes from 30 to 70 at the time, b, when the body is moved inside the store. This suggests we use the unit step function, $u(t)$, to construct $T_a(t)$, which is 30 for $0 \le t < b$ and 70 for $b \le t$; that is, $T_a(t) = 30 + 40u(t - b)$. Thus, b represents the number of hours the body was outside. We return to our explicit solution in (11.70) and perform the indicated integration for this form of $T_a(t)$:

$$T(t) = T(0)e^{kt} - ke^{kt} \int_0^t [30 + 40u(t - b)]\, e^{-kz}\, dz.$$

For $b \le t$ this becomes

$$T(t) = T(0)e^{kt} - ke^{kt} \left(\int_0^b 30e^{-kz}\, dz + \int_b^t 70e^{-kz}\, dz \right),$$

or

$$T(t) = T(0)e^{kt} - ke^{kt} \left[-\frac{1}{k} \left(30e^{-kb} - 30\right) - \frac{1}{k} \left(70e^{-kt} - 70e^{-kb}\right) \right].$$

We can rewrite this as

$$T(t) = e^{kt} \left[T(0) - 40e^{-kb} - 30 \right] + 70, \tag{11.73}$$

where we know $T(0) = 98.6$.

There are two unknowns in (11.73): k and b. But we also have two conditions to be met: $T(7) = 72.5$, and $T(8) = 72$. When we substitute these conditions into (11.73), we find

$$\begin{aligned} 2.5 &= e^{7k} \left(68.6 - 40e^{-kb}\right), \\ 2 &= e^{8k} \left(68.6 - 40e^{-kb}\right). \end{aligned} \tag{11.74}$$

Dividing the lower equation by the upper gives $e^k = 2/2.5$, or $k = \ln 0.8 \approx -0.223$. This is exactly the same value for k as we found in the previous example. This is not surprising, because k is the cooling constant for the body. For the same body it should have the same value. To find the value of b, we use this value of k in the upper equation of (11.74),

$$2.5 = e^{7 \ln 0.8} \left[68.6 - 40e^{-(\ln 0.8)b}\right],$$

or, solving for b,

$$b = -\frac{1}{\ln 0.8} \ln \left(\frac{68.6 - 2.5e^{-7 \ln 0.8}}{40} \right) \approx 2.15.$$

Thus, the body was moved at about 2:09 A.M. Figure 11.22 shows the graph of the temperature of the body.

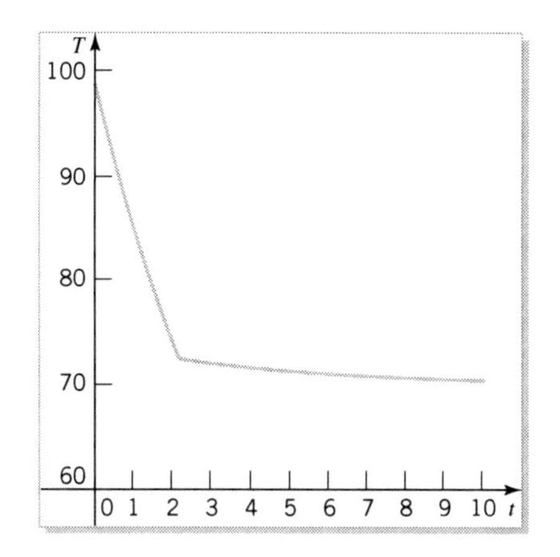

FIGURE 11.22 The temperature of the victim's body from midnight

General Initial Value Problem

We have used the Laplace transform to solve some first order linear differential equations with constant coefficients, so now we look at the general initial value problem associated with such differential equations; namely,

$$y' + ky = f(t), \qquad y(0) = y_0, \tag{11.75}$$

where $f(t)$ is a given function, and k and y_0 are constants. If we follow the same pattern that we used in the previous examples — take the Laplace transform of (11.75) and solve for $Y(s) = \mathcal{L}\{y(t)\}$ — we obtain

$$Y(s) = \frac{y_0}{s+k} + \frac{1}{s+k}F(s). \tag{11.76}$$

We apply the inverse Laplace transform to this equation and then use the Convolution Theorem to find

$$y(t) = y_0 e^{-kt} + \int_0^t e^{-k(t-z)} f(z)\, dz,$$

or

$$y(t) = y_0 e^{-kt} + e^{-kt} \int_0^t e^{kz} f(z)\, dz. \tag{11.77}$$

This is the general solution of the initial value problem (11.75). Notice in (11.77) that the first term on the right-hand side is a solution of the associated homogeneous equation, and that the second term is a particular solution of (11.75). Hence, we can associate these two terms with their counterparts in the transformed equation (11.76).

EXERCISES

1. Find $\mathcal{L}^{-1}\{1/[s^2(s-k)]\}$ by
   (a) using (11.43) and the fact that $\mathcal{L}\{(e^{kt}-1)/k\} = 1/[s(s-k)]$.
   (b) using partial fractions.

2. Derive (11.77) using reduction of order and the fact that $y(t) = e^{-kt}$ satisfies $y' + ky = 0$. (See Section 8.2.)

3. Solve the equation $y' + y = f(t)$ for the following functions $f(t)$. Compare your answers with those you found for Exercise 13 in Section 5.2.
   (a) $f(t) = \begin{cases} 2 \text{ if } 0 \le t < 1, \\ 1 \text{ if } 1 \le t, \end{cases} \quad y(0) = 0$
   (b) $f(t) = \begin{cases} 2 \text{ if } 0 \le t < 1, \\ 0 \text{ if } 1 \le t, \end{cases} \quad y(0) = 0$
   (c) $f(t) = \begin{cases} 5 \text{ if } 0 \le t < 10, \\ 1 \text{ if } 10 \le t, \end{cases} \quad y(0) = 6$
   (d) $f(t) = \begin{cases} e^{-t} \text{ if } 0 \le t < 2, \\ e^{-2} \text{ if } 2 \le t, \end{cases} \quad y(0) = y_0$

4. Solve the following initial value problems. In part (e), $f(t)$ is a given function, and a is a constant.
   (a) $y' + 4y = e^{-t} + u(t-2)e^{-t+2} \qquad y(0) = 1$
   (b) $y' - y = [1 - u(t-2)]e^t \qquad y(0) = 4$
   (c) $y' - y = [u(t-1) - u(t-3)]e^t \qquad y(0) = 2$
   (d) $y' + y = [1 - u(t-\pi)]\sin t \qquad y(0) = 1$
   (e) $y' + 4y = f(t) \qquad y(0) = a$

5. The differential equation for the current, $I(t)$, in a series circuit with inductance L and resistance R is
$$LI' + RI = E(t), \tag{11.78}$$
   where $E(t)$ is the applied voltage. Solve (11.78) subject to the initial condition $I(0) = 0$ for the following situations. (E_0 is a given constant.)
   (a) $E(t) = E_0[u(t-1) - u(t-2)]$ (a square pulse).
   (b) $E(t) = E_0[1 - u(t-\pi)]\sin t$ (a single pulse of a sine wave).
   (c) $E(t) = E_0 f(t)$, where $f(t)$ is the full square wave in Example 11.20 on page 538.
   (d) $E(t) = E_0 g(t)$, where $g(t)$ is the full-wave rectification of $\sin t$ from Exercise 3 on page 543.

6. The differential equation $Vy' = at[u(t) - u(t-T)] - by$ may be used to describe a compartmental model of the absorption of a drug by a body organ of volume V. The function $y(t)$ is the concentration of the drug in the organ's fluid at time t, a and b are the respective rates of fluid into and out of the organ, and T is the time period over which the drug is administered.
   (a) Find $y(t)$ if $y(0) = 0$ and $a = b$.
   (b) Find $y(t)$ if $y(0) = 1$ and $a \ne b$.

7. A pumpkin pie recipe says to place the ingredients in a preheated oven at 425°F for 15 minutes, then to turn the thermostat to 350°F and continue baking for an additional 45 minutes. Assume that the temperature of the oven decreases linearly from 425°F to 350°F in 10 minutes after the thermostat is changed, and that Newton's law of heating, $T' = k[T - T_a(t)]$, applies to this situation.
   (a) Find an expression for the oven temperature, $T_a(t)$, in terms of the unit step function.
   (b) If the initial temperature of the uncooked pie is 70°F, find the temperature of the pie as a function of t.
   (c) For $T(0) = 70$, find the temperature of the pie as a function of t if the temperature of the oven changes instantly from 425°F to 350°F at 15 minutes.
   (d) For $k = -0.09$ compare the graphs of your two solutions in parts (b) and (c), and comment on their similarities and differences.

8. A cup of coffee at 180°F is placed in a room where the air temperature is 70°F. After 30 minutes the temperature of the coffee is 120°F. At this time the coffee is placed in an oven, preheated to 400°F, until the coffee is once more at 180°F. Assume that Newton's law of cooling (heating) applies to both time periods.
   (a) Construct a separate phase line for each of these two time periods and then sketch the temperature of the coffee as a function of time for the entire time period.
   (b) Find the analytical solution to this problem using Laplace transforms, and compare this solution with your sketch.
   (c) At what time does the temperature of the coffee in the oven reach 180°F?

11.6 MODELS INVOLVING HIGHER ORDER LINEAR DIFFERENTIAL EQUATIONS

We now focus our attention on solving second order linear differential equations with constant coefficients using Laplace transforms. Here we find that the only change from the previous section is that we need to use the result for $\mathcal{L}\{y''\}$ in addition to that for $\mathcal{L}\{y'\}$. We revisit the application areas covered in Chapter 7 — namely electrical circuits, spring-mass systems, and the linear pendulum.

EXAMPLE 11.26 *A Spring-Mass System*

The differential equation that describes a spring-mass system that is subjected to an external forcing function is given by $mx'' + bx' + kx = f(t)$, where m is the mass, b is a damping coefficient, k is the spring constant, and $f(t)$ is the forcing function.

We consider the situation in which m, b, k, and $f(t)$ are chosen so the differential equation has the form

$$x'' + 4x' + 13x = 9e^{-2t}, \tag{11.79}$$

and where initially the mass is pulled down 3 units from equilibrium and released from rest. This means that our initial conditions are $x(0) = 3$, $x'(0) = 0$.

Take the Laplace transform

We start by taking the Laplace transform of (11.79) to obtain

$$\mathcal{L}\{x'' + 4x' + 13x\} = \mathcal{L}\{9e^{-2t}\}. \tag{11.80}$$

Using the properties of Laplace transforms from Tables 11.3 and 11.4 and our initial conditions, we can reduce (11.80) to

$$s^2\mathcal{L}\{x\} - 3s + 4(s\mathcal{L}\{x\} - 3) + 13\mathcal{L}\{x\} = \frac{9}{s+2}.$$

Solve for $\mathcal{L}\{x\}$

Solving for $\mathcal{L}\{x\}$, we find that

$$\mathcal{L}\{x\} = \frac{3s+12}{s^2+4s+13} + \frac{9}{(s+2)(s^2+4s+13)}. \tag{11.81}$$

Because the quadratic in the denominator may be written as $s^2 + 4s + 13 = (s+2)^2 + 9$, and because from Table 11.6 we have

$$\mathcal{L}^{-1}\left\{\frac{s+2}{(s+2)^2+9}\right\} = e^{-2t}\cos 3t$$

and

$$\mathcal{L}^{-1}\left\{\frac{3}{(s+2)^2+9}\right\} = e^{-2t}\sin 3t,$$

we rewrite $3s + 12$ as $3(s+2) + 2 \times 3$. Then we can recognize the inverse Laplace transform of the first term on the right-hand side of (11.81) as $e^{-2t}(3\cos 3t + 2\sin 3t)$. To evaluate the inverse Laplace transform of the remaining term, we could use either partial fractions or the Convolution Theorem. We use the latter with $G(s) = 1/(s+2)$, so $g(t) = e^{-2t}$, and $F(s) = 1/(s^2+4s+13)$, so $f(t) = (1/3)e^{-2t}\sin 3t$. This gives us

$$\mathcal{L}^{-1}\left\{\frac{9}{(s+2)(s^2+4s+13)}\right\} = 9\int_0^t \frac{1}{3}e^{-2z}\sin 3z\, e^{-2(t-z)}\,dz,$$

or

$$\mathcal{L}^{-1}\left\{\frac{9}{(s+2)(s^2+4s+13)}\right\} = 3e^{-2t}\int_0^t \sin 3z\,dz = 3e^{-2t}\left(-\frac{1}{3}\cos 3z\Big|_0^t\right).$$

Thus, we have

$$\mathcal{L}^{-1}\left\{\frac{9}{(s+2)(s^2+4s+13)}\right\} = e^{-2t}\left(-\cos 3t + 1\right). \tag{11.82}$$

Combining all our inverse transforms gives our solution of the initial value problem as

Explicit solution

$$x(t) = e^{-2t}\left(3\cos 3t + 2\sin 3t\right) + e^{-2t}(-\cos 3t + 1) = e^{-2t}(2\cos 3t + 2\sin 3t + 1). \tag{11.83}$$

Notice that had we used our earlier method of finding the general solution of the associated homogeneous differential equation and adding to this a particular solution, we would have derived the characteristic equation $r^2 + 4r + 13 = 0$. This gives us the same roots for r as we found for s when factoring the quadratic expression in the denominator of (11.81). Thus, some of the details in solving initial value problems using Laplace transforms are the same as those we encountered with methods used earlier.

One advantage of using Laplace transforms is the fact that the initial conditions are satisfied automatically. Another advantage is in treating forcing functions that are discontinuous or periodic, as we consider in the next two examples.

EXAMPLE 11.27 *Series RLC Circuit*

Recall that a series RLC circuit is governed by the differential equation $Lq'' + Rq' + \frac{1}{C}q = E(t)$, where L is the inductance, R the resistance, and C the capacitance. The charge at any time t is $q(t)$, and the applied voltage is $E(t)$.

We want to solve the initial value problem associated with this circuit where $L = 1$, $R = 6$, $C = 1/10$, $E(t) = 2[u(t-1) - u(t-2)]$, $q(0) = 0$, and $q'(0) = 1$. This gives our initial value problem as

$$q'' + 6q' + 10q = 2[u(t-1) - u(t-2)], \qquad q(0) = 0, \qquad q'(0) = 1.$$

The graph of the applied voltage is shown in Figure 11.23. (What phenomenon in the electrical circuit does $2[u(t-1) - u(t-2)]$ represent?)

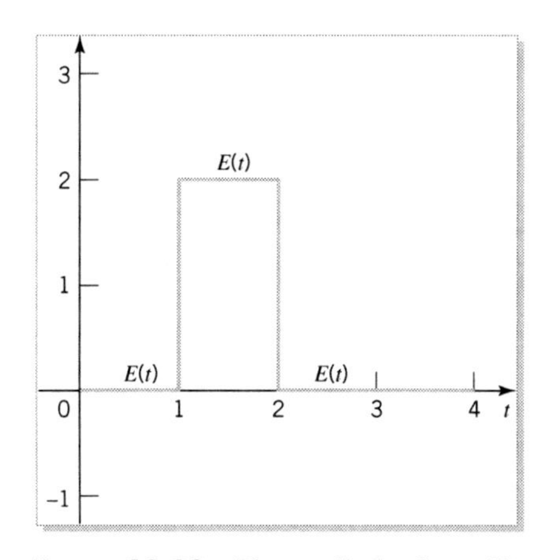

FIGURE 11.23 The applied voltage $E(t)$

Take the Laplace transform

If we take the Laplace transform, we obtain

$$\mathcal{L}\left\{q'' + 6q' + 10q\right\} = \mathcal{L}\left\{2[u(t-1) - u(t-2)]\right\}.$$

Using some properties of Laplace transforms changes this equation to

$$s^2\mathcal{L}\{q\} - 1 + 6s\mathcal{L}\{q\} + 10\mathcal{L}\{q\} = 2\left(\frac{e^{-s}}{s} - \frac{e^{-2s}}{s}\right).$$

Solve for $\mathcal{L}\{q\}$

Solving this equation for $\mathcal{L}\{q\}$ gives

$$\mathcal{L}\{q\} = \frac{1}{s^2 + 6s + 10} + \frac{1}{s^2 + 6s + 10}\frac{2}{s}\left(e^{-s} - e^{-2s}\right). \tag{11.84}$$

We now examine the first term on the right-hand side of (11.84) and express it as $1/(s^2 + 6s + 10) = 1/\left[(s+3)^2 + 1\right]$, so

$$\mathcal{L}^{-1}\left\{\frac{1}{s^2 + 6s + 10}\right\} = e^{-3t}\sin t.$$

In the second term on the right-hand side of (11.84) we have $[1/(s^2 + 6s + 10)]$ $[2/s]$, for which we may use the special case of the Convolution Theorem to find

$$\mathcal{L}^{-1}\left\{\frac{1}{s^2 + 6s + 10}\frac{2}{s}\right\} = 2\int_0^t e^{-3z}\sin z\,dz.$$

We designate this function by $h(t)$ and use a table of integrals, or integration by parts (twice), to find that

$$h(t) = 2\int_0^t e^{-3z}\sin z\,dz = -0.2e^{-3t}(3\sin t + \cos t) + 0.2.$$

Explicit solution

Combining these results gives the solution of our initial value problem as

$$q(t) = e^{-3t}\sin t + h(t-1)u(t-1) - h(t-2)u(t-2). \tag{11.85}$$

This can be expressed in the form

$$q(t) = \begin{cases} e^{-3t}\sin t & \text{if } t < 1, \\ e^{-3t}\sin t + h(t-1) & \text{if } 1 \le t < 2, \\ e^{-3t}\sin t + h(t-1) - h(t-2) & \text{if } 2 \le t. \end{cases}$$

The graph of $q(t)$ for $0 \le t \le 4$ is shown in Figure 11.24.

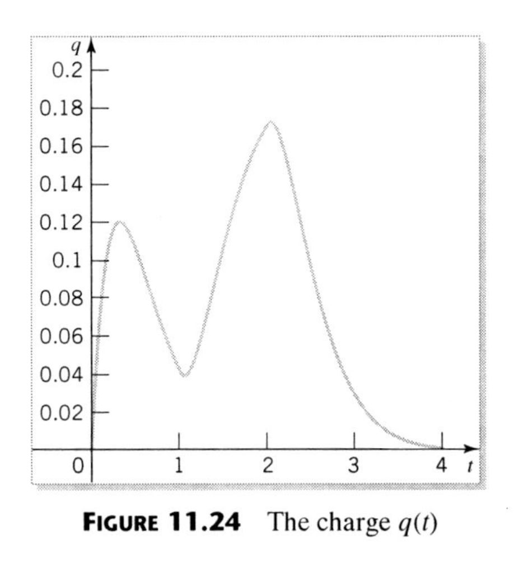

FIGURE 11.24 The charge $q(t)$

EXAMPLE 11.28

Here we consider the solution of the initial value problem

$$x'' + x = f(t), \qquad x(0) = 1, \qquad x'(0) = 1,$$

where $f(t)$ is the full square wave from Example 11.20 on page 538. This initial value problem models undamped oscillations of a forced

$$\left\{ \begin{array}{c} \text{spring-mass system} \\ \text{LC electrical circuit} \\ \text{linear pendulum} \end{array} \right\}$$

depending on the interpretation of $x, f(t), x(0)$, and $x'(0)$. Regardless of the physical situation, we will use the Laplace transform to solve this problem.

Take the Laplace transform

Taking the Laplace transform of our differential equation gives

$$\mathcal{L}\left\{x'' + x\right\} = \mathcal{L}\left\{f(t)\right\}.$$

Using some properties of Laplace transforms changes this equation to

$$s^2\mathcal{L}\{x\} - s - 1 + \mathcal{L}\{x\} = \mathcal{L}\{f(t)\}.$$

Solve for $\mathcal{L}\{x\}$ Solving this equation for $\mathcal{L}\{x\}$, we get

$$\mathcal{L}\{x\} = \frac{s}{s^2+1} + \frac{1}{s^2+1} + \frac{1}{s^2+1}\mathcal{L}\{f(t)\}. \tag{11.86}$$

The inverse Laplace transforms of the first two terms in (11.86) are in Table 11.6, whereas $\mathcal{L}\{f(t)\}$ is given by (11.52) as $(1/s)\sum_{k=0}^{\infty}(-1)^k e^{-sk}$. Because we know that $\mathcal{L}^{-1}\left\{e^{-as}F(s)\right\} = f(t-a)u(t-a)$, we use the special case of the Convolution Theorem to evaluate $\mathcal{L}^{-1}\left\{1/[s(s^2+1)]\right\}$ as

$$\mathcal{L}^{-1}\left\{\frac{1}{s\left(s^2+1\right)}\right\} = \int_0^t \sin z\, dz = 1 - \cos t.$$

Explicit solution Thus, we may write our solution of this initial value problem as

$$x(t) = \cos t + \sin t + \sum_{k=0}^{\infty}(-1)^k\left[1 - \cos(t-k)\right]u(t-k).$$

Figure 11.25 shows the effect of the square wave forcing function. Without the forcing function, we have oscillatory motion as given by the bottom curve in this figure.

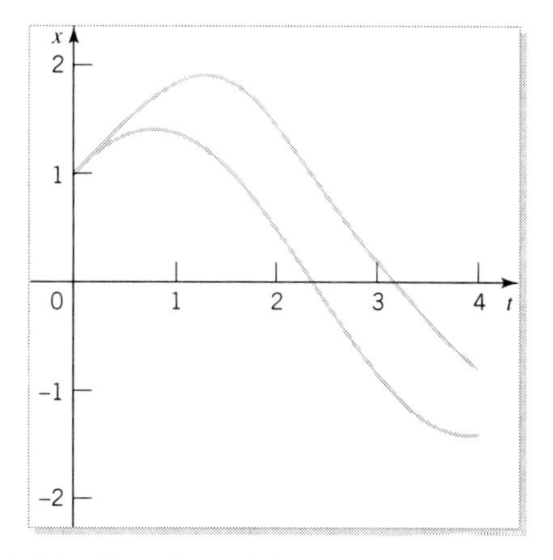

FIGURE 11.25 The effect of the square wave forcing function

Notice that in Examples 11.27 and 11.28 on pages 554 and 556, we had forcing functions that were discontinuous, but solutions that were not only continuous but also differentiable — see Figures 11.24 and 11.25. (How can you tell from the figures that the solutions are differentiable?) We will return to this comment in Section 11.8, where we will discuss the circumstances under which this occurs.

General Initial Value Problem

We have used the Laplace transform to solve some second order linear differential equations with constant coefficients, so now we look at the general initial value problem associated with such differential equations; namely,

$$ay'' + by' + cy = f(t), \qquad y(0) = y_0, \qquad y'(0) = y_0^*, \tag{11.87}$$

where $f(t)$ is a given function, and a, b, c, y_0, and y_0^* are constants. If we follow the same pattern that we used in the previous examples – take the Laplace transform of (11.87) and solve for $Y(s) = \mathcal{L}\{y(t)\}$ – we obtain

Take the Laplace transform

Solve for $\mathcal{L}\{y(t)\}$

$$Y(s) = \frac{asy_0 + ay_0^* + by_0}{as^2 + bs + c} + \frac{1}{as^2 + bs + c}F(s). \tag{11.88}$$

Next we introduce the two functions

$$h(t) = \mathcal{L}^{-1}\left\{\frac{1}{as^2 + bs + c}\right\}$$

and

$$g(t) = \mathcal{L}^{-1}\left\{\frac{s}{as^2 + bs + c}\right\}.$$

We then apply the inverse Laplace transform to (11.88) and use the Convolution Theorem to find

General solution

$$y(t) = ay_0 g(t) + \left(ay_0^* + by_0\right) h(t) + \int_0^t h(t - z) f(z)\, dz. \tag{11.89}$$

This is the general solution of the nonhomogeneous second order linear differential equation with constant coefficients. Notice that the functions $h(t)$ and $g(t)$ are solutions of the associated homogeneous equation, and that they correspond to the first term on the right-hand side of (11.88). The second term on the right-hand side of (11.88) corresponds to the particular solution $\int_0^t h(t - z) f(z)\, dz$.

In addition to the function f, the explicit solution (11.89) will depend on the factors of the quadratic expression $as^2 + bs + c$. There are three possibilities.

1. The quadratic expression $as^2 + bs + c$ has two real distinct factors, r_1 and r_2, so that $as^2 + bs + c = a\left(s - r_1\right)\left(s - r_2\right)$. In this case we have

$$h(t) = \mathcal{L}^{-1}\left\{\frac{1}{as^2 + bs + c}\right\} = \mathcal{L}^{-1}\left\{\frac{1}{a\left(s - r_1\right)\left(s - r_2\right)}\right\} = \frac{1}{a}\frac{1}{r_1 - r_2}\left(e^{r_1 t} - e^{r_2 t}\right)$$

and

$$g(t) = \mathcal{L}^{-1}\left\{\frac{s}{as^2 + bs + c}\right\} = \mathcal{L}^{-1}\left\{\frac{s}{a\left(s - r_1\right)\left(s - r_2\right)}\right\} = \frac{1}{a}\frac{1}{r_1 - r_2}\left(r_1 e^{r_1 t} - r_2 e^{r_2 t}\right).$$

The linear combination of $h(t)$ and $g(t)$ — written as $C_1 e^{r_1 t} + C_2 e^{r_2 t}$ — is the general solution of the associated homogeneous solution. This is consistent with the results we found in Chapter 6.

2. The quadratic expression $as^2 + bs + c$ has one real repeated factor, r, so that $as^2 + bs + c = a(s-r)^2$. In this case we have

$$h(t) = \mathcal{L}^{-1}\left\{\frac{1}{as^2+bs+c}\right\} = \mathcal{L}^{-1}\left\{\frac{1}{a(s-r)^2}\right\} = \frac{1}{a}te^{rt}$$

and

$$g(t) = \mathcal{L}^{-1}\left\{\frac{s}{as^2+bs+c}\right\} = \mathcal{L}^{-1}\left\{\frac{s}{a(s-r)^2}\right\} = \frac{1}{a}\left(e^{rt} + rte^{rt}\right).$$

The linear combination of $h(t)$ and $g(t)$ — written as $C_1 e^{rt} + C_2 te^{rt}$ — is the general solution of the associated homogeneous solution. This is consistent with the results we found in Chapter 6.

3. The quadratic expression $as^2 + bs + c$ has two complex factors, $\alpha \pm i\beta$, so that $as^2 + bs + c = a\left[(s-\alpha)^2 + \beta^2\right]$. In this case we have

$$h(t) = \mathcal{L}^{-1}\left\{\frac{1}{as^2+bs+c}\right\} = \mathcal{L}^{-1}\left\{\frac{1}{a\left[(s-\alpha)^2+\beta^2\right]}\right\} = \frac{1}{a\beta}e^{\alpha t}\sin\beta t$$

and

$$g(t) = \mathcal{L}^{-1}\left\{\frac{s}{as^2+bs+c}\right\} = \mathcal{L}^{-1}\left\{\frac{s}{a\left[(s-\alpha)^2+\beta^2\right]}\right\} = \frac{1}{a}e^{\alpha t}\left(\frac{\alpha}{\beta}\sin\beta t + \cos\beta t\right).$$

The linear combination of $h(t)$ and $g(t)$ — written as $e^{\alpha t}\left(C_1\cos\beta t + C_2\sin\beta t\right)$ — is the general solution of the associated homogeneous solution. This is consistent with the results we found in Chapter 6.

EXERCISES

1. Solve the following initial value problems using Laplace transforms. Then compare your answers with the ones you found in Exercise 3, Section 6.3.

(a) $y'' + y' - 2y = 0$ $\quad y(0) = 0 \quad y'(0) = 3$
(b) $y'' + 2y' - 10y = 0$ $\quad y(0) = 0 \quad y'(0) = 4$
(c) $y'' + 6y' + 9y = 0$ $\quad y(0) = 2 \quad y'(0) = 0$
(d) $12y'' + y' - y = 0$ $\quad y(0) = 4 \quad y'(0) = 0$
(e) $y'' + 3y' = 0$ $\quad y(0) = 4 \quad y'(0) = 3$
(f) $y'' - 2\pi y' + 2\pi^2 y = 0$ $\quad y(0) = 0 \quad y'(0) = -2\pi$
(g) $y'' + 10y' + 100y = 0$ $\quad y(0) = 15 \quad y'(0) = 4$
(h) $y'' + 10y' + 100y = 0$ $\quad y(1) = 0 \quad y'(1) = 0$

2. Create a *How to Solve Second Order Linear Differential Equations Using Laplace Transforms.* Add statements under the headings Purpose, Process, and Comments that summarize what you discovered in this section.

3. Solve the following initial value problems using Laplace transforms. Then compare your answers with the ones you found in Exercise 6, Section 7.2.

(a) $y'' + y' - 12y = 8e^{3t}$ $\quad y(0) = 0 \quad y'(0) = 1$
(b) $y'' + 6y' + 9y = e^{3t}$ $\quad y(0) = 0 \quad y'(0) = 6$
(c) $y'' - 5y' + 6y = 12te^{-t} - 7e^{-t}$ $\quad y(0) = 0 \quad y'(0) = 0$
(d) $y'' + 4y = 8\sin 2t + 8\cos 2t$ $\quad y(\pi) = 2\pi \; y'(\pi) = 2\pi$

4. Solve the differential equation (11.79), $x'' + 4x' + 13x = 9e^{-2t}$, subject to $x(0) = 3$, $x'(0) = 0$, using the methods of Chapter 7. Compare your answer with (11.83) on page 554.

5. Solve the following initial value problems. Here ω is a constant, $f(t)$ is any given function, and $g(t)$ is the full-wave rectification of $\sin t$ from Exercise 3 on page 543.

(a) $x'' + x = e^{-2t}\sin t \quad x(0) = 0 \quad x'(0) = 0$
(b) $x'' + 2x' + 2x = tu(t-1) \quad x(0) = 1 \quad x'(0) = 0$
(c) $x'' + 4x' + 4x = e^{-3t} \quad x(0) = 0 \quad x'(0) = 0$
(d) $x'' + x' - 2x = e^{-t}\sin t \quad x(0) = 0 \quad x'(0) = 0$
(e) $x'' + 3x' + 2x = t^2[u(t) - u(t-1)]$
$x(0) = 0 \quad x'(0) = 0$
(f) $x'' + 4x' + 4x = 4\cos 2t \quad x(0) = 1 \quad x'(0) = 1$
(g) $x'' + 4x' + 4x = f(t) \quad x(0) = 1 \quad x'(0) = 0$
(h) $x'' + \omega^2 x = f(t) \quad x(0) = 1 \quad x'(0) = 0$
(i) $x'' + \omega^2 x = f(t) \quad x(0) = 0 \quad x'(0) = 1$
(j) $x'' + \omega^2 x = u(t-2) - u(t-5)$
$x(0) = 1 \quad x'(0) = 1$
(k) $x'' + \omega^2 x = u(t-1) - u(t-2)$
$x(0) = 0 \quad x'(0) = 1$
(l) $x'' + \omega^2 x = g(t) \quad x(0) = 0 \quad x'(0) = 0$

6. A particular forced vibration of a mass m at the end of a vertical spring, is described by $mx''(t) + kx(t) = f(t)$, where $f(t) = 1 + u(t-1) - 2u(t-2)$, and k is the spring constant.

(a) Explain what $f(t)$ represents concerning the motion of the top of the spring.
(b) Solve this differential equation for $m = 1$, $k = 4$, $x(0) = x'(0) = 0$.
(c) Plot your solution for the interval $0 \le t \le 3$, as well as the graph of $f(t)$. What do you observe?

7. An initial value problem for a series RLC circuit consists of the differential equation $Lq'' + Rq' + \frac{1}{C}q = E(t)$, with initial conditions $q(0) = 1$, $q'(0) = 0$. Solve this initial value problem for $t \ge 0$ with $L = 1$ for the following situations.

(a) $R = 4 \quad 1/C = 20 \quad E(t) = 10e^{-t/2}$
(b) $R = 4 \quad 1/C = 20 \quad E(t) = 8\sin 4t$
(c) $R = 5 \quad 1/C = 6 \quad E(t) = 2[1 - u(t-2)]$
(d) $R = 5 \quad 1/C = 6$
$E(t) = 2[1 - u(t-2)] + 4u(t-4)$
(e) $R = 5 \quad 1/C = 6$
$E(t) = 2[1 - u(t-2)]$, $E(t+4) = E(t)$

11.7 APPLICATIONS TO SYSTEMS OF LINEAR DIFFERENTIAL EQUATIONS

Initial value problems involving systems of linear differential equations with constant coefficients may also be solved using Laplace transforms. The procedure is similar to that used many times in this chapter. Thus, we use the Laplace transform to transform our initial value problem for a system of differential equations into a system of algebraic equations. After solving this system of algebraic equations for the Laplace transform of our dependent variables, we use tables of inverse Laplace transforms to obtain our explicit solution. The advantages of the Laplace transform here include having the initial conditions satisfied automatically, and avoiding the need to find particular solutions. The procedure is illustrated with three examples.

EXAMPLE 11.29

Solve the system of differential equations

$$\begin{cases} x' = x - 2y, \\ y' = 5x - y, \end{cases} \tag{11.90}$$

subject to the initial conditions $x(0) = -1$, $y(0) = 2$.

Take the Laplace transform

If we apply the Laplace transform to both of these differential equations, we obtain

$$\begin{aligned} \mathcal{L}\{x'\} &= \mathcal{L}\{x - 2y\}, \\ \mathcal{L}\{y'\} &= \mathcal{L}\{5x - y\}. \end{aligned}$$

Using the results in Table 11.4, we can transform this system into

$$\begin{aligned} sX(s) + 1 &= X(s) - 2Y(s), \\ sY(s) - 2 &= 5X(s) - Y(s), \end{aligned} \tag{11.91}$$

where $X(s) = \mathcal{L}\{x\}$, and $Y(s) = \mathcal{L}\{y\}$. We may solve for $Y(s)$ from the top equation, $Y(s) = (1/2)[(1-s)X(s) + 1]$, and substitute the result into the bottom equation to find

Solve for $X(s)$

$$X(s) = -\frac{s+5}{s^2+9}. \tag{11.92}$$

If we rewrite this expression as

$$X(s) = -\frac{s}{s^2+9} - \frac{5}{3}\frac{3}{s^2+9},$$

we see that $x(t)$ is

$$x(t) = -\cos 3t - \frac{5}{3}\sin 3t. \tag{11.93}$$

We have two choices for determining $y(t)$. One is to substitute the expression for $X(s)$ from (11.92) into one of the equations in (11.91) to obtain an expression for $Y(s)$, and then find its inverse Laplace transform. However, here it is easier simply to use the explicit solution for $x(t)$ from (11.93) in the first of our original differential equations in (11.90) to obtain

$$y(t) = \frac{1}{2}\left(x - x'\right) = \frac{1}{2}\left(-\cos 3t - \frac{5}{3}\sin 3t - 3\sin 3t + 5\cos 3t\right),$$

Explicit solution

or

$$y(t) = 2\cos 3t - \frac{7}{3}\sin 3t.$$

EXAMPLE 11.30

Solve the initial value problem

$$\begin{cases} x' = -x + y + 25\sin t, \\ y' = -x - 3y, \end{cases} \tag{11.94}$$

where $x(0) = -1$, $y(0) = -1$.

Take the Laplace transform

If we apply the Laplace transform to both of these differential equations, we obtain

$$\mathcal{L}\{x'\} = \mathcal{L}\{-x + y + 25\sin t\},$$
$$\mathcal{L}\{y'\} = \mathcal{L}\{-x - 3y\}.$$

Using some results in Tables 11.3 and 11.4 transforms this to

$$sX(s) + 1 = -X(s) + Y(s) + 25/(s^2 + 1),$$
$$sY(s) + 1 = -X(s) - 3Y(s).$$

We may solve for $X(s)$ from the bottom equation, $X(s) = -(3 + s)Y(s) - 1$, and substitute the result into the top equation to find

$$(s + 1)[-(3 + s)Y(s) - 1] + 1 = Y(s) + \frac{25}{s^2 + 1}.$$

We may rewrite this expression as

$$-[(s + 1)(3 + s) + 1]Y(s) = s + \frac{25}{s^2 + 1},$$

Solve for $Y(s)$

or, solving for $Y(s)$,

$$Y(s) = -\frac{s}{(s + 2)^2} - \frac{25}{(s + 2)^2 (s^2 + 1)}.$$

Partial fractions

Here the inverse Laplace transform of the first term is $e^{-2t}(2t - 1)$. The second may be calculated in several ways. We will use partial fractions and write

$$\frac{25}{(s + 2)^2 (s^2 + 1)} = \frac{A}{s + 2} + \frac{B}{(s + 2)^2} + \frac{Cs + D}{(s^2 + 1)}$$
$$= \frac{A(s + 2)(s^2 + 1) + B(s^2 + 1) + (Cs + D)(s + 2)^2}{(s + 2)^2 (s^2 + 1)}.$$

This gives

$$25 = A(s + 2)(s^2 + 1) + B(s^2 + 1) + (Cs + D)(s + 2)^2,$$

or

$$25 = (A + C)s^3 + (2A + B + 4C + D)s^2 + (A + 4C + 4D)s + (2A + B + 4D).$$

This must be true for all values of s, so we equate the coefficients of like powers of s to find the four equations

$$\begin{aligned} A + C &= 0, \\ 2A + B + 4C + D &= 0, \\ A + 4C + 4D &= 0, \\ 2A + B + 4D &= 25. \end{aligned}$$

From the first equation we find $A = -C$, which when substituted into the third equation gives $D = -3C/4$. When we substitute these equations for A and D into the third equation, we find $B = 5C + 25$. We now use these equations for A, B, and D in the second equation, obtaining $C = -4$. Thus, the remaining constants are $A = 4$, $B = 5$, and $D = 3$. This gives us

$$\mathcal{L}^{-1}\left\{\frac{25}{(s+2)^2(s^2+1)}\right\} = 4e^{-2t} + 5te^{-2t} - 4\cos t + 3\sin t,$$

and the explicit form of $y(t)$ is

$$y(t) = -e^{-2t}(3t+5) + 4\cos t - 3\sin t.$$

Explicit solution

To find $x(t)$ we return to (11.94) and solve for x as $x = -3y - y'$. Using the value of y and its derivative gives

$$x(t) = e^{-2t}(3t+8) - 9\cos t + 13\sin t.$$

The graphs of $x(t)$ and $y(t)$ appear in Figure 11.26, where steady state motion is quickly achieved.

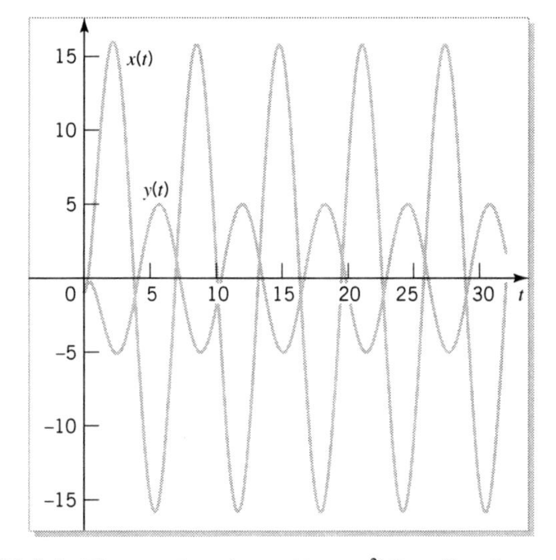

FIGURE 11.26 The two functions $x(t) = e^{-2t}(3t+8) - 9\cos t + 13\sin t$ and $y(t) = -e^{-2t}(3t+5) + 4\cos t - 3\sin t$

EXAMPLE 11.31

As our final example, we will find the solution of the nonhomogeneous system of differential equations

$$\begin{cases} x' = -2x + y + u(t-1), \\ y' = 4x - 2y + e^{2t}, \end{cases} \tag{11.95}$$

Take the Laplace transform

subject to the initial conditions $x(0) = x_0$, $y(0) = y_0$. Taking the Laplace transform and using our familiar properties gives

$$\begin{aligned} sX(s) - x_0 &= -2X(s) + Y(s) + e^{-s}/s, \\ sY(s) - y_0 &= 4X(s) - 2Y(s) + 1/(s-2). \end{aligned} \tag{11.96}$$

We may solve for $Y(s)$ from the top equation in (11.96) as

$$Y(s) = (s+2)X(s) - x_0 - \frac{1}{s}e^{-s}$$

and substitute this expression into the bottom equation to obtain

$$(s+2)\left[(s+2)X(s) - x_0 - \frac{1}{s}e^{-s}\right] - y_0 = 4X(s) + \frac{1}{s-2}.$$

Solve for $X(s)$

Solving this equation for $X(s)$ gives

$$X(s) = \frac{1}{s(s+4)}\left[(s+2)x_0 + y_0 + \frac{s+2}{s}e^{-s} + \frac{1}{s-2}\right].$$

Thus, the functions for which we need inverse Laplace transforms are

$$\frac{s+2}{s(s+4)}x_0, \quad \frac{1}{s(s+4)}y_0, \quad \frac{s+2}{s^2(s+4)}e^{-s}, \quad \text{and} \quad \frac{1}{s(s+4)(s-2)}.$$

The first two terms in this list may be written as

$$\frac{s+2}{s(s+4)}x_0 + \frac{1}{s(s+4)}y_0 = \frac{1}{s+4}x_0 + \frac{1}{s(s+4)}(2x_0 + y_0),$$

both of which occur in our list of inverse Laplace transforms in Table 11.6. Thus,

$$\mathcal{L}^{-1}\left\{\frac{s+2}{s(s+4)}x_0 + \frac{1}{s(s+4)}y_0\right\} = x_0 e^{-4t} + \frac{1}{4}(2x_0 + y_0)\left(1 - e^{-4t}\right).$$

Partial fractions

The last term in our list may be evaluated using a partial fraction decomposition as

$$\frac{1}{s(s+4)(s-2)} = \frac{A}{s} + \frac{B}{s+4} + \frac{C}{s-2} = \frac{A(s+4)(s-2) + Bs(s-2) + Cs(s+4)}{s(s+4)(s-2)}.$$

We then equate the numerators in the first and last terms in this continued equality to obtain

$$1 = A(s+4)(s-2) + Bs(s-2) + Cs(s+4).$$

Because this is to be an identity in s, it must be true for all values of s. Thus, it is true for the particular values of 0, 2, and -4. Using these three values in turn allows us to find $A = -1/8$, $B = 1/24$, and $C = 1/12$, so

$$\mathcal{L}^{-1}\left\{\frac{1}{s(s+4)(s-2)}\right\} = -\frac{1}{8} + \frac{1}{24}e^{-4t} + \frac{1}{12}e^{2t}.$$

The remaining term in our list has e^{-s} times a rational function of s. We first find the inverse Laplace transform of the rational function and then use a result involving the unit step function. We write

$$\frac{s+2}{s^2(s+4)} = \frac{s}{s^2(s+4)} + \frac{2}{s^2(s+4)} = \frac{1}{s(s+4)} + \frac{2}{s^2(s+4)}.$$

The first of these expressions occurs in Table 11.6, so

$$\mathcal{L}^{-1}\left\{\frac{1}{s(s+4)}\right\} = \frac{1}{4}\left(1 - e^{-4t}\right).$$

We now use the special case of the Convolution Theorem to write

$$\mathcal{L}^{-1}\left\{\frac{2}{s^2(s+4)}\right\} = \mathcal{L}^{-1}\left\{\frac{2}{s(s+4)}\frac{1}{s}\right\} = \int_0^t \frac{1}{4}\left(1 - e^{-4z}\right) dz.$$

If we denote this integral by $h(t)$, we have

$$h(t) = \int_0^t \frac{1}{4}\left(1 - e^{-4z}\right) dz = \frac{1}{8}\left(4t + e^{-4t} - 1\right).$$

(See Exercise 1 for an alternative way of computing this inverse Laplace transform.)

Finally, we collect our results to find our explicit solution for $x(t)$ as

$$x(t) = x_0 e^{-4t} + \frac{1}{4}\left(2x_0 + y_0\right)\left(1 - e^{-4t}\right) - \frac{1}{8} + \frac{1}{24}e^{-4t} + \frac{1}{12}e^{2t} + h(t-1)u(t-1). \quad (11.97)$$

To find $y(t)$, we return to the first equation in (11.95), where we have that $y = x' + 2x - u(t-1)$. Differentiating $x(t)$ in (11.97) and making the appropriate substitutions gives

Explicit solution

$$y(t) = \left(\frac{1}{2}y_0 - x_0 - \frac{1}{12}\right)e^{-4t} + x_0 + \frac{1}{2}y_0 - \frac{1}{4} + \frac{1}{3}e^{2t} + \left[h'(t-1) + 2h(t-1) - 1\right]u(t-1).$$ ▪

EXERCISES

1. Use partial fractions to find

$$\mathcal{L}^{-1}\left\{\frac{2}{s^2(s+4)}\right\}.$$

2. Create a *How to Solve Systems of Linear Differential Equations Using Laplace Transforms*. Add statements under the headings Purpose, Process, and Comments that summarize what you discovered in this section.

3. Use Laplace transforms to solve the following initial value problems. Then compare your answers with the ones you found for Exercise 4, Section 6.6.

(a) $x' = 2x - y \quad x(0) = 2$
$y' = -x + 2y \quad y(0) = 0$

(b) $x' = 2x - 2y \quad x(0) = 0$
$y' = 3x - 2y \quad y(0) = 2$

(c) $x' = 3x - 2y \quad x(0) = 2$
$y' = 2x - 2y \quad y(0) = 1$

(d) $x' = 5x - 2y \quad x(0) = 0$
$y' = 4x - 2y \quad y(0) = 6$

4. The differential equations that describe currents I_1 and I_2 in the electrical circuit in Figure 11.27 are

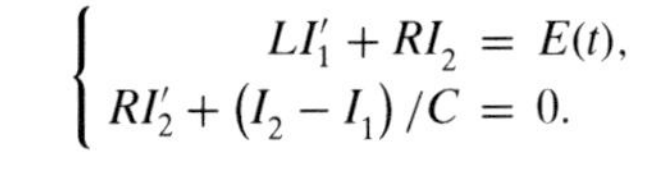

$$\begin{cases} LI_1' + RI_2 = E(t), \\ RI_2' + (I_2 - I_1)/C = 0. \end{cases}$$

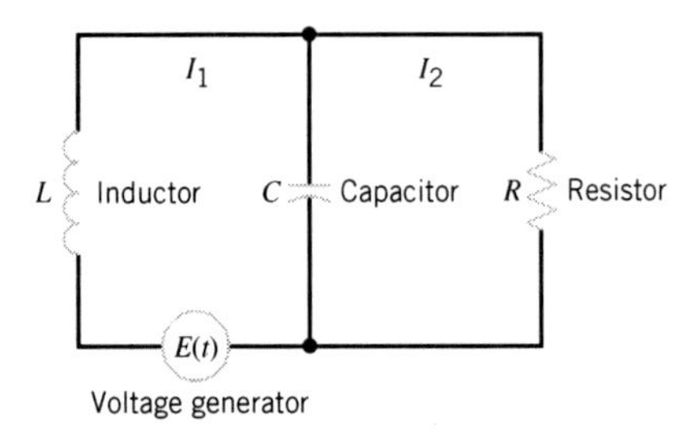

FIGURE 11.27 Electric circuit

(a) Solve these equations if $I_1(0) = I_2(0) = 0$, $E(t) = \sin \omega t$, $L = 1$, $R = 5$, and $C = 1/20$.

(b) Repeat part (a) if the initial conditions are changed to $I_1(0) = 0$, $I_2(0) = 1$.

5. Use Laplace transforms to solve the following initial value problems. Here $f(t)$ is the full square wave from Exercise 11.20 on page 538.

(a) $x' = -2x + 4y + 7e^{3t}$ $x(0) = 1$
$y' = x + y - 2e^{3t}$ $y(0) = 0$

(b) $x' = 2x + y + 3e^t$ $x(0) = 1$
$y' = -3x - 2y + e^{-t}$ $y(0) = 2$

(c) $x' = 2x + y + \cos t$ $x(0) = 1$
$y' = -3x - 2y$ $y(0) = 0$

(d) $x' = -2x + 4y + e^{2t}$ $x(0) = 0$
$y' = x + y - e^{2t}$ $y(0) = 1$

(e) $x' = -2x + 4y + u(t-1) - t(t-3)$ $x(0) = 0$
$y' = x + y$ $y(0) = 1$

(f) $x' = 2x + y + f(t)$ $x(0) = 1$
$y' = -3x - 2y$ $y(0) = 0$

6. Another advantage in using the Laplace transform is that the system of equations need not be in the form

$$\begin{cases} x' = ax + by + f(t), \\ y' = cx + dy + g(t). \end{cases}$$

(a) Show that if $X(s) = \mathcal{L}\{x(t)\}$ and $Y(s) = \mathcal{L}\{y(t)\}$, using the Laplace transform converts the initial value problem

$$\begin{cases} 2x' + y' - y = t, \\ x' + y' = t^2, \end{cases} \qquad x(0) = 1, \qquad y(0) = 0,$$

into the algebraic system

$$\begin{aligned} 2sX(s) + (s-1)\,Y(s) &= 2 + 1/s^2, \\ sX(s) + sY(s) &= 1 + 2/s^3. \end{aligned}$$

(b) Solve the equations in part (a) for $Y(s)$ as

$$Y(s) = \frac{4-s}{s^3(s+1)} = \frac{4}{s^3(s+1)} - \frac{1}{s^2(s+1)},$$

and use partial fractions or the Convolution Theorem to show that

$$y(t) = 5 - 5t + 2t^2 - 5e^{-t}.$$

(c) Show that $X(s) + Y(s) = 1/s + 2/s^4$, and solve for $x(t)$.

7. Use Laplace transforms to solve the following initial value problems.

(a) $2x' - 2x + y' = 1$ $x(0) = 4$
$x' - 3x + y' - 3y = 2$ $y(0) = -3$

(b) $x' + 2x + y' = 16e^{-2t}$ $x(0) = 0$
$2x' + 3y' + 5y = 15$ $y(0) = 0$

(c) $x' - y' = -e^t$ $x(0) = -1$
$2x' - 2y' - y = 8$ $y(0) = -10$

(d) $x' - y' - 6y = 0$ $x(0) = 2$
$x' - 3x + 2y' = 0$ $y(0) = 3$

(e) $x'' + 2x - 4y' = 0$ $x(0) = -4$ $x'(0) = 8$
$x' + y'' - 4y = 0$ $y(0) = 1$ $y'(0) = 2$

11.8 WHEN DO LAPLACE TRANSFORMS EXIST?

We now turn to the all-important question: What conditions must functions satisfy in order to guarantee the existence of their Laplace transform and the validity of the theorems that we have been using so frequently? In Section 11.1 we made the following working assumptions:

1. $\int_0^\infty e^{-st} y\, dt$ exists.

2. $\lim_{t\to\infty} e^{-st} y(t) = 0$.

3. $$\text{If } \int_0^\infty e^{-st} f(t)\,dt = \int_0^\infty e^{-st} g(t)\,dt, \text{ then } f(t) = g(t).$$

These led to a formal process that seemed to work very well for the examples we considered.[8] However, not everything is as straightforward as it looks. We will demonstrate this with an example.

Example 11.32 *The Inverse Laplace Transform of 1*

An obvious omission from Table 11.5 is a function $f(t)$ for which $F(s) = 1$. Let us try to find the function $f(t)$ for which $\mathcal{L}\{f(t)\} = 1$ — that is, find $\mathcal{L}^{-1}\{1\}$.

If we assume that such a function exists, then, because $F(s) = 1$, we also have $F(s-a) = 1$. From the Shifting Theorem — which states that if $\mathcal{L}\{f(t)\} = F(s)$, then $\mathcal{L}\{e^{at} f(t)\} = F(s-a)$ — we find $\mathcal{L}\{f(t)\} = \mathcal{L}\{e^{at} f(t)\}$. This leads to $f(t) = e^{at} f(t)$, for every $a > 0$, which implies that $f(t) = 0$. Thus, we have $\mathcal{L}\{0\} = 1$. However, we already know that $\mathcal{L}\{0\} = 0$, so $0 = 1$. ■

Clearly, something has gone wrong with our formal analysis. The mistake we made in this example was assuming that the Laplace transform exists. In fact, the correct logic to use in the previous example is to start with two possibilities. Either the Laplace transform exists or it does not. If it exists, then the previous example leads to a contradiction. Therefore, the first possibility cannot happen, and so we are led inevitably to the second — the Laplace transform of a function that gives us 1 does not exist.

So we are forced to look at conditions under which Laplace transforms exist. Let's look at the assumption $\lim_{t\to\infty} e^{-st} y(t) = 0$ and think of functions $y(t)$ for which this is true. In fact, most of the functions we work with, such as t^n, $\sin at$, $\cos at$, e^{at}, for $s > a$, the unit step function $u(t)$, and so on, have this property. Of these, the function e^{at} grows the most rapidly as $t \to \infty$. However, a little thought shows that there are functions that will violate this condition, such as e^{t^2}. Thus, we are forced to restrict ourselves to functions that don't grow more rapidly than exponentials. These functions are given a special name.

◆ *Definition 11.4:* **The function $f(t)$ is said to be of EXPONENTIAL ORDER α as $t \to \infty$ if there exists a constant M such that $|f(t)| \le Me^{\alpha t}$ for all t greater than some T.** ◆

Comments about Functions of Exponential Order α

- It is common to omit the phrase "as $t \to \infty$" when referring to a function being "of exponential order α as $t \to \infty$." We usually say the function is "of exponential order α," or just "of exponential order."
- This definition essentially says that a function is of exponential order α if it grows no faster than $e^{\alpha t}$.

---

[8] Mathematicians use the word formal to mean "we hope it works, and so far everything seems to be OK, but before we trust it we need a proof."

- This definition is equivalent to the statement that a function is of exponential order α if $e^{-\alpha t}|f(t)|$ is bounded for $t > T$.
- The functions $\sin at$ and $\cos at$ are of exponential order 0, because $|\sin at| \leq 1 = e^{0t}$ and $|\cos at| \leq 1 = e^{0t}$ for all t.
- The unit step function $u(t)$ is of exponential order 0, because $|u(t)| \leq 1 = e^{0t}$ for all t.
- The function t is of exponential order 1 because $|t| \leq e^{t} = e^{1t}$ for $t > 0$. The function t^n is also of exponential order 1.
- The function e^{at} is of exponential order a because $\left|e^{at}\right| \leq e^{at}$ for all t.
- The function e^{t^2} is not of exponential order because $e^{t^2}/e^{\alpha t} \to \infty$ as $t \to \infty$ for every α.
- When talking about a function being of exponential order, the exact values of α, M, and T are usually not relevant.
- If $y(t)$ is of exponential order α, then $\lim_{t\to\infty} e^{-st}y(t) = 0$, for $s > \alpha$.

We have calculated many Laplace transforms, not all of them of continuous functions. The unit step function $u(t)$ is an example of a discontinuous function that has a Laplace transform. Discontinuities can come in various forms, ranging from mild ones where the function jumps from one finite value to another, to extreme ones where the function goes to infinity at a finite value of t. It is common to distinguish between these types of discontinuities.

◆ *Definition 11.5:* **A function $f(t)$ is said to have a JUMP DISCONTINUITY AT $t = a$ if $\lim_{t\to a^-} f(t)$ and $\lim_{t\to a^+} f(t)$ exist, but $\lim_{t\to a^-} f(t) \neq \lim_{t\to a^+} f(t)$.** ◆

Comments about Jump Discontinuities

- This definition says that if $f(t)$ goes to a finite value as we approach a from the left of a, and if $f(t)$ goes to a finite value as we approach a from the right of a, but these finite values are not the same, then $f(t)$ has a jump discontinuity at $t = a$.
- The unit step function $u(t)$ has a jump discontinuity at $t = 0$ because $\lim_{t\to 0^-} u(t) = 0$ and $\lim_{t\to 0^+} u(t) = 1$.
- The function $1/t$ has a discontinuity at $t = 0$, but it is not a jump discontinuity.

In order to accommodate the possibility of a function having a reasonable number of jump discontinuities, we introduce another definition.

◆ *Definition 11.6:* **A bounded function $f(t)$ is said to be PIECEWISE CONTINUOUS ON THE INTERVAL $a \leq t \leq b$ if it is continuous except for a finite number of jump discontinuities.** ◆

Comments about Piecewise Continuous Functions

- The idea behind this definition is as follows: If we can find a finite number of values of t — say, $a_1, a_2, \cdots, a_n$ (where $a < a_1 < a_2 < \cdots < a_n < b$) — at which $f(t)$ has jump discontinuities and $f(t)$ is continuous within each of these subintervals, then $f(t)$ is piecewise continuous on $a \leq t \leq b$.

- If a function is continuous on the interval $a \leq t \leq b$, then it is piecewise continuous on the interval $a \leq t \leq b$.
- Every function we have considered in this chapter is piecewise continuous on any finite interval $a \leq t \leq b$. In particular, the square wave is piecewise continuous on every finite interval.
- If f is piecewise continuous in every finite interval, then $\int_0^t f(z)\,dz$ is a continuous function of t in every finite interval. (Why?)

The main result concerning the existence of the Laplace transform is as follows.[9]

▶ **Theorem 11.7: The Existence Theorem for Laplace Transforms** *Let $f(t)$ be a piecewise continuous function on every finite subinterval for $t \geq 0$ and be of exponential order. Then the Laplace transform of $f(t)$, $\mathcal{L}\{f(t)\} = F(s)$, exists.* ◀

Comments about the Existence Theorem for Laplace Transforms

- All the functions in this chapter for which we calculated the Laplace transform are piecewise continuous on every finite subinterval of $t \geq 0$. (Most of them are continuous.) Also, all these functions are of exponential order. Thus, this theorem guarantees that the Laplace transforms exist and are as calculated.
- It can be shown (see Exercise 8 on page 572) that if the conditions of this theorem are met, then $\lim_{s\to\infty} F(s) = 0$. Thus, many functions $F(s)$ do not have an inverse Laplace transform. We had already discovered this for the function $F(s) = 1$. If you evaluate a Laplace transform, and it has the property that $F(s)$ does not satisfy $\lim_{s\to\infty} F(s) = 0$, then you have made a mistake.
- The previous results justify the first two working assumptions listed on page 566.

Assuming we are dealing with functions that satisfy the conditions of the Existence Theorem for Laplace Transforms, we will turn to the third working assumption we made:

$$\text{if } \int_0^\infty e^{-st} f(t)\,dt = \int_0^\infty e^{-st} g(t)\,dt, \text{ then } f(t) = g(t).$$

If we consider the two functions $f(t) = 1$, and

$$g(t) = \begin{cases} 1 & \text{if } t \neq 1, \\ 0 & \text{if } t = 1, \end{cases}$$

[9] A proof of this theorem can be found in *The Use of Integral Transforms* by I.N. Sneddon, McGraw-Hill, 1972, page 138.

then they are both piecewise continuous on every finite subinterval for $t \geq 0$, they are both of exponential order, and

$$\int_0^\infty e^{-st} f(t)\,dt = \int_0^\infty e^{-st} g(t)\,dt = \frac{1}{s}.$$

However, $f(t) \neq g(t)$. Thus, the assertion that

$$\text{if } \int_0^\infty e^{-st} f(t)\,dt = \int_0^\infty e^{-st} g(t)\,dt, \text{ then } f(t) = g(t),$$

needs some qualifications. Notice in this example that $f(t) = g(t)$ at all points where $f(t)$ and $g(t)$ are continuous.[10]

▶ *Theorem 11.8: The Existence Theorem for Inverse Laplace Transforms* *If* $\int_0^\infty e^{-st} f(t)\,dt = \int_0^\infty e^{-st} g(t)\,dt$, *then* $f(t) = g(t)$ *at all points where* $f(t)$ *and* $g(t)$ *are continuous.* ◀

Comments about the Existence Theorem for Inverse Laplace Transforms

- This theorem guarantees the existence of inverse Laplace transforms.

We are now in a position to comment on those situations in which we found that piecewise continuous forcing functions gave rise to continuous solutions. In the first order case we found that the general solution of $y' + ky = f(t)$, $y(0) = y_0$, was (11.77); namely,

$$y(t) = y_0 e^{-kt} + e^{-kt} \int_0^t e^{kz} f(z)\,dz.$$

The only possible discontinuities in this solution could occur in the integral. However, $e^{kz} f(z)$ is piecewise continuous, and because the integral of a piecewise continuous function is continuous, the solution is continuous.

The second order case is a little more complicated. We found that the general solution of $ay'' + by' + cy = f(t)$, $y(0) = y_0$, $y'(0) = y_0^*$ was (11.89); namely,

$$y(t) = ay_0 g(t) + \left(ay_0^* + by_0\right) h(t) + \int_0^t h(t-z) f(z)\,dz,$$

where $h(t)$ is one of $\left(e^{r_1 t} - e^{r_2 t}\right) / \left[a\left(r_1 - r_2\right)\right]$, te^{rt}/a, or $e^{\alpha t} \sin \beta t/(a\beta)$. As in the first order case, the only possible discontinuities in this solution could occur in the integral. However, $h(t-z)f(z)$ is piecewise continuous, and because the integral of a piecewise continuous function is continuous, the solution is continuous.

---

[10] A proof of this theorem can be found in *The Use of Integral Transforms* by I. N. Sneddon, McGraw-Hill, 1972, page 141.

We now investigate the derivative of the solution. Again, the only possible discontinuities in the derivative of this solution could occur because of the integral. If we concentrate on the particular solution,

$$y_p(t) = \int_0^t h(t-z)f(z)\,dz,$$

and differentiate it,[11] we find

$$y_p'(t) = h(0)f(t) + \int_0^t h'(t-z)f(z)\,dz.$$

However, all three possibilities for $h(t)$ share the common property that $h(0) = 0$, so we have

$$y_p'(t) = \int_0^t h'(t-z)f(z)\,dz.$$

Again, all three possibilities for $h(t)$ share the common property that h'' exists; this requires that h' be continuous. Thus, $h'(t-z)f(z)$ is piecewise continuous, and because the integral of a piecewise continuous function is continuous, the derivative of the solution is continuous. We have just shown that $y_p'(t)$ is also continuous. Thus, $y(t)$ is differentiable everywhere and cannot have any "corners."

EXERCISES

1. Let $f(t)$ and $g(t)$ be of exponential order as $t \to \infty$. Show that their sum, $f(t) + g(t)$, is of exponential order as $t \to \infty$.

2. Let $f(t)$ and $g(t)$ be piecewise continuous on $a \le t \le b$. Show that their sum, $f(t) + g(t)$, is piecewise continuous on $a \le t \le b$.

3. Using the results of Exercises 1 and 2, show that $\mathcal{L}\{f(t) + g(t)\} = \mathcal{L}\{f(t)\} + \mathcal{L}\{g(t)\}$.

4. Which of the following functions are piecewise continuous for $0 \le t \le 100$?

   (a) $f(t) = t^{-1}$
   (b) $f(t) = (t+1)^{-1}$
   (c) $f(t) = (t-2)^{-1}$
   (d) $f(t) = \begin{cases} t^2 & \text{if } 0 \le t \le 1, \\ 2t^{-1} & \text{if } 1 < t \le 100 \end{cases}$
   (e) $f(t) = (\cos t + 1)^{-1}$
   (f) $f(t) = (\cos t - 2)^{-1}$
   (g) $f(t) = \tan t$
   (h) $f(t) = \begin{cases} \tan t & \text{if } 0 \le t \le 1, \\ (2t-1)^{-1} & \text{if } 1 < t \le 100 \end{cases}$

5. Find the values of α, M, and T from Definition 11.4 on page 567 for the following functions.

   (a) $f(t) = e^{3t}\cos 2t$
   (b) $f(t) = \begin{cases} 16 & \text{if } 0 \le t \le 10, \\ 100t^{-1} & \text{if } 10 < t \end{cases}$
   (c) $f(t) = 2t$
   (d) $f(t) = e^{7t}$

6. Show that e^{t^2} is not of exponential order.

7. Show that the function $f(t) = 3\sin(e^{t^2})$ is of exponential order but that its derivative, $f'(t)$, is not. List two other functions with this same property.

[11] If $y(t) = \int_0^t g(t,z)\,dz$, then $\dfrac{dy}{dt} = g(t,t) + \int_0^t \dfrac{\partial g(t,z)}{\partial t}\,dz$.

8. Prove that if $f(t)$ is piecewise continuous and of exponential order, then $\lim_{s\to\infty} F(s) = 0$. [Hint: Write

$$\mathcal{L}\{f(t)\} = \int_0^T e^{-st} f(t)\, dt + \int_T^\infty e^{-st} f(t)\, dt,$$

where T is taken from Definition 11.4, and use the facts that $|e^{-st}f(t)|$ is bounded for $t > T$, and that the absolute value of an integral is less than or equal to the integral of the absolute value of the integrand.]

9. The solution of $ay'' + by' + cy = f(t)$, where $f(t)$ is continuous everywhere except where it has jump discontinuities at $t = 1$ and $t = 3$, is allegedly represented by Figure 11.28. Comment on this claim.

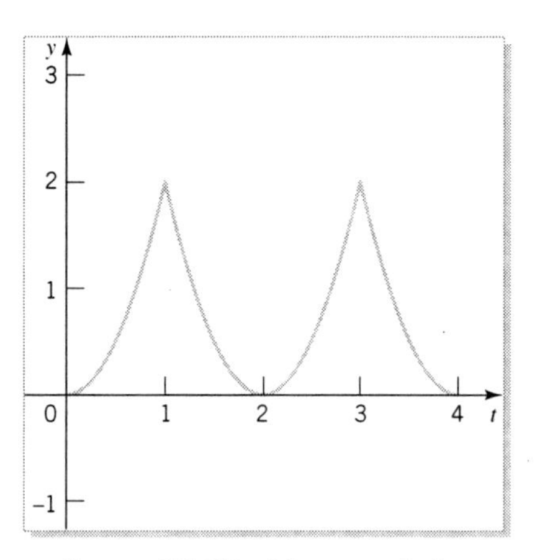

FIGURE 11.28 Mystery solution

10. Solve the initial value problem,

$$f''(t) + tf'(t) - 2f(t) = 1,$$
$$f(0) = f'(0) = 0, \tag{11.98}$$

in the following manner.

(a) Show that $\mathcal{L}\{tf'(t)\} = -sF'(s) - F(s)$.

(b) Show that taking the Laplace transform of (11.98) and using the initial conditions and the results of part (a) gives

$$F'(s) + \left(\frac{3}{s} - s\right) F(s) = -\frac{1}{s^2}.$$

(c) Find the integrating factor for this first order linear differential equation, and show that it has the solution

$$F(s) = \frac{1}{s^3} + C\frac{1}{s^3}e^{s^2/2}.$$

(d) Evaluate the arbitrary constant in part (b) by using the result of Exercise 8, and find the resulting solution $f(t)$. Verify that your answer is indeed the solution.

11. Solve the initial value problem $f''(t) - 2tf'(t) + 2f(t) = 0, f(0) = 0, f'(0) = 1$.

12. For what value of the constant a will the differential equation resulting from taking the Laplace transform of $t^2 f''(t) - atf'(t) + 3f(t) = 0$ be of the same form as the original differential equation?

What Have We Learned?

Main Ideas

In this chapter we have seen how Laplace transforms are used in solving a variety of equations.

- We can use Laplace transforms to solve linear differential equations with constant coefficients, of both first and second order. See *How to Solve First Order Linear Differential Equations Using Laplace Transforms* on page 514 and Exercise 2 on page 559.

- Laplace transforms also solve systems of linear differential equations with constant coefficients. See Exercise 2 on page 565.

- These transformations can also solve a special type of integral equation. See Exercise 5 on page 530.

There are four main advantages of using Laplace transforms to solve differential equations.

- They reduce differential equations to algebraic equations.
- They easily handle discontinuous forcing functions, or ones that have a different representation on different parts of the domain.
- They easily handle periodic forcing functions.
- They satisfy initial conditions automatically.

CHAPTER 12

USING POWER SERIES

Where We Are Going – and Why

At this point we have systematic techniques for solving the initial value problem

$$a_2(x)y'' + a_1(x)y' + a_0(x)y = 0, \qquad y(x_0) = y_0, \qquad y'(x_0) = y_0^*, \tag{12.1}$$

when the second order differential equation has the form of a Cauchy-Euler equation or constant coefficients. In this chapter we develop techniques to find solutions of (8.1) when $a_2(x)$, $a_1(x)$, and $a_0(x)$ are arbitrary functions of x. We do this by using Taylor series and base it on the techniques introduced in Chapters 1 and 3.

We state theorems that show when solutions of (8.1) exist in the form of Taylor series $y(x) = \sum_{n=0}^{\infty} c_n (x - x_0)^n$, where the coefficients c_n are to be determined. Because we often cannot find a general expression for the coefficients in this series solution, we introduce a theorem that determines the region of convergence of the infinite series solution.

If any of $a_2(x)$, $a_1(x)$, and $a_0(x)$ are not continuous, or if $a_2(x) = 0$ at some point, a Taylor series solution may not exist. However, with appropriate restrictions on the behavior of a_1/a_2 and a_0/a_2 at x_0, we can modify our techniques to find general solutions of the form $y(x) = (x - x_0)^s \sum_{n=0}^{\infty} c_n (x - x_0)^n$, where the exponent s and the coefficients c_n are to be determined. We give theorems regarding the convergence of these series solutions, so the region of convergence is known before we find the specific coefficients.

A knowledge of the first few terms in a series solution is useful in determining the behavior of the solution near a given initial value.

Occasionally the infinite series solution is the series expansion of a familiar function. When this occurs we can write such solutions in simpler, and more common, terms.

The functions with which we are familiar are usually defined by a simple formula. The techniques presented in this chapter are important because they allow us to introduce and define new functions – which occur naturally in many fields – as infinite series. This gives us a significant way to enlarge the family of familiar functions to include many important functions that cannot be defined by simple formulas. These functions are often designated Special Functions, and entire books are devoted to this subject.

12.1 SOLUTIONS USING TAYLOR SERIES

Thus far we have mapped out a strategy for solving two types of homogeneous linear second order differential equations — namely, equations with constant coefficients, and Cauchy-Euler equations. We now develop another technique that will also apply to linear differential equations of other types, as well as nonlinear ones. Recall that on several previous occasions we had no choice other than to leave our solution in terms of indefinite or definite integrals. At times we then used a Taylor series expansion of the known function that occurred in the integrand. This finally resulted in our solution taking the form of an infinite series. We now see how we can obtain such solutions directly.

Three Examples

EXAMPLE 12.1 *The Linearized Pendulum Again*

To consider a specific example, we return to the solution of the linearized pendulum equation

$$\frac{d^2y}{dx^2} + y = 0, \tag{12.2}$$

subject to

$$y(0) = 0.2, \qquad y'(0) = 0.15. \tag{12.3}$$

Here y is the angle, x is the time, and the gravitational constant divided by the pendulum length is 1. Note that these initial conditions correspond to the situation where the pendulum is released at an angle of 0.2 radian with an angular velocity of 0.15 radian per second, which means that initially the angle is going to increase. Although we can show that the solution for this problem is

$$y(x) = 0.2\cos x + 0.15\sin x, \tag{12.4}$$

(do this) let's look at the same problem again through different eyes and act as though we were unaware of this solution.

We note that the initial conditions in (12.3) give us the first two terms in a Taylor series expansion of our solution about $x = 0$,

$$y(x) = \sum_{n=0}^{\infty} \frac{1}{n!} y^{(n)}(0)x^n = y(0) + y'(0)x + \frac{1}{2!}y''(0)x^2 + \frac{1}{3!}y'''(0)x^3 + \cdots.$$

Thus, an approximation to our solution using just the first two terms of the Taylor series would be the Taylor polynomial of degree 1, $y(x) = 0.2 + 0.15x$. [Notice that this agrees with (12.4) if we use one-term approximations for $\cos x \approx 1$ and $\sin x \approx x$.] If we could find more terms in this Taylor series, then we would expect to have a better approximation to our solution. This means we should try to find $y''(0)$, $y'''(0)$, $y^{iv}(0)$, $\cdots$. How can we do this?

If we write (12.2) in the form

$$y''(x) = -y(x), \tag{12.5}$$

we can obtain the value of the second derivative at $x = 0$ as $y''(0) = -y(0) = -0.2$. Thus, an approximation to our solution using the first three terms of the Taylor series would be the Taylor polynomial of degree 2,

$$y(x) = 0.2 + 0.15x + \frac{1}{2!}(-0.2)x^2 = 0.2 + 0.15x - 0.1x^2.$$

More terms in this expansion can be obtained after differentiating (12.5). For example,

$$y''' = -y' \tag{12.6}$$

gives $y'''(0) = -y'(0) = -0.15$, and the first four terms in a Taylor series for the solution of (12.2) subject to (12.3) is the Taylor polynomial of degree 3,

$$y(x) = 0.2 + 0.15x + \frac{1}{2!}(-0.2)x^2 + \frac{1}{3!}(-0.15)\,x^3 = 0.2 + 0.15x - 0.1x^2 - 0.025x^3.$$

We can continue with this process. For example, in the next step we differentiate (12.6) and evaluate the result at $x = 0$ to find the Taylor polynomial of degree 4 as

$$y(x) = 0.2 + 0.15x - 0.1x^2 - 0.025x^3 + \frac{1}{120}x^4. \tag{12.7}$$

Following this same routine allows us to compute as many terms in a Taylor series expansion of a solution of such initial value problems as needed to achieve the desired accuracy. However, unless we can obtain the explicit form of the expression for the nth derivative of our solution evaluated at 0 from this procedure, we cannot determine for which values of x (if any) the series converges. In general, we cannot obtain this form for the nth derivative, although in the particular case of (12.2) we can (see Exercise 7 on page 581).

We can compare (12.7) with the Taylor series of the exact solution (12.4) by substituting the Taylor series for the $\cos x$ and $\sin x$ terms and keeping all terms up to x^4. This gives

$$\begin{aligned} y(x) &= 0.2[1 - x^2/2 + x^4/24 - \cdots] + 0.15[x - x^3/6 + \cdots] \\ &= 0.2 + 0.15x - 0.1x^2 - 0.025x^3 + (1/120)x^4 + \cdots, \end{aligned} \tag{12.8}$$

which is in exact agreement with (12.7). Figure 12.1 shows the Taylor polynomials of degrees 1 through 4 (previously obtained). Figure 12.2 shows the Taylor polynomials of degrees 5 through 8. Notice how we can begin to see the solution emerging as the common points between the seventh and eighth degree polynomials.

From the Taylor series in Figure 12.3 for the eighth degree polynomial, we can predict the maximum angle that the pendulum attains (before swinging back) and when this first occurs, given approximately by $y = 0.25$, $x = 0.64$. We can also find the first time that the pendulum is vertical, given approximately by $x = 2.21$.

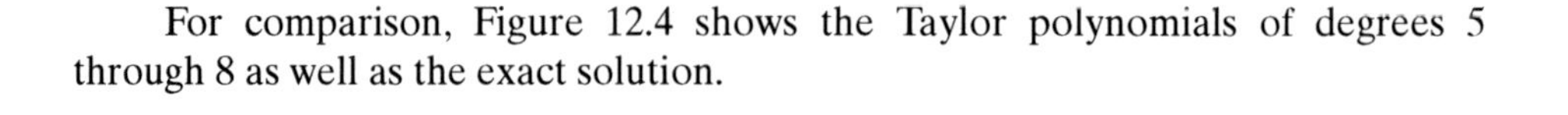

For comparison, Figure 12.4 shows the Taylor polynomials of degrees 5 through 8 as well as the exact solution.

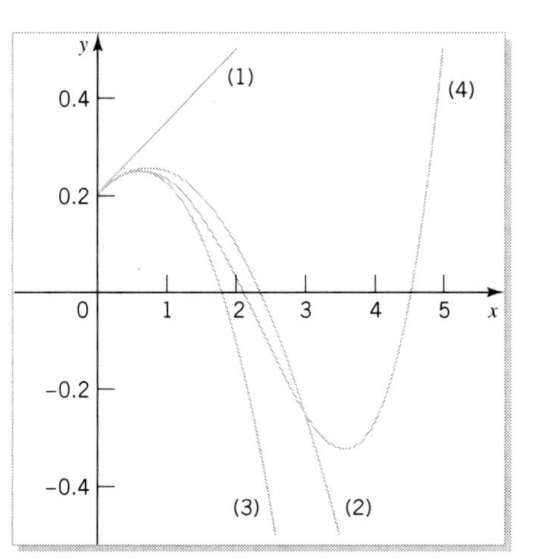

FIGURE 12.1 Taylor polynomials of degrees 1 through 4 for $y'' = -y$, $y(0) = 0.2$, $y'(0) = 0.15$

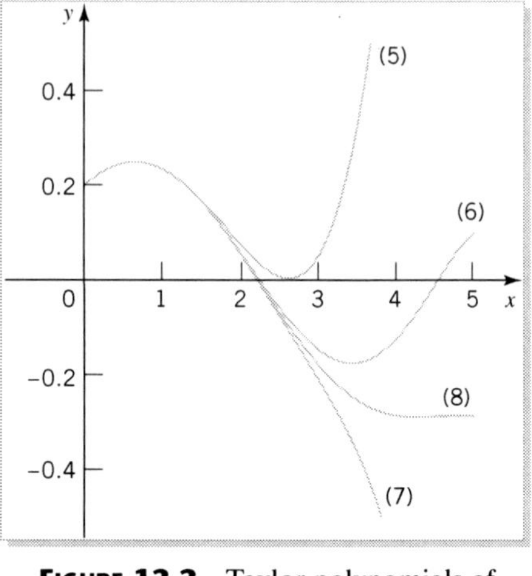

FIGURE 12.2 Taylor polynomials of degrees 5 through 8 for $y'' = -y$, $y(0) = 0.2$, $y'(0) = 0.15$

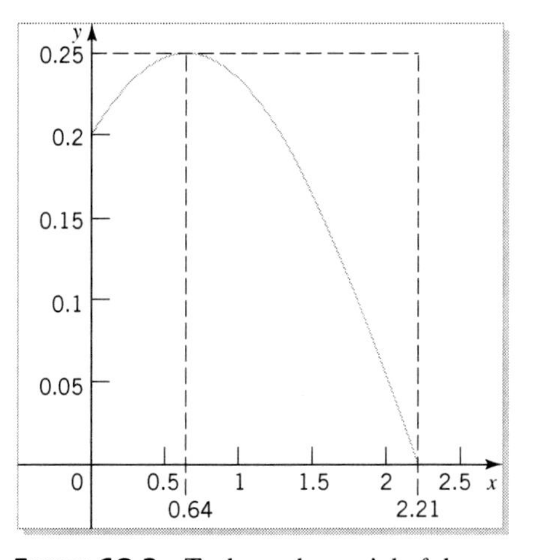

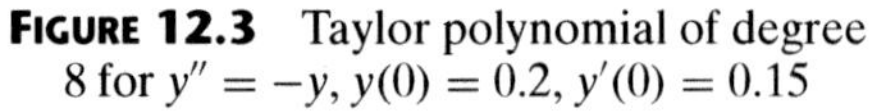

FIGURE 12.3 Taylor polynomial of degree 8 for $y'' = -y$, $y(0) = 0.2$, $y'(0) = 0.15$

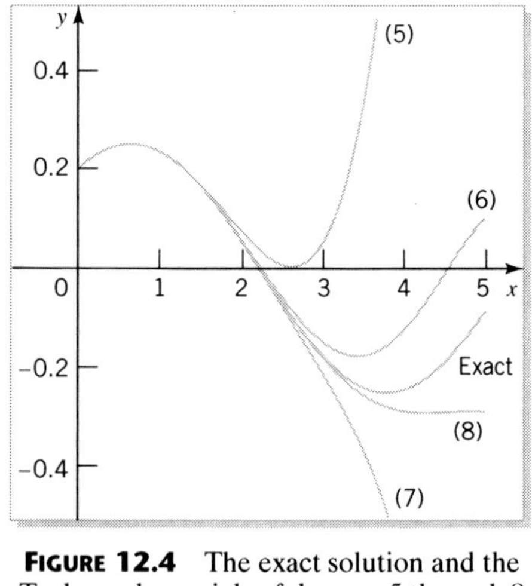

FIGURE 12.4 The exact solution and the Taylor polynomials of degrees 5 through 8 for $y'' = -y$, $y(0) = 0.2$, $y'(0) = 0.15$

EXAMPLE 12.2 *The Pendulum Again*

Here we want to compare the Taylor series of the solution of the initial value problem that governs the nonlinear motion of a pendulum,

$$y'' + \sin y = 0, \tag{12.9}$$

$$y(0) = 0.2, \qquad y'(0) = 0.15, \tag{12.10}$$

with that of the linearized problem. Remember that we cannot find a solution for (12.9) in terms of familiar functions.

The values of the solution and its first derivative are given by the initial conditions (12.10), and the value of the second derivative at $x = 0$ is obtained from (12.9) as $y''(0) = -\sin 0.2 \approx -0.1987$. To find values of the third and fourth derivatives at $x = 0$, we return to (12.9) and differentiate to obtain

$$y''' = -(\cos y)y' \qquad \text{and} \qquad y^{iv} = (\sin y)(y')^2 - (\cos y)y''.$$

Using the previously obtained values of y and its derivatives at $x = 0$ gives $y'''(0) = -(\cos 0.2)(0.15) \approx -0.1470$, and $y^{iv}(0) = (\sin 0.2)(0.15)^2 - (\cos 0.2)(-0.1987) \approx 0.1992$. Thus, we have the first five terms in a Taylor expansion of our original initial value problem (12.9) and (12.10) as

$$\begin{aligned} y(x) &\approx 0.2 + 0.15x - (0.1987/2!)x^2 - (0.1470/3!)x^3 + (0.1992/4!)x^4 \\ &\approx 0.2 + 0.15x - 0.0994x^2 - 0.0245x^3 + 0.0083x^4. \end{aligned} \tag{12.11}$$

Note that there is very little difference in the two expressions in (12.11) and (12.8) ($1/120 \approx 0.0083$). Figure 12.5 shows the Taylor polynomials of degrees 1 through 4, from (12.11), and Figure 12.6 compares the polynomials of degree 4 for the linearized and nonlinear pendulums.

Figure 12.7 shows the Taylor polynomials of degrees 5 through 8 for the solution of (12.9). Notice how we can begin to see the solution emerging as the common points between the seventh and eighth degree polynomials.

From the Taylor series in Figure 12.8 for the eighth degree polynomial, we can predict the maximum angle that the pendulum attains (before swinging back) and when this first occurs, given approximately by $y = 0.25$, $x = 0.65$. We can also find the first time that the pendulum is vertical, given approximately by $x = 2.24$.

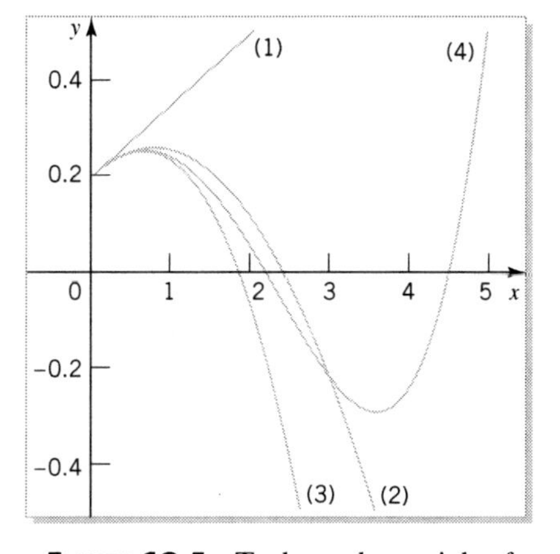

FIGURE 12.5 Taylor polynomials of degrees 1 through 4 for $y'' = -\sin y$, $y(0) = 0.2$, $y'(0) = 0.15$

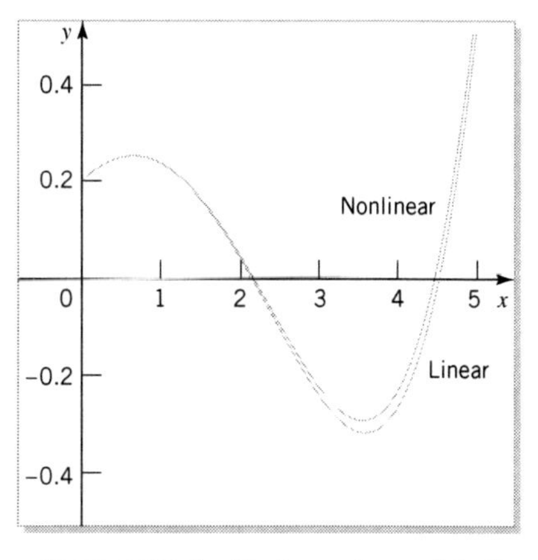

FIGURE 12.6 Taylor polynomial of degree 4 for $y'' = -\sin y$ and $y'' = -y$

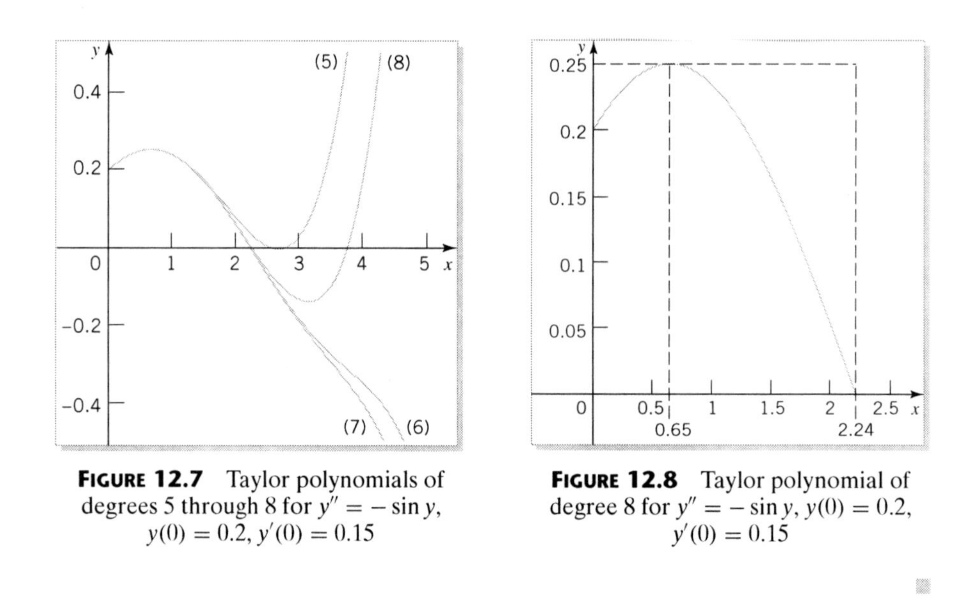

FIGURE 12.7 Taylor polynomials of degrees 5 through 8 for $y'' = -\sin y$, $y(0) = 0.2$, $y'(0) = 0.15$

FIGURE 12.8 Taylor polynomial of degree 8 for $y'' = -\sin y$, $y(0) = 0.2$, $y'(0) = 0.15$

EXAMPLE 12.3

Consider the second order linear equation

$$y'' + 2xy' + 2y = 0, \tag{12.12}$$

subject to the initial conditions

$$y(0) = y_0, \qquad y'(0) = y_0^*. \tag{12.13}$$

Because all the coefficients in (12.12) are continuous on all finite intervals, and the coefficient of the second derivative is never zero, the Existence-Uniqueness Theorem for Second Order Linear Differential Equations of Section 8.1 applies. Thus, the initial value problem (12.12) and (12.13) has a unique solution.

If we solve (12.12) for the second derivative, we have

$$y''(x) = -2xy'(x) - 2y(x), \tag{12.14}$$

from which we can obtain the value of the second derivative at $x = 0$ as $y''(0) = -2 \cdot 0 \cdot y'(0) - 2y(0) = -2y_0$. More terms in this expansion can be obtained by differentiating (12.14). For example,

$$y'''(x) = -2xy'' - 2y' - 2y' = -2xy''(x) - 4y'(x),$$

giving

$$y'''(0) = -2 \cdot 0 \cdot y''(0) - 4y'(0) = -4y_0^*.$$

This gives us the first four terms in a Taylor series for the solution of (12.12), subject to (12.13), as

$$y(x) = y_0 + y_0^* x - \frac{2}{2!} y_0 x^2 + \frac{1}{3!}\left(-4y_0^*\right) x^3. \tag{12.15}$$

Following this same routine allows us to compute as many terms in a Taylor series solution of such initial value problems as we desire. However, again, if we cannot obtain the general expression for the nth derivative evaluated at 0 from this procedure, we cannot determine the interval over which our series converges. However, we may use the Oscillation and Relation Theorems of Section 8.1 to discover whether this solution oscillates. (Does it?) ■

EXERCISES

1. Find the first four terms in the Taylor series expansion of the solution to $y' = \sqrt{x^2 + y^2}$, $y(0) = 1$.

2. Find the first four terms in the Taylor series expansion of the solution to $y' = e^x y^2 + 3\sin x$, $y(0) = 1$.

3. Find the first five terms in the Taylor series expansion of the solution to $y'' = -\sin y$, $y(0) = 1$, $y'(0) = 0$.

4. Find the first five terms in the Taylor series expansion of the solution to $y' = x^2 + y^2$, $y(-1) = -1$.

5. Find the Taylor series solution to $y' = 2xy$, $y(0) = 1$, and show that it may be expressed as e^{x^2}.

6. Create a *How to Find a Taylor Series Solution for the Initial Value Problem* $y'' = f(x, y, y')$, $y(0) = y_0$, $y'(0) = y_0^*$. Add statements under Purpose, Process, and Comments that summarize what is discussed in this section.

7. Solve $y'' + y = 0$, subject to $y(0) = y_0$, $y'(0) = y_0^*$, using Taylor series. Compare the answer with the one obtained by treating the equation as a linear differential equation with constant coefficients.

8. Solve $y'' - y = 0$, subject to $y(0) = y_0$, $y'(0) = y_0^*$, using Taylor series. Compare the answer with the one obtained by treating the equation as a linear differential equation with constant coefficients.

12.2 SOLUTIONS USING POWER SERIES

There is an alternative way of finding coefficients in a Taylor series expansion of a solution of a differential equation. We can simply substitute the power series

$$y(x) = \sum_{n=0}^{\infty} c_n x^n = c_0 + c_1 x + c_2 x^2 + c_3 x^3 + c_4 x^4 + \cdots \tag{12.16}$$

[and its derivatives $y'(x) = \sum_{n=0}^{\infty} c_n n x^{n-1} = \sum_{n=1}^{\infty} c_n n x^{n-1}$, and $y''(x) = \sum_{n=0}^{\infty} c_n n(n-1)x^{n-2} = \sum_{n=2}^{\infty} c_n n(n-1)x^{n-2}$] into the differential equation and try to evaluate the coefficients $c_0, c_1, c_2, \cdots, c_n$. For linear differential equations, this operation yields linear algebraic equations. For nonlinear differential equations, it leads to nonlinear algebraic equations.

The Main Idea

EXAMPLE 12.4 *The Linearized Pendulum Revisited Once More*

To illustrate this technique, we return to the linearized pendulum equation

$$y'' + y = 0. \tag{12.17}$$

Trial solution

If we substitute (12.16) into (12.17), we find

$$2c_2 + 6c_3x + 12c_4x^2 + 20c_5x^3 + 30c_6x^4 + \cdots + c_0 + c_1x + c_2x^2 + c_3x^3 + c_4x^4 + \cdots = 0.$$

If we rearrange these terms in increasing powers of x, we obtain[1]

$$(2c_2 + c_0) + (6c_3 + c_1)x + (12c_4 + c_2)x^2 + (20c_5 + c_3)x^3 + (30c_6 + c_4)x^4 + \cdots = 0.$$

Because we want this equation to be satisfied for all values of x, we set the coefficient of each power of x to zero to obtain

$$2c_2 + c_0 = 0, \quad 6c_3 + c_1 = 0, \quad 12c_4 + c_2 = 0, \quad 20c_5 + c_3 = 0, \quad 30c_6 + c_4 = 0,$$

from which we find

$$c_2 = -\frac{1}{2}c_0, \quad c_3 = -\frac{1}{6}c_1, \quad c_4 = -\frac{1}{12}c_2, \quad c_5 = -\frac{1}{20}c_3, \quad c_6 = -\frac{1}{30}c_4.$$

The first gives us c_2 in terms of c_0, whereas the second gives c_3 in terms of c_1. The third gives c_4 in terms of c_2, but we already know c_2 in terms of c_0. Thus, we find

$$c_4 = -\frac{1}{12}c_2 = -\frac{1}{12}\left(-\frac{1}{2}c_0\right) = \frac{1}{24}c_0.$$

In the same way, we have

$$c_5 = \frac{1}{120}c_1, \quad c_6 = -\frac{1}{720}c_0.$$

General solution

If we substitute these values for c_2 to c_6 into (12.16) and collect together terms involving c_0 and c_1, we find

$$y(x) = c_0\left(1 - \frac{1}{2}x^2 + \frac{1}{24}x^4 - \frac{1}{720}x^6 + \cdots\right) + c_1\left(x - \frac{1}{6}x^3 + \frac{1}{120}x^5 + \cdots\right).$$

If we impose the initial conditions $y(0) = 0.2$, $y'(0) = 0.15$, we see that $c_0 = 0.2$ and $c_1 = 0.15$, in agreement with (12.7).

While this method allows us to find as many terms in our solution as we desire, the calculations may become very tedious after a while. Our work would be greatly simplified if we could find a pattern for the values of c_n. To do this, we again substitute

[1] Recall that power series may be rearranged without changing their interval of convergence.

this series expression in (12.16) into the original differential equation (12.17), but this time we use the summation notation. This gives

$$\sum_{m=2}^{\infty} c_m m(m-1)x^{m-2} + \sum_{m=0}^{\infty} c_m x^m = 0.$$

In order to combine terms in this equation, we need the same power on x in the two series, as well as the same limits of summation. To accomplish this, we change the summation index in the first series by increasing it by 2 (letting $m - 2 = n$) while simply replacing m by n in the second series. These operations give

$$\sum_{n=0}^{\infty} c_{n+2}(n+2)(n+1)x^n + \sum_{n=0}^{\infty} c_n x^n = \sum_{n=0}^{\infty} \left[(n+2)(n+1)c_{n+2} + c_n\right] x^n = 0.$$

Because the differential equation is to be satisfied for all values of x, we may equate each of the coefficients of x^n to zero, giving

$$(n+2)(n+1)c_{n+2} + c_n = 0, \qquad n = 0, 1, 2, \cdots. \tag{12.18}$$

Because $(n+2)(n+1) \neq 0$ for $n = 0, 1, 2, \cdots$, we may rearrange (12.18) as

$$c_{n+2} = -\frac{c_n}{(n+2)(n+1)}, \qquad n = 0, 1, 2, \cdots. \tag{12.19}$$

If we now write down this equation for the first few values of n, we see that

$$c_2 = -\frac{1}{2 \cdot 1}c_0, \quad c_3 = -\frac{1}{3 \cdot 2}c_1, \quad c_4 = -\frac{1}{4 \cdot 3}c_2, \quad c_5 = -\frac{1}{5 \cdot 4}c_3.$$

Looking at these equations, we observe that coefficients with even subscripts are related to coefficients with even subscripts, and coefficients with odd subscripts are related to coefficients with odd subscripts. If we look at (12.19), we discover that it expresses a relationship between coefficients whose subscripts differ by 2. Equation (12.19) is an example of a RECURRENCE RELATION.

Recurrence relation

Even though (12.19) gives us a recipe for finding one coefficient in terms of the coefficient whose subscript is 2 less, it would be more convenient to have an explicit formula for c_n. Computing the first few coefficients with even subscripts, $n = 0, 2,$ and 4, gives

$$c_2 = -\frac{1}{2}c_0, \; c_4 = -\frac{1}{4 \cdot 3}c_2 = (-1)^2\frac{1}{4 \cdot 3 \cdot 2}c_0, \; c_6 = -\frac{1}{6 \cdot 5}c_4 = (-1)^3\frac{1}{6 \cdot 5 \cdot 4 \cdot 3 \cdot 2}c_0,$$

where we observe that all these coefficients with even subscripts are written as multiples of c_0 and there are no requirements on c_0. This means that c_0 is an arbitrary constant. To obtain a general formula for coefficients with even subscripts, we return to the recurrence relation (12.19) and let $n + 2 = 2m$ — that is, $n = 2m - 2$ — obtaining

$$c_{2m} = (-1)\,\frac{1}{(2m)\,(2m-1)}c_{2m-2}.$$

In this equation we now replace the term on the right-hand side by what we obtain by letting $n = 2m - 4$ in (12.19), namely

$$c_{2m-2} = (-1)\frac{1}{(2m-2)(2m-3)}c_{2m-4}.$$

This gives

$$c_{2m} = (-1)^2\frac{1}{2m(2m-1)(2m-2)(2m-3)}c_{2m-4}.$$

Continuing in this manner gives us

$$\begin{aligned} c_{2m} &= (-1)\frac{1}{(2m)(2m-1)}c_{2m-2} \\ &= (-1)^2\frac{1}{2m(2m-1)(2m-2)(2m-3)}c_{2m-4} \\ &= \cdots \\ &= (-1)^m\frac{1}{2m(2m-1)(2m-2)(2m-3)\cdots 3\cdot 2\cdot 1}c_0, \end{aligned}$$

or[2]

$$c_{2m} = (-1)^m\frac{1}{(2m)!}c_0, \qquad m = 0, 1, 2, \cdots. \tag{12.20}$$

This is an explicit formula for all the even coefficients, c_{2m}, in terms of the arbitrary constant c_0.

The recurrence relation for the odd subscripts is obtained from (12.19) by setting $n + 2 = 2m + 1$ — that is, $n = 2m - 1$ — as

$$c_{2m+1} = -\frac{1}{(2m+1)\,2m}c_{2m-1}, \qquad m = 1, 2, 3, \cdots,$$

which leads to

$$\begin{aligned} c_3 = -\frac{1}{3\cdot 2}c_1, \quad c_5 &= -\frac{1}{5\cdot 4}c_3 = (-1)^2\frac{1}{5\cdot 4\cdot 3\cdot 2}c_1, \\ c_7 &= -\frac{1}{7\cdot 6}c_5 = (-1)^3\frac{1}{7\cdot 6\cdot 5\cdot 4\cdot 3\cdot 2}c_1. \end{aligned}$$

Here we see that all of the coefficients with odd subscripts are written in terms of c_1. Because there are no conditions on c_1, it is also an arbitrary constant.

[2]It is wise to substitute what we think is the solution of a recurrence relation back into the recurrence relation to confirm that it is a solution.

For the general case for odd subscripts, we have

$$\begin{aligned} c_{2m+1} &= (-1)\frac{1}{(2m+1)\,2m}c_{2m-1} \\ &= (-1)^2\frac{1}{(2m+1)\,2m\,(2m-1)\,(2m-2)}c_{2m-3} \\ &= \cdots \\ &= (-1)^m\frac{1}{(2m+1)\,2m\,(2m-1)\,(2m-2)\cdots 4\cdot 3\cdot 2}c_1, \end{aligned}$$

or

$$c_{2m+1} = (-1)^m\frac{1}{(2m+1)!}c_1, \qquad m = 0, 1, 2, \cdots. \tag{12.21}$$

Thus, we can write our solution of (12.17) in the form

$$y(x) = \sum_{n=0}^{\infty} c_n x^n = \sum_{m=0}^{\infty} c_{2m}x^{2m} + \sum_{m=0}^{\infty} c_{2m+1}x^{2m+1},$$

General solution

or, using (12.20) and (12.21),

$$y(x) = c_0\sum_{m=0}^{\infty}(-1)^m\frac{x^{2m}}{(2m)!} + c_1\sum_{m=0}^{\infty}(-1)^m\frac{x^{2m+1}}{(2m+1)!}, \tag{12.22}$$

where c_0 and c_1 are arbitrary constants. This is the solution of (12.17) in terms of power series. Can we express these power series in terms of familiar functions? If we write out the first few terms of the coefficient of c_0, we find

$$\sum_{m=0}^{\infty}(-1)^m\frac{x^{2m}}{(2m)!} = 1 - \frac{x^2}{2!} + \frac{x^4}{4!} - \frac{x^6}{6!} + \cdots,$$

which we recognize as the Taylor series expansion for $\cos x$. Treating the coefficient of c_1 in the same way, we recognize

$$\sum_{m=0}^{\infty}(-1)^m\frac{x^{2m+1}}{(2m+1)!} = x - \frac{x^3}{3!} + \frac{x^5}{5!} - \frac{x^7}{7!} + \cdots$$

as the Taylor series expansion for $\sin x$. Both these series converge for all values of x, so the solution (12.22) can be written $y(x) = c_0 \cos x + c_1 \sin x$, valid for all x.

If we use the initial conditions $y(0) = 0.2$, $y'(0) = 0.15$ in (12.22), we find $c_0 = y(0) = 0.2$, and $c_1 = y'(0) = 0.15$, which agrees with (12.4). ■

◆ Definition 12.1: **A recurrence relation among c_0, c_1, c_2, $\cdots$, is an equation of the type**

$$c_{n+m} = f(c_n, c_{n+1}, \cdots, c_{n+m-1});$$

that is, c_{n+m} is determined by its previous m terms. ◆

Comments about Recurrence Relations

- Recurrence relations are also called finite difference equations.
- The positive integer m is the ORDER of the recurrence relation. For example, the recurrence relation (12.19),

$$c_{n+2} = -\frac{c_n}{(n+2)(n+1)}, \qquad n = 0, 1, 2, \cdots,$$

is a second order recurrence relation.

- There is no general way to solve recurrence relations. However, a simple technique sometimes works by a clever rewriting of the recurrence relation. This technique is based on the fact that the recurrence relation $a_n = a_{n-1}$ $(n = 0, 1, 2, \cdots)$ can be reapplied to give $a_n = a_{n-1} = a_{n-2} = \cdots = a_2 = a_1 = a_0$, and so has the solution $a_n = a_0$. For example,
  - the recurrence relation $c_n = -c_{n-1}$ $(n = 1, 2, 3, \cdots)$ can be rewritten in the form $(-1)^n c_n = (-1)^{n-1} c_{n-1}$, and so $(-1)^n c_n = (-1)^{n-1} c_{n-1} = (-1)^{n-2} c_{n-2} = \cdots = -c_1 = c_0$, giving $(-1)^n c_n = c_0$, so $c_n = (-1)^n c_0$.
  - the recurrence relation $c_n = 2c_{n-1}$ $(n = 1, 2, 3, \cdots)$ can be rewritten in the form $c_n/2^n = c_{n-1}/2^{n-1}$, and so $c_n/2^n = c_{n-1}/2^{n-1} = c_{n-2}/2^{n-2} = \cdots = c_1/2 = c_0$, giving $c_n/2^n = c_0$, so $c_n = 2^n c_0$.
  - the recurrence relation $nc_n = c_{n-1}$ $(n = 1, 2, 3, \cdots)$ can be rewritten in the form $n!c_n = (n-1)!c_{n-1}$, and so $n!c_n = (n-1)!c_{n-1} = (n-2)!c_{n-2} = \cdots = c_0$, giving $n!c_n = c_0$, so $c_n = c_0/n!$.
  - the recurrence relation $c_n = (n+1)\, c_{n-1}$ $(n = 1, 2, 3, \cdots)$ can be rewritten in the form $c_n/(n+1)! = c_{n-1}/n!$, and so $c_n/(n+1)! = c_{n-1}/n! = c_{n-2}/(n-1)! = \cdots = c_0$, giving $c_n/(n+1)! = c_0$, so $c_n = (n+1)!c_0$.
  - the recurrence relation (12.18) — namely, $(n+2)(n+1)\, c_{n+2} = -c_n$ — can be solved in this way. If n is even, so $n = 2m$, the recurrence relation can be rewritten as $(-1)^{m+1}(2m+2)!c_{2m+2} = (-1)^m (2m)!c_{2m}$, so

  $$(-1)^{m+1}(2m+2)!c_{2m+2} = (-1)^m (2m)!c_{2m} = (-1)^{m-1}(2m-2)!c_{m-2} = \cdots = c_0;$$

  that is, $(-1)^m (2m)!c_{2m} = c_0$, so $c_{2m} = (-1)^m c_0/(2m)!$, which agrees with (12.20). If n is odd, so $n = 2m+1$, the recurrence relation can be rewritten as $(-1)^{m+1}(2m+3)!c_{2m+3} = (-1)^m (2m+1)!c_{2m+1}$, so

  $$(-1)^{m+1}(2m+3)!c_{2m+3} = (-1)^m (2m+1)!c_{2m+1} = \cdots = c_1;$$

  that is, $(-1)^m (2m+1)!c_{2m+1} = c_1$, so $c_{2m+1} = (-1)^m c_1/(2m+1)!$, which agrees with (12.21).

An Important Theorem

In the previous example we were fortunate in two respects. First, we were able to solve the recurrence relation to find an explicit formula for c_n as a function of n.

Second, we were able to recognize the series we obtained as ones that converged to familiar functions. This meant that we did not have to worry about convergence. But how do we know that in general our power series solutions will converge? For linear differential equations with polynomial coefficients, the following theorem answers that question.[3]

► **Theorem 12.1: The First Power Series Theorem** *Consider*

$$a_2(x)y'' + a_1(x)y' + a_0(x)y = 0, \tag{12.23}$$

where $a_2(x)$, $a_1(x)$, and $a_0(x)$ are the polynomials in x remaining after all factors in common to all three terms have been removed. If $a_2(x_0) \neq 0$, the general solution of (12.23) may be obtained as the power series about x_0, $y(x) = \sum_{n=0}^{\infty} c_n(x - x_0)^n = C_1y_1(x) + C_2y_2(x)$, where $y_1(x)$ and $y_2(x)$ are linearly independent functions and C_1 and C_2 are arbitrary constants. This series converges at least for $|x - x_0| < R$, where R is the distance between x_0 and the zero of $a_2(x)$ that is closest to x_0. In considering the zeros of $a_2(x)$, we must include both real and complex zeros.[4] ◄

Comments about the First Power Series Theorem

- There are usually many values of x_0 that could be used in this theorem. If initial conditions are given for a particular value of x, then that is the value to use for x_0.
- In our last example — namely, $y'' + y = 0$ — we are assured that the two series we obtained converge for all values of x because $a_2(x) = 1$, $a_1(x) = 0$, and $a_0(x) = 1$ are all polynomials, and because $a_2(x)$ is never zero. The distance between $x_0 = 0$ and any roots of $a_2(x)$ is infinite, so $R = \infty$.
- If the hypotheses of the theorem are not satisfied, the convergence of each series must be examined separately.

How to Find a Power Series Solution of a Linear Differential Equation

Purpose To find a power series solution of

$$a_2(x)y'' + a_1(x)y' + a_0(x)y = 0, \tag{12.24}$$

about 0, where $a_2(x)$, $a_1(x)$, and $a_0(x)$ are polynomials in x and $a_2(0) \neq 0$.

---

[3] A proof of this theorem, and others in this chapter, can be found in *Ordinary Differential Equations* by E.L. Ince, Dover, 1956.

[4] The distance between x_0 and a complex zero, $z_1 = x_1 + iy_1$, is $\sqrt{(x_1 - x_0)^2 + y_1^2}$.

Process

1. **1.** Factor out any common factors in the three polynomials $a_2(x)$, $a_1(x)$, and $a_0(x)$, and assume a power series solution of the form $y(x) = \sum_{n=0}^{\infty} c_n x^n$.
2. **2.** Substitute this series into (12.24), after computing the appropriate derivatives.
3. **3.** Combine terms and obtain the same form of exponent in each series, which may mean changing indices of summation.
4. **4.** Have the lower limits of all the series start with the same integer, placing all the extra terms together.
5. **5.** Set the coefficient of each power of x equal to zero, obtaining a recurrence relation.
6. **6.** Find as many terms in the series solution as desired, or, if possible, find a general expression for c_n.
7. **7.** Check convergence using the First Power Series Theorem or the ratio test.[5]

Comments about Power Series Solutions of a Linear Differential Equation

- The form of the solution will be $y(x) = C_1 y_1(x) + C_2 y_2(x)$, where $y_1(x)$ and $y_2(x)$ are linearly independent functions and C_1 and C_2 are arbitrary constants.
- We may not be able to find a general expression for c_n.
- If $y(x) = \sum_{n=0}^{\infty} c_n x^n$ with initial conditions $y(0) = y_0$, $y'(0) = y_0^*$, we have $c_0 = y_0$, and $c_1 = y_0^*$.
- If initial values are given for $x_0 \neq 0$, the series expansion to use will be $y(x) = \sum_{n=0}^{\infty} c_n (x - x_0)^n$. There are two different ways to proceed in this case.
  - Follow the preceding technique, using $y(x) = \sum_{n=0}^{\infty} c_n (x - x_0)^n$ in place of $y(x) = \sum_{n=0}^{\infty} c_n x^n$. In this case it is useful to express the coefficients $a_2(x)$, $a_1(x)$, and $a_0(x)$ as polynomials in $x - x_0$. [For example, write $x^2 = (x - x_0)^2 + 2x_0(x - x_0) + x_0^2$.]
  - Change the independent variable from x to X, where $X = x - x_0$. In this way the point $x = x_0$ becomes the point $X = 0$, and $y(x) = \sum_{n=0}^{\infty} c_n (x - x_0)^n$ becomes $y(X) = \sum_{n=0}^{\infty} c_n X^n$, which is a power series about the origin of X. Now use the preceding process.
- This method will work on nonhomogeneous linear differential equations $a_2(x)y'' + a_1(x)y' + a_0(x)y = h(x)$, if $h(x)$ is first expanded in a power series. See Exercise 9 on page 598.

---

[5]See Section A.3 on infinite series in the Appendix.

- This method may also work with nonlinear differential equations; the fact that we will have nonlinear algebraic equations for the coefficients may present an additional complication. See Exercise 14 on page 598.

Three Examples

We demonstrate this procedure on three examples.

EXAMPLE 12.5

Find the power series solution of

$$y'' + 2xy' + 2y = 0. \tag{12.25}$$

(This is the differential equation from Example 12.3 on page 580.)

We are assured by the First Power Series Theorem that the series we obtain will converge for all values of x because $a_2(x) = 1$, $a_1(x) = 2x$, and $a_0(x) = 2$ are all polynomials, and $a_2(x)$ is never zero. The distance between $x_0 = 0$ and any zeros of $a_2(x)$ is infinite, so $R = \infty$.

Substituting the series $y(x) = \sum_{n=0}^{\infty} c_n x^n$ into the differential equation (12.25), we obtain

$$\sum_{n=2}^{\infty} c_n n(n-1)x^{n-2} + 2x \sum_{n=1}^{\infty} c_n n x^{n-1} + 2\sum_{n=0}^{\infty} c_n x^n = 0.$$

In order to combine terms in this equation, we need the same power on x in the three series, as well as the same limits of summation. To accomplish this, we combine the x multiplying the second series with the x^{n-1} inside the summation sign and change the summation index in the first series by raising it by 2. Also note that lowering the limit on the middle series to zero does not contribute anything because of the multiplicative factor of n. These operations give

$$\sum_{n=0}^{\infty} c_{n+2}(n+2)(n+1)x^n + \sum_{n=0}^{\infty} 2c_n n x^n + \sum_{n=0}^{\infty} 2c_n x^n$$

$$= \sum_{n=0}^{\infty} \left[(n+2)(n+1)c_{n+2} + 2(n+1)c_n\right] x^n = 0.$$

Recurrence relation

Because the differential equation is to be satisfied for all values of x, we may equate each of the coefficients to zero, giving the second order recurrence relation

$$(n+2)(n+1)c_{n+2} + 2(n+1)c_n = 0, \qquad n = 0, 1, 2, \cdots. \tag{12.26}$$

Rearranging (12.26) as

$$(n+2)c_{n+2} = -2c_n \text{ or } c_{n+2} = -\frac{2c_n}{n+2}, \qquad n = 0, 1, 2, \cdots, \tag{12.27}$$

shows that we have no requirements on c_1 and c_0, and that all the other coefficients are given by (12.27) in terms of c_1 and c_0. We note that all the coefficients with even subscripts will be given in terms of c_0, whereas all the coefficients with odd subscripts will be given in terms of c_1. We also note that c_0 and c_1 are arbitrary constants, which agrees with the number of arbitrary constants we expect for solutions of second order linear differential equations.

If n is even, so $n = 2m$, the recurrence relation can be rewritten as $(m+1)\,c_{2m+2} = -c_{2m}$, or $(-1)^{m+1}(m+1)!c_{2m+2} = (-1)^m m! c_{2m}$, so

$$(-1)^{m+1}(m+1)!c_{2m+2} = (-1)^m m! c_{2m} = (-1)^{m-1}(m-1)!c_{2m-2} = \cdots = c_0;$$

that is, $(-1)^m m! c_{2m} = c_0$, so

$$c_{2m} = (-1)^m \frac{1}{m!} c_0, \qquad m = 0, 1, 2, \cdots.$$

If n is odd, so $n = 2m+1$, the recurrence relation can be rewritten as $(2m+3)\,c_{2m+3} = -2c_{2m+1}$, or

$$\begin{aligned}(-1)^{m+1}\frac{1}{2^{m+1}}&[(2m+3)(2m+1)\cdots 5\cdot 3\cdot 1]\,c_{2m+3}\\ &= (-1)^m \frac{1}{2^m}[(2m+1)\cdots 5\cdot 3\cdot 1]\,c_{2m+1},\end{aligned}$$

so,

$$(-1)^m \frac{1}{2^m}[(2m+1)\cdots 5\cdot 3\cdot 1]\,c_{2m+1} = c_1;$$

that is,

$$c_{2m+1} = (-1)^m \frac{2^m}{1\cdot 3\cdot 5\cdots(2m+1)} c_1, \qquad m = 0, 1, 2, \cdots.$$

Thus, we have the solution of (12.25) in the form

$$y(x) = \sum_{m=0}^{\infty} c_{2m} x^{2m} + \sum_{m=0}^{\infty} c_{2m+1} x^{2m+1},$$

General solution

which can be written

$$y(x) = c_0 \sum_{m=0}^{\infty} (-1)^m \frac{x^{2m}}{m!} + c_1 \sum_{m=0}^{\infty} (-1)^m \frac{2^m}{1\cdot 3\cdot 5\cdots(2m+1)} x^{2m+1}.$$

We recognize that the first series converges for all values of x to

$$c_0 \sum_{m=0}^{\infty} (-1)^m \frac{(x^2)^m}{m!} = c_0 \sum_{m=0}^{\infty} \frac{1}{m!}\left(-x^2\right)^m = c_0 e^{-x^2},$$

but the second series does not lead to a familiar function (see Exercises 5 and 6 on pages 597 and 598.).

Thus, the general solution of (12.25), valid for all x, is given by

$$y(x) = c_0 e^{-x^2} + c_1 \sum_{m=0}^{\infty} (-1)^m \frac{2^m}{1 \cdot 3 \cdot 5 \cdots (2m+1)} x^{2m+1}. \tag{12.28}$$

If we use the initial conditions $y(0) = y_0$, $y'(0) = y_0^*$ in (12.28), we have $c_0 = y(0) = y_0$ and $c_1 = y'(0) = y_0^*$. If we write out the first few terms of each of the series in (12.28), we find (12.15).

We know what the function e^{-x^2} behaves like (see Figure 12.9), but what about the function multiplying c_1? We know from the Oscillation Theorem and Relation Theorem of Section 8.1 that this solution cannot oscillate. (Why?) Furthermore, if we impose the initial conditions $y(0) = 0$, $y'(0) = 1$, then (12.28) reduces to the solution

$$y(x) = \sum_{m=0}^{\infty} (-1)^m \frac{2^m}{1 \cdot 3 \cdot 5 \cdots (2m+1)} x^{2m+1}. \tag{12.29}$$

Figure 12.10 shows the power series polynomials of degrees 1 through 10 for (12.29). Notice how we can begin to see the solution emerging as the common points between the ninth and tenth degree polynomials. However, in spite of appearances, we know that this solution cannot cross the positive x-axis. If it did, it would then have two zeros ($x = 0$ being the other one), and so e^{-x^2} would have to vanish between these two zeros, which it does not. It is possible to confirm these observations in other ways (see Exercise 5, page 597).

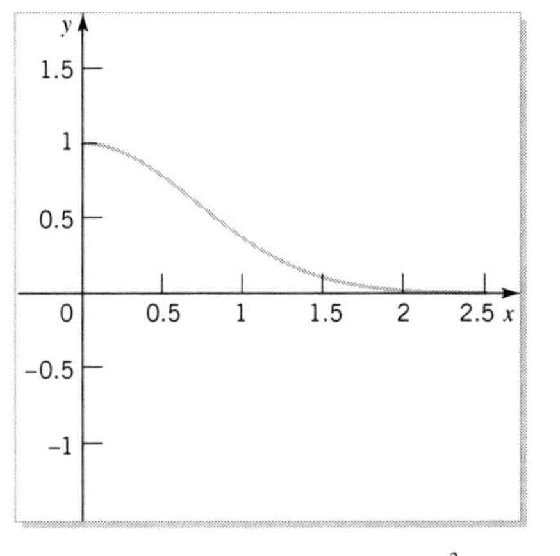

FIGURE 12.9 The solution e^{-x^2} of $y'' + 2xy' + 2y = 0$

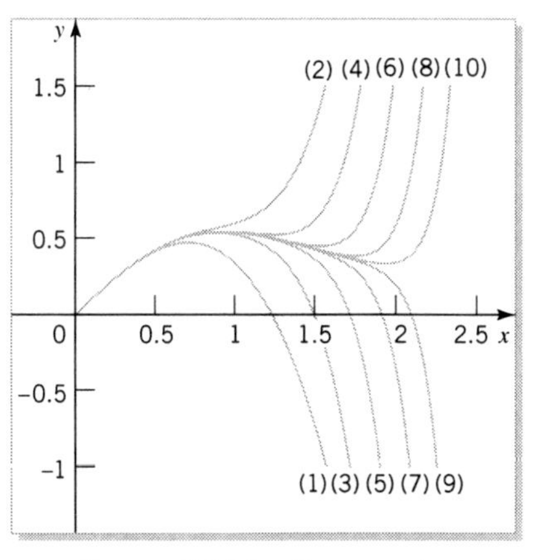

FIGURE 12.10 Power series polynomials of degrees 1 through 10 for $y'' + 2xy' + 2y = 0$, $y(0) = 0$, $y'(0) = 1$

EXAMPLE 12.6 *Airy's Equation*

Let us return to Airy's equation that we discussed in Section 8.1; that is,

$$y'' - xy = 0. \tag{12.30}$$

We already know from the Oscillation Theorem of Section 8.1 that every nontrivial solution of (12.30) oscillates for $x < 0$ and has, at most, one zero for $x > 0$. Now we wish to find a power series solution for this equation. The First Power Series Theorem guarantees that the series we obtain will converge for all values of x. Substituting the series $y(x) = \sum_{n=0}^{\infty} c_n x^n$ into the differential equation (12.30), we find $c_2 = 0$, and the third order recurrence relation

Recurrence relation

$$c_{n+2} = \frac{1}{(n+2)(n+1)} c_{n-1}, \qquad n = 1, 2, 3, \cdots.$$

After several calculations, we find

$$\begin{aligned}
c_{3m} &= (3m-2)\cdot(3m-5)\cdots 7\cdot 4\cdot 1\cdot c_0/(3m)!, && m = 1, 2, 3, \cdots, \\
c_{3m+1} &= (3m-1)\cdot(3m-4)\cdots 8\cdot 5\cdot 2\cdot c_1/(3m+1)!, && m = 1, 2, 3, \cdots, \\
c_{3m+2} &= 0, && m = 0, 1, 2, \cdots.
\end{aligned}$$

General solution

The general solution of (12.30), valid for all x, is therefore

$$\begin{aligned}
y(x) = c_0 + c_0 \sum_{m=1}^{\infty} \frac{(3m-2)\cdot(3m-5)\cdots 7\cdot 4\cdot 1}{(3m)!} x^{3m} + \\
+ c_1 x + c_1 \sum_{m=1}^{\infty} \frac{(3m-1)\cdot(3m-4)\cdots 8\cdot 5\cdot 2}{(3m+1)!} x^{3m+1}.
\end{aligned} \tag{12.31}$$

Figure 12.11 shows the power series polynomials of degrees 6 through 27 for the coefficient of c_0 in (12.31). Notice that it corresponds to the initial condition $y(0) = 1$, $y'(0) = 0$, and that we can begin to see the solution emerging as the common points between the 24th and 27th degree polynomials. We also see that for $x > 0$, there are no roots, whereas for $x < 0$, the solution begins to show oscillatory behavior. These observations are consistent with the previous results on oscillations and zeros.

Figure 12.12 shows the power series polynomials of degrees 7 through 28 for the coefficient of c_1 in (12.31). Notice that it corresponds to the initial condition $y(0) = 0$, $y'(0) = 1$, and that we can begin to see the solution emerging as the common points between the 25th and 28th degree polynomials. We also see that for $x > 0$, there are no roots, whereas for $x < 0$, the solution oscillates. These observations are consistent with the previous results on oscillations and zeros.

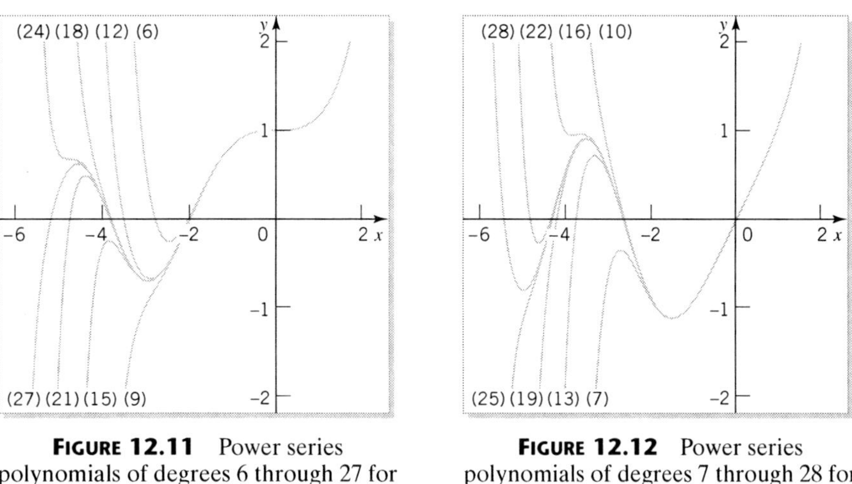

FIGURE 12.11 Power series polynomials of degrees 6 through 27 for Airy's equation with $c_0 = 1, c_1 = 0$

FIGURE 12.12 Power series polynomials of degrees 7 through 28 for Airy's equation with $c_0 = 0, c_1 = 1$

EXAMPLE 12.7

Solve

$$y'' - 2(x-1)y' + 2y = 0 \tag{12.32}$$

by using a power series solution about $x = 1$.

The coefficients of y'', y', and y are polynomials, so we know that there is a solution of (12.32) of the form

$$y(x) = \sum_{n=0}^{\infty} c_n (x-1)^n .$$

There are two different ways to find the coefficients in this expansion.

We could substitute $y(x) = \sum_{n=0}^{\infty} c_n (x-1)^n$ into (12.32), collect like powers of $x - 1$ together, equate their coefficients to zero, and proceed as usual (see Exercise 7 on page 598).

Change variable

A second way is to make the change of variable $X = x - 1$, so that the point $x = 1$ becomes the point $X = 0$, and the differential equation (12.32) becomes the differential equation

$$\frac{d^2y}{dX^2} - 2X\frac{dy}{dX} + 2y = 0. \tag{12.33}$$

Thus, the original problem of finding a series solution of (12.32) about $x = 1$ is equivalent to finding a power series solution of (12.33) about the point $X = 0$; that is,

$$y(X) = \sum_{n=0}^{\infty} c_n X^n.$$

Recurrence relation In Exercise 8 on page 598 you will find that the recurrence relation in this case becomes

$$c_{n+2} = \frac{2\,(n-1)}{(n+2)\,(n+1)}c_n, \qquad n = 0, 1, 2, \cdots,$$

which leads to the general solution

$$y(X) = c_0\left(1 - X^2 - \frac{1}{6}X^4 - \frac{1}{30}X^6 + \cdots\right) + c_1 X.$$

General solution In terms of the original variable x, the solution of (12.32) is

$$y(x) = c_0\left[1 - (x-1)^2 - \frac{1}{6}(x-1)^4 - \frac{1}{30}(x-1)^6 + \cdots\right] + c_1(x-1).$$

Another Important Theorem

The First Power Series Theorem deals with the second order linear differential equation $a_2(x)y'' + a_1(x)y' + a_0(x)y = 0$, where $a_2(x), a_1(x)$, and $a_0(x)$ are polynomials in x with $a_0(x_0) \neq 0$. But what happens if this is not the case? If we divide by the coefficient of the second derivative, we obtain

$$y'' + p(x)y' + q(x)y = 0, \tag{12.34}$$

which we use as our standard form. In (12.34) we have $p(x) = a_1(x)/a_2(x)$, and $q(x) = a_0(x)/a_2(x)$.

The success of the power series method is ensured by the following theorem.[6]

► **Theorem 12.2: The Second Power Series Theorem** *If both $p(x)$ and $q(x)$ are analytic[7] at $x = x_0$, then the general solution of $y'' + p(x)y' + q(x)y = 0$ may be obtained in the form of a power series about $x = x_0$. The radius of convergence of this series is at least as large as the minimum of the radii of convergence of the series expansions for $p(x)$ and $q(x)$ about $x = x_0$.* ◄

Comments about the Second Power Series Theorem

- If the hypothesis of the Second Power Series Theorem is satisfied, we may confidently proceed with the calculation of a power series solution knowing the interval of convergence even before we start. We could either use the power series method or implicitly differentiate the differential equation as we did in the previous section.

[6] A proof of this theorem can be found in *Ordinary Differential Equations* by E.L. Ince, Dover, 1956.

[7] A function is analytic at a point if it has a Taylor series that converges in some interval about the point.

- If both $p(x)$ and $q(x)$ are analytic at the point $x = x_0$, then the point $x = x_0$ is called an ORDINARY POINT of (12.34). If $x = x_0$ is not an ordinary point of (12.34), the point $x = x_0$ is a SINGULAR POINT of the differential equation, and the function with the singular point is said to be SINGULAR and have a SINGULARITY AT $x = x_0$. However, it does not follow that a differential equation with a singular point has a singular solution. The Cauchy-Euler differential equation $y'' - \frac{2}{x}y' + \frac{2}{x^2}y = 0$ has $x = 0$ as a singular point, and the nonsingular function $y(x) = C_1x + C_2x^2$ as its solution.
- This theorem can be extended to nonhomogeneous differential equations if the radius of convergence of the forcing term is taken into account.

EXAMPLE 12.8 *Legendre's Differential Equation*

Exercise 13 on page 598 is concerned with LEGENDRE'S DIFFERENTIAL EQUATION, which has the form

$$(1 - x^2)y'' - 2xy' + \lambda y = 0,$$

where λ is a specified constant.

Singular point

Legendre's differential equation has singular points at 1 and -1, with all other points being ordinary points. Thus, according to either the First or Second Power Series Theorems, a power series for the solution about $x = 0$ will converge at least for $-1 < x < 1$. ■

EXAMPLE 12.9

Find a power series up to degree 5 for the solution of

$$xy'' + (\sin x)\,y' + 2xy = 0 \tag{12.35}$$

about $x = 0$, and check its convergence.

To check the hypothesis of the Second Power Series Theorem, we divide (12.35) by x to put it in the standard form of

$$y'' + \frac{\sin x}{x}y' + 2y = 0. \tag{12.36}$$

The coefficients

$$p(x) = \frac{\sin x}{x} = \frac{1}{x}\sum_{m=0}^{\infty}(-1)^m\frac{x^{2m+1}}{(2m+1)!} = \sum_{m=0}^{\infty}(-1)^m\frac{x^{2m}}{(2m+1)!}$$

and $q(x) = 2$ are analytic at $x = 0$, so the hypothesis of the Second Power Series Theorem is satisfied with $x_0 = 0$. Because the radii of convergence of $p(x)$ and $q(x)$ are infinite, we are assured that the power series solution of (12.35) converges for all values of x.

If we substitute the series $y(x) = \sum_{n=0}^{\infty} c_n x^n$ into (12.36), we obtain

$$\sum_{n=2}^{\infty} n(n-1)c_n x^{n-2} + \left[\sum_{m=0}^{\infty}(-1)^m \frac{x^{2m}}{(2m+1)!}\right]\sum_{n=1}^{\infty} nc_n x^{n-1} + \sum_{n=0}^{\infty} 2c_n x^n = 0.$$

If we want to find the power series up to degree 5, we truncate all these series to obtain

$$\begin{gathered} 2c_2 + 6c_3x + 12c_4x^2 + 20c_5x^3 + 30c_6x^4 + \\ + \left[1 - (1/6)\,x^2 + (1/120)\,x^4\right](c_1 + 2c_2x + 3c_3x^2 + 4c_4x^3 + 5c_5x^4) + \\ +2c_0 + 2c_1x + 2c_2x^2 + 2c_3x^3 + 2c_4x^4 + 2c_5x^5 = 0. \end{gathered}$$

Equating to zero the coefficients of powers of x^n for $n = 0$, 1, 2, and 3 gives the following set of homogeneous algebraic equations:

$$\begin{aligned} 2c_2 + c_1 + 2c_0 &= 0, \\ 6c_3 + 2c_2 + 2c_1 &= 0, \\ 12c_4 + 3c_3 - (1/6)\,c_1 + 2c_2 &= 0, \\ 20c_5 + 4c_4 - (1/3)\,c_2 + 2c_3 &= 0. \end{aligned}$$

Note that this system of four equations with six unknowns is a dependent system. As expected, we may choose c_0 and c_1 arbitrarily, and the solution of this system of equations becomes

$$\begin{aligned} c_2 &= -(2c_0 + c_1)/2, \\ c_3 &= (2c_0 - c_1)/6, \\ c_4 &= (3c_0 + 5c_1)/36, \\ c_5 &= -(24c_0 + 7c_1)/360. \end{aligned}$$

This means that the solution in the form of a power series, including terms of degree 5 or less, is given by

$$\begin{aligned} y(x) = c_0 &\left[1 - x^2 + \frac{1}{3}x^3 + \frac{1}{12}x^4 - \frac{1}{15}x^5 + \cdots\right] + \\ &+c_1\left[x - \frac{1}{2}x^2 - \frac{1}{6}x^3 + \frac{5}{36}x^4 - \frac{7}{360}x^5 + \cdots\right]. \end{aligned} \tag{12.37}$$

Figure 12.13 shows the power series polynomials of degrees 2 through 5 for the coefficient of c_0 in (12.37). Notice that it corresponds to the initial condition $y(0) = 1$, $y'(0) = 0$, and that we can begin to see the solution emerging as the common points between the fourth and fifth degree polynomials. Figure 12.14 shows the power series polynomials of degrees 2 through 5 for the coefficient of c_1 in (12.37). It corresponds to the initial condition $y(0) = 0$, $y'(0) = 1$. Again we can begin to see the solution emerging as the common points between the fourth and fifth degree polynomials. However, from this analysis we cannot tell whether these solutions oscillate. In fact they do oscillate (see Exercise 10 on page 598).

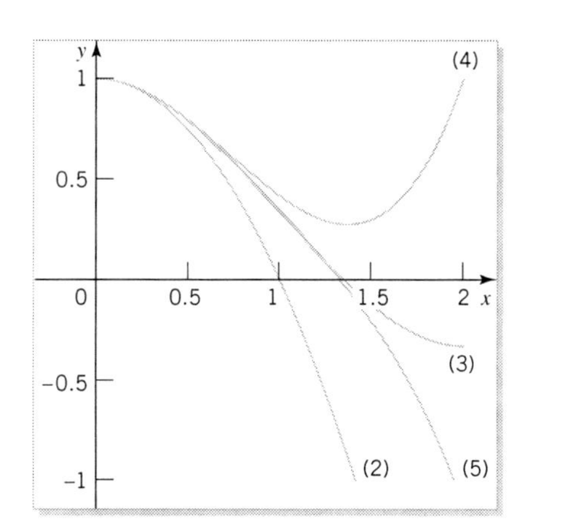

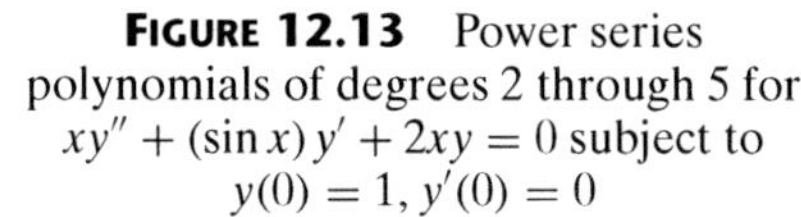

FIGURE 12.13 Power series polynomials of degrees 2 through 5 for $xy'' + (\sin x)\,y' + 2xy = 0$ subject to $y(0) = 1$, $y'(0) = 0$

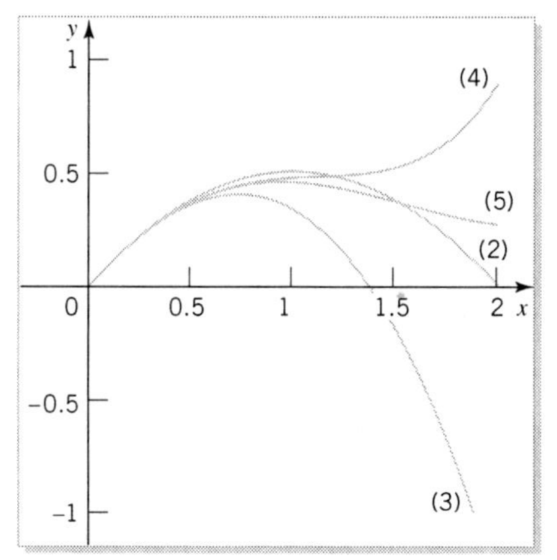

FIGURE 12.14 Power series polynomials of degrees 2 through 5 for $xy'' + (\sin x)\,y' + 2xy = 0$ subject to $y(0) = 0$, $y'(0) = 1$

EXERCISES

1. Make a change in the index of summation to show the validity of the following equations.

(a) $\sum_{k=0}^{\infty} c_k x^{k+2} = \sum_{m=2}^{\infty} c_{m-2} x^m$

(b) $\sum_{k=4}^{\infty} c_{k-1} x^{k-2} = \sum_{m=2}^{\infty} c_{m+1} x^m$

(c) $\sum_{m=n}^{\infty} c_m x^m = \sum_{k=0}^{\infty} c_{k+n} x^{k+n}$ (That is, if we reduce the lower limit of the summation by n, we increase the index to the right of the summation sign by n.)

(d) $\sum_{m=n}^{\infty} c_m x^{m-p} = \sum_{k=n-p}^{\infty} c_{k+p} x^k$ (That is, if we increase the exponent by p, we increase the corresponding subscript by p and reduce the lower limit of summation by p.)

2. Show that

(a) $\sum_{k=2}^{\infty} c_k + A\sum_{k=0}^{\infty} a_k = A(a_0 + a_1) + \sum_{k=2}^{\infty} (c_k + Aa_k)$.

(b) $\sum_{k=0}^{\infty} c_k = \sum_{k=0}^{n} c_k + \sum_{k=n+1}^{\infty} c_k$ for any positive integer n.

3. Use the power series method to solve $y'' - xy' + 3y = 0$, $y(0) = 2$, $y'(0) = 0$.

4. Use the power series method to solve $y'' + 2x^2y' + xy = 0$, $y(0) = 0$, $y'(0) = 3$.

5. Make the change of dependent variable $y(x) = z(x)e^{-x^2}$ in (12.25) — that is, $y'' + 2xy' + 2y = 0$ — to obtain a second solution in a different form from that given in (12.28); namely, $y(x) = e^{-x^2}\int_0^x e^{t^2}\,dt$.

(a) Show that this solution cannot vanish for $x > 0$ in agreement with the observation made following (12.29).

(b) Show that this solution has an extreme value at the point $(x_m, 1/(2x_m))$ where x_m satisfies $y(x_m) = 1/(2x_m)$. Show that at this extreme value, we have $y''(x_m) = -2y(x_m)$. Explain why this requires that there is a maximum at $(x_m, 1/(2x_m))$. Can this solution have a minimum? Can it have a second maximum? Find a numerical value for x_m, and compare your answer with Figure 12.10.

(c) Show that this solution has a point of inflection at the point $\left(x_i, x_i/\left(2x_i^2 - 1\right)\right)$, where x_i satisfies $y(x_i) = x_i/\left(2x_i^2 - 1\right)$. Find a numerical value for x_i, and compare your answer with Figure 12.10. Can this solution have a second point of inflection?

(d) Use local linearity or L'Hôpital's rule to show that this solution has the property that $y(x) \to 0$ as $x \to \infty$.

(e) Use local linearity or L'Hôpital's rule to show that this solution has the property that $2xy(x) \to 1$ as $x \to \infty$, so that $y(x) \to 1/(2x)$ as $x \to \infty$.

(f) Use the information from parts (a) through (e) to sketch this solution for $x > 0$.

(g) Show that the solution satisfies $y(x) = -y(-x)$. Use this information to sketch this solution for all x.

6. Show that the substitution $u(x) = y' + 2xy$ converts the differential equation (12.25) — that is, $y'' + 2xy' + 2y = 0$ — into $u' = 0$. Use this fact to find the general solution of (12.25). Explain how this idea can be used to solve differential equations of the form $y'' + p(x)y' + p'(x)y = 0$.

7. Use the power series method about $x = 1$ to find the general solution of $y'' - 2(x-1)y' + 2y = 0$.

8. Use the power series method about $x = 0$ to find the general solution of $y'' - 2xy' + 2y = 0$.

9. Find the power series solution, up to terms of degree 4, of the following initial value problems, and give a value of R such that the series solution converges for $|x - x_0| < R$.

(a) $y'' + x^2y' + (\sin x)\, y = 3 \quad y(0) = 0 \quad y'(0) = 3$
(b) $y'' + 3xy' + (\cos x)\, y = x^2 \quad y(0) = 1 \quad y'(0) = 0$
(c) $y'' - 2xy' + e^x y = e^x \quad y(0) = 0 \quad y'(0) = -2$
(d) $y'' + (\sin x)\, y' + xy = 0 \quad y(0) = 3 \quad y'(0) = 0$
(e) $y'' - xy' + (\ln x)\, y = 0 \quad y(1) = 0 \quad y'(1) = 2$
(f) $y'' - xy' + y/(1-2x) = 0 \quad y(0) = 1 \quad y'(0) = 0$

10. Show that the differential equation (12.36) — namely, $y'' + \frac{\sin x}{x} y' + 2y = 0$ — has, from the Oscillation Theorem and Relation Theorem of Section 8.1, a $Q(x)$ given by $Q(x) = q(x) - \frac{1}{2}p'(x) - \frac{1}{4}p^2(x)$, which can be written in the form

$$Q(x) = \frac{3}{2} + \frac{1}{4x^2}\left[(x - \cos x)^2 + x^2 + 2\sin x - 1\right].$$

Show that $x^2 + 2\sin x - 1 \geq x^2 - 3$, so that if $x \geq \sqrt{3}$, then $Q(x) \geq 3/2$. Explain how this guarantees that nontrivial solutions of (12.36) must oscillate for $x \geq \sqrt{3}$.

11. Find the general solution of the following differential equations as a power series about $x_0 = 0$, and give the radius of convergence.

(a) $y'' + x^2y = 0$
(b) $y'' - 3xy' + y = 0$
(c) $\left(1 - x^2\right) y'' + 2xy' + 5y = 0$

12. Find the first five nonzero terms in a power series solution about an appropriate point of the following initial value problems. Give the radius of convergence.

(a) $xy'' + x^2y' + y = 0 \quad y(2) = 0 \quad y'(2) = 1$
(b) $x^2y'' + y' + 2y = 0 \quad y(1) = 0 \quad y'(1) = 1$
(c) $\left(x^2 + 1\right) y'' + xy' + 3y = 0 \quad y(2) = 0 \quad y'(2) = 3$

13. Legendre Polynomials. Legendre polynomials are solutions of Legendre's differential equation $\left(1 - x^2\right) y'' - 2xy' + \lambda y = 0$ for special values of λ.

(a) Show that Legendre's differential equation will have a polynomial solution whenever λ is the product of two consecutive nonnegative integers; that is, $\lambda = m(m+1)$.

(b) Find the first five Legendre polynomials [denoted by $P_0(x)$, $P_1(x)$, $P_2(x)$, $P_3(x)$, and $P_4(x)$] by choosing the arbitrary constants in the solutions for $m = 0, 1, 2, 3$, and 4, so that $P_m(1) = 1$.

(c) Define $w(x) = \left(x^2 - 1\right)^m$ and show that $\left(1 - x^2\right) w' + 2mxw = 0$. Differentiate this $m + 1$ times to show that

$$\left(1 - x^2\right) \frac{d^{m+2}w}{dx^{m+2}} - 2x\frac{d^{m+1}w}{dx^{m+1}} + \lambda\frac{d^m w}{dx^m} = 0.$$

[Hint: Leibnitz' Rule, $\frac{d^n}{dx^n}\left[f(x)\, g(x)\right] = \sum_{k=0}^{n} \frac{n!}{(n-k)!k!}\left(\frac{d^{n-k}}{dx^{n-k}}f\right)\left(\frac{d^k}{dx^k}g\right)$, may be useful.] Explain why this implies that $P_m(x) = \frac{1}{2^m m!}\frac{d^m}{dx^m}\left(x^2 - 1\right)^m$ is a solution of Legendre's differential equation.

(d) Show that if $w(x) = \left(x^2 - 1\right)^m$, then w' must have at least one root between -1 and 1. [Hint: Evaluate $w(-1)$ and $w(1)$ and use the Mean Value Theorem.] Continue in this way to show that $P_m(x)$ has at least m roots between -1 and 1.

14. Find the first six terms in a power series solution of the nonlinear differential equations

(a) $y'' + \left(y'\right)^2 = e^x y$, $y(0) = 0$, $y'(0) = 1$.

(b) $y' = 2x^2 - y^2$, $y(0) = 1$. Now make the change of variable $y = u'/u$, and solve $u'' = 2x^2u, u'(0) = u(0)$. Comment on this form of your answer.

15. List the singular points for the following differential equations.

(a) $(1-x)y'' + \frac{3x}{x+2}y' + \frac{(1-x)^2}{x+3}y = 0$
(b) $(x^2 + x)y'' + \frac{x^3}{x-1}y' + \frac{x^4+3x}{x+2}y = 0$
(c) $\frac{1}{x}y'' + \frac{3(x-4)}{x+6}y' + \frac{x^2(x-2)}{x-1}y = 0$
(d) $(x^2 + 3x + 2)y'' + \frac{x+2}{x-1}y' + \frac{x-2}{x}y = 0$

(e) $(x^2+9)y'' + \frac{x-2}{x+7}y' + \frac{x^2+3}{2}y = 0$

(f) $e^x y'' + \frac{3x-4}{x+4}y' + \frac{x}{x-4}y = 0$

(g) $\sin x\, y'' + \frac{x^3}{x-1}y' + \frac{x^4+3x}{x+2}y = 0$

16. Use the First and Second Power Series Theorems to determine the interval of convergence for each of the differential equations in Exercise 15 for the following values of x_0.

(a) 1 (b) 4 (c) -1 (d) 2

(e) 0 (f) 0 (g) 3

12.3 WHAT TO DO WHEN POWER SERIES FAIL

In Section 12.2 we used a power series method to find series solutions of differential equations about ordinary points. If we use that technique to find a power series solution about the point $x = 0$ for the linear differential equation

$$x^2y'' - 2y = 0, \tag{12.38}$$

we find that

$$x^2(2c_2 + 6c_3x + 12c_4x^2 + \cdots) - 2(c_0 + c_1x + c_2x^2 + c_3x^3 + c_4x^4 + \cdots) = 0.$$

Combining like terms gives

$$-2c_0 - 2c_1x + 4c_3x^3 + 10c_4x^4 + \cdots = 0,$$

which requires that $c_0 = c_1 = c_3 = c_4 = \cdots = 0$. This gives us no condition on c_2, so c_2x^2 is a solution, where c_2 is an arbitrary constant. However, we do not get a second solution by this method.

If we try this same method on a similar differential equation,

$$4x^2y'' - 3y = 0, \tag{12.39}$$

we find that all of the coefficients in $\sum_{n=0}^{\infty} c_nx^n$ must equal zero. (Do this!)

What is happening? If we look at the Existence-Uniqueness Theorem for Second Order Linear Differential Equations of Section 8.1, we notice that the hypothesis that $a_2(0) \neq 0$ is violated in both these examples, so that theorem does not apply here. Also, if we look at the Second Power Series Theorem on page 594, we notice that the hypothesis that $q(x)$ be analytic at $x = 0$ is violated in both these examples, because $q(x) = -2/x^2$, and $q(x) = -3/\left(4x^2\right)$, respectively. So the point $x = 0$ is not an ordinary point, but a singular point.

So what can we try if $x = 0$ is not an ordinary point? We note that both (12.38) and (12.39) are Cauchy-Euler differential equations with general solutions $C_1x^2 + C_2x^{-1}$, and $C_1x^{3/2} + C_2x^{-1/2}$, respectively. Because the exponents in a power series must be nonnegative integers, only one of the four functions in these general solutions is a power series expansion about $x = 0$; but all these solutions have the form of x to a power. Maybe we should try a solution of the form of a power series multiplied by x to a power — that is, $y(x) = x^s \sum_{n=0}^{\infty} c_nx^n$, where s is a constant.

The Indicial Equation

EXAMPLE 12.10

We now try this form of a series; namely,[8]

$$y(x) = x^s \sum_{n=0}^{\infty} c_n x^n, \tag{12.40}$$

in the first differential equation in this section, (12.38), $x^2 y'' - 2y = 0$. In this series, s, as well as the coefficients c_n, $n = 0, 1, 2, \cdots$, are to be found in order that (12.40) satisfies the differential equation (12.38). To simplify subsequent derivatives, we move the x^s term inside the series as

$$y(x) = \sum_{n=0}^{\infty} c_n x^{n+s} = c_0 x^s + c_1 x^{s+1} + c_2 x^{s+2} + c_3 x^{s+3} + \cdots. \tag{12.41}$$

Because the series must start somewhere, we **always** assume that c_0 is the coefficient of the smallest surviving power of x, and so $c_0 \neq 0$. We substitute (12.41) into $x^2 y'' - 2y = 0$ to obtain

$$x^2 \left[c_0 s(s-1)x^{s-2} + c_1(s+1)sx^{s-1} + c_2(s+2)(s+1)x^s + c_3(s+3)(s+2)x^{s+1} + \cdots \right] + \\ -2\left[c_0 x^s + c_1 x^{s+1} + c_2 x^{s+2} + c_3 x^{s+3} + \cdots \right] = 0.$$

If we combine like powers of x, this reduces to

$$[s(s-1) - 2]c_0 x^s + [(s+1)s - 2]c_1 x^{s+1} + [(s+2)(s+1) - 2]c_2 x^{s+2} + \\ + [(s+3)(s+2) - 2]c_3 x^{s+3} + \cdots = 0.$$

Equating coefficients of each power of x to zero, we find

$$\begin{aligned} [s(s-1) - 2]c_0 &= 0, \\ [(s+1)s - 2]c_1 &= 0, \\ [(s+2)(s+1) - 2]c_2 &= 0, \\ [(s+3)(s+2) - 2]c_3 &= 0. \end{aligned} \tag{12.42}$$

Because $c_0 \neq 0$, the first equation gives $s(s-1) - 2 = 0$, or

$$(s-2)(s+1) = 0, \tag{12.43}$$

so the possible choices for s are 2 and -1. We now show how these choices lead to our previous general solution of $C_1 x^2 + C_2 x^{-1}$.

1. With $s = 2$ in (12.42), we find $c_1 = c_2 = c_3 = \cdots = 0$, so in this case the series (12.40) reduces to $c_0 x^2$, where c_0 is arbitrary. Thus, $C_1 y_1(x)$, where $y_1(x) = x^2$, is one solution of (12.38).

---

[8] Here s need not be an integer.

2. With $s = -1$ in (12.42), we also find $c_1 = c_2 = c_3 = \cdots = 0$, so in this case the series (12.40) reduces to $c_0 x^{-1}$, where c_0 is arbitrary. Thus, $C_2 y_2(x)$, where $y_2(x) = x^{-1}$, is another solution of (12.38).

General solution

Now $y_1(x)$ and $y_2(x)$ are linearly independent, so the general solution of (12.38) is

$$y(x) = C_1 y_1(x) + C_2 y_2(x) = C_1 x^2 + C_2 \frac{1}{x},$$

in agreement with our previous observation. ■

The quadratic equation that determines the possible values for s — in this case, (12.43) — is given a special name.

◆ *Definition 12.2:* **If $x = x_0$ is a singular point of the differential equation $a_2(x)y'' + a_1(x)y' + a_0(x)y = 0$, and we substitute $y(x) = (x - x_0)^s \sum_{n=0}^{\infty} c_n (x - x_0)^n$ into the differential equation, collect like powers of $x - x_0$, and then set the coefficient of the lowest power of $x - x_0$ equal to zero, the equation so obtained is called the INDICIAL EQUATION. It is a quadratic equation in s.** ◆

EXAMPLE 12.11

We apply the technique used in Example 12.10 to the differential equation given in (12.39),

$$4x^2 y'' - 3y = 0, \tag{12.44}$$

which has a singular point at $x = 0$.

We assume a solution of the form

$$y(x) = x^s \sum_{n=0}^{\infty} c_n x^n = \sum_{n=0}^{\infty} c_n x^{n+s} = c_0 x^s + c_1 x^{s+1} + c_2 x^{s+2} + c_3 x^{s+3} + \cdots, \tag{12.45}$$

where $c_0 \neq 0$, and substitute (12.45) into (12.44) to obtain

$$4x^2 \left[c_0 s(s-1)x^{s-2} + c_1 (s+1)sx^{s-1} + c_2(s+2)(s+1)x^s + c_3(s+3)(s+2)x^{s+1} + \cdots\right] + \\ -3\left[c_0 x^s + c_1 x^{s+1} + c_2 x^{s+2} + c_3 x^{s+3} + \cdots\right] = 0.$$

If we combine like powers of x, this reduces to

$$[4s(s-1) - 3]c_0 x^s + [4(s+1)s - 3]c_1 x^{s+1} + [4(s+2)(s+1) - 3]c_2 x^{s+2} + \\ + [4(s+3)(s+2) - 3]c_3 x^{s+3} + \cdots = 0.$$

Indicial equation

The lowest power of x is x^s, so the indicial equation is $[4s(s-1) - 3]c_0 = 0$. Because $c_0 \neq 0$, we have the quadratic equation

$$4s^2 - 4s - 3 = 0,$$

with roots 3/2 and −1/2. Equating the coefficients of the next powers of x to zero, we find

$$\begin{aligned} [4(s+1)s-3]c_1 &= 0, \\ [4(s+2)(s+1)-3]c_2 &= 0, \\ [4(s+3)(s+2)-3]c_3 &= 0. \end{aligned} \tag{12.46}$$

From the indicial equation, the possible choices for s are 3/2 and −1/2. We now use each of these choices in turn.

1. With $s = 3/2$ in (12.46), we find $c_1 = c_2 = c_3 = \cdots = 0$, so in this case the series (12.45) reduces to $c_0 x^{3/2}$, where c_0 is arbitrary. Thus, $C_1 y_1(x)$, where $y_1(x) = x^{3/2}$, is one solution of (12.44).
2. With $s = -1/2$ in (12.46), we also find $c_1 = c_2 = c_3 = \cdots = 0$, so in this case the series (12.45) reduces to $c_0 x^{-1/2}$, where c_0 is arbitrary. Thus, $C_2 y_2(x)$, where $y_2(x) = x^{-1/2}$, is another solution of (12.44).

General solution

Now $y_1(x)$ and $y_2(x)$ are linearly independent, so the general solution of (12.44) is

$$y(x) = C_1 y_1(x) + C_2 y_2(x) = C_1 x^{3/2} + C_2 x^{-1/2},$$

again in agreement with our previous observation.

Motivating Regular Singular Points

EXAMPLE 12.12

If we use the preceding technique on the linear differential equation

$$x^4 y'' + 2x^3 y' - y = 0, \tag{12.47}$$

which has a singular point at $x = 0$, we find that

$$\begin{aligned} x^4 &\left[c_0 s(s-1)x^{s-2} + c_1(s+1)s x^{s-1} + c_2(s+2)(s+1)x^s + \cdots\right] + \\ &+2x^3\left[c_0 s x^{s-1} + c_1(s+1)x^s + c_2(s+2)x^{s+1} + \cdots\right] + \\ &-\left[c_0 x^s + c_1 x^{s+1} + c_2 x^{s+2} + c_3 x^{s+3} + \cdots\right] = 0. \end{aligned}$$

Combining like terms gives

$$-c_0 x^s - c_1 x^{s+1} + [c_0 s(s-1) + 2c_0 s + c_2]x^{s+2} + \cdots = 0.$$

The coefficient of the lowest power of x does not depend on s, and the only way this coefficient can equal zero is for $c_0 = 0$. However, $c_0 \neq 0$, so there is no solution of (12.47) of the form $y(x) = x^s \sum_{n=0}^{\infty} c_n x^n$. In fact, the solution of (12.47) is

$$y(x) = C_1 e^{1/x} + C_2 e^{-1/x}$$

(see Exercise 2 on page 608), which is of the form $x^s \sum_{n=0}^{\infty} c_n x^{-n}$. Figure 12.15 shows the graphs of $e^{1/x}$ and $e^{-1/x}$.

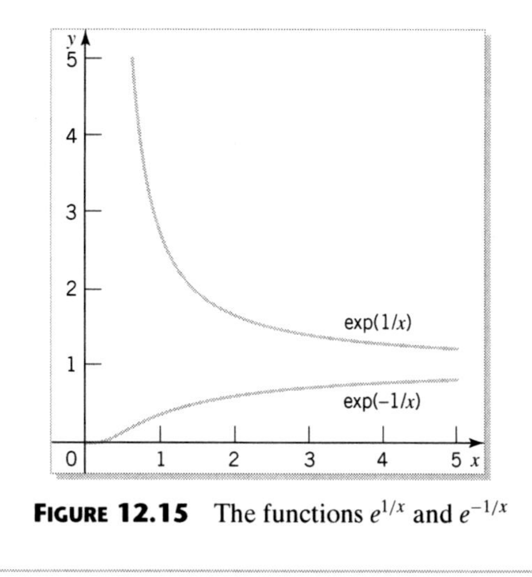

FIGURE 12.15 The functions $e^{1/x}$ and $e^{-1/x}$

Why did this technique work for the first two differential equations with singular points at $x = 0$, but not for the third in (12.47)? To answer that question, we recall that, in Exercise 12 of Section 8.4, we showed that if $y_1(x)$ and $y_2(x)$ are linearly independent solutions of a second order linear differential equation $y'' + p(x)y' + q(x)y = 0$, then[9]

$$p(x) = -\frac{y_1 y_2'' - y_1'' y_2}{y_1 y_2' - y_1' y_2}, \qquad \text{and} \qquad q(x) = \frac{y_1' y_2'' - y_1'' y_2'}{y_1 y_2' - y_1' y_2}.$$

If we look for solutions of the form $y(x) = x^s \sum_{n=0}^{\infty} c_n x^n$ — that is, if $y_1(x) = x^{s_1} \sum_{n=0}^{\infty} c_n x^n$ and $y_2(x) = x^{s_2} \sum_{n=0}^{\infty} d_n x^n$ — then

$$\begin{aligned}
y_1(x) &= c_0 x^{s_1} + \cdots \\
y_1'(x) &= c_0 s_1 x^{s_1 - 1} + \cdots \\
y_1''(x) &= c_0 s_1 (s_1 - 1) x^{s_1 - 2} + \cdots \\
y_2(x) &= d_0 x^{s_2} + \cdots \\
y_2'(x) &= d_0 s_2 x^{s_2 - 1} + \cdots \\
y_2''(x) &= d_0 s_2 (s_2 - 1) x^{s_2 - 2} + \cdots,
\end{aligned}$$

so,

$$\begin{aligned}
y_1 y_2' - y_1' y_2 &= c_0 d_0 (s_2 - s_1) x^{s_1 + s_2 - 1} + \cdots \\
y_1 y_2'' - y_1'' y_2 &= c_0 d_0 (s_2 - s_1)(s_1 + s_2 - 1) x^{s_1 + s_2 - 2} + \cdots \\
y_1' y_2'' - y_1'' y_2' &= c_0 d_0 s_1 s_2 (s_2 - s_1) x^{s_1 + s_2 - 3} + \cdots,
\end{aligned}$$

[9] Based on "An elementary analysis for motivating the definition of the regular singular point of the ordinary linear differential equation" by S. Scott, *Int. J. Math. Educ. Sci. Technol.* 17, 1986, pages 39–40.

giving

$$p(x) = (s_1 + s_2 - 1)x^{-1} + \cdots$$
$$q(x) = (s_2 - s_1)x^{-2} + \cdots.$$

This suggests that for this technique to work, the structure for $p(x)$ must be

$$p(x) = \frac{a_{-1}}{x} + a_0 + a_1 x + \cdots,$$

while

$$q(x) = \frac{b_{-2}}{x^2} + \frac{b_{-1}}{x} + b_0 + b_1 x + \cdots.$$

This requires that $xp(x)$ and $x^2q(x)$ must both be analytic about $x = 0$.

Thus, we need to subdivide singular points into two types: one type is called regular, where $xp(x)$ and $x^2q(x)$ are both analytic at $x = 0$, and the other type is called irregular, where they are not. This leads to the following definition.

◆ *Definition 12.3:* **Consider a differential equation in the form**

$$y'' + p(x)y' + q(x)y = 0, \tag{12.48}$$

which has $x = x_0$ as a singular point. If the functions

$$(x - x_0)p(x) \qquad \text{and} \qquad (x - x_0)^2 q(x) \tag{12.49}$$

are both analytic at x_0, then x_0 is called a REGULAR SINGULAR POINT of (12.48). If at least one of the products in (12.49) is not analytic at x_0, then x_0 is called an IRREGULAR SINGULAR POINT of (12.48). ◆

Notice that in (12.38) and (12.39), $x = 0$ is a regular singular point, whereas $x = 0$ is an irregular singular point of (12.47).

EXAMPLE 12.13

In Example 12.8 on page 595 we noted that Legendre's differential equation,

$$(1 - x^2)y'' - 2xy' + \lambda y = 0, \tag{12.50}$$

has singular points at 1 and −1. Let us check to see if these singular points are regular.

We first put (12.50) in standard form as

$$y'' - \frac{2x}{1 - x^2}y' + \frac{\lambda}{1 - x^2}y = 0.$$

Regular singular point

Because $(x - 1)p(x) = 2x/(1 + x)$ and $(x - 1)^2 q(x) = -\lambda(x - 1)/(x + 1)$ are both analytic at $x = 1$, then $x = 1$ is a regular singular point. A similar situation holds true for $x = -1$. [Don't forget in that case we would consider $(x + 1)p(x)$ and $(x + 1)^2 q(x)$.] ■

EXAMPLE 12.14

As an example with coefficients that are not rational functions of x, we consider

$$(\sin x)\,y'' + 3(x+1)y' - \frac{2}{x^2(x-4)^2}y = 0.$$

We want to find and classify the singular points of this differential equation.

In terms of the differential equation in standard form, we have $p(x) = 3(x+1)/\sin x$, and $q(x) = -2/[x^2(x-4)^2 \sin x]$. The singular points occur at $x = 0, x = 4$, and places where $\sin x = 0$; namely, $x = \pm k\pi, k = 0, 1, 2, \cdots$.

Singular point

Because $x^2 q(x) = -2/[(x-4)^2 \sin x]$ is not analytic at $x = 0, x = 0$ is an irregular singular point. Because $(x-4)p(x) = 3(x-4)(x+1)/\sin x$ and $(x-4)^2 q(x) = -2/(x^2 \sin x)$ are both analytic at $x = 4, x = 4$ is a regular singular point.

To determine the nature of the remaining singularities at $x = \pm k\pi$, $k = 1, 2, 3, \cdots$, we examine $(x - k\pi)p(x) = (x-k\pi)3(x+1)/\sin x$ and $(x-k\pi)^2 q(x) = (x-k\pi)^2(-2)/[x^2(x-4)^2 \sin x]$ near $x = k\pi$ (a similar analysis works if x is near $x = -k\pi$). We know that the quotient of analytic functions is analytic at all points, except possibly where the denominator is zero, so we need only check the behavior of both of these expressions near $x = k\pi, k = 1, 2, 3, \cdots$. By using either the local linearity property of the sine function near $x = k\pi$ or L'Hôpital's rule, we see that the limits as $x \to k\pi$ of both $(x-k\pi)p(x)$ and $(x-k\pi)^2 q(x)$ exist, being $3(k\pi+1)/(-1)^k$, and 0, respectively. Thus, we have regular singular points at $x = \pm k\pi, k = 1, 2, 3, \cdots$. ■

The Regular Singular Point Theorem

Knowing the nature of the singular point of a differential equation is important because of the following theorem.

▶ **Theorem 12.3: The Regular Singular Point Theorem** *Consider the second order linear differential equation in the standard form*

$$y'' + p(x)y' + q(x)y = 0, \tag{12.51}$$

where $x = x_0$ is a regular singular point. Let R denote the smallest radius of convergence of the functions $(x-x_0)p(x)$ and $(x-x_0)^2 q(x)$, and let s denote the larger root of the indicial equation. Then (12.51) has a solution of the form $y(x) = (x-x_0)^s \sum_{n=0}^{\infty} c_n (x-x_0)^n$, which converges for $0 < |x - x_0| < R$. ◀

Comments about the Regular Singular Point Theorem

- The indicial equation is a quadratic that will have two solutions, s_1 and s_2, where $s_1 \geq s_2$. This theorem says that if we use $s = s_1$ we are sure of one solution. Sometimes the other choice for s — that is, s_2 — generates a second linearly independent solution.
- Whether or not the choice s_2 for s produces a second linearly independent solution, we can always use the reduction of order technique from Section 8.2 to generate a second solution from the first.

- In general, the interval of convergence $0 < |x - x_0| < R$ does not contain $x = x_0$, so there are usually two disconnected intervals of convergence, $x_0 < x < x_0 + R$, and $x_0 - R < x < x_0$. If $x_0 = 0$, then these intervals are $0 < x < R$ and $-R < x < 0$. In all examples, we will concentrate on the interval $x_0 < x < x_0 + R$, which, in the case of $x_0 = 0$, reduces to $0 < x < R$.
- For cases where the roots of the indicial equation are not integers or ratios of odd integers, we need to replace $(x - x_0)^s$ in our solution with $|x - x_0|^s$.
- This theorem assumes that the roots of the indicial equation are real numbers, and in this book we restrict our attention to this assumption. However, these ideas can be extended to the case in which the indicial equation has complex roots.

Because the behavior of the series solution near a regular singular point depends so heavily on the root of the indicial equation, we seek ways of determining these roots without having to go through the details of trying a series solution. We list the main result here as a theorem; another technique is given as Exercise 4 on page 609.

Theorem 12.4: The Indicial Equation Theorem *If $x = x_0$ is a regular singular point of the second order linear differential equation*

$$a_2(x)y'' + a_1(x)y' + a_0(x)y = 0, \tag{12.52}$$

then the indicial equation is obtained by equating to zero the lowest power of $x - x_0$ that occurs when $(x - x_0)^s$ is substituted into (12.52).

Although we won't prove the Indicial Equation Theorem, we note that because x_0 is a regular singular point, $a_1(x)/a_2(x)$ has, at most, a factor of $1/(x - x_0)$, and $a_0(x)/a_2(x)$ has, at most, a factor of $1/(x - x_0)^2$. Thus, the indicial equation associated with a regular singular point will be a quadratic equation in s.

EXAMPLE 12.15

Find the roots of the indicial equation associated with a series solution about the regular singular point of

$$2xy'' + 6y' - 9xy = 0,$$

and find the radius of convergence of this series solution.

We can see that $x = 0$ is the only singular point for this differential equation. Because $xp(x) = 3$ and $x^2q(x) = -9x^2/2$ are both analytic at $x = 0$, we have a regular singular point there. To find the indicial equation, we substitute x^s into the differential equation to obtain

Regular singular point

$$2xs(s-1)x^{s-2} + 6sx^{s-1} - 9xx^s = 2s(s-1)x^{s-1} + 6sx^{s-1} - 9x^{s+1}.$$

Indicial equation

Here the lowest power of x is x^{s-1}, so the indicial equation is given by

$$2s(s-1) + 6s = 2s^2 + 4s = 0,$$

with roots of 0 and $-1/2$. From the Regular Singular Point Theorem, we have that the accompanying series solution for the larger root, $s = 0$, will converge for all values of x. This theorem tells us nothing about the series solution associated with the smaller root, $s = -1/2$. The way to find out if we have a solution for $s = -1/2$ is to try to compute the coefficients. This we do in the next section. ∎

EXAMPLE 12.16

Find the roots of the indicial equation associated with a series solution about the regular singular point $x = 0$ of

$$3x^2y'' + xy' - \frac{2}{2-x}y = 0, \tag{12.53}$$

and find the radius of convergence of this series solution.

Regular singular point

The singular point $x = 0$ is a regular singular point because both $xp(x) = 1/3$ and $x^2q(x) = -2/[3(2-x)]$ are analytic at $x = 0$. (Are there any other singular points?) To find the indicial equation, we substitute x^s into (12.53) and use the expansion

$$\frac{1}{1-a} = 1 + a + a^2 + \cdots$$

to find that

$$\frac{2}{2-x} = \frac{2}{2\left(1-\frac{x}{2}\right)} = 1 + \frac{x}{2} + \left(\frac{x}{2}\right)^2 + \cdots.$$

This gives

$$3x^2s(s-1)x^{s-2} + xsx^{s-1} - \left[1 + \frac{x}{2} + \left(\frac{x}{2}\right)^2 + \cdots\right]x^s,$$

Indicial equation

so the indicial equation is

$$3s(s-1) + s - 1 = 3s^2 - 2s - 1 = 0.$$

Region of convergence

The roots of this indicial equation are 1 and $-1/3$, and from the Regular Singular Point Theorem, the resulting series solution corresponding to $s = 1$ converges at least for the interval $0 < |x| < 2$. (Where does this interval come from?) ∎

EXAMPLE 12.17

Find the roots of the indicial equation associated with a series solution about the regular singular point of

$$(x-3)\,y'' + 2xy' + 5y = 0, \tag{12.54}$$

and find the radius of convergence of this series solution.

The only singular point in this differential equation is $x = 3$. We see that it is a regular singular point because both $(x-3)\,p(x) = 2x$ and $(x-3)^2\,q(x) = 5\,(x-3)$ are analytic at $x = 3$. To find the indicial equation, we substitute $(x-3)^s$ into (12.54) and obtain

Regular singular point

$$(x-3)\,s\,(s-1)\,(x-3)^{s-2} + (2x)\,s\,(x-3)^{s-1} + 5\,(x-3)^s\,. \tag{12.55}$$

To find the coefficient of the lowest power of $(x-3)$, we must express each term in (12.55) as a polynomial in $(x-3)$. Thus, we write $x = (x-3)+3$, and the middle term in (12.55) in the form

$$(2x)\,s\,(x-3)^{s-1} = 2\,[(x-3)+3]\,s\,(x-3)^{s-1} = 2s\,(x-3)^s + 6x\,(x-3)^{s-1}\,.$$

This means that (12.55) may be rewritten in the form

$$s\,(s-1)\,(x-3)^{s-1} + 2s\,(x-3)^s + 6x\,(x-3)^{s-1} + 5\,(x-3)^s\,,$$

Indicial equation

so the indicial equation is

$$s\,(s-1) + 6s = s^2 + 5s = 0.$$

Region of convergence

Here the roots of the indicial equation are 0 and −5. From the Regular Singular Point Theorem, we have that the series associated with $s = 0$ will converge for all values of x, except possibly $x = 3$. ■

In the next two sections we will use this technique to solve differential equations that have regular singular points.

EXERCISES

1. Determine the regular and irregular singular points for the following differential equations.

   (a) $x^2\left(x^2-1\right)^2 y'' + \frac{x(x+1)}{x-4} y' + \frac{3(x-1)}{x^2-16} y = 0$

   (b) $x\left(x^2-3x-10\right) y'' + \frac{x+4}{x-2} y' + 16y = 0$

   (c) $x \sin x\, y'' + \frac{3(x-1)}{x+1} y' + \cos x\, y = 0$

   (d) $\sin x\, y'' + \frac{x\cos x}{x+1} y' - \frac{x^2}{x-2} y = 0$

   (e) $x^2\left(x^2-16\right) y'' + \frac{(x-2)}{x+2} y' + 32y = 0$

   (f) $e^x y'' + 3xy' + \frac{1}{1-e^x} y = 0$

   (g) $x\,(x-1)\, y'' + \frac{x+1}{(x-4)^2} y' + \frac{1}{x^2} y = 0$

   (h) $\left(x^2+x-6\right) y'' + \frac{14}{1-x^2} y' + \frac{12}{1+x^2} y = 0$

2. The purpose of this exercise is to find the general solution of (12.47); namely,

$$x^4 y'' + 2x^3 y' - y = 0. \tag{12.56}$$

   (a) Show that the change of independent variable from x to t, where $x = 1/t$, converts the differential equation (12.56) into $\frac{d^2y}{dt^2} - y = 0$.

   (b) Solve the differential equation in part (a), finding $y = y(t)$.

   (c) Substitute $t = 1/x$ into the solution $y = y(t)$ of part (b) to show that the solution of (12.56) is $y = C_1 e^{1/x} + C_2 e^{-1/x}$.

3. Find the indicial equation and the minimum interval of convergence of series solutions about the regular singular points of the following differential equations.

(a) $4xy'' + 2(1+x)y' + y = 0$
(b) $3xy'' + y' - y = 0$
(c) $4xy'' + 2y' + y = 0$
(d) $2x^2y'' + (2x^2 + x)y' + 2xy = 0$
(e) $2x^2y'' - xy' + (x-5)y = 0$
(f) $2(x-1)y'' - y' + e^x y = 0$
(g) $2(x+1)^2 y'' - 3(x^2 + 3x + 2)y' + (3x+5)y = 0$
(h) $xy'' + (x-1)y' - [2/(1-2x)]y = 0$
(i) $3(x-1)y'' + y' - y = 0$
(j) $(x+1)y'' - (x+4)y' + 2y = 0$
(k) $(x-1)^2y'' + (3x^2 - 4x + 1)y' - 2y = 0$
(l) $9x^2y'' + 9x^2y' + 2y = 0$

4. An alternative method of finding the indicial equation applies to the case in which we express our second order linear differential equation with a regular singular point at $x = x_0$ as $(x-x_0)^2 b_2(x)y'' + (x-x_0)b_1(x)y' + b_0(x)y = 0$, where $b_2(x_0) = \alpha \neq 0$. In this case the indicial equation is $\alpha s(s-1) + \beta s + \gamma = 0$, where $\beta = b_1(x_0)$ and $\gamma = b_0(x_0)$. Use this result to find the indicial equation for the differential equations in Exercise 3.

12.4 SOLUTIONS USING THE METHOD OF FROBENIUS

In this section we develop solutions of second order linear differential equations by using a series expression of the form

$$y(x) = (x-x_0)^s \sum_{n=0}^{\infty} c_n (x-x_0)^n, \tag{12.57}$$

where x_0 is a regular singular point of the differential equation. In the previous section we discovered how to obtain the indicial equation for the evaluation of the index s. In this section we determine the coefficients. The method of determining solutions of the form (12.57) is called the METHOD OF FROBENIUS.

Three Examples

EXAMPLE 12.18

Find the general solution of

$$4xy'' + 2y' + y = 0 \tag{12.58}$$

as a series about the origin using the method of Frobenius.

Regular singular point

We see that $x = 0$ is a regular singular point, and, using the Regular Singular Point Theorem, we see that the series solution we obtain will converge for all values of x except possibly $x = 0$. If we substitute the series in (12.57), with $x_0 = 0$, into this differential equation, we obtain

$$\sum_{n=0}^{\infty} 4(n+s)(n+s-1)c_n x^{n+s-1} + \sum_{n=0}^{\infty} 2(n+s)c_n x^{n+s-1} + \sum_{n=0}^{\infty} c_n x^{n+s} = 0.$$

We now raise the index of summation in the first two terms by 1 (that is, we let $n-1=m$) and find

$$4s(s-1)c_0x^{s-1}+2sc_0x^{s-1}+\sum_{m=0}^{\infty}[4(m+1+s)(m+s)c_{m+1}+2(m+1+s)c_{m+1}]x^{m+s}+$$
$$+\sum_{n=0}^{\infty}c_nx^{n+s}=0. \tag{12.59}$$

Indicial equation

The indicial equation is obtained by setting the coefficient of the lowest power of x to 0, giving

$$[4s(s-1)+2s]c_0=0.$$

Because $c_0 \neq 0$, we have

$$4s^2-2s=2s(2s-1)=0,$$

so this gives $1/2$ and 0 as our two values of s.

Recurrence relation

The recurrence relation is obtained by setting the coefficients of the remaining powers of x in (12.59) to 0. This gives the first order recurrence relation

$$[4(m+1+s)(m+s)+2(m+1+s)]c_{m+1}+c_m=0, \qquad m=0,1,2,\cdots. \tag{12.60}$$

If we use the larger value of s, $s=1/2$, in this recurrence relation, we obtain

$$(2m+3)(2m+2)c_{m+1}+c_m=0, \qquad m=0,1,2,\cdots,$$

or

$$(2m+3)\,(2m+2)\,c_{m+1}=-c_m.$$

This can be solved by rewriting it as (see page 586)

$$(-1)^{m+1}\,(2m+3)!c_{m+1}=(-1)^m\,(2m+1)!c_m,$$

so $(-1)^{m+1}\,(2m+3)!c_{m+1}=c_0$, or

$$c_{m+1}=\frac{(-1)^{m+1}}{(2m+3)!}c_0, \qquad m=0,1,2,\cdots.$$

One solution

This gives the solution of (12.58) corresponding to $s=1/2$ as

$$y_1(x)=\sqrt{x}\left[c_0+\sum_{m=0}^{\infty}\frac{(-1)^{m+1}}{(2m+3)!}x^{m+1}c_0\right]=c_0\sqrt{x}\sum_{m=0}^{\infty}\frac{(-1)^m}{(2m+1)!}x^m$$
$$=c_0\sum_{m=0}^{\infty}\frac{(-1)^m}{(2m+1)!}x^{m+1/2},$$

where we have restricted ourselves to the case $x > 0$. This series can be written in the form

$$y_1(x) = c_0 \sum_{m=0}^{\infty} \frac{(-1)^m}{(2m+1)!} \left(x^{1/2}\right)^{2m+1},$$

which we recognize converges to $y_1(x) = c_0 \sin \sqrt{x}$.

Reduction of order

Because we have found one solution, we can now use reduction of order techniques to obtain a second solution. Recall from Section 8.2 that when we know one solution; say, $y_1(x)$ with $c_0 = 1$, of $a_2(x)y'' + a_1(x)y' + a_0(x)y = 0$, then the second one is

$$y_2(x) = y_1 \int \left[\frac{1}{y_1^2} \exp\left(-\int \frac{a_1}{a_2} dx \right) \right] dx.$$

In our example, $a_1(x)/a_2(x) = 1/(2x)$, so $\int [a_1(x)/a_2(x)]\, dx = (\ln x)/2$. This means that

$$y_2(x) = y_1 \int \frac{1}{y_1^2} \exp\left(-\frac{1}{2} \ln x \right) dx$$
$$= y_1(x) \int \frac{1}{\sqrt{x} y_1^2} dx,$$

or

$$y_2(x) = \sin \sqrt{x} \int \frac{1}{\sqrt{x} \sin^2 \sqrt{x}} dx.$$

By using the substitution $u = \sqrt{x}$, this integral can be evaluated, giving rise to $y_2(x) = -2\cos \sqrt{x}$. Thus, the general solution may be written

General solution

$$y(x) = C_1 \sin \sqrt{x} + C_2 \cos \sqrt{x}.$$

Although this is the solution we seek, let's see what happens if we use the other solution of the indicial equation ($s = 0$) in the recurrence relation of (12.60). Doing so gives

$$[4(m+1)(m) + 2(m+1)]c_{m+1} + c_m = (2m+2)(2m+1)c_{m+1} + c_m = 0, \quad m = 0, 1, 2, \cdots.$$

Recurrence relation

This recurrence relation,

$$(2m+2)(2m+1)\, c_{m+1} = -c_m,$$

can be solved by rewriting it as

$$(-1)^{m+1} (2m+2)! c_{m+1} = (-1)^m (2m)! c_m,$$

so

$$c_{m+1} = \frac{(-1)^{m+1}}{(2m+2)!} c_0, \qquad m = 0, 1, 2, \cdots.$$

Thus, we have the series solution corresponding to $s = 0$ as

$$c_0 + \sum_{m=0}^{\infty} \frac{(-1)^{m+1}}{(2m+2)!} x^{m+1} c_0 = c_0 \sum_{m=0}^{\infty} \frac{(-1)^m}{(2m)!} x^m,$$

which we recognize as the series expansion for the function $c_0 \cos \sqrt{x}$.

Figure 12.16 shows the graphs of $\sin \sqrt{x}$ and $\cos \sqrt{x}$. Notice how the roots are interlaced.

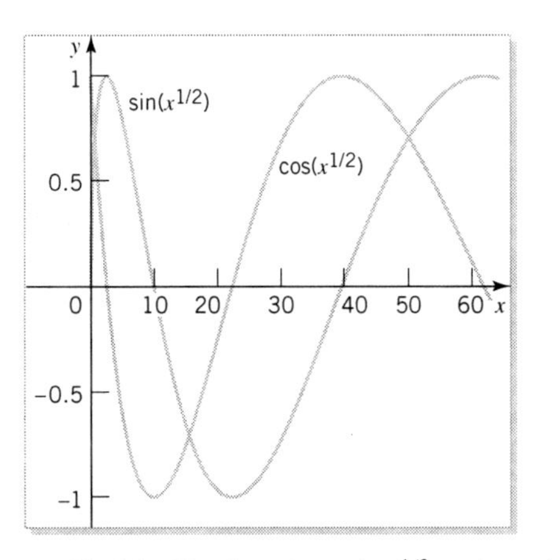

FIGURE 12.16 The functions $\sin x^{1/2}$ and $\cos x^{1/2}$

Comments about Using Both Roots of the Indicial Equation

- This example, and the previous ones, suggest that we can always obtain the general solution of a differential equation by using both roots of the indicial equation. Unfortunately, this is not always true even if the roots are distinct, because sometimes the smaller root does not yield a series solution different from the first. For example, $2xy'' + 6y' - 9xy = 0$ has two distinct roots of the indicial equation (0 and -2), and each gives the same solution (see Exercise 2 on page 628).

EXAMPLE 12.19 *Bessel's Differential Equation of Order 1/2*

Bessel's differential equation occurs in many applications, such as vibrations of circular membranes, and radiation from right circular cylinders, and is given by

$$x^2 y'' + xy' + (x^2 - \mu^2) y = 0, \tag{12.61}$$

where μ is a constant. On physical grounds we are usually interested only in $x > 0$.

Solutions of this differential equation are called Bessel functions of order μ. It is straightforward to verify that $x = 0$ is a regular singular point of (12.61). In this example, we solve (12.61) for the special case where $\mu = 1/2$, namely,

Regular singular point

$$x^2 y'' + xy' + \left(x^2 - \frac{1}{4}\right) y = 0. \tag{12.62}$$

Because $x = 0$ is a regular singular point of (12.62), we substitute the proper form of a series solution, $y(x) = x^s \sum_{n=0}^{n=\infty} c_n x^n$, into (12.62) and obtain

$$\sum_{n=0}^{\infty}(n+s)(n+s-1)c_n x^{n+s} + \sum_{n=0}^{\infty}(n+s)c_n x^{n+s} + \sum_{n=0}^{\infty} c_n x^{n+s+2} - \frac{1}{4}\sum_{n=0}^{\infty} c_n x^{n+s} = 0. \tag{12.63}$$

Indicial equation

The indicial equation is found by setting the coefficient of the lowest power of x — namely, x^s — to zero, and is given by

$$s(s-1) + s - \frac{1}{4} = s^2 - \frac{1}{4} = 0.$$

This gives $1/2$ and $-1/2$ as roots of the indicial equation.

In (12.63) we see that three of the series have the same power of x, so we lower the index of summation in the remaining series to obtain

$$\left(s^2 - \frac{1}{4}\right) c_0 x^s + \left[(s+1)^2 - \frac{1}{4}\right] c_1 x^{s+1} + \sum_{n=2}^{\infty}\left[(n+s)^2 - \frac{1}{4}\right] c_n x^{n+s} + \sum_{n=2}^{\infty} c_{n-2} x^{n+s} = 0.$$

If we now consider the smaller root of the indicial equation — namely, $s = -1/2$ — we see that the term multiplying c_0 is zero. (Explain why this must be.) The term multiplying c_1 is also 0, giving both c_0 and c_1 as arbitrary constants. The remaining coefficients are obtained from the second order recurrence relation,

Recurrence relation

$$c_n = \frac{-1}{(n-1/2)^2 - 1/4} c_{n-2} = \frac{-1}{n(n-1)} c_{n-2}, \qquad n = 2, 3, 4, \cdots.$$

or

$$n\,(n-1)\, c_n = -c_{n-2}, \qquad n = 2, 3, 4, \cdots.$$

This has already been solved in (12.18) once we realize that it is equivalent to

$$(n+2)\,(n+1)\, c_{n+2} = -c_n, \qquad n = 0, 1, 2, \cdots,$$

and we have

$$c_{2m} = \frac{(-1)^m}{(2m)!} c_0, \text{ and } c_{2m+1} = \frac{(-1)^m}{(2m+1)!} c_1.$$

General solution

Thus, the general solution of (12.62) may be written as

$$y(x) = x^{-1/2}[c_0 + c_0 \sum_{m=1}^{\infty} \frac{(-1)^m}{(2m)!} x^{2m} + c_1 x + c_1 \sum_{m=1}^{\infty} \frac{(-1)^m}{(2m+1)!} x^{2m+1}].$$

A more compact way of writing these series is

$$y(x) = c_0 x^{-1/2} \sum_{m=0}^{\infty} \frac{(-1)^m}{(2m)!} x^{2m} + c_1 x^{-1/2} \sum_{m=0}^{\infty} \frac{(-1)^m}{(2m+1)!} x^{2m+1}. \qquad (12.64)$$

We note that the two series in (12.64) seem familiar. In fact, they are the Taylor expansions for the cosine and sine functions, so we may write (12.64) as

$$y(x) = c_0 \frac{1}{\sqrt{x}} \cos x + c_1 \frac{1}{\sqrt{x}} \sin x.$$

Because c_0 and c_1 are undetermined, and because $x^{-1/2} \cos x$ and $x^{-1/2} \sin x$ are linearly independent, this is the general solution.

Solutions of Bessel's differential equation with parameter $\mu = 1/2 + m$, where m is an integer, and with specific values for the constants c_0 and c_1, are called SPHERICAL BESSEL FUNCTIONS. Choosing $c_0 = \sqrt{2/\pi}$ and $c_1 = 0$, and then $c_0 = 0$ and $c_1 = \sqrt{2/\pi}$, gives the spherical Bessel functions

$$J_{-1/2}(x) = \sqrt{\frac{2}{\pi x}} \cos x \text{ and } J_{1/2}(x) = \sqrt{\frac{2}{\pi x}} \sin x.$$

These functions are plotted in Figure 12.17. Notice how the roots are interlaced as predicted by the Oscillation Theorem and Relation Theorem (see Exercise 3 of Section 8.1).

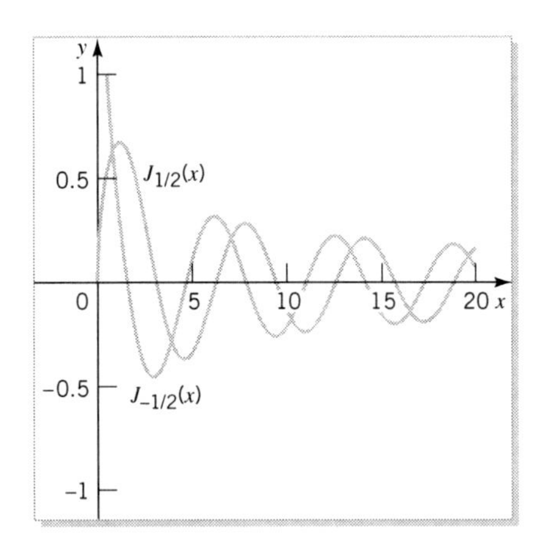

FIGURE 12.17 The Bessel functions of order 1/2

Comments about Using One Root of the Indicial Equation

- It may have come as a surprise that we obtained the general solution of our differential equation by considering only one root of the indicial equation. It turns out that in cases where the two roots of the indicial equation differ by an integer, choosing the smaller root of the indicial equation sometimes gives the general solution. For example, we consider the case where the two roots of the indicial equation are -1 and 1. Taking the smaller root, our solution would have the form

$$c_0 x^{-1} + c_1 + c_2 x + c_3 x^2 + c_4 x^3 + \cdots.$$

If, from the recurrence relation associated with $s = -1$, we find that c_1 is determined in terms of c_0, and that c_3, c_4, and so on are determined in terms of c_2, then c_0 and c_2 are the arbitrary constants, and we have found the general solution. The series that is the coefficient of c_2 would have been the same series we would have obtained if we had used the larger root of the indicial equation $s = 1$. Unfortunately, this situation does not always occur.

EXAMPLE 12.20

Find the first few terms in the series solution about 0 of the differential equation from Example 12.16 on page 607,

$$3x^2 y'' + xy' - \frac{2}{2-x} y = 0. \tag{12.65}$$

Regular singular point

In Example 12.16 we discovered that $x = 0$ is a regular singular point, and that a series solution will converge at least for $-2 < x < 2$, except possibly at $x = 0$. We also determined roots of the indicial equation as 1 and $-1/3$.

Let us now find the recurrence relation by substituting the series expression $x^s \sum_{n=0}^{\infty} c_n x^n$ into (12.65) and using the expansion

$$\frac{2}{2-x} = \frac{2}{2(1-x/2)} = 1 + \frac{x}{2} + \left(\frac{x}{2}\right)^2 + \cdots.$$

This gives

$$\sum_{n=0}^{\infty} 3(n+s)(n+s-1)c_n x^{n+s} + \sum_{n=0}^{\infty} (n+s)c_n x^{n+s} - \left[1 + \frac{x}{2} + \left(\frac{x}{2}\right)^2 + \cdots\right] \sum_{n=0}^{\infty} c_n x^{n+s} = 0,$$

or

$$\sum_{n=0}^{\infty} (n+s)(3n+3s-2)c_n x^{n+s} - \left[1 + \frac{x}{2} + \left(\frac{x}{2}\right)^2 + \cdots\right] \sum_{n=0}^{\infty} c_n x^{n+s} = 0.$$

Indicial equation

We see that the coefficient of the lowest power of x, $s(3s-2) - 1$, equals zero for $s = -1/3$ and $s = 1$. (Explain why.) Equating the coefficients of the next few powers of x to zero gives

$$(1+s)(3+3s-2)c_1 - \left(c_1 + \frac{1}{2}c_0\right) = 0,$$

$$(2+s)(6+3s-2)c_2 - \left(c_2 + \frac{1}{2}c_1 + \frac{1}{4}c_0\right) = 0,$$

$$(3+s)(9+3s-2)c_3 - \left(c_3 + \frac{1}{2}c_2 + \frac{1}{4}c_1 + \frac{1}{8}c_0\right) = 0.$$

For both values of s, there will no condition specified for c_0, so it is an arbitrary constant, and all the other constants will be obtained in terms of c_0. For the larger root, $s = 1$, we have

$$c_1 = \frac{1}{14}c_0,\ c_2 = \frac{1}{5 \cdot 14}c_0,\ c_3 = \frac{1}{260}c_0,$$

so the first four terms in this part of our solution are

$$x\left(1 + \frac{1}{14}x + \frac{1}{70}x^2 + \frac{1}{260}x^3\right)c_0.$$

Similar calculations for the smaller root, $s = -1/3$, lead to the solution of the form

$$x^{-1/3}\left(1 - \frac{1}{2}x\right)c_0.$$

These two series are not proportional, and so they must be linearly independent.

General solution

Thus, the general solution of (12.65) has the form

$$y(x) = C_1 x\left(1 + \frac{1}{14}x + \frac{1}{70}x^2 + \frac{1}{260}x^3 + \cdots\right) + C_2 x^{-1/3}\left(1 - \frac{1}{2}x\right).$$

In most of the examples we have looked at in this section, using the two roots of the indicial equation gave series solutions in the form of two linearly independent functions. However, this is not always the case. We now consider such an example.

An Example

EXAMPLE 12.21

Find the general solution of

$$x(x-1)y'' + (5x-2)y' + 4y = 0 \tag{12.66}$$

as a series about the origin using the method of Frobenius.

If we divide this expression by $x(x-1)$, we obtain

$$y'' + \frac{5x-2}{x(x-1)}y' + \frac{4}{x(x-1)}y = 0.$$

Regular singular point

Here we see that both $xp(x) = x(5x-2)/\left[x(x-1)\right] = (5x-2)/(x-1)$ and $x^2q(x) = x^2 4/\left[x(x-1)\right] = 4x/(x-1)$ are analytic at $x = 0$, giving the origin as a regular singular point. Because the radius of convergence for both analytic functions is $R = 1$, we know our solution will be valid for $0 < x < 1$.

To find the indicial equation, we substitute $y = x^s$ into the left-hand side of (12.66) and obtain $(x^2 - x)s(s-1)x^{s-2} + (5x - 2)sx^{s-1} + 4x^s = s(s-1)x^s - s(s-1)x^{s-1} + 5sx^s - 2sx^{s-1} + 4x^s$.

Indicial equation

From this equation we see that the indicial equation (obtained by setting the coefficient of the lowest power of s to 0) is

$$-s(s-1) - 2s = -s^2 - s = -s(s+1) = 0.$$

Thus, we have $s = 0$ and $s = -1$ as the roots, and the method of Frobenius guarantees us one solution if we use the larger root $s = 0$.

We substitute the series $y(x) = x^0 \sum_{n=0}^{\infty} c_n x^n = \sum_{n=0}^{\infty} c_n x^n$ into (12.66) and find that

$$(x^2 - x) \sum_{n=0}^{\infty} n(n-1)c_n x^{n-2} + (5x - 2) \sum_{n=0}^{\infty} nc_n x^{n-1} + 4 \sum_{n=0}^{\infty} c_n x^n = 0,$$

or

$$\sum_{n=0}^{\infty} n(n-1)c_n x^n - \sum_{n=0}^{\infty} n(n-1)c_n x^{n-1} + \sum_{n=0}^{\infty} 5nc_n x^n - \sum_{n=0}^{\infty} 2nc_n x^{n-1} + \sum_{n=0}^{\infty} 4c_n x^n = 0.$$

If we now combine series with like exponents, we obtain

$$\sum_{n=0}^{\infty} [n(n-1) + 5n + 4]c_n x^n - \sum_{n=0}^{\infty} [n(n-1) + 2n]c_n x^{n-1}$$
$$= \sum_{n=0}^{\infty} (n+2)^2 c_n x^n - \sum_{n=0}^{\infty} n(n+1) c_n x^{n-1} = 0.$$

We now change the dummy index in the second series and obtain

$$\sum_{n=0}^{\infty} (n+2)^2 c_n x^n - \sum_{n=0}^{\infty} (n+1)(n+2) c_{n+1} x^n$$
$$= \sum_{n=0}^{\infty} (n+2) [(n+2)c_n - (n+1) c_{n+1}] x^n = 0.$$

Recurrence relation

This gives the first order recurrence relation as

$$(n+1)c_{n+1} - (n+2) c_n = 0, \qquad n = 0, 1, 2, \cdots.$$

This recurrence relation can be solved if we write it in the form

$$\frac{1}{(n+2)} c_{n+1} = \frac{1}{(n+1)} c_n,$$

which gives

$$\frac{1}{(n+2)}c_{n+1} = \frac{1}{(n+1)}c_n = \frac{1}{n}c_{n-1} = \cdots = c_0,$$

One solution

so $c_{n+1} = (n+2)\,c_0$, and our solution is

$$y_1(x) = c_0\left(1 + 2x + 3x^2 + 4x^3 + \cdots\right).$$

It is possible to express this series in terms of familiar functions, as follows. Because

$$\frac{1}{1-x} = 1 + x + x^2 + x^3 + x^4 + \cdots, \qquad \text{for } |x| < 1,$$

by differentiating we obtain

$$\frac{1}{(1-x)^2} = 1 + 2x + 3x^2 + 4x^3 + \cdots,$$

so our first solution is simply

$$y_1(x) = \frac{1}{(1-x)^2}.$$

Exercise 9 on page 629 shows that the smaller root of the indicial equation, $s = -1$, does not give a second solution, but gives us the same $y_1(x)$ again. However, because we have obtained one solution, we now use reduction of order techniques to obtain a second solution. Recall from Section 8.2 that when we know one solution — say, $y_1(x)$ — of $a_2(x)y'' + a_1(x)y' + a_0(x)y = 0$, then the second one is

Reduction of order

$$y_2(x) = y_1 \int \left[\frac{1}{y_1^2}\exp\left(-\int \frac{a_1}{a_2}\,dx\right)\right] dx.$$

In our example, $a_1(x)/a_2(x) = (5x-2)\,/\,[x\,(x-1)]$, so

$$\begin{aligned}\int \frac{a_1}{a_2}\,dx &= \int \frac{5x-2}{x\,(x-1)}\,dx \\ &= \int \frac{2}{x}\,dx - \int \frac{3}{1-x}\,dx \\ &= 2\ln x + 3\ln(1-x), \qquad \text{if } 0 < x < 1, \\ &= \ln\left[x^2(1-x)^3\right].\end{aligned}$$

This means that

$$\begin{aligned}y_2(x) &= y_1(x)\int \frac{1}{y_1^2}\exp\left(-\ln\left[x^2(1-x)^3\right]\right) dx \\ &= \frac{1}{(1-x)^2}\int \frac{(1-x)^4}{x^2(1-x)^3}\,dx = \frac{1}{(1-x)^2}\int \frac{1-x}{x^2}\,dx,\end{aligned}$$

or

$$y_2(x) = \frac{1}{(1-x)^2}\left(-\frac{1}{x} - \ln x\right). \tag{12.67}$$

General solution Thus, the general solution of (12.66) is $y(x) = C_1y_1(x) + C_2y_2(x)$, so that

$$y(x) = C_1\frac{1}{(1-x)^2} - C_2\left[\frac{1}{(1-x)^2}\ln x + \frac{1}{x}\frac{1}{(1-x)^2}\right]. \tag{12.68}$$

The functions $y_1(x)$ and $-y_2(x)$ are plotted in Figure 12.18.

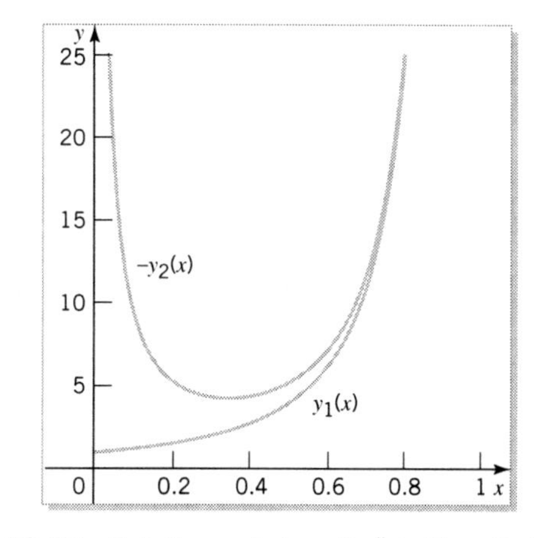

FIGURE 12.18 Solutions of $x(x-1)y'' + (5x-2)y' + 4y = 0$

We used reduction of order because we could not obtain a second series solution from the second root. However, because both of these solutions usually will be series, we expand $y_2(x)$ so we can see its form as an infinite series. From our previous calculations, we have

$$\frac{1}{(1-x)^2} = 1 + 2x + 3x^2 + 4x^3 + \cdots,$$

so we can write our solution as

$$y(x) = C_1y_1(x) - C_2\left[y_1(x)\ln x + x^{s_2}\left(1 + 2x + 3x^2 + 4x^3 + \cdots\right)\right], \tag{12.69}$$

where s_2 is the smaller root of the indicial equation, and $y_1(x)$ is the solution corresponding to the larger root, s_1. ■

A similar situation, in which we obtain only one solution from $y(x) = x^s\sum_{n=0}^{\infty} c_nx^n$, ***Indicial equation*** arises if the indicial equation has equal roots. For example (see Exercise 11 on page 629), the differential equation $x^2y'' + (x^2 - 3x)y' + (-2x+4)y = 0$ has $s_1 =$

$s_2 = 2$ as the repeated root of the indicial equation associated with the regular singular point $x_0 = 0$. Using $s = 2$, we find one solution as $y_1(x) = x^2$. The second solution can be obtained by reduction of order, and we have

Reduction of order

$$y_2(x) = x^2 \ln x + x^2 \sum_{n=1}^{\infty} \frac{(-1)^n}{n!n} x^n.$$

This general solution can be written in a form similar to (12.69); namely,

$$y(x) = C_1 y_1(x) + C_2 \left[y_1(x) \ln x + x^s \sum_{n=1}^{\infty} \frac{(-1)^n}{n!n} x^n \right].$$

The Frobenius Theorem

The reduction of order procedure will yield a second solution as long as we can find one solution of the differential equation. The last two examples gave simple forms of the second solution because the simple forms of $y_1(x)$; namely, $1/(x-1)^2$, and x^2, made it very easy to divide by $[y_1(x)]^2$ before integrating. In general, when we use the method of Frobenius, our first solution will be in the form of an infinite series that we cannot express as a familiar function. Although we can divide by an infinite series to find the second solution by reduction of order, it is not the simplest of operations. What we can do is rely on the following theorem, which gives an alternative to using reduction of order to obtain the second solution.

▶ Theorem 12.5: The Frobenius Theorem *Consider the second order differential equation*

$$a_2(x)y'' + a_1(x)y' + a_0(x)y = 0, \tag{12.70}$$

with $x = 0$ as a regular singular point. Let s_1 and s_2 be solutions of the indicial equation, with $s_1 \geq s_2$. Then there are two linearly independent solutions of (12.70) — namely, $y_1(x)$ and $y_2(x)$ — whose series converge for at least $0 < x < R$, where R is the smaller radius of convergence for $xa_1(x)/a_2(x)$ and $x^2a_0(x)/a_2(x)$.[10] *For all cases,*

$$y_1(x) = x^{s_1} \sum_{n=0}^{\infty} c_n x^n,$$

while the form for $y_2(x)$ depends on the relationship between the roots of the indicial equation, as follows.
(a) If $s_1 - s_2 \neq$ integer, then

$$y_2(x) = x^{s_2} \sum_{n=0}^{\infty} c_n x^n. \tag{12.71}$$

[10] Similar results hold for $-R < x < 0$.

(b) If $s_1 = s_2$, then

$$y_2(x) = y_1(x)\ln x + x^{s_1}\sum_{n=1}^{\infty} c_n x^n. \tag{12.72}$$

(c) If $s_1 - s_2 =$ integer, then

$$y_2(x) = Cy_1(x)\ln x + x^{s_2}(1 + \sum_{n=1}^{\infty} c_n x^n), \tag{12.73}$$

where, depending on the differential equation, the constant C may or may not be zero. ◀

Comments about the Frobenius Theorem

- If the regular singular point is at x_0, use the translation $X = x - x_0$ to move the singularity to the origin of the X-axis.
- Note that for the Bessel functions in Example 12.19 on page 612, the indicial equation had roots $1/2$ and $-1/2$, which differ by an integer. Thus, the second solution, corresponding to $s = -1/2$, has the form of (12.73) with $C = 0$.
- Note that the structure of the second solution in (12.68) has the form of (12.73), with $C = 1$.
- It is worthwhile for you to return to all the examples in Sections 12.3 and 12.4 to see exactly to which case each example belongs.

How to Use the Method of Frobenius

Purpose To find a series solution of

$$a_2(x)y'' + a_1(x)y' + a_0(x)y = 0 \tag{12.74}$$

of the form $x^s\sum_{n=0}^{\infty} c_n x^n$, where $x = 0$ is a regular singular point of (12.74).

Process

1. Verify that 0 is a regular singular point of (12.74), and use the Regular Singular Point Theorem to determine the minimum radius of convergence of the series solution we are seeking.
2. Determine the indicial equation and solve for its two roots, s_1 and s_2, where $s_1 \geq s_2$. The indicial equation may be determined by using the Indicial Equation Theorem, or Exercise 4 on page 609, or by equating the lowest power of x to zero in the result of step 3 following.
3. Determine the recurrence relation. This is obtained by substituting the series expression $x^{s_1}\sum_{n=0}^{\infty} c_n x^n$ into (12.74) and equating the coefficient of each power of x to zero.

4. Use s_1 in the recurrence relation to find a general expression for c_n, or as many terms as desired, giving $y_1(x)$.
5. To find the second solution corresponding to s_2, there are three cases.
   (a) If $s_1 - s_2 \neq$ integer, then assume a solution of the form (12.71) and return to steps 3 and 4, using (12.71) in place of $x^{s_1} \sum_{n=0}^{\infty} c_n x^n$ and s_2 in place of s_1.
   (b) If $s_1 = s_2$, then assume a solution of the form (12.72) and return to steps 3 and 4, using (12.72) in place of $x^{s_1} \sum_{n=0}^{\infty} c_n x^n$.
   (c) If $s_1 - s_2 =$ integer, then assume a solution of the form (12.73) and return to steps 3 and 4, using (12.73) in place of $x^{s_1} \sum_{n=0}^{\infty} c_n x^n$. Depending on the differential equation, the constant C may or may not be zero.

Two Examples

In two previous examples we found a second solution to a differential equation using reduction of order. In our next example, we use the preceding results to find the second solution.

EXAMPLE 12.22

Find the general solution of

$$xy'' + y' + 4xy = 0 \tag{12.75}$$

as a series about the origin, using the method of Frobenius.

If we divide this expression by x, we obtain $y'' + \frac{1}{x}y' + 4y = 0$. Here we see that both $xp(x) = 1$ and $x^2 q(x) = 4x^2$ are analytic at $x = 0$, making the origin a regular singular point.

Regular singular point

To find the indicial equation, we substitute $y = x^s$ into the left-hand side of (12.75) and obtain

$$xs(s-1)x^{s-2} + sx^{s-1} + (4x)x^s = s(s-1)x^{s-1} + sx^{s-1} + 4x^{s+1}.$$

Indicial equation

From here we see that the indicial equation (obtained by setting the coefficient of the lowest power of x to 0) is

$$s(s-1) + s = s^2 = 0.$$

Thus, we have $s = 0$ as a double root, and we proceed to find the associated solution.

We substitute the series $y = \sum_{n=0}^{\infty} c_n x^n$ into (12.75) and find that

$$x \sum_{n=0}^{\infty} n(n-1)c_n x^{n-2} + \sum_{n=0}^{\infty} nc_n x^{n-1} + 4x \sum_{n=0}^{\infty} c_n x^n$$
$$= \sum_{n=0}^{\infty} n(n-1)c_n x^{n-1} + \sum_{n=0}^{\infty} nc_n x^{n-1} + \sum_{n=0}^{\infty} 4c_n x^{n+1} = 0.$$

If we now combine series with like exponents, we obtain

$$\sum_{n=0}^{\infty} n^2 c_n x^{n-1} + \sum_{n=0}^{\infty} 4c_n x^{n+1} = 0,$$

or

$$\sum_{n=1}^{\infty} n^2 c_n x^{n-1} + \sum_{n=0}^{\infty} 4c_n x^{n+1} = 0.$$

We now separate out the first term in the first series, and lower by 2 the index of summation in the second series, to obtain

$$c_1 x^0 + \sum_{n=2}^{\infty} n^2 c_n x^{n-1} + \sum_{n=2}^{\infty} 4c_{n-2} x^{n-1} = 0.$$

Recurrence relation

Because this equation must be an identity in x, we see that we must have $c_1 = 0$, and the recurrence relation becomes

$$n^2 c_n + 4c_{n-2} = 0, \qquad n = 2, 3, 4, \cdots.$$

We now rearrange the recurrence relation as

$$c_n = -\frac{4}{n^2} c_{n-2}, \qquad n = 2, 3, 4, \cdots.$$

Because $c_1 = 0$, all the odd coefficients equal 0. The even ones are obtained from

$$c_{2m} = \frac{-4}{(2m)^2} c_{2m-2}, \qquad m = 1, 2, 3, \cdots;$$

that is,

$$(2m)^2 c_{2m} = -4c_{2m-2}.$$

This can be solved if we write it in the form $m^2 c_{2m} = -c_{2m-2}$, or

$$(-1)^m (m!)^2 c_{2m} = (-1)^{m-1} [(m-1)!]^2 c_{2m-2},$$

which gives

$$(-1)^m (m!)^2 c_{2m} = (-1)^{m-1} [(m-1)!]^2 c_{2m-2} = (-1)^{m-2} [(m-2)!]^2 c_{2m-4} = \cdots = c_0,$$

so $(-1)^m (m!)^2 c_{2m} = c_0$, or

$$c_{2m} = \frac{(-1)^m}{(m!)^2} c_0, \qquad m = 1, 2, 3, \cdots.$$

One solution This gives one solution of our original differential equation as

$$y_1(x) = 1 + \sum_{m=1}^{\infty} \frac{(-1)^m}{(m!)^2} x^{2m} = \sum_{m=0}^{\infty} \frac{(-1)^m}{(m!)^2} x^{2m}, \tag{12.76}$$

where we have set $c_0 = 1$.

According to the Frobenius Theorem, our second solution will have the form

$$y_2(x) = y_1(x) \ln x + \sum_{n=1}^{\infty} c_n x^n, \tag{12.77}$$

where $y_1(x)$ is given by (12.76) [see (12.72)]. Because the exact form of $y_1(x)$ is not needed for the first steps in this procedure, we will not use its series expansion until necessary. We first differentiate (12.77) to obtain

$$y_2'(x) = y_1'(x) \ln x + \frac{1}{x} y_1(x) + \sum_{n=1}^{\infty} n c_n x^{n-1},$$

$$y_2''(x) = y_1''(x) \ln x + \frac{1}{x} 2y_1'(x) - \frac{1}{x^2} y_1(x) + \sum_{n=1}^{\infty} n(n-1) c_n x^{n-2}.$$

Substituting these values into (12.75) and rearranging terms gives

$$\left(xy_1'' + y_1' + 4xy_1\right) \ln x + 2y_1' + \sum_{n=1}^{\infty} n(n-1) c_n x^{n-1} + \sum_{n=1}^{\infty} n c_n x^{n-1} + \sum_{n=1}^{\infty} 4c_n x^{n+1} = 0.$$

Because $y_1(x)$ is a solution of (12.75), the coefficient of $\ln x$ is 0, and the rest of the expression is

$$2y_1' + \sum_{n=1}^{\infty} n^2 c_n x^{n-1} + \sum_{n=1}^{\infty} 4c_n x^{n+1} = 2y_1' + c_1 + 4c_2 x + \sum_{n=3}^{\infty} \left(n^2 c_n + 4c_{n-2}\right) x^{n-1} = 0,$$

where we combined the last two series as before.

We now differentiate our series expression for y_1 from (12.76) and use the result in the preceding equation to obtain

$$\sum_{m=1}^{\infty} (-1)^m \frac{4m}{(m!)^2} x^{2m-1} + c_1 + 4c_2 x + \sum_{n=3}^{\infty} \left(n^2 c_n + 4c_{n-2}\right) x^{n-1} = 0,$$

or

$$-4x + \sum_{m=2}^{\infty} (-1)^m \frac{4m}{(m!)^2} x^{2m-1} + c_1 + 4c_2 x + \sum_{n=3}^{\infty} \left(n^2 c_n + 4c_{n-2}\right) x^{n-1} = 0.$$

Recurrence relation Equating coefficients of like powers of x to zero gives us

$$c_1 = 0, \qquad -4 + 4c_2 = 0, \qquad n^2 c_n + 4c_{n-2} = 0, \quad n = 3, 5, 7, \cdots,$$

and

$$(-1)^m \frac{4m}{(m!)^2} + (2m)^2 c_{2m} + 4c_{2m-2} = 0, \quad m = 2, 3, 4, \cdots.$$

Thus, all the odd coefficients are equal to zero, and the first few even coefficients are given by

$$\begin{aligned} c_2 &= 1, \\ c_4 &= \frac{1}{16}\left(-4c_2 - \frac{8}{(2!)^2}\right) = -\frac{3}{8}, \\ c_6 &= \frac{1}{36}\left(-4c_4 + \frac{12}{(3!)^2}\right) = \frac{11}{216}, \\ c_8 &= \frac{1}{64}\left(-4c_6 - \frac{16}{(4!)^2}\right) = -\frac{25}{6912}. \end{aligned}$$

This gives our second solution the form

$$y_2(x) = y_1(x)\ln x + x^2 - \frac{3}{8}x^4 + \frac{11}{216}x^6 - \frac{25}{6912}x^8 + \cdots. \tag{12.78}$$

General solution The general solution of (12.75) is therefore given by $y(x) = C_1 y_1(x) + C_2 y_2(x)$, where C_1 and C_2 are arbitrary constants, and $y_1(x)$ and $y_2(x)$ are given by (12.76) and (12.78).

Because $xa_1(x)/a_2(x) = 1$ and $x^2 a_0 x/a_2(x) = 4x^2$ are both polynomials, both series in our solution converge for $0 < x < \infty$. Using the Oscillation Theorem and Relation Theorem of Section 8.1 with $Q(x) = 4 - 1/2(-1/x)^2 - 1/4(1/x)^2 = 4 + 1/(4x^2) > 0$, we know our solution oscillates and is bounded for both positive and negative values of x, something not possible to tell from the series solution. The first few terms in our solution for $y_1(x)$ are shown in Figure 12.19. Notice how the oscillatory behavior is beginning to emerge.

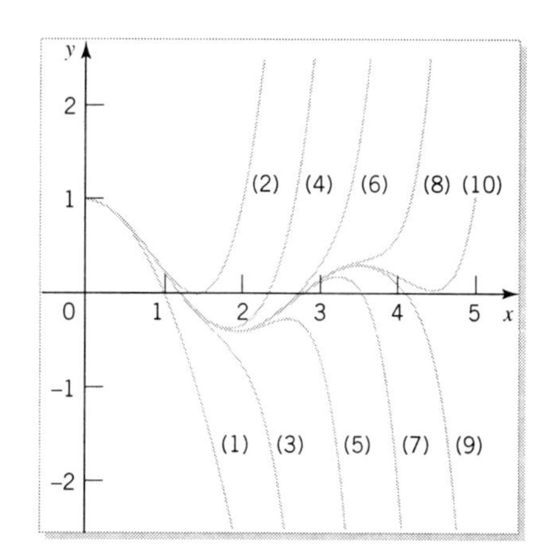

FIGURE 12.19 Approximations to $y_1(x)$, a solution of $xy'' + y' + 4xy = 0$

■

We conclude this section with an example in which the roots of the indicial equation differ by a positive integer, and in which the C in (12.73) of the Frobenius Theorem is not zero.

EXAMPLE 12.23

Find the general solution of

$$xy'' + (x-1)y' - 2y = 0 \tag{12.79}$$

as a series about $x = 0$.

Regular singular point

We first note that $x = 0$ is a regular singular point. Also, because $xp(x) = x - 1$ and $x^2q(x) = -2x$ are both polynomials, the series solutions we obtain will converge for $x > 0$.

We now substitute x^s into (12.79) and obtain

$$xs(s-1)x^{s-2} + (x-1)sx^{s-1} - 2x^s = s(s-1)x^{s-1} + sx^s - sx^{s-1} - 2x^s.$$

Indicial equation

Thus, we have the indicial equation given as

$$s^2 - s - s = s(s-2) = 0,$$

with roots of 0 and 2. If we let $s = 2$, and substitute the series $\sum_{n=0}^{\infty} c_n x^{n+2}$ into (12.79), we obtain

$$x\sum_{n=0}^{\infty}(n+2)(n+1)c_n x^n + x\sum_{n=0}^{\infty}(n+2)c_n x^{n+1} - \sum_{n=0}^{\infty}(n+2)c_n x^{n+1} - 2\sum_{n=0}^{\infty}c_n x^{n+2} = 0,$$

or

$$\sum_{n=0}^{\infty}(n+2)(n+1)c_n x^{n+1} + \sum_{n=0}^{\infty}(n+2)c_n x^{n+2} - \sum_{n=0}^{\infty}(n+2)c_n x^{n+1} - \sum_{n=0}^{\infty}2c_n x^{n+2} = 0.$$

If we now combine the terms with like powers of x, we obtain

$$\sum_{n=0}^{\infty}(n+2)nc_n x^{n+1} + \sum_{n=0}^{\infty}nc_n x^{n+2} = 0.$$

To combine the two series, we lower the index of summation on the second series to obtain

$$\sum_{n=0}^{\infty}(n+2)nc_n x^{n+1} + \sum_{n=1}^{\infty}(n-1)c_{n-1} x^{n+1} = 0,$$

so we may combine the two series into one as

$$\sum_{n=1}^{\infty}\left[(n+2)nc_n + (n-1)c_{n-1}\right]x^{n+1} = 0.$$

Recurrence relation This gives the first order recurrence relation as

$$c_n = -\frac{n-1}{(n+2)n}c_{n-1}, \qquad n = 1, 2, 3, \cdots.$$

Now c_0 is arbitrary, and from the recurrence relation we have that $c_1 = 0$, and all the remaining coefficients are also zero. If we choose $c_0 = 1$, we obtain our first solution of (12.79) as $y_1(x) = x^2$.

One solution

The proper form for our second solution is found in (12.73), namely,

$$y_2(x) = Cx^2 \ln x + x^0 \left(1 + \sum_{n=1}^{\infty} c_n x^n\right).$$

Differentiation of this expression gives

$$y_2'(x) = C(2x \ln x + x) + \sum_{n=1}^{\infty} nc_n x^{n-1},$$

$$y_2''(x) = C(2 \ln x + 3) + \sum_{n=2}^{\infty} n(n-1)c_n x^{n-2},$$

and substitution of these results into (12.79) gives

$$x\left[C(2\ln x + 3) + \sum_{n=2}^{\infty} n(n-1)c_n x^{n-2}\right] + (x-1)\left[C(2x\ln x + x) + \sum_{n=1}^{\infty} nc_n x^{n-1}\right] +$$

$$-2\left[Cx^2 \ln x + 1 + \sum_{n=1}^{\infty} c_n x^n\right] = 0.$$

Combining terms gives

$$-2 + 2Cx + Cx^2 + \sum_{n=2}^{\infty} n(n-1)c_n x^{n-1} + \sum_{n=1}^{\infty} nc_n x^n - \sum_{n=1}^{\infty} nc_n x^{n-1} - 2\sum_{n=1}^{\infty} c_n x^n = 0$$

or

$$-2 + 2Cx + Cx^2 + \sum_{n=1}^{\infty} n(n-2)c_n x^{n-1} + \sum_{n=1}^{\infty} (n-2)c_n x^n = 0,$$

which can be written as

$$-2 + 2Cx + Cx^2 + \sum_{n=1}^{\infty} n(n-2)c_n x^{n-1} + \sum_{n=2}^{\infty} (n-3)c_{n-1} x^{n-1} = 0.$$

We now set the coefficients of like powers of x to zero to obtain

$$\begin{aligned} -2 - c_1 &= 0, \\ 2C - c_1 &= 0, \\ C + 3c_3 &= 0, \\ n(n-2)c_n &= -(n-3)c_{n-1}, \qquad n = 4, 5, 6, \cdots. \end{aligned}$$

Recurrence relation

This gives the coefficients as $c_1 = -2, C = -1, c_2$ is arbitrary, $c_3 = 1/3$. The recurrence relation $n(n-2)c_n = -(n-3)c_{n-1}$ can be solved. We write it in the form

$$(-1)^n n!\,(n-2)\,c_n = (-1)^{n-1}\,(n-1)!\,(n-3)\,c_{n-1}, \qquad n = 4, 5, 6, \cdots,$$

or,

$$(-1)^n n!\,(n-2)\,c_n = -3!c_3,$$

which gives

$$c_n = (-1)^{n+1}\frac{2}{n!\,(n-2)}.$$

Our second solution may now be expressed as

$$y_2(x) = -x^2\ln x + 1 - 2x + c_2x^2 + \frac{1}{3}x^3 + 2\sum_{n=4}^{\infty}\frac{(-1)^{n+1}}{n!(n-2)}x^n,$$

General solution

where we notice that c_2 is arbitrary. Because $y_1(x) = x^2$, the term c_2x^2 can be combined with the arbitrary constant multiplying $y_1(x)$ in our general solution, which has the form $C_1x^2 + C_2y_2(x)$, where C_1 and C_2 are arbitrary constants. Another way would be to set $c_2 = 0$ in the preceding expression for $y_2(x)$.

Because one of our solutions is x^2, we know from the Oscillation Theorem and Relation Theorem of Section 8.1 that $y_2(x)$ will not oscillate. ■

EXERCISES

1. Use the method of Frobenius to find general solutions to the following differential equations. For series for which the formula for the nth term is not apparent, find terms in the series expansion for the solution up to x^4.

(a) $xy'' - (3+x)y' + 2y = 0$
(b) $4xy'' + 2(1+x)y' + y = 0$
(c) $xy'' + (x^3 - 1)y' + x^2y = 0$
(d) $3xy'' + y' - y = 0$
(e) $2xy'' + (2x+1)y' + 2y = 0$
(f) $4x^2y'' - 4xy' + (3 - 4x^2)y = 0$
(g) $2x^2y'' - xy' + (x-5)y = 0$
(h) $4x^2y'' + (4x - 2x^2)y' - (25 + 3x)y = 0$

2. Concerning the differential equation $2xy'' + 6y' - 9xy = 0$, do the following:

(a) Show that $x = 0$ is a regular singular point.

(b) Find the indicial equation associated with a series solution using the method of Frobenius.

(c) Find the series solution associated with the larger root of the indicial equation.

(d) Show that this method for the smaller root of the indicial equation gives the same solution as in part (c).

(e) Use the Frobenius Theorem to find the second solution.

(f) Show that any nontrivial solution of this differential equation has, at most, one zero for $x > 0$.

3. Find the general solution of the following differential equations using the method of Frobenius with expansions about the given point.

(a) $2(x-1)^2y'' + (5x-5)y' + xy = 0 \quad x_0 = 1$
(b) $2(x-1)y'' - y' + e^xy = 0 \quad x_0 = 1$
(c) $x(x+1)^2y'' - (x^2 + 3x + 2)y' + 9y = 0$
$x_0 = -1$
(d) $2(x+1)^2y'' - 3(x^2 + 3x + 2)y' + (3x+5)y = 0$
$x_0 = -1$
(e) $(x-1)^2y'' - (x-1)(x^2 - 2x)y' + (x^2 - 2x)y = 0$
$x_0 = 1$

4. Create a *How to Use the Method of Frobenius* that encompasses the use of the Regular Singular Point Theorem by adding statements under Purpose, Process, and Comments that summarize what you discovered in this section.

5. Find the general solution to the following differential equations for $x > 0$ using the method of Frobenius. For series for which the formula for the nth term is not apparent, find terms in the series up to x^4. Also find R such that the series converge for $0 < x < R$.

(a) $xy'' + y' - 4xy = 0$
(b) $xy'' + y' - xy = 0$
(c) $x^2y'' + 2xy' + xy = 0$
(d) $(x^2 - x)y'' + 3y' - 2y = 0$
(e) $xy'' + (x - 1)y' - y = 0$
(f) $2xy'' + (1 + x)y' - 2y = 0$
(g) $x^2y'' + (x^2 - x)y' - (x - 1)y = 0$
(h) $x^2y'' + (x^2 - x)y' + y = 0$
(i) $xy'' + xy' + y = 0$
(j) $(x - x^2)y'' - 3xy' - y = 0$

6. Show that one solution of Exercise 5(h) is $y(x) = xe^{-x}$, and use the reduction of order technique to find the second solution.

7. Show that the answer to Exercise 5(j) may also be written as

$$y(x) = c_0\frac{x}{(1-x)^2} + c_1\left[\frac{x}{(1-x)^2}\ln x + \frac{1}{(1-x)^2}\right].$$

8. Show that using $s = 0$ in a Frobenius series solution of (12.79) on page 626, $xy'' + (x - 1)y' - 2y = 0$, gives the same solution as $s = 2$.

9. Show that trying a Frobenius-type solution with $s = -1$ in (12.66), $x(x - 1)y'' + (5x - 2)y' + 4y = 0$, yields the same solution as that for $s = 0$.

10. Use the method of Frobenius to obtain the second solution of Example 12.21 on page 616, $x(x - 1)y'' + (5x - 2)y' + 4y = 0$, and compare your answer with (12.68).

11. Show that the indicial equation obtained by substituting $y(x) = x^s\sum_{n=0}^{\infty} c_n x^n$ in $x^2y'' + (x^2 - 3x)y' + (-2x + 4)y = 0$ gives rise to a double root of $s = 2$ and a solution of $y_1(x) = x^2$. Use reduction of order to find the second solution in the form $y_2(x) = x^2\int \frac{1}{x}e^{-x}\,dx$. By expanding e^{-x} in a Taylor series about the origin and integrating term by term, show that

$$y_2(x) = x^2\ln x + x^2\sum_{n=1}^{\infty}\frac{(-1)^n}{n!n}x^n.$$

What Have We Learned?

Main Ideas

The following comments pertain to the linear differential equation

$$a_2(x)y'' + a_1(x)y' + a_0(x)y = 0. \qquad (12.80)$$

- See page 595 for the definition of ordinary and singular points.
- If $x = x_0$ is an ordinary point of (12.80), then the solution of (12.80), subject to the initial conditions $y(x_0) = y_0$ and $y'(x_0) = y_0^*$, may be obtained as a power series about $x = x_0$. The answer to Exercise 6 on page 581 explains how to find a Taylor series solution. Also see *How to Find a Power Series Solution of a Linear Differential Equation* on page 587.
- A regular singular point is a point $x = x_0$ where at least one of a_1/a_2 and a_0/a_2 is not analytic, but both $(x - x_0)a_1/a_2$ and $(x - x_0)^2a_0/a_2$ are analytic. An irregular singular point is a singular point that is not regular. See page 604.
- The indicial equation needed to find a solution of (12.80) using the method of Frobenius may be found in three different ways. See pages 601 and 606 and Exercise 4 on page 609.

- If x_0 is a regular singular point of (12.80), the method of Frobenius always gives one solution of the form

$$y(x) = (x - x_0)^{s_1} \sum_{n=0}^{\infty} c_n (x - x_0)^n ,$$

where s_1 is the larger root of the indicial equation. See the Regular Singular Point Theorem on page 605.

- If x_0 is a regular singular point of (12.80), the second solution may have the form

$$y_2(x) = y_1(x) \ln (x - x_0) + (x - x_0)^{s_1} \sum_{n=1}^{\infty} c_n (x - x_0)^n , \qquad (12.81)$$

or

$$y_2(x) = C y_1(x) \ln (x - x_0) + (x - x_0)^{s_2} \left[1 + \sum_{n=1}^{\infty} c_n^n (x - x_0)^n \right] , \qquad (12.82)$$

where (12.81) works for repeated roots of the indicial equation, and (12.82) works for all other cases. It is possible that in some cases the constant C in (12.82) will be zero. See the Frobenius Theorem on page 620.

- The radius of convergence of the series obtained by using the method of Frobenius is at least as large as the radii of convergence of the two functions $(x - x_0)a_1/a_2$ and $(x - x_0)^2 a_0/a_2$. See the Regular Singular Point Theorem and the Frobenius Theorems, which have $x = 0$ as a regular singular point.

APPENDICES

You should remember the following piece of advice. There is so little that is true in mathematics, that anything you make up is likely to be wrong.

There are seven appendices.

1. **Background Material.** The material in this section is critical for success in any ordinary differential equations course. You will have seen all of this material in previous courses, but, if you are a typical student, you will have forgotten much of it. In fact, you might even think that you haven't seen some of it before.
2. **Partial Fractions.** This material should also be familiar to you. You will use it extensively when integrating and when using Laplace transforms.
3. **Infinite Series, Power Series, and Taylor Series.** This material should also be familiar to you, but most students don't really master series until they use it in differential equations.
4. **Complex Numbers.** Much of this material will be new to most students.
5. **Elementary Matrix Operations.** Most students will have seen special cases of these results.
6. **Least Squares Approximation.** This will be new to most students. It shows how to find the best straight-line approximation, $y = mx + b$, to a data set consisting of n points.
7. **Proofs of the Oscillation Theorems.** This contains the proofs of the theorems stated in Chapter 8.

A.1 BACKGROUND MATERIAL

Solving Quadratic Equations

The solutions of the quadratic equation $ax^2 + bx + c = 0$ are

$$x = \frac{-b \pm \sqrt{b^2 - 4ac}}{2a}.$$

If the discriminant, that is, $b^2 - 4ac$, is positive, the equation has distinct real roots. If $b^2 - 4ac = 0$, the roots are real, but repeated. If $b^2 - 4ac < 0$, the roots are complex, and are complex conjugates.

Perpendicular Lines

The two nonvertical straight lines, $y = m_1x + b_1$ and $y = m_2x + b_2$, are perpendicular if $m_1m_2 = -1$.

Properties of Trigonometric Functions

All angles are measured in radians where $180° = \pi$ radians. Thus, $90° = \pi/2$ radians, $60° = \pi/3$ radians, $45° = \pi/4$ radians, and $30° = \pi/6$ radians. An easy way to remember the values of the trig functions at frequently used angles is to use the following table.

$$\begin{array}{rcccccc} & & 0° & 30° & 45° & 60° & 90° \\ x & = & 0 & \pi/6 & \pi/4 & \pi/3 & \pi/2 \\ \sin x & = & \sqrt{0}/2 & \sqrt{1}/2 & \sqrt{2}/2 & \sqrt{3}/2 & \sqrt{4}/2 \\ \cos x & = & \sqrt{4}/2 & \sqrt{3}/2 & \sqrt{2}/2 & \sqrt{1}/2 & \sqrt{0}/2 \end{array},$$

which simplifies to

$$\begin{array}{rcccccc} & & 0° & 30° & 45° & 60° & 90° \\ x & = & 0 & \pi/6 & \pi/4 & \pi/3 & \pi/2 \\ \sin x & = & 0 & 1/2 & 1/\sqrt{2} & \sqrt{3}/2 & 1 \\ \cos x & = & 1 & \sqrt{3}/2 & 1/\sqrt{2} & 1/2 & 0 \end{array}.$$

The graphs of $\sin x$ and $\cos x$ are shown in Figure A.1.

$$\begin{aligned} \sin(-x) &= -\sin x \\ \cos(-x) &= \cos x \\ \sin^2 x + \cos^2 x &= 1 \\ \sin(x+y) &= \sin x \cos y + \cos x \sin y \\ \cos(x+y) &= \cos x \cos y - \sin x \sin y \end{aligned}$$

$$\tan x = \frac{\sin x}{\cos x}$$

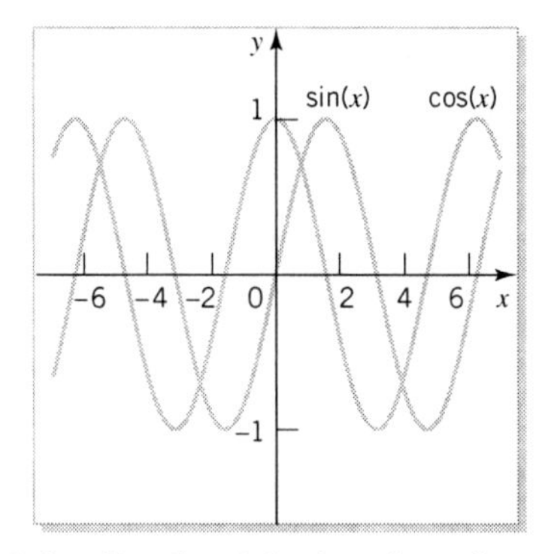

FIGURE A.1 Graphs of the functions $\sin x$ and $\cos x$

$$\tan(x+y) = \frac{\tan x + \tan y}{1 - \tan x \tan y}$$

$$\begin{aligned}\tfrac{d}{dx}\sin x &= \cos x \\ \tfrac{d}{dx}\cos x &= -\sin x \\ \tfrac{d}{dx}\tan x &= \sec^2 x\end{aligned}$$

$$\begin{aligned}\textstyle\int \sin x\,dx &= -\cos x + C \\ \textstyle\int \cos x\,dx &= \sin x + C \\ \textstyle\int \tan x\,dx &= -\ln|\cos x| + C\end{aligned}$$

$$\sin x = x - \frac{x^3}{3!} + \frac{x^5}{5!} - \cdots, \text{ for } -\infty < x < \infty$$

$$\cos x = 1 - \frac{x^2}{2!} + \frac{x^4}{4!} - \cdots, \text{ for } -\infty < x < \infty$$

The Inverse Trigonometric Functions

If $x = \sin\theta$, and $-\pi/2 \le \theta \le \pi/2$, then $\theta = \arcsin x$. The function $\arcsin x$ is sometimes written $\sin^{-1} x$. If $x = \tan\theta$, and $-\pi/2 < \theta < \pi/2$, then $\theta = \arctan x$. The function $\arctan x$ is sometimes written $\tan^{-1} x$.

$$\begin{aligned}\tfrac{d}{dx}\arcsin x &= 1/\sqrt{1-x^2} \\ \tfrac{d}{dx}\arctan x &= 1/\left(1+x^2\right)\end{aligned}$$

Properties of Exponential Functions

$$\begin{aligned}e^0 &= 1 \\ e^{x+y} &= e^x e^y \\ e^{x-y} &= e^x e^{-y} \\ e^{-x} &= 1/e^x \\ e^{ax} &= (e^a)^x \\ a^x &= e^{x\ln a}, \text{ for } a > 0\end{aligned}$$

$$\frac{d}{dx}e^{ax} = ae^{ax}$$

$$\int e^{ax}\,dx = \frac{1}{a}e^{ax} + C$$

$$e^x = 1 + x + \frac{x^2}{2!} + \frac{x^3}{3!} + \cdots, \text{ for } -\infty < x < \infty$$

The graphs of e^x and e^{-x} are shown in Figure A.2.

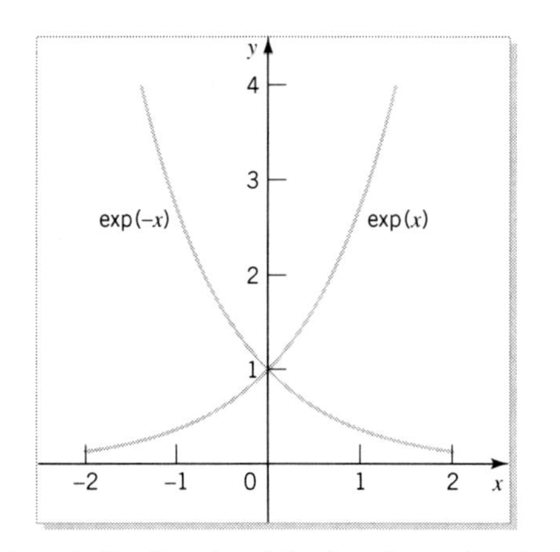

FIGURE A.2 Graphs of the functions e^x and e^{-x}

Properties of Logarithmic Functions

$\ln x$ is defined only for $x > 0$

$$
\begin{aligned}
\ln 1 &= 0 \\
\ln e &= 1 \\
\ln (xy) &= \ln x + \ln y \\
\ln (x/y) &= \ln x - \ln y \\
\ln x^n &= n \ln x \\
\ln x^{-1} &= -\ln x \\
e^{\ln x} &= x, \text{ if } x > 0 \\
\ln e^x &= x \\
\tfrac{d}{dx} \ln x &= \tfrac{1}{x} \\
\int \tfrac{1}{x}\, dx &= \ln |x| + C \\
\int \ln x \, dx &= x \ln x - x + C
\end{aligned}
$$

There are **NO** general formulas that simplify either $\ln(x + y)$ or $\ln(x - y)$.

WRONG: $\ln(x + y) = \ln x + \ln y$
WRONG: $\ln(x - y) = \ln x - \ln y$

The graphs of e^x and $\ln x$ are shown in Figure A.3.

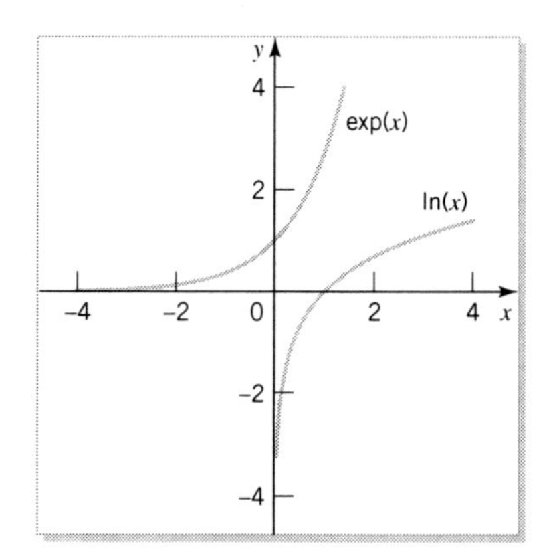

FIGURE A.3 Graphs of the functions e^x and $\ln x$

The Hyperbolic Functions

$$\sinh x = \frac{1}{2}\left(e^x - e^{-x}\right)$$

$$\cosh x = \frac{1}{2}\left(e^x + e^{-x}\right)$$

$$\tanh x = \frac{\sinh x}{\cosh x} = \frac{e^x - e^{-x}}{e^x + e^{-x}}$$

$$\begin{aligned} \sinh 0 &= 0 \\ \cosh 0 &= 1 \end{aligned}$$

$$\begin{aligned} \tfrac{d}{dx}\sinh x &= \cosh x \\ \tfrac{d}{dx}\cosh x &= \sinh x \end{aligned}$$

Properties of Derivatives

$$\frac{d}{dx}[cf(x)] = c\frac{d}{dx}f(x)$$

$$\frac{d}{dx}[f(x) \pm g(x)] = \frac{d}{dx}f(x) \pm \frac{d}{dx}g(x)$$

$$\frac{d}{dx}[f(x)g(x)] = \frac{d}{dx}[f(x)]\,g(x) + f(x)\frac{d}{dx}[g(x)]$$

$$\frac{d}{dx}\left[\frac{f(x)}{g(x)}\right] = \frac{f'(x)g(x) - f(x)g'(x)}{g^2(x)}$$

$$\text{If } y = f(u(x)), \text{ then } \frac{dy}{dx} = \frac{df}{du}\frac{du}{dx}.$$

$$\frac{d}{dx}\int_a^x f(t)\,dt = f(x), \text{ if } f(t) \text{ is continuous at } t = x.$$

If $f(x)$ is differentiable at $x = a$, then $f(x)$ is continuous at $x = a$.
If $f'(x) > 0$ in the interval $a < x < b$, then $f(x)$ is increasing in that interval.
If $f'(x) < 0$ in the interval $a < x < b$, then $f(x)$ is decreasing in that interval.
If $f''(x) > 0$ in the interval $a < x < b$, then $f(x)$ is concave up in that interval.
If $f''(x) < 0$ in the interval $a < x < b$, then $f(x)$ is concave down in that interval.
The function $f(x)$ has a local, or relative, maximum at x_0,
  if $f(x_0) \geq f(x)$ for all x near x_0.
The function $f(x)$ has a local, or relative, minimum at x_0,
  if $f(x_0) \leq f(x)$ for all x near x_0.

Table A.1 Table of derivatives

$f(x)$	$f'(x)$
c	0
x^n	nx^{n-1}
e^x	e^x
$\sin x$	$\cos x$
$\cos x$	$-\sin x$
$\tan x$	$\sec^2 x = 1/\cos^2 x$
$\cot x$	$-\csc^2 x = -1/\sin^2 x$
$\sec x = 1/\cos x$	$\sec x \tan x = \sin x/\cos^2 x$
$\csc x = 1/\sin x$	$-\csc x \cot x = -\cos x/\sin^2 x$
$\ln \lvert x \rvert$	$1/x$
$\sinh x = (e^x - e^{-x})/2$	$\cosh x$
$\cosh x = (e^x + e^{-x})/2$	$\sinh x$
$\arcsin x$	$1/\sqrt{1-x^2}$
$\arctan x$	$1/(1+x^2)$

Properties of Integrals

$$\int cf(x)\,dx = c\int f(x)\,dx$$

$$\int [f(x) \pm g(x)]\,dx = \int f(x)\,dx \pm \int g(x)\,dx$$

$$\int f(x)g'(x)\,dx = f(x)g(x) - \int f'(x)g(x)\,dx$$

$$\int u\,dv = uv - \int v\,du$$

$$\int_a^b f(x)g'(x)\,dx = f(x)g(x)|_a^b - \int_a^b f'(x)g(x)\,dx$$

$$\int f(g(x))g'(x)\,dx = \int f(u)\,du \text{ where } u = g(x)$$

There are **NO** general formulas that simplify either $\int[f(x)g(x)]\,dx$ or $\int[f(x)/g(x)]\,dx$.

WRONG: $$\int f(x)g(x)\,dx = \int f(x)\,dx \int g(x)\,dx$$

WRONG: $$\int \frac{f(x)}{g(x)}\,dx = \frac{\int f(x)\,dx}{\int g(x)\,dx}$$

Table A.2 **Table of integrals**

$f(x)$	$\int f(x)\,dx$
x^n	$x^{n+1}/(n+1)+C,\ n\neq -1$
$1/x$	$\ln\|x\|+C$
e^x	e^x+C
$\sin x$	$-\cos x+C$
$\cos x$	$\sin x+C$
$\tan x$	$-\ln\|\cos x\|+C=\ln\|\sec x\|+C$
$\cot x$	$\ln\|\sin x\|+C$
$\sec x=1/\cos x$	$\ln\|\sec x+\tan x\|+C$
$\csc x=1/\sin x$	$\ln\|\csc x-\cot x\|+C$
$\sec^2 x$	$\tan x+C$
$\csc^2 x$	$-\cot x+C$
$\ln x$	$x\ln x-x+C$
$\sinh x=(e^x-e^{-x})/2$	$\cosh x+C$
$\cosh x=(e^x+e^{-x})/2$	$\sinh x+C$
$1/[(x-a)(x-b)]$	$(\ln\|x-a\|-\ln\|x-b\|)/(a-b)+C,\ a\neq b$
$1/(1+x^2)$	$\arctan x+C$
$1/\sqrt{1+x^2}$	$\ln(x+\sqrt{x^2+1})+C$
$1/\sqrt{1-x^2}$	$\arcsin x+C$
$1/\sqrt{x^2-1}$	$\ln(x+\sqrt{x^2-1})+C$
$e^{ax}\sin bx$	$e^{ax}(a\sin bx-b\cos bx)/(a^2+b^2)+C$
$e^{ax}\cos bx$	$e^{ax}(b\sin bx+a\cos bx)/(a^2+b^2)+C$

A.2 PARTIAL FRACTIONS

We sometimes need to express a rational polynomial — that is, a function of the type

$$R(x)=\frac{P(x)}{Q(x)}, \tag{A.1}$$

where $P(x)$ and $Q(x)$ are polynomials — in an alternative form. The standard technique, known as partial fractions, goes as follows. (We should point out that this general explanation is much more involved than doing a particular example.)

1. If the degree of the polynomial $Q(x)$ is less than or equal to the degree of $P(x)$, then divide $Q(x)$ into $P(x)$, obtaining a polynomial plus a term similar to $R(x)$ in

(A.1), but where the degree of $Q(x)$ is greater than the degree of the new $P(x)$. From now on we concentrate on this new $R(x)$.

2. Factor $Q(x)$ into linear factors and quadratic factors (that cannot be written as the product of linear factors with real coefficients), so that

$$Q(x) = (x - r_1)^{n_1} \cdots \left(x - r_p\right)^{n_p} \left(a_1 x^2 + b_1 x + c_1\right)^{m_1} \cdots \left(a_q x^2 + b_q x + c_q\right)_{mq},$$

where n_1 through n_p and m_1 through m_q are positive integers, and $a_1 x^2 + b_1 x + c_1$, and so on, have no real roots. For example, if $Q(x) = x^3 - x$, then $Q(x) = x(x-1)(x+1)$, whereas, if $Q(x) = x^3 + x$, then $Q(x) = x(x^2+1)$.

3. For each linear factor of $Q(x)$ of degree n — say, $(x-r)^n$ — write down a contribution to $R(x)$ that is an expansion with n terms; namely,

$$\frac{A_1}{x-r} + \frac{A_2}{(x-r)^2} + \cdots + \frac{A_n}{(x-r)^n},$$

where A_1 through A_n are constants to be determined.

4. For each quadratic factor of $Q(x)$ of degree m — say, $\left(ax^2 + bx + c\right)^m$ — write down a contribution to $R(x)$ that is an expansion with m terms; namely,

$$\frac{B_1 x + C_1}{ax^2 + bx + c} + \frac{B_2 x + C_2}{\left(ax^2 + bx + c\right)^2} + \cdots + \frac{B_m x + C_m}{\left(ax^2 + bx + c\right)^m},$$

where B_1 through B_m and C_1 through C_m are constants to be determined.

5. Add the contributions to $R(x)$ from all the terms in $Q(x)$ and set them equal to $R(x)$. Now cross-multiply this identity in x by $Q(x)$ to evaluate the constants.

EXAMPLE A.1

Write $1/\left(x^2 - 1\right)$ as a partial fraction.

Here $P(x) = 1$, and $Q(x) = x^2 - 1$. The degree of $Q(x)$ — two — exceeds that of $P(x)$ — one — so we do not divide $P(x)$ by $Q(x)$ but use $R(x) = 1/\left(x^2 - 1\right)$. Now $Q(x) = (x-1)(x+1)$, so we have two linear roots, each of degree one. The contribution from $(x-1)$ is $A/(x-1)$, and the contribution from $(x+1)$ is $B/(x+1)$. Thus the total contribution to $R(x)$ is

$$\frac{1}{(x-1)(x+1)} = \frac{A}{x-1} + \frac{B}{x+1},$$

where A and B are constants to be determined by making this last equation an identity. Cross-multiplying by $(x-1)(x+1)$ gives $1 = A(x+1) + B(x-1)$, or

$$1 = (A+B)x + (A-B).$$

For this to be true for all x, we must have

$$\begin{aligned} A+B &= 0 \\ A-B &= 1' \end{aligned}$$

which can be solved to give $A = 1/2$ and $B = -1/2$. Thus, the partial fraction form of $1/\left(x^2-1\right)$ is

$$\frac{1}{x^2-1} = \frac{1/2}{x-1} - \frac{1/2}{x+1}.$$

EXAMPLE A.2

Write $x/\,(x-1)^2$ as a partial fraction.

Here $Q(x) = (x-1)^2$ has a linear factor of degree two, so we try

$$\frac{x}{(x-1)^2} = \frac{A}{x-1} + \frac{B}{(x-1)^2}.$$

This gives $x = A(x-1) + B$, or

$$x = Ax - A + B.$$

Thus $A = 1$, and $-A + B = 0$, so $B = 1$, giving

$$\frac{x}{(x-1)^2} = \frac{1}{x-1} + \frac{1}{(x-1)^2}.$$

EXAMPLE A.3

Write $(x+1)\,/[x(x^2+1)]$ as a partial fraction.

The contribution from x will be A/x, and from $\left(x^2+1\right)$ will be $(Bx+C)/(x^2+1)$. Thus, we try

$$\frac{x+1}{x\left(x^2+1\right)} = \frac{A}{x} + \frac{Bx+C}{x^2+1}.$$

Cross-multiplying by $x(x^2+1)$ gives $x+1 = A(x^2+1) + (Bx+C)\,x$, or

$$x+1 = (A+B)\,x^2 + Cx + A.$$

Thus $A = 1$, $B = -A = -1$, and $C = 1$, so we have

$$\frac{x+1}{x\left(x^2+1\right)} = \frac{1}{x} + \frac{-x+1}{x^2+1}.$$

A.3 INFINITE SERIES, POWER SERIES, AND TAYLOR SERIES

An infinite series, $\sum_{k=0}^{\infty} a_k$, either **converges** or **diverges**.

Convergent series are divided into two groups — **absolutely convergent** and **conditionally convergent**.

- If $\sum_{k=0}^{\infty} a_k$ and $\sum_{k=0}^{\infty} |a_k|$ are both convergent, then $\sum_{k=0}^{\infty} a_k$ is absolutely convergent.
- If $\sum_{k=0}^{\infty} a_k$ is convergent, but $\sum_{k=0}^{\infty} |a_k|$ is divergent, then $\sum_{k=0}^{\infty} a_k$ is conditionally convergent.
- The third possibility — that $\sum_{k=0}^{\infty} a_k$ is divergent, but $\sum_{k=0}^{\infty} |a_k|$ is convergent — cannot occur because if $\sum_{k=0}^{\infty} |a_k|$ is convergent, so is $\sum_{k=0}^{\infty} a_k$.

The reason that it is important to distinguish between absolutely and conditionally convergent series is Riemann's rearrangement theorem.

▶ **Theorem A.1:** *The terms of an absolutely convergent series may be rearranged in any order without affecting the convergence of the series. In particular, its sum is unchanged. Rearranging the terms of a conditionally convergent series may change its sum.* ◀

There are a variety of tests to decide whether a series converges or diverges, but, for ordinary differential equations, the most important is the RATIO TEST.

▶ **Theorem A.2: The Ratio Test** *If the series $\sum_{k=0}^{\infty} a_k$ has nonzero terms $a_0, a_1, a_2, \cdots$, then calculate*

$$\rho = \lim_{k\to\infty} \left| \frac{a_{k+1}}{a_k} \right| .$$

If $\rho < 1$, then the infinite series is absolutely convergent. If $\rho > 1$ (including ∞), then the infinite series is divergent. If $\rho = 1$, then the test fails and the series may converge or diverge. ◀

Change of index. The following equations should be studied carefully. This technique is used heavily when using series in differential equations.

$$\sum_{k=0}^{\infty} a_k = \sum_{j=0}^{\infty} a_j = \sum_{h=r}^{\infty} a_{h-r}$$

In differential equations, the most important series are power series.

◆ **Definition A.1:** **If $b_0, b_1, b_2, \cdots$, are constants, then**

$$\sum_{k=0}^{\infty} b_k (x-a)^k$$

is a POWER SERIES about the point $x = a$. ◆

For a power series, exactly one of the following is true.

1. The series converges only for $x = a$. The series is said to have a radius of convergence of zero.
2. The series converges absolutely for all x. The series is said to have an infinite radius of convergence.
3. There is a positive number, R, for which the series converges absolutely for $|x - a| < R$, but diverges for $|x - a| > R$. The series is said to have a radius of convergence R.

To find the radius of convergence, use the ratio test.

Properties of Power Series

- If

$$\sum_{k=0}^{\infty} b_k (x-a)^k = \sum_{k=0}^{\infty} c_k (x-a)^k,$$

then

$$b_k = c_k.$$

- If $\sum_{k=0}^{\infty} b_k (x-a)^k$ converges to $f(x)$ with radius of convergence either $R > 0$ or ∞, then

$$f(x) = \sum_{k=0}^{\infty} b_k (x-a)^k.$$

$$\frac{d}{dx} f(x) = \sum_{k=0}^{\infty} k b_k (x-a)^{k-1}.$$

$$\int f(x)\,dx = \sum_{k=0}^{\infty} \frac{1}{k+1} b_k (x-a)^{k+1} + C.$$

When

$$b_k = \frac{1}{k!} f^{(k)}(a),$$

the power series is called the TAYLOR SERIES of $f(x)$ about $x = a$.

- The Taylor series expansions about $x = 0$, and regions of convergence for $\sin x$, $\cos x$, e^x, $1/(1-x)$, and $\ln(1-x)$, are

$$\sin x = x - \frac{x^3}{3!} + \frac{x^5}{5!} - \cdots, \text{ for } -\infty < x < \infty,$$

$$\cos x = 1 - \frac{x^2}{2!} + \frac{x^4}{4!} - \cdots, \text{ for } -\infty < x < \infty,$$

$$e^x = 1 + x + \frac{x^2}{2!} + \frac{x^3}{3!} + \cdots, \text{ for } -\infty < x < \infty,$$

$$\frac{1}{1-x} = 1 + x + x^2 + x^3 + x^4 + \cdots, \text{ for } -1 < x < 1,$$

$$\ln(1-x) = x + \frac{1}{2}x^2 + \frac{1}{3}x^3 + \frac{1}{4}x^4 + \cdots, \text{ for } -1 < x < 1.$$

A.4 Complex Numbers

Let x and y be real numbers and consider the complex number

$$z = x + iy, \tag{A.2}$$

where i has the property that

$$i^2 = -1. \tag{A.3}$$

In (A.2), x is called the **real** part of z, and y is called the **imaginary** part of z. The complex number 0 is defined by $0 = 0 + i0$.

If $z_1 = x_1 + iy_1$ and $z_2 = x_2 + iy_2$ are two complex numbers, then

$$z_1 = z_2$$

is equivalent to the two equations

$$\begin{aligned} x_1 &= x_2, \\ y_1 &= y_2. \end{aligned}$$

In other words, two complex numbers are equal if and only if their real parts are equal and their imaginary parts are equal. Consequently, if $x + iy = 0$, then $x = 0$ and $y = 0$.

The complex conjugate $\tilde{z}$ of a complex number $z = x + iy$ is defined by

$$\tilde{z} = x - iy. \tag{A.4}$$

We treat complex numbers like real numbers when we do addition, subtraction, multiplication, and division, except that whenever we see i^2 we replace it by -1. For example,

$$(2 + 3i)(3 - 4i) = 6 - 8i + 9i - 12i^2 = 6 - 8i + 9i + 12 = 18 + i,$$

$$z\tilde{z} = (x + iy)(x - iy) = x^2 - (iy)^2 = x^2 + y^2.$$

The modulus, $|z|$, of a complex number is defined by

$$|z| = \sqrt{z\tilde{z}} = \sqrt{x^2 + y^2},$$

which is always a real number. This is useful because it allows us to convert division by complex numbers into division by real numbers, by realizing that

$$\frac{1}{z} = \frac{\tilde{z}}{z\tilde{z}} = \frac{\tilde{z}}{|z|^2} = \frac{\tilde{z}}{x^2 + y^2}.$$

For example,

$$\frac{1}{2 + 3i} = \frac{2 - 3i}{(2 + 3i)(2 - 3i)} = \frac{2 - 3i}{4 + 9} = \frac{2}{13} - \frac{3}{13}i.$$

We can think of the complex number $z = x + iy$ geometrically as the point (x, y) in the xy-plane. If we switch to polar coordinates (r, θ), where

$$\begin{aligned} x &= r\cos\theta \\ y &= r\sin\theta \end{aligned},$$

then

$$z = x + iy = r(\cos\theta + i\sin\theta).$$

This suggests that the expression

$$E(\theta) = \cos\theta + i\sin\theta$$

might be important. Let us differentiate $E(\theta)$ with respect to θ:

$$E'(\theta) = -\sin\theta + i\cos\theta = i^2\sin\theta + i\cos\theta = i(\cos\theta + i\sin\theta) = iE(\theta).$$

Because $E(0) = 1$, this suggests that we denote $E(\theta)$ by $e^{i\theta}$. We therefore define

$$e^{ix} = \cos x + i\sin x. \tag{A.5}$$

If we then define

$$e^z = e^{x+iy} = e^x e^{iy},$$

we find that e^z has all the usual properties of exponentials, such as

$$e^{0+i0} = 1,$$

$$e^{z_1+z_2} = e^{z_1}e^{z_2}. \tag{A.6}$$

From (A.5) we have

$$e^{-ix} = \cos x - i\sin x. \tag{A.7}$$

If we solve (A.5) and (A.7) for $\sin x$ and $\cos x$, we find

$$\sin x = \frac{e^{ix} - e^{-ix}}{2i}, \tag{A.8}$$

and

$$\cos x = \frac{e^{ix} + e^{-ix}}{2}. \tag{A.9}$$

Consider the following identity,

$$e^{2ix} = e^{ix}e^{ix}.$$

This can be rewritten in the form

$$\cos 2x + i\sin 2x = (\cos x + i\sin x)^2,$$

or

$$\cos 2x + i \sin 2x = \cos^2 x - \sin^2 x + i(2 \cos x \sin x),$$

from which we have the trig identities

$$\cos 2x = \cos^2 x - \sin^2 x,$$

and

$$\sin 2x = 2 \cos x \sin x.$$

[Use this idea to find trig identities for $\sin(x + y)$, and $\cos 3x$.]

We also have

$$\frac{d}{dx} e^{iax} = iae^{iax}, \tag{A.10}$$

and

$$\int e^{iax}\, dx = \frac{1}{ia} e^{iax} + C. \tag{A.11}$$

Let's consider the problem of evaluating

$$\int e^{ax} e^{ibx}\, dx = \int e^{(a+ib)x}\, dx = \frac{1}{a + ib} e^{(a+ib)x} + C = \frac{a - ib}{a^2 + b^2} e^{ax} e^{ibx} + C. \tag{A.12}$$

The right-hand side of (A.12) can be written as

$$\frac{a - ib}{a^2 + b^2} e^{ax} e^{ibx} = \frac{e^{ax}}{a^2 + b^2} (a - ib)(\cos bx + i \sin bx),$$

or

$$\frac{a - ib}{a^2 + b^2} e^{ax} e^{ibx} = \frac{e^{ax}}{a^2 + b^2} \left[a \cos bx + b \sin bx + i\,(a \sin bx - b \cos bx)\right]. \tag{A.13}$$

From (A.12) and (A.13) we thus have

$$\int e^{ax} \cos bx\, dx = \frac{e^{ax}}{a^2 + b^2} (a \cos bx + b \sin bx) + C, \tag{A.14}$$

and

$$\int e^{ax} \sin bx\, dx = \frac{e^{ax}}{a^2 + b^2} (a \sin bx - b \cos bx) + C. \tag{A.15}$$

It is a good exercise to start from the left-hand side of (A.14) or (A.15) and use (A.11), (A.8), and (A.9) to verify (A.14) or (A.15).

A.5 ELEMENTARY MATRIX OPERATIONS

An m by n matrix, $\mathbf{A}$, is an array of mn real numbers a_{ij}, where $1 \le i \le m, 1 \le j \le n$:

$$\mathbf{A} = \begin{bmatrix} a_{11} & a_{12} & \cdots & a_{1n} \\ a_{21} & a_{22} & \cdots & a_{2n} \\ \vdots & \vdots & \ddots & \vdots \\ a_{m1} & a_{m2} & \cdots & a_{mn} \end{bmatrix}.$$

This m by n matrix has m rows and n columns, with entries, elements, or components a_{ij}.

Another notation that is often used to represent the matrix $\mathbf{A}$ is

$$\mathbf{A} = \{a_{ij}\}, \qquad 1 \le i \le m, 1 \le j \le n.$$

Vectors are special cases of matrices where either $m = 1$, in which case we have

$$\mathbf{A} = \begin{bmatrix} a_{11} & a_{12} & \cdots & a_{1n} \end{bmatrix},$$

or $n = 1$, giving

$$\mathbf{A} = \begin{bmatrix} a_{11} \\ a_{21} \\ \vdots \\ a_{m1} \end{bmatrix}.$$

Two particularly important matrices are the zero matrix $\mathbf{0}$, and the identity matrix $\mathbf{I}$, where

$$\mathbf{0} = \begin{bmatrix} 0 & 0 & \cdots & 0 \\ 0 & 0 & \cdots & 0 \\ \vdots & \vdots & \ddots & \vdots \\ 0 & 0 & \cdots & 0 \end{bmatrix},$$

and

$$\mathbf{I} = \begin{bmatrix} 1 & 0 & \cdots & 0 \\ 0 & 1 & \cdots & 0 \\ \vdots & \vdots & \ddots & \vdots \\ 0 & 0 & \cdots & 1 \end{bmatrix}.$$

The zero matrix has 0 in every entry, and is defined for any m and n. The identity matrix has 0 in every entry except the main diagonal, where its entries are 1. The identity matrix is therefore defined only if $m = n$; that is, for **square** matrices where the number of rows is the same as the number of columns.

Two matrices, $\mathbf{A} = \{a_{ij}\}$ and $\mathbf{B} = \{b_{ij}\}$, are **equal** — that is, $\mathbf{A} = \mathbf{B}$ — if

$$a_{ij} = b_{ij}, \qquad \text{for every } i, j.$$

Thus, equality of matrices is defined only between matrices that have the same number of rows and the same number of columns. Also, if $\mathbf{A} = \mathbf{0}$, then $a_{ij} = 0$ for every i, j.

The **sum** $\mathbf{C}$ of two matrices $\mathbf{A} = \{a_{ij}\}$ and $\mathbf{B} = \{b_{ij}\}$ – that is, $\mathbf{C} = \mathbf{A} + \mathbf{B}$ – is defined by the matrix $\mathbf{C} = \{c_{ij}\}$, where

$$c_{ij} = a_{ij} + b_{ij}, \qquad \text{for every } i, j.$$

Thus, the sum of matrices is defined only between matrices that have the same number of rows and the same number of columns.

The **scalar multiple** $\mathbf{C}$ of a real number r (also called a scalar) and a matrix $\mathbf{A} = \{a_{ij}\}$ – that is, $\mathbf{C} = r\mathbf{A}$ – is defined by the matrix $\mathbf{C} = \{c_{ij}\}$, where

$$c_{ij} = ra_{ij}, \qquad \text{for every } i, j.$$

Thus, every entry of the matrix $\mathbf{A}$ is multiplied by the real number r.

The **product** $\mathbf{C}$ of two matrices $\mathbf{A} = \{a_{ij}\}$ and $\mathbf{B} = \{b_{ij}\}$ – that is, $\mathbf{C} = \mathbf{AB}$ – is defined by the matrix $\mathbf{C} = \{c_{ij}\}$, where

$$c_{ij} = \sum_{k=1}^{n} a_{ik} b_{kj}, \qquad \text{for every } i, j.$$

Thus, the product of matrices is defined only if the number of columns of $\mathbf{A}$ is the same as the number of rows of $\mathbf{B}$. So if $\mathbf{A}$ is an m by n matrix, then $\mathbf{B}$ must be n by p. The product $\mathbf{C}$ is an m by p matrix. In general, if $\mathbf{AB}$ is defined, there is no guarantee that $\mathbf{BA}$ is defined, so, in general, $\mathbf{AB} \neq \mathbf{BA}$.

Some properties of matrices, assuming everything is defined, are

$$\begin{aligned}
\mathbf{A} + \mathbf{B} &= \mathbf{B} + \mathbf{A} \\
(\mathbf{A} + \mathbf{B}) + \mathbf{C} &= \mathbf{A} + (\mathbf{B} + \mathbf{C}) \\
\mathbf{A} + \mathbf{0} &= \mathbf{A} \\
r(\mathbf{A} + \mathbf{B}) &= r\mathbf{A} + r\mathbf{B} \\
(r + s)\mathbf{A} &= r\mathbf{A} + s\mathbf{A} \\
\mathbf{AI} &= \mathbf{IA} = \mathbf{A} \text{ (if } \mathbf{A} \text{ is square)} \\
\mathbf{A}(\mathbf{BC}) &= (\mathbf{AB})\mathbf{C} \\
\mathbf{A}(\mathbf{B} + \mathbf{C}) &= \mathbf{AB} + \mathbf{AC} \\
(\mathbf{A} + \mathbf{B})\mathbf{C} &= \mathbf{AC} + \mathbf{BC} \\
r(\mathbf{AB}) &= (r\mathbf{A})\mathbf{B} \\
r(\mathbf{AB}) &= \mathbf{A}(r\mathbf{B})
\end{aligned}$$

The system of m equations in the n unknowns x_1 through x_n,

$$\begin{aligned}
a_{11}x_1 + a_{12}x_2 + \cdots + a_{1n}x_n &= b_1 \\
a_{21}x_1 + a_{22}x_2 + \cdots + a_{2n}x_n &= b_2 \\
\vdots \qquad\qquad &\vdots \quad \vdots \\
a_{m1}x_1 + a_{m2}x_2 + \cdots + a_{mn}x_n &= b_m,
\end{aligned}$$

can be written in matrix form

$$\begin{bmatrix} a_{11} & a_{12} & \cdots & a_{1n} \\ a_{21} & a_{22} & \cdots & a_{2n} \\ \vdots & \vdots & \ddots & \vdots \\ a_{m1} & a_{m2} & \cdots & a_{mn} \end{bmatrix} \begin{bmatrix} x_1 \\ x_2 \\ \vdots \\ x_n \end{bmatrix} = \begin{bmatrix} b_1 \\ b_2 \\ \vdots \\ b_m \end{bmatrix},$$

or

$$\mathbf{AX} = \mathbf{B},$$

where

$$\mathbf{A} = \begin{bmatrix} a_{11} & a_{12} & \cdots & a_{1n} \\ a_{21} & a_{22} & \cdots & a_{2n} \\ \vdots & \vdots & \ddots & \vdots \\ a_{m1} & a_{m2} & \cdots & a_{mn} \end{bmatrix}, \quad \mathbf{X} = \begin{bmatrix} x_1 \\ x_2 \\ \vdots \\ x_n \end{bmatrix}, \quad \mathbf{B} = \begin{bmatrix} b_1 \\ b_2 \\ \vdots \\ b_m \end{bmatrix}.$$

With every square matrix $\mathbf{A}$, we can associate a single real number called its **determinant**, denoted by $|\mathbf{A}|$.

If $n = 2$, then

$$|\mathbf{A}| = \begin{vmatrix} a_{11} & a_{12} \\ a_{21} & a_{22} \end{vmatrix} = a_{11}a_{22} - a_{12}a_{21}.$$

If $n = 3$, then

$$|\mathbf{A}| = \begin{vmatrix} a_{11} & a_{12} & a_{13} \\ a_{21} & a_{22} & a_{23} \\ a_{31} & a_{32} & a_{33} \end{vmatrix} = a_{11} \begin{vmatrix} a_{22} & a_{23} \\ a_{32} & a_{33} \end{vmatrix} - a_{12} \begin{vmatrix} a_{21} & a_{23} \\ a_{31} & a_{33} \end{vmatrix} + a_{13} \begin{vmatrix} a_{21} & a_{22} \\ a_{31} & a_{32} \end{vmatrix}.$$

If $n = 4$, then

$$|\mathbf{A}| = \begin{vmatrix} a_{11} & a_{12} & a_{13} & a_{14} \\ a_{21} & a_{22} & a_{23} & a_{24} \\ a_{31} & a_{32} & a_{33} & a_{34} \\ a_{41} & a_{42} & a_{43} & a_{44} \end{vmatrix} = a_{11} \begin{vmatrix} a_{22} & a_{23} & a_{24} \\ a_{32} & a_{33} & a_{34} \\ a_{42} & a_{43} & a_{44} \end{vmatrix} - a_{12} \begin{vmatrix} a_{21} & a_{23} & a_{24} \\ a_{31} & a_{33} & a_{34} \\ a_{41} & a_{43} & a_{44} \end{vmatrix} +$$

$$+a_{13} \begin{vmatrix} a_{21} & a_{22} & a_{24} \\ a_{31} & a_{32} & a_{34} \\ a_{41} & a_{42} & a_{44} \end{vmatrix} - a_{14} \begin{vmatrix} a_{21} & a_{22} & a_{23} \\ a_{31} & a_{32} & a_{33} \\ a_{41} & a_{42} & a_{43} \end{vmatrix}.$$

Follow this pattern for $n \geq 5$.

A square matrix $\mathbf{A}$ is called **singular** if its determinant is zero; that is, $|\mathbf{A}| = 0$. Otherwise it is called **nonsingular**.

An important result involving determinants is that the n by n system of equations $\mathbf{AX} = \mathbf{B}$, where $\mathbf{A}$ is a given n by n matrix, has a unique solution if and only if $|\mathbf{A}| \neq 0$; that is, $\mathbf{A}$ is nonsingular. A consequence of this is that if $|\mathbf{A}| \neq 0$, then the only solution of the n by n system of equations $\mathbf{AX} = \mathbf{0}$ is $\mathbf{X} = \mathbf{0}$.

Assume we are given an n by n matrix $\mathbf{A}$. If we can find a second n by n matrix that, when multiplied by $\mathbf{A}$, results in the n by n identity matrix $\mathbf{I}$, then the second matrix is called the **inverse** of $\mathbf{A}$, and is denoted by $\mathbf{A}^{-1}$. Thus, $\mathbf{A}^{-1}$ satisfies the equation

$$\mathbf{A}\mathbf{A}^{-1} = \mathbf{I}.$$

If we think of this as an equation of the form $\mathbf{AX} = \mathbf{I}$, where we seek $\mathbf{X}$ (which will be $\mathbf{A}^{-1}$), it can be shown from the comments in the previous paragraph that $\mathbf{A}^{-1}$ exists if and only if $|\mathbf{A}| \neq 0$. So to find $\mathbf{A}^{-1}$, we first confirm that $|\mathbf{A}| \neq 0$, and then solve $\mathbf{AX} = \mathbf{I}$ for $\mathbf{X} = \mathbf{A}^{-1}$. It can also be shown that if $\mathbf{A}^{-1}$ exists, then the inverse is unique and $\mathbf{A}^{-1}\mathbf{A} = \mathbf{I}$.

We outline an efficient method for finding $\mathbf{A}^{-1}$ using a 2 by 2 matrix $\mathbf{A}$ as an example. Assume that

$$\mathbf{A} = \begin{bmatrix} a & b \\ c & d \end{bmatrix}$$

is the given matrix for which we seek $\mathbf{A}^{-1}$. We first check that $|\mathbf{A}| \neq 0$ and then construct the matrix

$$\begin{bmatrix} a & b & 1 & 0 \\ c & d & 0 & 1 \end{bmatrix}$$

by adjoining the identity matrix to $\mathbf{A}$. The method for finding $\mathbf{A}^{-1}$ depends on using elementary row operations (described later) on this augmented matrix to convert it to the form

$$\begin{bmatrix} 1 & 0 & e & f \\ 0 & 1 & g & h \end{bmatrix}.$$

After this is done, the inverse of $\mathbf{A}$ can be read off from the last matrix. It is

$$\mathbf{A}^{-1} = \begin{bmatrix} e & f \\ g & h \end{bmatrix}.$$

The elementary row operations referred to are:

1. Interchanging two rows.
2. Replacing a row by a nonzero multiple of that row.
3. Replacing a row by that row plus a multiple of another row.

Applying this method to

$$\mathbf{A} = \begin{bmatrix} a & b \\ c & d \end{bmatrix}$$

gives

$$\mathbf{A}^{-1} = \frac{1}{ad - bc} \begin{bmatrix} d & -b \\ -c & a \end{bmatrix},$$

provided that

$$|\mathbf{A}| = ad - bc \neq 0.$$

We demonstrate this method on a specific 3 by 3 matrix.

EXAMPLE A.4

Find the inverse of

$$\mathbf{A} = \begin{bmatrix} 0 & 3e^{-t} & 3e^{5t} \\ 3e^{3t} & 0 & 0 \\ 0 & -e^{-t} & e^{5t} \end{bmatrix}.$$

We first evaluate the determinant of **A**, which is

$$|\mathbf{A}| = \begin{vmatrix} 0 & 3e^{-t} & 3e^{5t} \\ 3e^{3t} & 0 & 0 \\ 0 & -e^{-t} & e^{5t} \end{vmatrix} = 0 \begin{vmatrix} 0 & 0 \\ -e^{-t} & e^{5t} \end{vmatrix} - 3e^{-t} \begin{vmatrix} 3e^{3t} & 0 \\ 0 & e^{5t} \end{vmatrix} + 3e^{5t} \begin{vmatrix} 3e^{3t} & 0 \\ 0 & -e^{-t} \end{vmatrix},$$

or

$$|\mathbf{A}| = 0 - 3e^{-t}3e^{3t}e^{5t} + 3e^{5t}3e^{3t}\left(-e^{-t}\right) = -18e^{7t}.$$

Because $|\mathbf{A}| \neq 0$, we know that $\mathbf{A}^{-1}$ exists, so we construct the augmented matrix

$$\begin{bmatrix} 0 & 3e^{-t} & 3e^{5t} & 1 & 0 & 0 \\ 3e^{3t} & 0 & 0 & 0 & 1 & 0 \\ 0 & -e^{-t} & e^{5t} & 0 & 0 & 1 \end{bmatrix}.$$

We now use the elementary row operations to rewrite this in a form where the entries of the identity matrix occur in the first 3 by 3 block. Using the first elementary row operation, we interchange the first and second rows, obtaining

$$\begin{bmatrix} 3e^{3t} & 0 & 0 & 0 & 1 & 0 \\ 0 & 3e^{-t} & 3e^{5t} & 1 & 0 & 0 \\ 0 & -e^{-t} & e^{5t} & 0 & 0 & 1 \end{bmatrix}.$$

The second elementary row operation allows us to replace the first row by $e^{-3t}/3$ times the first row, so we find

$$\begin{bmatrix} 1 & 0 & 0 & 0 & e^{-3t}/3 & 0 \\ 0 & 3e^{-t} & 3e^{5t} & 1 & 0 & 0 \\ 0 & -e^{-t} & e^{5t} & 0 & 0 & 1 \end{bmatrix}.$$

The beginnings of the entries of a 3 by 3 identity matrix are starting to show up on the left-hand side. We want to put 0 where $-e^{-t}$ is at present. This we can do by using the third of the elementary row operations. We replace the third row by the third row plus 1/3 times the second row, obtaining

$$\begin{bmatrix} 1 & 0 & 0 & 0 & e^{-3t}/3 & 0 \\ 0 & 3e^{-t} & 3e^{5t} & 1 & 0 & 0 \\ 0 & 0 & 2e^{5t} & 1/3 & 0 & 1 \end{bmatrix}.$$

We can eliminate the $3e^{5t}$ term by again using the third elementary row operation, but this time we replace the second row by the second row plus $-3/2$ times the third row. In this way we find

$$\begin{bmatrix} 1 & 0 & 0 & 0 & e^{-3t}/3 & 0 \\ 0 & 3e^{-t} & 0 & 1/2 & 0 & -3/2 \\ 0 & 0 & 2e^{5t} & 1/3 & 0 & 1 \end{bmatrix}.$$

We now use the second of the elementary row operations twice. We replace the second row by $e^t/3$ times the second row, and we replace the third row by $e^{-5t}/2$ times the third row, giving

$$\begin{bmatrix} 1 & 0 & 0 & 0 & e^{-3t}/3 & 0 \\ 0 & 1 & 0 & e^t/6 & 0 & -e^t/2 \\ 0 & 0 & 1 & e^{-5t}/6 & 0 & e^{-5t}/2 \end{bmatrix}.$$

Thus,

$$\mathbf{A}^{-1} = \begin{bmatrix} 0 & e^{-3t}/3 & 0 \\ e^t/6 & 0 & -e^t/2 \\ e^{-5t}/6 & 0 & e^{-5t}/2 \end{bmatrix}.$$

You should check that $\mathbf{AA}^{-1} = \mathbf{I}$. ■

Finally, we sometimes need the derivative of a determinant whose entries depend on a variable — that is, given $|\mathbf{A}|$ find $|\mathbf{A}|'$. We could do this by first evaluating $|\mathbf{A}|$ and then taking its derivative. However, there is another way, which uses the following result. If

$$\mathbf{A} = \begin{bmatrix} a_{11} & a_{12} & \cdots & a_{1n} \\ a_{21} & a_{22} & \cdots & a_{2n} \\ \vdots & \vdots & \ddots & \vdots \\ a_{n1} & a_{n2} & \cdots & a_{nn} \end{bmatrix},$$

then

$$|\mathbf{A}|' = \begin{vmatrix} a'_{11} & a_{12} & \cdots & a_{1n} \\ a'_{21} & a_{22} & \cdots & a_{2n} \\ \vdots & \vdots & \ddots & \vdots \\ a'_{n1} & a_{n2} & \cdots & a_{nn} \end{vmatrix} + \begin{vmatrix} a_{11} & a'_{12} & \cdots & a_{1n} \\ a_{21} & a'_{22} & \cdots & a_{2n} \\ \vdots & \vdots & \ddots & \vdots \\ a_{n1} & a'_{n2} & \cdots & a_{nn} \end{vmatrix} + \cdots + \begin{vmatrix} a_{11} & a_{12} & \cdots & a'_{1n} \\ a_{21} & a_{22} & \cdots & a'_{2n} \\ \vdots & \vdots & \ddots & \vdots \\ a_{n1} & a_{n2} & \cdots & a'_{nn} \end{vmatrix},$$

or

$$|\mathbf{A}|' = \begin{vmatrix} a'_{11} & a'_{12} & \cdots & a'_{1n} \\ a_{21} & a_{22} & \cdots & a_{2n} \\ \vdots & \vdots & \ddots & \vdots \\ a_{n1} & a_{n2} & \cdots & a_{nn} \end{vmatrix} + \begin{vmatrix} a_{11} & a_{12} & \cdots & a_{1n} \\ a'_{21} & a'_{22} & \cdots & a'_{2n} \\ \vdots & \vdots & \ddots & \vdots \\ a_{n1} & a_{n2} & \cdots & a_{nn} \end{vmatrix} + \cdots + \begin{vmatrix} a_{11} & a_{12} & \cdots & a_{1n} \\ a_{21} & a_{22} & \cdots & a_{2n} \\ \vdots & \vdots & \ddots & \vdots \\ a'_{n1} & a'_{n2} & \cdots & a'_{nn} \end{vmatrix}.$$

A.6 LEAST SQUARES APPROXIMATION

We are given n data points, (x_1, y_1), (x_2, y_2), (x_3, y_3), $\cdots$, (x_n, y_n), and want to find constants m and b of the straight line $y(x) = mx + b$, which make the quantity S, where

$$S = \sum_{k=1}^{n} \left[y_k - y(x_k)\right]^2,$$

a minimum. Thus, we want to find constants m and b that make

$$S = \sum_{k=1}^{n} \left(y_k - mx_k - b\right)^2$$

a minimum. The function $y = mx + b$ is then called the line of best-fit or the least-squares line. In this appendix we show how to find b and m using the fact that a quadratic $a_2 t^2 + a_1 t + a_0$ has a minimum at $t = -a_1/(2a_2)$ if $a_2 > 0$.

Now, a typical member of S is $\left(y_k - mx_k - b\right)^2$, and

$$\left(y_k - mx_k - b\right)^2 = y_k^2 + m^2 x_k^2 + b^2 - 2my_k x_k - 2by_k + 2mbx_k,$$

so applying this to each of the terms in S gives

$$S = \sum_{k=1}^{n} y_k^2 + m^2 \sum_{k=1}^{n} x_k^2 + nb^2 - 2m \sum_{k=1}^{n} x_k y_k - 2b \sum_{k=1}^{n} y_k + 2mb \sum_{k=1}^{n} x_k.$$

To simplify S we define

$$A = \sum_{k=1}^{n} y_k^2, \quad B = \sum_{k=1}^{n} x_k^2, \quad C = \sum_{k=1}^{n} x_k y_k, \quad \bar{x} = \frac{1}{n} \sum_{k=1}^{n} x_k, \quad \bar{y} = \frac{1}{n} \sum_{k=1}^{n} y_k,$$

so

$$S = A + m^2 B + nb^2 - 2mC - 2bn\bar{y} + 2mbn\bar{x}.$$

Here the quantities A, B, C, $\bar{x}$, and $\bar{y}$ are known constants computed from the data points (x_1, y_1), (x_2, y_2), (x_3, y_3), $\cdots$, (x_n, y_n). The quantities $\bar{x}$ and $\bar{y}$ are their means.

First, thinking of S as a quadratic in m; that is,

$$S = Bm^2 + 2\left(bn\bar{x} - C\right) m + \left(A + nb^2 - 2bn\bar{y}\right),$$

we see that S has a minimum when

$$m = -\frac{bn\bar{x} - C}{B},$$

so that

$$mB + bn\bar{x} = C.$$

Second, thinking of S as a quadratic in b; that is,

$$S = nb^2 + 2(m\bar{x} - \bar{y})\,nb + \left(A + m^2B - 2mC\right),$$

we see that S has a minimum when $b = -m\bar{x} + \bar{y}$, so that

$$m\bar{x} + b = \bar{y}.$$

Solving $mB + bn\bar{x} = C$ and $m\bar{x} + b = \bar{y}$ for m and b gives

$$m = \frac{C - n\bar{x}\bar{y}}{B - n\bar{x}^2}$$

and

$$b = \frac{B\bar{y} - C\bar{x}}{B - n\bar{x}^2},$$

or

$$m = \frac{\sum_{k=1}^{n} x_k y_k - n\bar{x}\bar{y}}{\sum_{k=1}^{n} x_k^2 - n\bar{x}^2}$$

and

$$b = \frac{\bar{y}\sum_{k=1}^{n} x_k^2 - \bar{x}\sum_{k=1}^{n} x_k y_k}{\sum_{k=1}^{n} x_k^2 - n\bar{x}^2}.$$

These are the values for b and m that give the line of best-fit.

Further Comments

From

$$m = \frac{\sum_{k=1}^{n} x_k y_k - n\bar{x}\bar{y}}{\sum_{k=1}^{n} x_k^2 - n\bar{x}^2}$$

we find

$$-m\bar{x} = \frac{-\bar{x}\sum_{k=1}^{n} x_k y_k + n\bar{x}^2\bar{y}}{\sum_{k=1}^{n} x_k^2 - n\bar{x}^2}.$$

By adding and subtracting $\bar{y}\sum_{k=1}^{n} x_k^2$ to the numerator, we have

$$-m\bar{x} = \frac{\bar{y}\sum_{k=1}^{n} x_k^2 - \bar{x}\sum_{k=1}^{n} x_k y_k - \bar{y}\sum_{k=1}^{n} x_k^2 + n\bar{x}^2\bar{y}}{\sum_{k=1}^{n} x_k^2 - n\bar{x}^2} = b - \bar{y},$$

so the equation for the least-squares line can be written

$$y = mx + b = m(x - \bar{x}) + \bar{y}.$$

Thus, the point $(\bar{x}, \bar{y})$ lies on the least-squares line. The point $(\bar{x}, \bar{y})$ is the center of mass of the data points.

We now show that requiring that a straight line pass through the point $(\bar{x}, \bar{y})$ is equivalent to requiring that the sum of the vertical distance between the data points and that line is zero; that is,

$$\sum_{k=1}^{n} (y_k - y(x_k)) = 0.$$

To show this, we rewrite the left-hand side as

$$\sum_{k=1}^{n} (y_k - y(x_k)) = \sum_{k=1}^{n} (y_k - mx_k - b),$$

so

$$\sum_{k=1}^{n} (y_k - y(x_k)) = \sum_{k=1}^{n} y_k - m\sum_{k=1}^{n} x_k - nb = n(\bar{y} - m\bar{x} - b),$$

which is zero if and only if $\bar{y} - m\bar{x} - b = 0$, which is exactly the condition that needs to be satisfied to make $y = mx + b$ pass through the point $(\bar{x}, \bar{y})$.

We have seen that the line of best-fit passes through the point $(\bar{x}, \bar{y})$. Thus, if we know the slope of the line, then we know the line. There is a way to interpret the slope m of the line by rewriting

$$m = \frac{\sum_{k=1}^{n} x_k y_k - n\bar{x}\bar{y}}{\sum_{k=1}^{n} x_k^2 - n\bar{x}^2}$$

in a different way. First we notice that the quantity N defined by

$$N = \sum_{k=1}^{n} (x_k - \bar{x})(y_k - \bar{y})$$

has a typical term, $(x_k - \bar{x})(y_k - \bar{y}) = x_k y_k - \bar{x} y_k - \bar{y} y_k + \bar{x}\bar{y}$, so that

$$N = \sum_{k=1}^{n} (x_k y_k - \bar{x} y_k - \bar{y} y_k + \bar{x}\bar{y}),$$

which can be rewritten as

$$N = \sum_{k=1}^{n} x_k y_k - \bar{x}\sum_{k=1}^{n} y_k - \bar{y}\sum_{k=1}^{n} x_k + n\bar{x}\bar{y},$$

or, using the definitions of $\bar{x}$ and $\bar{y}$,

$$N = \sum_{k=1}^{n} x_k y_k - n\bar{x}\bar{y},$$

which is the numerator of m. Second, a similar calculation shows that

$$\sum_{k=1}^{n} (x_k - \bar{x})^2 = \sum_{k=1}^{n} x_k^2 - n\bar{x}^2,$$

the denominator of m. Thus, m can be written in the form

$$m = \frac{\sum_{k=1}^{n} (x_k - \bar{x})(y_k - \bar{y})}{\sum_{k=1}^{n} (x_k - \bar{x})^2}.$$

In spite of its appearance, we can give a geometrical interpretation to this form of m. Think about a typical data point — say, (x_k, y_k) — and the straight line joining this point to the point $(\bar{x}, \bar{y})$. This line will have slope s_k, where

$$s_k = \frac{y_k - \bar{y}}{x_k - \bar{x}},$$

so $y_k - \bar{y} = s_k (x_k - \bar{x})$, and m can be written as

$$m = \frac{\sum_{k=1}^{n} (x_k - \bar{x})^2 s_k}{\sum_{k=1}^{n} (x_k - \bar{x})^2}.$$

Thus, m is the weighted average of the slopes of the straight lines joining the data points to the center of mass $(\bar{x}, \bar{y})$. This is easily seen if we introduce the nonnegative quantities

$$w_1 = (x_1 - \bar{x})^2, \qquad w_2 = (x_2 - \bar{x})^2, \ \cdots, \ w_n = (x_n - \bar{x})^2,$$

so that

$$m = \frac{\sum_{k=1}^{n} w_k s_k}{\sum_{k=1}^{n} w_k}.$$

Notice that, if $x_1 < x_2 < \cdots < x_n$, the points x_1 and x_n are the farthest from $\bar{x}$, so the terms $w_1 = (x_1 - \bar{x})^2$ and $w_n = (x_n - \bar{x})^2$ will be the dominant terms. Thus, in general, the points near the ends have more influence on the slope than the points near the center of mass.

A.7 PROOFS OF THE OSCILLATION THEOREMS

Theorem A.3: *Consider the two differential equations,*

$$u'' + Q_1(t)u = 0 \tag{A.16}$$

and

$$v'' + Q_2(t)v = 0, \tag{A.17}$$

where $Q_1(t)$ and $Q_2(t)$ are continuous for $t \geq t_0$, and $Q_1(t) \leq Q_2(t)$ for $t \geq t_0$. If a nontrivial solution, $u(t)$, of (A.16) has successive zeros at t_1 and t_2, where $t_0 \leq t_1 < t_2$, and there is at least one point where $Q_1(t) < Q_2(t)$ for $t_1 < t < t_2$, then every nontrivial solution, $v(t)$, of (A.17) has a zero between t_1 and t_2.

Proof The continuity of $Q_1(t)$ and $Q_2(t)$ for $t \geq t_0$ guarantees that (A.16) and (A.17) have unique solutions.

Let us assume that $v(t)$ does not vanish between t_1 and t_2; that is, $v(t) \neq 0$ for $t_1 < t < t_2$. We want to show that this leads to a contradiction.

We are given $u(t_1) = u(t_2) = 0$, and $u(t) \neq 0$ for $t_1 < t < t_2$. Let us define the quantity

$$W(t) = u(t)v'(t) - v(t)u'(t), \tag{A.18}$$

which, when differentiated, gives

$$W'(t) = u(t)v''(t) - v(t)u''(t) = -\left[Q_2(t) - Q_1(t)\right]u(t)v(t),$$

where we have used (A.16) and (A.17). Because $u(t)$ and $v(t)$ are both nonzero for $t_1 < t < t_2$, and $Q_2(t) - Q_1(t) \geq 0$ [where for at least one point $Q_2(t) - Q_1(t) > 0$], the sign of W' is determined by the sign of $u(t)v(t)$, and it does not change for $t_1 < t < t_2$.

To fix ideas, let's consider the case where $u(t) > 0$ and $v(t) > 0$ for $t_1 < t < t_2$. This means that $W'(t) \leq 0$ [where for at least one point $W'(t) < 0$] for $t_1 < t < t_2$, so W is a decreasing function and hence $W(t_1) > W(t_2)$.

If we evaluate $W(t)$ at t_1 and at t_2, we find

$$W(t_1) = -v(t_1)u'(t_1), \text{ and } W(t_2) = -v(t_2)u'(t_2).$$

Because $u(t_1) = u(t_2) = 0$ and $u(t) > 0$ for $t_1 < t < t_2$, we must have $u'(t_1) > 0$ and $u'(t_2) < 0$. Thus, $W(t_1) < 0$ and $W(t_2) > 0$, which contradicts $W(t_1) > W(t_2)$. [The same analysis follows for the other three choices for the signs of $u(t)$ and $v(t)$.] This establishes the theorem. ◀

First, consider what Theorem A.3 tells us in the special case where $Q_2(t) = 0$. In this case, $v(t) = 1$ is a solution of (A.17) and it has no zeros. This means that no nontrivial solution of (A.16) for $Q_1(t) \leq 0$ can have two zeros; otherwise, $v(t)$ would have to vanish between them. This gives rise to the following result.

▶ **Theorem A.4:** *Consider the differential equation*

$$\frac{d^2x}{dt^2} + Q(t)x = 0, \tag{A.19}$$

where $Q(t)$ is continuous for $t \geq t_0$, and $Q(t) \leq 0$ for $t \geq t_0$. If $x(t)$ is a nontrivial solution of (A.19), then it has, at most, one zero for $t > t_0$. ◀

Second, consider what Theorem A.3 tells us in the special case when $Q_1(t) = k^2 > 0$. In this case, $u(t) = \sin kt$ is a solution of (A.16) and has an infinite number of zeros (at $t = n\pi/k$, $n = 0, \pm 1, \pm 2, \cdots$). This means that every nontrivial solution of (A.17) for $Q_2(t) \geq k^2$ has a zero between each successive pair of the zeros of $u(t)$, so $v(t)$ has an infinite number of solutions. This gives rise to the following result.

▶ **Theorem A.5:** *Consider the differential equation*

$$\frac{d^2x}{dt^2} + Q(t)x = 0, \tag{A.20}$$

where $Q(t)$ is continuous for $t \geq t_0$, and $Q(t) \geq k^2 > 0$ (k a constant) for $t \geq t_0$. If $x(t)$ is a nontrivial solution of (A.20), then it has an infinite number of zeros for $t \geq t_0$.

Third, the argument used to prove Theorem A.3 can be extended to prove the following result.

Theorem A.6: *Consider the differential equation*

$$\frac{d^2x}{dt^2} + Q(t)x = 0, \tag{A.21}$$

where $Q(t)$ is continuous for $t \geq t_0$, $x = x_1(t)$ and $x = x_2(t)$ are nontrivial solutions of (A.21) for $t \geq t_0$, and there is no constant c for which $x_1(t) = cx_2(t)$. Between two consecutive zeros of $x_1(t)$ there is a zero of $x_2(t)$, and vice versa. Furthermore, if $x_1(t) = 0$ has an infinite number of roots, so does $x_2(t) = 0$, but if $x_1(t) = 0$ has a finite number of roots, so does $x_2(t) = 0$.

There are results that guarantee that these oscillations are bounded. We consider the differential equation

$$x'' + Q(t)x = 0, \tag{A.22}$$

where $Q(t)$ is continuous and $Q(t) \geq k^2 > 0$ for $t \geq t_0$. By the preceding theorems, we know that all nontrivial solutions of (A.22) oscillate. Here we show that if $Q'(t)$ is continuous and monotonic for $t \geq t_0$, then these oscillations are bounded. We prove two theorems.

Theorem A.7: *Let $Q(t) \geq k^2 > 0$, and let $Q'(t)$ be continuous for $t \geq t_0$. The amplitude of the oscillations of any nontrivial solution of (A.22) is nonincreasing if $Q'(t) \geq 0$ for $t \geq t_0$, and nondecreasing if $Q'(t) \leq 0$ for $t \geq t_0$.*

Proof We restrict ourselves to $t \geq t_0$. The proof of this theorem hinges on introducing the function[1]

$$h(t) = x^2(t) + \frac{1}{Q(t)}\left[x'(t)\right]^2,$$

which is well-defined because $Q(t) > 0$. If we differentiate $h(t)$ with respect to t, we find

$$h'(t) = 2xx' + \frac{1}{Q}2x'x'' - \frac{Q'}{Q^2}\left(x'\right)^2,$$

which, by (A.22), reduces to

$$h'(t) = -\frac{Q'}{Q^2}\left(x'\right)^2.$$

[1] We would like to thank our colleague, Maciej Wojtkowski, for suggesting this proof.

Thus, $h(t)$ is nonincreasing if $Q'(t) \geq 0$, and nondecreasing if $Q'(t) \leq 0$. From the definition of $h(t)$, we see that if $t_1 < t_2$, then

$$x^2(t_1) + \frac{1}{Q(t_1)}\left[x'(t_1)\right]^2 \geq x^2(t_2) + \frac{1}{Q(t_2)}\left[x'(t_2)\right]^2 \qquad \text{if } Q'(t) \geq 0,$$

and

$$x^2(t_1) + \frac{1}{Q(t_1)}\left[x'(t_1)\right]^2 \leq x^2(t_2) + \frac{1}{Q(t_2)}\left[x'(t_2)\right]^2 \qquad \text{if } Q'(t) \leq 0.$$

We now apply these results to particular values of t_1 and t_2. Because $x(t)$ oscillates for $t \geq t_0$, it will have successive local extrema. We consider two of these — at t_1 and t_2 — where $t_0 < t_1 < t_2$. We thus have $x'(t_1) = x'(t_2) = 0$, and so the amplitudes $x(t_1)$ and $x(t_2)$ satisfy

$$x^2(t_1) \geq x^2(t_2) \qquad \text{if } Q'(t) \geq 0$$

and

$$x^2(t_1) \leq x^2(t_2) \qquad \text{if } Q'(t) \leq 0,$$

or

$$|x(t_1)| \geq |x(t_2)| \qquad \text{if } Q'(t) \geq 0$$

and

$$|x(t_1)| \leq |x(t_2)| \qquad \text{if } Q'(t) \leq 0.$$

Thus, the amplitudes are nonincreasing if $Q'(t) \geq 0$, and nondecreasing if $Q'(t) \leq 0$. ◀

▶ **Theorem A.8:** *Under the conditions of Theorem A.7, the oscillations are bounded.* ◀

Proof If $Q'(t) \geq 0$, then Theorem A.7 guarantees that the amplitude is nonincreasing and the oscillations are bounded.

To show that $x(t)$ is bounded if $Q'(t) \leq 0$, we will show that $x^2(t) \leq a/k^2$, where a is a constant, for $t \geq t_0$. The proof of this depends on multiplying (A.22) by x', namely,

$$x'x'' + Qxx' = 0,$$

and then rewriting it in the form

$$\left[(x')^2 + Qx^2\right]' = Q'x^2. \tag{A.23}$$

Because $Q'(t) \leq 0$, (A.23) implies that

$$\left[(x')^2 + Qx^2\right]' \leq 0,$$

so that $(x')^2 + Qx^2$ is a nonincreasing function of t. Thus, its value at any $t \geq t_0$ can be no larger that its value at t_0, from which we have

$$[x'(t)]^2 + Q(t)x^2(t) \leq [x'(t_0)]^2 + Q(t_0)x^2(t_0),$$

for $t \geq t_0$. We note that the right-hand side is a positive constant — which we denote by a — while the left-hand side satisfies

$$k^2x^2(t) \leq (x'(t))^2 + Q(t)x^2(t).$$

Thus, $k^2x^2(t) \leq a$.

ANSWERS

CHAPTER 1

Section 1.1

1. (a) $y(x) = \frac{1}{4}x^4 + \frac{3}{4}$. (c) $y(x) = \sin x$.
(e) $y(x) = -e^{-x} + 2$. (g) $y(x) = \ln(-x) + 1$.
(i) $y(x) = \ln[\frac{x}{2(x-1)}] + 1, x > 1$.
(k) $y(x) = -x^2e^{-x} - 2xe^{-x} - 2e^{-x} + 3$.

3. (a) $S(0) = 0$.
(b) $S(-x) = -S(x)$.

	x	$y(x)$	x	$y(x)$
(c–e)	0.5	0.033	3.0	0.617
	1.0	0.248	3.5	0.394
	1.5	0.621	4.0	0.596
	2.0	0.642	4.5	0.483
	2.5	0.344	5.0	0.421

(f) As $x \to \infty$, $S(x) \to 0.5$.

5. $y(x) = e^x + C_1$ and $y(x) = e^{-x} + C_2$. The angle of intersection is 90°.

Section 1.2

1. Uniqueness assures us that the curves cannot cross.
(a) Increasing when $x > 0$. Decreasing when $x < 0$. Concave up for all x, except for $x = 0$. Symmetric across the y-axis. No singularities.
(c) Increasing when $(4n-1)\frac{\pi}{2} < x < (4n+1)\frac{\pi}{2}$ for $n = 0, \pm1, \pm2, \cdots$. Decreasing when $(4n+1)\frac{\pi}{2} < x < (4n+3)\frac{\pi}{2}$ for $n = 0, \pm1, \pm2, \cdots$. Concave up when $(2n-1)\pi < x < 2n\pi$ for $n = 0, \pm1, \pm2, \cdots$. Concave down when $2n\pi < x < (2n+1)\pi$ for $n = 0, \pm1, \pm2, \cdots$. Inflection points when $x = n\pi$ for $n = 0, \pm1, \pm2, \cdots$. Symmetric about the origin. No singularities.
(e) Increasing for all x. Concave down for all x. No symmetry. No singularities.
(g) Increasing when $x > 0$. Decreasing when $x < 0$. Concave down for all x, except for $x = 0$. Symmetric across the y-axis. Singularity at $x = 0$.
(i) Increasing when $0 < x < 1$. Decreasing when $x < 0$ or $1 < x$. Concave up when $x > 1/2$. Concave down when $x < 1/2$. Inflection point $x = 1/2$. No symmetry. Singularities at $x = 0$ and $x = 1$.
(k) Increasing for all x except $x = 0$. Concave up when $0 < x < 2$. Concave down when $x < 0$ or $x > 2$. No symmetry. No singularities.

3. (a) $y(x) = \ln[(x-4)/(5x)], x < 0$.
(b) $y(x) = \ln[(4-x)/(3x)], 0 < x < 4$.
(c) $y(x) = \ln[5(x-4)/x], x > 4$.

5. Increasing when $\sqrt{2n\pi} < x < \sqrt{(2n+1)\pi}$, $n = 0, 1, 2, \cdots$. Decreasing when $\sqrt{(2n+1)\pi} < x < \sqrt{(2n+2)\pi}$, $n = 0, 1, 2, \cdots$. Concave up when $0 < x < \pi/2$ or $\sqrt{-\pi/2 + 2n\pi} < x < \sqrt{\pi/2 + 2n\pi}$, $n = 1, 2, 3, \cdots$. Concave down when $\sqrt{\pi/2 + 2n\pi} < x < \sqrt{3\pi/2 + 2n\pi}$, $n = 0, 1, 2, \cdots$. Symmetric about the origin. No singularities. As $x \to \infty$, $y \to 0.5$.

Section 1.3

1. (a) Isoclines are given by $x = m^{1/3}$.
(c) Isoclines are given by $x = \pm\arccos m + 2n\pi$, $n = 0, \pm1, \pm2, \cdots$.
(e) Isoclines are given by $x = \ln\frac{1}{m}$.
(g) Isoclines are given by $x = \frac{1}{m}$.
(i) Isoclines are given by $x = \frac{1}{2} \pm \sqrt{\frac{m-4}{4m}}$.
(k) Isoclines are given by $x^2e^{-x} = m$.

3. Isoclines are given by $\sqrt{\frac{2}{\pi}}\sin x^2 = m$. As $x \to \infty$, if $y(0) = 0$, then $y(x) \to 1$.

5. (b) No.

7. (a) Figure 1.20: $y' = \ln|x|$.

9. (b) $y(x) = 49x + 245e^{-0.2x} - 245$. Hits the ground after 107.04 seconds.

Section 1.4

1. $\sqrt{\frac{2}{\pi}}\left(\frac{x^3}{3\cdot 1!} - \frac{x^7}{7\cdot 3!} + \frac{x^{11}}{11\cdot 5!} - \frac{x^{15}}{15\cdot 7!} + \cdots\right)$.

CHAPTER 2

Section 2.1

1. (a) ii. $y(x) = [3(-x + C)]^{1/3}$.
(c) ii. $y(x) = 0$ and $y(x) = \frac{-1}{x+C}$.
(e) ii. $y(x) = 0$ and $y(x) = \left(-\frac{2}{x+C}\right)^2$ for $x + C < 0$.

3. (a) $y(x) = 1 - e^{-x}$. (c) $y(x) = 1 + e^{-x}$.

5. (a) Figure 2.11: $y' = \frac{y^2-1}{y^2+1}$. Figure 2.12: $y' = -\frac{y^2+1}{y^2-1}$.

7. No.

9. $y(x) = 0$ and $y(x) = 1/(x - C)$.

Section 2.2

3. $\ln 2/k \approx 131$ years.

5. $t = -10\ln 0.1/\ln 2 \approx 33.2$ hours.

7. $P(t) = 18.632\exp(0.039t)$. $t \approx 346$ years.

9. 8.19×10^{-4} sec. The bullet can never come to rest in finite time.

Section 2.3

1. 3.02 days.

5. The curve labelled A.

7. (a) $y(x) = (e^x - 1)/(e^x + 1)$.
(b) $y(x) = 1$. (c) $y(x) = -1$.

9. (a) $y(x) = 0$, $y(x) = \frac{\alpha}{\beta}$, $y(x) = \frac{\alpha y_0}{(\alpha - \beta y_0)e^{-\alpha x} + \beta y_0}$.
(b) $y(x) = 0$, $y(x) = y_0 e^{\alpha x}$.
(c) $y(x) = 0$, $y(x) = \frac{y_0}{\beta y_0 x + 1}$.
(d) $y(x) = 0$, $y(x) = \frac{\alpha}{\beta}$.

11. (a) The logistic equation for $y > 0$.
(b) Increasing when $y < b$. Decreasing when $y > b$. Concave up when $y < be^{-1}$ or $y > b$. Concave down when $be^{-1} < y < b$. Points of inflection $y = be^{-1}$.
(c) $y(t) = b$.
(d) $y(t) = b\left(\frac{y(0)}{b}\right)^{e^{-kt}}$.
(e) $y(0) > be^{-1}$.

15. (a) $-1/0.087\ln(0.0498/990) \approx 87.3$ days.
(c) $2\ln 99 \approx 9.2$ years.

Section 2.4

1. $y(x) = \frac{2}{3}$. $y(x_0) = \frac{2}{3}$.

3. $y(x) = 0$ for all x; $y(x) = 0$ for $x \le 0$, $y(x) = \left(\frac{2}{3}x\right)^{3/2}$ for $x > 0$; and $y(x) = 0$ for $x \le 0$, $y(x) = -\left(\frac{2}{3}x\right)^{3/2}$ for $x > 0$.

5. No.

7. (c) If $C > 1$ there are no equilibrium solutions. If $C = 1$ there is one (repeated) equilibrium solution. If $C < 1$ there are two equilibrium solutions.
(d) $C = 1$, semistable. $C < 1$, $r_1 = 1 - \sqrt{1 - C}$, $r_2 = 1 + \sqrt{1 - C}$. $Y = r_2$ is a stable equilibrium solution. $Y = r_1$ is an unstable equilibrium solution.
(f) $C = 1$, $Y(0) < 1$. $C < 1$, $Y(0) < r_1$.

9. (a) No. (b) No.
(c) Yes. One example is $y' = C(1 - y)(2 - y)^2(3 - y)^2$, where $C > 0$.

11. $P(t) = 1/[C - (n - 1)rt]^{1/(n-1)}$.

Section 2.5

1. The logistics equation.

3. The equilibrium solutions are $P(t) = 0$, $P(t) = b$, and $P(t) = c$. The solutions 0 and c are stable, while b is unstable.

5. (a) Yes.
(c) i. $y(x) = 1$ is unstable. ii. $y(x) = -1$ is unstable. $y(x) = 1$ is semistable. iii. $y(x) = -1$ and $y(x) = 1$ are unstable, but $y(x) = 0$ is stable.

11. A possible differential equation is $y' = a(b - y)(c - y)$. $a = 0.03$, $b = 276$, and $c = 211$, give the ultimate world record as 211 seconds.

13. (a) $y(x) = 0$ and $y(x) = 4$. They are both stable.

(b) $y(x) = 2 + \sqrt{4 + (y_0^2 - 4y_0)e^{-x}}$ if $y_0 > 2$.

$y(x) = 2 - \sqrt{4 + (y_0^2 - 4y_0)e^{-x}}$ if $y_0 < 2$.
As $x \to \infty$, $y(x) \to 4$ if $y_0 > 2$ and $y(x) \to 0$ if $y_0 < 2$.

CHAPTER 3

Section 3.1

3. (a) $n = 3$. (c) $n = 32$. (e) $n = 1$.

5. (a) C may be any real number.

(c) $C = -1/4$.

7. (a) Symmetric across the x- and y-axes, and about origin.

(b) $x = 0$, $y = \frac{x}{m}$.

(d) Increasing in the first and third quadrants. Decreasing in the second and fourth quadrants. Concave up when $y > 0$ and $|y| > |x|$, or $y < 0$ and $|y| < |x|$. Concave down when $y < 0$ and $|y| > |x|$, or $y > 0$ and $|y| < |x|$.

13. (a) Figure 3.16: $y' = -(y^2 + 1)/(x^2 + 1)$.

(b) Figure 3.17: $y' = (x^2 + 1)/(y^2 + 1)$.

Section 3.2

1. Symmetric across the x- and y-axes, and about origin.

5. The solution curve is an even function.

9. (a) $g(y)$ is an odd function.

(c) $f(-x)g(-y) = f(x)g(y)$.

11. Every member of the family of solutions differs by a vertical translation.

Section 3.3

1. (a) $y' = x^3$ subject to $y(1) = 1$.

i. Exact solution: $y(x) = \frac{1}{4}x^4 + \frac{3}{4}$.
$y(x_1) = 1.1160$, $y(x_2) = 1.2684$, $y(x_3) = 1.4640$, $y(x_4) = 1.7104$

ii. $y(x_1) = 1.1160$, $y(x_2) = 1.2684$, $y(x_3) = 1.4640$, $y(x_4) = 1.7104$

iii. $y(x_1) = 1.1000$, $y(x_2) = 1.2331$, $y(x_3) = 1.4059$, $y(x_4) = 1.6256$

iv. $y(x_1) = 1.0500$, $y(x_2) = 1.1079$, $y(x_3) = 1.1744$, $y(x_4) = 1.2505$, $y(x_5) = 1.3369$, $y(x_6) = 1.4345$, $y(x_7) = 1.5444$, $y(x_8) = 1.6674$

(c) $y' = e^{-x}$ subject to $y(0) = 1$.

i. Exact solution: $y(x) = -e^{-x} + 2$.
$y(x_1) = 1.0952$, $y(x_2) = 1.1813$, $y(x_3) = 1.2592$, $y(x_4) = 1.3297$

ii. $y(x_1) = 1.0952$, $y(x_2) = 1.1813$, $y(x_3) = 1.2592$, $y(x_4) = 1.3297$

iii. $y(x_1) = 1.1000$, $y(x_2) = 1.1905$, $y(x_3) = 1.2724$, $y(x_4) = 1.3464$

iv. $y(x_1) = 1.0500$, $y(x_2) = 1.0976$, $y(x_3) = 1.1428$, $y(x_4) = 1.1858$, $y(x_5) = 1.2268$, $y(x_6) = 1.2657$, $y(x_7) = 1.3028$, $y(x_8) = 1.3380$

(e) $y' = \frac{1}{1+x^2}$ subject to $y(1) = \frac{\pi}{4}$.

i. Exact solution: $y(x) = \arctan x$.
$y(x_1) = 0.8330$, $y(x_2) = 0.8761$, $y(x_3) = 0.9151$, $y(x_4) = 0.9505$

ii. $y(x_1) = 0.8330$, $y(x_2) = 0.8761$, $y(x_3) = 0.9151$, $y(x_4) = 0.9505$

iii. $y(x_1) = 0.8354$, $y(x_2) = 0.8806$, $y(x_3) = 0.9216$, $y(x_4) = 0.9588$

iv. $y(x_1) = 0.8104$, $y(x_2) = 0.8342$, $y(x_3) = 0.8568$, $y(x_4) = 0.8783$, $y(x_5) = 0.8988$, $y(x_6) = 0.9183$, $y(x_7) = 0.9369$, $y(x_8) = 0.9546$

7. (a) $y(x) = e - e^x$, and $y(1) = 0$.

(b) With step-sizes of 0.1, 0.05, 0.25, and 0.125 we find $y(1) \approx 0.124539$, 0.064984, 0.033218, and 0.016797.

(c) With step-sizes of 0.1, 0.05, 0.25, and 0.125 we find $y(1) \approx 0.004201$, 0.001091, 0.000278, and 0.000070.

(d) With step-sizes of 0.1, 0.05, 0.25, and 0.125 we find $y(1) \approx 0.0000020843$, 0.0000001358, 0.0000000087, and 0.0000000005.

11. For $h = 0.18$ to $h \approx 0.282$ we find period 1 solutions. For $h \approx 0.282$ to $h \approx 0.301$ we find period 2 solutions. For $h \approx 0.301$ to $h \approx 0.310$ we find period 4 solutions.

13. (a) Somewhere between 0.8862 and 0.8863.

(e) If $y_0 - \frac{\sqrt{\pi}}{2} > 0$ then $y(x) \to \infty$, as $x \to \infty$.
If $y_0 - \frac{\sqrt{\pi}}{2} < 0$ then $y(x) \to -\infty$, as $x \to \infty$.
If $y_0 - \frac{\sqrt{\pi}}{2} = 0$ then $y(x) \to 0$, as $x \to \infty$.

Section 3.5

1. (a) $c_0\left(1 - x + \frac{1}{2!}x^2 - \frac{1}{3!}x^3 + \frac{1}{4!}x^4 + \cdots\right) = c_0 e^{-x}$.
(c) $c_0\left(1 + 2x + x^2\right) = c_0(1+x)^2$.

3. $c_0\left(1 - x^2 + \frac{1}{2!}x^4 - \frac{1}{3!}x^6 + \cdots\right) = c_0 e^{-x^2}$.

CHAPTER 4

Section 4.1

1. (a) Separable (c) Neither (e) Neither
(g) Autonomous and Separable

3. $y(x) = \pm\sqrt{x^2 + 2C}$.

5. (a) $y(x) = x/e$, $x > 0$.
(c) $y(x) = 2\sin\left(\ln\frac{x}{8} + \frac{\pi}{6}\right)$, where $-\frac{\pi}{2} < \ln\frac{x}{8} + \frac{\pi}{6} < \frac{\pi}{2}$.
(e) $y(x) = -2$.
(g) $\frac{y^2}{2} + \ln y = \frac{3}{2} - e^{-\cos x}$, $y > 0$.
(i) $y(x) = \left(\int_0^x \sin t^2\, dt + 27\right)^{1/3}$.
(k) $y(x) = 2(1 - x^2)$, where $-1 < x < 1$.

7. (a) $600\left(1 - e^{-x/50}\right)$.
(b) $50 \ln 3 \approx 54.9306$ minutes.
(c) 3.

9. $P(T) = \exp(-a/T + C)$.
$P(T) = \exp(-4458.965/T + 18.276)$.

13. (b) An oval region.

Section 4.2

5. (a) $x^2 y + y^3 = \frac{1}{3}A$.
(c) $\ln|y| - \arctan\left(\frac{y}{x}\right) = C$.
(e) $y = x(\ln|y| + C)$. (g) $y = \frac{x}{\ln|x| + C}$.
(i) $x^3 + 3xy^2 = c$. (k) $y = xe^{cx}$.

7. (a) A homogeneous function of degree 2.
(c) A homogeneous function of degree 1.
(e) A homogeneous function of degree 0.

9. (a) $y(x) = x(1 \pm \sqrt{2}) + 4 \pm \sqrt{2}$, and
$(x+1)^2 + 2(x+1)(y-3) - (y-3)^2 = c$.
(c) $(x - 2y)^2 + 2x - 6y = C$.

Section 4.3

1. (a) $\alpha \approx 0.0875$, $\beta \approx -2.465$. $T(t) = \alpha t^2/2 + \beta t + c$.
$c = 82.3$. As $t \to \infty$, $T \to \infty$.
(b) $a \approx -0.0434$, $b \approx 1.066$. $T(t) = (Ce^{at} - b)/a$.
$C = b + 82.3a = -2.506$. As $t \to \infty$, $T \to -\frac{b}{a}$.
(c) $T(t) = (C - b)/a + Ct + Cat^2/2 +$
$(C/a)(at)^3/3! + \cdots$.

3. (b) $70 + 110e^{-0.0263t}$ for $0 < t < 30$.
$400 - 280e^{-0.0263(t-30)}$ for $t > 30$.
(c) 39.18 minutes after noon.

7. (a) $L(t) = -a/k$. L_e is the maximum length of the fish while a must be a positive number.
(b) Newton's Law of Heating.
(c) $L(t) = 212 - \exp(5.246 - 0.054t)$.

9. (a) $T(t) = T_a$ is stable.
(b) $T(t) = T_a + [(-n+1)(kt + C)]^{1/(1-n)}$.
$T(t) = T_a + kc/\left(e^{-kt} - ch\right)$.

11. $k < \ln(0.72/26.6)/24 \approx 0.1504$.

Section 4.4

1. $(1/\alpha)\ln(19) \approx 8.23$ sec. 835 feet.

3. (a) Note the phase line for the first time period has a stable equilibrium at $v = 180$, while for the second time period this stable equilibrium is at $v = 22$.
(b) $v(t) = V(e^{\alpha t} - 1)/(e^{\alpha t} + 1)$.
$x(t) = (2V/\alpha)\ln[(e^{\alpha t} + 1)/2] - tV$, where $V = 180$, and $\alpha = 2(32.2)/180 \approx 0.358$.
48.22 seconds. 180 ft/sec.
(c) $v(t) = 22\left[0.7822e^{2.927(t-48.22)} - 1\right]/$
$[0.7822e^{2.927(t-48.22)} + 1]$.
$x(t) = 9000 + 15.03\ln[(e^{2.927(t-48.22)} - 0.7822)/$
$0.2178] - 22(t - 48.22)$. 22 ft/sec.

7. $x(t) = Vt - V^2\left(1 - e^{-gt/V}\right)/g$.

9. $x(t) = -gt^2/2 + v_0 t$. $t = v_0/g$. Total time in the air is $2v_0/g$. The time interval of v_0/g is 50% of the total time in the air.

Section 4.5

1. (a) $y = x/2 + C$.
   (c) $y^2 = -2x^2 + C$.
   (e) $y^2 = -x^2 + 32\ln|x| + C$.
   (g) $y(x) = x^a C$.
   (i) $y(x) = \pm\sqrt{2C - 2x}$.
   (k) $y(x) = \pm\sqrt{2C - 2x/a}$.
   (m) $y = -\ln[C - 1/(3x)]$.
   (o) $\ln|\sin y| = -x + C$.

3. (a) They are either parallel or perpendicular.

5. Orthogonal trajectories are $y^2/2 - x^2/2 = C$.

7. (b) $r(\theta) = c\sin\theta$.

9. $r(\theta) = c(1 + \sin\theta)$.

11. $r(\theta) = ce^{\theta^2}$.

Section 4.6

1. (a) $v(t) = 180(e^{0.354t} - 1)/(e^{0.354t} + 1), 0 < t < 6$.
   $v(t) = 22 + 121e^{-1.464(t-6)}, t > 6$.

3. As N approaches infinity, both terms approach a constant value, with their difference being $-c$.

5. The concentration increases without bound as t increases.

7. $V(t) = 0$ for $0 < t < T_1$. $V(t) = Ee^{-(t-T_1)/(RC)}$ for $T_1 < t < T_1 + T_2$. $V(t) = 0$ for $T_1 + T_2 < t < 2T_1 + T_2$. $V(t) = Ee^{-(t-2T_1-T_2)/(RC)}$ for $2T_1 + T_2 < t < 2T_1 + 2T_2$, and so on.

9. (a) If y is differentiable, then $\lim_{x\to k^-} y(x) = \lim_{x\to k^+} [b - y(x)]$.
   (b) Approach $y(x) = b$. b would be the carrying capacity.
   (c) For $0 < x < k$, $y(x) = y_0 e^x$. $k = \ln(b/(2y_0))$. For $x > k$, $y(x) = b(1 - 0.5e^{k-x})$.
   (d) $x = k$, $y = b/2$.
   (e) i. $a = 1/b$. $y(x) = by_0/\left[(b - y_0)e^{-x} + y_0\right]$
   ii. $a = -[1/(bk)]\ln[y_0/(b - y_0)]$.

CHAPTER 5

Section 5.1

1. (a) $y(x) = x - 1 + Ce^{-x}$. Valid for all x.
   (c) $y(x) = \frac{1}{2}e^x + Ce^{-x}$. Valid for all x.
   (e) $y(x) = x^3e^{-3x} + Ce^{-3x}$. Valid for all x.
   (g) $y(x) = \frac{1}{x} - \frac{\sqrt{2}}{x^2}\arctan\frac{x}{\sqrt{2}} + \frac{C}{x^2}$. Valid for all $x \neq 0$.
   (i) $y(x) = \frac{\sin x + C}{x+1}$. Valid for all $x \neq -1$.
   (k) $y(x) = x^2 - 1 + Ce^{-x^2}$. Valid for all x.
   (m) $y(x) = 4x + C|x|^{1/2}$. Valid for all $x \neq 0$.
   (o) $y(x) = -\frac{1}{2} + Ce^{-2/x}$. Valid for all $x \neq 0$.
   (q) $y(x) = xe^{-2x} + \frac{C}{x}e^{-2x}$. Valid for all $x \neq 0$.
   (s) $y(x) = (x^2 + 1)/2 + C(x^2 + 1)^{-2}$. Valid for all x.

3. (a) Separable
   (c) Separable
   (e) Linear
   (g) None of these

5. $P(20) = 2.14$ million.

7. $y(x) = (x + C)/\sqrt{x^2 + 1}$. As $x \to \infty$, $y(x) \to 1$ for any C.

Section 5.2

1. (a) $y(t) = 2e^{-t} + Ce^{-2t}$.
   (c) $y(t) = 2 + e^{-t^2}C$.
   (e) $y(t) = [-\cos t + C]/t^2$.
   (g) $P(t) = (2\ln t - 1)/(4t) + C/t^3$.
   (i) $P(t) = 4\cos t + 4 + C\exp(\cos t)$.

3. (a) $y(t) = -2 + 3e^{3t}$.
   (c) $y(t) = -2t - \frac{2}{7} + \frac{2}{7}e^{7t}$.
   (e) $y(t) = \frac{1}{2} + \frac{3}{2}e^{-t^2}$.

5. (a) $x(t) = 3(200 - t) - 3(200 - t)^5/(200)^4$. 200 minutes.
   (b) $200 - 200/3^{1/4} \approx 48$ minutes.

7. $T(t) = 45 + 600[(12/5)\sin(\pi t/12) - \pi\cos(\pi t/12)]/[144 + 25\pi^2] + \left[25 + 600\pi/\left(144 + 25\pi^2\right)\right]e^{-0.2t}$
   The plants would not be safe.

9. $I(t) = \frac{3}{901}(30\sin 2t - \cos 2t) + \frac{3}{901}e^{-60t}$. Steady state solution $\frac{3}{901}(30\sin 2t - \cos 2t)$. Transient $\frac{3}{901}e^{-60t}$. No equilibrium solutions.

11. $q(t) = 0.12 + 4.88e^{-10t}$. 0.12 is the steady state solution. $q(t) = 0.12$ is also the equilibrium solution.

13. (a) For $0 \le t < 1$, $y(t) = 2(1 - e^{-t})$.
For $t \ge 1$, $y(t) = 1 + (e - 2)e^{-t}$.
(c) For $0 \le t < 10$, $y(t) = 5 + e^{-t}$.
For $t \ge 10$, $y(t) = e^{-t}\left(4e^{10} + 1\right) + 1$.

15. (a) $x(t) = tV - 2\sqrt{Vx_0 t}$.
(b) $t = 4x_0/V$.

17. $e^{k\sqrt{x}}\left|3 - k\sqrt{x}\right|^3/27 = e^{-tk^2/(50\pi)}$. $(3/k)^2 < 10$.

19. If $x(t)$ is the depth of the water $V' = (\pi/4)x^2 dx/dt = 3 - kx$. $3/k < 10$.

21. $t \approx 71.4$ minutes. The container has not overflowed.

23. $k = (Cr - R)/x_\infty$.

25. $x_i \approx 1.502$.

Section 5.3

1. (a) $y^2 = 1 + Ce^{-x^2}$.
(c) $y(x) = xe^x/(-e^x + C)$.
(e) $y^4 = 1/\left(2x^{10} + Cx^8\right)$.
(g) $y(x) = (2/3 + Ce^{-x^{3/2}})^{2/3}$.
(i) $y^2 = e^{2x}/\left(x^2 + C\right)$.

5. (a) $y(x) = 0$ (stable) and $y(x) = 1/m$ (unstable).
(b) $y(x) = 0$ (stable), $y(x) = 1/m$ (unstable), and $y(x) = -1/m$ (unstable).
(c) $y(x) = 0$ (stable), $y(x) = 1/m$ (unstable), and $y(x) = -1/m$ (unstable).

7. $P(t) = \mu(t)/[\int \mu(t)k(t)/b(t)dt + C]$, $\mu(t) = \exp\left(\int k(t)\,dt\right)$.

9. (a) Separable
(c) None of these
(e) Separable
(g) Bernoulli, homogeneous coefficients
(i) Linear

CHAPTER 6

Section 6.1

5. $\frac{1}{2}\ln\left[5\left(\frac{y}{x}\right)^2 - 2\frac{y}{x} + 10\right] - \frac{1}{7}\arctan\left(\frac{5y}{7x} - \frac{1}{7}\right) = -\ln|x| + C$.

Section 6.2

1. (a) Linear, constant coefficients, nonhomogeneous, nonautonomous.
(c) Linear, homogeneous, nonautonomous.
(e) Nonlinear, autonomous.

15. (a) $y'' + 4y' + y = 0$.
(c) $-2y'' - \frac{7}{2}y' - 2y - 16 = 2e^{2t}$.
(e) $\left(\frac{1}{2}t + 1\right)y'' + \left(\frac{1}{2}t + 1\right)y' + y = \sin t + t\cos t$.

Section 6.3

1. (a) $x(t) = C_1e^{-2t} + C_2e^{3t}$.
(c) $x(t) = C_1e^{t} + C_2te^{t}$.
(e) $x(t) = C_1e^{-4t} + C_2e^{2t}$.
(g) $x(t) = C_1e^{-3t} + C_2e^{4t}$.
(i) $x(t) = C_1e^{2t} + C_2e^{3t}$.
(k) $x(t) = C_1\cos 4t + C_2\sin 4t$.
(m) $x(t) = C_1e^{-t/2} + C_2e^{t/3}$.
(o) $x(t) = e^{-t}(C_1\cos\sqrt{3}t + C_2\sin\sqrt{3}t)$.

3. (a) $x(t) = e^{t} - e^{-2t}$.
(c) $x(t) = 2e^{-3t} + 6te^{-3t}$.
(e) $x(t) = 5 - e^{-3t}$.
(g) $x(t) = 15e^{-5t}\cos 5\sqrt{3}t + \frac{79}{15}\sqrt{3}e^{-5t}\sin 5\sqrt{3}t$.

5. $x(t) = C_1e^{-0.2t}\cos(1.4t) + C_2e^{-0.2t}\sin(1.4t)$.

9. (a) Amplitude $= \sqrt{8}$, period $= 2\pi/3$, phase angle $= 3\pi/4$.

11. (i) $x(t) = e^{\alpha t}\left(C_1\cos\beta t + C_2\sin\beta t\right)$, an infinite number of times.
(ii) $x(t) = \left(C_1 + C_2t\right)e^{rt}$, once or no times.
(iii) $x(t) = C_1e^{r_1t} + C_2e^{r_2t}$, once or no times.

13. $c = 2\sqrt{2}$ is the smallest value of c that satisfies the conditions.

Section 6.4

1. (a) $x(t) = (8/\sqrt{255})e^{-t/16}\sin(\sqrt{255}t/16)$.
   (c) $x(t) = e^{-t/8}\cos t/2 + (1/4)e^{-t/8}\sin t/2$.
   (e) $x(t) = 0$.

3. (a) overdamped if $b > 12$.
   (b) underdamped if $0 < b < 12$.
   (c) critically damped if $b = 12$.

13. Shorten the pendulum.

15. (a) $\pm Ae^{\alpha t}$.
   (d) The curve that oscillates is $Ae^{\alpha t}\cos(\beta t + \phi)$. Dotted curves are the envelopes $\pm Ae^{\alpha t}$. The other two curves are the curves $\pm Ae^{\alpha t}\cos(\arctan(\alpha/\beta))$.

19. (a) $X(t) = At + B$. $Y(t) = C_1\cos\omega t + C_2\sin\omega t$. X is the center of mass of x and y.
   (c) $x(t) = At + B - M/(m+M)(C_1\cos\omega t + C_2\sin\omega t)$
   $y(t) = At + B + m/(m+M)(C_1\cos\omega t + C_2\sin\omega t)$.

23. (a) $3 > R > 0$. (b) $R > 3$. (c) $R = 3$.

25. ≈ 2.149.

Section 6.5

1. (a) $\frac{dy}{dx} = \frac{-4x}{y}$. (b) $x = 2, y = 0$.
   (c) $4x^2 + y^2 = 16$. (d) $x(t) = 2\cos 2t$.

3. (a) $E = 74$.
   (b) At the highest and lowest points on the ellipse.

7. Second order, nonlinear, autonomous equation. A nonlinear, autonomous system of equations.

9. (a) $x(t) = x_0(1 - rt)e^{rt}$, where $r = -\frac{R}{2L}$.
   (b) ≈ 3.319.

11. (i) $-\ln|x| = \frac{1}{2}\ln\left[\left(\frac{y}{x} + \frac{1}{2}\right)^2 + 2\right] - \frac{1}{2\sqrt{2}}\arctan\left(\left(\frac{y}{x} + \frac{1}{2}\right)/\sqrt{2}\right) + C$.
   (ii) $-\ln|x| = \ln\left|\frac{y}{x} + \frac{3}{2}\right| + \frac{3}{2}\left(\frac{y}{x} + \frac{3}{2}\right)^{-1} + C$.
   (iii) $-\ln|x| = -\frac{1}{8}\ln\left|\frac{y}{x} + \frac{1}{2}\right| + \frac{9}{8}\ln\left|\frac{y}{x} + \frac{9}{2}\right| + C$.

Section 6.6

1. (a) The equilibrium point is $(0, 0)$.
   (c) The equilibrium point is $(3, 9)$.
   (e) This is not an autonomous system of differential equations.

3. (a) $(6, 3)$.
   (c) $(2 + 3y, y)$ for every y.

5. (i) a. $x'' - 4x' + 3x = 0$. $x(t) = C_1e^{3t} + C_2e^{t}$.
   b. $x(t) = e^{3t} + e^{t}$ $y(t) = -e^{3t} + e^{t}$.
   c. $x - y = 2(x - y)^3$.
   (iii) a. $x'' - x' - 2x = 0$. $x(t) = C_1e^{2t} + C_2e^{-t}$.
   b. $x(t) = 2e^{2t}$ $y(t) = e^{2t}$.
   c. $x = 2y$.

Section 6.7

1. (a) There are an infinite number, although only three occur on Figure 6.43.
   (b) There are an infinite number, although only four occur on Figure 6.43.

3. (a) $0 \le \beta^2 \le 4$. (b) $\beta^2 > 4$ (c) No.

5. (a) $dy/dx = \lambda^2(x^3/6 - x)/y$.
   $(0, 0)$, $(\sqrt{6}, 0)$, and $(-\sqrt{6}, 0)$.
   (b) $y^2/2 = \lambda^2(x^4/24 - x^2/2) + C$.
   (c) $(0, 0)$, $(\sqrt{6}, 0)$, and $(-\sqrt{6}, 0)$.

7. (a) The period is unchanged.
   (b) The period gets larger.
   (c) The period gets smaller.

CHAPTER 7

Section 7.1

1. (a) $x(t) = C_1e^{t} + C_2e^{-t} - 2 - t^2$.
   (c) $x(t) = C_1e^{-2t} + C_2e^{-t} + te^{-t}$.
   (e) $x(t) = C_1e^{2t} + C_2te^{2t} + 2t^3e^{2t}$.

3. (a) $x(t) = C_1 + C_2e^{-t} + t$.
   (b) $x(t) = C_1 + C_2e^{-t} + t + 2$.

5. (b) $x(t)/(20e^{3t}) \to 1/20$ as $t \to \infty$.
   (c) $x(t)/e^{-t} \to \infty$ (like $t + 1$) as $t \to \infty$.
   (e) $x(t)/\left(12te^{2t}\right) \to \infty$ [like $\left(1 + 2t^2\right)/12$] as $t \to \infty$.

Section 7.2

1. (a) $x(t) = C_1 e^t + C_2 e^{-t} - t^2 - 2$.
(c) $x(t) = C_1 e^{-2t} + C_2 e^{-t} + te^{-t}$.
(e) $x(t) = C_1 e^{2t} + C_2 te^{2t} + 2t^3 e^{2t}$.

3. (a) $x(t) = C_1 e^{2t} + C_2 e^{-3t} + 2e^{4t}$.
(c) $x(t) = C_1 e^{2t} + C_2 e^{-3t} + 8/5te^{2t}$.
(e) $\sin t^2$ is not in Table 7.1.

5. (a) $x(t) = -\frac{1}{36} - \frac{1}{6}t + C_1 e^{2t} + C_2 e^{-3t}$.
(c) $x(t) = -\frac{7}{50}\sin t - \frac{1}{50}\cos t + C_1 e^{2t} + C_2 e^{-3t}$.
(e) $x(t) = -6te^{-t} + C_1 + C_2 e^{-t}$.
(g) $x(t) = -2\sin 3t + C_1 \cos 2t + C_2 \sin 2t$.
(i) $x(t) = -\frac{1}{4}t\cos 2t + C_1 \cos 2t + C_2 \sin 2t$.
(k) $x(t) = t^3 e^{\frac{3}{2}t} + C_1 e^{\frac{3}{2}t} + C_2 te^{\frac{3}{2}t}$.

7. (a) $x(t) = -\frac{1}{49}e^{3t} + \frac{8}{7}te^{3t} + \frac{1}{49}e^{-4t}$.
(c) $x(t) = -e^{3t} + e^{2t} + te^{-t}$.

11. (a) $x(t) = C_1 + C_2 e^t - t^3/3 - t^2 - 2t$.
(b) $x(t) = -t^3/3 - t^2 - 2t + C_1 e^t + C_2$.

15. $x(t) = C_1 \cos t + C_2 \sin t + \sin t \ln|\csc t - \cot t|$.

19. $a_2(x_1 + x_2)'' + a_1(x_1 + x_2)' + a_0(x_1 + x_2) = f_1(t) + f_2(t)$. $x_1(t_0) + x_2(t_0) = h_1 + h_2$ and $x_1'(t_0) + x_2'(t_0) = k_1 + k_2$.

Section 7.3

1. $x(t) = e^{-t/4}[-B\cos(3t/4) + (-B/3 - 4\omega A/3)\sin(3t/4)] + A\sin\omega t + B\cos\omega t$, where $A = (5 - 8\omega^2)/\alpha$, $B = -4\omega/\alpha$, and $\alpha = 2[64\omega^2 - 64\omega - 9]$.

3. $19t\sin 10t/(20)$.

5. (a) $x(t) = \frac{1}{4}(-B - A\omega)e^{-t}\sin 4t - Be^{-t}\cos 4t + A\sin\omega t + B\cos\omega t$, where $A = -2(\omega^2 - 17)\alpha$, $B = -4\omega\alpha$, and $\alpha = 1/[(\omega^2 - 17)^2 + 4\omega^2]$.
(b) Steady state solution is $A\sin\omega t + B\cos\omega t$. Transient solution is $\frac{1}{4}(-B - A\omega)e^{-t}\sin 4t - Be^{-t}\cos 4t$.

7. $\omega = 0$ is a local minimum while $\omega = \pm 2$ are global maxima.

11. (a) ii. 1 unit of time to reach maximum height. It takes as long to go up as it does to come down.
(b) ii. $\ln 2 \approx 0.6931$ units of time to reach maximum height. It takes longer to come down than to go up.

13. (b) As $c \to 2$ the amplitude increases. At $c = 2$ the solution becomes unbounded as $t \to \infty$. As c increases for $c > 2$, the amplitude decreases.

Chapter 8

Section 8.1

1. All nontrivial solutions oscillate.

3. All nontrivial solutions oscillate and are bounded. $x(t) = \frac{1}{\sqrt{t}}(C_1 \sin t + C_2 \cos t)$.

9. The first equation.

13. (a) $x(t) = e^{-t^2/2}(C_1 + C_2 t)$.
(c) $x(t) = \frac{1}{t}(C_1 \cos 2t + C_2 \sin 2t)$.

15. This means that all solutions are concave up everywhere, and so they cannot oscillate.

Section 8.2

1. (a) $x(t) = e^{2t}(C_1 + C_2 t)$.
(c) $x(t) = t(C_1 \ln t + C_2)$.
(e) $x(t) = -C_1 t\cos t + C_2 t\sin t$.
(g) $x(t) = t(C_1 + C_2 e^t)$.

3. (a) $x(t) = (-2\ln t + C_1 t + C_2)e^{-t}$.
(c) $x(t) = t\sin t + \cos t\ln\cos t + C_1 \sin t + C_2 \cos t$.

5. (a) $-2\ln t$. (c) $t\tan t + \ln\cos t$.

7. (b) $x(t) = \frac{1}{2}(-t - 2) - C_1\frac{1}{t} + C_2 e^t$.

Section 8.3

3. (a) $x(t) = -2e^{-t}\ln t - 2e^{-t} + C_1 e^{-t} + C_2 te^{-t}$.
(c) $x(t) = (\ln\cos t + C_1)\cos t + (t + C_2)\sin t$.

Section 8.4

3. (a) $k \neq 0$. (c) $k \neq \frac{1}{2}$. (e) $k \neq 0$.
(g) $k \neq 0$. (i) All k. (k) All k.

9. $x(0) = x_0$ and $x'(0) = x_0^*$.

11. (a) $t^2x'' - 2tx' + 2x = 0$.

(c) $(t-1)x'' - tx' + x = 0$.

(e) $x'' + 2x' + x = 0$.

(g) $x'' + \beta^2 x = 0$.

(i) $(\sin^2 t \cos t)x'' - (2\sin t\cos^2 t - \sin^3 t)x' + (2\cos t - 2\cos t\sin^2 t)x = 0$.

(k) $(\cos t - \sin t)x'' + 2(\sin t)x' - (\sin t + \cos t)x = 0$.

Section 8.5

1. (a) $y(x) = C_1/x + C_2x^2$.

(c) $y(x) = C_1 + C_2 \ln x$.

(e) $y(x) = C_1\cos(3\ln x) + C_2\sin(3\ln x)$.

(g) $y(x) = \dfrac{C_1}{x^2} + \dfrac{C_2}{x^2}\ln x$.

(i) $y(x) = C_1x + C_2x^5$.

(k) $y(x) = C_1x + C_2x\ln x$.

3. (a) $y(x) = -\frac{1}{2}\left(\ln x - \frac{1}{2}\right) + C_1/x + C_2x^2$.

(c) $y(x) = 2 - \ln x/x + C_1/x + C_2/x^5$.

Section 8.6

3. (a) $x(t) = 4\cos 10t + 500\sin 10t$.

(c) $x(t) = 0$.

5. (a) The trivial solution is always a solution.

(b) $b \neq n\pi/4,\ n = 0, \pm1, \pm2, \cdots$.

(c) $x(t) = C_2\sin 4t$, for any C_2.

7. $\lambda = (2n+1)/2,\ n = 0, \pm1, \pm2, \cdots$.
$x(t) = C_2\sin[(2n+1)t/2]$.

11. $x_0^* \approx 1.1.\ x_0^* = 0$.

Section 8.7

1. (a) $x(t) = C_1 + C_2\cos t + C_3\sin t$.

(c) $x(t) = C_1\cos\sqrt{2}t + C_2\sin\sqrt{2}t + C_3t\cos\sqrt{2}t + C_4t\sin\sqrt{2}t$.

(e) $x(t) = C_1\cos 2t + C_2\sin 2t + C_3e^{\sqrt{2}t} + C_4e^{-\sqrt{2}t}$.

(g) $x(t) = C_1e^t + C_2e^{2t} + C_3e^{-t}$.

(i) $x(t) = C_1e^t + C_2e^{-t}\cos 2t + C_3e^{-t}\sin 2t$.

3. (a) $x(t) = -6 + 4t + 8e^{-t} - 2e^{-2t}$.

(c) $x(t) = e^t - e^{-3t}$.

5. (a) Linearly independent.

(c) Linearly dependent.

(e) Linearly independent.

(g) Linearly independent.

11. $x(t) = At + B + C_1\cos\omega t + C_2\sin\omega t$.
$y(t) = At + B - (m/M)C_1\cos\omega t - (m/M)C_2\sin\omega t$.

Section 8.8

1. (a) $x(t) = \frac{1}{6}\cos 2t + C_1 + C_2\cos t + C_3\sin t$.

(c) $x(t) = -\frac{1}{2}t\cos t + C_1 + C_2\cos t + C_3\sin t$.

(e) $x(t) = \frac{1}{9}e^{-2t} + C_1\cos\sqrt{2}t + C_2\sin\sqrt{2}t + C_3t\cos\sqrt{2}t + C_4t\sin\sqrt{2}t$.

(g) $x(t) = te^{-t} + \frac{1}{6}e^t + C_1 + C_2t + C_3e^{-t} + C_4e^{-2t}$.

(i) $x(t) = \frac{1}{144}e^{-3t} + \frac{1}{4}\cos t + C_1\cos\sqrt{3}t + C_2\sin\sqrt{3}t + C_3t\cos\sqrt{3}t + C_4t\sin\sqrt{3}t$.

3. $x(t) = gt^2/2 + C_3t + C_4 + C_1\cos\omega t + C_2\sin\omega t$.
$y(t) = gt^2/2 + C_3t + C_4 - (m/M)C_1\cos\omega t - (m/M)C_2\sin\omega t + l$.

CHAPTER 9

Section 9.1

1. (a) $x(t) = C_1e^t + C_2e^{7t}$.
$y(t) = -\frac{1}{5}C_1e^t + C_2e^{7t}$.

(c) $x(t) = C_1e^{2t} + C_2e^{7t}$.
$y(t) = -\frac{1}{2}C_1e^{2t} + \frac{1}{3}C_2e^{7t}$.

(e) $x(t) = C_1e^{-t} + C_2te^{-t}$.
$y(t) = \frac{1}{9}(-3C_1 + C_2)e^{-t} - \frac{1}{3}C_2te^{-t}$.

(g) $x(t) = C_1e^t\sin t + C_2e^t\cos t$.
$y(t) = \frac{1}{2}(3C_1 + C_2)e^t\sin t - \frac{1}{2}(C_1 - 3C_2)e^t\cos t$.

(i) $x(t) = C_1e^{3t}\sin 5t + C_2e^{3t}\cos 5t$.
$y(t) = -C_2e^{3t}\sin 5t + C_1e^{3t}\cos 5t$.

(k) $x(t) = (C_1 + C_2t)e^{2t}$.
$y(t) = (C_1 + C_2 + C_2t)e^{2t}$.

3. (a) $x(t) = 16 + C_1e^t + C_2e^{7t}$.
$y(t) = -5 - \frac{1}{5}C_1e^t + C_2e^{7t}$.

(c) $x(t) = 1 + C_1e^{3t} + C_2te^{3t}$.
$y(t) = 1 - \frac{1}{2}(2C_1 + C_2)e^{3t} - C_2te^{3t}$.

(e) $x(t) = -1 + e^{3t}(C_1 \cos\sqrt{2}t + C_2 \sin\sqrt{2}t)$.
$y(t) = 1 + e^{3t}(\sqrt{2}C_1 + 2C_2)\sin\sqrt{2}t +$
$e^{3t}(-\sqrt{2}C_2 + 2C_1)\cos\sqrt{2}t$.

5. (a) $x(t) = 400 - 300e^{-t/50} - 100e^{-3t/50}$.
$y(t) = 400 - 600e^{-t/50} + 200e^{-3t/50}$.

(b) $x(t) = 400 - 225e^{-t/50} - 175e^{-3t/50}$.
$y(t) = 400 - 450e^{-t/50} + 350e^{-3t/50}$.

Section 9.2

1. (a) Unstable node. (c) Unstable node.
(e) Stable node. (g) Unstable focus.
(i) Unstable focus. (k) Unstable node.

Section 9.3

1. (a) $y = x, y = -\frac{1}{5}x$. (c) $y = \frac{1}{3}x, y = -\frac{1}{2}x$.
(e) $y = -\frac{1}{3}x$. (g) None.
(i) None. (k) $y = x$.

3. (a) Figure 9.18. (b) Figure 9.20.
(c) Figure 9.19.

Section 9.4

1. (a) Unstable node. (c) Unstable node.
(e) Inconclusive. (g) Inconclusive.
(i) Inconclusive. (k) Inconclusive.

Section 9.5

5. (a) $\mathbf{X}(t) = \begin{bmatrix} e^{4t} & e^{-t} \\ -2e^{4t} & 3e^{-t} \end{bmatrix} \begin{bmatrix} C_1 \\ C_2 \end{bmatrix}$.
$y = 3x, y = -2x$.

(c) $\mathbf{X}(t) =$
$$\begin{bmatrix} 3e^{(2+2\sqrt{3})t} & 3e^{(2-2\sqrt{3})t} \\ (3-2\sqrt{3})e^{(2+2\sqrt{3})t} & (3+2\sqrt{3})e^{(2-2\sqrt{3})t} \end{bmatrix} \begin{bmatrix} C_1 \\ C_2 \end{bmatrix}.$$
$y = (3 \pm 2\sqrt{3})x/3$.

(e) $\mathbf{X}(t) = \begin{bmatrix} -3e^{-t} & (-3t-1)e^{-t} \\ e^{-t} & te^{-t} \end{bmatrix} \begin{bmatrix} C_1 \\ C_2 \end{bmatrix}$.
$y = -x/3$.

(g) $\mathbf{X}(t) = \begin{bmatrix} e^{2t}\cos 2t & e^{2t}\sin 2t \\ -2e^{2t}\sin 2t & 2e^{2t}\cos 2t \end{bmatrix} \begin{bmatrix} C_1 \\ C_2 \end{bmatrix}$.
No straight-line orbits.

(i) $\mathbf{X}(t) = \begin{bmatrix} e^{3t}\cos 5t & e^{3t}\sin 5t \\ -e^{3t}\sin 5t & e^{3t}\cos 5t \end{bmatrix} \begin{bmatrix} C_1 \\ C_2 \end{bmatrix}$.
No straight-line orbits.

7. (a) $25\ln 3 \approx 27.5$ min. (b) $\mathbf{X}(25) \approx \begin{bmatrix} 4.1 \\ 3.8 \end{bmatrix}$ lbs.
(c) 0.

9. A case (c). B case (a). C case (b).

Section 9.6

1. (b) $y(t) = -vx(t)/V + C$.
If $y > x$, the orbits will move to the right. If $y < x$, the orbits will move to the left.

(c) $x(t) = C_1 + C_2 e^{-(w_1+w_2)t}$.
$y(t) = C_1 - \frac{w_2}{w_1} C_2 e^{-(w_1+w_2)t}$.
$w_1 = a/v$, $w_2 = a/V$.

CHAPTER 10

Section 10.1

1. $\frac{dy}{dx} = \frac{y(y-1)}{x}$.
Equilibrium solutions $y(x) = 0$ and $y(x) = 1$.
Nonequilibrium solutions $y = \frac{cx}{1-cx}$.

3. (a) $x = 0$ and $y = 0$.
(b) (i) $x(t) = K_1 e^{-t}$. $y(t) = \frac{1}{3}K_1^2 e^{-2t} + K_2 e^t$.
(c) As $t \to \infty$, $x(t) \to 0$, $y(t) \to \infty$ if $K_2 \neq 0$ and $y(t) \to 0$ if $K_2 = 0$.
(d) (i) $y = \frac{1}{3}x^2 + K_2K_1/x$, if $K_1 \neq 0$. $x = 0$ and $y = K_2 e^t$.
(f) $x \to 0$ and $y \to \pm\infty$, except that there appears to be an orbit from the right that approaches $(0, 0)$, and a similar one from the left.
(g) Separatrix $y = \frac{1}{3}x^2$. There is no basin of attraction.

5. (a) $(0, 0)$ and $(2, 0)$.
(b) $\frac{dy}{dx} = \frac{2x-x^2}{y}$. $y^2 = 2x^2 - \frac{2}{3}x^3 + C$.
(d) Orbits near $(2, 0)$ rotate around $(2, 0)$. Almost all other orbits have the property that $x \to -\infty$, and $y \to -\infty$. The orbit $y = \pm\sqrt{2x^2 - \frac{2}{3}x^3}$ approaches $(0, 0)$ as t increases.
(e) Separatrix $y^2 = 2x^2 - \frac{2}{3}x^3$. No basin of attraction for the equilibrium point $(0, 0)$. The equilibrium point $(2, 0)$ attracts all curves that lie inside the separatrix.

7. (a) $(0, 0)$.

(b) $\frac{dy}{dx} = \frac{y(y-2x)}{x(x-2y)}$. $y = 0$, $y = x$ and $y = \frac{1}{2}(x \pm \sqrt{x^2 - 4c/x})$.

(d) Orbits are (i) asymptotic to $x = 0$ for $y > 0$ and $y > x$, (ii) asymptotic to $y = 0$ for $x > 0$ and $y < x$, (iii) asymptotic to $y = x$ for $x < 0$ and $y < 0$, or (iv) approach $(0, 0)$ along $y = 0$ for $x < 0$, along $x = 0$ for $y < 0$, and along $y = x$ for $x > 0$.

(e) Separatrices (i) $y = 0$ for $x < 0$, (ii) $x = 0$ for $y < 0$, and (iii) $y = x$ for $x > 0$. No basin of attraction.

9. (b) A possible shape of the wire is $y(x) = (x + 2)(x + 1)x(x - 1)(x - 2)$.

Section 10.2

1. (a) The equilibrium point $(0, 0)$ behaves like an unstable focus, while $(1, -1)$ behaves like a saddle point.

(c) A nullcline analysis is inconclusive as far as $(-1, 0)$ is concerned. The equilibrium point $(1, 0)$ behaves like a saddle point.

3. (a) The equilibrium point $(0, 0)$ behaves like a saddle point.

(c) The nullcline analysis is inconclusive as far as $(2, 0)$ is concerned. The equilibrium point $(0, 0)$ behaves like a saddle point.

(e) The equilibrium point $(0, 0)$ behaves like a saddle point.

Section 10.3

5. (a) At $(0, 0)$, $D = -1$ and $T = 0$, characteristic of a saddle point.

(c) At $(0, 0)$, $D = -2$ and $T = 0$, characteristic of a saddle point.
At $(2, 0)$, $D = 2$ and $T = 0$, which is inconclusive.

(e) At $(0, 0)$, $D = 0$ and $T = 0$, which is inconclusive.

Section 10.4

1. (a) $(0, 0)$, $(1/\sqrt{-2b}, 0)$, and $(-1/\sqrt{-2b}, 0)$.

(b) For the hard spring, the nullcline analysis is inconclusive.
For the soft spring, the equilibrium points $(1/\sqrt{-2b}, 0)$ and $(-1/\sqrt{-2b}, 0)$ behave like saddle points. The nullcline analysis is inconclusive for $(0, 0)$.

(c) At $(0, 0)$, $D = \lambda^2$ and $T = 0$, inconclusive for both the hard and soft springs.
For the soft spring, at $(\pm 1/\sqrt{-2b}, 0)$, $D = -2\lambda^2$ and $T = 0$, which characterize saddle points.

(e) Figure 10.37, hard spring. Figure 10.38, soft spring.

(f) The motion is periodic.

3. (a) $(0, 0)$.

(b) All orbits seem to be approaching a limit cycle with "center" $(0, 0)$ and "radius" approximately 2.

(c) Nullcline analysis is inconclusive.

(d) At $(0, 0)$, $D = 1$ and $T = 1$. Unstable focus.

5. If a solution does not start at $(0, 0)$ or $(0, 6)$, all solutions tend to $(9, 0)$ — that is, the y population becomes extinct.

Section 10.6

1. $x(t) = t^2 + C_1 \ln|t| + C_2$.

3. (a) $x(t) = \pm t + C$ and $x(t) = t - 2C_1 \ln|t - C_1| + C_2$.

(c) $x^2 = 2C_1 t + C_2$.

(e) $\arctan x = C_1 t + C_2$.

5. $(y - a)^2 + (x - b)^2 = \frac{1}{\kappa^2}$, where $b = -C/\kappa$.

7. $x(t) = 1/\left(C_1 e^t + C_2 e^{-t}\right)$.

9. $x(t) = \left(C_2 + 2C_3 t\right) / \left(2C_1 + 2C_2 t + 2C_3 t^2\right)$.

CHAPTER 11

Section 11.1

3. (a) $\mathcal{L}\{\sin at\} = a/(s^2 + a^2)$.

(c) $\mathcal{L}\{\sinh at\} = a/(s^2 - a^2)$.

5. (a) $\exp(2t) + \exp(-3t)$.

(c) $\exp(-2t) - \exp(-3t)$.

(e) $2\exp(2t) - 2\sinh t - 4\cosh t$.

7. (a) $y(t) = \exp 2t - \exp(-3t)$.

(c) $y(t) = \exp(-2t) - \cos t + 2\sin t$.

(e) $y(t) = [\exp(2t) - \exp(-3t)]/5 +$
$[1 - \exp(10)][\exp(-3t)]/5$.

Section 11.2

3. (a) $2s/(s^2 - 1)^2$.

5. (a) $a/[s^2 - 2sa]$. (c) $2/(s - a)^3$.

7. (a) $(2s^3 - 6a^2s)/(s^2 + a^2)^3$.
(c) $(s^2 + a^2)/(s^2 - a^2)^2$.

Section 11.3

1. (a) $\exp 2t - \exp t$.
(c) $-(1/3)\cosh t + 2/3 \sinh t + (1/3)\exp(-2t)$.
(e) $1/(a^2+1)\exp t - 1/(a^2+1)\cos at + a/(a2+1)\sin at$.
(g) $(1/39)(\exp(-2t)[-2\sin 3t - 3\cos 3t] + 3)$.
(i) $(a\exp t - \sin at - a\cos at)/(1+a^2)$.
(k) This is listed in Table 11.6.
(m) This is listed in Table 11.6.

5. (a) $y(t) = 2t - 3/2 - (3/2)\cos(\sqrt{2}t) + \sqrt{2}\sin(\sqrt{2}t)$.
(c) $y(t) = 3t\exp(-t)$. (e) $y(t) = t - t^2/2$.

Section 11.4

1. (a) $u(t) - u(t-4)$.
(c) $t[u(t) - u(t-2)] + 2[u(t-2) - u(t-4)] + (6-t)[u(t-4) - u(t-6)]$.
(e) $[u(t) - u(t-a)] - [u(t-2a) - u(t-3a)]$.

3. $f(t) = \sin t[u(t) - u(t-\pi)]$.
$\mathcal{L}\{f(t)\} = [1/(1-\exp(-\pi s))][\exp(-s\pi)+1]/[1+s^2]$.

5. (a) $[1-\exp(-2s)]/s^2 - 2\exp(-2s)/s$.
(c) $1/s^2 - (2/s)/(\exp 2s - 1)$.

7. (a) $(t-1)\exp(-\{t-1\})u(t-1)$.
(c) $[1-\exp(-\{t-3\})]u(t-3)$.
(e) $(t-1)\exp(4[t-1])u(t-1) - (t-2)\exp(4[t-2])u(t-2)$.
(g) $(1/4)\exp(-3[t-3])\sin[4(t-3)]\,u(t-3)$.

Section 11.5

1. (a) $(\exp kt - 1)/k^2 - t/k$.

3. (a) $2[1-\exp(-t)] - [1-\exp\{-(t-1)\}]u(t-1)$.
(c) $6\exp(-t) + 5[1-\exp(-t)] - 4[1-\exp(-\{t-10\})]u(t-10)$.

5. (a) $[1-\exp(-R[t-1]/L)]u(t-1)/R - [1-\exp(-R[t-2]/L)]u(t-2)/R$.
(c) $\sum_{k=0}^{\infty}\{(-1)^k[1-\exp(-R[t-k]/L)]u(t-k)/R\}$.

7. (a) $425u(t) - 7.5(t-15)u(t-15) + 7.5(t-25)u(t-25)$.
(b) $425 - 355e^{kt} + 7.5[f(t-15)u(t-15) - f(t-25)u(t-25)]$, where $f(t) = (e^{kt} - 1 - kt)/k$.
(c) $425 - 355e^{kt} + 75\left[e^{k(t-15)} - 1\right]u(t-15)$.

Section 11.6

1. (a) $\exp t - \exp(-2t)$.
(c) $2\exp(-3t) + 6t\exp(-3t)$.
(e) $5 - \exp(-3t)$.
(g) $15e^{-5t}\cos(5\sqrt{3}t) + (79\sqrt{3}/15)e^{-5t}\sin(5\sqrt{3}t)$.

3. (a) $(8t/7)\exp 3t + (1/49)[\exp(-4t) - \exp(3t)]$.
(c) $t\exp(-t) + \exp(2t) - \exp(3t)$.

5. (a) $\exp(-2t)[\cos t + \sin t]/8 + [\sin t - \cos t]/8$.
(c) $\exp(-3t) - \exp(-2t) + t\exp(-2t)$.
(e) If $0 < t < 1$,
$x(t) = t^2 - 2t + 2 - 2e^{-t} + (2t^2 - 2t + 1 - e^{-2t})/4$.
If $1 < t$, $x(t) = e^{-t}(e-2) + e^{-2t}\left(e^2 - 1\right)/4$.
(g) $\int_0^t z\exp(-2z)f(t-z)\,dz + \exp(-2t) + 2t\exp(-2t)$.
(i) $\int_0^t [\sin(\omega z)/\omega]f(t-z)\,dz + \sin(\omega t)/\omega$.
(k) If $0 < t < 1$, $x(t) = \sin(\omega t)/\omega$.
If $1 < t < 2$,
$x(t) = \sin(\omega t)/\omega + [1 - \cos(\omega(t-1))]/\omega$.
If $2 < t$,
$x(t) = \sin(\omega t)/\omega + [\cos(\omega(t-2)) - \cos\omega]/\omega$.

7. (a) $(10/73)\exp(-2t)[-3/2\sin 4t - 4\cos 4t] + (40/73)\exp(-t/2) + \exp(-2t)[\cos(4t) + \sin(4t)/2]$.
(c) $3\exp(-2t) - 2\exp(-3t) + f(t) - f(t-2)u(t-2)$.
(e) $3\exp(-2t) - 2\exp(-3t) + \sum_{n=0}^{\infty}(-1)^n f(t-2n)u(t-2n)$.

Section 11.7

3. (a) $x(t) = \exp 3t + \exp t$, $y(t) = \exp t - \exp 3t$.
(c) $x(t) = 2\exp 2t$, $y(t) = \exp 2t$.

5. (a) $x(t) = -(2/5)\exp(-3t) + (2/5)\exp 2t + \exp 3t$.
$y(t) = (1/10)\exp(-3t) + (2/5)\exp 2t - (1/2)\exp 3t$.
(c) $x(t) = (9/4)\exp t - (1/4)\exp(-t) + (1/2)\sin t - \cos t$.
$y(t) = (-9/4)\exp t + (3/4)\exp(-t) + (3/2)\cos t$.
(e) $x(t) = (4/5)[\exp(2t) - \exp(-3t)] + g(t-1)u(t-1) - g(t-3)u(t-3)$, where
$g(t) = 1/6 - (4/15)\exp(-3t) + (1/10)\exp(2t)$.
$y(t) = (4/5)\exp(2t) + (1/5)\exp(-3t) + f(t-1)u(t-1) - f(t-3)u(t-3)$, where
$f(t) = -1/6 + (1/15)\exp(-3t) + (1/10)\exp(2t)$.

7. (a) $x(t) = -1/2 + (19/2)\exp(2t) - 5\exp(3t)$.
$y(t) = -1/6 - (19/2)\exp(2t) + (20/3)\exp(3t)$.

(c) $x(t) = 2 - 3\exp t$. $y(t) = -8 - 2\exp t$.

(e) $x(t) = 4\sin 2t - 4\cos 2t$. $y(t) = \sin 2t + \cos 2t$.

Section 11.8

5. (a) $\alpha = 3$, $M = 1$, and $T = 0$.

(c) $\alpha = 2$, $M = 1$, and $T = 0$.

11. $f(t) = t$.

CHAPTER 12

Section 12.1

1. $1 + x + \frac{1}{2}x^2 + \frac{1}{3}x^3 + \cdots$.

3. $1 - \left(\frac{1}{2}\sin 1\right)x^2 + \left(\frac{1}{24}\sin 1 \cos 1\right)x^4 + \cdots$.

5. $1 + x^2 + \frac{1}{2!}x^4 + \frac{1}{3!}x^6 + \cdots$. These are the first 4 terms of the Taylor series for $\exp(x^2)$.

7. $y_0\left(1 - \frac{1}{2!}x^2 + \frac{1}{4!}x^4 + \cdots\right) + y_0^*\left(x - \frac{1}{3!}x^3 + \frac{1}{5!}x^5 + \cdots\right) = y_0\cos x + y_0^*\sin x$.

Section 12.2

3. $2\left(1 - \frac{3}{2!}x^2 + \frac{3}{4!}x^4 + \frac{3}{6!}x^6 + \cdots\right)$.

7. $c_0[1 - (x-1)^2 - \frac{2}{4\cdot 3}(x-1)^4 - \frac{2^2 3}{6\cdot 5\cdot 4\cdot 3}(x-1)^6 + \cdots] + c_1(x-1)$.

9. (a) $3x + \frac{3}{2}x^2 - \frac{1}{2}x^4 + \cdots$. $R = \infty$.

(c) $-2x + \frac{1}{2}x^2 - \frac{1}{6}x^3 + \frac{1}{3}x^4 + \cdots$. $R = \infty$.

(e) $2(x-1) + (x-1)^2 + \frac{2}{3}(x-1)^3 + \frac{1}{6}(x-1)^4 + \cdots$. $R = 1$.

11. (a) $c_0\left[1 - \frac{1}{4\cdot 3}x^4 + (-1)^2\frac{1}{8\cdot 7\cdot 4\cdot 3}x^8 + \cdots\right] + c_1\left[x - \frac{1}{5\cdot 4}x^5 + (-1)^2\frac{1}{9\cdot 8\cdot 5\cdot 4}x^9 + \cdots\right]$. $R = \infty$.

(c) $c_0 + c_1 x - \frac{5}{2}c_0 x^2 - \frac{7}{6}c_1 x^3 + \frac{35}{24}c_0 x^4 + \frac{7}{24}c_1 x^5 + \cdots$. $R = 1$.

15. (a) $1, -2, -3$. (c) $-6, 1$.

(e) $3i, -3i, -7$.

(g) $1, -2, \pm n\pi$, $n = 0, 1, 2, \cdots$.

Section 12.3

1. (a) Regular $0, -1, 4, -4$. Irregular 1.

(c) Regular -1, $\pm n\pi$, $n = 1, 2, 3, \cdots$. Irregular 0.

(e) Regular $-2, 4, -4$. Irregular 0.

(g) Regular 1. Irregular $0, 4$.

3. (a) Converges for all x. $4s^2 - 2s = 0$.

(c) Converges for all x. $4s^2 - 2s = 0$.

(e) Converges for all x. $2s^2 - 3s - 5 = 0$.

(g) Converges for all x. $2s(s-1) - 3s + 2 = 0$.

(i) Converges for all x. $3s(s-1) + s = 0$.

(k) Converges for all x. $s(s-1) + 2s - 2 = 0$.

Section 12.4

1. (a) $c_0\left(1 + \frac{2}{3}x + \frac{1}{6}x^2\right) + c_4\sum_{n=4}^{\infty}\frac{(n-3)4!}{n!}x^n$.

(c) $c_0\sum_{n=0}^{\infty}\frac{(-1)^n}{3^n n!}x^{3n} + c_2[x^2 + \sum_{n=2}^{\infty}\frac{(-1)^{n+1}}{(3n-1)(3n-4)\cdots(8)(5)}x^{3n-1}]$.

(e) $c_0[1 + \sum_{n=1}^{\infty}\frac{(-1)^n(2x)^n}{(2n-1)(2n-3)\cdots(3)(1)}] + c_0^* x^{1/2}\sum_{n=0}^{\infty}\frac{(-1)^n x^n}{n!}$.

(g) $c_0 x^{-1}[1 + \sum_{n=1}^{\infty}\frac{(-1)^n x^n}{n!(2n-7)(2n-9)\cdots(-3)(-5)}] + c_0^* x^{5/2}[1 + \sum_{n=1}^{\infty}\frac{(-1)^n x^n}{n!(2n+7)(2n+5)\cdots(11)(9)}]$.

3. (a) $y(x) = y_1(x) + y_2(x)$, where

$y_1(x) = c_0(x-1)^{-1/2}[1 + \sum_{n=1}^{\infty}\frac{(-1)^n(x-1)^n}{n!(2n+1)(2n-1)\cdots(5)(3)}]$

$y_2(x) = c_0^*(x-1)^{-1}[1 + \sum_{n=1}^{\infty}\frac{(-1)^n(x-1)^n}{n!(2n-1)(2n-3)\cdots(3)(1)}]$.

(c) $c_0(x+1)^{-3}[1 - 3(x+1) + 3(x+1)^2 - (x+1)^3] + c_6(x+1)^{-3}\sum_{n=6}^{\infty}\frac{(n-4)!6!}{n!2!}(x+1)^n$.

(e) $c_0(x-1) + c_0^*(x-1)[\ln|x-1| + \sum_{n=1}^{\infty}\frac{(-1)^n}{n!2^n 2n}(x-1)^{2n}]$.

5. (a) $y(x) = c_1 y_1(x) + c_2\left[y_1(x)\ln x - x^2 + \sum_{n=2}^{\infty}c_{2n}^* x^{2n}\right]$, where $y_1(x) = \sum_{n=0}^{\infty}\frac{1}{(n!)^2}x^{2n}$. $R = \infty$.

(c) $y(x) = C_1 y_1(x) + C_2[-y_1(x)\ln x + x^{-1}\left(1 - \frac{3}{2}x + \frac{5}{72}x^3 - \frac{17}{1728}x^4 + \cdots\right)]$, where $y_1(x) - c_0[\sum_{n=0}^{\infty}\frac{(-1)^n}{(n+1)!n!}x^n]$. $R = \infty$.

(e) $c_0(1-x) + 2c_2\left(e^{-x} + x - 1\right)$. $R = \infty$.

(g) $c_0 x + c_0^*[x\ln x + \sum_{n=1}^{\infty}\frac{(-1)^n}{nn!}x^{n+1}]$. $R = \infty$.

(i) $y(x) = y_1(x) + y_2(x)$, where $y_1(x) = c_0 x e^{-x}$, and $y_2(x) = \left(x - x^2 + \frac{1}{2}x^3 - \frac{1}{6}x^4 + \cdots\right)\ln|x| + x - \frac{1}{12}x^3 + \frac{5}{36}x^4$. $R = \infty$.

LaVergne, TN USA
25 February 2011
218023LV00001B/4/P